AF803683

The Minerals, Metals & Materials Series

Tao Jiang · Jiann-Yang Hwang ·
Dean Gregurek · Zhiwei Peng ·
Jerome P. Downey · Baojun Zhao ·
Onuralp Yücel · Ender Keskinkilic ·
Rafael Padilla
Editors

10th International Symposium on High-Temperature Metallurgical Processing

TMS

Springer

Editors
Tao Jiang
Central South University
Changsha, China

Jiann-Yang Hwang
Michigan Technological University
Houghton, MI, USA

Dean Gregurek
RHI Magnesita
Leoben, Austria

Zhiwei Peng
Central South University
Changsha, China

Jerome P. Downey
Montana Tech
Butte, MT, USA

Baojun Zhao
The University of Queensland
Brisbane, QLD, Australia

Onuralp Yücel
Istanbul Technical University
Istanbul, Turkey

Ender Keskinkilic
Atilim University
Ankara, Turkey

Rafael Padilla
University of Concepción
Concepción, Chile

ISSN 2367-1181 ISSN 2367-1696 (electronic)
The Minerals, Metals & Materials Series
ISBN 978-3-030-05954-5 ISBN 978-3-030-05955-2 (eBook)
https://doi.org/10.1007/978-3-030-05955-2

Library of Congress Control Number: 2018964681

© The Minerals, Metals & Materials Society 2019
This work is subject to copyright. All rights are solely and exclusively licensed by the Publisher, whether the whole or part of the material is concerned, specifically the rights of translation, reprinting, reuse of illustrations, recitation, broadcasting, reproduction on microfilms or in any other physical way, and transmission or information storage and retrieval, electronic adaptation, computer software, or by similar or dissimilar methodology now known or hereafter developed.
The use of general descriptive names, registered names, trademarks, service marks, etc. in this publication does not imply, even in the absence of a specific statement, that such names are exempt from the relevant protective laws and regulations and therefore free for general use.
The publisher, the authors and the editors are safe to assume that the advice and information in this book are believed to be true and accurate at the date of publication. Neither the publisher nor the authors or the editors give a warranty, express or implied, with respect to the material contained herein or for any errors or omissions that may have been made. The publisher remains neutral with regard to jurisdictional claims in published maps and institutional affiliations.

This Springer imprint is published by the registered company Springer Nature Switzerland AG
The registered company address is: Gewerbestrasse 11, 6330 Cham, Switzerland

Preface

This book presents selected papers submitted for the 10th International Symposium on High-Temperature Metallurgical Processing organized in conjunction with the TMS 2019 Annual Meeting and Exhibition in San Antonio, Texas, USA. More than 120 abstracts were received. Among them, 90 were selected for oral presentation and 34 were provided with a poster presentation opportunity. After reviewing the submitted manuscripts, 75 were accepted for publication in this book.

As the title of symposium suggests, the interest of the symposium is on thermal processing of minerals, metals and materials that intends to promote physical and chemical transformations of materials to enable the extraction and production of valuable materials such as metals, alloys, ceramics and compounds. The symposium was open to participants from both industry and academia and focused on innovative high-temperature technologies including those based on non-traditional heating methods as well as their environmental aspects such as handling and treatment of emission gases and by-products. Because high-temperature processes require high energy input to sustain the temperature at which the processes take place, the symposium addressed the needs for sustainable technologies with reduced energy consumption and reduced emission of pollutants. The symposium also welcomed contributions on thermodynamics and kinetics of chemical reactions, phase transformations that take place at elevated temperatures, as well as simulation of high-temperature metallurgical processes. We hope the book will serve as a reference for both new and experienced metallurgists, particularly those who are actively engaged in exploring innovative technologies and routes that lead to more energy-efficient and environmentally sustainable solutions.

We would like to acknowledge the contributions from the authors of included papers, time and effort that reviewers dedicated to the manuscripts and help from the publisher. We also want to thank Dr. Elsa Olivetti and Dr. Camille Fleuriault for their help in identifying and screening manuscripts, and Dr. Mingjun Rao for his assistance in collating the submitted abstracts and manuscripts.

Tao Jiang
Jiann-Yang Hwang
Dean Gregurek
Zhiwei Peng
Jerome P. Downey
Baojun Zhao
Onuralp Yücel
Ender Keskinkilic
Rafael Padilla

Contents

Part I Simulation of High-Temperature Processes

A Mathematical Model for Carbon Loss of Blast Furnace Based on Traditional Engineering Method 3
Shun Yao, Sheng-li Wu, Bo Song, Ming-yin Kou and Heng Zhou

Study on Alkali Circulation Process and Its Influence on Coke Ratio in Blast Furnace 15
Haokun Li, Yijie Wang, Kexin Jiao, Jianliang Zhang, Rong Zhu and Hanjie Guo

Fuzzy Gray Relational Analysis for Electromagnetic Parameters of Induction Heating Process 25
Pei Fu, Ping Zhou, Tianyang Zhao, Zhuo Chen and Chenn Q. Zhou

Numerical Simulation of Inclusion Removal in a Novel Tundish with Swirl Flow 39
Jianchuan Yan, Tao Li and Jun Liu

Numerical Simulation Study on Design Optimization of Inner Cavity Dimensions of Large-Capacity Tundish 51
Yong Zhong, Mingmei Zhu, Bing Huang and AiPing Zhang

Part II Energy Efficient Clean Metallurgical Technologies

Preparation of High-Carbon Metallic Briquette for Coke Saving in Blast Furnace 65
Huiqing Tang, Shihong Liu and Kai Fan

Study of Siderite Fluidized Magnetization Roasting and Magnetic Separation 75
Qiang Zhao and Jilai Xue

Emission Profile of PM_{10} and $PM_{2.5}$ in Iron Ore Sintering Process and Control Technology 85
Zhiyun Ji, Xiaohui Fan, Min Gan, Xuling Chen, Wei Lv, Guojing Wong and Tao Jiang

Viscosity Properties of Mould Flux Under Low-Frequency Electromagnetic Field 95
Wei Qian, Yu Wang and Lu-ming Zhao

Part III Fundamentals of Metallurgical Processes

Effects of Cr_2O_3, FeO and CaO/SiO_2 Ratio on the Apparent Viscosity of $CaO\text{-}SiO_2\text{-}MgO\text{-}MnO\text{-}FeO\text{-}Cr_2O_3$ Slags 109
Bing Huang, Mingmei Zhu, Yong Zhong, Aiping Zhang and Tiancheng Lin

Thermodynamic Analysis of Carbothermic Reduction of Electric Arc Furnace Dust 117
Qing Ye, Zhiwei Peng, Lei Ye, Liancheng Wang, Robin Augustine, Joonho Lee, Yong Liu, Mudan Liu, Mingjun Rao, Guanghui Li and Tao Jiang

Influence of Cr_2O_3 Content on Slag Viscosity Under Different Melting States and Temperature Programs 125
Fang Yuan, Tuo Wu, Yanling Zhang and Zheng Zhao

Influence of Mould Slag Compositions with Different Reactivity on the Erosion Rate of ZrO_2-C Bearing Submergence Entry Nozzle 137
Xue-Si Wang, Qian Wang, Chang-Ping Zeng and Hua-Zhi Yuan

A New Method for Determining High-Temperature Wettability of Bonding Phase 145
Yijia Dong, Guanghui Li, Chen Liu, Qiang Zhong, Hu Sun, Jun Luo and Tao Jiang

Part IV High-Temperature Processing

Research on Dezincification of Ironmaking and Steelmaking Ashes 157
Yan Zhang, Xiaojiang Wu, Meng Xu, Gele Qing, Haoyu Cai, Wenbin Huang, Yunqing Tian and Wenwang Liu

Effects of Temperature and Alkali Carbonates on Graphitization and Metallurgical Properties of Coke 165
Rongjin Zhu, Shengfu Zhang, Guangsheng Suo, Yue Wu, Xiaohu Zhou and Shuxing Qiu

Field-Assisted Sintering of Nickel-Based Superalloy Powder for High-Temperature Hybrid Turbine Disc Applications 175
Charis I. Lin, Sebastian Niuman, Namiko Yamamoto, Anil Kulkarni and Jogender Singh

Sintering Test Research of High Proportion Limonite 189
Qiang Zhao and Jilai Xue

A Machine Learning Method for State Identification of Superheat Degree with Flame Interference 199
Shiwei Zhao, Yongfang Xie, Weichao Yue and Xiaofang Chen

Part V Extraction and Recovery of Metals

Study on Volatizing Tin from Tin-Bearing Middling by Carbothermic Reduction in Rotary Kiln 211
Jianfa Jing, Yufeng Guo, Feng Chen, Fuqiang Zheng and Lingzhi Yang

Isothermal Sulphation Roasting of Nickel Sulphide Minerals in a Static Air Atmosphere 221
Lizhen Wei, Caixiang Yu, Guangshi Li, Xiaolu Xiong, Hongwei Cheng, Qian Xu and Xionggang Lu

Thermal Transformations of Main Components in Molybdenite Concentrates Under SO_2-Containing Atmosphere 231
Hu Sun, Guanghui Li, Junjie Yu, Jun Luo, Mingjun Rao and Tao Jiang

A Novel Method of Recovering Rare Earths from Bayan Obo Rare-Earth Concentrate Under Super-Gravity Field 243
Xi Lan, Jintao Gao and Zhancheng Guo

Part VI Ironmaking and Steelmaking

Sintering Characteristic and Consolidation Behaviour of Chromite Fines 257
Xiaohui Fan, Guojing Wong, Min Gan, Xuling Chen, Zhiyun Ji, Xunwei Zhou and Wei Lv

Construction and Practice on Energy Flow Network of New Generation Recyclable Iron and Steel Manufacturing Process 269
Fuming Zhang

Dependency of Microstructure and Inclusions on the Different Growth Rate for Directionally Solidified Non-quenched and Tempered Steel 279
Hui Liu, Jianbo Xie, Honggang Zhong, Qijie Zhai and Jianxun Fu

Development of Offshore Steel for High Heat Input Welding 289
Xiaodong Ma, Peng Zhang, Tingliang Dong, Feng Wang and Baojun Zhao

Slag Basicity: What Does It Mean? 297
G. A. Brooks, M. M. Hasan and M. A. Rhamdhani

Investigation on Clogging of Submerged Entry Nozzles for GCr15 Bearing Steels 309
Gong Cheng, Lifeng Zhang, Wenbo Wang, Qiangqiang Wang and Piotr Roman Scheller

Part VII Preparation of Alloys and Materials I

Effects of Electrolytic Parameters on the Preparation of Al–Sc Master Alloy in Na_3AlF_6-K_3AlF_6-AlF_3 Melt 321
Kai Yang, Zhongliang Tian, Xun Hu, Yanqing Lai and Jie Li

Numerical Simulation Study on the Position Layout of the Permeable Brick at the Bottom of 300t Reblown Converter 329
Yun Huang, Chengbo Wu, Yong Zhong, Haitao Zhang and Gaopeng Zhang

Optimization of Zn-Al-Fe Alloy Vacuum Distillation Experiments by Response Surface Methodology 341
Zhenghao Pu, Yifu Li, Bin Yang and Huan Zhang

Review on Preparation of Medium- and Low-carbon Ferrochrome Alloys 349
Ting Hu, Hua Liu, Bingguo Liu, Linqing Dai, Libo Zhang and Shenghui Guo

Study on Separation of Sn-Sb Alloy by Vacuum Distillation 361
Yanjun You, Zhenghao Pu, Yifu Li, Bin Yang and Junjie Xu

Statistical Optimization of Tungsten Carbide Synthesis Parameters 371
Grant C. Wallace, Jerome P. Downey, Jannette Chorney, Katie Schumacher and Trenin Bayless

Part VIII Preparation of Alloys and Materials II

Effects of CeO_2 on Melting Temperature, Viscosity and Structure of Mold Fluxes for Casting Rare Earths Alloyed Steels 381
Zeyun Cai, Bo Song, Longfei Li and Zhen Liu

Experimental Study on the Mechanism of Lead Vapor Condensation Under Vacuum 393
Huan Zhang, Zhenghao Pu, Yifu Li, Junjie Xu, Baoqiang Xu and Bin Yang

Effect of Al on the Formation of IAF in Al–Ti–Mg Deoxidized and RE-Treated Steel 405
Xiaokang Cui, Bo Song, Zhen Liu and Longfei Li

Effect of Ce Treatment on the Composition of Nucleation Inclusion in Ti–Mg Complex Deoxidized C–Mn Steel 415
Zhen Liu, Bo Song, Longfei Li, Zeyun Cai and Xiaokang Cui

Effect of La Addition on Inclusions, Microstructures, and High-Temperature Mechanical Properties of As-Cast FeCrAl Alloys 427
Yang He, Jianhua Liu, Yindong Yang and Alex McLean

Removal of Copper from Fe–Cu Alloy by Using Iodine 435
Yuichi Takamatsu and Takashi Nagai

Part IX Treatment and Recycling of Wastes

Effect of Chemical Composition on the Crystallization Behaviour of Rare Earth Phase in Slag 443
Tengfei Ma, Fu Feng, Xuefeng She, Jingsong Wang and Qingguo Xue

Effects of Steel Scrap Oxidation on Scrap Preheating Process in an Electric Arc Furnace 453
Guangwu Tang, Yuchao Chen, Armin K. Silaen, Yury Krotov and Chenn Q. Zhou

Enriching and Separating Iron Impurity from Galvanizing Dross by Super-Gravity Technology 467
Anjun Shi, Zhe Wang, Lei Guo, Ning Zhang and Zhancheng Guo

Parameters of the Metallic Calcium Reduction from Magnesium Production Residues 479
K. C. Tasyurek, M. Bugdayci and O. Yucel

Smelting Studies for Recovery of Iron from Red Mud 489
Ender Keskinkilic, Saeid Pournaderi, Ahmet Geveci and Yavuz A. Topkaya

Optimization on Drying of Acid Leaching Slag by Microwave Heating Using Response Surface Methodology 501
Xuemei Zheng, Aiyuan Ma, Hairong Gao, Xiaoling Li, Xianzhu He, Ming Sun and Fengjiao Gu

Part X Utilization of Complex Ores

Effect of Core Diameter on the Compressive Strength and Porosity of Itakpe Iron Ore Pellets 511
JU Odo and VU Nwoke

Correlation Between Reduction Degree and Softening and Melting Properties of Pellets 523
Yu-Zhu Pan, Ai-Jun Zhang, Lin Lin, Jing-Song Wang, Huai-Xuan Feng and Qing-Shan Lin

Effect of TiO_2 on the Viscous Behaviour of the CaO–SiO_2–14 Mass% Al_2O_3–8 Mass% MgO–TiO_2 Slag 531
Zhengde Pang, Yuyang Jiang, Xuewei Lv, Zhiming Yan and Wenchao He

Granulation of Semisteel by Rotary Disc Atomizer 539
Wenchao He, Xuewei Lv, Feifei Pan, Xueqin Li, Zhiming Yan and Zhengde Pang

Dissolution Kinetics of Titanium in Carbon-Saturated Iron 545
Leizhang Gao, Tongxiang Ma, Zhiming Yan and Meilong Hu

Research on Mineral Structure and Compositions of Peru Raw Ore 553
Wen Pan, Lei Liu, Ya-peng Zhang, Xia Zhao and Zhi-xing Zhao

Part XI Poster Session

A Literature Review of Heat Capacity Measurement Methods 569
Guishang Pei, Junyi Xiang, Gang Li, Shanshan Wu, Feifei Pan and Xuewei Lv

Application of Offgas Analysis on Predicting Carbon Content at End Point During Steelmaking Process 579
Rong Cheng, Jiongming Zhang, Liangjin Zhang and Haitao Ma

Calcination of Strontium Carbonate in Rotary Kiln Furnace 591
Raşit Sezer, Emre Yilmaz, Selim Ertürk and Cüneyt Arslan

Determination of Effect of Li_2O on the Structure of CaO-Al_2O_3-Based Slag by Molecular Dynamics Simulation and Raman Spectrum 599
Sai Wang, Bo Ran Jia, Sheng Ping He and Qian Wang

Dissolution Reaction of Earthy Graphite in Liquid Steel 609
Hongyan Yan, Xiaojun Hu, Chao Luo, Jinglong Liang and KuoChih Chou

Distribution Behaviour of Metals in Copper Alloy Under Super-Gravity Field 619
Long Meng, Zhe Wang, Yiwei Zhong, Kuiyuan Chen and Zhancheng Guo

Effect of H_2/CO Ratio on Gas Consumption and Energy Utilization Rate of Gas-Based Direct Reduction Process 631
C. Y. Xu, A. Y. Zheng, J. L. Zhang, R. R. Wang, Y. Li, Y. Z. Wang and Z. J. Liu

Effects of Particle Size of Coke on Iron Ore Sintering Process 649
Huaiying Ma, Wen Pan, Lei Liu, Zhidong Zhang and Chunlai Wang

Experimental Study on Dechlorination of Cold-Rolling Sludge at High-Temperature Roasting 657
Yi Li, Hongwei Cheng, Guangshi Li, Xiaoyong Mei, Xionggang Lu and Qian Xu

Extraction Process of Antimony from Stibnite by Electrothermal Volatilization 665
Dongbo Li and Xiaohua Yang

Influence Factors Analysis on Scavenging of Chlorine Impurity from Crude Titanium Sponge 681
Liang Li, Fuxing Zhu, Dachun Liu, Kaihua Li, Zhuo Sheng and Baoqiang Xu

Kinetic Study on Decomposition and Dissolution of Limestone in Converter Slag 693
Haohua Deng, Nan Wang, Min Chen and Guangzong Zhang

Low-Grade Phosphorus-Containing Iron Ore for the Removal of Cu(II) Ion from Wastewater 705
Xiaoli Yuan, Dongshan Zhou, Wentang Xia and Qingyun Huang

Mechanism of the Chlorination Roasting of Nickel Sulphide Concentrate with Ammonium Chloride 713
Xiaoyong Mei, Hongwei Cheng, Cong Xu, Guangshi Li, Xionggang Lu and Qian Xu

Numerical Simulation Investigation on the Flow and Temperature Fields in Tundish with Gas Injection into Ladle Shroud 723
Wang Zhou, Tao Zhang and San-Xing Chen

Recovery Indium from Indium-Enriched Cementation and Co-precipitation Simultaneously Fe-As 733
Hui Huang, Zhigan Deng, Chang Wei, Xingbin Li, Gang Fan, Cunxiong Li and Minting Li

Research on Comprehensive Recovery and Harmless Treatment Process of Copper Smelting Slag 741
Dongbo Li, Yaguang Guo, Shuaibiao Liang and Deng Ma

Structure–Property Correlations of $Al_2O_3 \leftrightarrow SiO_2$ Substitution in Blast Furnace Slag 759
Zhiming Yan, Xue-wei Lv, Ramana G. Reddy, Zhengde Pang and Wenchao He

The Study of Hot Metal Dephosphorization by Replacing Part of Lime with Limestone 769
Haohua Deng, Min Chen, Nan Wang and Guangzong Zhang

Study on Energy Utilization of High Phosphorus Oolitic Haematite by Gas-Based Shaft Furnace Reduction and Electric Furnace Smelting Process 777
Hui Sun and Miaolian Bian

Study on the Effect of Different CO_2–O_2 Mixture Gas Blowing Modes on Vanadium Oxidation 787
Zheng-Lei Guo, Yu Wang, Qi Lu and Shu-Chao Wang

Thermodynamic Analysis of Precipitation of La-O-S-As Inclusions in Steel 795
Congxiao Li, Hongpo Wang, Bin Bai and Lei Zhang

Author Index 805

Subject Index 809

About the Editors

Tao Jiang received his M.S. in 1986 and Ph.D. in 1990, both from Central South University of Technology. Then he joined the university and served as an assistant professor (1990–1992) and full professor (1992–2000). From 2000 to 2003, he was a Visiting Scientist in the Department of Metallurgical Engineering at the University of Utah. Since 2003, Dr. Jiang has been a Professor in the School of Minerals Processing and Bioengineering at Central South University. He was elected as Specially Appointed Professor of Chang Jiang Scholar Program of China in 2008 and has been the dean of the school since 2010.

His research interests include sintering, pelletizing and non coke ironmaking of iron ores, and extraction of refractory gold ores. He has completed more than 50 projects from government and industry, including the National Science Fund for Distinguished Young Scholars Program. He and co-workers invented the direct reduction process of composite binder pellets, and three plants were set up in China based on the invention. He proposed the innovative composite agglomeration process of iron ore fines, which was put into production in Baotou Steel Company, China. He has been actively involved in the areas of utilization of non-traditional ferrous resources such as complex ores and various solid wastes. Dr. Jiang has published 320 technical papers, and 6 books including *Direct Reduction of Composite Binder Pellets and Use of DRI*, *Principle & Technology of Agglomeration of Iron Ores*, *Chemistry of Extractive Metallurgy of Gold*, and

Electrochemistry and Technology of Catalytical Leaching of Gold. He holds 42 patents and has more than 40 conference presentations.

Jiann-Yang Hwang is a Professor in the Department of Materials Science and Engineering at Michigan Technological University. He is also the Chief Energy and Environment Advisor at the Wuhan Iron and Steel Group Company, a Fortune Global 500 company. He has been the editor-in-chief of the *Journal of Minerals and Materials Characterization and Engineering* since 2002. He has founded several enterprises in areas including water desalination and treatment equipment, microwave steel production, chemicals, fly ash processing, antimicrobial materials and plating wastes treatment. Several universities have honoured him as a Guest Professor, including the Central South University, University of Science and Technology Beijing, Chongqing University, Kunming University of Science and Technology and Hebei United University.

Dr. Hwang received his B.S. from National Cheng Kung University in 1974, M.S. in 1980 and Ph.D. in 1982, both from Purdue University. He joined Michigan Technological University in 1984 and served as its Director of the Institute of Materials Processing from 1992 to 2011 and the Chair of Mining Engineering Department in 1995. He has been a TMS member since 1985. His research interests include the characterization and processing of materials and their applications. He has been actively involved in the areas of separation technologies, pyrometallurgy, microwaves, hydrogen storage, ceramics, recycling, water treatment, environmental protection, biomaterials, and energy and fuels. He has more than 28 patents and has published more than 200 papers. He has chaired the Materials Characterization Committee and the Pyrometallurgy Committee in TMS and has organized several symposia. He is the recipient of TMS Technology Award and the Michigan Tech's Bhakta Rath Research Award.

Dean Gregurek has been a Senior Mineralogist in the RHI Magnesita Technology Center, Leoben, Austria, since 2001. Dr. Gregurek received his M.Sc. degree at the University of Graz in 1995 and his doctorate degree in Applied Mineralogy from the University of Leoben in 1999. Prior to RHI Magnesita, he worked two years for Luzenac Europe in the talc business. His current research interests and technical expertise are focused on chemical and mineralogical studies related to interactions between refractories, and molten metals and slags from pyrometallurgical furnaces. Dr. Gregurek has been a TMS member since 2012, was *JOM* advisor (2014–2017) and has been Chair of the Pyrometallurgy Committee and a co-organizer for the 7–10th International Symposium on High-Temperature Metallurgical Processing (TMS Annual Meetings 2016–2019).

Zhiwei Peng is an Associate Professor in the School of Minerals Processing and Bioengineering at Central South University. He received his B.E. and M.S. degrees from Central South University in 2005 and 2008, respectively, and his Ph.D. degree in Materials Science and Engineering from Michigan Technological University in 2012. His research interests include heat transfer in microwave heating, dielectric characterization of materials, non-thermal microwave effects, extractive metallurgy, computational electromagnetics, microwave absorbing materials and biomaterials.

Dr. Peng has published 113 papers, including 80 peer-reviewed articles in multiple journals. He has served as an editor for *Cogent Chemistry* since 2018, a guest editor for *JOM* from 2013 to 2018, a member of editorial board of the *Journal of Minerals and Materials Characterization and Engineering* since 2012, and a reviewer for 41 journals. He received a TMS Travel Grant Award for the 141st TMS Annual Meeting and Exhibition, the Doctoral Finishing Fellowship and Dean's Award for Outstanding Scholarship of Michigan Technological University in 2012, and the Bhakta Rath Research Award of Michigan Technological University in 2013.

Dr. Peng is an active member of The Minerals, Metals and Materials Society (TMS). He has co-organized 8 TMS symposia (Characterization of

Minerals, Metals and Materials in 2013–2018, and the 9 and 10th International Symposia on High-Temperature Metallurgical Processing in 2018 and 2019) and co-chaired 16 TMS symposia sessions since 2012. He is a member of the Pyrometallurgy Committee and Materials Characterization Committee and is the Chair of the Continuing Education Sub-Committee of the Materials Characterization Committee. He was a winner of the TMS EPD Young Leaders Professional Development Award in 2014.

Jerome P. Downey earned his Ph.D. in Metallurgical and Materials Engineering at Colorado School of Mines and his B.S. and M.S. degrees in Metallurgical Engineering at Montana Tech. Dr. Downey is a registered professional engineer with active licences in Colorado and Montana. He has over 40 years of professional experience that includes industrial operations, applied process research and development, and corporate management. His technical expertise includes chemical and metallurgical thermodynamics, thermal processing, materials synthesis and processing, and hazardous materials treatment.

Dr. Downey is presently the Goldcorp Professor of Extractive Metallurgy at Montana Tech. He also serves as the Campus Director of the Montana University System Materials Science Ph.D. programme. Dr. Downey's research efforts are currently focused on the study of fundamental properties of slags, molten salts, and glasses; vapour phase extraction and refining of rare earth elements; and synthesis of non-oxide ceramic compounds for energy applications. Other research interests include pyro- and hydrometallurgical process development, polymer and biomass gasification, and wastewater treatment.

Baojun Zhao is Codelco-Fangyuan Professor in the School of Chemical Engineering at The University of Queensland, Brisbane, Australia. His primary fields of research are fundamental and applied investigations relevant to high-temperature processing of metals and materials. He has developed a number of novel research techniques to enable high-quality research to be carried out. He has received a number of international awards to demonstrate his leading research achievements. He has long-term collaborations with many international companies on metallurgy, materials and resources.

Onuralp Yücel completed his technical education with a Ph.D. in Metallurgical Engineering from Istanbul Technical University (ITU) where he currently holds the post of Professor since 2002. He was a Visiting Scientist at Berlin Technical University between 1987 and 1988. He carried out postdoctoral studies at New Mexico Institute of Mining and Technology, Socorro, USA, between 1993 and 1994. He has as many as 330 publications/presentations to his credit, which include topics such as technological developments in the production of a wide range of metals, ferroalloys, advanced ceramic powders, and application of carbothermic and metallothermic processes among others. He was the Vice Chairman of ITU, Metallurgical and Materials Engineering Department (MMED) between 2004 and 2007, and Director of ITU, Applied Research Center of Material Science and Production Technologies between 2006 and 2012. He has been the Chairman of ITU, MMED, since 2016. He is a member of the International Advisory Board of the International Symposium on Boron, Borides and Related Materials; International Symposium on Self-Propagating High-Temperature Synthesis; and International Metallurgy and Materials Congress (IMMC).

Dr. Yücel's areas of interest include:

Pyrometallurgy: Pretreatment of concentrates (production of WO_3, Sb_2O_3, As_2O_3, MoO_3), smelting and reduction of slags, production ferroalloys, alloys and metals carbothermic and metallothermic processes in EAF or in ladle (copper, cobalt, vanadium, chromium, ferroboron, cobalt boron, nickel boron, ferromolybdenum, ferromanganese, silicomanganese, ferrovanadium, ferrotungsten, ferrochromium, nickel–chromium molybdenum–iron and aluminium–titanium–boron alloys)

Ceramic Powder Production and Processing: Production of carbide, nitride, boride powders and their processing by explosive consolidation or sintering techniques. (B_4C, TiB_2, ZrB_2, SiC, CrB_2).

Beneficiation of Industrial Wastes: Production of metals and compounds from galvanizing ash, brass production wastes and vanadium sludges produced aluminum production. Grit production from aluminium, copper and steel slags.

Ender Keskinkilic earned his undergraduate degree from the Department of Metallurgical and Materials Engineering of Middle East Technical University (METU), Ankara (the capital city of Turkey), in 1999. He continued M.S. and Ph.D. studies in the same department. He worked as a Research Assistant in METU between 1999 and 2003. After receiving the master's degree in 2001, he progressed further in the field of extractive metallurgy. During the Ph.D. period, he moved to Eregli-Zonguldak in 2003 and worked in the Quality Metallurgy and RD Department of Eregli Iron and Steel Works Co. (ERDEMIR), which is the leading steel company in Turkey regarding the qualities produced and the production capacity. After earning his Ph.D. degree in 2007, he returned to university and to work in the Department of Metallurgical and Materials Engineering of Atilim University, Ankara, in 2008. He has been working as a faculty in Atilim since then. He was assistant professor between 2009 and 2014. He has been working as an associate professor since 2014. His primary field of interest is extractive metallurgy and more specifically pyrometallurgical processes such as iron- and steelmaking, ladle metallurgy, ferroalloy production and non-ferrous extractive metallurgy. He has been acting as the Chairman of the Department of Metallurgical and Materials Engineering of Atilim University since July 2018.

Rafael Padilla received his Ph.D. and M.Sc. degrees in Metallurgy from the University of Utah in 1984 and 1977, respectively, and Professional Engineering Title as Metallurgical Engineer from the Technical University of Oruro, Bolivia, in 1975. He joined the Department of Metallurgical Engineering, University of Concepcion, Chile, in 1986, where he currently holds the rank of Full Professor. His present research interest is on the vaporization of minor elements such as arsenic, antimony and bismuth from sulphide concentrates at roasting and smelting temperatures, and the removal and recovery of impurities from refractory chalcopyrite–enargite concentrates by pressure and atmospheric leaching.

Part I
Simulation of High-Temperature Processes

A Mathematical Model for Carbon Loss of Blast Furnace Based on Traditional Engineering Method

Shun Yao, Sheng-li Wu, Bo Song, Ming-yin Kou and Heng Zhou

Abstract A carbon loss mathematical model was established based on the Fe–O–C balance and traditional engineering method. The coke carbon loss and carbon loss of coupled direct reduction are calculated based on this model and actual production data of B# blast furnace in Bayisteel. And then, the model is applied to research the effect of coal ratio, burden metallization rate, blast temperature, oxygen enrichment rate and blast humidity on carbon loss of direct reduction and carbon loss of coupled direct reduction. The results show that burden metallization rate and oxygen enrichment rate have significant influences on carbon loss of direct reduction. The burden metallization rate, oxygen enrichment rate and coal ratio have significant influences on carbon loss of coupled direct reduction. In addition, comparing with 4# blast furnace in Baosteel, the high-quality coke can observably reduce carbon loss of coupled direct reduction.

Keywords Blast furnace · Coke · Solution loss · Direct reduction
Carbon loss

Introduction

Metallurgical coke is the most important and inevitable raw material in the blast furnace. Metallurgical coke is a fuel, a reductant, a carburetant, a permeable grid in the blast furnace. And coke is indispensable material in the blast furnace, because it sustains the passages of liquid metal and slag towards the lower part and of high-temperature reducing gas towards the upper part, and it is the only material that remains a solid phase in the high-temperature zones adjacent to the blast furnace raceways. Coke quality affects the gas permeability, liquid permeability of the burden and state of hearth [1].

S. Yao (✉) · S. Wu · B. Song · M. Kou · H. Zhou
School of Metallurgical and Ecological Engineering, University of Science and Technology Beijing, No. 30, Xueyuan Road, Haidian District, Beijing 100083, China
e-mail: yaoshunapp@163.com

© The Minerals, Metals & Materials Society 2019

T. Jiang et al. (eds.), *10th International Symposium on High-Temperature Metallurgical Processing*, The Minerals, Metals & Materials Series,
https://doi.org/10.1007/978-3-030-05955-2_1

Due to the increasing of blast furnace volume, increasing environmental concerns and increasing shortage of coke, there is an increasing trend to substitute coke by injecting pulverized coal in order to decrease reliance on coke. With the coke rate decreasing, the ratio of ore to coke is increased in the burden, so less amount of coke has to maintain sufficient permeability and smooth state of blast furnace. And it is prolonged that residence time of coke in the blast furnace. Coke is deteriorating as the coke moves towards lower zone of blast furnace, as a result of mechanical, thermal and chemical effects. As the residence time of coke is gradually extended, coke suffers more mechanical, thermal and chemical effects than before, leading to the increased deterioration of coke. The reaction of coke gasification by CO_2 and H_2O in the furnace shaft is called solution loss reaction. In the blast furnace, in addition to the decrease of coke strength caused by inevitable mechanical and thermal effects, solution loss reaction is a major factor causes the decrease of coke strength [2–6].

There are extensive researches on the solution loss reaction and deterioration of coke. For example, Wang et al. [5] studied the effect of solution loss reaction on CSR (coke strength after reaction) and coke degradation. The results showed that solution loss reduces CSR and accelerates deterioration of coke, and the reaction rate of solution loss is decreased with increase of ore pre-reduction rate. Haapakangas et al. [6] investigated that the effects of H_2 and H_2O on coke reactivity in a range of temperatures, the effects of H_2 and H_2O on threshold temperature of gasification, and calculating the activation energies of CO_2 and H_2O in the homogenous model and shrinking core model. Xing et al. [7] studied the deterioration behaviour of coke under the conditions of simulated blast furnace temperature and $CO–CO_2–N_2$ atmosphere, and found that both gasification and annealing decrease the coke strength. Compared with annealing at 1673 K, gasification at the same temperature caused larger degradation of all three cokes, and the effect was more significant on the more reactive coke.

Fang et al. [8] studied the effect of different temperatures on high-temperature compressive strength of coke. It was provided that the high-temperature compression strength of coke is in a linear relationship with carbon loss rate of coke under the actual conditions of blast furnace. And the influence of carbon loss rate of coke on the compressive strength is reduced with the increasing temperature. Carbon loss rate of coke is about 30% in stock column. Cui et al. [9] experimentally researched the influence of temperature and solution loss reaction to the high-temperature compressive strength of coke under different gas composition conditions. The results showed that with the increase of temperature, CO_2 concentration and solution loss reaction extent, the high-temperature compressive strength of coke decreases. Guo et al. [10] investigated that the variation of pore structures of coke after gasification with CO_2 and steam. In addition, the influences of solution loss rate, temperature, and pore structure on the high-temperature compressive strength of coke were studied. The results showed that compared with that in CO_2 gasification, the average pore diameter of coke becomes smaller, and the specific surface and the quantity of pore under 100 μm increase in steam gasification. At the same solution loss, the decrease of coke strength was lower for steam gasified coke than CO_2 gasified coke. After reacting with CO_2 and steam, the high-temperature compressive strength of coke will decrease with the increasing of solution loss or temperature. Guo et al. [11] researched the solution loss kinetics behaviour on coke strength after reaction. The results showed that the temperature of gradient reaction brings the most serious degradation to three cokes are different due to different reactivity, and the temperature of the high reactive coke is about 1100 °C which is lower than another two cokes. Liu et al. [12] investigated kinetics of coke gasification with CO_2 by non-isothermal thermogravimetry. The results showed that when the gasification temperature is higher than 1200 K, both of the gasification reaction rate and carbon conversion rate increase significantly. When the gasification temperature reaches about 1500 K, the gasification reaction rate reaches the maximum. NOMURA et al. [13] investigated the strength of catalyst-added highly reactive coke after solution loss reaction. This result showed that the adding method and the types of catalyst have a great influence on the post-reaction strength of coke.

These researches have important guiding significance to ironmaking process, but there is few information about the quantization of coke solution loss in the blast furnace. In this study, a carbon loss calculation model is established based on traditional engineering method and Fe–O–C balance. The coke carbon loss of direct reduction and carbon loss of coupled direct reduction in B# blast furnace of Bayisteel are calculated based on actual production data. Discussions were made on the influences of different operating factors to coke carbon loss.

Modelling

There are some shortcomings in traditional engineering method, such as using the assumed direct reduction degree, unchangeable empirical specific heat capacity and ignoring the effects of blast humidity and pulverized coal. Therefore, the shortcomings of traditional engineering method have been corrected in this work. Firstly, carbon loss of direct reduction is calculated by this model based on conservation of carbon. Then the amount of carbon loss of coupled direct reduction can be calculated by the result of carbon loss of direct reduction. This model assumes the following conditions: all pulverized coal is burned with oxygen; all combustion of carbon and oxygen is incomplete combustion; it is 0.999:0.001 that the distribution ratio of iron to hot metal and slag.

Calculation of Direct Reduction Degree

According to the principles of ironmaking, the formula of direct reduction degree is shown as:

$$r_{\mathrm{d}} = \frac{R_{\mathrm{Fe(DR)}}}{R_{\mathrm{Fe(pig)}}} \quad (1)$$

where r_{d} represents the direct reduction degree; $R_{\mathrm{Fe(DR)}}$ represents the amount of Fe that is generated by direct reduction, kg/t; $R_{\mathrm{Fe(pig)}}$ represents the amount of Fe in hot metal, kg/t.

According to Fe–O–C balance, the amount of Fe element produced by direct reduction could be calculated by carbon loss of direct reduction. The calculating formula is shown as follows:

$$R_{\mathrm{Fe(DR)}} = \frac{56}{12} \times R_{\mathrm{C(DR)}} \quad (2)$$

where $R_{\mathrm{Fe(DR)}}$ is the amount of Fe element produced by direct reduction, kg/t; $R_{\mathrm{C(DR)}}$ is the coke carbon loss of direct reduction, kg/t.

According to the conservation of carbon, the formula of coke carbon loss of direct reduction is shown as:

$$R_{C(DR)} = R_{C(total)} - R_{C([C])} - R_{C(Vad)} - R_{C(XO)} - R_{C(combustion)} - R_{C(H_2O)} - R_{C(dust)} \quad (3)$$

where $R_{\mathrm{C(total)}}$ is the total amount of carbon in coke, kg/t; $R_{\mathrm{C([C])}}$ is the amount of carbon consumed by hot metal carburization, kg/t; $R_{\mathrm{C(Vad)}}$ is the amount of carbon in volatiles of coke, kg/t; $R_{\mathrm{C(XO)}}$ is the amount of carbon consumed by non-ferrous oxide direct reduction, kg/t; $R_{\mathrm{C(combustion)}}$ is the amount of carbon consumed by combustion of coke in raceway, kg/t; $R_{\mathrm{C(H_2O)}}$ is the amount of carbon consumed by oxygen of blast moisture, kg/t; $R_{\mathrm{C(dust)}}$ is the amount of carbon in dust, kg/t.

Calculation of Carbon Loss

In blast furnace, carbon loss of direct reduction reaction can be divided into coupled direct reduction carbon loss reaction and molten direct reduction carbon loss reaction according to different reaction modes. The direct reduction reaction is coupled by solution loss reaction and indirect reduction reaction, and this direct reduction reaction is regarded as coupled direct reduction reaction. Therefore, the amount of carbon consumed by solution loss reaction is same to coupled direct reduction carbon loss reaction. Besides, molten direct reduction carbon loss reaction is caused by direct reduction of coke and molten slag.

In this model, the reduction zone of iron oxide in blast furnace is divided into 3 parts as follows:

(1) Indirect reduction zone. In this zone, the direct reduction reaction does not occur. The indirect reduction reaction equations as follows:

$$FeO(s) + CO(g) = Fe(s) + CO_2(g) \tag{4}$$

$$FeO(s) + H_2(g) = Fe(s) + H_2O(g) \tag{5}$$

(2) Coupled direct reduction zone. In this zone, indirect reduction reaction and solution loss reaction can be coupled into direct reduction reaction. The solution loss reaction equations are shown as:

$$C(s) + CO_2(g) = 2CO(g) \tag{6}$$

$$C(s) + H_2O(g) = CO(g) + H_2(g) \tag{7}$$

Eqn (4) + Eq. (6) and Eq. (5) + Eq. (7) can obtain direct reduction reaction equation as follows:

$$FeO(s) + C(s) = Fe(s) + CO(g) \tag{8}$$

(3) Molten direct reduction zone. In this zone, the indirect reduction reaction does not occur. And coke is the only material that remains a solid phase, the direct reduction reaction occurs between coke and molten slag. The molten direct reduction reaction equation is shown as:

$$FeO(l) + C(s) = Fe(l) + CO(g) \tag{9}$$

The coke consumption in BF is shown in the following ways: iron oxide direct reduction; hot metal carburization; volatile matter of coke volatile; non-ferrous oxide direct reduction; combustion in raceway; coke fines are discharged with gas into dust. The iron oxide direct reduction includes coupled direct reduction and molten direct reduction, and the coke carbon loss includes carbon loss of coupled direct reduction and carbon loss of molten direct reduction.

The following assumptions are made for the calculation of coupled direct reduction carbon loss: the coke carbon loss rate is 30% before coke falling into the raceway [9, 14]; before coke falling into raceway, the coke carbon loss is caused by coupled direct reduction, hot metal carburization, coke volatiles and coke fines that are discharged into dust; the carbon loss of molten direct reduction is ignored before coke drops into raceway; the hot metal carburization can be completed 80% before hot metal droplets drop into raceway [15].

The formula for carbon loss of coupled direct reduction is as follows:

$$R_{C(\mathrm{CDR})} = \lambda_{LC} \times R_{C(\mathrm{total})} - R_{C(\mathrm{Vad})} - R_{C(\mathrm{dust})} - \lambda_{CC} \times R_{C([\mathrm{C}])} \tag{10}$$

where $R_{C(\mathrm{CDR})}$ represents the amount of carbon loss of coupled direct reduction, kg/t; λ_{LC} represents the coke carbon loss ratio, it is 0.3 in assumptions; λ_{CC} represents the completed ratio of hot metal carburization, it is 0.8 in assumptions.

The formula for carbon loss of molten direct reduction is as follows:

$$R_{C(MR)} = R_{C(DR)} - R_{C(CDR)} \quad (11)$$

where $R_{C(MR)}$ represents the amount of carbon loss of molten direct reduction, kg/t.

Results and Analysis

The Influence of Different Factors on Carbon Loss

Based on the carbon loss model, it is investigated that the influence of burden metallization rate, oxygen enrichment rate and blast humidity on coke carbon loss and carbon loss of coupled direct reduction. The calculation results are shown in Fig. 1 and Fig. 2, respectively.

The meanings of these symbols in Fig. 1 are as follows. A represents the effect of coal ratio on carbon loss of direct reduction. It shows that with the increasing 10 kg/t of the coal ratio, the carbon loss of direct reduction is 0.45 kg/t. The effect of blast temperature on carbon loss of direct reduction is shown in B. The carbon loss of direct reduction increases 0.14 kg/t when the blast temperature increasing 50 °C.

The effects of burden metallization rate on carbon loss of direct reduction are shown in C and D. C represents the carbon loss of direct reduction decreases 1.69 kg/t with the increasing 2% of the burden metallization rate by adding scrap iron. It can be seen from D that using the pre-reduced burden to increase burden metallization rate by 2% will reduce the carbon loss of direct reduction by about 7.16 kg/t. The effects of oxygen enrichment rate on carbon loss of direct reduction are shown in E and F. E represents that an increment in oxygen enrichment rate of

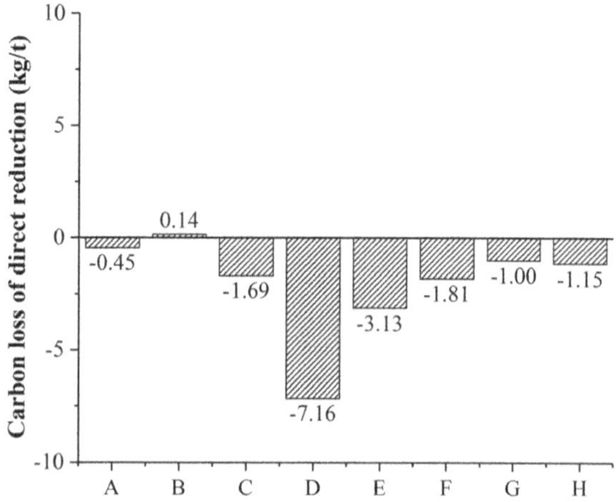

Fig. 1 Influence of different factors on carbon loss of direct reduction

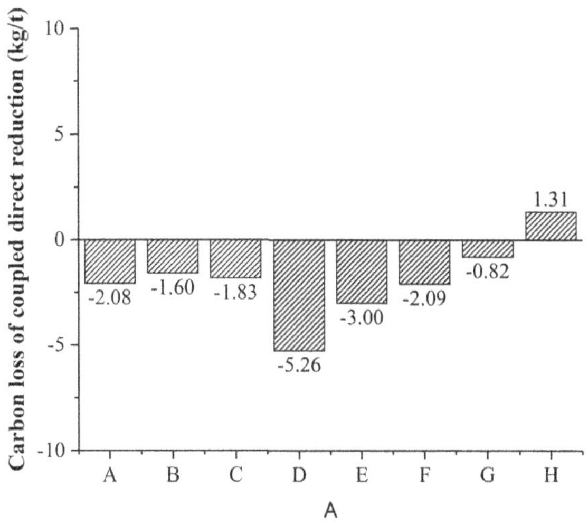

Fig. 2 Influence of different factors on carbon loss of coupled direct reduction

1% under the condition that blast volume is constant will reduce the carbon loss of direct reduction by about 3.13 kg/t. F represents that an increment in oxygen enrichment rate of 1% under the condition that bosh gas volume is constant will reduce the carbon loss of direct reduction by about 1.81 kg/t. The effects of blast humidity on carbon loss of direct reduction are shown in G and H. G represents that in order to ensure sufficient heat in the hearth, the blast temperature is increased during the process of increasing the blast humidity. It can be seen from G that as the blast humidity increases 5 g/m^3, the carbon loss of direct reduction is decreased by 1.15 kg/t. H represents that in order to ensure sufficient heat in the hearth, the amount of pulverized coal injection is reduced during the process of increasing the blast humidity. It can be seen from H that with the blast humidity increasing 5 g/m^3, the carbon loss of direct reduction is reduced by 1.15 kg/t.

The meanings of these symbols in Fig. 2 are as follows. A represents the effect of coal ratio on carbon loss of coupled direct reduction. It shows that with the increasing 10 kg/t of the coal ratio, the carbon loss of coupled direct reduction is 2.08 kg/t. The effect of blast temperature on carbon loss of coupled direct reduction is shown in B. The carbon loss of coupled direct reduction decreases 1.6 kg/t when the blast temperature increasing 50 °C. The effects of burden metallization rate on carbon loss of coupled direct reduction are shown in C and D. C represents the carbon loss of coupled direct reduction decreases 1.83 kg/t with the increasing 2% of the burden metallization rate by adding scrap iron. It can be seen from D that using the pre-reduced burden to increase burden metallization rate by 2% will reduce the carbon loss of coupled direct reduction by about 5.26 kg/t. The effects of oxygen enrichment rate on carbon loss of coupled direct reduction are shown in E and F. E represents that an increment in oxygen enrichment rate of 1% under the condition that blast volume is constant will reduce the carbon loss of coupled direct reduction by about 3.00 kg/t. F represents that an increment in oxygen enrichment rate of 1% under the condition that bosh gas volume is constant will reduce the carbon loss of coupled direct reduction by about 2.09 kg/t. The effects of blast humidity on carbon loss of coupled direct reduction are shown in G and H. G represents that in order to ensure sufficient heat in the hearth, the blast temperature is increased during the process of increasing the blast humidity. It can be seen from G that as the blast humidity increases 5 g/m^3, the carbon loss of coupled direct reduction is decreased by 0.82 kg/t. H represents that in order to ensure sufficient heat in the hearth, the amount of pulverized coal injection is reduced during the process of increasing the blast humidity. It can be seen from H that with the blast humidity increasing 5 g/m^3, the carbon loss of coupled direct reduction is increased by 1.31 kg/t.

The Influence of Coke Quality on Carbon Loss

The coke quality of B# BF in Bayisteel and 4# BF in Baosteel are shown in Table 1.

As shown in Fig. 3, the amounts of carbon loss in 4# BF are smaller than those in B# BF. And the carbon loss of coupled direct reduction in 4# BF is significantly smaller than that in B# BF. Accordingly, improving the quality of coke can reduce carbon loss of direct reduction and carbon loss of coupled direct reduction.

Table 1 Comparison of coke quality

	Mad	Fcad	Ad	Vdaf	CRI	CSR
B#	0.05	87.82	11.53	1.21	36.86	43.43
4#	0	86.23	12.74	1.04	24.00	70.80

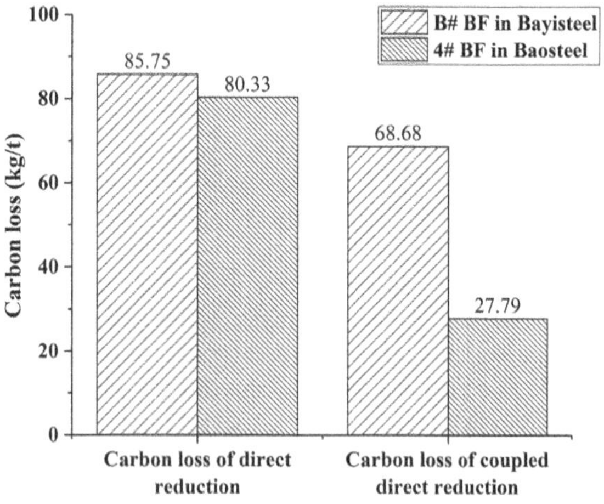

Fig. 3 Comparisons of carbon loss

Measures to Reduce Carbon Loss of Coupled Direct Reduction

According to the calculation results, the CRI and CSR significantly affect the carbon loss of coupled direct reduction. And, using low CRI and high CSR coke can effectively reduce the carbon loss of coupled direct reduction. In actual production, coke quality has great influence on the gas permeability, liquid permeability of the burden and state of hearth, and coke quality directly affects the smooth operation. Therefore, the carbon loss of coupled direct reduction can show the coke quality, and carbon loss of coupled direct reduction characterize the state of blast furnace to a certain extent.

From the calculation results of the model, it can be known that the method of increasing the burden metallization rate by using the pre-reduced burden can significantly reduce the carbon loss of coupled direct reduction, because this method can greatly reduce the amount of oxygen element brought into the blast furnace by burden, thereby reducing carbon loss of coupled direct reduction. Although the pre-reduction of burden consumes fuel, the total energy consumption may not be increased by using non-coking coal in the production of pre-reduced burden [16]. Using scrap iron to increase the burden metallization rate can decrease carbon loss of coupled direct reduction, but the reduction extent is relatively small. Although the scrap iron does not participate in the oxidation-reduction reaction in the blast furnace, the melting of the scrap iron will absorb a lot of heat, so the amount of scrap iron should not be too large. In terms of increasing the oxygen enrichment rate, the method of constant blast volume has a lower carbon loss of coupled direct reduction than the method of constant bosh gas volume. Because the method of increasing the oxygen enrichment rate by constant blast volume can inject more coal, reduce more coke and produce more bosh gas. Therefore, the specific surface area of the reaction between coke and carbon dioxide is reduced, resulting in a deterioration of reaction conditions between C and CO_2, so that the carbon loss of coupled direct reduction is relatively small. For the method of increasing the oxygen enrichment rate by constant blast volume, if the oxygen enrichment rate is too high, the pressure loss of the blast furnace will be too large, and then the blast furnace operation cannot be stable, so the oxygen enrichment rate should not be too high. The method of increasing the oxygen enrichment rate by using a constant bosh gas volume may cause the theoretical combustion temperature to be too high, resulting in a decrease in the life of tuyere. The method of increasing the blast temperature to increase the blast humidity can slightly reduce the carbon loss of coupled direct reduction. Moreover, in actual production, the improvement of blast temperature is difficult. The method of reducing the amount of coal injection to increase the blast humidity will increase the carbon loss of coupled direct reduction. Therefore, the dehumidified blast technology could be adopted to reduce carbon loss of coupled direct reduction.

Conclusions

A carbon loss mathematical model was established based on the Fe–O–C balance and traditional engineering method. The coke carbon loss and carbon loss of coupled direct reduction are calculated based on this model and actual production data of B# blast furnace in Bayisteel. And then, the results show that the carbon loss of direct reduction is 85.75 kg/t and carbon loss of coupled direct reduction is 68.68 kg/t. For carbon loss of direct reduction, the burden metallization rate (using pre-reduced burden) affects the most, followed by oxygen enrichment rate (constant blast volume) and oxygen enrichment rate (constant bosh gas volume). The burden metallization rate (using pre-reduced burden) affects the most, followed by oxygen enrichment rate (constant blast volume), oxygen enrichment rate (constant bosh gas volume) and coal ratio on carbon loss of coupled direct reduction. In addition, comparing with 4# blast furnace in Baosteel, the high-quality coke can observably reduce carbon loss of direct reduction and carbon loss of coupled direct reduction.

References

1. Xu W, Wu X, Chen J et al (2003) Influence of high PCR operation on coke degradation in blast furnace. Iron Steel 38(3):4–7,24 (in Chinese)
2. Cao J, Zhang Q, Wu X (2003) Size decrease of baosteel coke in blast furnace. Baosteel Technol (1):39–43 (in Chinese)
3. Ren R, Fang J, Zhang J, Zhang H (2007) Strength distribution of coke in blast furnace. J Hebei Inst Technol 29(1):21–26 (in Chinese)
4. Liu P, Fang J, Guo L, Gong R (2006) Research on the distribution of coke decarbonization. Henan Metall 14(1):9–16 (in Chinese)
5. Wang P, Li J, Zhou L (2005) Fundamental study on coke degradation in BF with oxygen enriched blast and high hydrogen atmosphere. J Iron Steel Res 12(3):1–4, 10
6. Juho H, Hannu S, Mikko I et al (2016) Coke reactivity in simulated blast furnace shaft conditions. Metall Mater Trans B 47(4):2357–2370
7. Xing Xing, Rogers Harold, Guangqing Zhang et al (2016) Coke degradation under simulated blast furnace conditions. ISIJ Int 56(5):786–793
8. Fang J, Wang X, Guo L, Shao J (2006) Research on high temperature compressive strength of coke. Iron Steel 41(5):20–23
9. Cui P, Wang W, Wang Z (2014) Study on high temperature compressive strength of coke in the presence of CO_2. Fuel Chem Process 45(4):1–7
10. Guo W, Wang J, She X, Xue Q, Guo Z (2015) Pore structure and high temperature compressive strength of gasified coke with CO_2 and steam. J Fuel Chem Technol 43(6):654–662
11. Guo R, Wang Q, Zhang S (2012) Influence of solution loss reaction on post reaction strength of coke. Coal Convers 35:12–16
12. Liu Z, Wang Q, Zou Z, Tan G (2010) Non-isothermal thermogravimetric investigation on kinetics of coke gasification with CO_2. Res Iron Steel 38(6):1–3, 43 (in Chinese)
13. Nomura S, Naito M, Yamaguchi K (2007) Post-reaction strength of catalyst-added highly reactive coke. ISIJ Int 47(6):831–839

14. Goleczka J, Tucker J (1985) Coke quality and its assessment in the CRE laboratory. Ironmaking Conf Proc 44:217–232
15. Zhang Z, Zhang J, Jiao K, Liu Z (2015) Research progress of iron carburization in blast furnace. TMS (The Minerals, Metals & Materials Society), pp 627–633
16. Liu Z, Wang J (2010) Discussion on measures and potentials of coke-rate reduction of blast furnace. J Hunan Univ Technol 24(5):1–4 (in Chinese)

Study on Alkali Circulation Process and Its Influence on Coke Ratio in Blast Furnace

Haokun Li, Yijie Wang, Kexin Jiao, Jianliang Zhang, Rong Zhu and Hanjie Guo

Abstract Alkali metal can be circulated and accumulated in blast furnace. In this work, the distribution of alkali in blast furnace is determined by the calculation of alkali balance. Combined with the cyclic of alkali in the blast furnace, the influence of alkali on coke ratio is calculated by using blast furnace operation line. The results show that the circulation of alkali in the blast furnace increases the coke ratio.

Keywords Alkali · Circulation · Thermodynamic behaviour · Blast furnace operation line · Coke ratio

Introduction

Blast furnace ironmaking plays a very important role in the steel production process. The energy consumption of ironmaking accounts for more than 70% of the total energy consumption of steel production, and the cost of ironmaking accounts for about 75% of the cost of steel production [1–3]. To this end, steel companies reduce cost pressures by purchasing low-cost raw materials. However, the content of low cost iron ore alkali is higher, so that the alkali load is increased.

In the blast furnace ironmaking process, the circulation of alkali in the blast furnace is extremely harmful to the blast furnace production. It can not only reduce the softening temperature of the ore, cause abnormal expansion of the pellets, but also strengthen the gasification reaction capacity of the coke, so that the strength of the coke reaction is drastically reduced and powdering occurs, resulting in poor gas permeability of the column. At the same time, the alkali adheres to the furnace wall, which can cause scaffolding in blast furnace and directly destroy the brick wall refractory material [4–9]. Therefore, steel companies reduce the hazards of alkali to the blast furnace by controlling the load of alkali.

H. Li · Y. Wang · K. Jiao · J. Zhang · R. Zhu · H. Guo (✉)
School of Metallurgical and Ecological Engineering, University of Science and Technology Beijing, Beijing, China
e-mail: guohanjie@ustb.edu.cn

© The Minerals, Metals & Materials Society 2019

T. Jiang et al. (eds.), *10th International Symposium on High-Temperature Metallurgical Processing*, The Minerals, Metals & Materials Series,
https://doi.org/10.1007/978-3-030-05955-2_2

However, the effect of alkali on the blast furnace coke ratio has not been clarified, and it is necessary to quantitatively study the influence of harmful elements on the coke ratio of blast furnace. In this paper, the distribution of alkali in blast furnace is determined by the calculation of alkali balance. Combined with the cyclic of alkali in the blast furnace, the influence of alkali on coke ratio is calculated by using blast furnace operation line.

Alkali Circulation Process in Blast Furnace

Calculation of Alkali Balance in Blast Furnace

At present, the traditional research process on the circulation and distribution of alkali in blast furnace is initially based on the chemical analysis of alkalis in input–output items, then proceeding with balance calculations according to the quantity of input consumption and output production.

In this paper, the samples, which include sinter, pellet, coke, pulverized coal and pig iron, slag and dust, were taken from a commercial blast furnace. The balance of input and output items for production of 1t pig iron are listed in Table 1. The alkalis content of input and output for blast furnace are listed in Table 2.

It is indicated that sinter, which containing 0.2% alkalis, dominates alkalis content for raw materials, followed by pellets, coke and coal. For output items, alkalis mainly concentrate on slag and dust.

Table 1 Balance sheet of input and output of a blast furnace for production of 1t pig iron

	Items	Quantity (kg/tHM)
Input	Sinter	1080
	Pellet	500
	Coke	350
	Coal	168
Output	Slag	285
	Dust	17
	Pig iron	1000

Table 2 Content of alkali of the samples

Items	K_2O (%)	Na_2O (%)
Sinter	0.08	0.12
Pellet	0.02	0.02
coke	0.01	0.02
Coal	0.01	0.02
Slag	0.37	0.49
Dust	0.18	0.12
Pig iron	0	0

Table 3 Quantity of alkalis in input–output items for a blast furnace

	Items	K_2O (kg/tHM)	Na_2O (kg/tHM)	$K_2O + Na_2O$ (kg/tHM)
Input	Sinter	0.86	1.30	2.16
	Pellet	0.10	0.10	0.20
	coke	0.04	0.07	0.11
	Coal	0.02	0.03	0.05
	Total	1.02	1.5	2.52
Output	Slag	1.05	1.40	2.45
	Dust	0.03	0.02	0.05
	Pig iron	0	0	0
	Total	1.08	1.42	2.50

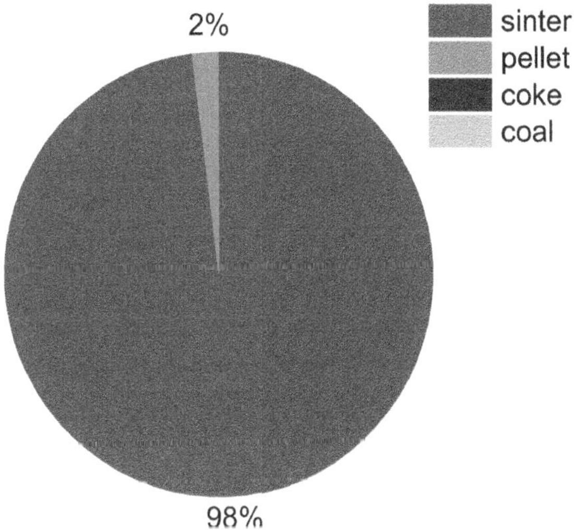

Fig. 1 Proportion of alkalis in input items for blast furnace

Based on the results of Tables 1 and 2, the quantity and proportion of alkalis in input–output items are listed in Table 3, Figs. 1 and 2 by simple calculation.

It is indicated that the total input of alkalis for a blast furnace reaching 2.52 kg/tHM. Sinter, pellet, coke and coal contribute 85.7, 7.9, 4.4 and 2.0% of alkalis, respectively. The output of alkalis, which discharging from blast furnace in the form of slag and dust, total 2.5 kg/tHM and 98% of which concentrates in slag.

The Thermodynamic Behaviour of Alkalis in Blast Furnace

Alkalis mainly exist in the form of complex silicates such as mica feldspar and silicate, and these alkali minerals are eventually reduced in the form of alkali silicate [10, 11]. The chemical equation for the reduction of alkali silicates is as follows:

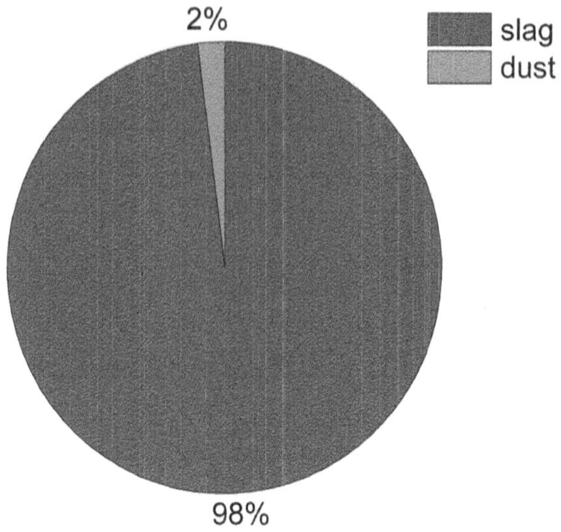

Fig. 2 Proportion of alkalis in output items for blast furnace

$$K_2SiO_3 + CO = 2K(g) + SiO_2 + CO_2(g) \quad (1)$$

$$Na_2SiO_3 + CO = 2Na(g) + SiO_2 + CO_2(g) \quad (2)$$

The alkali, which descending to bosh and hearth area with temperature being 1300–1600 °C and higher than that of alkali boiling point, always exists in the form of elemental vapour. However, when the alkali vapour, which ascending upward with blast furnace gas and reaching over belly area, will react with SiO_2 and CO_2 and produce alkali silicate as follows:

$$2K(g) + SiO_2 + CO_2 = K_2SiO_3 + CO(g) \quad (3)$$

$$2Na(g) + SiO_2 + CO_2 = Na_2SiO_3 + CO(g) \quad (4)$$

The newly formed alkali silicate, which descending to bosh and hearth area, will again be reduced to elemental vapour, finally, a micro-cycle between alkali silicate and elemental vapour will be formed between bosh and hearth area.

C and N_2, which sharing high partial pressure in hearth area, will react with alkali vapour and result in the formation of cyanide in bosh and hearth area, the reaction equation is as follows:

$$2K(g) + 2C + N_2 = 2KCN \quad (5)$$

$$2Na(g) + 2C + N_2 = 2NaCN \quad (6)$$

Since negligible cyanide is produced, alkali mainly exists in the form of simple vapour, and a few exist in the form of cyanide vapour.

$$2K(g) + CO_2 = K_2O + CO \tag{7}$$

$$2Na(g) + CO_2 = Na_2O + CO \tag{8}$$

The elemental vapour, which combined with alkali cyanide ascending to bosh and belly area, will be oxidized and produce alkali oxides or alkali carbonates, the reaction equation is as follows:

$$2K(g) + CO_2 = K_2O + CO \tag{9}$$

$$2Na(g) + CO_2 = Na_2O + CO \tag{10}$$

Under the actual production conditions of blast furnaces, it is difficult to form alkali oxides at the lower part of the shaft, belly and bosh area, while a small amount of alkali oxides can be formed in the middle and upper part of the shaft. Most alkali elemental vapours and alkali cyanide are oxidized by CO_2 in the furnace gas to form alkali carbonate in the middle and upper part of the blast furnace. The reaction equation is as follows:

$$2K(g) + 2CO_2 = K_2CO_3 + CO \tag{11}$$

$$2Na(g) + 2CO_2 = Na_2CO_3 + CO \tag{12}$$

$$2KCN + 4CO_2 = K_2CO_3 + 5CO + N_2 \tag{13}$$

$$2NaCN + 4CO_2 = Na_2CO_3 + 5CO + N_2 \tag{14}$$

The alkali carbonates formed in the middle and upper part of the furnace shaft will deposit on the furnace material and descend along with the material. During the heating process, they may be reduced by CO to form elemental alkali again. The calculation shows that alkali carbonates can be reduced to alkali vapour before reaching to the soft melting zone. Therefore, alkali carbonates form cyclic enrichment in the upper part of blast furnace.

Effect of Alkali on Coke Ratio

Operating Line of Blast Furnace

The operation line of blast furnace is to study the change and transfer of Fe–O–C elements in blast furnace based on carbon balance, oxygen balance and heat balance. The operating line of blast furnace reveals the essence and law of blast furnace smelting process and can be used to discuss the effect of alkalis on coke ratio. In this paper, the mathematical relation between alkali and coke ratio in the process of blast furnace smelting is calculated and analysed.

Fig. 3 Operating line of blast furnace

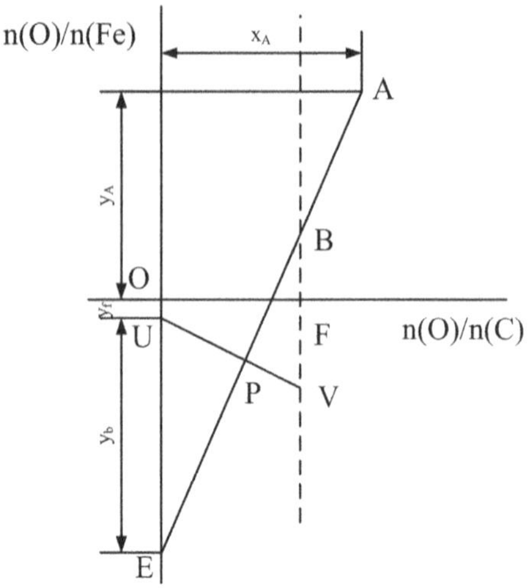

As shown in Fig. 3, the x-coordinate is the ratio of oxygen atom to carbon atom, and the y-coordinate is the ratio of oxygen atom to iron atom. AE represents the operating line of blast furnace under certain operating conditions. The coordinates of points A and E in the figure can be calculated according to the data related to material balance and regional heat balance in the blast furnace, such as raw fuel composition, top gas composition, pig iron and slag composition, so as to determine the operating line equation of the blast furnace.

$$y_A = \frac{w(Fe_2O_3) \times 3/160 + w(FeO) \times 1/72}{w(TFe)/56} \quad (15)$$

$$x_A = 1 + \frac{\varphi(CO_2)}{\varphi(CO_2) + \varphi(CO)} \quad (16)$$

$$y_E = -(y_f + y_b) \quad (17)$$

where $w(Fe_2O_3)$ is the mass fraction of Fe_2O_3 in the ore; $w(FeO)$ is the mass fraction of FeO in the ore; $w(TFe)$ is the grade of iron ore; $\varphi(CO)$ is the CO contents of the top gas; $\varphi(CO_2)$ is the CO_2 contents of the top gas; y_f is the oxygen combined with trace elements; y_b is the oxygen consumed by carbon combustion.

In addition, the operational line AE should pass through the point P determined by the heat balance in the high-temperature zone of the blast furnace. The thermal equilibrium in the high-temperature zone is:

$$y_b \cdot q_b = y_d \cdot q_d + Q \quad (18)$$

where y_b is the degree of direct reduction; q_d is the heat consumed by reducing FeO; Q is other effective heat consumption; q_b is the teat generated by carbon combustion.

According to the heat balance equation of the high-temperature zone, the coordinates of point P can be determined:

$$x_P = \frac{q_d}{q_d + q_b} \tag{19}$$

$$y_P = y_f + x_P\left(\frac{Q}{q_d} - y_f\right) \tag{20}$$

The cyclic accumulation of alkalis in the blast furnace leads to changes in the smelting process of the blast furnace and changes in the operating line of the blast furnace.

$$x_{A'} = 1 + \frac{V(CO_2) - \frac{X_i Y_i}{M_i} V_m}{V(CO_2) + V(CO)} \tag{21}$$

$$y_{P'} = y_f + x_P\left(\frac{Q + \frac{X_i \cdot Y_i \cdot \Delta H_i^{\ominus}/M_i}{1000w[Fe]/56}}{q_d} - y_f\right) \tag{22}$$

where X_i is the load of alkali; Y_i is the accumulation times of alkali; M_i is the molar mass of specie i.

The coke ratio can be calculated according to the operating line inclination:

$$K = \left(\tan\alpha \times 12 \times \frac{1000w[Fe]}{56} + 1000w[C] - M \cdot w(C)_M\right) \cdot \frac{1}{w(C)_K} \tag{23}$$

where M is the coal injection rate; $w(C)_K$ is the carbon content of coke; $w(C)_M$ is the carbon content of coal; $w[C]$ is the carbon content of hot metal.

Operation Line Calculation Results

According to the above calculation method, the production data of a commercial blast furnace is selected for calculation. Some production data of blast furnace is listed in Table 4.

Figure 4 shows the calculation results of the blast furnace operation line. From the diagram, the coke ratio increased significantly with the increase of alkali load. When the load of K rises from 1 to 2.5, the coke ratio rises from 334 to 379 kg/t; when the load of Na rises from 1 to 2.5 kg/t, the coke ratio rises from 311 to 384 kg/t. There are two reasons for the increase of coke ratio caused by circulation

Table 4 Production data of a blast furnace

	Parameter	Result
Production	Productivity, t/m^3 d	2.40
Fuel	Coke ratio, kg/t	350.51
	PCI ratio, kg/t	167.57
	Coke's carbon content, wt%	87.19
	Fuel's carbon content, wt%	69.90
Raw material	TFe, wt%	57.52
	FeO, wt%	6.33
	Fe_2O_3, wt%	71.87
Pig Iron	Pig iron temperature, °C	1516
	[C], %	4.76
	[Si], %	0.48
	[Mn], %	0.037
	[P], %	0.07
	[S], %	0.024
Slag	CaO, %	42.41
	SiO_2, %	35.52
	Al_2O_3, %	11.66
	MgO, %	7.04
	S, %	1.03
	Slag ratio, kg/t	285.57
Blast parameter	Blast volume, m^3/min	6110
	Blast temperature, °C	1265
	Blast humidity, g/m^3	9.56
	Oxygen enrichment rate, %	5.62
Top gas	ηCO, %	48.60
	CO_2, %	23.55
	CO, %	24.94

of alkali in blast furnace: (1) thermodynamic behaviour of alkali in blast furnace shows that alkali circulation and enrichment in the blast furnace leads to a decrease in gas utilization rate, resulting in an increase in coke ratio; (2) the reduction of alkali metals in the blast furnace requires heat consumption, resulting in an increase in coke ratio. It is noteworthy that the influence of Na on coke ratio is greater than that of K. For each 0.5 kg/t increase in the load of K, the coke ratio is increased by approximately 15 kg/t. When the load of Na increases by 0.5 kg/t, the coke ratio increases by about 25 kg/t. Therefore, in actual production, it is necessary to control the load of K and Na separately, so as to reduce the coke ratio of the blast furnace.

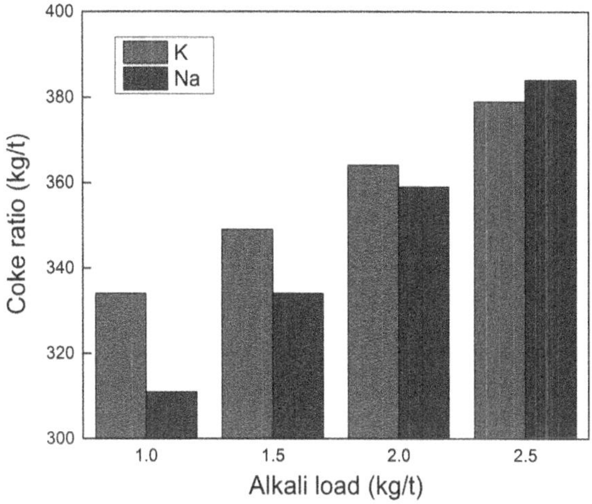

Fig. 4 Effect of alkali on coke ratio of blast furnace

Conclusions

(1) The blast furnace alkali metal balance was calculated. The results show that the total input of alkalis for a blast furnace is 2.52 kg/tHM, and the output of alkalis is 2.5 kg/tHM. Therefore, every ton of molten iron produced will accumulate 0.02 kg of alkali in the blast furnace.
(2) Alkali metal can be circulated and accumulated in blast furnace. In the upper part of blast furnace, the alkali circulates in the form of carbonate; in the lower part of the blast furnace, alkali metal circulates in the form of silicate.
(3) The circulation of alkali in the blast furnace increases the coke ratio. For each 0.5 kg/t increase in the load of K, the coke ratio is increased by approximately 15 kg/t. When the load of Na increases 0.5 kg/t, the coke ratio increases by about 25 kg/t.

References

1. Van Laar R, Van Stein Callenfels E, Geerdes M (2003) Blast furnace hearth management for safe and long campaigns. Iron Steelmaker 30(8):123–130
2. Jiao KX, Zhang JL, Hou QF, et al (2017) Analysis of the relationship between productivity and hearth wall temperature of a commercial blast furnace and model prediction. Steel Res Int
3. Zhang FM (2013) Design and operation control for long campaign life of blast furnaces. J Iron Steel Res (Int) 09:53–60
4. Qi CL, Zhang JL, Lin CC, Yin J, Chang J and Lu WJ (2011) Erosion of carbon bricks at the hearth sidewall by harmful elements in a blast furnace. J Univ Sci Technol Beijing 33(4):491–498
5. Li KJ, Zhang JL, Barati M et al (2015) Influence of alkaline (Na, K) vapors on carbon and mineral behavior in blast furnace cokes. Fuel:202–213

6. Van Niekerk W, Dippenaar R, Kotze D (1986) The influence of potassium on the reactivity and strength of coke, with special reference to the role of coke ash. J S Afr Inst Min Metall 86 (1):25–29
7. Kapteijn F, Jurriaans J, Moulijn JA (1983) Formation of intercalate-like structures by heat treatment of K_2CO_3-carbon in an inert atmosphere. Fuel 62(2):249–251
8. Yang YY, Gao ZK (1983) Study on the mechanism of scaffolding caused by alkali cycling and accumulation and promoted by the presence of fluoring in the blast furnaces of Baotou Iron and Steel Co. and measures to prevent scab-formation. Iron Steel (12):36–42
9. Li KJ, Zhang JL, Liu ZJ et al (2014) Gasification of graphite and coke in carbon-carbon dioxide-sodium or potassium carbonates system. Ind Eng Chem Res 53(14):5737–5748
10. Bai L, Zhang JL, Guo H, Zhang XS, Zhang XD (2008) Circulation and enrichment of alkali metal in blast furnace. J Iron Steel Res Int 20(9):5–8
11. Zhao HB, Cheng SS (2012) New cognition on coke degradation by potassium and sodium in alkali enriched regions and quantificational control model for BF. J Univ Sci Technol Beijing 34(3):333–341

Fuzzy Gray Relational Analysis for Electromagnetic Parameters of Induction Heating Process

Pei Fu, Ping Zhou, Tianyang Zhao, Zhuo Chen and Chenn Q. Zhou

Abstract Induction heating is widely used for heat treatment, providing fast and precise heating effect. A wide range of electromagnetic parameters, such as the structure parameters of coil and the electrical operating parameters, have significant influences on the temperature distribution of the workpiece in induction heating process, which is important for the subsequent heat treatment process. In this work, the main factors including exciting current, power frequency, coil inner diameter and coil spacing are chosen to be studied by numerical simulation. Meanwhile the single-factor experimental design and the Fuzzy Gray Relational Analysis are combined to investigate the impacts of the four factors on the temperature distribution, providing great reference value for further research of induction heating. The result shows that, for axial temperature difference of specimen, the impacts of the four factors are ranked from the most important to the least important as coil inner diameter, coil spacing, power frequency and exciting current. While for radial temperature difference, the ranking list of importance becomes exciting current, power frequency, coil inner diameter and coil spacing.

Keywords Heat treatment · Induction heating · Temperature distribution · Numerical simulation · Fuzzy gray relational analysis

P. Fu · P. Zhou (✉) · T. Zhao · Z. Chen
School of Energy Science and Engineering, Central South University, Changsha 410083, People's Republic of China
e-mail: zhoup@csu.edu.cn

C. Q. Zhou
Center for Innovation Through Visualization and Simulation, Purdue University Calumet, Hammond, IN 46323, USA
e-mail: chenn.q.zhou@gmail.com

© The Minerals, Metals & Materials Society 2019

T. Jiang et al. (eds.), *10th International Symposium on High-Temperature Metallurgical Processing*, The Minerals, Metals & Materials Series,
https://doi.org/10.1007/978-3-030-05955-2_3

Introduction

Induction heating is one of the preferred heating technologies in industrial, domestic and medical applications due to its advantages compared with other classical heating techniques such as flame heating, resistance heating, and traditional ovens or furnaces [1]. It provides contactless, controlled, efficient and clean heating of conductive materials, therefore, it is currently used in smelting, forging and heat treatment, promoting the development of advanced material and metallurgy. During the heat treatment of superalloy, the initial temperature distribution is absolutely necessary, and the temperature distribution and temperature gradient play vital roles for microstructure and mechanical properties [2]. However, the limited current penetration depth and various influence factors lead to nonuniform temperature distribution of workpiece, making it difficult to control the induction heating process and the subsequent heat treatment process. Meanwhile, it is impossible to get the overall temperature distribution of workpiece because of the limited means of measurement. The numerical simulation technique is an efficient way to obtain the temperature distribution which is difficult through the traditional measurement method, making it convenient for the parameter control and optimization [3].

Induction heating is a complex process including electromagnetic, thermal and metallurgic phenomena. In this process, an alternating electric current induces electromagnetic field, which in turn induces eddy currents in the workpiece [4]. The induced eddy currents release energy in the form of heat, which is then distributed throughout the workpiece. It is difficult to solve the electromagnetic thermal coupling problem during induction heating. Based on lumped parameter method, models like equivalent circuit model of power supply and empirical formula of experimental modification are put forward to solve simple induction heating problems [5, 6]. However, the oversimplified models and excessive auxiliary parameters cause computational errors and less understanding of induction heating. Then, distributed parameter method like finite element analysis is adopted to analyse the induction heating [7, 8].

Besides the coupling problem, it is difficult to choose the right shape and position of the induction coil and adjust the electric current properties to attain a desired temperature profile in the workpiece. Huang [9] studied the induction heating process of steel tube and compared the temperature field caused by different power frequency, finally an optimum frequency combination was selected. Shokouhmand [10] analysed the process of moving induction heat treatment, and the effect of velocity, initial position of inductor and inner to outer radius ratio on temperature distribution are investigated. Hammi [11] developed a 2D axisymmetric model of induction heating process to study the parametric and sensitivity effect of machine parameters including current density, heating time, frequency and some dimensional factors. To sum up, many investigations have been done to study the influencing factors in order to obtain an ideal induction heating effect. However, few of them focused on the importance extent of the influencing factors. As there is a wide range of influence factors, it is crucial to compare the different influencing factors and find the main factors to make a better understanding and more precise control of induction heating.

Be different from previous work, this work aims to determine the primary factors that affect the temperature distribution in induction heating process through the numerical simulation. Previously mentioned, the structure parameters of coil (inner and outside diameter, number of turns and coil spacing) and the electrical operating parameters (exciting current, power frequency and heating time) all affect the temperature distribution of workpiece obviously, which are too many to make a comprehensive investigation. In this work, main factors including exciting current, power frequency, coil inner diameter and coil spacing are chosen to be investigated.

In addition, Gray Relational Analysis (GRA) is a useful method to evaluate the impacts of various factors without knowing the mathematical relationship among the investigated factors and the results. Meanwhile the single-factor experimental design and Fuzzy Gray Relational Analysis (FGRA) are combined to investigate the impacts of various factors on the temperature distribution, providing great reference value for further research of induction heating.

Evaluating Method

Single-Factor Experimental Design

The Single-factor experimental design is a popular method to deal with the test, including multiple factors and levels. It has been widely applied to many fields since it is simple and distinct [12]. In this work, four factors are investigated, including exciting current, power frequency, coil inner diameter and coil spacing.

Fuzzy Gray Relational Analysis

Gray relational analysis (GRA) is an effective statistical method for measuring the degree of approximation among the sequences using a gray relational grade. It was developed by Deng [13] and has been successfully applied in other fields [14–17]. In this work, this method is improved and employed to evaluate the impact of various factors on the temperature uniformity of the induction heating process. The steps are as follows:

Step 1: List the reference matrix and comparison matrix. The reference matrix y and the comparison matrix x are expressed as follows:

$$y = [y(1)y(2)\ldots y(n)] \tag{1}$$

$$x = \begin{bmatrix} x_1 \\ x_2 \\ \ldots \\ x_m \end{bmatrix} = \begin{bmatrix} x_1(1) & x_1(2) & \cdots & x_1(n) \\ x_2(1) & x_2(2) & \cdots & x_2(n) \\ \vdots & \vdots & \vdots & \vdots \\ x_m(1) & x_m(2) & \cdots & x_m(n) \end{bmatrix} \tag{2}$$

where m is the number of investigated factors and n is number of investigated conditions.

Step 2: Nondimensionalize original matrix. The original matrix should be nondimensionalized because of the different dimensions of the investigated factors and the reference variable by the following equation:

$$x_i(k)' = \frac{x_i(k) - \min x_i(k)}{max x_i(k) - \min x_i(k)} \tag{3}$$

Step 3: Calculate the cosine value of fuzzy membership. The similarity of the two factors is related to the include dangle cosine of the two factors, which is expressed as follows:

$$r_1 = \frac{\sum_{k=1}^{n} y(k)x(k)}{\sqrt{\sum_{k=1}^{n} y(k)^2}\sqrt{\sum_{k=1}^{n} x(k)^2}} \tag{4}$$

Step 4: Calculate the gray relational grade.

$$\xi_i(k) = \frac{\Delta min + l\Delta \max}{\Delta k + l\Delta \max} \tag{5}$$

where l is the resolution coefficient, which represents the weight of the maximum absolute difference and is used to satisfy the integrity and anti-interference of the relational grade. Absolute difference $\bar{\Delta}$ is determined as follows:

$$\bar{\Delta} = \frac{1}{m \cdot n} \sum_{j=1}^{m} \sum_{k=1}^{n} |y_t(k) - x_{tj}(k)| \tag{6}$$

The resolution coefficient is determined as follows:

$$1 \in \begin{cases} [c, 1.5c] & c < 1/3 \\ (1.5c, 2c] & c \geq 1/3 \end{cases} \tag{7}$$

where ratio $c = \bar{\Delta}/\Delta\max$. If $c < 1/3,\ 1 = 1.5c$, else if $c \geq 1/3,\ 1 = 1.75c$.

Step 5: Calculate the Euclidean gray relational grade. The difference between the reference matrix and the comparison matrix can be evaluated by the Euclidean distance, which improves the evaluation accuracy. The Euclidean gray relational grade r_2 is expressed as follows:

$$r_2 = 1 - 2\sqrt{\sum_{k=1}^{n} [w(1 - \xi_i(k))]^2} \quad (8)$$

Where w is weight factor of different factors in the comparison matrix.

Step 6: Calculate the fuzzy gray relational grade. The fuzzy relational grade is a combination of the fuzzy membership coefficient and the Euclidean gray relational grade, which is calculated by the following formula:

$$r = \sqrt{\frac{r_1^2 + r_2^2}{2}} \quad (9)$$

Step 7: Rank. Based on the value of the fuzzy gray relational grade, impacts of the investigated factors are ranked.

Investigated Model

Geometric Model

In this work, the computational domain includes specimen, insulated cotton, inductance coil, cooling water and ambient air. The model is simplified in the following ways:

a. The model can be simplified to be axisymmetric since the cylindrical specimen and the inductance coil are coaxial;
b. The 3D model can be simplified to be 2D as the materials are all isotropic.

The schematic diagram of the induction heating process is presented in Fig. 1. The specimen is 12.5 mm wide and 100 mm long, while the air section is 400 mm wide and 800 mm long. The material of the specimen is FGH96 and the physical properties are determined by former experiment. Point A is set as temperature control point, which is the mid-point of the side surface of specimen. While point B is set as temperature measure point, which is 10 mm from the top of the side surface.

Mathematical Model

The mathematical model of induction heating process consists of the electromagnetic field equation and temperature field equation.

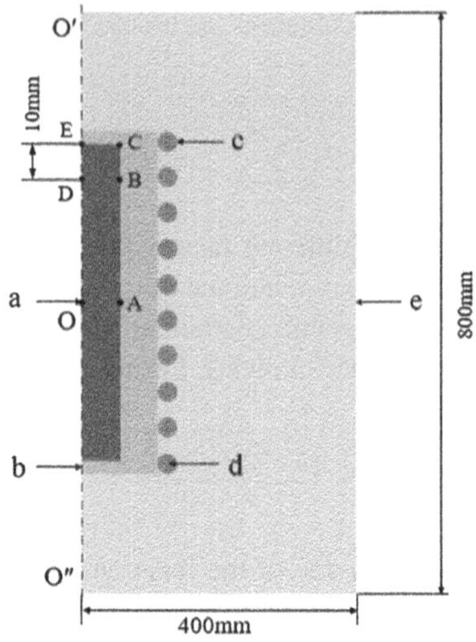

Fig. 1 Schematic diagram of the induction heating process **a**—specimen; **b**—insulated cotton; **c**—inductance coil; **d**—cooling water; **e**—ambient air

For the electromagnetic field, combined with Coulomb gauge and auxiliary quantity, like magnetic vector potential $\vec{A}$, and electric scalar potential $\phi^*(\phi = \int \phi^* dt)$, the Maxwell's equations are described as:

$$\vec{\boldsymbol{J}} = -\sigma\left(\frac{\partial \vec{\boldsymbol{A}}}{\partial t} + \nabla\frac{\partial \phi}{\partial t}\right) + \overrightarrow{\boldsymbol{J_e}} \tag{10}$$

$$\nabla \times \frac{1}{\mu}\nabla \times \vec{\boldsymbol{A}} - \nabla\frac{1}{\mu}\left(\nabla \cdot \vec{\boldsymbol{A}}\right) + \sigma\left(\frac{\partial \vec{\boldsymbol{A}}}{\partial t} + \nabla\frac{\partial \phi}{\partial t}\right) = \overrightarrow{\boldsymbol{J_e}} \tag{11}$$

$$-\nabla \cdot \sigma\left(\frac{\partial \vec{\boldsymbol{A}}}{\partial t} + \nabla\frac{\partial \phi}{\partial t}\right) = 0 \tag{12}$$

where $\vec{\boldsymbol{J}}$, σ, ∇, $\overrightarrow{\boldsymbol{J_e}}$ are the total current density, total conductivity, Hamilton operator and current density of external exciting source respectively.

For the temperature field, the Joule heat source caused by induced eddy current can be determined according to Joule's law. The heat conduction equation is:

$$\frac{\partial}{\partial t}\left(\rho C_p T\right) = \frac{J^2}{\sigma} + \nabla \cdot (\lambda \nabla T) \tag{13}$$

where ρ, C_p, λ is density, specific heat capacity and thermal conductivity of the specimen, respectively.

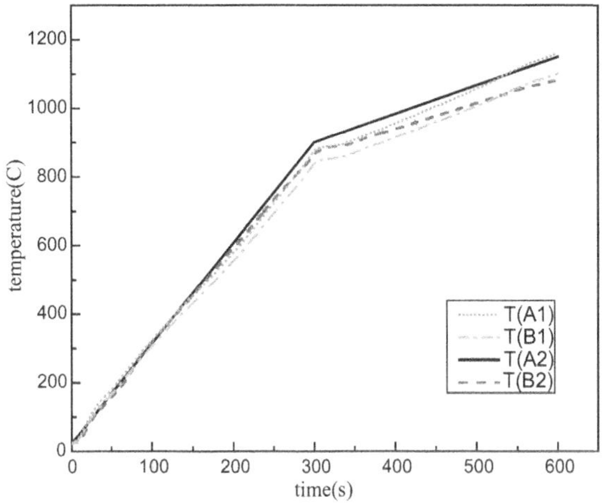

Fig. 2 Comparison between numerical and experimental results of point A and B

Based on the geometric model and the equations above, the numerical computation is solved using the COMSOL Multiphysics 5.2a. The boundary conditions in the numerical model are listed as follows: the temperature of cooling water is 20 °C, the air is at room temperature and the side surface of the specimen is adiabatic as the specimen is wrapped in insulation cotton.

Model Verification

To verify the numerical model, the simulation results (line A1, line B1) and the experimental results (line A2, line B2) of specimen are shown in Fig. 2. Overall, the simulated results agree with experimental results very well in terms of the magnitude and the trend. The maximum relative error is less than 3%, which is acceptable. As a result, the numerical model is adopted to investigate the effects of electromagnetic parameters on the temperature distribution of induction heating process.

Results and Discussion

Results Based on Single-Factor Experimental Design

The investigated factors are exciting current, power frequency, coil inner diameter and coil spacing, and they all have five levels, respectively. In this work, distance between adjacent turns from middle to the end is designed as arithmetic progression, and the coil spacing is described as the common difference of the arithmetic

Table 1 Factors and levels

Levels	Factors			
	Exciting current (A)	Power frequency (kHz)	Coil inner diameter (mm)	Coil spacing (mm)
1	100	50	40	−4
2	150	75	45	−2
3	200	100	50	0
4	250	125	55	2
5	300	150	60	4

progression. The investigated factors and their levels are presented in Table 1. On basic condition, exciting current is 200 A, power frequency is 100 kHz, coil inner diameter is 50 mm and coil spacing is 0. Temperature difference equals to the temperature in the middle minus the temperature at the end of the specimen. The simulation conditions and results of test cases are shown in Table 2.

Table 2 Simulation conditions and results of test cases

Case	Factors				Results	
	Exciting current (A)	Power frequency (kHz)	Inner diameter of coil (mm)	Coil spacing (mm)	Axial temperature difference (K)	Radial temperature difference (K)
1	100	100	50	0	95	4.5
2	200	100	50	0	90	11.5
3	300	100	50	0	85	21.5
4	400	100	50	0	82	34
5	500	100	50	0	78	50
6	200	50	50	0	95	13
7	200	75	50	0	90	17
8	200	100	50	0	85	21
9	200	125	50	0	82	26
10	200	150	50	0	79	29
11	200	100	40	0	70	23
12	200	100	45	0	80	22
13	200	100	50	0	85	21
14	200	100	55	0	90	20
15	200	100	60	0	93	19
16	200	100	50	−4	525	65
17	200	100	50	−2	180	25
18	200	100	50	0	90	22
19	200	100	50	2	14	20
20	200	100	50	4	−130	17

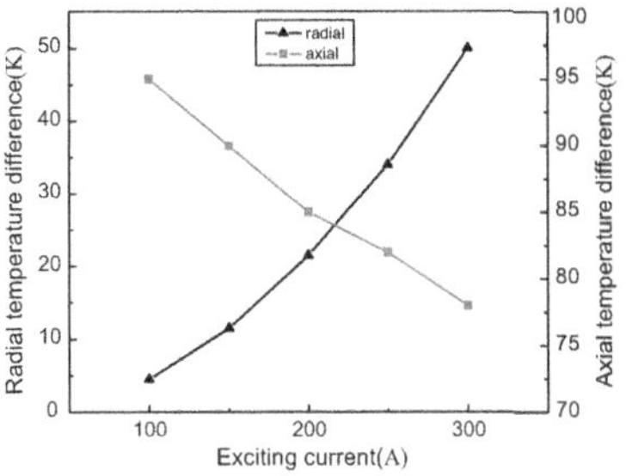

a. Temperature difference with exciting current

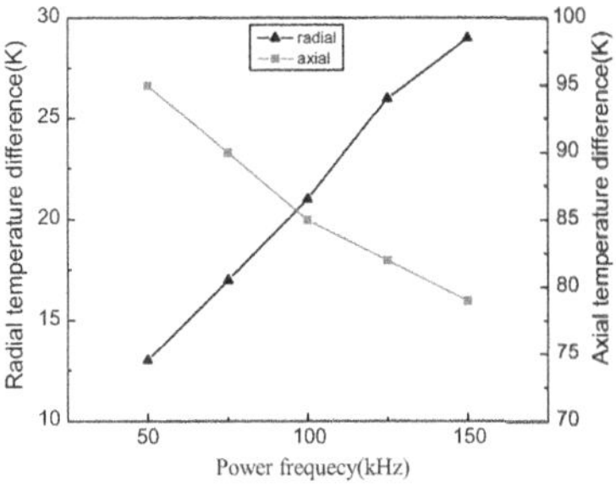

b. Temperature difference with power frequency

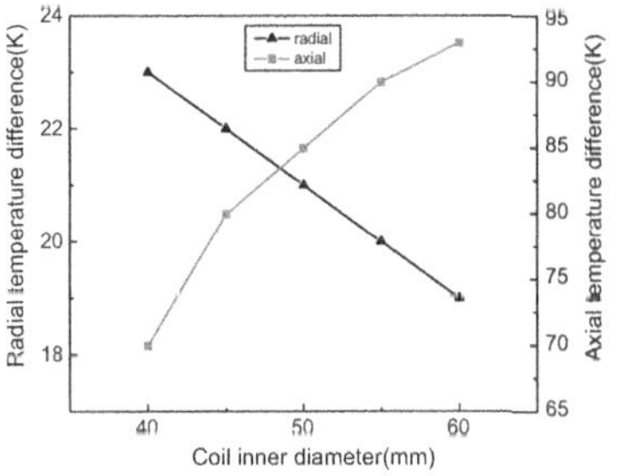

c. Temperature difference with coil inner diameter

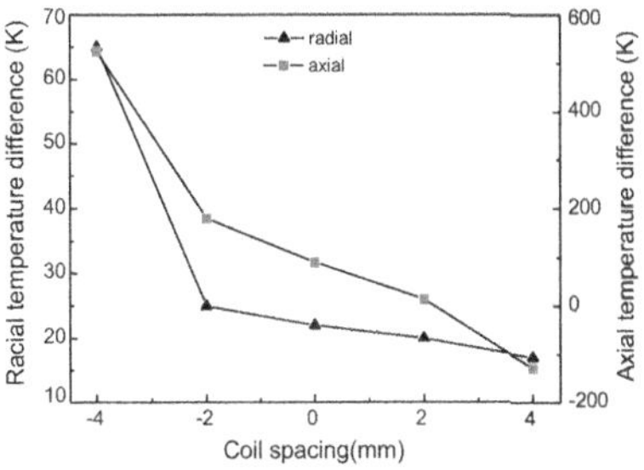

d. Temperature difference with coil spacing

Fig. 3 Radial and axial temperature difference of the specimen with different factors

Through various investigations, the temperature difference of the 20 cases is presented in Fig. 3. It can be clearly observed that all the four factors have obvious influences on the radial and axial temperature difference of the specimen. The tendency of radial and axial temperature difference is opposite for the exciting current, power frequency and coil inner diameter. The larger exciting current, higher power frequency and smaller inner distance of coil lead to shorter heating time, causing shorter depth of penetration and smaller heat loss at the end, which means radial temperature difference gets larger and axial temperature difference gets smaller. When the coil spacing is sparse in the middle and dense at the end, it is beneficial to decrease the temperature difference due to the less heat loss at the end. However, if the coil spacing is over sparse in the middle, the temperature difference becomes negative, meaning the temperature in the middle is lower than that at the end, which has a negative effect on the temperature uniformity.

Results Based on Fuzzy Gray Relational Analysis

The results above have shown that the investigated factors have various impacts on temperature difference. As a result, Fuzzy Gray Relational Analysis, an improved method based on the gray relational theory, is adopted in this work.

Firstly, the axial and radial temperature difference under various conditions is considered as reference matrix $y_1(k)$ and $y_2(k)$. Exciting current, power frequency, coil inner diameter and coil spacing are taken to be the element $x_1(k)$, $x_2(k)$, $x_3(k)$ and $x_4(k)$ in comparison matrix, respectively. The reference matrix and comparison matrix are processed by (3).

Secondly, the cosine value of fuzzy membership is calculated by (4). Figure 4 illustrates the fuzzy membership grades of four factors to the temperature difference. It indicates that the fuzzy membership grades of the four factors have a huge difference. For axial temperature difference, the fuzzy membership grade of coil inner diameter is the largest, while the coil spacing is the smallest. Based on the analysis of (4), the higher the fuzzy membership grade, the better the similarity of the changing trends of the investigated factors and axial temperature difference, which means the coil inner diameter has the most noticeable effect on axial temperature difference, while the coil spacing has the lowest effect. Similarly, the exciting current has the most remarkable effect on radial temperature difference, while the coil spacing has the lowest effect.

Thirdly, the Euclidean gray relational grade is computed by (5)–(8). In (8), the weights are considered to be equal due to the mutual independence of the cases. Figure 5 depicts the Euclidean gray relational grades of the four factors to the temperature difference. It shows that the Euclidean gray relational grades of the four factors have a small difference. Based on the analysis of (5)–(8), it still can be attributed to that the coil inner diameter has great effect on axial and radial temperature difference.

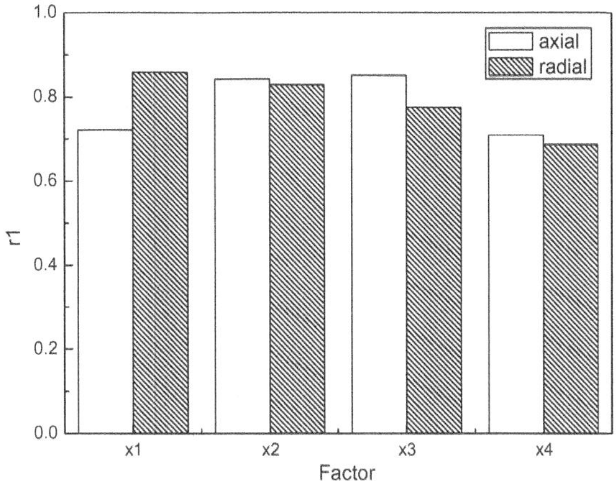

Fig. 4 Fuzzy membership grades

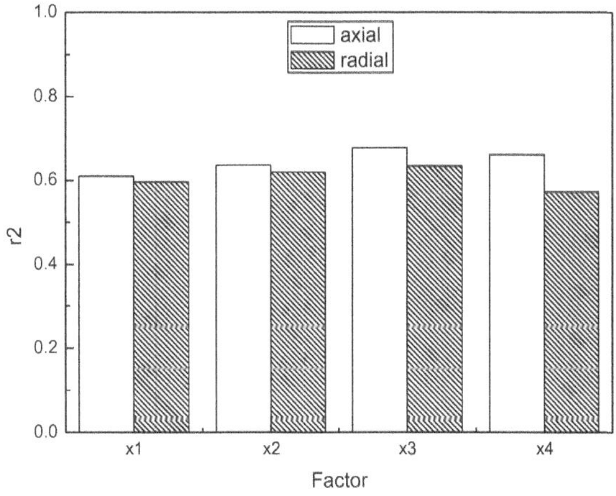

Fig. 5 Euclidean gray relational grades

Finally, the fuzzy gray relational grade is computed by (9), as shown in Fig. 6. It offers us comprehensive consideration for evaluating impacts of the four factors. It shows that, for axial temperature difference, the fuzzy gray relational grades of the four factors, exciting current, power frequency, coil inner diameter and coil spacing, are 0.6684, 0.7464, 0.7691 and 0.6853, respectively, meaning that the impacts of the four factors are ranked from the most important to the least important as coil inner diameter, coil spacing, power frequency and exciting current. While the ranking list of importance becomes exciting current, power frequency, coil inner diameter and coil spacing for radial temperature difference, since the fuzzy gray relational grades are 0.7389, 0.7311, 0.7078 and 0.6315, respectively. As a result,

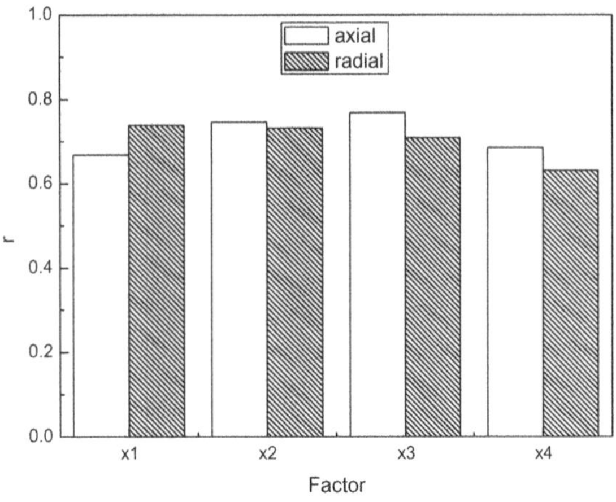

Fig. 6 Fuzzy gray relational grades

coil inner diameter has the most important effect on the axial temperature difference. The exciting current has the most important effect on the radial temperature difference, while it is not important for the axial temperature difference. This work provides us great reference value for optimizing the temperature uniformity of the specimen in induction heating process. As few studies investigated the importance extent of the influencing factors before, the results will be instructive and provide great reference for further research of induction heating.

Conclusions

In this work, numerical investigation of induction heating process was conducted. The single-factor experimental design and the Fuzzy Gray Relational Analysis (FGRA) were employed to evaluate the impacts of four electromagnetic parameters on temperature uniformity. As a result, the major conclusions are summarized as follows:

(1) The larger exciting current, the higher power frequency and the smaller inner distance of coil led to larger radial temperature difference and smaller axial temperature difference. Also, it was beneficial to the temperature uniformity of specimen to make the coil spacing sparser in the middle and denser at the end properly.
(2) The fuzzy membership grades and the Euclidean gray relational grades both showed that coil inner diameter had the most noticeable effect on axial temperature difference, while coil spacing had the lowest effect on radial temperature difference.

(3) For axial temperature difference of specimen, the impacts of the four factors were ranked from the most important to the least important as coil inner diameter, coil spacing, power frequency and exciting current, since the fuzzy gray relational grades were 0.7691, 0.7464, 0.6853 and 0.6684, respectively. While for radial temperature difference, the ranking list of importance became exciting current, power frequency, coil inner diameter and coil spacing, since the fuzzy gray relational grades were 0.7389, 0.7311, 0.7078 and 0.6315, respectively

Acknowledgements The authors would like to express their appreciation to the financial support from the National Key Research and Development Program of China (2016YFB0700304). Special thanks to Professor L. Jiang, we are honoured to have the privilege to learn from such a leader full of inspiration to the scientific research work.

References

1. Lucia O, Maussion P, Dede EJ (2013) Induction heating technology and its applications: past developments, current technology, and future challenges. IEEE Trans Ind Electron 61 (5):2509–2520
2. Zhang D, Wen Z, Yue Z (2015) Effects of strain rate and temperature on mechanical property of nickel-based super alloy GH3230. Rare Metal Mater Eng 44(11):2601–2606
3. Lei W, Mao X, Lu Y (2011) Numerical simulation of temperature field and thermal stress field in quenching process of TC21 titanium alloy forging. Rare Metal Mater Eng 40 (10):1721–1726
4. Kranjc M, Zupanic A, Miklavcic D et al (2010) Numerical analysis and thermographic investigation of induction heating. Int J Heat Mass Transf 53(17):3585–3591
5. Matloubieh AY, Roemer RB, Cetas TC (1984) Numerical simulation of magnetic induction heating of tumors with ferromagnetic seed implants. IEEE Trans Biomed Eng 31(2):227
6. Ismail KS, Marzouk RA (1996) Iterative hybrid finite element-boundary element method for the analysis of induction heating system with nonlinear charge. IEEE Trans Magn 32 (4):3212–3218
7. Chaboudez C, Clain S, Glardon R et al (1997) Numerical modeling in induction heating for axisymmetric geometries. IEEE Trans Magn 33(1):739–745
8. Bermudez A, Reales C, Rodriguez R et al (2010) Numerical analysis of a finite-element method for the axisymmetric eddy current model of an induction furnace. IMA J Numer Anal 30(3):654–676
9. Huang J, Wu WF, Wang BF et al (2012) Numerical simulation of steel tube induction heating during quenching process. J Iron Steel Res 19(1):187–190
10. Shokouhmand H, Ghaffari S (2012) Thermal analysis of moving induction heating of a hollow cylinder with subsequent spray cooling: effect of velocity, initial position of coil, and geometry. Appl Math Model 36(9):4304–4323
11. Hammi H, Barka N, Ouafi AE (2015) Effects of induction heating process parameters on hardness profile of 4340 steel bearing shoulder using 2D axisymmetric model. In: International symposium on aircraft materials, Marrakech, Maroc, April 2014
12. Sun Y, Xiao SL, Feng L et al (2014) Effect of single factor experiment design on foam wood residual fiber cushion packaging material. Forest Eng 9(30):50–52
13. Deng JL (1989) Introduction to grey system theory, 1. Sci-Tech Information Services, China

14. Kadier A, Abdeshahian P, Simayi Y et al (2015) Grey relational analysis for comparative assessment of different cathode materials in microbial electrolysis cells. Energy 90(2):1556–1562
15. Wang Z, Lei T, Chang X et al (2015) Optimization of a biomass briquette fuel system based on grey relational analysis and analytic hierarchy process: a study using cornstalks in China. Appl Energy 157:523–532
16. Zhou S, Zhou L, Yu L et al (2016) Monitoring chip fatigue in an IGBT module based on grey relational analysis. Microelectron Reliab 56:49–52
17. Zuo W, E JQ, Liu X et al (2016) Orthogonal experimental design and fuzzy grey relational analysis for emitter efficiency of the micro-cylindrical combustor with a step. Appl Therm Eng 103:945–951

Numerical Simulation of Inclusion Removal in a Novel Tundish with Swirl Flow

Jianchuan Yan, Tao Li and Jun Liu

Abstract It is of great importance to remove the inclusions as much as possible in the molten steel. The inclusion collision rate is significantly improved by generating centrifugal flow in the tundish with an electromagnetic device, which induces the inclusions to grow and float. However, the electromagnetic device requires an external equipment component that is difficult to maintain. Additionally, a turbulence inhibitor is commonly used to optimize the turbulence flow in the tundish. This paper describes a novel tundish design with a turbulence inhibitor that generates swirl flow in a cylindrical zone, which produced an effect similar to the electromagnetic device. The gravitational potential energy of the molten steel from the nozzle is converted into kinetic energy of the swirling flow in the tundish. The inclusion removal rates with various nozzle diameters were investigated, and the optimal turbulence inhibitor size and structure were identified by numerical simulation.

Keywords Tundish · Inclusion removal · Turbulence inhibitor
Numerical simulation · Centrifugal flow

Introduction

In steelmaking, nonmetallic inclusions are mainly generated during deoxidization process or originate with the refractory materials; these inclusions are of great harm to the quality and mechanical properties of the steel products [1]. To improve the quality of the molten steel, many methods to reduce inclusion formation and

J. Yan · T. Li (✉) · J. Liu
College of Materials Science and Engineering, Chongqing University,
Chongqing 400044, People's Republic of China
e-mail: litao0518@hotmail.com

T. Li
Chongqing Key Laboratory of Vanadium-Titanium Metallurgy and Advanced Materials,
Chongqing University, Chongqing 400044, People's Republic of China

© The Minerals, Metals & Materials Society 2019

T. Jiang et al. (eds.), *10th International Symposium on High-Temperature Metallurgical Processing*, The Minerals, Metals & Materials Series,
https://doi.org/10.1007/978-3-030-05955-2_4

remove inclusions have been investigated [2]. In the continuous casting process, a tundish acts not only as a buffer and distributor between ladle and casting mold, but also plays an important role in the removal of inclusions [3]. Studies of the flow characteristics of molten steel in the tundish are of great significance to improve the quality and productivity of steel products. Inclusion separation is abetted by increased residence time, elimination of dead zones and smooth molten steel flow fields. Various flow control devices have been put into application, including bubble curtains, weirs, dams, baffles and turbulence inhibitors [4]. Kawasaki Steel proposed a Centrifugal Flow Tundish (CFT) [5], which could generate centrifugal flow in the rotating chamber of the tundish driven by an external electromagnetic field. The CFT method has proved to effectively eliminate inclusions; inclusion removal is accomplished by increasing residence time and the collision rate of inclusion particles in the centrifugal flow [6]. However, the CFT is not widely applied due to its requirements of the external electromagnetic device and energy consumption.

In this paper, a novel tundish is proposed based on the idea of a CFT installed with an innovative turbulence inhibitor to generate swirling flow. The turbulence inhibitor could obtain the similar effect to the electromagnetic device in the CFT without the requirements of external device and electricity consumption. Furthermore, it is cost efficient and easy to be maintained and replaced. The size and structure of the novel tundish are optimized through computational fluid dynamics (CFD) simulation implemented by ANSYS Fluent in the current work.

Model Description

Tundish Model

The novel tundish differs from tradition by employing the turbulence inhibitor that is designed as a hollow cylinder with an inlet from its top surface and two tangential outlets located symmetrically around the side wall of the cylinder (See Fig. 1b). The tangential outlets generate swirling flow in the rotating chamber (See Fig. 1a) of the tundish. The gravitational potential energy of the molten steel from the nozzle is converted to kinetic energy of the swirling flow in the rotating chamber. The turbulence inhibitor will be called the "swirling flow generator" in the following text. The components and planes of the tundish are labelled in Fig. 1. These cross sections Plane 1 and Plane 2 are parallel to the bottom surface of the tundish. Plane 1 is located at 125 mm above the bottom of the tundish, which is going through the middle of the outlet of the swirling flow generator. While Plane 2 is located at 275 mm above the bottom of the tundish, which is between the upper surface of the swirling flow generator and the nozzle outlet.

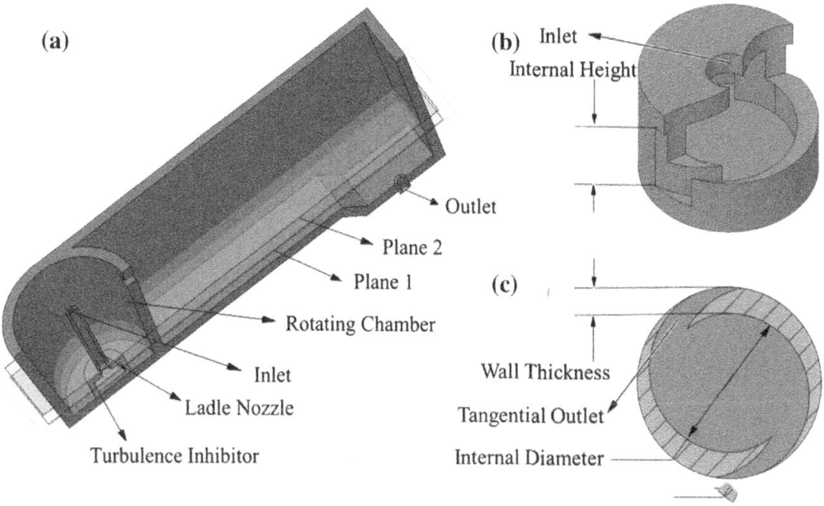

Fig. 1 Structure and components of the tundish. **a** Components of the tundish. **b** and **c** Structure of the swirling flow generator

The three-dimensional (3D) model of a simplified tundish installed with a swirling flow generator is established. The detailed geometry and parameters are shown in Fig. 2, in which the internal diameter of the swirling flow generator is variously changed and optimized in the following discussions.

Numerical Simulation

Numerical simulations were carried out by using the commercial software ANSYS FLUENT 18.0®, which was widely applied in the field of the numerical simulation of the flow fields in the tundishes [7–10]. In this study, the meshes of the 3D tundish models are produced by ICEM with the same operating process for each tundish model. The total quantity of the grids is approximately 1.7 million. The calculations are considered to have converged as the residual of all the variables are below 1×10^{-3}. The numerical simulations were carried out based on the following consumptions to simplify the simulation model:

(1) The molten steel is a Newtonian incompressible fluid. The density and viscosity of the molten steel are constants, which are 7080 kg/m^3 and 0.0055 kg/(m s), respectively.
(2) The temperature and compositions of steel within the tundish are assumed to be homogenously distributed.

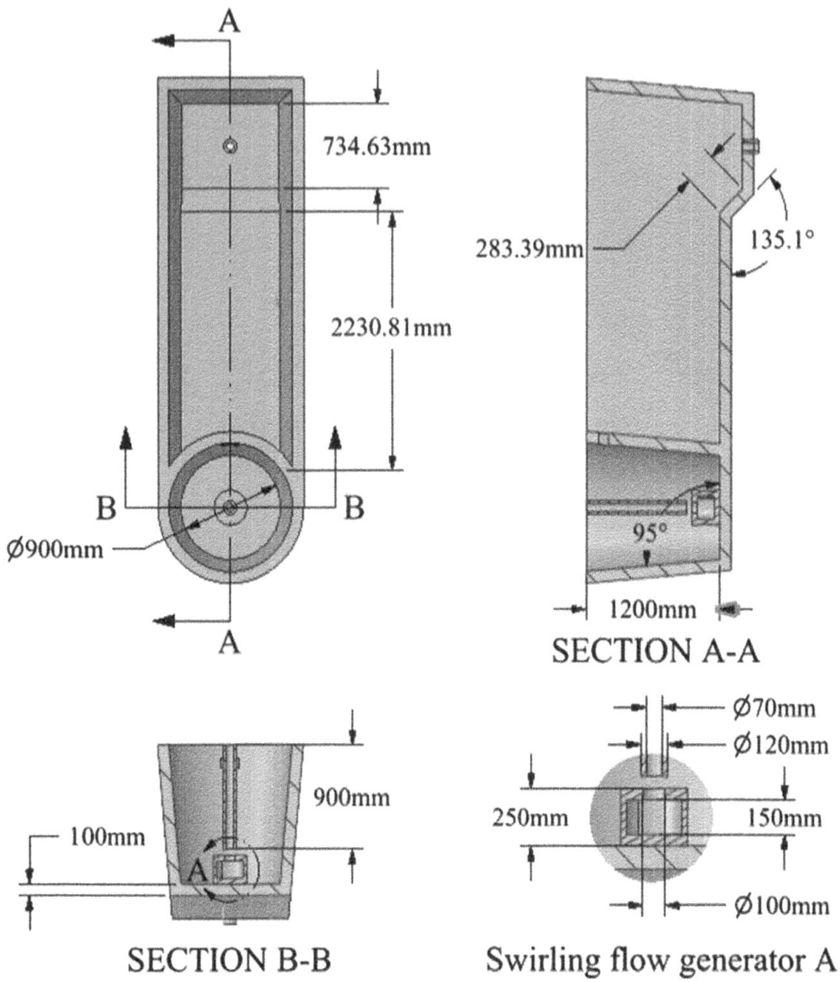

Fig. 2 Dimensions of the three-dimensional model

(3) The walls of the tundish are considered as smooth wall.
(4) Regardless of the influence of the slag phase, the top surface of the molten steel is set to be specified shear wall boundary condition.
(5) The inclusion particles are assumed to be spherical with the density of 3000 kg/m^3.

Turbulence Model

The renormalization group (RNG) k-ε two-equation turbulence model is adopted in this work based on the previous study [11]. The governing equations for the steady flow are solved within a finite control volume are presented as follows.

The continuity equation of the fluid is,

$$\frac{\partial u_i}{\partial x_i} = 0 \tag{1}$$

The momentum conservation equation is given by,

$$\rho \frac{\partial(u_i u_j)}{\partial x_j} = -\frac{\partial p}{\partial x_i} + \frac{\partial}{\partial x_j}\left[\mu_{eff}\left(\frac{\partial u_i}{\partial x_j} + \frac{\partial u_j}{\partial x_i}\right)\right] \tag{2}$$

The turbulence kinetic energy ε can be derived from the following equation,

$$\rho\left(\frac{\partial}{\partial x_j}\right)\left(u_j k - \frac{u_{eff}}{\sigma_k} \times \frac{\partial k}{\partial x_j}\right) = G - \rho\varepsilon \tag{3}$$

The dissipation rate k is calculated by the following formula,

$$\rho \frac{\partial(u_j \varepsilon)}{\partial x_j} = \frac{\partial}{\partial x_j}\left(\frac{u_{eff}}{\sigma_\varepsilon} \times \frac{\partial \varepsilon}{\partial x_j}\right) + \frac{(C_1 G\varepsilon - C_2 \rho \varepsilon^2)}{k} \tag{4}$$

where, C_1, C_2, σ_ε and σ_k are constants recommended by Launder and Spalding [12]; and G is given by,

$$G = \mu_t \frac{\partial u_j}{\partial x_i}\left(\frac{\partial u_i}{\partial x_j} + \frac{\partial u_j}{\partial x_i}\right) \tag{5}$$

The effective viscosity is modified by,

$$\mu_{eff} = \mu\left(1 + \sqrt{\frac{C_\mu}{\mu}\frac{k}{\sqrt{\varepsilon}}}\right)^2 \tag{6}$$

Inclusions Tracking Model

The capacity to separate inclusion particles from molten steel is one of the most important criteria in the optimization of the tundish structure. The discrete phase model (DPM) is introduced to simulate the motion of the inclusion particles in the phase of the molten steel, and thus calculate the removal rate of the inclusion particles. In the discrete phase model, the boundary condition of the outlet of the tundish is set to be "escape", which means that the inclusion particles could exit from the outlet with the molten steel. The boundary condition of the top surface of the tundish is set to be "trap", where the inclusion particles become trapped. The boundary conditions at all of the other tundish walls are set to be "reflect", where the inclusion particles are reflected back into the molten steel after they collide with those walls.

Results and Discussion

Comparison Between Novel and Traditional Tundishes

The molten steel flow in the tundishes was simulated with the swirling flow generator and the traditional turbulence inhibitor (200 mm internal diameters in both devices) to illustrate the function of swirling flow generator. Figure 3 compares the molten steel streamlines in the two models. With the swirling flow generator, an obvious circular motion of the molten steel is found in the cylindrical rotating chamber. The velocity contours of the cross section A-A (See Fig. 2) are shown in Fig. 4. The velocity of the molten steel in the tundish with the swirling flow generator is much more symmetric than that with the traditional turbulence inhibitor. It is clear that a swirling flow is created in the rotating chamber by guiding the molten steel from the nozzle to flow out of the tangential outlets.

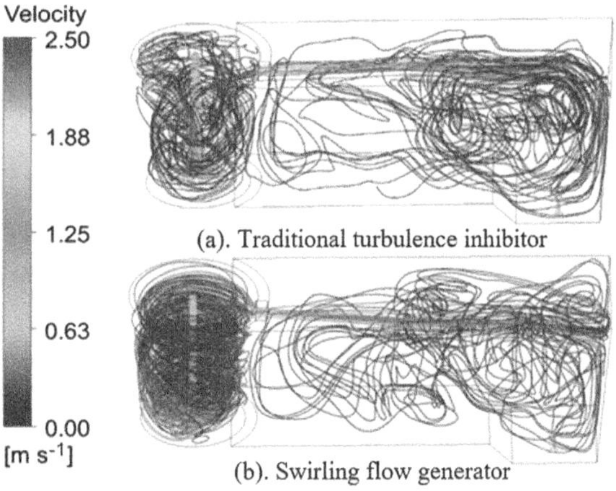

Fig. 3 Streamlines of tundishes with traditional turbulence inhibitor and swirling flow generator

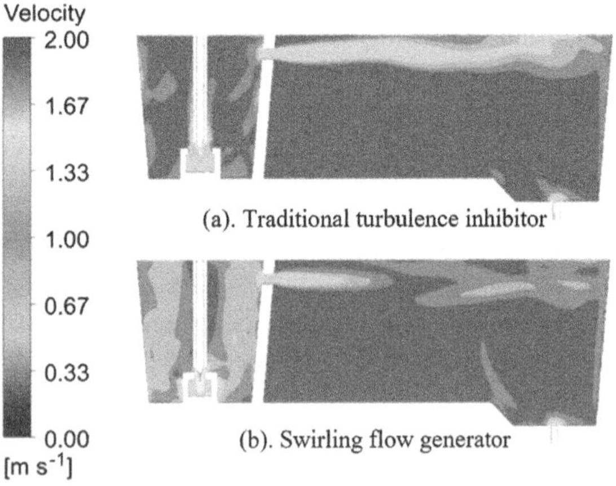

Fig. 4 Velocity contours in section A-A of tundishes with traditional turbulence inhibitor and swirling flow generator

Influence of Nozzle Length

Two tundish models are established to clarify the influence of the immersing depth of the 250 mm internal diameter nozzle. The model of case (a) shown in Fig. 5a is designed exactly following the parameters shown in Fig. 2. The immersion depth of the nozzle is 900 mm, which is positioned 50 mm above the inlet of the swirling flow generator. The immersion depth is increased to 970 mm in case (b) shown in Fig. 5, in which the nozzle is put into the swirling flow generator. The streamlines and the velocity contours of the molten steel are shown in Figs. 5 and 6. The streamlines of the molten steel in the tundish become more disordered and asymmetrical when the nozzle immersion depth is extended from 900 to 970 mm. Meanwhile, the velocity contours of the two tundish models indicate that a stronger swirling flow is generated in the rotating chamber when the outlet of the nozzle has a distance of 50 mm above the inlet of the swirling flow generator.

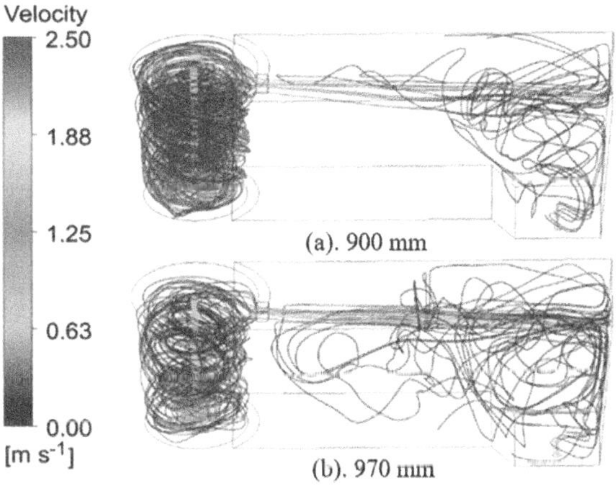

Fig. 5 Streamlines in the tundishes with different immersing depth

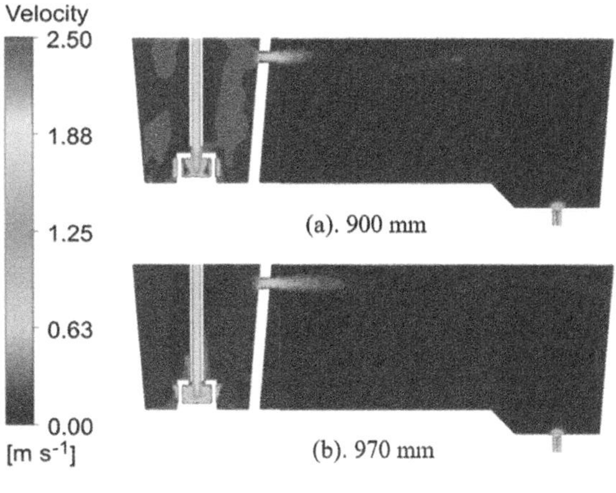

Fig. 6 Velocity contours in section A-A with different immersing depth

Influence of Internal Diameter

In order to determine the optimum structure, the internal diameters of the swirling flow generator were varied, (150, 200, 250 and 300 mm). The streamlines of the molten steel of each model are presented in Fig. 7. The poorest tundish performance appears when the internal diameter is 150 mm, indicating that the internal diameter is too small for the swirling flow generator to generate a sufficient swirling flow in the rotating chamber. With internal diameters of 200, 250 and 300 mm, the swirling flow generator produced a similar swirling flow in the rotating chamber. With the increase of the internal diameter, the molten steel is more likely to flow along the wall of the rotating chamber.

The velocity contours of the rotating chamber area in the position of Plane 1 and Plane 2 (See Fig. 1) are shown in Figs. 8 and 9. In Fig. 8a, the molten steel flowing

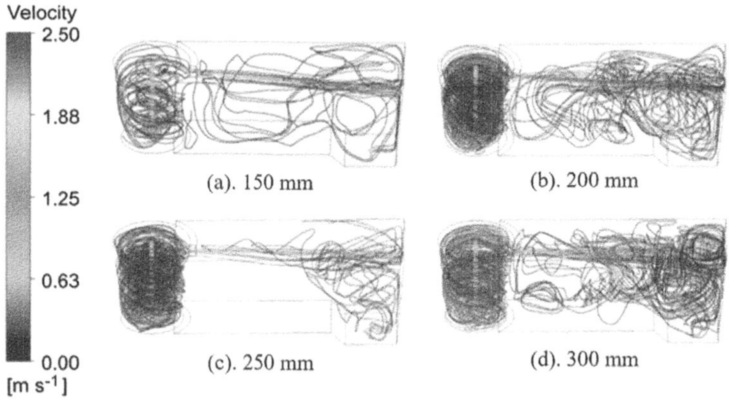

Fig. 7 Streamlines of tundishes with various sizes of swirling flow generators

Fig. 8 Velocity contours in Plane 1 in rotating chamber with different internal diameters of swirling flow generator

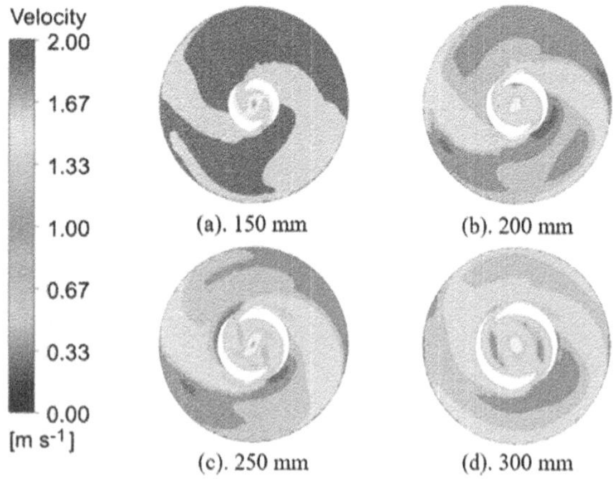

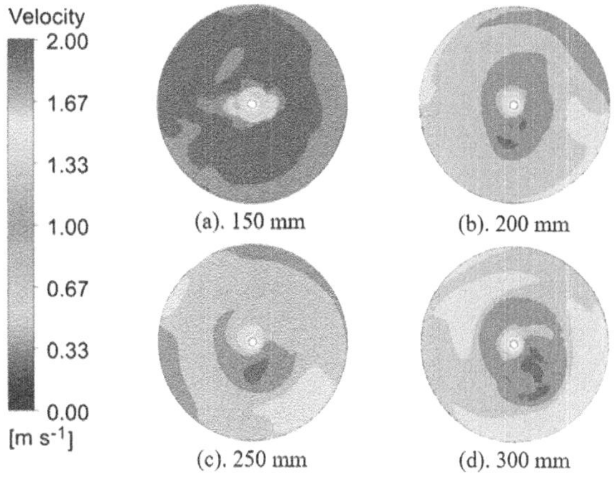

Fig. 9 Velocity contours in Plane 2 in rotating chamber with different internal diameters of swirling flow generator

out of the tangential outlets is shown to go straight to the wall of the rotating chamber when the internal diameter of the swirling flow generator is 150 mm. Furthermore, the velocity contour at Plane 2 in Fig. 9a indicates that the kinetic energy of the molten steel is insufficient to generate a well-developed swirling flow at the upper part of the rotating chamber. By comparison with the other velocity contours, the swirling flow is well developed and symmetric in Fig. 8c; while the swirling flow is not well developed in Fig. 8b and not symmetric in Fig. 8d. Therefore, it may be concluded that the swirling flow generator with an internal diameter of 250 mm generated a relatively ideal flow field for inclusion removal.

Removal of Inclusion Particles

The inclusion removal rate must be considered to find the optimum structure. Inclusion particles with diameters of 20, 50 and 80 μm were tracked and the results are presented at Fig. 10. In all models, the removal rate increased with the increasing particle size because larger inclusions are easier to float and be removed. Additionally, when the internal diameter of the swirling flow generator is less than 250 mm, the inclusion removal rate increases with increased internal diameter; while the inclusion removal decreases with the increasing of the internal diameter of the swirling flow generator when its diameter is larger than 250 mm. The model with the internal diameter of 250 mm achieves a higher removal rate than the others regardless of the inclusion size. The highest removal rate of inclusions reaches 91.9% as the inclusion size is 80 μm and the internal diameter of the swirling flow generator is 250 mm, which indicates that the optimal internal diameter is 250 mm.

The comparison of inclusion removal rates between swirling flow generator and traditional turbulence inhibitor with the internal diameter of 250 mm was investigated. The sizes of the inclusion particles are set at 20 μm 50 and 80 μm, the results

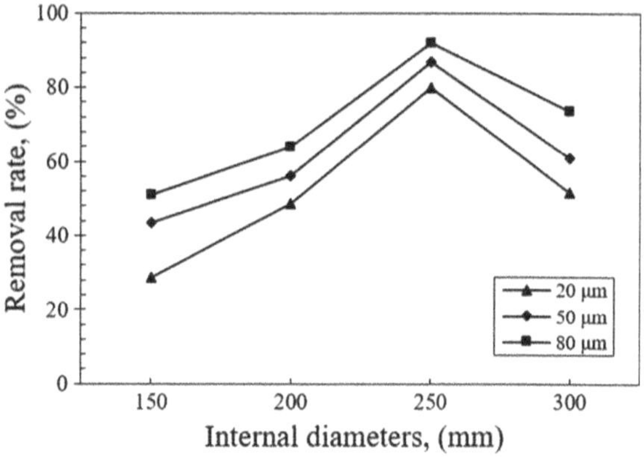

Fig. 10 Removal rates of different inclusions

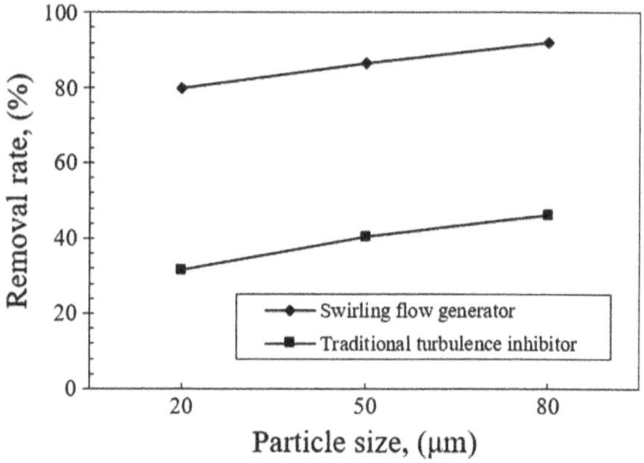

Fig. 11 Comparsion of inclusion removal rates

of which are shown in Fig. 11. The inclusion removal rates increase with increasing particle size for both the tundish applied with swirling flow generator and traditional turbulence inhibitor. However, the inclusion removal rate of the tundish with swirling flow generator is double that of the traditional one for any size of inclusion particles. The application of the swirling flow generator is feasible to significantly strengthen the removal of inclusions.

Conclusions

A novel tundish design incorporates a swirling flow generator. The swirling flow generator was optimized by numerical simulation. The following conclusions are drawn from this study:

(1) By applying the swirling flow generator, a swirling flow of the molten steel is successfully generated in the rotating chamber in the tundish. The idea of swirling flow generator takes advantage of the gravitational potential energy of the molten steel to generate a swirling flow that increases the collision rate of the inclusion.
(2) The influence of the nozzle immersion depth was investigated. The patterns of streamlines indicate that the swirling flow is much stronger when the immersing depth is 900 mm that is 50 mm above the inlet of the swirling flow generator.
(3) The flow characteristics of the molten steel in the tundish with various internal diameters of the swirling flow generator were studied. The optimized internal diameter of the swirling flow generator is found to be 250 mm, in which a steady and symmetrical swirling flow is generated in rotating chamber.
(4) Inclusion particles of various size were tracked in the tundish installed a swirling flow generator with a series of internal diameters. The highest inclusion removal rate reaches 91.9% when the particle size is 80 μm and the internal diameter of the swirling flow generator is 250 mm. By comparison with traditional tundish, the swirling flow generator has the potential to significantly improve inclusion removal efficiency.

Acknowledgements The authors are especially grateful to the grants from the National Natural Science Foundation of China (Grant No. 51704052 and No. 51874061) for their financial support.

References

1. Ånmark N, Karasev A, Jönsson PG (2017) The influence of microstructure and non-metallic inclusions on the machinability of clean steels. Steel Res Int 88(1):1–10
2. Ramesha DK, Hosur KS (2013) Numerical investigation of steel grade change using ANSYS. Int J Innovative Res Sci Eng Technol 2(10):5601–5606
3. Sahai Y, Emi T (2007) Tundish technology for clean steel production. World Sci:8–11
4. Solorio-Díaz G, Morales RD, Ramos-Banderas A (2005) Effect of a swirling ladle shroud on fluid flow and mass transfer. Int J Heat Mass Transf 48(17):3574–3590
5. Miki Y, Ogura S, Fujii T (1996) Separation of inclusions from molten steel in a Tundish by use of a rotating electromagnetic field. Kawasaki Steel Tech Rep 35:67–73
6. Miki Y, Kitaoka H, Sakuraya T, Fujii T (2009) Mechanism of separation of inclusions from molten steel stirred with rotating electro-magnetic field. Tetsu- to- Hagane 78(3):431–438
7. Ni P, Josson L, Ersson M, Jösson PG (2017) Application of a swirling flow producer in a conventional Tundish. ISIJ Int 57(12):2175–2184
8. Tripathi A, Ajmani SK (2011) Effect of shape and flow control devices on the fluid flow characteristics in three different industrial six strand billet caster Tundish. ISIJ Int 51 (10):1647–1656
9. Hou Q, Yue Q, Wang H, Zou Z, Yu A (2008) Modelling of inclusion motion and flow patterns in swirling flow tundishes with symmetrical and asymmetrical structures. ISIJ Int 48 (6):787–792

10. Aguilar-Rodriguez CE, Ramos-Banderas JA, Torres-Alonso E, Solorio-Diaz G, Hernández-Bocanegra CA (2018) Flow characterization and inclusions removal in a slab Tundish equipped with bottom argon gas feeding. Metallurgist 61(11–12):1055–1066
11. Hou Q, Zou Z (2005) Comparison between standard and renormalization group k-ε models in numerical simulation of swirling flow Tundish. ISIJ Int 45(3):325–330
12. Launder BE, Spalding DB (1974) The numerical computational of turbulent flows. Comput Methods Appl Mech Eng 3(2):269–289

Numerical Simulation Study on Design Optimization of Inner Cavity Dimensions of Large-Capacity Tundish

Yong Zhong, Mingmei Zhu, Bing Huang and AiPing Zhang

Abstract In this paper, the design and optimization of tundish size for single-strand slab are studied. A three-dimensional mathematical model is established for calculating the flow field and temperature field of molten steel in the tundish with different length to height (L/H) ratios, width to height ratios (W/H), and angles of edges. The concentration distribution curves of the tracer at the outlet are calculated by using component transfer equation. On this basis, the average residence times and flow patterns of the molten steel can be obtained. The trajectories and the removal ratios of the inclusions are calculated by using Lagrange particles tracking model and DPM model. The validity of the mathematical model is verified by the results of water model experiment. The optimized L/H ratio and W/H ratio are obtained from the numerical simulation results, and it can provide a theoretical basis for the tundish used in steel mill.

Keywords Tundish · Structure size · Numerical simulation · Residence time distribution curve

Introduction

The tundish plays an important role as a refining container in the continuous casting process. The larger the volume of the tundish, the deeper the molten pool, the longer the average residence time of the molten steel. The probability of inclusions being lifted up is greater, so the tundish also develops towards a larger capacity. Its shape and geometrical dimensions play an important role in the flow of molten steel in the tundish, and the flow of molten steel directly affects the removal of inclusions, which in turn affects the purity of the cast. Therefore, it is important to optimize the size of the cavity of the large-capacity tundish [1–3]. The methods for

Y. Zhong · M. Zhu (✉) · B. Huang · A. Zhang
College of Materials Science and Engineering,
Chongqing University, Chongqing 400044, China
e-mail: zhumingmei@cqu.edu.cn

© The Minerals, Metals & Materials Society 2019

T. Jiang et al. (eds.), *10th International Symposium on High-Temperature Metallurgical Processing*, The Minerals, Metals & Materials Series,
https://doi.org/10.1007/978-3-030-05955-2_5

studying the tundish structure mainly include two methods: physical simulation and numerical simulation.

However, the flow of molten steel in the actual tundish is a very complicated transmission process and many factors affecting the flow of molten steel. In order to save costs, a three-dimensional numerical model method of tundish is used to analyse the flow field, temperature field, residence time, and inclusion trajectory and removal rate in the tundish [4, 5]. Optimize the aspect ratio and aspect ratio of the inner cavity of the single-flow slab, increase the retention time while increasing the inclusion removal rate. The re-use model is used to verify the feasibility, ensure the accuracy of the model, and provide a theoretical basis for actual production.

Numerical Simulation Research Method

Model Hypothesis

In the numerical simulation, all the influencing factors in actual production cannot be considered, and it is impossible to completely simulate all the behaviours of molten steel flow. So in the simulation, some minor factors affecting the flow should be simplified. At the same time, it is necessary to ensure that the flow characteristics of the molten steel in the simulation do not change, so it is necessary to make some assumptions about the flow of molten steel in the tundish [6]:

(1) It is considered that the tundish molten steel is a viscous, incompressible Newtonian fluid;
(2) The thermophysical properties of the molten steel such as density, viscosity, specific heat capacity are constant;
(3) The heat loss of the tundish is mainly caused by heat radiation and heat conduction, and the heat transfer process is a three-dimensional steady state process;
(4) It is assumed that the inclusions are all spherical particles, and the movement of the inclusions does not affect the flow state of the tundish steel;
(5) Assume that the flow rate is the same across the cross section at the exit of the ladle.

Basic Equation

The process of describing the physical phenomenon of molten steel in the tundish has the following equation. Describe the turbulence process using the standard k-ε equation [7, 8].

(1) Continuity equation

$$\frac{\partial(\rho\mu_i)}{\partial x_i} = 0$$

(2) Momentum equation (N-S equation)

$$\frac{\partial(\rho\mu_i\mu_j)}{\partial x_j} = -\frac{\partial P}{\partial x_j} + \frac{\partial}{\partial x_j}\left(\mu_{eff}\frac{\partial\mu_i}{\partial x_j}\right) + \frac{\partial}{\partial x_i}\left(\mu_{eff}\frac{\partial\mu_j}{\partial x_i}\right) + \rho g_i$$

(3) Kinetic energy (K) equation

$$\frac{\partial(\rho\mu_i)}{\partial x_i} = \frac{\partial}{\partial x_i}\left[\left(\mu_{eff} + \frac{\mu_t}{\sigma_k}\right)\frac{\partial k}{\partial x_i}\right] + G - \rho\varepsilon$$

(4) Turbulent energy dissipation rate (ε) equation

$$\frac{\partial(\rho\mu_i\varepsilon)}{\partial x_i} = \frac{\partial}{\partial x_i}\left[\left(\mu_{eff} + \frac{\mu_t}{\sigma_\varepsilon}\right)\frac{\partial\varepsilon}{\partial x_i}\right] + c_1\frac{\varepsilon}{k}G - c_2\frac{\varepsilon^2}{k}\rho$$

among them $G = \mu_t\frac{\partial\mu_j}{\partial x_i}\left(\frac{\partial\mu_j}{\partial x_j} + \frac{\partial\mu_j}{\partial x_j}\right)$

$$\mu_{eff} = \mu + \mu_t = \mu + c_\mu\rho\frac{k^2}{\varepsilon}$$

(5) Heat transfer equation

$$\frac{\partial T}{\partial t} + \mu\frac{\partial T}{\partial x} + v\frac{\partial T}{\partial y} + w\frac{\partial T}{\partial z} = a\left(\frac{\partial^2}{\partial x^2} + \frac{\partial^2 T}{\partial y^2} + \frac{\partial^2 T}{\partial z^2}\right) + S$$

where μ_i and μ_j represent velocity vectors, x_i and x_j represent direction vectors, ρ is the fluid density, μ is laminar viscosity, μ_t is the turbulent viscosity, and μ_{eff} is the effective viscosity of the fluid.

Common values for the coefficients appearing in the above equations are shown in Table 1.

Table 1 Common values in the $k-\varepsilon$ model

Parameter		c_1	c_2	c_μ	σ_k	σ_ε
Value		1.43	1.92	0.09	1.0	1.0

Calculation Method

(1) In this paper, the velocity inlet is used for the ladle nozzle, the outlet is the pressure outlet, and the velocity value is 0 on the solid wall, and the condition of no slip boundary is adopted.
(2) In this paper, the SIMPLE pressure coupling method is adopted, and the convergence criterion is that the residual value is less than 0.001.
(3) After adding the tracer at the nozzle, the concentration response curve at the outlet of the tundish is obtained using the component transfer equation.
(4) The inclusion particles are regarded as discrete phases, and the Lagrange particle tracking model and the DPM model are used to calculate the trajectory and removal rate of the inclusions in the tundish. The calculation parameters are shown in Table 2.

Simulation Scheme

According to literature research, the capacity range of the tundish is recommended to be within 50–80t, and the working level of the tundish is controlled at 1100–1400 mm. The ratio of the bottom length to the height of the working level (L/H) should be greater than 3.5, and the ratio of the bottom width to the working level (W/H) is in the range of 0.8–1.0. This shape of the tundish facilitates the removal of inclusions.

In order to facilitate the research, the working liquid level in this paper is fixed at 1300 mm, the height of the tundish is unchanged at 1550 mm, the long side inclination angle is 7°, and the wide side inclination angle is 10°. A simulation study was carried out by changing the ratio of the base length to the working level height (L/H) and the ratio of the bottom width to the working level height (W/H).

Table 2 Tundish numerical simulation basic parameters

Parameter	Data	Parameter	Data
Steel liquid boundary layer viscosity	0.0067 kg m^{-1} s	Liquid steel thermal conductivity	41 w m^{-1} K^{-1}
Molar mass of molten steel	55.85 kg mol^{-1}	Steel liquid expansion coefficient	1.0×10^{-4} K^{-1}
Specific heat capacity of molten steel	750 J kg^{-1} K^{-1}	Free surface heat flow	15 kw m^{-2}
Steel liquid reference density	7000 kg m^{-3}	Bottom wall heat loss	1.4 kw m^{-2}
Inlet kinetic energy	$0.01V_{in}^2$ m^2/s^2	Long wall heat loss	3.2 kw m^{-2}
Inlet turbulent dissipation rate	$33.33K_{in}^{1.5}$ m^2/s^3	Short wall heat loss	3.8 kw m^{-2}
Inlet temperature	1823 K	Export conditions	Pressure outlet

Table 3 Tundish simulation scheme

Case	L/H	Bottom length/mm	Top length/mm	W/H	Bottom width/mm	Top width/mm	Working capacity/t
1	3.5	4550	5096	1	1300	1681	66
2	3.6	4680	5226	1	1300	1681	67
3	3.7	4810	5356	1	1300	1681	69
4	3.8	4940	5486	1	1300	1681	71
5	3.9	5070	5616	1	1300	1681	73
6	3.5	4550	5096	0.95	1235	1616	63
7	3.5	4550	5096	0.9	1170	1551	60
8	3.5	4550	5096	0.85	1105	1486	57
9	3.5	4550	5096	0.8	1040	1421	54

For high-speed single-flow slab tundish, nine different schemes were designed for the tundish in the range of 50–75t. The effects of L/H and W/H on the flow field and residence time in the tundish were studied. The specific scheme is shown in Table 3.

Model Verification

The accuracy of the mathematical model establishment has a great impact on the final simulation results, and it is related to whether the real flow phenomenon can be reproduced. The three-dimensional tundish flow field was simulated by Fluent software, and the results were compared with the stimulus response method in the physical experiment to verify the correctness of the mathematical model.

The residence time distribution curve (RTD) of any flow control device in the tundish with L/H of 3.5 and W/H of 1 was measured by water simulation experiment. Comparing the numerical simulation with the physical simulation, it can be found that the trend of the residence time distribution curve in the air condition in the tundish is basically consistent, indicating that the established mathematical model is in line with the actual situation and the model is established reasonably. The curve is shown below (Fig. 1).

Simulation Results and Discussion

Flow Field in the Tundish

After the simulation is completed, the horizontal cross-sectional flow field at the symmetrical longitudinal section and the long nozzle outlet in the tundish is obtained, as shown in the following figures (Figs. 2, 3, 4, 5, 6 and 7).

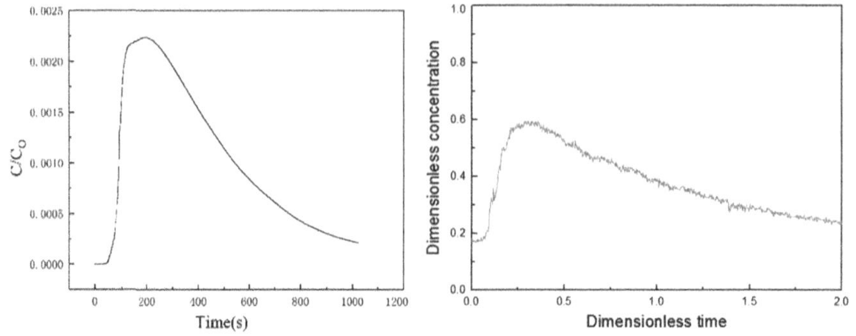

Fig. 1 RTD curve for tundish numerical simulation and water simulation with L/H of 3.5 and W/H of 1

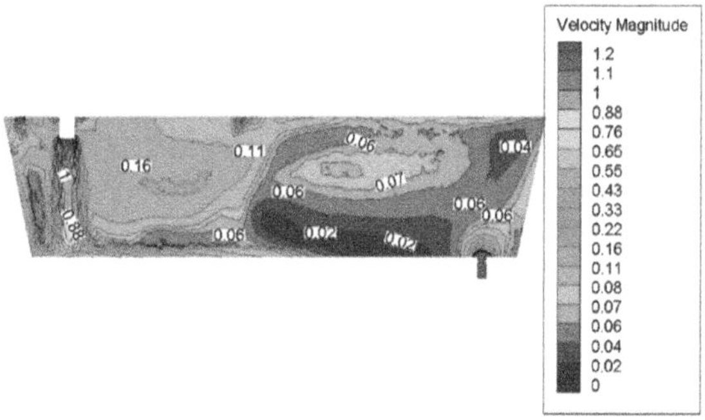

Fig. 2 Case 5 flow field longitudinal section

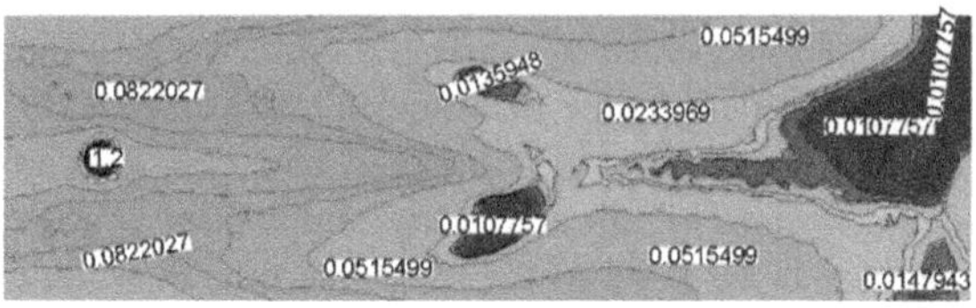

Fig. 3 Case 5 flow field cross section

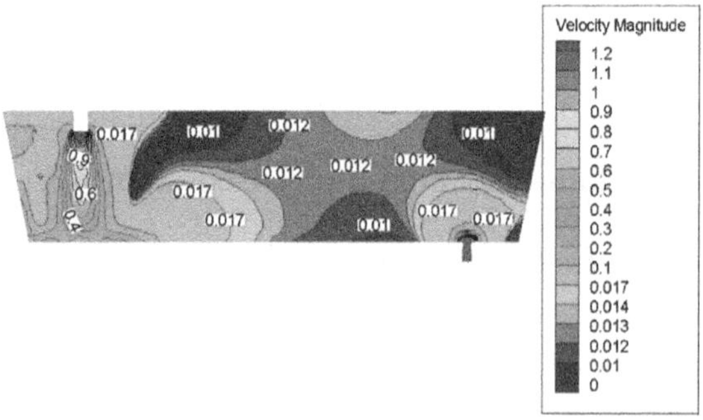

Fig. 4 Case 1 flow field longitudinal section

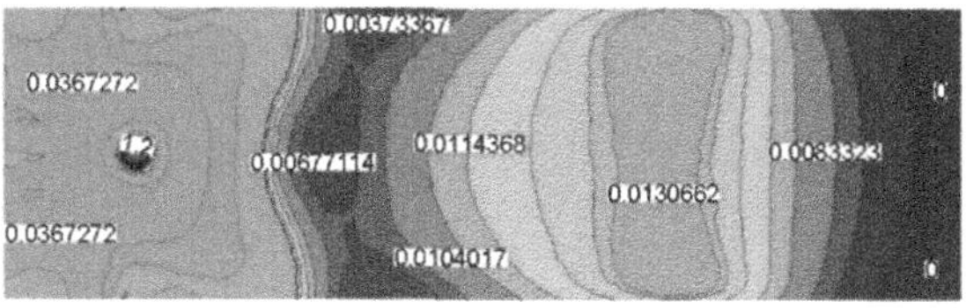

Fig. 5 Case 1 flow field cross section

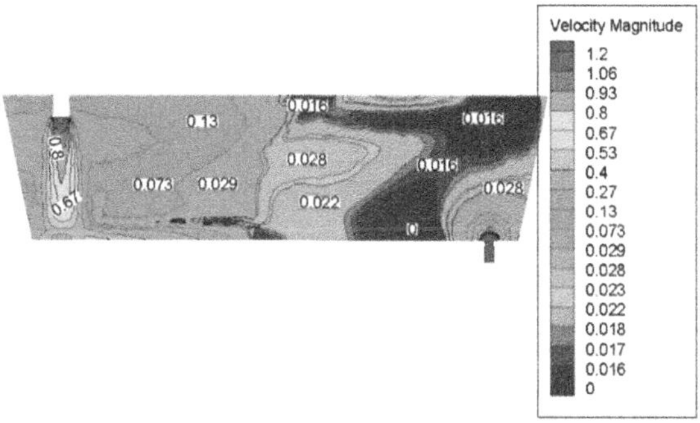

Fig. 6 Case 8 flow field longitudinal section

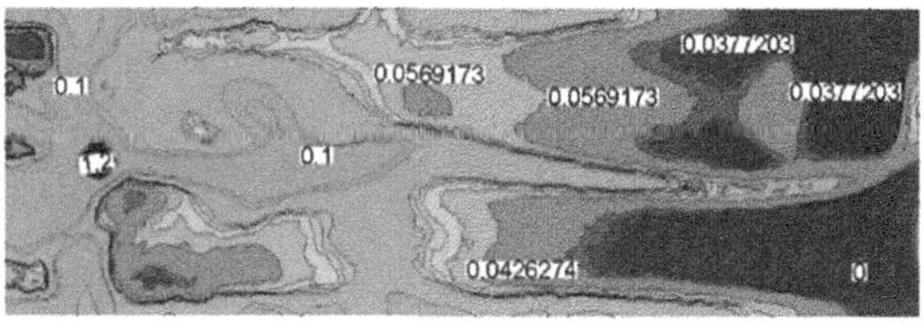

Fig. 7 Case 8 flow field cross section

From the simulation results of the flow field, the flow pattern of the Case 5 is the most reasonable, the dead volume is smaller than the other case, and the average speed is the largest. Conducive to the removal of non-metallic inclusions in molten steel to the surface.

When the base length is less than 3.9, the dead volume begins to increase, which is not conducive to the inclusions floating up. When the bottom width ratio working level is lower than 1, the dead volume is larger than the case where the aspect ratio is 1. Therefore, when L/H is 3.9 and W/H is 1, the tundish flow mode is the most reasonable.

Temperature Field in the Tundish

The temperature difference between different schemes is shown in Table 4. From the temperature field, the temperature difference of each case is not very large. The smallest temperature difference is the group of Case 9; the minimum temperature difference is 10 °C. The maximum temperature difference is that Case 5 is 21 °C.

Table 4 Temperature difference between different schemes

Case	Temperature difference/°C
1	12
2	14
3	17
4	20
5	21
6	17
7	16
8	13
9	10

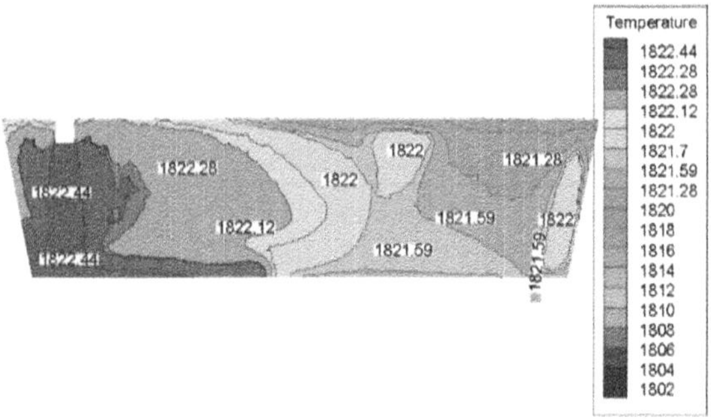

Fig. 8 Case 5 temperature field

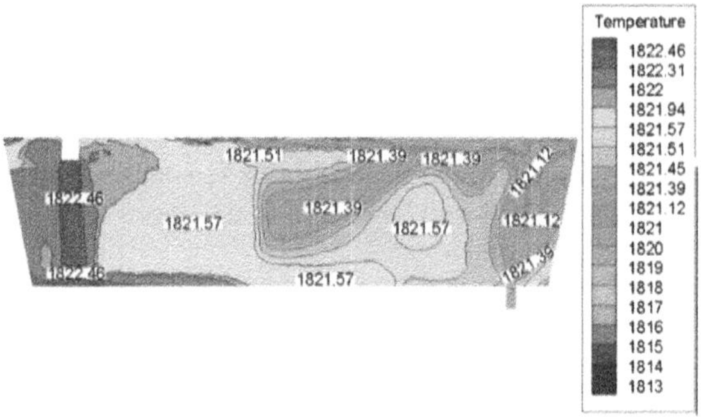

Fig. 9 Case 9 temperature field

Here is a picture of two sets of simulation results for the minimum temperature difference and the maximum temperature difference (Figs. 8 and 9).

Inclusion Trajectory

From the trajectory of inclusions, the ratio of the bottom width to the working level is 1 compared to the scheme below 1. The number of times the inclusions are in

Fig. 10 Case 5 inclusion trajectory

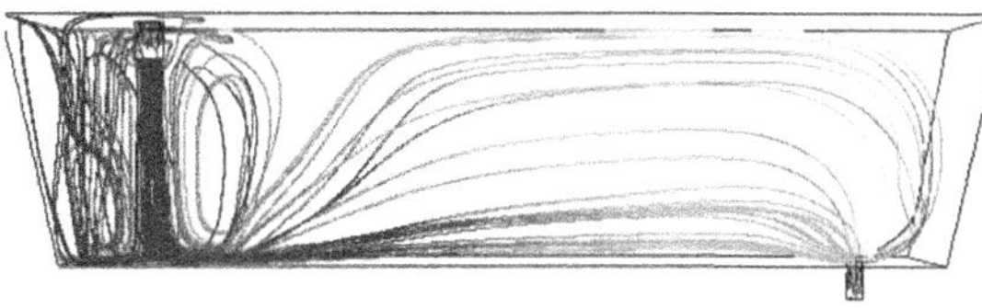

Fig. 11 Case 7 inclusion trajectory

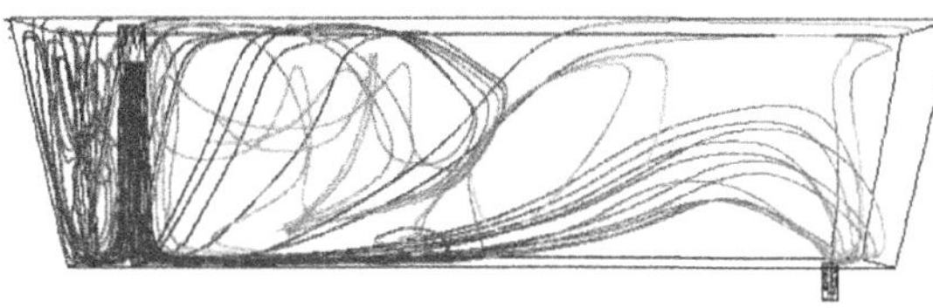

Fig. 12 Case 8 inclusion trajectory

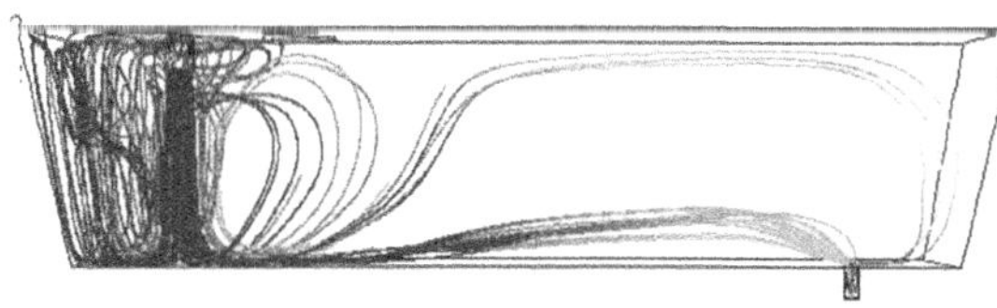

Fig. 13 Case 1 inclusion trajectory

contact with the surface of the fresh liquid is large, and the probability of inclusion removal is large. Below 1 solution, the inclusion trajectory is lower.

In the case where the bottom width and the working level ratio are both, the three groups of the ratio of the bottom length to the working level height are less than 3.9, and the inclusion circulation process is weak, which is not favourable for the floating contact, so the aspect ratio is 3.9. The aspect ratio is reasonable (Figs. 10, 11, 12 and 13).

Average Residence Time and Inclusion Removal Rate

As can be seen from Table 5, the average residence time is the group of Case 5. The longer the residence time, the higher the removal rate of inclusions. It can also be seen that in the range of the base width to the liquid level ratio of 0.8–1, the aspect ratio (W/H) is lowered, the residence time is small, and the inclusion removal rate is low. Not conducive to the purification of molten steel. Therefore, it is reasonable to choose 1 for the aspect ratio (W/H).

Table 4 Average residence time and inclusion removal rate for each solution

Case	V_p/%	V_d/%	Vm/%	V_p/V_d	Average residence time/s	Inclusion removal rate/%
1	28	10	61	2.8	378	66
2	29	9	60	3.2	382	68
3	29	8	64	3.6	387	69.2
4	30	8	62	3.75	388	71
5	32	7	61	4.6	391	71.9
6	28	8	62	3.5	370	65
7	27	9	62	3.0	355	67
8	26.5	11	60	2.4	350	69.8
9	26	12	61	2.2	336	68

The dead volume of Case 5 is smaller than that of the other 8 Cases, and the volume of plug flow and full mixed flow is relatively large, so this situation is better than the other 8 schemes. In addition, the residence time of the scheme 5 is longer, which provides sufficient time for the inclusions to float, and the corresponding inclusion removal rate is also the highest, which is beneficial to improve the quality of the slab.

Conclusion

This paper mainly uses fluent to establish a three-dimensional numerical simulation method to analyse the flow field, temperature field, inclusion motion trajectory, average residence time, and inclusion removal rate in large-capacity tundish. The results show that when the ratio of the bottom length of the tundish to the tundish is 3.9 and the bottom width ratio is 1, the flow mode in the tundish is the most reasonable, the removal rate of the inclusion rate is relatively high, and the average residence time is longer. The design of the capacity tundish provides a theoretical basis.

References

1. Lei H, Wang L, Wu Z et al (2002) Collision and coalescence of alumina particles in the vertical bending continuous caster. ISIJ Int 42(7):717–725
2. Zhang L, Taniguchi S, Cai K (2000) Fluid flow and inclusion removal in continuous casting tundish. Metall Mater Trans B 31B:253–266 (2000)
3. Hou Q, Yue Q, Wang H (2008) Modeling of inclusion motion and flow patterns in swirling flow tundishes with symmetrical and asymmetrical structures. ISIJ Int 48(6):87–792
4. Palai P, Sahoo PP, Dey A et al (2013) Constitutional segregation of Al_2O_3 in mold slag and its impact on steel cleanliness during continuous casting. Metall Mater Trans B 44(5):1185–1189

5. Chattopadhyay K, Isac M, Guthrie RIL (2010) Physical and mathematical modelling of steelmaking tundish operations: a review of the last decade(1999–2009). ISIJ Int 50(3):331
6. Kumar A, Mazumdar D, Koria SC (2008) Modeling of fluid flow and residence time distribution in a four-strand tundish for enhancing inclusion removal. ISIJ Int 48(1):38
7. Zong JH, Yi KW, Yoon JK (1999) Residence time distribution analysis by the modified combined model for the design of continuous refining vessel. ISIJ International 39(2):139–148
8. Xu K, Thomas BG (2012) Particle-size-grouping model of precipitation kinetics in microalloyed steel. Metall Mater Trans A 43A:1079–1096

Part II
Energy Efficient Clean Metallurgical Technologies

Preparation of High-Carbon Metallic Briquette for Coke Saving in Blast Furnace

Huiqing Tang, Shihong Liu and Kai Fan

Abstract Developments in blast furnace (BF) sector have long focused on coke saving in ironmaking. In this research, high-carbon metallic briquette was proposed as a novel BF feeding material for coke saving in BF. The preparing condition of the proposed high-carbon metallic briquette was optimized regarding its carbon content, cold strength and crushing strength after the reaction. Gasification behaviour of the optimally prepared briquette was examined using a custom-built thermalgravimetric analysis device. Results showed that the optimal preparation condition was $m_{hematite}/m_{coal} = 2.0$. The activation energy of carbon gasification in the briquette was 166 kJ/mol; analysis of the effective application of the briquette indicated that a suitable addition level of high-carbon metallic briquette in the ore layer in BF could suppress the lump coke gasification.

Keywords High-carbon metallic briquette · Blast furnace · Gasification · Coke saving

Introduction

Blast furnace(BF) ironmaking is the major hot metal making process in the world and will remain the same for the foreseeable future. The most important raw material fed into BF is coke, which constitutes a major portion of the production costs of hot metal. Due to a low availability of coking coal resources and the environmental problems in the coke production [1, 2], developments in the BF sector have long focused on coke-saving technologies.

Most of the coke-saving technologies are concentrated on reducing coke consumption below the cohesive zone in the BF. The major development in this aspect is the pulverized coal injection (PCI) [3, 4]. Some attention [5, 6] has also been paid

H. Tang (✉) · S. Liu · K. Fan
State Key Laboratory of Advance Metallurgy, University of Science and Technology Beijing, 30, Xueyuan Rd, Beijing 100083, China
e-mail: hqtang@ustb.edu.cn

© The Minerals, Metals & Materials Society 2019
T. Jiang et al. (eds.), *10th International Symposium on High-Temperature Metallurgical Processing*, The Minerals, Metals & Materials Series,
https://doi.org/10.1007/978-3-030-05955-2_6

to reduce coke consumption in the region of the BF shaft with a temperature of 1073 K through 1473 K as approximately 10% of the coke is gasified there by the carbon solution-loss reaction. Metallic briquette (direct reduced iron and hot briquette iron) have partially replaced the traditional BF iron ore burden in some countries for improving the productivity of the BF. Metallic briquette is usually produced from iron ore–carbon agglomerates, and various carbonaceous materials (coal, biochar and BF flue dust) could be introduced as a solid reducing agent in the agglomerates. Presently, the carbon content in the obtained metallic briquettes is approximately 2.0–3.0 wt%. If the metallic briquette could be prepared with a high carbon content without deteriorating its strength, carbon in the metallic briquette could also be gasified where the carbon solution-loss reaction occurs. This, in turn, would alter the BF gas composition and may realize a partial substitution of coke by the carbon in the metallic briquette.

In this study, the high-carbon metallic briquette was prepared by roasting cold-bonded iron-oxide–coal agglomerates. The preparing condition of high-carbon metallic briquette was optimized regarding its carbon content, cold strength, and crushing strength after the reaction. Thereafter, the gasification behaviour of the optimally prepared briquette was examined and the effective application of the briquette in BF was discussed at the same time.

Materials and Methods

Preparation of High-Carbon Metallic Briquette

Non-coking coal and pure iron oxide were used as raw materials. The non-coking coal powders for PCI operation are supplied by Jianbang Iron & Steel company (Shanxi, China); its proximate analysis is volatile matter: 20.3wt%, fixed carbon: 69.97 wt%, ash: 9.09 wt% and moisture: 0.91 wt%. The Fe_2O_3 powders (analytical reagent grade) were purchased from Sinopharm Chemical Reagent Co.(Shanghai, China). Mean diameters of Fe_2O_3 powders and of coal powders were 2.85 and 55.39 μm, respectively.

The iron oxide and coal powders were thoroughly mixed with an addition of 10% distilled water and 2% organic binder (cellulose). In the mixture, the mass ratio of the iron oxide fines to the coal fines ($m_{hematite}/m_{coal}$) was predetermined. These moistened fines were pressed into iron-oxide–coal agglomerates using a die under a pressure of 30 MPA. After being air-dried for 24 h, followed by drying in an electric oven at 383 K for 2 h, the dried agglomerates were subjected to roasting in an electric furnace under N_2 atmosphere. During the roasting process, the furnace was heated from room temperature to 1273 K at a rate of 5 K/min; after maintaining at 1273 K for 30 min, the furnace was allowed to cool down naturally. The prepared briquettes were with a cylindrical shape (diameter = 15 mm and height = 15 mm). The mass of each briquette was approximately 6.0 g. In the present research, six briquette samples were prepared and their preparing conditions are listed in Table 1.

Table 1 Preparing conditions of different briquette samples

Sample	A	B	C	D	E	F
$(m_{hematite}/m_{coal})$/-	0	0.5	1.0	1.5	2.0	2.5

Strength Tests

The cold strength of the samples was tested using a MJDW-10B electronic universal testing machine (Maijie Co, Jinan, China). The load speed of the jig was set at 20 mm/min. The arithmetic mean of five tests for each sample was calculated as the final value. Method for testing the crushing strength after the reaction of the briquette samples was the same for the cold strength. Reaction conditions for the crushing strength tests were determined with a reference to the BF in-furnace conditions at 1373 K. The briquettes reacted with a gas mixture of 80 vol.% CO–20 vol.% CO_2 under 1373 K in an alumina tube reactor (diameter = 45 mm) for 60 min. The gas flow rate into the reactor was 2000 ml/min (standard pressure and temperature (STP)).

Gasification Tests

Gasification tests of the briquette sample were conducted using a custom-built thermalgravimetric analysis device. A detailed description of the device is given in Ref. [7]. The furnace was firstly preheated to the desired temperature and stabilized for 30 min under N_2. One briquette was loaded into the sample holder, preheated at 773 K for 5 min in the upper part of the reaction tube, then introduced into the hot zone. At this time, the N_2 feed was replaced with a CO–CO_2 gas mixture and the mass of the briquette was measured by the electronic scale and recorded by a computer every 2 s. After the predetermined time, the briquette was withdrawn and cooled under N_2. During the test, the gas flow rate into the reaction tube was maintained at 3000 ml/min (STP).

Analysis and Characterization

Selected briquettes were analyzed and characterized using the following techniques. Carbon content was measured using a CS-2800 infrared carbon–sulphur analyzer (NCS Co., China). Phase examination was conducted by X-ray diffraction (XRD) using a DMAX-RB X-ray diffractometer (Rigaku Co., Japan). Microstructure analysis was conducted by SEM using a Quanta-250 scanning electron microscope (FEI Co., US).

Results and Discussion

Assessment of Briquette Quality

Carbon contents of the six samples are listed in Table 2. Carbon content of the briquette samples is in a wide range. Sample A, which was prepared using the pure coal fines, was with a carbon content of 91.0 wt%, and sample F, which was the briquette prepared under $m_{hematite}/m_{coal}$ of 2.5, was with a carbon content of 14.9 wt%.

Cold strength of the samples is shown in Fig. 1. It could be seen that the samples A and F were with a high strength, samples B and E with a medium strength and samples C and D with a low strength.

Crushing strength after reaction of the samples is shown in Fig. 2. As can be seen in Fig. 2, after reacting under BF in-furnace conditions of 1373 K, samples A, B, C and D showed a strength of lower than 1000 N/briquette; however, samples E and F exhibited a strength of more than 2500 N/briquette.

During BF operation, normal iron-ore–carbon composite briquettes are usually required to have a strength of more than 1200 N/briquette. With respect to cold strength, crushing strength after reaction, and carbon content, sample E is considered to be qualified for the purpose of the present research.

The strength of the briquette is closely related to its microstructure. The iron oxide powders had been completely reduced in the roasting process. Figure 3 shows the cross-section SEM images of sample E. Figure 3a shows that metallic iron grains are highly dispersed in the briquette and Fig. 3b shows that the iron particles are with a size of less than 15 μm and with various shapes, and they are

Table 2 Carbon contents of different samples

Sample	A	B	C	D	E	F
Carbon content(wt%)	91.0	65.5	45.7	33.4	25.6	14.9

Fig. 1 Cold compressive strength of different briquette samples

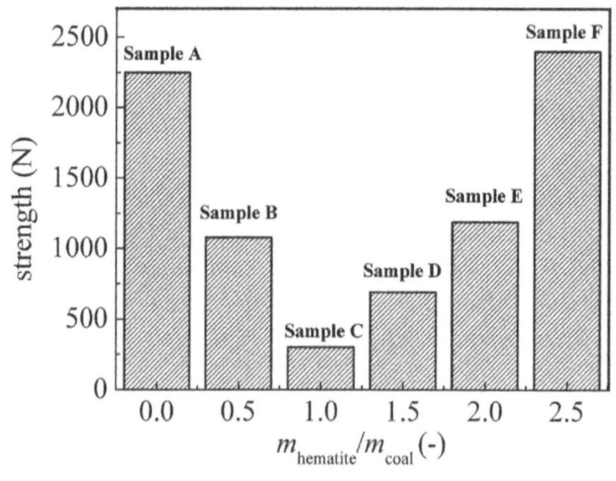

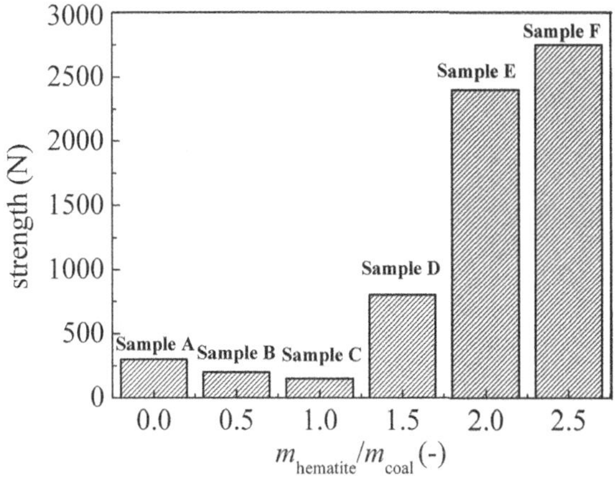

Fig. 2 Crushing strength after reaction of different samples

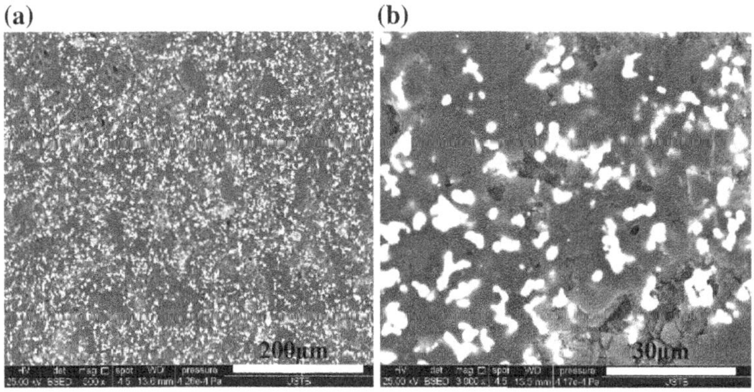

Fig. 3 SEM images in the cross section of Sample E: **a** morphology of the briquette; **b** morphology of the iron grains (white) and of the carbon particles (grey)

intimately adhered to the surfaces of the carbon particles forming a network. Thus, it is considered that the bonding force of sample E was originated from the formation of a primary network by the tiny iron fibres.

Figure 4 shows the cross-section SEM images of sample E after the reaction. Figure 4a shows that metallic iron grains become larger. Figure 4b shows that some iron grains partially coalesced around the carbon particle, indicating that the iron network structure had been strengthened in the briquette; therefore, the strength of the briquette got increased after the reaction.

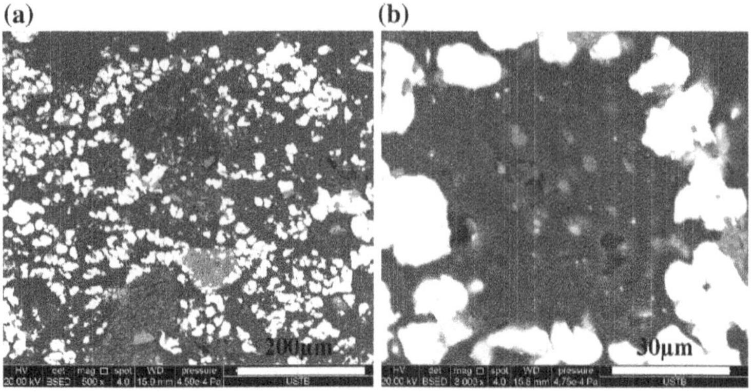

Fig. 4 SEM images in the cross section of sample E after the reaction: **a** morphology of the briquette; **b** morphology of the iron grains (white) and of the carbon particles (grey)

Gasification of High-Carbon Metallic Briquette

Calculation of activation energy is useful to estimate the reactivity of the briquette. Reaction kinetics of the briquette gasification were studied under temperature ranging from 1273 K to 1373 K in a gas atmosphere of 80 vol.% CO–20 vol.% CO_2. Under this condition, iron oxidation is not able to take place. Plots of carbon gasification fraction(f) against time(t) under different temperatures are depicted in Fig. 5. In Fig. 5, carbon gasification fraction was calculated by Eq. (1)

$$f = -1.0 \times \Delta m/(m_0[\%C]) \tag{1}$$

where [%C] is the carbon content in sample E and is available in Table 2.

Fig. 5 Gasification of sample E under different temperatures

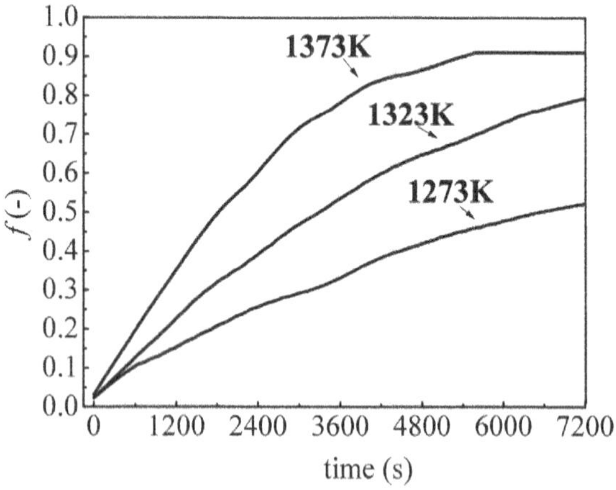

For a kinetic study, it is important to know the controlling step of the briquette gasification rate. Under the experimental temperature range, increasing the gas flow rate had a negligible effect on the carbon gasification rate; therefore, it was concluded that the briquette gasification was not controlled by the external gas diffusion. Diameter and height of the briquette were nearly equal, so its shape was considered to be spherical. Thus, carbon gasification in the briquette can then be described using the unreacted shrinking core model. The high-carbon metallic briquette is porous. The model is thus formed on the basis of control by chemical reaction and is described by Eq. (2).

$$1-(1-f)^{\frac{1}{3}}=kt \tag{2}$$

where k is the temperature-dependent rate constant(1/s).

Figure 6 shows the results of the reaction-controlling mechanism estimation under different temperatures. The lines in Fig. 6 are regressed using the least square method. The correlation coefficient (adj. R-square) under each temperature is larger than 0.99, indicating that the dominant mechanism of the briquette gasification under the tested temperatures is the chemical reaction.

The gasification kinetics of the briquette can be reasonably described using Eq. (2). In Fig. 6, the value of k under each temperature was calculated using the slope of the fitting line and they were 3.303×10^{-5}, 5.93×10^{-5} and 1.05×10^{-4} 1/s under 1273, 1323 and 1373 K, respectively. It was found that $ln\ k$ had a linear relationship with 10,000/T as depicted in Fig. 7, so it can be stated that the gasification of the briquette followed the Arrhenius law expressed as Eq. (3).

$$k=k_0e^{-E/RT} \tag{3}$$

where k_0 is the pre-exponential factor(1/s). E is the activation energy(J/mol) and T is the temperature(K), respectively.

From the slope of the fitting line in Fig. 7, the activation energy of the briquette is calculated to be 169 kJ/mol. Activation energy of conventional coke gasification is mostly more than 200 kJ/mol. Compared to the coke, CO_2 reactivity of the briquette is much higher. Iron matters are usually considered to have a catalytic effect on the carbon solution-loss reaction. In the briquette, the freshly generated

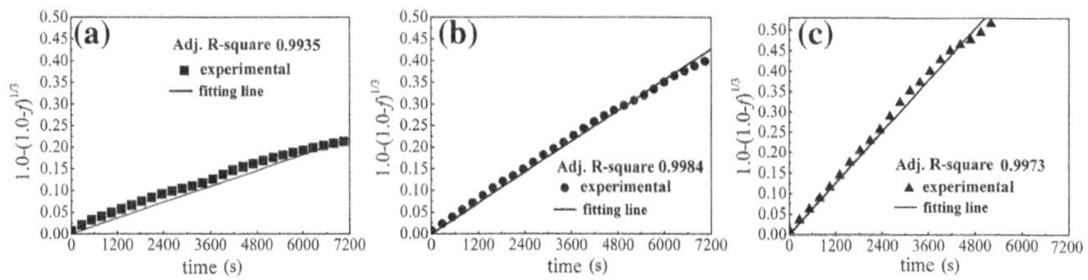

Fig. 6 Result of reaction-controlling mechanism identification by Eq. (3) under different temperatures: **a** 1273 K; **b** 1323 K; **c** 1373 K

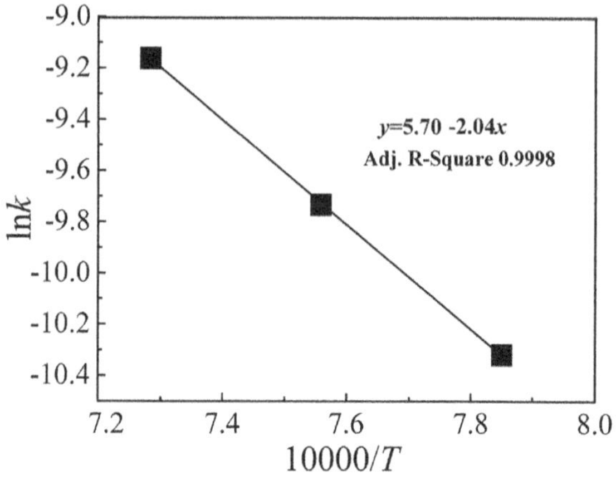

Fig. 7 Arrhenius plot calculated using Eq. (3)

iron grains were intimately adhered to the surfaces of the carbon particles, forming many activated centres. Accordingly, the carbon solution-loss reaction could be catalyzed.

Effective Application of High-Carbon Metallic Briquettes to BF

In BF operation, coke and iron ore materials are fed alternatively at the top of the furnace using the charging device. Layered structures of coke and ore are formed in the shaft. Coke is gasified in the BF shaft by CO_2 following the carbon solution-loss reaction. The carbon gasification is controlled by the partial pressure of CO_2 in BF. The high-carbon metallic briquette could be charged with a suitable additional level in the ore layer in BF. Due to the briquettes with a high CO_2 reactivity, the carbon in briquette is preferentially consumed and CO_2 from iron oxide reduction could be converted to CO immediately. Therefore, CO_2 concentration in the coke layer drops. When the briquettes reach the cohesive zone, carbon content in the briquette becomes low. The analysis of the effective application of the briquette indicates that the presence of high-carbon metallic briquette in the BF burden could give a positive effect on suppressing coke gasification and would not influence the permeability of the lower part of BF.

Conclusions

(1) The high-carbon metallic briquette can be optimally prepared with $m_{hematite}/m_{coal}$ of 2.0. The briquette is with a carbon content of 25 wt% and with both cold strength and crushing strength after reaction of more than 1200 N/briquette. Its quality meets the requirement of the BF process.

(2) The activation energy of the briquette is 169 kJ/mol, compared to the conventional coke. Its CO_2 reactivity is much higher.

(3) The briquette can be charged into BF with the ore layer. A suitable mixing level of the briquette could increase the partial pressure of CO in BF gas, suppressing coke gasification. Moreover, it would not influence the permeability of the lower part of BF.

References

1. Guo ZC, Fu ZX (2010) Current situation of energy consumption and measures taken for energy saving in the iron and steel industry in China. Energy 35(11):4356–4360
2. Li Y, Zhu L (2014) Cost of energy saving and CO_2 emissions reduction in China's iron and steel sector. Appl Energ 130:603–616
3. Mathieson JG, Truelove JS, Rogers H (2005) Toward an understanding of coal combustion in blast furnace tuyere injection. Fuel 84:1229–1237
4. Nomura S, Callcott TG (2011) Maximum rates of pulverized coal injection in ironmaking blast furnaces. ISIJ Int 51(7):1033–1043
5. Zhao H, Bai H, Cheng Y (2013) Effect of coke reaction index on reduction and permeability of ore layer in blast furnace lumpy zone under non-isothermal condition. J Iron Steel Res Int 20 (4):6–10
6. Li P, Liu W, Zhang H (2017) Impact of potassium on gasification reaction and post-reaction strength of ferro-coke. ISIJ Int 57(11):1947–1954
7. Tang HQ, Yun ZW, Fu XF, Du S (2018) Modeling and experimental study of ore-carbon briquette reduction under $CO–CO_2$ atmosphere. Metals 8(4):205–215

Study of Siderite Fluidized Magnetization Roasting and Magnetic Separation

Qiang Zhao and Jilai Xue

Abstract The development and utilization of siderite is worldwide difficult problem. The conventional beneficiation methods only can get iron concentrate with iron grade 45–50% and recovery 40–55%; the resource utilization is low. The research creatively applies fluidization technology in aluminium industry to refractory iron ore roasting, aiming at siderite with TFe35.19% and iron carbonate content 91.28%, in the best fluidized magnetization roasting conditions, by one magnetic separation; the magnetic concentrate with iron grade 59.17% and recovery rate 73.74% was obtained; and the index is good. Fluidized magnetization roasting is the proper process for refractory siderite resource; it is of great significance to enhance the competitiveness of domestic iron ore concentrate and ensure the safety of China's iron ore supply.

Keywords Siderite · Fluidization · Magnetizing roasting · Magnetic separation

Introduction

In China, the reserves of iron ore are 85.08 billion tons, because the refractory iron ore characteristics are poor, thin, scattered and miscellaneous, the utilization cost is high. The beneficiation problem of refractory iron ore such as siderite was unresolved; the iron ore concentrate with 45–60% recovery and 45–55% iron grade was got by conventional physical beneficiation method [1–3]. Studies showed that magnetization roasting-weak magnetic separation was the best way to deal with this kind of ore [4, 5]; magnetization roasting which had been used to industry in the early stage was rotary kiln and shaft furnace. However, the existing problems

Q. Zhao (✉) · J. Xue
School of Metallurgical and Ecological Engineering, University of Science and Technology Beijing, Beijing 100083, China
e-mail: 13319532520@163.com

J. Xue
e-mail: jx@ustb.edu.cn

© The Minerals, Metals & Materials Society 2019

T. Jiang et al. (eds.), *10th International Symposium on High-Temperature Metallurgical Processing*, The Minerals, Metals & Materials Series,
https://doi.org/10.1007/978-3-030-05955-2_7

included high heat consumption, long roasting time, low iron recovery, low equipment operating rate and uneven quality of roasted ore; the large-scale industrial application was seriously restricted. The fluidized magnetization roasting technology was the most promising technology to treat siderite resources at the present stage [6–11].

In this study, a detailed fluidized magnetization roasting-magnetic separation experiment was carried out for a siderite from Guizhou, China, expecting excellent test indicators. The factors such as fluidizing velocity, roasting temperature, roasting time and grinding fineness were explored. The raw ore was characterized by X-ray diffraction (XRD), elemental composition analysis, scanning electron microscopy (SEM) and optical microscopy. Characterization of the roasted ore was detected by SEM and XRD; the iron concentrate was estimated by elemental XRD and composition analysis.

Raw Material

The chemical compositions of the raw siderite ore sample are shown in Table 1. The iron content was 35.19%, while the main impurities include SiO_2, CaO, Al_2O_3 and the loss on ignition (LOI) reached up to 25.74%. The chemical phases and distribution of iron in siderite ore are shown in Table 2. XRD analyses of siderite ore are shown in Fig. 1. It was noted that most of iron existed in siderite phase, which accounted for a proportion of 91.28% in the total iron.

Table 1 Chemical compositions of siderite ore sample (wt%)

TFe	**Fe_2O_3**	**FeO**	**SiO_2**	**CaO**	**MgO**	**Al_2O_3**
35.19	2.58	40.89	15.33	2.93	1.61	3.14
MnO	**K_2O**	**Na_2O**	**As_2O_3**	**P**	**S**	**LOI**
0.32	0.080	0.024	0.012	0.01	0.08	25.74

Table 2 Chemical phases and distribution of iron in siderite ore (wt%)

Minerals	Siderite	Haematite	Magnetite	Ferrocite	Iron sulphide	Total
Content	32.12	1.39	0.45	0.99	0.24	35.19
Fraction	91.28	3.95	1.28	2.81	0.68	100.00

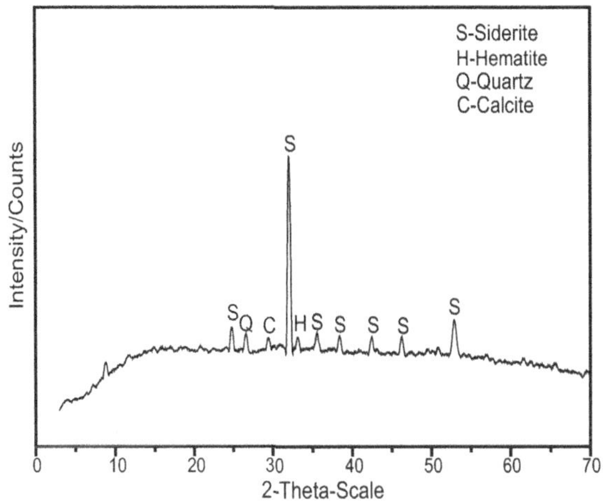

Fig. 1 XRD analyses of siderite ore

Experiment Methods and Evaluation Standard

Experiment Methods

The test installation diagram was shown in Fig. 2. First, the furnace tube was cleaned by the inert gas(N_2). Then, the raw ore materials were added to the furnace from charging hopper for fluidized magnetization roasting at designed temperature and duration. When the magnetization roasting test was finished, the roasted ore was blown out by water cooling. Next, the wet roasted product was dried at 100 °C for at least 2 h in the oven. And then the dried ore was dressed by wet magnetic separation for 15 min in an XCGS-73 Davis magnetic tube at a magnetic field strength of 0.15 T. Finally, both dry concentrate and tailings were chemically analysed.

Evaluation Standard

At home and abroad, some researchers had used the magnetization rate or conversion rate to characterize the effect of magnetization roasting; some researchers had used the iron grade and recovery rate of magnetic concentrate to evaluate it. This study mainly adopted the latter as the evaluation standard.

Results and Discussion

The flash magnetization roasting test was carried out in the self-developed roasting equipment; the effects of fluidization velocity, roasting time, roasting temperature and fineness of grinding were investigated.

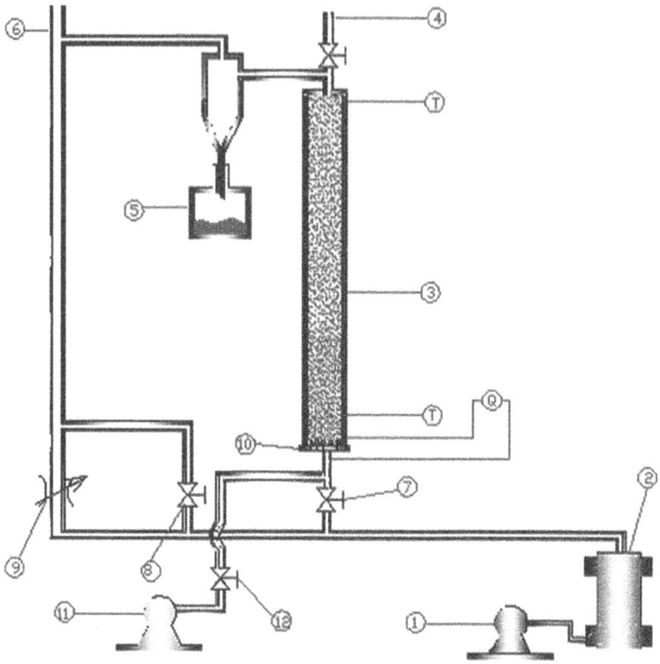

Fig. 2 Test installation diagram of fluidized magnetizing roasting unit(1—draught fan, 2—gas generator, 3—reacting furnace, 4—charging hole, 5—materials receiver, 6—exhaust pipe, 7—changeover valve, 8—changeover valve, 9—regulating valve, 10—wind-distribution plate, 11—draught fan, 12—control valve, T—temperature measuring port, andQ–gas flow measurement points)

Fluidizing Velocity Test

The test conditions were as follows: the feed ore fineness was −0.074 mm 55.63%, roasting temperature was 800 °C, roasting atmosphere was 100% N_2, and roasting time was 3 min; after roasting, roasted ore was directly disposed by weak magnetic separation without ball milling, magnetic field intensity was 0.25 T, and the influence of gas fluidization velocity on the magnetic concentration was investigated, which was presented in Fig. 3.

The experimental tests results showed that in the case of low gas fluidizing velocity(0.20–0.30 m/s), the magnetic concentration index was better, the iron grade of magnetic concentrate was about 57%, and the recovery rate was about 78%. When the fluidizing velocity increased to 0.6 m/s, the iron grade changed a little, but the recovery decreased to 68.76%, and the decrease was significant. With the increase of fluidization velocity, the recovery rate decreased, this indicated that the gas fluidization velocity was too fast and was not conducive for magnetization. This was because as the gas fluidization speed increased, the

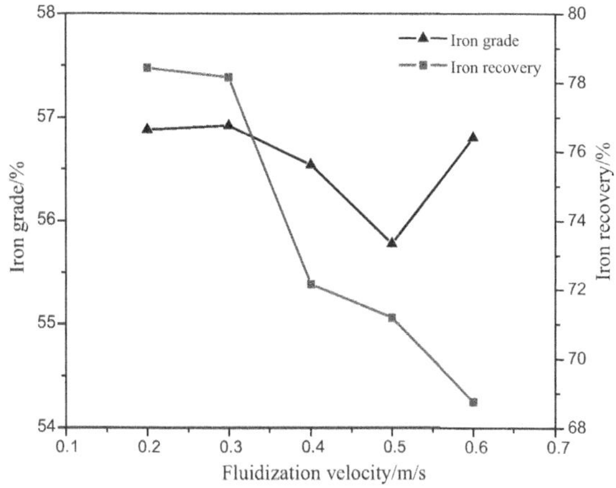

Fig. 3 Influence of gas fluidizing velocity on the magnetic concentration index

material would also be sprayed to a higher height, the existence form was too dispersed, the gas-solid contact probability decreased, and the magnetization effect became worse. Therefore, the selected gas fluidization velocity was 0.30 m/s for the follow-up study.

Roasting Time Tests

The test conditions were as follows: the feed ore fineness was −0.074 mm 55.63%, roasting temperature was 800 °C, roasting atmosphere was 100% N_2, and gas fluidization velocity was 0.3 m/s; after roasting, roasted ore was directly disposed by weak magnetic separation without ball milling, magnetic field intensity was 0.25 T, and the influence of roasting time on the magnetic concentration was investigated, which was presented in Fig. 4.

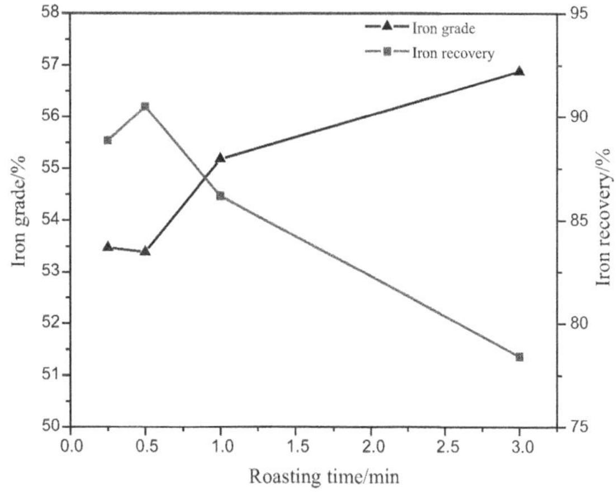

Fig. 4 Influence of roasting time on the magnetic concentration index

The experimental tests results presented that when the roasting temperature was 800 °C, the flash roasting time was extended from 0.25 to 3 min, the iron grade of magnetic separation concentrates increased from 53.47 to 56.88%, while the recovery decreased from 88.84 to 78.43%. The longer the flash roasting time, the higher the iron grade and the lower the recovery. Considering the double index of iron grade and recovery, this study selected roasting time 0.50 min for follow-up study; at this time, the iron grade was 53.39% and the recovery was 90.50%.

Roasting Temperature Tests

The test conditions were as follows: the feed ore fineness was −0.074mm 55.63%, roasting time was 0.5 min, roasting atmosphere was 100% N_2, and gas fluidization velocity was 0.3 m/s; after roasting, roasted ore was directly disposed by weak magnetic separation without ball milling, magnetic field intensity was 0.25 T, and the influence of roasting temperature on the magnetic concentration was investigated, which was presented in Fig. 5.

It was noted that when the flash roasting time was 0.5 min, the roasting temperature increased from 700 to 900 °C, the iron grade increased from 50.92 to 57.33%, while the recovery rate increased from 81.09 to 90.78% and then fell to 68.17%. It was indicated that with the increase of roasting temperature, the high temperature decomposition and reduction of siderite was increased. When the roasting temperature was in the range of 750–800 °C, the recovery was about 90%. The roasting temperature continued to rise, but the recovery rate of iron was decreasing; it was demonstrated that there was a reduction phenomenon, and the FeO was generated, the magnetic drop of roasted ore led to decrease of recovery. Therefore, the roasting temperature 750 °C was chosen to carry out the subsequent experiments.

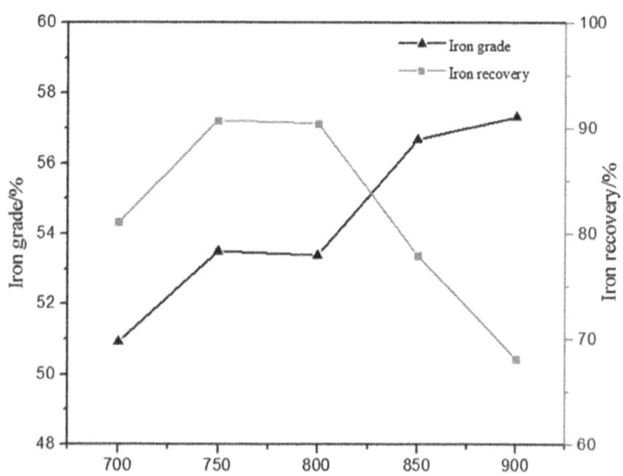

Fig. 5 Influence of roasting temperature on the magnetic concentration index

Grinding Fineness Tests

The test conditions were as follows: roasting temperature was 750 °C, roasting time was 0.5 min, roasting atmosphere was 100% N_2, and gas fluidization velocity was 0.3 m/s; after roasting, roasted ore was disposed by ball milling, magnetic field intensity was 0.25 T, and the influence of grinding fineness on the magnetic concentration was investigated, which was presented in Fig. 6.

The roasted ore structure of siderite after magnetization roasting was relatively loose, which was beneficial to improve the grinding effect and reduce the grinding cost. The results showed that with the extension of grinding time, the size of roasted ore became finer, the iron grade increased from 53.51 to 61.30%, while the recovery was reduced from 90.78 to 66.80%. Obviously, the improvement of grinding fineness was good for increasing iron grade, but it had great influence on iron recovery. After taking into account all the factors, the grinding fineness −0.045 mm 95.3% was selected, this time the iron grade was 59.17%, and the recovery was 73.74%.

Comprehensive Process Conditions

The optimum technological condition for the siderite fluidized magnetization roasting-magnetic separation was as follows: the feed ore fineness was −0.074 mm 55.63%, roasting temperature was 750 °C, roasting time was 0.5 min, roasting atmosphere was 100% N_2, gas fluidization velocity was 0.3 m/s, grinding fineness was −0.045 mm 95.3%, and magnetic field intensity was 0.25 T. The iron concentration with 59.17% iron grade, and 73.74% recovery was obtained. Table 3 was the chemical composition of magnetic concentrate obtained under this condition.

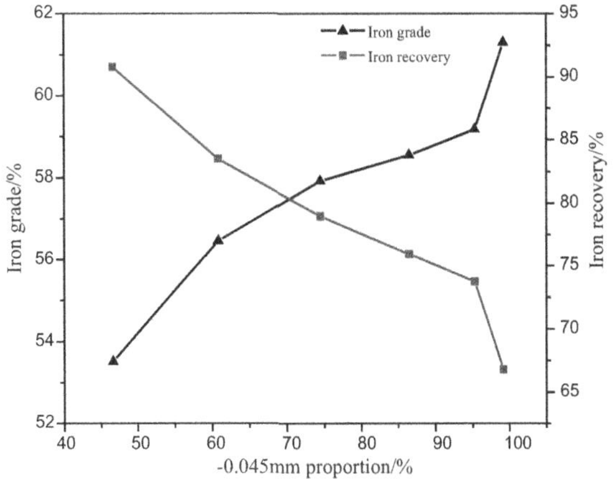

Fig. 6 Influence of grinding fineness on the magnetic concentration index

Table 3 Chemical composition of magnetic concentrate %

Tfe	FeO	Fe_2O_3	SiO_2	Al_2O_3	CaO
59.17	29.81	51.94	8.95	2.17	2.11
MgO	MnO	K_2O	Na_2O	P	S
1.61	1.12	0.063	0.089	0.21	0.065

Conclusions

(1) The development and utilization of siderite is worldwide difficult problem; the conventional beneficiation methods only can get iron concentrate with iron grade 45–50% and recovery 40–55%; and the resource utilization is low. Studies showed that magnetization roasting-weak magnetic separation was the best way to deal with this kind of ore
(2) The optimum technological condition for the siderite fluidized magnetization roasting-magnetic separation was as follows: the feed ore fineness was −0.074 mm 55.63%, roasting temperature was 750 °C, roasting time was 0.5 min, roasting atmosphere was 100% N_2, gas fluidization velocity was 0.3 m/s, grinding fineness was −0.045 mm 95.3%, and magnetic field intensity was 0.25 T. The iron concentration with 59.17% iron grade and 73.74% recovery was obtained. It was important to show that fluidized magnetization roasting was the appropriate process for the treatment of refractory siderite resources, and it was of great significance to improve the competitiveness of domestic iron concentrate and guarantee the safe supply of iron ore in China.

References

1. Liu XH (2013) Research on beneficiation process for low-grade limonite Ore. Min Metall Eng 4:70–72
2. Xie XZ, Wang YH (2010) Current situation and thinking of mineral processing technique on limonite ore. Met Mine 1:6–9
3. Yu YF (2006) Development trends of iron mines and status and problems of beneficiation technology in China. Min Metall Eng 1:21–23
4. Zhao Q (2010) Study of new technique and mechanism of strengthening limonite's magnetizing roasting. Central South University, MS.D. thesis, pp 25–33
5. Yu YF (2016) Research of limonite and siderite type refractory ore flash magnetizing roasting and engineering transformation. Min Metall Eng 8:1–9
6. Ren YF (2006) Study on flash magnetization roasting technology for refractory low-grade red iron ores. Wuhan University of Technology, MS.D.thesis, pp 33–52
7. Feng ZL, Yu YF, Liu GF (2011) Kinetics of the thermal decomposition of wangjiatan siderite. J Wuhan Univ Technol 3:523–526
8. Liao ZH (2015) Research on siderite flash magnetizing roast mechanism and reasonable size limit. MS.D.thesis, Changsha Research Institute of Mining and Metallurgy Co., LTD:25–53

9. Liu XY, Yu YF, Chen W (2009) Research on flash magnetizing roasting-magnetic separation for daxigou siderite. Met Mine 10:84–85
10. Zhang HQ, Yu YF, Peng ZY (2009) Study on flash magnetizing roasting of Huangmei limonitie ore. Iron Steel 7:11–14
11. Chen W (2017) Six hundred thousand t/a refractory siderite flash magnetizing roasting complete sets technique and equipment. Met Mine 3:54–56

Emission Profile of PM_{10} and $PM_{2.5}$ in Iron Ore Sintering Process and Control Technology

Zhiyun Ji, Xiaohui Fan, Min Gan, Xuling Chen, Wei Lv, Guojing Wong and Tao Jiang

Abstract Controlling the emission of ultra-fine particulate matters (PM_{10} and $PM_{2.5}$) in sintering process is significant for green iron & steel manufacturing. In this investigation, the emission profile and control technique of PM_{10} and $PM_{2.5}$ during sintering process were studied. PM_{10} and $PM_{2.5}$ characterized high emission concentration in special sintering areas, which were sintering stage-4 to stage-6, and sintering stage-4 to stage-5, respectively. The emission load of PM_{10} and $PM_{2.5}$ in special areas accounted for about 63.5 and 47.0% of the total, respectively. Based on these properties, spraying organic binder solution on granules during granulation process was found to improve the absorption efficiency of sintering bed to PM_{10} and $PM_{2.5}$, which made them intensively emitted within sintering stage-5 and stage-6, with accounting for 74.8 and 74.2% of the total. The research findings were helpful to realize the efficient and economic control of PM_{10} and $PM_{2.5}$ in practical sintering plants.

Keywords Iron ore sintering · PM_{10} and $PM_{2.5}$ · Emission profile · Efficient control

Introduction

Iron ore sintering process is the main source of PM_{10} and $PM_{2.5}$ throughout the whole steelmaking chain, accounting for 40% of the total amount. As PM_{10} and $PM_{2.5}$ characterized huge specific surface area and high surface activity, they tended to be the carrier of heavy metals, alkali metals and persistent organic pollutants generated or removed from high-temperature process [1, 2]. After being released into the atmosphere, PM_{10} and $PM_{2.5}$ formed aerosol, which acted as the inducement of hazy weather, smog episode and ozone depletion, etc. The environmental

Z. Ji (✉) · X. Fan · M. Gan · X. Chen · W. Lv · G. Wong · T. Jiang
School of Minerals Processing & Bioengineering, Central South University, South Lushan Road, No.932, Changsha 410083, Hunan, People's Republic of China
e-mail: zhiyunji@sina.com

© The Minerals, Metals & Materials Society 2019

T. Jiang et al. (eds.), *10th International Symposium on High-Temperature Metallurgical Processing*, The Minerals, Metals & Materials Series,
https://doi.org/10.1007/978-3-030-05955-2_8

problems caused by PM_{10} and $PM_{2.5}$ have been exerting serious threat to human health [3]. Therefore, controlling the emission of PM_{10} and $PM_{2.5}$ during sintering process contributed to reducing the emission of ultra-fine particulate matters from the entire iron & steel industry, which is of great significance for green iron & steel manufacturing.

Unfortunately, current research about PM_{10} and $PM_{2.5}$ mainly focused on coal combustion area, and the characteristics, formation mechanism and controlling techniques were systematically studied or developed [4, 5]. However, the research in sintering area is just the beginning, the reports gave the basic physicochemical properties and emission property of dust emitted from the stack, while there was little research about the emission rule, physicochemical property of PM_{10} and $PM_{2.5}$ during sintering process [6, 7]. According to our knowledge, there are even no direct controlling techniques for reducing the emissions of PM_{10} and $PM_{2.5}$ from sintering flue gas. In this investigation, the emission rule of PM_{10} and $PM_{2.5}$ was examined, and a controlling technique aimed at regulating the intensive emission of PM_{10} and $PM_{2.5}$ during sintering process was studied as well, which could guide the efficient and economic control of ultra-fine particulate matters.

Materials and Methods

Raw Materials

The raw materials used in this investigation included iron ores, fluxes, return fines and solid fuel, the chemical composition of which can be found in Table 1. For meeting the requirements of blast furnace for ironmaking, the basicity (mass ratio of CaO/SiO_2), and contents of SiO_2 and MgO were kept at 1.8, 5 and 2.2%, respectively.

Experimental Methods

A laboratory-scale sinter pot with size of Φ180 mm *700 mm was used to simulate the practical sintering process. A natural gas-fuelled ignition hood was used to

Table 1 Chemical composition of raw materials

Material types	Chemical compositions%						LOI%
	TFe	FeO	SiO_2	CaO	MgO	Al_2O_3	
Iron ore mix	58.79	4.50	5.49	2.44	0.87	1.88	4.97
Quicklime	0.20	–	0.62	77.8	0.7	0.70	15.45
Dolomite	0.19	–	0.36	32.07	18.38	0.03	45.79
Coke breeze	0.84	0.14	7.05	1.01	0.19	4.68	84.88
Return fines	54.17	6.18	5.22	10.16	2.29	1.80	'1.91

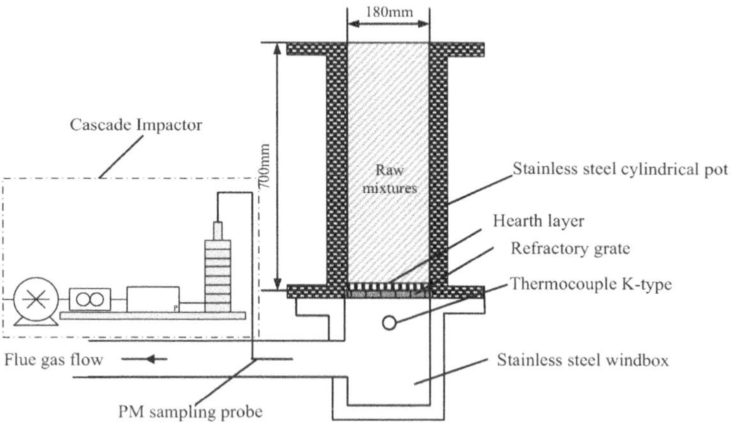

Fig. 1 Schematic diagram for PM sampling from sintering flue gas

ignite the coke breeze in the surface mixture, under ignition temperature of 1100 ± 50 °C for 1 min, heat preservation temperature of about 950 °C for 0.5 min, and ignition pressure of 5 kPa. While the pressure was adjusted to 10 kPa during the sintering process with the support of a downdraught system, the whole sintering process started from the ignition to the burning through point at which the sinter waste gas had reached its maximum temperature, during which a K-type thermocouple was used to record the change of flue gas temperature (FGT). For each individual sampling case, a low-pressure cascade impactor (Model WY-1, China) was used to collect particulate matter from flue gas, as shown in Fig. 1.

During the sampling process, PM samples were collected by Whatman QMA Quartz fibre filters with the help of the low-pressure cascade impactor. To elucidate the PM emission properties during an integrated sintering process, six typical sintering stages have been divided according to the FGT curve (Fig. 2). Among them, stage-1 referred to the ignition period, stage-2 showed the gradually stabilizing process of FGT, stage-3 indicated the stable process of FGT, stage-4 was the short period before the rise of FGT, stage-5 and stage-6 comprised the first half and the second half of FGT rising process.

After the particle-bearing flue gas being sucked into the impactor through a sampling probe, PM was automatically segregated into 0.7–1.4, 1.4–2.5, 2.5–3.6, 3.6–5.0, 5.0–6.9, 6.9–10 μm according to their aerodynamic diameters. The mass of PM collected in each case was determined by weighing the substrates before and after the sampling with a Sartorius BT125D microbalance (accuracy: 0.01 mg). Relevant equation for calculating the mass concentration of PM is given in Eq. (1).

$$C = \frac{m_1 - m_0}{\int_0^t S v_t dt} \tag{1}$$

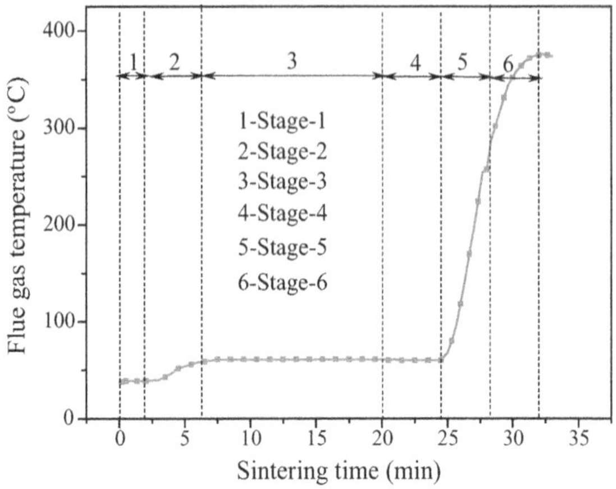

Fig. 2 Typical sampling stage division of sintering process

Where C indicates the mass concentration of PM of each sintering stage (mg/m^3); m_1 and m_0 indicate the mass of filter after and before sampling (mg); t indicates the sampling time (s); S indicates the cross-sectional area of the sampling probe (m^2); v_t indicates the real-time linear velocity of the flue gas (m/s).

For guaranteeing the accuracy of PM quantitative analysis, isokinetic sampling is the prerequisite. To achieve this goal, the real-time linear velocity of flue gas is calculated on the basis of real-time flue gas flow and the dimension of the flue. This information serves as the guidance to adjust the linear velocity of sampling, thereby making linear velocity of gas flue and sampling equal. Besides, the sampling time was properly set so that the mass of PM collected in each stage was less than 10 mg to avoid particles bounce off from the substrates. The PM sampling was repeated at least three times for each combustion test case to ensure satisfactory reproducibility and accuracy. After that, the upper part of particles was removed from the filter using a scalpel for further characterization.

An absorption layer consisted of wet mixture was employed to examine the absorption efficiency to PM_{10} and $PM_{2.5}$. During the research, a kind of polymer organic binder solution was sprayed onto the surface of granules with its purpose to change the absorption behaviour to particulate matters. Equation (2) was used to calculate the absorption efficiency of wet mixture layer to PM_{10} and $PM_{2.5}$.

$$\eta = \frac{(C_O - C_L)}{C_L} \times 100\% \tag{2}$$

Where η is the absorption efficiency of wet mixture layer; C_O is the emission concentration of particulate matter before passing the wet mixture layer, mg/m^3; C_L is the emission concentration of particulate matter before passing the wet mixture layer, mg/m^3

Results and Discussion

Emission Property of PM_{10} and $PM_{2.5}$

With the sintering process proceeding, the emission property of PM_{10} and $PM_{2.5}$ from different sintering stages can be found in Figs. 3 and 4, respectively. It can be found that the emission concentration of PM_{10} and $PM_{2.5}$ was the lowest in stage-1, while their emission concentration showed ever-increasing tendency from stage-2 to stage-4. After that, the emission concentration of PM_{10} and $PM_{2.5}$ in stage-4 was considerably higher than them in stage-1 to stage-3. The emission concentration of PM_{10} and $PM_{2.5}$ reached the peak value in stage-5, while a drop again can be

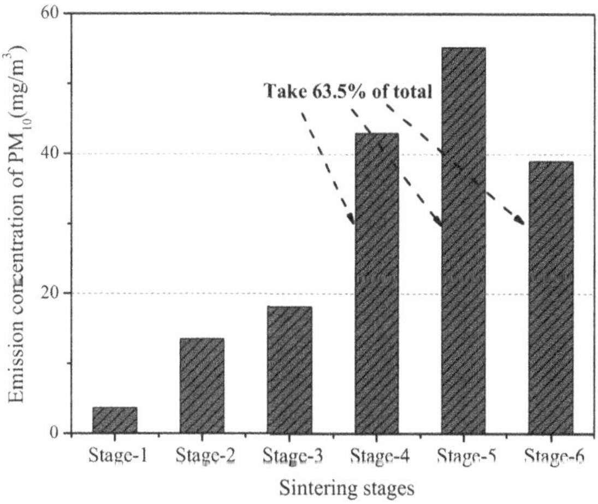

Fig. 3 Emission property of PM_{10} during sintering

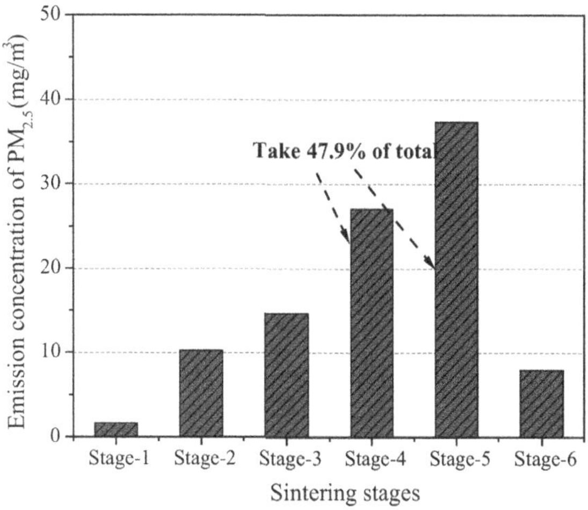

Fig. 4 Emission property of $PM_{2.5}$ during sintering

observed in stage-6. Consequently, the emission property of PM_{10} and $PM_{2.5}$ was similar, which characterized high-proportion emission area, with stage-4 and stage-5 for $PM_{2.5}$, and stage-4 to stage-5 for PM_{10}. According to the accumulation of emission concentration of PM_{10} and $PM_{2.5}$ from the whole sintering process, the proportion emitted from high-proportion emission areas reached 63.5 and 47.9% of the total for PM_{10} and $PM_{2.5}$, respectively.

Figure 5 gives the proportion of $PM_{2.5}$ in PM_{10} during the whole sintering process for laboratory test. It can be found that the ratio of $PM_{2.5}$ to PM_{10} was relatively lower at stage-1 and stage-6, which indicated that more bigger particles emitted from these stages. Moreover, the ratio reached the highest level at stage-2 and stage-3, which implied that $PM_{2.5}$ was the main emitted particles in these stages. The ratio of $PM_{2.5}$ to PM_{10} was similar in stage-4 and stage-5, while it was lower than that in stage-2 and stage-3.

Figure 6 shows the emission concentration of particles with different diameters during sintering process. It can be found that the emission concentration of particles with size range of 6.9–10 and 0.7–1.4 μm was obviously higher than others in sintering stage-1. Moreover, the emission concentration of particulate matters in stage-2 and stage-3 exhibited an increasing tendency with size reducing, and a great increase can be found for the particles within 0.7–1.4 μm. This phenomenon well agreed with the high $PM_{2.5}/PM_{10}$ in these stages. The emission concentration of particles with size range of 6.9–10 and 0.7–1.4 μm were considerably higher than other particles in stage-4. The particles range from 0.7 to 1.4 μm was much higher than other bigger particles in stage-5. Unlikely, the emission concentration of particles with size range of 5.1–6.9 and 6.9–10 μm was much higher than other particles, which resulted in the low proportion of $PM_{2.5}$ in PM_{10}.

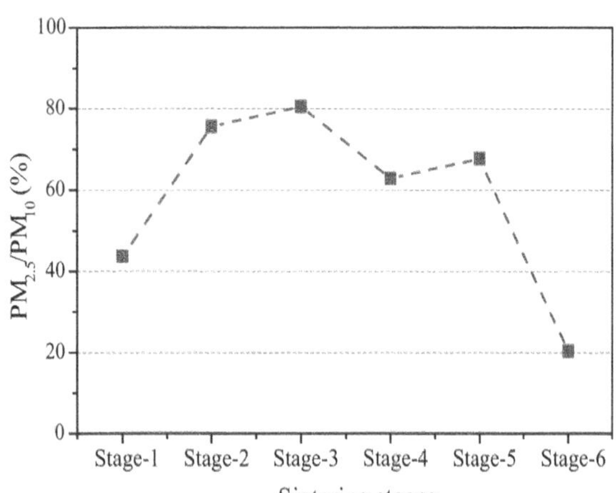

Fig. 5 Proportion of $PM_{2.5}$ in PM_{10} from different sintering stages

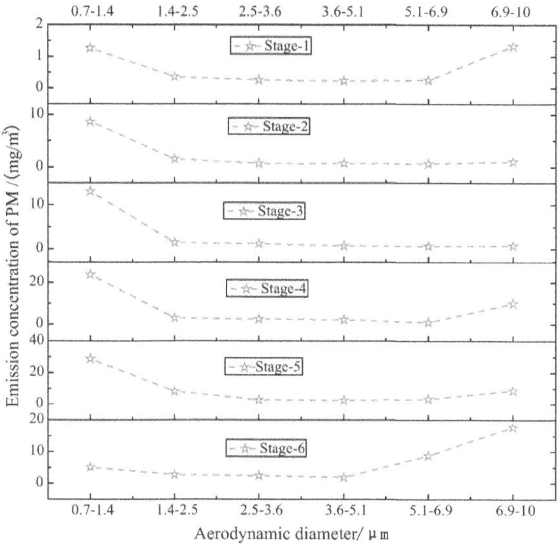

Fig. 6 Emission concentrations of PM with different sizes from different sintering process

Influences of Adding Organic Binder on Emission Property of PM_{10} and $PM_{2.5}$

According to the research above, the emission of PM_{10} and $PM_{2.5}$ characterized high-concentration area, the reason can be probably attributed to the scrubbing function of wet mixture, where similar effect has been found for coarser particles. Moreover, the absorbed particles would be released when the high-temperature layer approached the bottom layer. [6, 7] For changing the absorption capacity of wet mixture layer, a kind of polymer binder (mainly consisted of C, H, and O) solution was sprayed onto the surface of granules, with its purpose to improve the absorption capacity.

Figure 7 gives the influences of adding polymer organic binder on emission concentration of PM_{10} and $PM_{2.5}$. It can be found that with the concentration of organic binder increasing, the emission concentration of PM_{10} and $PM_{2.5}$ showed decreasing tendency, which indicated that adding organic binder improved the absorption capacity of wet mixture layer. Compared with the case without adding organic binder, the emission concentration of PM_{10} and $PM_{2.5}$ for the case adding organic binder with concentration of 0.5% dropped from 8.87 and 13.73 mg/m^3 to 5.15 and 8.56 mg/m^3. Accordingly, the absorption efficiency of wet mixture layer increased from 59.97 and 64.23% to 76.74 and 77.70%. The reason for this phenomenon can be described that when spraying binder onto granules, the bed exhibited stronger absorption capacity with the help of adhesive force of binders, which made PM more efficiently captured and temporarily retained by the bed.

Based on the above research, the influence of adding polymer organic binder on the emission property of PM_{10} and $PM_{2.5}$ during the sintering process were investigated, and the results were shown in Tables 2 and 3. It can be found that the emission concentration of both PM_{10} and $PM_{2.5}$ were decreased to low level from sintering stage-2 to stage-4 after adding organic binder. Especially for the PM_{10} and $PM_{2.5}$ emitted from stage-4, which presented an obvious drop. Compared with the

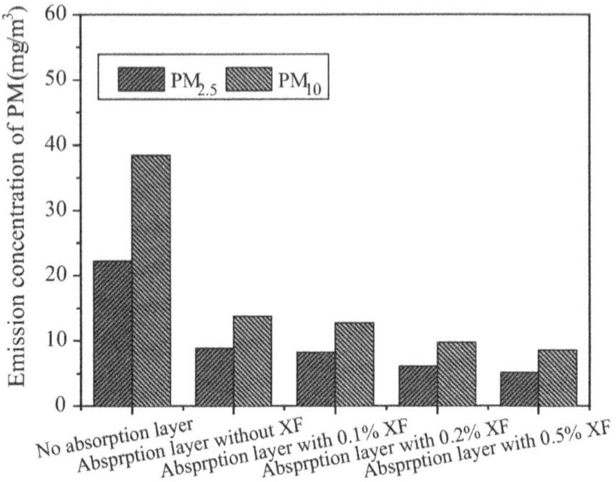

Fig. 7 Influences of XF concentration on emission of PM_{10} and $PM_{2.5}$

Table 2 Influence of strengthening sintering layer absorption on PM_{10} emissions

Sintering stages	PM_{10}/(mg/m^3)		Proportion of $PM_{2.5}$ emitted from high-concentration area%	
	Before strengthening	After strengthening	Before strengthening	After strengthening
Stage-1	3.7	3.3	–	–
Stage-2	13.5	6.5	–	–
Stage-3	18.2	9.0	–	–
Stage-4	42.9	15.7	63.5	–
Stage-5	55.2	105.3		74.7
Stage-5	38.9	59.0		

Table 3 Influence of strengthening sintering layer absorption on $PM_{2.5}$ emissions

Sintering stages	$PM_{2.5}$(mg/m^3)		Proportion of $PM_{2.5}$ emitted from high-concentration area%	
	Before strengthening	After strengthening	Before strengthening	After strengthening
Stage-1	1.6	1.6	–	–
Stage-2	10.2	3.7	–	–
Stage-3	14.6	5.6	–	–
Stage-4	27.0	8.7	47.9	–
Stage-5	37.4	66.9		74.2
Stage-6	7.9	27.7	–	

case no binder added, the emission concentration of PM_{10} from the stage-2 to stage-4 dropped from 13.5, 18.2, and 42.9 mg/m^3 to 6.5, 9.0, and 15.7 mg/m^3, while that of $PM_{2.5}$ dropped from 10.2, 14.6 and 27.0 mg/m^3 to 3.7, 5.6 and 8.7 mg/m^3. Moreover, the emission concentration of PM_{10} and $PM_{2.5}$ emitted from stage-5 and stage-6 for the case with organic binder added were higher that no binder added. Especially for particles emitted from stage-5, the emission concentration increased considerably from 55.2 to 105.3 mg/m^3 for PM_{10}, and from 37.4 to 66.9 mg/m^3 for $PM_{2.5}$. Accordingly, the accumulated emission load of PM_{10} and $PM_{2.5}$ during the high-concentration area increased from 63.5 and 47.9% to 74.8 and 74.2% after adding binder.

Consequently, the absorption capacity of wet mixture to PM_{10} and $PM_{2.5}$ from stage-2 to stage-4 was effectively improved after adding organic binder, which contributed to decreasing their emission concentration to a lower level. However, the particles absorbed by wet mixture layer were rereleased into the flue gas from stage-5 and stage-6, where no wet mixture layer existed. The narrowed high-concentration emission property paved the way the intensive control of PM_{10} and $PM_{2.5}$ within special sintering stages, which facilitated the efficient and economic reduction of PM_{10} and $PM_{2.5}$.

Conclusions

(1) PM_{10} and $PM_{2.5}$ characterized high emission concentration in special sintering areas, and they were sintering stage-4 to stage-6, and sintering stage-4 to stage-5, respectively. The emission load of PM_{10} and $PM_{2.5}$ in special areas accounted for about 63.5 and 47.0% of the total, respectively.

(2) Spraying polymer organic binder solution on the surface of granules helps to improve the absorption efficiency of wet mixture layer to PM_{10} and $PM_{2.5}$, which made them intensively emitted within sintering stage-5 and stage-6, with accounting for 74.8 and 74.2% of the total. The research findings pave the way for realizing the efficient and economic control of PM_{10} and $PM_{2.5}$ in practical sintering plants.

Acknowledgements This work was supported by the National Natural Science Foundation of China (Grant No. 51804347, Fund No. U1660206), and Hunan Provincial Co-Innovation Center for Clean and Efficient Utilization of Strategic Metal Mineral Resources.

References

1. Hleis D, Fernández-Olmo I, Ledoux F, Kfoury A, Courcot L, Desmonts T, Courcot D (2013) Chemical profile identification of fugitive and confined particle emissions from an integrated iron and steelmaking plant. J Hazard Mater 250(2):246–255
2. Tsai JH, Lin LK, Chen CY, Ding JY, Choa CG, Chiang HL (2017) Chemical constituents in particulate emissions from an integrated iron and steel facility. J Hazard Mater 147(1–2):11–119

3. Guerriero E, Guarnieri A, Mosca S, Rossetti G, Rotatori M (2009) PCDD/Fs removal efficiency by electrostatic precipitator and wetfine scrubber in an iron ore sintering plant. J Hazard Mater 172(2–3):1498–1504
4. Baxter LL (1992) Char fragmentation and fly ash formation during pulverized-coal combustion. Combust Flame 90(2):174–184
5. Mcnallan MJ, Yurek GJ, Elliott JF (1981) The formation of inorganic particulates by homogeneous nucleation in gases produced by the combustion of coal. Combust Flame 42 (81):45–60
6. Nakano M, Okazaki J (2007) Influence of operational conditions on dust emission from sintering bed. ISIJ Int 47(2):240–244
7. Debrincat D, Eng LC (2007) Factors influencing particulate emissions during iron ore sintering. ISIJ Int 47(5):652–658

Viscosity Properties of Mould Flux Under Low-Frequency Electromagnetic Field

Wei Qian, Yu Wang and Lu-ming Zhao

Abstract The effects of magnetic field intensity, magnetic field frequency and Al_2O_3 content on the viscosity of molten slag in a low-frequency magnetic field were studied in this paper. The flow field of mould flux under low-frequency electromagnetic field was simulated by the finite element software. The experimental results showed that under the condition of this experiment, the magnetic field intensity increased from 0 to 30mT, and the viscosity of mould flux (1573 K) increased by 0.8–1.7 poise; the magnetic field frequency ranged from 6 to 12 Hz, and the viscosity(1573 K)increased by 0.9–1.6 poise; the content of Al_2O_3 increased from 3 to 7%, and the viscosity (1573 K) increased by 0.9–2.1 poise. Numerical simulation results showed that the flow field of mould flux is changed under low-frequency electromagnetic field, the amount of the slag in the disordered flow state increased and the viscosity increased.

Keywords Keywords · Low frequency · Viscosity

Introduction

Mould flux is a very important functional material in the process of continuous casting in modern iron and steel industry [1–3]. It plays an important role in the smooth progress of continuous casting process and the improvement of slab surface quality [4, 5]. Mould flux viscosity is a measure of the capacity of slag to flow into the slab and mould under certain temperature and shear force [6]. Therefore, the proper viscosity is the key to ensure that the mould flux can be smoothly filled into the channel between the mould and the billet, to ensure the thickness of the slag film, to ensure a reasonable heat transfer rate and lubrication of the billet.

W. Qian · Y. Wang (✉) · L. Zhao
College of Materials Science and Engineering, Chongqing University,
Chongqing 400044, China
e-mail: wangyu@cqu.edu.cn

© The Minerals, Metals & Materials Society 2019

T. Jiang et al. (eds.), *10th International Symposium on High-Temperature Metallurgical Processing*, The Minerals, Metals & Materials Series,
https://doi.org/10.1007/978-3-030-05955-2_9

At present, the research on the viscosity properties of electromagnetic continuous casting mould flux is limited to high-frequency magnetic field and static magnetic field. The frequency of some electromagnetic stirring is lower in actual continuous casting production. Therefore, it is of great significance to the study of the viscosity properties of electromagnetic continuous casting mould flux under low-frequency magnetic field.

The effects of magnetic field intensity, magnetic field frequency and Al_2O_3 content on the viscosity of molten slag in a low-frequency magnetic field were studied by a rotating viscosity tester in this paper. The flow properties of mould flux under low-frequency magnetic field were simulated by finite element software, and the mechanism of low-frequency electromagnetic field on the viscosity of mould flux was explored.

Experimental Procedure

Experimental Apparatus

SB6 rotary viscometer is adopted in this experiment, compared with other types of viscometers, it has the advantages of high temperature resistance, suitable for electromagnetic environment and accurate measurement. Its device is shown in Fig. 1. The relevant equipment parameters of the viscosity test system are shown in Table 1.

Experimental Method

In order to meet the design principle of the mould flux, the mould flux used in a factory is directly used as the base slag. Because the main inclusion absorbed by the mould flux is Al_2O_3, so the content of Al_2O_3 is changed and the alkalinity was kept unchanged, which meets the requirement of taking Al_2O_3 as the variable in this experiment. The mould flux used in the experiment is shown in Table 2.

Based on the actual continuous casting production and numerical simulation experiment, the magnetic field intensity of magnetic field to the viscosity of slag is 0–30 mt, and the magnetic field frequency is 6–12 Hz.

In this experiment, the alternating magnetic field is used. The magnetic field frequency parameters can be adjusted directly, but the magnetic field intensity needs to be changed by adjusting the current. The relationship between the magnetic field intensity and the current is shown in Table 3.

The single variable was studied by two groups of experiments, and the first one was to set the magnetic field intensity 8 Hz at 1573 K and to measure the viscosity

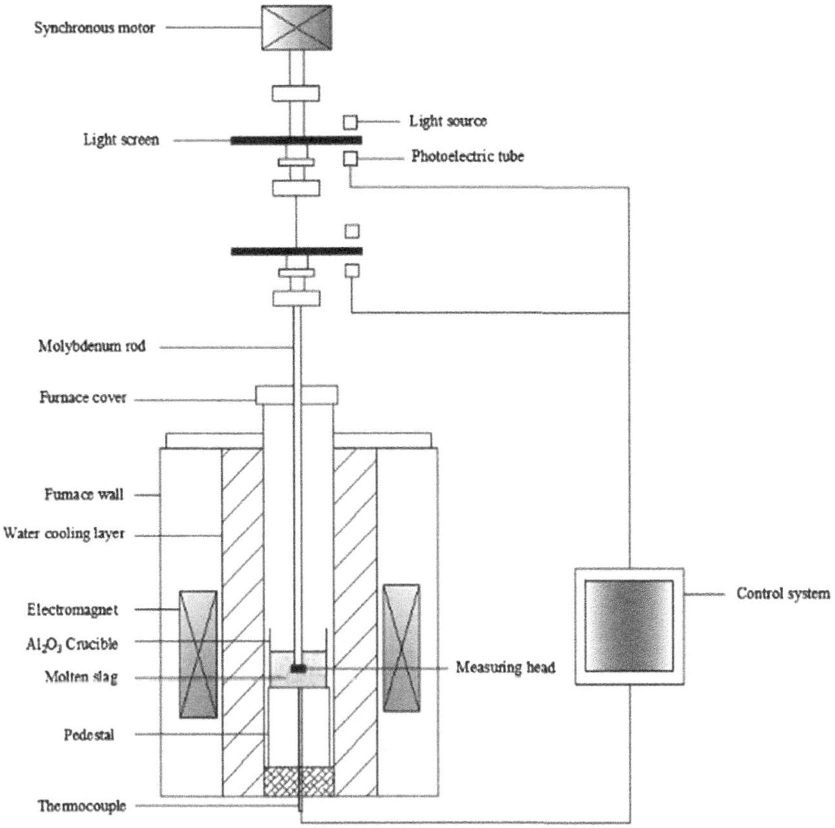

Fig. 1 A schematic diagram of a viscosity testing device under electromagnetic field

Table 1 Viscosity test device parameters

Property	Parameter	Property	Parameter
Sling material	Steel wire	Probe speed	12 r/min
Suspender material	Molybdenum	Viscosity range	$10^{-2}\sim30$ Pa s
Probe material	Corundum	Viscosity accuracy	±2.5% F.S

Table 2 Chemical composition of protective slag for experiment (mass%)

CaO	SiO_2	Al_2O_3	CaF_2	Na_2O	MnO	Sample
39.69	39.69	3.00	4.84	4.29	2.96	M (1)
38.69	38.69	5.00	4.84	4.29	2.96	M (2)
37.69	37.69	7.00	4.84	4.29	2.96	M (3)

Table 3 The correspondence between magnetic field intensity and current size

B/mT	0	10	20	30
f/Hz	0	2.2	4.4	6.6

Table 4 Experimental schemes 1

f/Hz	B/mT	Sample	f/Hz	B/mT	Sample
8	0	M-1	8	20	M-1
		M-2			M-2
		M-3			M-3
	10	M-1		30	M-1
		M-2			M-2
		M-3			M-3

Table 5 Experimental schemes 2

B/mT	f/Hz	Sample	B/mT	f/Hz	Sample
20	6	M-1	20	10	M-1
		M-2			M-2
		M-3			M-3
	8	M-1		12	M-1
		M-2			M-2
		M-3			M-3

of the slag by changing the frequency of the electromagnetic field and the content of the alumina component. As shown in Table 4

In the second experiment group, when the temperature is 1573 K, the magnetic field frequency is 20 mT. The viscosity of the powder is measured by changing the intensity of the electromagnetic field and the content of alumina (Table 5).

Results and Discussion

As shown in Fig. 2, the viscosity (1573 K) becomes larger as the magnetic field becomes larger. When the magnetic field intensity increased from 0 to 30 mT, viscosity (1573 K) increased by 0.8–1.7 poise. This is due to mould flux is mainly composed of SiO_4^-. When applying low-frequency magnetic field, it receives electromagnetic force with changing direction and magnitude, and polarizes repeatedly, which promotes the increase of covalent bond fraction. It forms more complex silica complex ions by attracting each other. Therefore, with the increase of magnetic field strength, the viscosity of mould flux increases.

As can be seen from Fig. 3, the viscosity (1573 K) of mould fluxes increases with the increase in magnetic field frequency. From 6 to 12 Hz, the viscosity (1573 K) increases by 0.9–1.6 poise. This is because the induced electric field in the slag increases with the increase in magnetic field frequency under the condition of

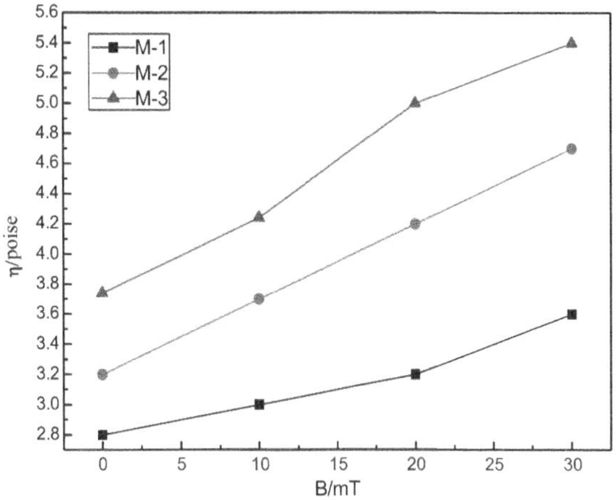

Fig. 2 f = 8 Hz, T = 1573 K Effects of magnetic field intensity on viscosity

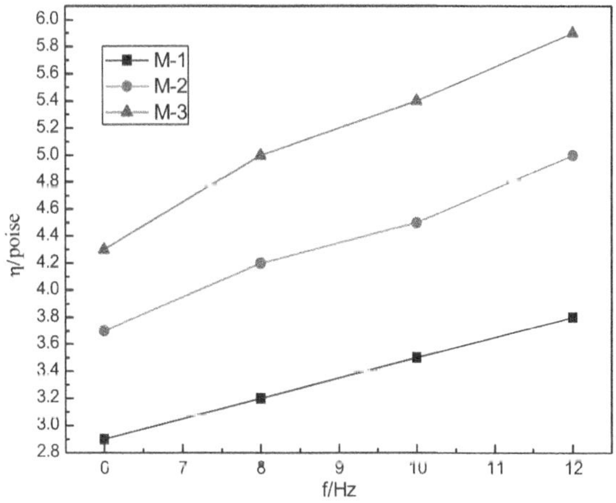

Fig. 3 B = 20 mT, T = 1573 K Effects of magnetic field frequency on viscosity

alternating magnetic field, and then the Lorentz force increases, the specific solutes migrate, the silica-oxygen ions aggregate to form more complex silica-oxygen complex ions, thus increasing the viscosity of mould flux.

From Fig.4, we can see that the viscosity increases with the increase in Al_2O_3 content. When Al_2O_3 content ranges from 3 to 7%, the viscosity (1573 K) is increased by 0.9–2.1 poise. This is because the content of Al_2O_3 increases, the ratio of silicon to oxygen decreases, so that the network structure of mould powder increases, resulting in an increase in viscosity.

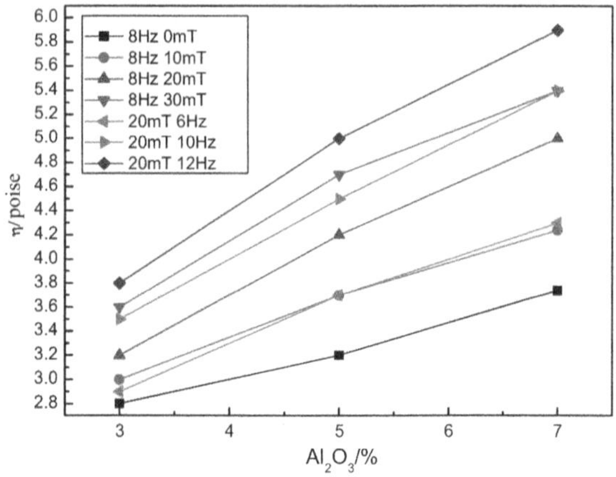

Fig. 4 Effects of Al_2O_3 content on viscosity (1573K)

Numerical Simulation

Basic Hypothesis

It is necessary to make reasonable simplification and necessary assumptions to study the flow behaviour of mould fluxes. Conditions set forth below.

(1) ignoring the influence of other factors on mould fluxes.
(2) the mould flux is an incompressible viscous fluid.
(3) the pressure on the mould powder is zero.
(4) the speed of free surface vertical downward is zero.
(5) the conductivity, density and mould flux viscosity of mould fluxes are constant.

Fundamental Equation

In the fluid under the action of electromagnetic force, the flow equation of the mould flux is different from that of the ordinary fluid. Due to the addition of electromagnetic force, an electromagnetic force needs to be added to the momentum equation as the momentum source. Since the problem of solidification of the mould flux is not considered, the control equation of the mould flux is as follows:

Continuity equation:

$$\frac{\partial(\rho u_i)}{\partial x_i} = 0 \tag{1}$$

Momentum equation:

$$\frac{\partial(\rho u_i u_j)}{\partial x_j} = -\frac{\partial p}{\partial x_i} + \frac{\partial}{\partial x_j}\left[\mu_{eff}\left(\frac{\partial u_i}{\partial x_j} + \frac{\partial u_j}{\partial x_i}\right)\right] + \rho g_i + F_{em} \tag{2}$$

Turbulent kinetic energy K equation:

$$\frac{\partial(\rho u_i k)}{\partial x_i} = \frac{\partial}{\partial x_i}\left(\frac{u_{eff}}{\sigma_k}\frac{\partial k}{\partial x_i}\right) + G - \rho\varepsilon \tag{3}$$

Turbulent energy dissipation rate equation ε:

$$\frac{\partial(\rho u_i \varepsilon)}{\partial x_i} = \frac{\partial}{\partial x_i}\left(\frac{u_{eff}}{\sigma_\varepsilon}\frac{\partial \varepsilon}{\partial x_i}\right) + C_1\frac{\varepsilon}{k}G - C_2\frac{\varepsilon^2}{K} \tag{4}$$

In the formula:

$$G = \mu_t \frac{\partial \mu_i}{\partial x_i}\left(\frac{\partial \mu_i}{\partial x_j} + \frac{\partial \mu_j}{\partial x_i}\right) \tag{5}$$

Effective viscosity is

$$\mu_{eff} = \mu_t + \mu \tag{6}$$

$$\mu_t = \rho C_\mu \frac{k^2}{\varepsilon} \tag{7}$$

ρ density, Kg/m^3;
F_{em} mean value of electromagnetic force density, N/m^3;
u_i, u_j time averaged velocity of turbulent flow, m/s;
u_{eff} effective viscosity coefficient, Pa s;
μ viscosity, Pa s;
μ_t turbulent momentum diffusion coefficient or turbulent viscosity coefficient, Pa s;
k turbulent kinetic energy, m^2/s^2;
σ_k, σ_ε the Prandtl number corresponding to turbulent kinetic energy k and dissipation rate ε
ε turbulent dissipation rate, m^2/s^3;
k_{eff} effective transmission coefficient, W m^{-1} k^{-1}.

Parameter Selection

This simulation selects a section of crucible under experimental conditions as a model for two-dimensional data analysis, the crucible diameter is 75 mm, the head is 25 mm, the rotating speed of the head is 12 r/min. Assuming the density of the mould flux is 2500 kg/m^3, the viscosity of the mould flux is set at 0.3 Pa s. This is the viscosity of the original slag used in the experiment. The specific parameters are shown in Table 6.

Analysis of Simulation Results of Mould Fluxes Flow Field Under Electromagnetic Field

(1) Distribution of magnetic field intensity in mould flux

Figure 5shows the distribution of electromagnetic force on the mould flux under the action of electromagnetic field. It can be seen from the figure that the magnetic field intensity is distributed in the range of 10mT to 20mT, indicating that this simulation coincides with the experiment.

(2) Flow field simulation of mould fluxes

It can be seen from Fig. 6 that the flow direction of mould flux is changed due to the effect of Lorentz force after the application of electromagnetic field, and the mould flux forms two swirls which flow inward. It shows that the effect of electromagnetic force is opposite to the static pressure of molten steel, thus improving the quality of casting slab. From Fig. 7, it is not difficult to find out that the flux flow is stable and the velocity is small without applying electromagnetic field. After applying electromagnetic field, the ordered structure state of the flux is disturbed, and the disordered ion group increases, which promotes the free flow of the flux and the ion flow rate increases and finally leads to the viscosity increase.

Table 6 Finite element simulation parameters

Project	Reference value
Power frequency	8 Hz
Effective value of current intensity	200 A
The number of turns	10
Copper coil resistivity	$3 \times 10^{-6}\Omega$.m
Vacuum permeability	$4\pi \times 10^{-7}$H/m
Vacuum dielectric constant	8.854×10^{-12}
Copper relative permeability	1
Relative permeability of mould fluxes	1
Relative permeability of air	1

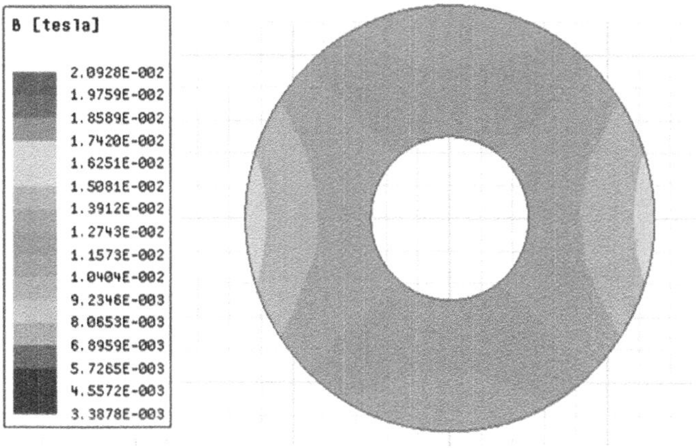

Fig. 5 Simulation results of mould fluxes flow field under electromagnetic field

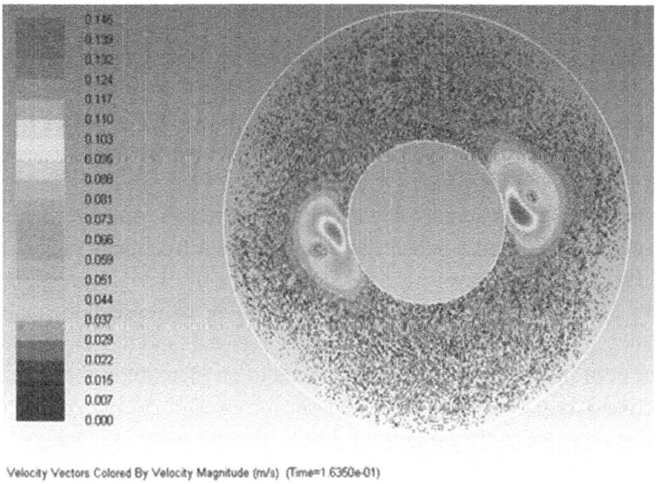

Fig. 6 Velocity vector diagram of mould fluxes under low-frequency electromagnetic field

(3) The two eddies in the mould flux correspond to the two positions with the highest magnetic field intensity in Fig. 5. This shows that the stronger the magnetic field intensity, the faster the flux flow rate, and the more the disordered ion clusters. Because these two parts are adjacent to the copper coils where the electromagnetic field is generated, the electromagnetic intensity is large. In addition, because the magnetic field strength of the mould flux is small, the flow direction and velocity of mould fluxes remain unchanged.

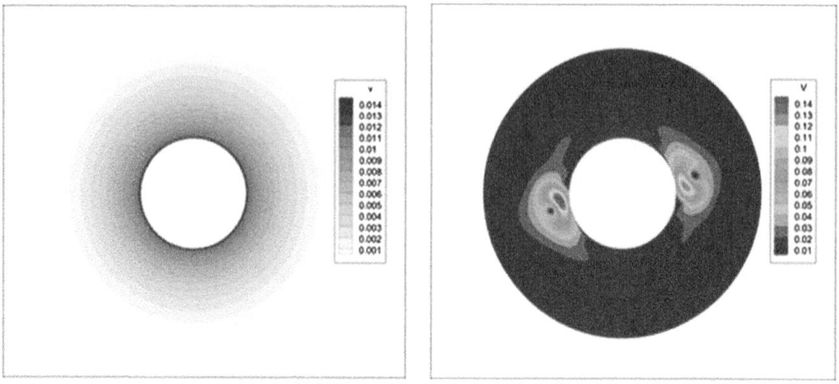

Fig. 7 No electromagnetic field applied (left)

Conclusions

(1) The viscosity of mould flux (1573 K) increases with the increase in magnetic field strength. When the magnetic field intensity is from 0 to 30 mT, the viscosity (1573 K) also increases by 0.8–1.7 pool. This is because the Lorentz force on the molten slag increases with the increase in magnetic field strength, which leads to the increase of viscosity.

(2) The viscosity of the mould flux (1573 K) increases with the increase in the magnetic field frequency when the magnetic field frequency ranges from 6 Hz to 12 Hz, and the viscosity (1573 K) increases by 0.9–1.6 poise, because the induced electric field in the slag increases with the increase in the magnetic field frequency. As a result, the Lorentz force increased, and the stirring force of the molten pool increased, which in turn leads to an increase in the viscosity of the protective slag.

(3) With the increase of the content of Al_2O_3, the viscosity (1573 K) of the mould flux increases. When the content of Al_2O_3 increases from 3 to 7%, the viscosity (1573 K) increases by 0.9–2.1 pool. This is because the increase of the content of Al_2O_3 leads to the increase of the oxygen-silicon ratio, and the network structure of the mould flux also increases. Finally, the viscosity of the molten slag increases

(4) Through the simulation of finite element software, it can be found that the direction and velocity of flux flow will change under the action of electromagnetic field, and then the whirlpool will be formed in the flux and finally the viscosity of flux will increase.

Acknowledgements This work was financially supported by National Natural Science Foundation of China (project No.51574051) and Sharing Fund of Large Scale Equipment, Chongqing University (project No. 201512150025).

References

1. Chi JH, Gan YN (1992) Mold Flux. Northeastern University Press, Shengyang
2. Pinheiro CA (1996) Mold Flux for CC of Steel. Iron Steelmaker 23(3):85
3. Xie B (1986) Translation of mold flux for casting steel. Chongqing University, Chongqing
4. Nakata H, Inoue T, Mori H et al (2002) Improvement of Billet Surface Quality by Ultra-High-Frequency Electromagnetic Casting. ISIJ Int 42(3):264
5. Wei ZW (2010) Basic research on the performance of molten slag under electric field. Chongqing University, Chongqing
6. Wang XJ, Zhu LG, Liu R, et al (2013) Viscosity characteristics of mold fluxes under high frequency electromagnetic field. Iron Steel Vanadium Titanium 34(2):79–83

Part III
Fundamentals of Metallurgical Processes

Effects of Cr_2O_3, FeO and CaO/SiO_2 Ratio on the Apparent Viscosity of CaO-SiO_2-MgO-MnO-FeO-Cr_2O_3 Slags

Bing Huang, Mingmei Zhu, Yong Zhong, Aiping Zhang and Tiancheng Lin

Abstract In this study, the effects of Cr_2O_3, FeO and CaO/SiO_2 ratio on the apparent viscosity of the CaO-SiO_2-MgO-MnO-FeO-Cr_2O_3 slags are investigated at 1500 °C. Apparent viscosity of the slags is measured using the rotating cylinder method and the liquid and solid phases are calculated using the thermodynamic calculation software FactSage. The experimental results showed that the apparent viscosity of slags increases with increasing CaO/SiO_2 ratio and Cr_2O_3 content and decreases with the FeO addition. The thermodynamic calculations results showed that Cr_2O_3 content has a minor effect on the mass fraction of liquid phases, nevertheless it can promote the formation of spinels resulting in increased amount of solid phases as well as, increased slag apparent viscosity. The liquid slag phases were increased with increasing of FeO content therefore, the apparent viscosity of the slag was reduced. However, the liquid slag phases were decreased with increasing of CaO/SiO_2 ratio, which resulted in increased slag apparent viscosity.

Keywords Steelmaking slag bearing chromium · Apparent viscosity
High-temperature experiment · Thermodynamic calculation

Introduction

Viscosity is one of the most important physical properties of slag, not only related to the smooth progress of the smelting process, but also affecting the heat and mass transfer of metallurgical process, and thus the rate of chemical reaction, the discharge of impurities in the molten pool, the loss of metal in slag and the life of furnace lining [1]. Chromium is an important alloy element in many different metals. In the process of smelting of chromium-containing metals, the chromium can easily be oxidized to Cr_2O_3. The Cr_2O_3 entering the slag will form with other

B. Huang · M. Zhu (✉) · Y. Zhong · A. Zhang · T. Lin
College of Materials Science and Engineering, Chongqing University,
Chongqing 400044, China
e-mail: zhuminmei@cqu.edu.cn

© The Minerals, Metals & Materials Society 2019

T. Jiang et al. (eds.), *10th International Symposium on High-Temperature Metallurgical Processing*, The Minerals, Metals & Materials Series,
https://doi.org/10.1007/978-3-030-05955-2_10

components of the slag high melting point spinel, such as $MgCr_2O_4$ and $FeCr_2O_4$, which significantly influence the viscosity of slag. Therefore, further studies to better understand of the viscosity behaviour of chromium-containing slag were necessary.

In recent years, many scholars have studied the effect of Cr_2O_3 on slag viscosity. Ostrovski [2] and Xu [3] investigated the effect of Cr_2O_3 on the viscosity of CaO-Cr_2O_3-CaF_2 slags and mould flux, respectively. Qiu [4], Huang [5] and Xu [6] have studied the effect of Cr_2O_3 on the viscosity of Ti-bearing slag. Furthermore, Li [7, 8] studied the viscosities of CaO-SiO_2-MgO-Al_2O_3-Cr_2O_3 and CaO-SiO_2-MgO-Na_2O-B_2O_3-Al_2O_3-Cr_2O_3 slags. Their research results showed that with the addition of Cr_2O_3, the viscosity of slag increases.

In this study, the effects of Cr_2O_3, FeO and CaO/SiO_2 ratio on the apparent viscosity of the CaO-SiO_2-MgO-MnO-FeO-Cr_2O_3 slags were investigated at 1500 °C. Apparent viscosity of the slags was measured using the rotating cylinder method and the liquid and solid phases were calculated using the thermodynamic calculation software FactSage.

Experimental

Samples Preparation

The CaO-SiO_2-FeO-MgO-MnO-Cr_2O_3 slags were selected for the conceptual design. The chemical composition of slags used in the experimental study was listed in Table 1. The compositions were obtained from analytical-grade reagents CaO, SiO_2, Ferrous oxalate, MgO, MnO and Cr_2O_3. The CaO, MnO and MgO were calcined for 10 h in a muffle furnace at 1000 °C to remove carbonates and hydroxides. In order to remove the moisture, samples were treated in the furnace at 110 °C for 12 h. According to the proportion of each test slag sample shown in Table 1, a 200 g of sample were mixed thoroughly in an agate mortar and placed into the grinding tool to make cylindrical sample. Samples were added to the MgO crucible and then placed together with the crucible in the dryer to prevent re-suction of moisture.

Viscosity Measurements

In this experiment, apparent viscosity measurements were carried out by a Brookfield digital viscometer and experimental apparatus, which is graphically sketched in Fig. 1. The main temperature control equipment used in the experiment is a high-temperature silicon-molybdenum furnace with the precision of temperature control ±2 °C. The crucibles used in the experiment are based on Al_2O_3 and

Table 1 Chemical composition of experimental slags (wt%)

No	CaO	FeO	MgO	Cr_2O_3	MnO	R
1	40.0	24.0	6.0	0.0	5.0	1.6
2	38.8	24.0	6.0	2.0	5.0	1.6
3	37.5	24.0	6.0	4.0	5.0	1.6
4	36.3	24.0	6.0	6.0	5.0	1.6
5	35.1	24.0	6.0	8.0	5.0	1.6
6	40.0	20.0	6.0	4.0	5.0	1.6
7	38.8	22.0	6.0	4.0	5.0	1.6
8	36.3	26.0	6.0	4.0	5.0	1.6
9	35.1	28.0	6.0	4.0	5.0	1.6
10	33.3	24.0	6.0	4.0	5.0	1.2
11	35.6	24.0	6.0	4.0	5.0	1.4
12	39.2	24.0	6.0	4.0	5.0	1.8
13	40.7	24.0	6.0	4.0	5.0	2.0

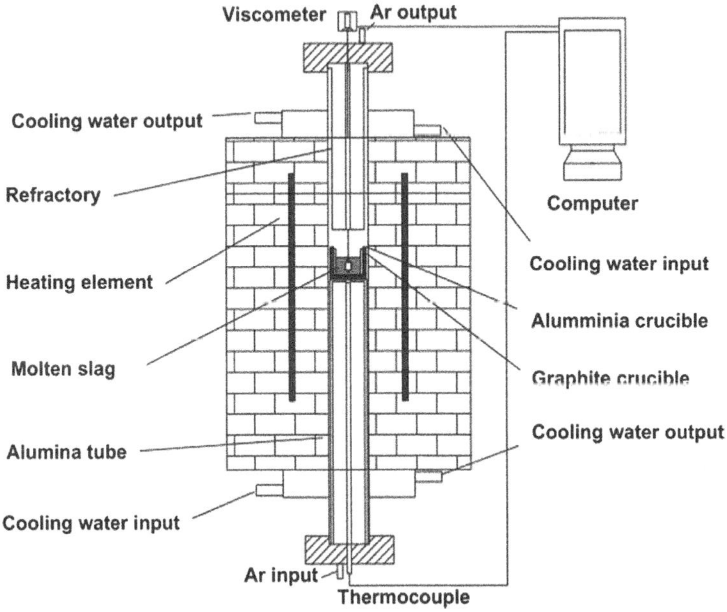

Fig. 1 Experimental apparatus for measurements of slag viscosity

graphite with an inner diameter of 60 and 75 mm, respectively. In order to reduce the experimental error, the viscometer was calibrated with three Brookfield standard oils of known viscosity in a water bath crucible at 25 °C.

The cylindrical sample was placed in the Al_2O_3 crucible and together with the graphite crucible inserted in the silicon-molybdenum furnace. The furnace was heated with 2 L/min of high purity nitrogen as a protective gas. When the furnace temperature reached 1550 °C, the heating was stopped and temperature kept on 1550 °C. In order to homogenize the slag to achieve better fluidity, the slag was heated for 1 h and stirred with molybdenum rod every 20 min. By reducing the furnace temperature to 1500 °C, the apparent viscosity of slags was measured at the rotational speeds of 12 rpm. The distance between the bottom of the probe and the bottom of the crucible was 10 mm. Due to a slight deviation from the axis the probe, the centre of the crucible and the viscometer were in the same line [9].

Thermodynamic Calculation

Thermodynamic calculation software FactSage 6.3 was used to calculate the multiphase equilibrium of CaO-SiO_2-FeO-MgO-MnO-Cr_2O_3 slags under different composition conditions. The calculations were performed at temperature of 1500 °C and pressure of 1 atm. Equilibrium data could be calculated for the liquid and solid phases at 1500 °C.

Results and Discussion

Effect of Cr_2O_3 on Apparent Viscosity of Slags

Figure 2 shows the apparent viscosity changes of CaO-SiO_2-FeO-MgO-MnO-Cr_2O_3 slags with different content of Cr_2O_3 at 1500 °C. The results showed that the apparent viscosity of slags increases gradually, namely from 0.48 Poise to 4.55 Poise when the Cr_2O_3 content was increased from 0 to 8%. Additionally when the

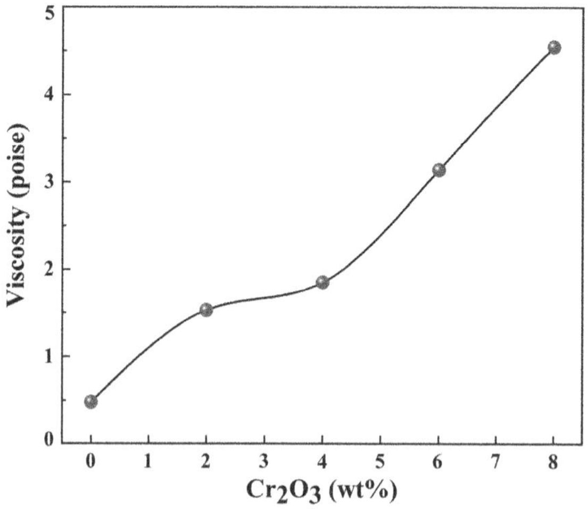

Fig. 2 Effect of Cr_2O_3 on apparent viscosity of slags at 1500 °C

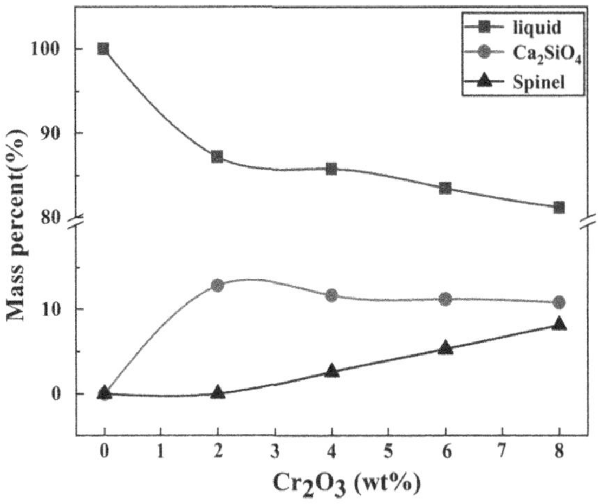

Fig. 3 Effect of Cr_2O_3 on phases of slags at 1500 °C

Cr_2O_3 content in the slag was increased to 4%, the apparent viscosity increases by 20.9%. After the Cr_2O_3 content in slag exceeds 4%, the apparent viscosity of slag increased almost linearly. This can be explained due to precipitation of solid phases that automatically caused increase in the slag viscosity.

Figure 3 shows the change of liquid and solid phases of CaO-SiO_2-FeO-MgO-MnO-Cr_2O_3 slags as a function of different content of Cr_2O_3 at a temperature of 1500 °C. As can be seen from Fig. 3, when Cr_2O_3 is not present in the slag, the slag is completely molten, (i.e., low slag viscosity). With the addition of 2% Cr_2O_3, the dicalcium silicate and spinel start to precipitate, the amount of liquid phase in the slag decreases with the increasing of Cr_2O_3 content. Additionally, the amount of solid phase in the slag increases with increasing Cr_2O_3 content from 0 to 8%. It can be also seen from Fig. 3 that dicalcium silicate in the solid phase increases gradually at higher content of Cr_2O_3. Thus, the higher slag apparent viscosity is mainly caused by the increase of dicalcium silicate. At increased Cr_2O_3 content in the slag, e.g. from 2 to 4%, the spinel begins to precipitate. The amount of dicalcium silicate does not change significantly, but the solid content in the slag increases.

Effect of FeO on Apparent Viscosity of Slags

Figure 4 shows the apparent viscosity changes of CaO-SiO_2-FeO-MgO-MnO-Cr_2O_3 slags with different content of FeO at the temperature of 1500 °C. As can be seen, the apparent viscosity of slags decreases rapidly with higher FeO content. At increased FeO content from 20 to 28%, the apparent viscosity of the slag decreases from 2.4 to 1.3 Poise. Oxygen dioxide reacts with complex silicate structures and reduces the polymerization degree of the slag. Therefore, FeO acts as a network modifier and reduces the viscosity of slag.

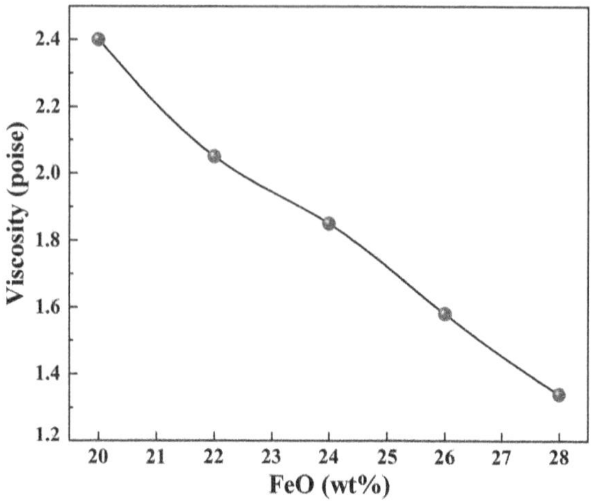

Fig. 4 Effect of FeO on apparent viscosity of slags at 1500 °C

Figure 5 indicates the transformation of liquid and solid phases of the CaO-SiO_2-FeO -MgO-MnO-Cr_2O_3 slags with different content of FeO at a temperature of 1500 °C. From Fig. 5, it can be seen that the amount of spinel in slag is almost unchanged when the content of FeO in slag was increased from 20 to 28%. Additionally, the amount of liquid phase and dicalcium silicate decreases. The liquid phase increased from 74.8 to 91.6% when FeO content was increased from 20 to 28%. On the other hand, the amount of dicalcium silicate is reduced because FeO tends to form low melting silicates like calcium iron olivine ($Ca_3Fe_2SiO_{12}$).

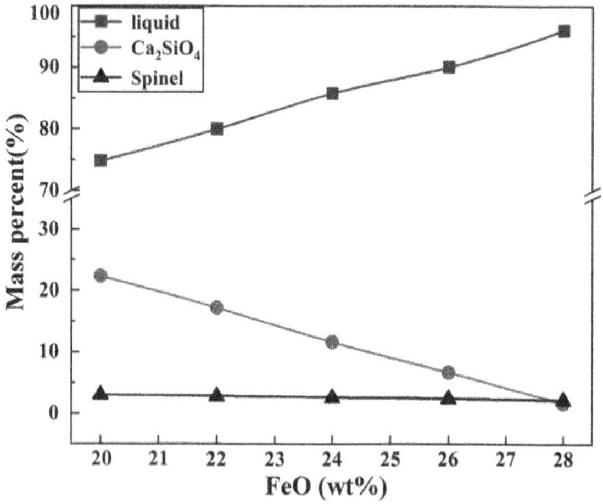

Fig. 5 Effect of FeO on phases of slags at 1500 °C

Effect of CaO/SiO_2 Ratio on Apparent Viscosity of Slags

Figure 6 represents the apparent viscosity changes of the $CaO\text{-}SiO_2\text{-}FeO\text{-}MgO\text{-}MnO\text{-}Cr_2O_3$ slags with different CaO/SiO_2 ratio at a temperature of 1500 °C. Results indicate that the apparent viscosity of slag increases as a function of increased CaO/SiO_2 ratio from 1.2 to 2.0. At increased CaO/SiO_2 ratio, the apparent viscosity of the slag increases from 0.68 to 11.3 Poise.

Figure 7 shows the change of liquid and solid phases of the $CaO\text{-}SiO_2\text{-}FeO\text{-}MgO\text{-}MnO\text{-}Cr_2O_3$ slags with different content of CaO/SiO_2 ratio at a temperature of 1500 °C. In the steelmaking process, the converter slag usually has a high basicity[10]. As observed by Seong, there is a high amount of Ca_2SiO_4 in the converter slag even at a temperature of 1600 °C [10]. Additionally, it can be seen that the change of CaO/SiO_2 ratio has a minor effect on the formation of spinel in the slag; nevertheless it has a great influence on the content of liquid phase and calcium silicate in the slag (see Fig. 7). When the CaO/SiO_2 ratio increases to 2.0,

Fig. 6 Effect of basicity on apparent viscosity at 1500 °C

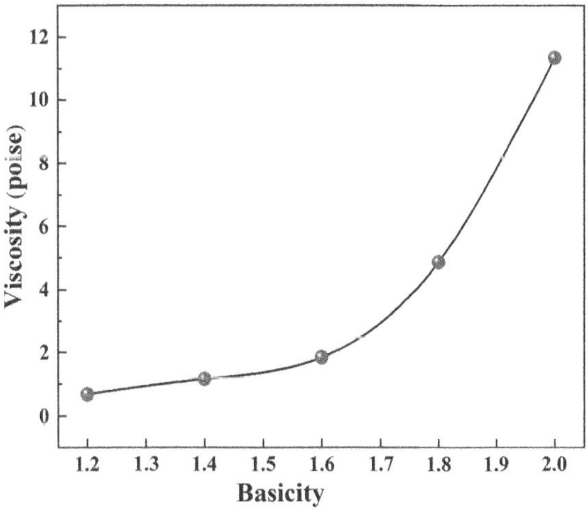

Fig. 7 Effect of basicity on phases of slags at 1500 °C

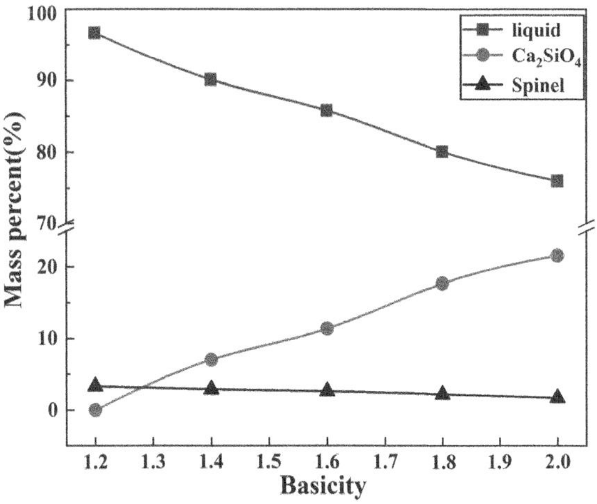

the liquid content of slag is ~76%, and the content of dicalcium silicate is ~21%. At the same time, the viscosity of slag reached 11.3 Poise. Therefore, the main reason for the increased slag apparent viscosity is the higher dicalcium silicate content formed in the slag.

Conclusions

The effects of Cr_2O_3, FeO and CaO/SiO_2 ratio on the apparent viscosity of the CaO-SiO- FeO-MgO-MnO-Cr_2O_3 slag were measured at 1500 °C using the rotating cylinder method. The liquid and solid phases of slags were calculated at the same temperature by thermodynamic calculation software FactSage. The main conclusions are as follows:

(1) The apparent viscosity of slag increases with the increased Cr_2O_3 content. This is mainly due to increase of Cr_2O_3 content resulting in a large quantity of solid phases that are formed in the slag.
(2) The apparent viscosity of slag decreases with increased FeO content, which is mainly caused by increased amount of liquid phase in the slag.
(3) The apparent viscosity of slag increases sharply with increased CaO/SiO_2 ratio. The latter phenomenon can be explained by higher amount of precipitated dicalcium silicate in the slag.

References

1. Huang XH (2014) Iron and steel metallurgy principle, 4th ed. Metallurgical industry press, metallurgical industry press, Beijing, p 316
2. Ostrovski OI, Utochkin YI, Pavlov AV, et al (2007) Viscosity of the CaO-CaF_2 system containing chromium oxide. ISIJ Int, 34(9):773–775
3. Xu C, Wang W, Zhou L, Xie S, Zhang C (2015) The effects of Cr_2O_3 on the melting, viscosity, heat transfer, and crystallization behaviors of mold flux used for the casting of Cr-bearing alloy steels. Metall Mater Trans B 46(2):882–892
4. Qiu G, Chen L, Zhu J et al (2015) Effect of Cr_2O_3 addition on viscosity and structure of ti-bearing Blast Furnace slag. ISIJ Int 55(7):1367–1376
5. Huang WJ, Zhao YH, et al (2016) Viscosity property and structure analysis of FeO-SiO_2-V_2O_3-TiO_2-Cr_2O_3 slags. ISIJ Int 56(4):594–601
6. Xu R, Zhang J, Fan X et al (2017) Effect of MnO on high-alumina slag viscosity and corrosion behavior of refractory in slags. ISIJ Int 57(11):1887–1894
7. Li Q, Gao J, Zhang Y et al (2017) Viscosity measurement and structure analysis of Cr_2O_3-bearing CaO-SiO_2-MgO-Al_2O_3 slags. Metall Mater Trans B 48(1):346–356
8. Li Q, Yang S, Zhang Y et al (2017) Effects of MgO, Na_2O, and B_2O_3 on the viscosity and structure of Cr_2O_3-bearing CaO-SiO_2-Al_2O_3 slags. ISIJ Int 57(4):689–696
9. Chen M, Raghunath S et al (2013) Viscosity measurements of "FeO"-SiO_2 Slag in equilibrium with metallic Fe. Metall Mater Trans B, 44(3):506–515
10. Seong SH, Sungmo J, Youngseok L, et al (2007) Viscosity of highly basic slags. ISIJ Int, 47(8):1090–1096

Thermodynamic Analysis of Carbothermic Reduction of Electric Arc Furnace Dust

Qing Ye, Zhiwei Peng, Lei Ye, Liancheng Wang, Robin Augustine, Joonho Lee, Yong Liu, Mudan Liu, Mingjun Rao, Guanghui Li and Tao Jiang

Abstract Electric arc furnace (EAF) dust is a kind of secondary resource which contains multiple metallic elements, including Fe, Mn and Cr. Pyrometallurgical processes for recovering metal elements from EAF dust have been investigated for many years although they are suffered from high energy consumption due to the spinel-structured components of EAF dust. In this study, the thermodynamic analysis of carbothermic reduction of EAF dust was performed. The main components of EAF dust were magnetite (Fe_3O_4), hausmannite (Mn_3O_4) and chromate spinel ($FeCr_2O_4$). The gangue minerals were mainly composed of magnesium silicates. The thermodynamic analysis indicated that magnetite and hausmannite can be reduced to metallic iron and MnO, respectively. Meanwhile, the chromate spinel will be reduced to chromium oxide and then to form $CaCr_2O_4$. The results also demonstrated that the gangue components can promote the separation of Fe and Cr, agreeing well with the experimental results.

Keywords Thermodynamic analysis · Carbothermic reduction · EAF dust · Ferrite

---

Q. Ye · Z. Peng (✉) · L. Ye · L. Wang · M. Rao · G. Li · T. Jiang
School of Minerals Processing and Bioengineering,
Central South University, Changsha, Hunan 410083, China
e-mail: zwpeng@csu.edu.cn

R. Augustine
Department of Engineering Sciences,
Uppsala University, Uppsala 75121, Sweden

J. Lee
Department of Materials Science and Engineering,
Korea University, Seoul 02841, South Korea

Y. Liu · M. Liu
Guangdong Provincial Key Laboratory of Development and Comprehensive
Utilization of Mineral Resources, Guangdong 510650, China

© The Minerals, Metals & Materials Society 2019

T. Jiang et al. (eds.), *10th International Symposium on High-Temperature Metallurgical Processing*, The Minerals, Metals & Materials Series,
https://doi.org/10.1007/978-3-030-05955-2_11

Introduction

According to the World Steel Association, electric arc furnaces produce about 29% of worldwide steel. It is estimated that along with production of 1 ton of molten metal about 11–20 kg of electric arc furnace (EAF) dust is generated [1, 2]. EAF dust is an important secondary resource which contains multiple metallic elements, such as iron, chromium, manganese and zinc. In general, it contains 3–40% zinc and up to 50% iron [3]. Its recycling is not only favorable to improving economic potential of the dust by recovering the valuable metals, but also of benefit to solving disposal and environmental problems caused by the heavy elements (e.g. lead and chromium) entrained in the dust [4, 5].

In recent years, many scholars have paid attention to recovering metal resources from EAF dust. The hydrometallurgical process is the earliest method applied for recycling of EAF dust, based on acidic or alkali leaching [6]. Inorganic acids such as sulfuric, nitric and hydrochloric acids have been widely used for selective dissolution of the metal elements from EAF dust. It was reported that the extraction ratios of zinc and cadmium reached up to 97 and 95%, respectively, using sulfuric acid [7]. However, this method often requires multiple steps to achieve high extractions [8]. Compared with acidic leaching, alkali leaching may enable selective leaching of EAF dust [9, 10]. It has the potential to obtain better results technically and economically, with less generation of residue. Nevertheless, high extractions of zinc and lead are still difficult.

Although the hydrometallurgical and other processes have been studied, pyrometallurgical routes are still considered to be the primary choice for processing of EAF dust because of its high potential metal recovery and relatively short flow sheet. The most frequently applied pyrometallurgical processes involve use of Waelz Kiln for recycling of EAF dust [11]. Coke and coal are the main reducing agents, and the iron metallization degree might reach 70–80% [12]. Actually, the stable spinel-structured components (e.g. $ZnFe_2O_4$, $MnFe_2O_4$ and $FeCr_2O_4$) of EAF dust often lead to difficult reduction, and high operating temperatures (1300–1650 °C) are required [13]. Most recently, a study of thermal stability of zinc ferrite indicated that adding calcium oxide would promote its decomposition. In order to decompose more than 90% of $ZnFe_2O_4$ in EAF dust, it was necessary to add more than 46% CaO to the dust [14]. However, the relevant thermodynamic analysis under reductive atmosphere receives little attention.

For understanding of the principle of pyrometallurgical processes of EAF dust based on carbothermic reduction, it is essential to investigate the reactions involved from the thermodynamic perspective to achieve efficient recovery of the entrained valuable elements. In this study, a thermodynamic analysis of carbothermic reduction of EAF dust was carried out and the results were verified experimentally by reduction of an EAF dust sample using biochar under microwave irradiation.

Experimental

Materials

The EAF dust, collected from Baosteel Group Corporation, Shanghai, China, was used for the analysis in this study. Table 1 shows the main components of the EAF dust sample. Its iron and zinc contents were 22.46 and 1.63%, respectively. The contents of other elements, including Mg, Al and Si, were 2.61, 10.59 and 8.25%, respectively. The biochar (charcoal) with the particle size between 0.045 and 0.053 mm was used as reducing agent. Its proximate analysis and ash composition are shown in Table 2.

The phase composition of EAF dust is shown in Fig. 1. The iron was primarily enriched in the dust in the forms of Fe_3O_4 and $ZnFe_2O_4$. The main phases of gangue components were $CaAl_8Fe_4O_{19}$ and $MgAl(SiO_4)_3$.

Method

The thermodynamic calculations of carbothermic reduction of EAF dust was performed using the software FactSage 7.0 (Thermfact/CRCT, Montreal, Quebec, Canada; GTT-Technologies, Herzogenrath, Germany). The results of the thermodynamic analysis were verified experimentally. The EAF dust and biochar were firstly pre-dried and then mixed with a mass ratio of 0.25 using a blender. The mixture was subsequently pressed at 50 MPa to prepare cylindrical briquettes with diameter of 20 mm and height of 10 mm. Lastly, the briquettes were charged in a microwave furnace for reduction at 1323 K for 15 min.

Table 1 Chemical composition of EAF dust (wt%)

Element	Fe	Ca	Zn	Pb	Mn	Cr	F	Cl
Content	22.46	21.03	1.63	0.049	3.50	0.23	0.59	0.18
Element	Si	Al	Mg	S	P	Na	K	
Content	8.25	10.59	2.61	0.71	0.039	1.19	0.31	

Table 2 Proximate analysis of biochar and its ash composition (wt%)

Proximate analysis			Main chemical composition of ash				
F_c	A_d	V_{daf}	Al_2O_3	SiO_2	Fe_2O_3	TiO_2	CaO
77.47	5.55	16.98	1.51	8.77	23.91	0.17	37.09

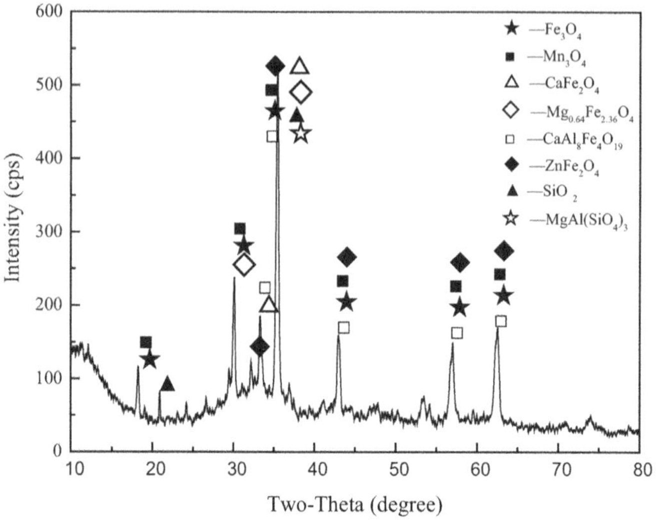

Fig. 1 XRD pattern of EAF dust

Results and Discussion

The Gibbs free energy changes of the reactions involved in reduction of main components of EAF dust were calculated, and the results are shown in Table 3 and Fig. 2. Note that $ZnFe_2O_4$ (franklinite) was the main zinc-bearing constituent of EAF dust, easily decomposed in reductive atmosphere. It can be reduced to ZnO and Fe_3O_4 at 800 K (see T_0 in Table 3). The reaction temperature of zinc oxide with carbon is 1229 K, which is lower than that of zinc oxide with carbon monoxide at 1591 K. The thermodynamic analysis indicated zinc oxide reacts with carbon easily. The stepwise reductions of iron oxide and zinc oxide take place in the temperature range of 800–1100 K. In addition, metallic zinc can evaporate above 1183 K. These results indicated that iron and zinc constituents can be easily separated in a reduction process.

In order to investigate the correlation between reduction reactions and reductive atmosphere, the relevant thermodynamic calculations based on determination of the relationship between thermodynamic equilibrium constant and temperature for some reactions are shown in Table 4. A corresponding plot of temperature versus gas equilibrium composition is shown in Fig. 3. The equilibrium curve of the Boudouard reaction, i.e. Eq. (14), and indirect reduction curves of iron oxide and zinc oxide intersects at T_a, T_b and T_c, respectively, indicating relatively easy reduction of the oxides. The thermodynamic analysis indicated that Mn_3O_4 can be easily reduced to MnO below 600 K, but further reduction of MnO demands high temperatures (>1650 K). MnO and metallic iron can be separated based on the

Table 3 Expressions for Gibbs free energy changes of main reactions involved in EAF dust reduction

No.	Reaction equation	$\Delta_r G_m^\theta$ (J·mol^{-1})	T_0 (K)
(1)	$Fe_3O_4(s) + 4C = 3Fe(s) + 4CO(g)$	684400-698T	980.5
(2)	$FeO(s) + C = Fe(s) + CO(g)$	158970-160.25T	992.0
(3)	$Cr_2O_3(s) + 3C(s) = 2Cr(s) + 3CO(g)$	819936-541.2T	1515.0
(4)	$Cr_2O_3(s) + 13/3C(s) = 2/3Cr_3C_2(s) + 3CO(g)$	755927-522.03T	1448.1
(5)	$Cr_2O_3(s) + 27/7C(s) = 2/7Cr_7C_2(s) + 3CO(g)$	766692-546.86T	1402.0
(6)	$FeO{\cdot}Cr_2O_3(s) + 4C(s) = 2Cr(s) + Fe(s) + 4CO(g)$	951540-653.61T	1455.8
(7)	$FeO{\cdot}Cr_2O_3(s) + C(s) = Cr_2O_3(s) + Fe(s) + 4CO(g)$	163830-138.43T	1183.5
(8)	$3ZnFe_2O_4(s) + C(s) = 3ZnO(s) + 2Fe_3O_4(s) + CO\ (g)$	159510-182.02T	876.3
(9)	$6ZnFe_2O_4(s) + C(s) = 6ZnO(s) + 4Fe_3O_4(s) + CO_2(g)$	144579-180.69T	800.15
(10)	$ZnFe_2O_4(s) + C(s) = ZnO(s) + 2FeO(s) + CO\ (g)$	192297-208.84T	920.79
(11)	$2ZnFe_2O_4(s) + C(s) = 2ZnO(s) + 4FeO(s) + CO_2(g)$	212742-241.93T	879.35
(12)	$ZnFe_2O_4(s) + 3C(s) = ZnO(s) + 2Fe(s) + CO\ (g)$	498437-515.35T	967.14
(13)	$ZnFe_2O_4(s) + 4C(s) = Zn\ (g) + 2Fe(s) + 4CO(g)$	746525-722.18T	1033.7

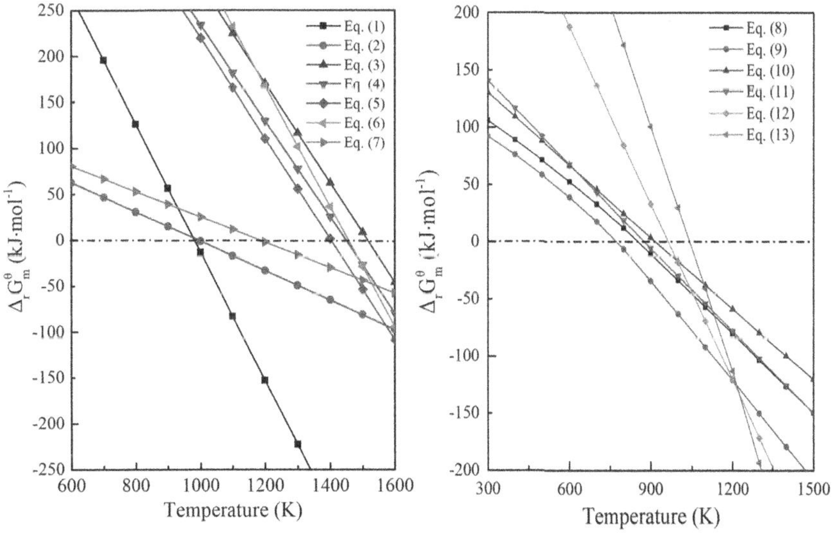

Fig. 2 Plot of Gibbs free energy changes of main reactions involved in EAF dust reduction versus temperature

discrepancy of their magnetisms. Hence, maintaining a temperature above 1183 K is sufficient for separation of Fe, Mn and Zn.

According to previous study, the transformation of spinel-structured components of EAF dust was the main reason that caused the high energy consumption by using the pyrometallurgical method. Based on the calculations of Gibbs free energy

Table 4 Relationship between thermodynamic equilibrium constant and temperature for reactions involved in reduction of EAF dust (CO as reducing agent)

No.	Reaction equation	$\ln K^\theta - T$
(14)	$C(s) + CO_2(g) = 2CO(g)$	$\ln K^\theta = -20{,}530.47/T + 20.99$
(15)	$Fe_2O_3(s) + 1/3CO(g) = 2/3Fe_3O_4(s) + 1/3CO_2(g)$	$\ln K^\theta = 6269.91/T + 4.93$
(16)	$Fe_3O_4(s) + CO(g) = 3FeO(s) + CO_2(g)$	$\ln K^\theta = -4255.23/T + 4.83$
(17)	$FeO(s) + CO(g) = Fe(s) + 3CO_2(g)$	$\ln K^\theta = 2742.21/T - 2.92$
(18)	$Fe_3O_4(s) + 4CO(g) = 3Fe(s) + 4CO_2(g)$	$\ln K^\theta = 1182.52/T - 1.03$
(19)	$ZnO(g) + CO = Zn(l) + CO_2(g)$	$\ln K^\theta = -16{,}551.27/T + 11.84$
(20)	$Mn_3O_4(s) + C(s) = 3MnO(g) + CO(g)$	$\ln K^\theta = -6175.25/T + 4.75$

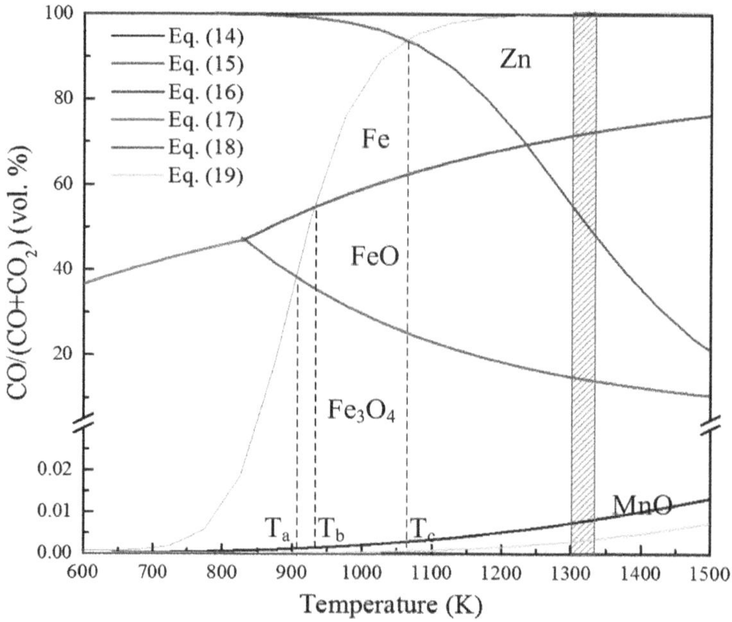

Fig. 3 Plot of gas equilibrium composition versus temperature concerning reduction of iron oxide and manganese oxide

changes of the reactions, $FeCr_2O_4$ can react with carbon to form Cr_2O_3 and metallic iron at 1183 K. Chromium oxide can be transformed to spinel phases (e.g., $CaCr_2O_4$) with the mass ratio of Cr_2O_3 to CaO less than 0.25 at 1350 K, as shown in Fig. 4. The results also indicated than Cr_2O_3 is converted to $Ca_3Cr_2Si_3O_{12}$ in the presence of excessive SiO_2. In addition, the components of MgO and Al_2O_3 will promote the formation of spinel phases (e.g., $MgCr_2O_4$). Therefore, when the temperature is higher than 1350 K, the gangue components (CaO, SiO_2, MgO and Al_2O_3) have potential to facilitate separation of Cr and Fe.

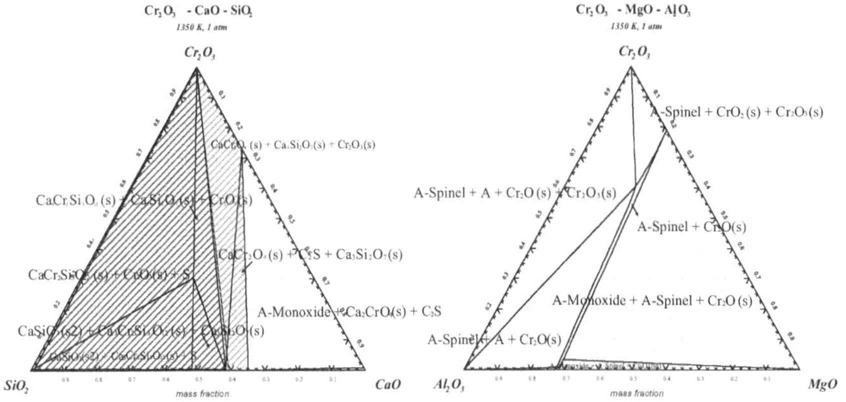

Fig. 4 Phase diagrams of Cr_2O_3-CaO-SiO_2 and Cr_2O_3-MgO-Al_2O_3

For verifying the above thermodynamic results, the reduction of the EAF dust sample was performed under the aforementioned conditions. It was found that after reduction at 1323 K for 15 min, the iron metallization degree of the reduced briquettes and the volatilization ratios of zinc and lead reached 96%, 99.6% and 80.2%, respectively, confirming the relevant reaction tendencies revealed by the thermodynamic analysis.

Conclusions

In this study, the thermodynamic analysis of the main reactions in the process of carbothermic reduction of EAF dust was reported. It was revealed that the stepwise reductions of iron oxide and zinc oxide take place in the temperature range of 800–1100 K. Meanwhile, the chromate spinel ($FeCr_2O_4$) will be reduced to chromium oxide and then to form $CaCr_2O_4$ in the presence of CaO. It was shown that the metal elements can be recovered through the reduction process. The thermodynamic results were also confirmed by the reduction of EAF dust using biochar under microwave irradiation. After reduction of the composite briquettes composed of EAF dust and biochar at 1323 K for 15 min, the iron metallization degree of the reduced briquettes and the volatilization ratios of zinc and lead reached 96, 99.6 and 80.2%, respectively.

Acknowledgements This work was partially supported by the National Natural Science Foundation of China under Grants 51774337, 51504297, 51811530108 and 51881340420, the Science and Technology Major Project of Gansu Province, China, under Grant 1602FKDC007, the Research Fund Program of Guangdong Provincial Key Laboratory of Development and Comprehensive Utilization of Mineral Resources under Grant SK-201801, the Innovation-Driven Program of Central South University under Grant 2016CXS021, the Shenghua Lieying Program of Central South University under Grant 502035001, the Natural Science Foundation of Hunan Province, China, under Grant 2017JJ3383, the Hunan Provincial Co-Innovation Centre for Clean and Efficient Utilization of Strategic Metal Mineral Resources under Grant 2014-405, the Fundamental Research Funds for the Central Universities of Central South University under Grant 2018zzts222, and the Open-End Fund for the Valuable and Precision Instruments of Central South University under Grant CSUZC201706.

References

1. Oustadakis P, Tsakiridis PE, Katsiapi A, Agatzini-Leonardou S (2010) Hydrometallurgical process for zinc recovery from electric arc furnace dust (EAFD) Part I: Characterization and leaching by diluted sulphuric acid. J Hazard Mater 179:1–7
2. Penterson RD, Twidwell LG (1996) Recycling of metals and engineered materials. JOM 48 (3):43–44
3. Orhan G (2005) Leaching and cementation of heavy metals from electric arc furnace dust in alkaline medium. Hydrometallurgy 78(3):236–245
4. Lin XL, Peng ZW, Yan JX, Li ZZ, Hwang JY, Zhang YB, Li GH, Jiang T (2017) Pyrometallurgical recycling of electric arc furnace. J Clean Prod 149(15):1079–1100
5. Peng ZW, Lin XL, Yan JX, Hwang JY, Zhang YB, Li GH, Jiang T (2017) Characterization of low-zinc electric arc furnace dust. Character Miner Metals Mater 2017:103–109
6. Kukurugya F, Vindt T, Havlik T (2015) Behavior of zinc, iron, and calcium from electric arc furnace (EAF) dust in hydrometallurgical processing in sulfuric acid solutions: thermodynamic and kinetic aspects. Hydrometallurgy 154:20–32
7. Montenegro V, Leonardou SA, Oustadakis P, Tsakiridis P (2016) Hydrometallurgical treatment of EAF Dust by direct sulphuric acid leaching at atmospheric pressure. Waste Biomass Valori 7(6):1531–1548
8. Halli P, Hamuyuni J, Revitzer H, Lundstrom M (2017) Selection of leaching media for metal dissolution form electric arc furnace dust. J Clean Prod 164(15):265–276
9. Bakkar A (2014) Recycling of electric arc furnace dust through dissolution in deep eutectic ionic liquids and electrowing. J Hazard Mater 280:191–199
10. Fujimoto RC, Maruyama K, Miki T, Nagasaka T (2016) The selective alkaline leaching of zinc oxide from electric arc furnace dust pre-treated with calcium oxide. Hydrometallurgy 159:120–125
11. Morcali M, Yucel O, Aydin A, Derin B (2012) Carbothermic reduction of electric arc furnace dust and calcination of Waelz oxide by semi-pilot scale Potary Furnace. J Min Metall Sect B-Metall 48(2):173–184
12. Leclerc N, Meux E, Lecuire JM (2002) Hydrometallurgical recovery of zinc and lead from electric arc furnace dust using mononitrilotriacetate anion and hexahydrated ferric chloride. J Hazard Mater 91(1–3):257–270
13. Zhang HN, Li JL, Xu AJ, Yang QX, He DF, Tian NY (2014) Carbothermic reduction of zinc and iron oxides in electric arc furnace. J Iron Steel Res Int 21(4):427–432
14. Yakornov S, Panshin A, Grudinsky P, Dyubanov V, Leontev L, Kozlov P, Ivakin D (2017) Thermodynamic analysis of zinc ferrite decomposition in electric arc furnace dust by lime. Russ J Non-ferrous Met 58(6):586–590

Influence of Cr_2O_3 Content on Slag Viscosity Under Different Melting States and Temperature Programs

Fang Yuan, Tuo Wu, Yanling Zhang and Zheng Zhao

Abstract Viscosities of Cr_2O_3-bearing slags were examined under different situations. The influence of Cr_2O_3 content on slag viscosity greatly depends on slag basicity and temperature. In $CaO–SiO_2–Cr_2O_3$ ternary slags (R = 1.2, 0.8, and 0.5) with a Cr_2O_3 content lower than 6%, viscosity decreases with an increase in Cr_2O_3 content; with increasing Cr_2O_3 content, the slag viscosity tends to decrease, especially in the cases of higher basicity. For $CaO–SiO_2–10\%Al_2O_3–Cr_2O_3$ (R = 1.2) quaternary slags, when holding at 1953 K for 2 h and measurement was performed during cooling down to 1773 K, viscosities of this slag gradually increased with decreasing temperature to a point and then a sharp increase on viscosity appeared with further decreasing temperature; above 1873 K, the viscosity decreased with increasing Cr_2O_3 content, while below 1853 K, a raise in Cr_2O_3 content increased the viscosity. However, in the case of measurement performed from 1813 to 1673 K, the viscosity increased with an increase in Cr_2O_3 content. It revealed that the slag viscosity showed greater dependency on the precipitation of Cr_2O_3 and their morphologies.

Keywords Cr-containing slag · Viscosity · Temperature programs · Morphology

Introduction

Slag viscosity is a critical physical property that affects the separation and transportation of elements between the metal and slag phases [1]. Extensive research has been conducted to examine viscosity in CrO_x-bearing slag systems [2–21]. However, results concerning the dependency of slag viscosity on Cr_2O_3 content have been inconsistent. Minami et al. [2] suggested that the slag viscosity increases with increasing Cr_2O_3 content, while both Forsbacka et al. [8, 9] and Wu [10]

F. Yuan · T. Wu · Y. Zhang (✉) · Z. Zhao
State Key Laboratory of Advanced Metallurgy, University of Science and Technology Beijing, Beijing 100083, People's Republic of China
e-mail: zhangyanling@metall.ustb.edu.cn

© The Minerals, Metals & Materials Society 2019

T. Jiang et al. (eds.), *10th International Symposium on High-Temperature Metallurgical Processing*, The Minerals, Metals & Materials Series,
https://doi.org/10.1007/978-3-030-05955-2_12

showed that slag viscosity decreased with increasing Cr_2O_3 content. This difference could be explained by the different fluid characteristics of Cr_2O_3-bearing slag under different conditions.

In this research, the effects of Cr_2O_3 content on viscosity of $CaO–SiO_2–Cr_2O_3$ ternary slags with varied basicity were initially examined. Subsequently, the viscosities of $CaO–SiO_2–10\%Al_2O_3–Cr_2O_3$(R = 1.2) quaternary slags were measured under two temperature regimes: A, wherein samples were held at 1953 K for 2 h and then the temperature dependency of viscosity was measured during cooling from 1953 to 1683 K; and B, wherein the holding temperature was changed to 1813 K and viscosity was measured during cooling from 1813 to 1683 K. The effects of Cr_2O_3 content on slag viscosity under these two temperature regimes were analyzed.

Experimental Methods

Apparatus

The configuration of the experimental apparatus is illustrated in Fig. 1. The system includes an electrical resistance ($MoSi_2$ elements) furnace with a maximum

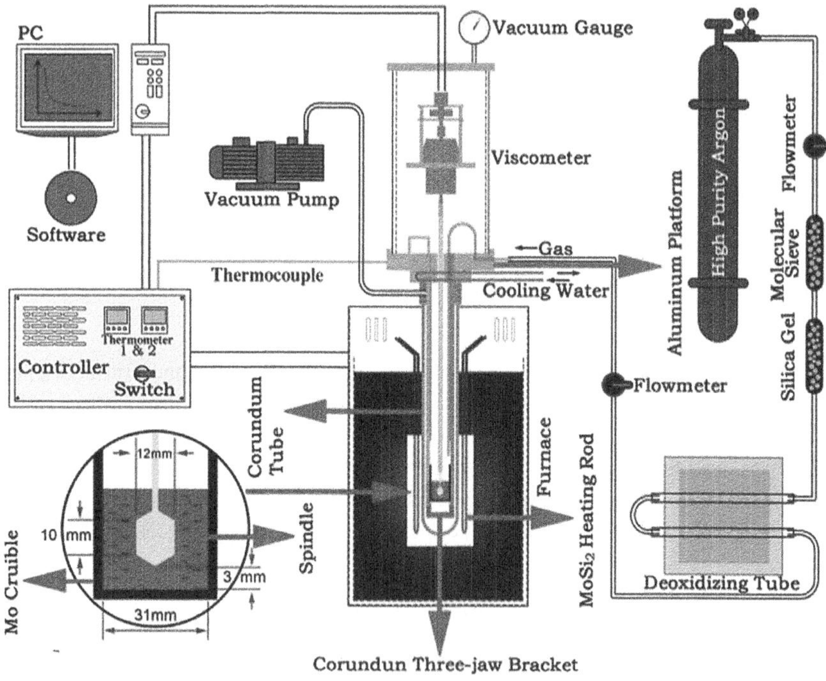

Fig. 1 Diagram of high-temperature viscometer and its auxiliary device

temperature of 1973 K (1700 °C), a rotary digital viscometer (Brookfield DV3T LV, Brookfield company, USA), a gas protection system (High purity Ar, ≥ 99.999%), and data collection software (Viscometer Test Software, Version Number HA1.10). The furnace and sample temperatures were measured by two B-type thermocouples (Omega, USA) with measurement accuracies of ±0.5%. The viscometer was protected by a transparent glass tube installed on an aluminum platform, which can be connected with the furnace tube to form a complete cavity. A vacuum gauge and vacuum pump was used to quickly evacuate the air inside the cavity. This setup ensures that the slag can be measured in a closed and controllable atmosphere at high temperatures. The constant used with the Brookfield viscometer is referred to as the spindle multiplier constant (SMC). The viscometer is calibrated using standard silicone oils (provided by Brookfield Company) with viscosities of 100, 500, and 5000 cP; the final SMC is the average of the instrument constants after calibration with different standard silicone oils. A similar calibration operation has been described in detail elsewhere [22].

Sample Preparation

The raw materials are described in Table 1. The 45 g slag quantity was based on an estimation of the slag density and the pre-experiment. Pretreated pure reagents (about 400 mesh) were carefully weighed, blended, and fully ground to produce the compositions presented in Table 2. Then, the mixed powder samples were pressed into a cylindrical block in a steel mould under 30 MPa, which facilitated fitting the samples into the molybdenum crucible and subsequent melting by bringing the oxide particles into close contact. Finally, the sample was placed in a dryer.

Table 1 Raw materials used for experiments

Materials	Purity	Pretreatment	Provider
$CaCO_3$	≥ 99.0%	Decomposed at 1473 K for 12 h, Verified by XRD	Sinopharm Chemical Reagent Co., Ltd.
SiO_2	AR	Dried at 1073 K for 8 h	
Al_2O_3	AR		
Cr_2O_3	≥ 99.0%		
Mo	≥ 99.7%	Crucible and Spindle were machined by CNC machine. Deviation: ±0.02 mm	Baoji Sheng Hua Non-ferrous Materials Co., Ltd., China
Ar	≥ 99.999%	Dehydrated by molecular sieve and silica gel Deoxidized by Cu pieces (5 N) and Mg stripes (>99%) at 773 K	Beijing Qianxi gas company, China

Table 2 Target composition of slags

No.	Target composition wt%					Temperature program
	CaO	SiO_2	Al_2O_3	Cr_2O_3	R_{target}[a]	
A0/B0	40	50	10	0	1.2	A/B
A5/B5	46.4	38.6	10	5	1.2	A/B
A7/B7	45.3	37.7	10	7	1.2	A/B
A10/B10	43.6	36.4	10	10	1.2	A/B
C0	33.3	66.7	0	0	0.5	A
C3	32.3	64.7	0	3	0.5	A
C6	31.3	62.7	0	6	0.5	A
C9	30.3	60.7	0	9	0.5	A
D0	44.4	55.6	0	0	0.8	A
D3	43.1	53.9	0	3	0.8	A
D6	41.8	52.2	0	6	0.8	A
D9	40.5	50.5	0	9	0.8	A
E0	54.5	45.5	0	0	1.2	A
E3	52.9	44.1	0	3	1.2	A
E6	51.3	42.7	0	6	1.2	A
E9	49.7	41.3	0	9	1.2	A

[a]R_{target} = (wt%CaO)/(wt%SiO_2)

Viscosity Measurements

The viscometer was mechanically zeroed. A hollow corundum tube equipped with a molybdenum rotor was suspended on the viscometer hook. The crucible containing the sample was fixed on the three-jaw bracket. The spindle was adjusted to coincide with the crucible centreline to attain full alignment. The aluminum platform was lowered to place the sample in the constant temperature zone of the furnace. The flange on the alumina tube was connected to the aluminum platform by screws. The air inside the tube was pumped out using the vacuum pump, and high-purity argon was used throughout the entire measurement process at a flow rate of 150 mL/min.

The temperature programs used during experiments are shown in Fig. 2. For program A, the furnace was programed to heat from room temperature to 1953 K (1680 °C), and samples were held at 1953 K for 2 h before the first measurement, to ensure complete melting of the Cr_2O_3-containing slags. The viscosity of the sample was then measured while cooling at 3 K/min. In program B, the sample was first heated to 1813 K (1540 °C), maintained at this temperature for 2 h, and then viscosity was measured during cooling at 3 K/min. Measurement was terminated when the viscosity of slag exceeded 3000 cP or if the viscosity–temperature curve showed obvious fluctuation. After measurement, the sample was again heated to the target holding temperature and maintained for 2 h to pull out the spindle. Finally, the crucible was quickly removed from the furnace and quenched rapidly in water to obtain glass samples for analysis.

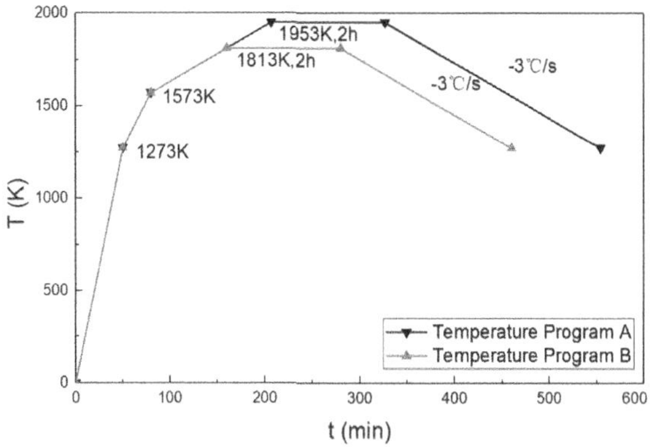

Fig. 2 Temperature programs A and B

The phase of the samples was analyzed by X-ray diffraction (XRD, Smart RIGAKU, Japan) and the morphological character of the samples was analyzed by scanning electron microscope (SEM, FEI MLA250, USA).

Results and Discussion

Effects of Cr_2O_3 on viscosity of CaO-SiO_2-Cr_2O_3 slag with basicities of R = 0.5, 0.8, and 1.2 were shown in Fig. 3a–c, respectively. The basicity dependency of the slag (3% Cr_2O_3) viscosity was illustrated in Fig. 3d. The viscosities of CaO-SiO_2-Cr_2O_3 slag significantly decreased with increasing basicity within the 1793–1953 K range. For three kinds of slag (Fig. 3a–c), when Cr_2O_3 content is lower than 6%, the viscosity decreases with an increase in Cr_2O_3 content. Moreover, the variation of viscosity tends to increase with decreasing basicity and decreasing temperature. For the acidic slags with R = 0.5 (Fig. 3a), when the Cr_2O_3 content increases from 0 to 6% at 1833 K (1560 °C), the slag viscosity decreases from 2.146 to 0.857 Pa S, representing a drop of 60%. While in the cases of R = 0.8(Fig. 3b) and R = 1.2 (Fig. 3c), the corresponding viscosity decreases from 0.355 to 0.240 Pa S and from 0.164 to 0.124 Pa S, a decrease of 32 and 24%, respectively. It suggested that the Cr_2O_3 (lower than 6%) could be totally dissolved into molten slag (or the amount of undissloved Cr_2O_3 is too little to change the rheology characteristic of slag flow) and tends to behave as basic ions [23, 24] and work as a Si-O network breaker [25] above 1793 K.

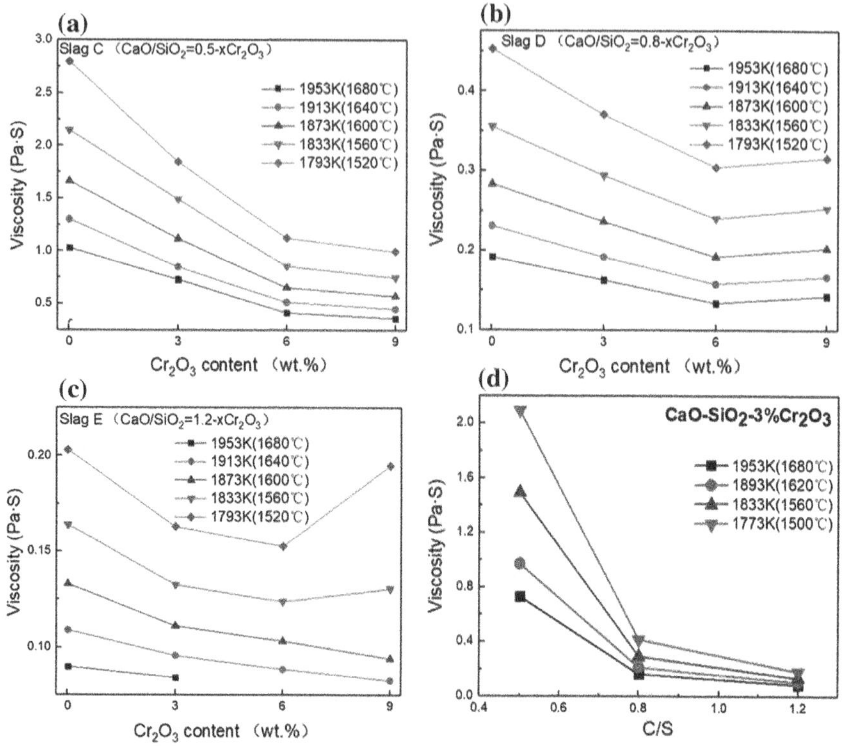

Fig. 3 Effects of Cr_2O_3 and basicity on the viscosity of $CaO–SiO_2–Cr_2O_3$ ternary slags

Effect of Cr_2O_3 Content on the Viscosity of $CaO\text{-}SiO_2\text{-}Cr_2O_3$ Slag Under Different Basicity

However, when the Cr_2O_3 content exceeds 6%, the viscosities of slag with higher basicity (R = 0.8 and 1.2) tend to increase with further increases in Cr_2O_3 content, as observed in Fig. 3b, c. This effect could be due to that in these cases some Cr_2O_3 remains undissolved or some Cr_2O_3-bearing phases precipitate at lower temperature, which causes slag to behave as a non-homogeneous liquid phase with suspended fine solid particles. Obviously, more undissolved Cr_2O_3 or precipitated Cr_2O_3-bearing solid particles will be present when the overall Cr_2O_3 content is elevated, and so the slag viscosities will be higher. Figure 4 shows the XRD patterns of quenched samples which Cr_2O_3 content is greater than 6% (C6, C9, D6, D9, E6, and E9, seen in Table 2). Some solid Cr_2O_3 or $CaCr_2O_4$ (formed in slag with higher basicity during decreasing temperature) remained in the testing samples as shown in Fig. 4, which adds credibility to the above offered explanations. In the case of slag with the basicity of 0.5 (Fig. 3a), even some Cr_2O_3 remained undissolved (C9 in Fig. 4), the viscosity of slag C still slightly decreased as the Cr_2O_3

Fig. 4 XRD pattern of the quenched samples

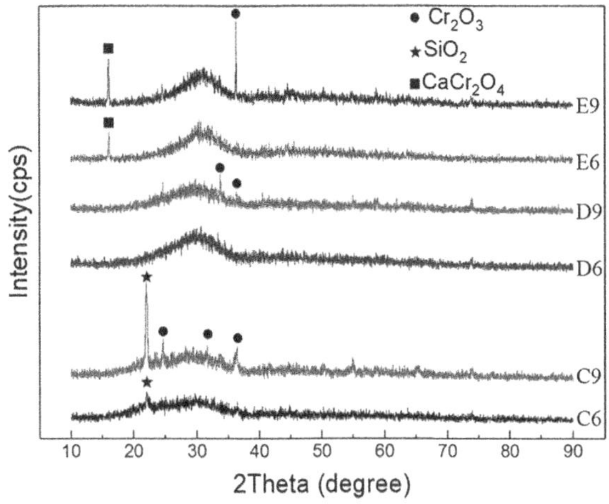

increased from 6 to 9%, but the decline has been significantly reduced (from 39 to 12%). These findings reveal that Cr_2O_3 tends to behave as basic materials and shows more effective function of decreasing viscosity especially in the much lower basicity slag.

Effect of Cr_2O_3 Content on the Viscosity of CaO-SiO_2-10% Al_2O_3-Cr_2O_3 Slag Under Different Temperature Program

The viscosities of CaO–SiO_2–10%Al_2O_3–Cr_2O_3(R = 1.2) slags as functions of temperature, under two temperature programs, are shown in Fig. 5. Slag viscosity

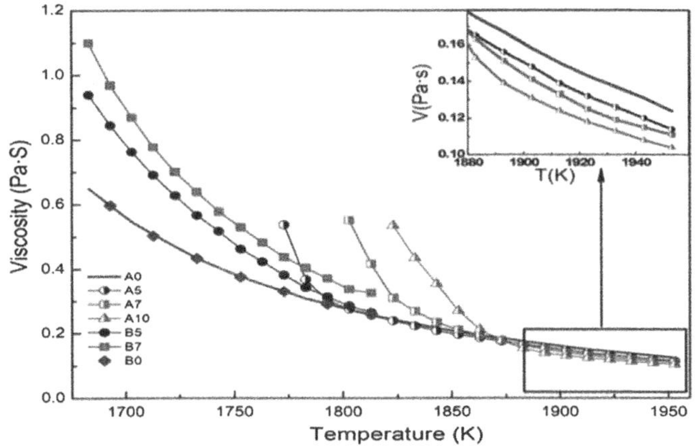

Fig. 5 Temperature dependence of slag viscosity under temperature programs A and B

generally decreased with increasing temperature. Moreover, for $CaO–SiO_2–10\%$ Al_2O_3 slags without Cr_2O_3 (samples A0 and B0) evaluated under two temperature programs, the measured viscosity data were in agreement and a smooth viscosity versus temperature curves were produced. This outcome suggests that the rheological behaviour of $CaO–SiO_2–10\%Al_2O_3$ slag without Cr_2O_3 tends to stay uniform over the whole temperature range (1953–1683 K). However, in the case of Cr_2O_3-bearing $CaO–SiO_2–10\%Al_2O_3–Cr_2O_3$(R = 1.2) samples (A5, A7, A10, B5, and B7), large fluctuations appeared on the viscosity–temperature curves even at same temperature range (1813–1793 K) under two temperature programs. For program A, the slag sample viscosities gradually increased as the temperature decreased from 1953 K to a certain point (i.e. 1783, 1813, and 1863 K for samples A5, A7, and A10, respectively), and then sharply increased with further decreases in temperature.

More importantly, for temperature program A, when the temperature was above 1873 K, viscosities of $CaO–SiO_2–10\%Al_2O_3–Cr_2O_3$(R = 1.2) slags decreased with increasing Cr_2O_3 content. This observation agrees well with the results of ternary slags in this study (Cr_2O_3 content < 6%) and the results obtained by Wu [10] and Forsbacka et al. [8, 9].The explanation is same as the ternary slags (Cr_2O_3 content < 6%) above. However, at temperatures below 1803 K, viscosity increased with increasing Cr_2O_3 content. Below 1803 K, some Cr-bearing solid phase tended to precipitate (as discussed in the next section) with decreasing temperature, which caused a sharp increase in viscosity of the slag. The slags containing higher Cr_2O_3 contents would have had more solid precipitation, thus resulting in higher viscosities.

Similarly, under temperature program B, when the sample was held at temperatures as low as 1813 K, Cr_2O_3 was not completely dissolved into the molten slag, the higher initial Cr_2O_3 content, the more Cr_2O_3 remained in solid particles, and then the larger viscosities. This tendency agrees reasonably well with previously reported data [2–7], wherein viscosity measurements were performed over a similar temperature range (1673–1793 K).

However, even over the same temperature range (1813–1793 K), the viscosities varied greatly between A5 and B5, and between A7 and B7. The data in Fig. 5 show that the viscosities of A5 and A7 were much higher than those of B5 and B7. The formation of different solid phases, along with the different solid compositions, under different temperature program was considered to be the main factors resulting in these differences [1, 26].

To verify whether the differences between the viscosities of slag samples A5 and B5 and of A7 and B7 were caused by different precipitation behaviours of solid phases, the morphologies of samples A5 and B5 (at the same temperature, 1813 K) were investigated by SEM. After viscosity measurements were taken, samples A5 and B5 were heated up again and maintained at 1953 K and 1813 K for 2 h, respectively. Sample A5 was then cooled at 3 K/min to 1813 K. Both samples were then quenched in water. SEM results for samples A5 and B5 are shown in Figs. 6 and 7, respectively, along with the corresponding XRD or EDS analysis result.

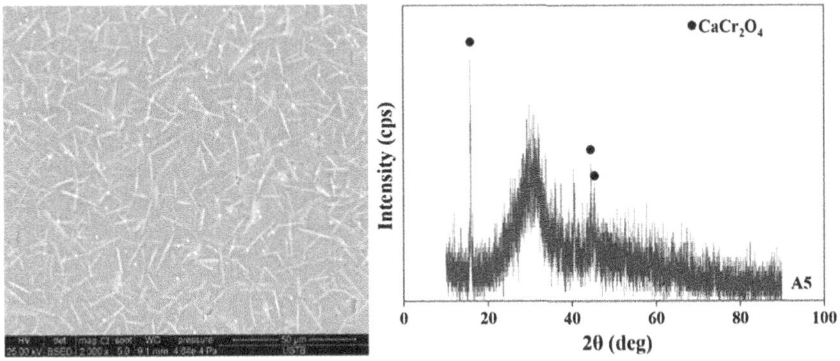

Fig. 6 Morphology and XRD results for slag A5

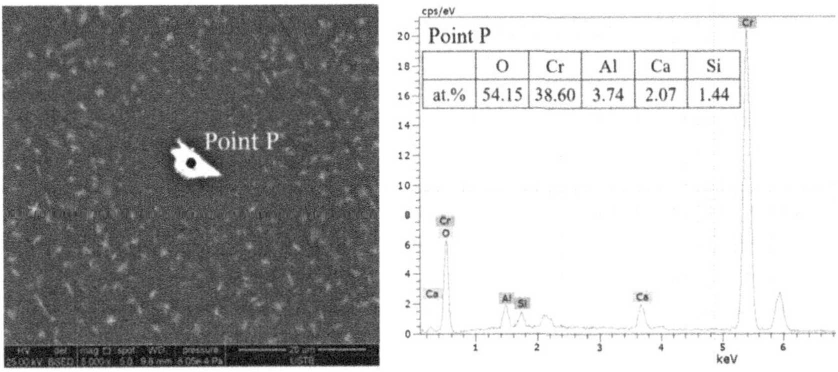

Fig. 7 Morphology and EDS results for slag B5

A large difference was observed between the morphologies of A5 and B5, even though they were obtained at the same temperature (1813 K) and had the same slag composition (Table 2). The solid phase in A5, white acicular materials uniformly distributed in the slag matrix (Fig. 6), was identified as $CaCr_2O_4$ by XRD. It is suggested that under temperature program A, during the cooling of the homogeneous liquid phase from 1953 to 1813 K, $CaCr_2O_4$ was formed and precipitated. This event would have led to a sharp increase in the solid fraction of the molten slag, explaining the obvious inflection point in the viscosity–temperature curve. For sample B5, while holding it at 1813 K, white spherical particles still existed, as shown in Fig. 7. According to EDS analysis, these particles are Cr_2O_3, indicating that the initial Cr_2O_3 was not completely dissolved in the molten slag at 1813 K. Obviously, the higher the initial Cr_2O_3 content in the samples (B0, B5, B7), the more particles remained in solid phase, resulting in higher apparent viscosities, which agrees reasonably well with the measurement results below 1813 K reported both herein and previously [2–7]. Moreover, the differences between the viscosities of A5 and B5, and between A7 and B7 below 1813 K are mainly caused by two phenomena. First, the solid fractions in A5 and A7 are higher than those in B5 and B7, since $CaCr_2O_4$ was formed through the reaction between CaO and Cr_2O_3 (the Cr_2O_3 contents were equivalent in A5 and B5 and in A7 and B7). Second, the network-like distribution of $CaCr_2O_4$ precipitated during solidification tends to generate a larger resistance against fluid of the liquid phase. Hence, higher viscosities were recorded for A5 and A7 than for B5 and B7, even at the same temperature.

Conclusions

(1) The effects of Cr_2O_3 content on viscosity of $CaO–SiO_2–Cr_2O_3$ ternary slags with varied basicity (R = 0.5, 0.8, and 1.2) were examined. The results showed that in the case of Cr_2O_3 content lower than 6%, the viscosity decreases with an increase in Cr_2O_3 content. And, the decreased viscosity scope tends to increase with decreasing basicity and decreasing temperature. While with further increasing Cr_2O_3 content the slag viscosity tends to decrease, especially in the cases of higher basicity.

(2) The viscosities of $CaO–SiO_2–10\%Al_2O_3–Cr_2O_3$ (R = 1.2) slags were measured under different temperature programs. For program A (in which samples were held at 1953 K for 2 h and viscosity was measured during cooling to 1813 K at 3 K/min), viscosity gradually increased with decreasing temperature to a certain point, after which a sharp increase in viscosity was observed upon further temperature decreases. When the temperature was above 1873 K, the viscosity decreased with increasing Cr_2O_3 content, while below 1853 K, an increase in Cr_2O_3 content led to increases in viscosity. In program B (in which samples were held at 1813 K for 2 h and the measurement was performed during cool-down to 1683 K at 3 K/min), viscosity increased with increasing Cr_2O_3 content. Even under the same temperature range (1813–1793 K), the viscosities varied greatly different under different temperature programs. That result could be explained by the change of solubility of Cr_2O_3 in slag and rheological characteristics of slag under different temperature program.

Acknowledgements The authors would like to express their appreciation to the National Natural Science Foundation (No. 51674022) for its financial support of this research work.

References

1. Miyamoto KI, Kato K, Yuki T (2002) Tetsu-to-Hagane 88:838–844
2. Minami E, Amatatsu M, Sano N (1987) Tetsu-to-Hagane 73:S871
3. Qiu G, Chen L, Zhu J, Lv X, Bai C (2015) ISIJ Int 55:1367–1376

4. Xu C, Wang W, Zhou L, Xie S, Zhang C (2015) Metall Mater Trans B 46B:882–892
5. Huang W, Zhao Y, Yu S, Zhang L, Ye Z, Wang N, Chen M (2016) ISIJ Int 56:594–601
6. Xu RZ, Zhang JL, Wang ZY, Jiao KX (2016) Steel Res Int 87:1–7
7. Li Q, Gao J, Zhang Y, An Z, Guo Z (2017) Metall Mater Trans B 48B:346–356
8. Forsbacka L, Holappa L (2004) In: VII International conference on molten slags, fluxes and salts, The South African Institute of Mining and Metallurgy, Johannesburg, pp 129–136
9. Forsbacka L, Holappa L, Kondratiev A, Jak E (2007) Steel Res Int 78:676–684
10. Wu T, Zhang Y, Yuan F, An Z. Metall Mater Trans B. Published online, https://link.springer.com/article/10.1007/s11663-018-1258-z
11. Holappa L, Xiao Y (2004) J S Afr I Min Metall 104:429–437
12. Behera RC, Mohanty UK (2001) ISIJ Int 41:834–843
13. Ostrovski OI, Utochkin YuI, Pavlov AV, Akberdin RA (1994) ISIJ Int 34:773–775
14. Nakamoto M, Forsbacka L, Holappa L (2007) In: XI International conference on innovations in the ferro alloy industry (INFACON XI), vol 1, pp 159–164
15. Kalicka Z, Kawecka-Cebula E, Pytel K (2009) Arch Metall Mater 54:179–187
16. Behera RC, Mohanty UK (2001) ISIJ Int 41:827–833
17. Forsbacka L, Holappa L, Iida T, Kita Y, Toda Y (2003) Scand J Metall 32:273–280
18. Forsbacka L, Holappa L (2004) Scand J Metall 33:61–68
19. Wang Z, Sun Y, Sridhar S, Zhang M, Guo M, Zhang Z (2015) Metall Mater Trans B 46B:537–541
20. Kim TS, Park JH (2014) ISIJ Int 54:2031–2038
21. Schumacher KJ, White JF, Downey JP (2015) Metall Mater Trans B 46B:119–124. Arrhenius S (1887). Zeitschrift für Physikalische Chemie 1:285–298
22. Forsbacka L. Experiences in slag viscosity measurement by rotation cylinder method. Research report, Laboratory of Metallurgy, Helsinki University of Technology
23. Mills K (2011) The estimation of slag properties. Short course presented as part of Southern African Pyrometallurgy 2011, 7 Mar 2011
24. Fincham CJB, Richardson FD (1952) Proc R Soc Lond 223:40–62
25. Mills KC, Yuan L, Li Z, Zhang GH, Chou KC (2012) High Temp Mater P 31:301–321
26. Verein Deutscher Eisenhüttenleute (VDEh) (1995) Slag atlas, 2nd edn. Verlag Stahleisen CmbH, Germany, pp 101–120

Influence of Mould Slag Compositions with Different Reactivity on the Erosion Rate of ZrO_2-C Bearing Submergence Entry Nozzle

Xue-Si Wang, Qian Wang, Chang-Ping Zeng and Hua-Zhi Yuan

Abstract Several slag compositions of differing reactivity were designed as a potential measure for reducing erosion on submergence entry nozzle during continuous casting of high-Mn high-Al steel. Reactivities were determined according to the SiO_2 content of the slag considering the reaction between [Al] and (SiO_2) at the steel–slag interface. The erosion rate was evaluated by means of the rotating cylinder method; a ZrO_2-C rod was immersed in the 1673 K slag sample for two hours, and the erosion rate was evaluated according to the reduction of rod diameter. The effect of slag components Na_2O, Li_2O, CaF_2 and B_2O_3 on the erosion rate was investigated. The main results indicate that slag with low SiO_2 content was more effective in decreasing the erosion rate than conventional high-SiO_2 slag. Listed in decreasing order of intensity, the effects of the individual components on the erosion rate are Na_2O, CaF_2, Li_2O and B_2O_3. The ultimate aim of the mould slag optimization effort is stabilizing the operation by extending the submergence entry nozzle service life.

Keywords ZrO_2-C refractory · Rotation speed · Slag composition · Dissolution rate

Introduction

Submerged entry nozzles (SEN) are extensively used in steel continuous casting to prevent re-oxidation of the molten steel and to improve its flow behaviour [1]. In general, SEN service life is limited by the erosion at the slag line zone. The slag line zone is usually strengthened with ZrO_2-C refractory because ZrO_2 withstands thermal shock and resists erosion from molten steel and slag to a higher degree than Al_2O_3. The stability of continuous casting and the quality of steel products depend

X.-S. Wang · Q. Wang (✉) · C.-P. Zeng · H.-Z. Yuan
College of Materials Science and Engineering, Chongqing University,
Chongqing 400044, China
e-mail: wangqian66@vip.sina.com

© The Minerals, Metals & Materials Society 2019

T. Jiang et al. (eds.), *10th International Symposium on High-Temperature Metallurgical Processing*, The Minerals, Metals & Materials Series,
https://doi.org/10.1007/978-3-030-05955-2_13

on SEN service life, and it is reported that the mould slag composition is a crucial factor in refractory erosion [1–3]. A number of SEN erosion studies have been previously performed [1–6]. For instance, Nakamura [3] concluded that slag with low viscosity or high $[F^-]$ content causes greater erosion of ZrO_2-C refractory. Cirilli [5] proposed that high SiO_2/low CaO activity slag causes high degree of zirconia degradation. Wioeniewska [2] concluded that the erosion of SEN is effectively suppressed by reducing the content of Li_2O, Na_2O and MgO in slag. Previous research mainly focused on the erosion of SEN by conventional CaO-SiO_2-based slag, and accounts of the effects of the low-reactive slag with high Al_2O_3 and low SiO_2 contents designed for continuous casting of high-Mn high-Al steel are rare. The current study was conducted to investigate the effect of slag for high manganese–high aluminium steel on erosion of ZrO_2-C refractory via the rotating cylinder method.

Experimental

Material Preparation

Zirconia rods 1.0×10^{-2} m in diameter and 3.5×10^{-2} m in length were made from the ZrO_2-C refractories at a pressure of 180 MPa. The sample material with granularity of ~0.5 mm was selected from a commercial refractory manufacturer, and the supplied ratio of chemical compositions is shown in Table 1. The slag samples were prepared with high-purity $CaCO_3$, SiO_2, MgO, Al_2O_3, CaF_2, Na_2CO_3, Li_2CO_3, $BaCO_3$ and B_2O_3 powders mixed in graphite crucible. The slag samples were heated in the graphite crucible for 1 h at 1573 K to ensure complete melting. The chemical compositions of the initial slag samples (after pre-melting) are listed in Table 2.

Table 1 Chemical composition of ZrO_2-C samples

ZrO_2 %	C (%)	CaO (%)	SiO_2 (%)	The others(%)
72.43	12	5.22	3.29	7.06

Table 2 Chemical composition of slag, wt%

Slag	CaO + BaO + MgO	Al_2O_3	$SiO_2 + B_2O_3$	CaF_2	$Li_2O + Na_2O$	Fe_2O_3
CaO-Al_2O_3 based slag	41	18	22	10	9	0
CaO-SiO_2 based slag	25.4	4.1	45.7	8.1	15.7	1

Experimental Method

A schematic diagram of the experimental apparatus is shown in Fig. 1. The dissolution experiments were carried out at constant temperature (1673 K) under an argon atmosphere in a resistance furnace with molybdenum disilicide elements. Each 270 g pre-melted slag sample was placed in a 50-mm internal diameter graphite crucible. When the slag was completely molten, the ZrO_2-C rod was lowered to a height of 10 cm above the slag surface and preheated for 10 min. After temperature stabilized, the rod was immersed in the molten slag about 15×10^{-3}m and rotated for 120 min at varying rotational speeds. At the end of each experiment, the ZrO_2-C rod was withdrawn and cooled to room temperature in air. The observation shows that the rods immersed in molten slag underwent different degrees of erosion, in which the area at the interface of the molten slag and gas phase was eroded severely, while the area that below the interface was not eroded significantly. Thus, the erosion rate was determined from the average reduction of the diameter of the rods in the area at the interface of the molten slag and gas phase.

Results and Discussion

Dissolution Rate of Zirconia Rod

The dissolution rate (cm s^{-1}) of the ZrO_2 rod is calculated by measuring the erosion depth of the rod at the air–slag interface (for dissolution in molten slag), as given by Eq. 1 [4]:

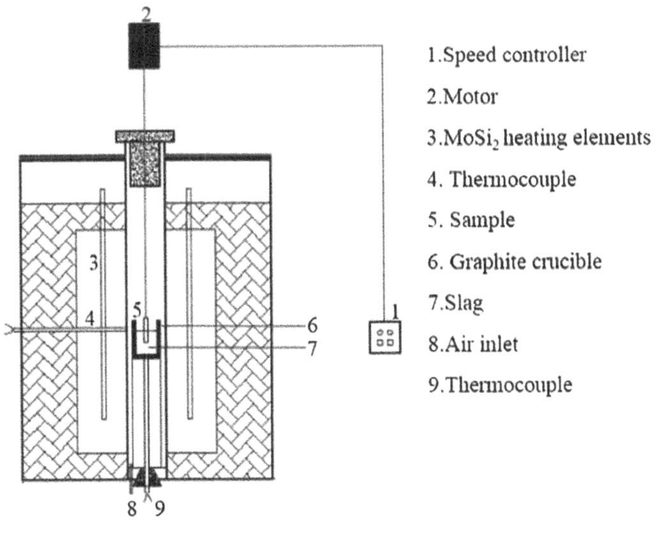

Fig. 1 Schematic drawing of experimental apparatus used to measure refractory erosion in molten slag

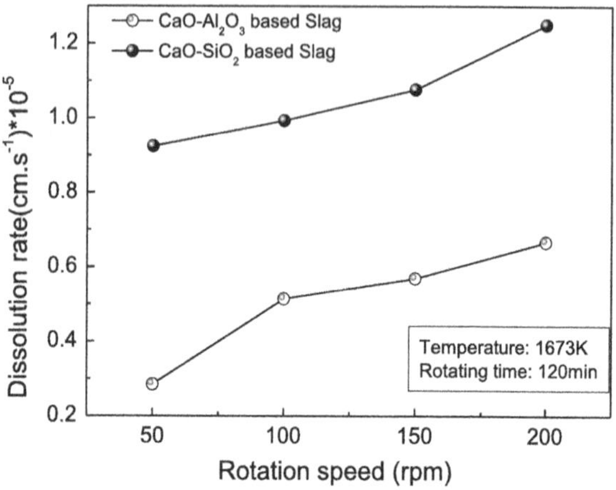

Fig. 2 Effect of slag composition on the ZrO_2-C rod dissolution rate at 1673 K

$$R_d = \frac{d_o - d_t}{2t} \tag{1}$$

where t is the erosion time(s), and d_0 and d_t are the diameter of the rods before and after the dissolution experiments (cm), respectively. The dissolution rate was evaluated according to the reduction of rod diameter after rods were immersed for 2 h.

In order to compare the effect of conventional high SiO_2 content slag with low-reactive slag, the dissolution rate of ZrO_2 rod was measured at 1673 K at different rotational speeds. The results in Fig. 2 show that the erosion rate of conventional slag is much higher than low-reactive slag. However, this study did not consider molten steel; thus, the dissolution rate is lower than previous studies that employed the rotational rod method [4]. Erosion of ZrO_2-C refractories was increased as the rotational speed of ZrO_2 rod increased. Higher rotational promotes the dissolution of oxide in molten slag and results in the aggravation of the erosion of ZrO_2-C refractories.

The effect of low-reactivity slag compositions on the dissolution rate of ZrO_2-C refractory was studied at 100 rpm and 1673 K. The data in Fig. 3 show the dissolution rate of zirconia rod increases with Na_2O, F^{-1}, Li_2O, B_2O_3 content and, listed in decreasing order of intensity, the following component intensifies the erosion rate: Na_2O, CaF_2, Li_2O and B_2O_3. The Na_2O and CaF_2 are network inhibitors that reduce slag viscosity by breaking up the silicate structure. The addition of B_2O_3 also reduces slag viscosity because the B-O bond is weaker than the Si-O bond. Reduction of slag viscosity accelerates the erosion of ZrO_2-C refractories.

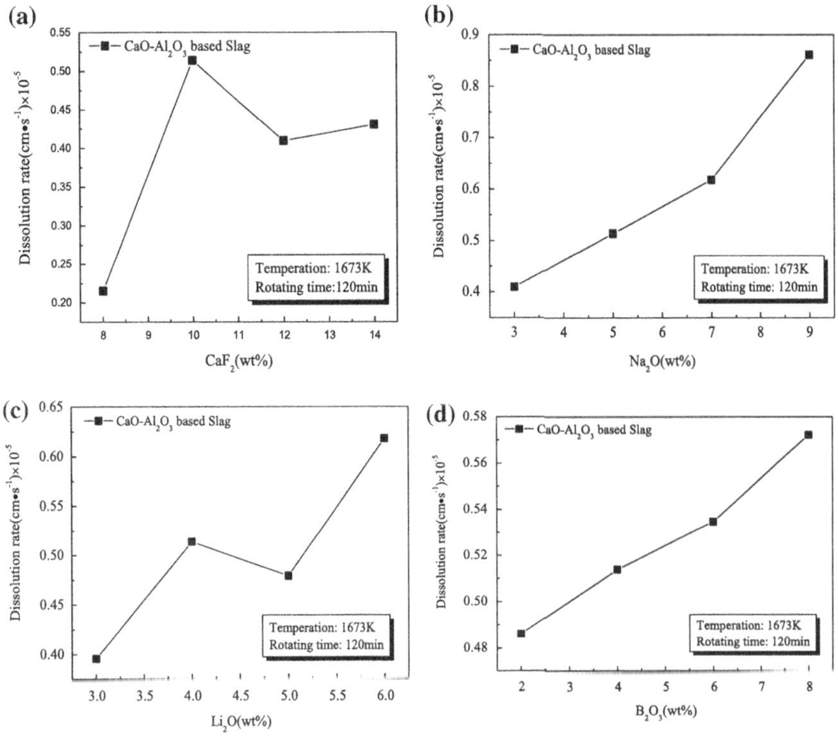

Fig. 3 Effect of individual slag components on the ZrO_2-C rod dissolution rates at 1673 K

Microscopic Examination of the Rod Surface

Some ZrO_2 rods that reacted with low-reactive slag were cut at the interface of slag–refractory–air phase. The interface was examined by scanning electron microscopy (SEM) and energy dispersive X-ray analysis (EDXA). The SEM micrograph and energy spectrum analysis are shown in Fig. 4 and Table 3.

In reference to Fig. 4, the dark material at point 6 is mainly graphite; points 1, 4 and 8 are zirconia containing partial CaO. Points 1 and 4 were located in the slag zone, which contained little CaO, so it is possible that the CaO in stabilized zirconia did not completely react or dissolve in the slag; it may be caused by the intrusion of CaO in the slag. The light black material is also carbon, which was not oxidized completely in the permeable zone.

As shown in Fig. 5, zirconia particles were eroded by slag penetration, which created numerous holes. This phenomenon is mainly due to the reaction of CaO with penetrated slag, which caused the instability of zirconia, resulting in the break-up of the zirconia grain.

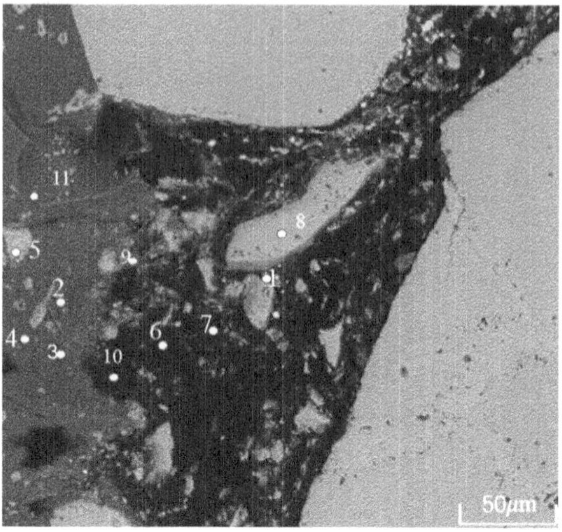

Fig. 4 Microstructure of low-reactive slag A after erosion (SEM × 1000)

Table 3 Local EDX analysis of the points labelled in Fig. 4

Element (wt%)	C	O	F	Na	Al	Si	Ca	Ba	Zr
1	–	32.72	–	–	–	–	1.90	–	65.38
2	–	37.80	13.69	2.97	8.26	8.37	15.04	8.32	5.55
3	46.31	20.28	6.47	1.82	5.30	5.26	9.03	5.53	–
4		36.61	–	–	–	–	2.28	–	61.11
5	36.26	–	–	–	1.25	–	3.01	–	59.48
6	100	–	–	–	–	–	–	–	–
7	–	36.81	13.07	3.22	9.22	11.24	17.70	8.74	–
8	–	24.90	–	–	–	–	2.12	–	72.98
9	–	31.14	–	–	–	–	–	–	68.86
10	7.38	36.53	13.60	2.81	5.95	8.18	12.60	7.41	5.55
11	32.45	35.30	5.41	2.01	1.77	1.88	5.30	–	15.89

Conclusion

The effect of slag composition for high manganese–high aluminium steel on erosion of ZrO_2–graphite refractory was investigated by rotating cylinder method. The erosion rate of ZrO_2-C refractory by low-reactivity slag was found to be lower than conventional high SiO_2 slag, and the dissolution rate increases with increased rotational speed of zirconia rod. Also, erosion of ZrO_2-C refractory was accelerated by increasing the Na_2O, CaF_2 and B_2O_3 content of the slag, and the degree of influence is $Na_2O > CaF_2 > Li_2O > B_2O_3$. These components reduce slag viscosity

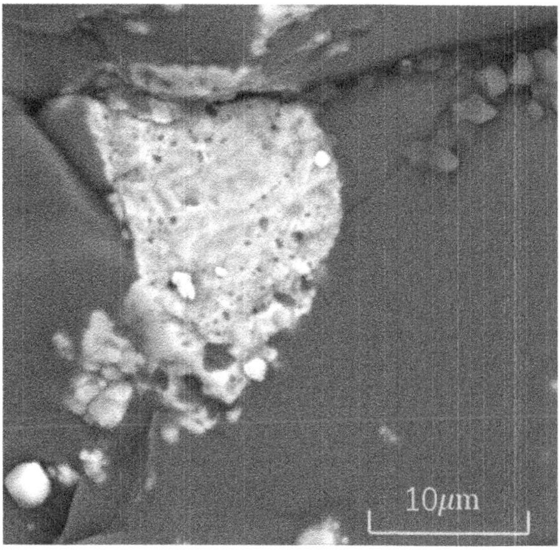

Fig. 5 Microstructure of low-reactive slag A after erosion (SEM × 4000)

and change the surface tension of the slag. The erosion of ZrO_2-C refractory is explained by dissolution of the graphite phase and zirconia instability.

Acknowledgements This work was supported by the Key Program of National Science Foundation of China (No. U1660204).

References

1. Dick AF et al (1997) Attack of submerged entry nozzles by mould flux and dissolution of refractory oxides in the flux. ISIJ Int 37(2):102–108
2. Imaoka T, Iwamoto Y (2012) Effect of mold powder properties on corrosion rate of SEN ZrO_2-C material. Shinagawa Tech Rep 55:21–30
3. Nakamura Y, Ando T, Kurata K (1986) Effect of chemical composition of mold powder on the erosion of submerged nozzles for continuous casting of steel. Trans. ISIJ 26:1052–1058
4. Bui A-H, Park S-C, Chung I-S (2006) Dissolution behavior of zirconia-refractories during continuous casting of steel. Met Mater Int 12(5):435–440
5. Cirilli F, Donato AD, Martini U (2008) Corrosion mechanisms of zirconia/carbon based refractory materials by slag in presence of steel La. Metallurgia Italiana 10:43–50
6. Suk MO, Park JH (2009) Corrosion behaviors of zirconia refractory by $CaO–SiO_2–MgO–CaF_2$ Slag. Am Ceramic Soc 92(3):717–723

A New Method for Determining High-Temperature Wettability of Bonding Phase

Yijia Dong, Guanghui Li, Chen Liu, Qiang Zhong, Hu Sun, Jun Luo and Tao Jiang

Abstract High-temperature wettability of bonding phase can be elucidated by its contact angle which is an important property reflecting the bonding behaviour between bonding phases and unmelted components. A new method for determining the high-temperature wettability of bonding phase was developed and applied to analyse the liquid phase generation capability, liquid quantity and bonding strength. Results showed that two cryogenic liquid generated areas of andradite and calcium ferrite formation appeared in Fe_2O_3–CaO–SiO_2 ternary system phase by thermodynamic analysis and the initial temperature of liquid phase formation was divided into three parts. The contact angle will decrease gradually with the increase of furnace temperature until a constant value was reached which measured by a professional software (PolyPro). A positive correlation has been found between contact angle and liquid phase generation ability. This method is meaningful for studying the high-temperature wettability of bonding phase.

Keywords High-temperature wettability · Iron ore · Phase diagram

Introduction

Sintering mineralization is a process to make part of the raw material melt after serial complex chemical reactions at high temperature. Based on the sintering process, whether binding phase coheres unfused ferrous minerals well,it will play a vital role to the formation and strength of sinter [1]. In the sintering process of iron ore, the low melting point compounds were generated by the solid phase reaction of various mineral in sinter mixture, then, the liquid formed by the action in high temperature to bond the surrounding unmelted containing iron content in molten cooling process and got the sinter finally [2]. According to the micro-structure of

Y. Dong · G. Li (✉) · C. Liu · Q. Zhong · H. Sun · J. Luo · T. Jiang
School of Minerals Processing & Bioengineering, Central South University, Changsha 410083, Hunan, China
e-mail: liguangh@csu.edu.cn

© The Minerals, Metals & Materials Society 2019
T. Jiang et al. (eds.), *10th International Symposium on High-Temperature Metallurgical Processing*, The Minerals, Metals & Materials Series,
https://doi.org/10.1007/978-3-030-05955-2_14

finished sinter, the mainly consist of sinter have two parts: iron-bearing minerals and bonding phase [3]. The strength of both iron-bearing minerals is very well, so the sinter strength depends on the characteristics of bonding phase intensity directly, which is the intensity of bonding phase and the bond strength between the bonding phase and the unmelted components. At present, most studies based on the intensity of bonding phase, the crushing strength and flexural strength of bonding phase sample have been measured to be an evaluation index [4, 5], but there is less research on bond strength of bonding phase.

Sintering is a high-temperature physicochemical process that depends on the formation and crystallization of a liquid phase to bond ore particles. Wettability, which can be characterized by the contact angle, is important for assimilation [6]. Nakashima et al. [7] have studied the wetting behaviour of CF slag with sintered haematite. They concluded that the wettability of a CF-based slag on a sintered haematite substrate can be improved by adding Al_2O_3 and that the spreading rate increases with increasing η/γ (viscosity/surface tension). Yu et al. [8] studied the influence of the different ratio of flux on bonding phase wetting behaviour on different substrate. They concluded that addition of certain amounts of MgO or SiO_2 or TiO_2 can improve the wettability of CF to the Al_2O_3 substrate. Adding MgO to the melt increases the spreading rate, and the addition of Al_2O_3 and SiO_2 reduces the spreading rate on the TiO_2 substrate. Zhu et al. [9] studied the bond strength of bonging phase, they determining the shear force between binding phase and unfused ferrous minerals as the bonding strength. Finally, they establishing characteristic number about bonding strength of binding phase, it can be chosen as a reference index to guide ore blending.

At present, most studies are focus on the bond strength of bonding phase or itself intensity of bonding phase. There are very few studies about the role of bonding phase plays in the process of sintering. In the paper, contact angle and bonding strength of bonding phase were studied to analyse the high-temperature wettability of bonding phase, and a method has been obtained to reveal its high-temperature wettability.

Materials and Methods

Materials

In this study, the ideal phase compositions of Fe_2O_3-CaO-SiO_2 (CFS) for liquid formation were confirmed by FactSage 7.1. The pure-grade reagents (Fe_2O_3, CaO, SiO_2) were utilized to prepare the cylindrical bulk samples φ 3 mm (CFS) in dimension and approximately 60 mg in weight for the wetting experiment. The samples were used for the wetting experiment.

Wetting Experiment

The pure-grade Fe_2O_3 was utilized to prepare the Fe_2O_3 substrate in the wetting experiment. The Fe_2O_3 reagent with 0.8 g weight was wetted by anhydrous ethanol and pressed into a cylindrical pellet (φ 10 mm) using a uniaxial hydraulic presser with 700 kN pressure. The Fe_2O_3 substrate was put into an electric resistance furnace of 1450 °C roasted for 2 h under air atmosphere, and then naturally cooled to room temperature in the furnace. After cooling, the Fe_2O_3 substrate was ground with an abrasive paper six times (100, 500, 1000, 1500, 2000, and 2500 CW successively) to obtain a smooth surface. The Fe_2O_3 substrate was cleaned with anhydrous ethanol using an ultrasonic wave cleaner for 10 min.

RDS—2010 automatic continuous casting slag melting point and melting speed meter were used in the wettability measurement. The schematic of this experiment is shown in Fig. 1. The experimental setup mainly includes a furnace, a heating control system, a cooling water circulation system, a sample introduction system, an atmosphere constant system, and an image-forming and data-processing system. In summary, the operation involves heating the furnace by resistance to the experiment temperature, pushing the sample into the high-temperature zone, and then capturing photographs with a camera. A strong light was used to irradiate the sample and reduce the strong background influence on the image. Photographs of test process are shown in Fig. 2. The samples with the substrate after the wetting experiment were embedded with resin and then vertically cut along the centre of the drop.

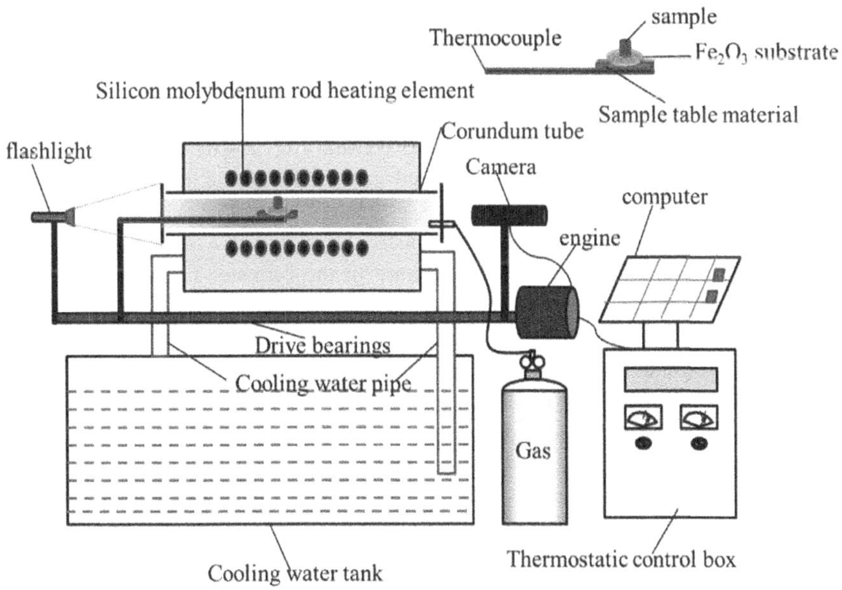

Fig. 1 Schematic of the wetting experiment

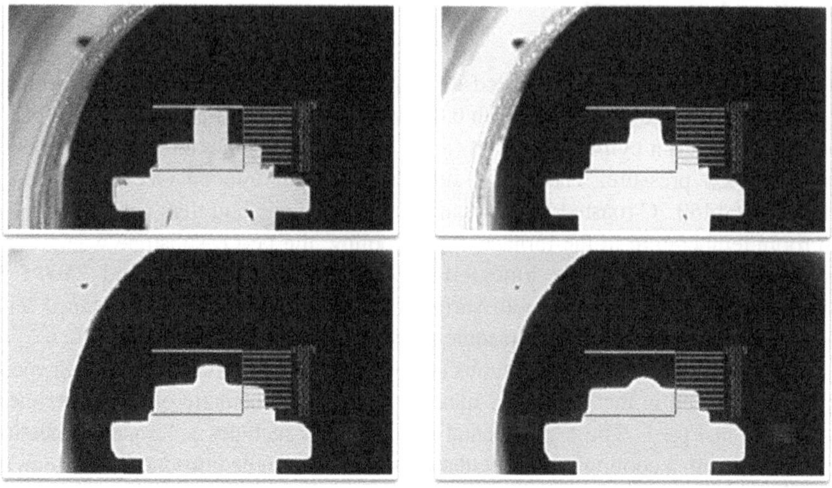

Fig. 2 The process of wetting experiment

The contact angle was measured by a professional software (PolyPro) from the photographs. The cross-section of the drop was observed by SEM-EDS, and the substrate was polished for SEM and EDS observations.

Bonding Strength Experiment

Determining the shear force between binding phase and unfused ferrous minerals is the bonding strength. The pure-grade Fe_2O_3 was used to prepare the Fe_2O_3 substrate (50 * 50 mm) for the shear force test. The Fe_2O_3 reagent with 25 g weight was wetted by deionized water and pressed into a quadrate pellet using a uniaxial hydraulic presser with 1500 kN pressure. The Fe_2O_3 substrate was loaded into an electric resistance furnace of 1350 °C roasted for 2 h under air atmosphere, and then naturally cooled to ambient temperature in the furnace. The quadrate Fe_2O_3 substrate had been cut into two kind pieces (40 * 6 mm and 20 * 6 mm). To obtain a smooth surface, the Fe_2O_3 substrate was ground with an abrasive paper three times (100, 500 and 1000 CW, successively).

The schematic diagram of shear force experiment is shown in Fig. 3. The cylindrical bulk samples with φ 3 mm (CFS) in dimension and approximately 60 mg in weight were used in the shear force experiment. Two pieces of shorter Fe_2O_3 substrates (20 * 6 mm) were put in the corundum crucible in a straight line, and they have a spacing distance about 30 mm. The sample was put on the Fe_2O_3 substrate in a range of 5 mm, and the longer Fe_2O_3 (40 * 6 mm) substrate was put on the sample horizontally. Then, the corundum crucible with the sample and substrates was put into horizontal furnace under nitrogen atmosphere. The samples

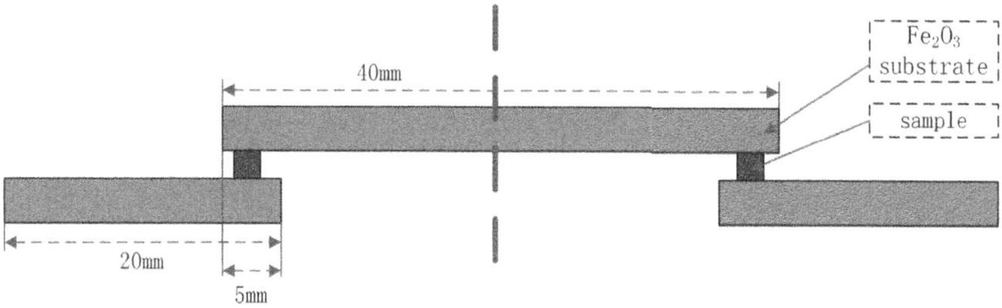

Fig. 3 Schematic diagram of shear force experiment

were roasted at constant temperature of 1400 °C for 3 min. When the sample had been sent into the highest temperature zone, it should be kept for 2 min, and then, the sample was pulled out for cooling down. After 3 min cooling, the sample was cut off at the middle and ground for the shear force test. Finally, the results of wetting experiment and bonding strength experiment are combined to research the high-temperature wettability of bonding phase.

Results and Discussion

Thermodynamic Calculation

With the change of the content of iron, calcium and silicon components as is show in Fig. 4, the main content of Fe_2O_3–CaO–SiO_2 ternary system phase includes tridymite (SiO_2), andradite ($Ca_3Fe_2Si_3O_{12}$), Ps-wollastoni ($CaSiO_3$), rankinite ($Ca_3Si_2O_7$), calcium ferrite ($CaFe_2O_4$), dicalcium ferrite ($Ca_2Fe_2O_5$)and calcium diferrite ($CaFe_4O_7$). Two cryogenic liquid generation areas of andradite and calcium ferrite formation area appear in Fe_2O_3–CaO–SiO_2 ternary system phase. The minimum generating temperature (1200.41 °C) is appeared at the cryogenic liquid generated point, which is formed by andradite, tridymite and rankinite and the components of ferric oxide, silicon oxide and calcium oxide content are 28.70, 50.55, 20.75 wt%, respectively. Because of the two cryogenic liquid generation areas have a same content of calcium oxide, about 20 wt%, so the contents of calcium oxide are determined to four scales of 15, 20, 25 and 30 wt% for researching the component variation of Fe_2O_3–CaO–SiO_2 ternary system at high temperature.

Figure 5 shows the ternary system phase diagram of Fe_2O_3-CaO-SiO_2 in the composition of calcium oxide is 15 wt%. The main mineral phases are changed from calcium ferrite to andradite with the increase of SiO_2 component (SiO_2/Fe_2O_3 + (SiO_2) (g/g)). The initial temperature of liquid phase formation is divided into three parts by two points of 0.15 and 0.19 SiO_2 components. The silica composition of the demarcation point increase gradually with the increase of the

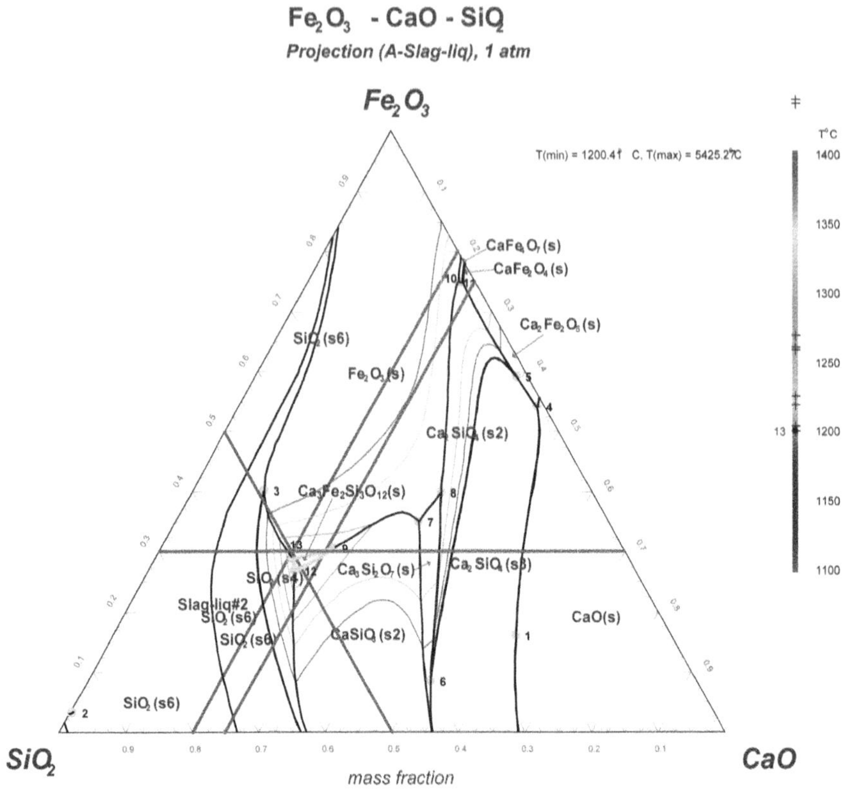

Fig. 4 Ternary system phase diagram of Fe_2O_3-CaO-SiO_2

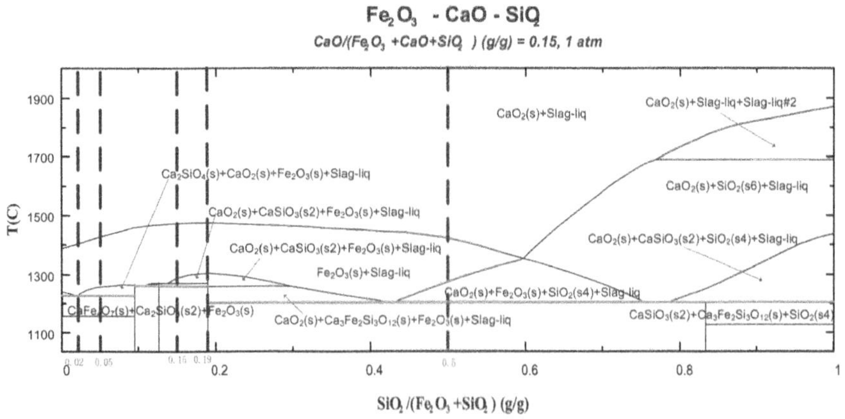

Fig. 5 Ternary system phase diagram of Fe_2O_3-CaO-SiO_2 under the condition of 15% CaO composition

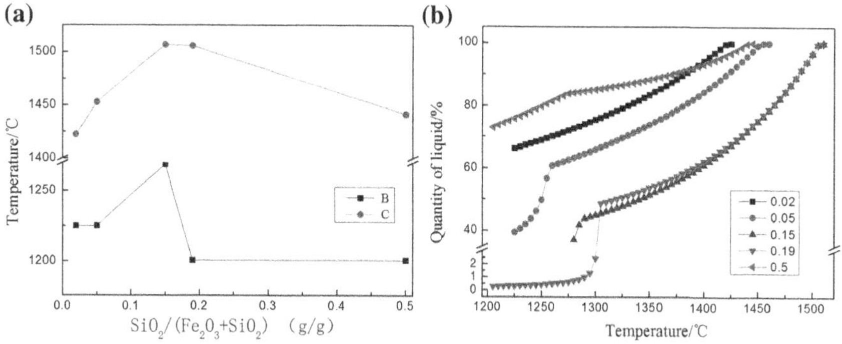

Fig. 6 Liquid phase generation capability and liquid quantity of Fe_2O_3–CaO–SiO_2 ternary system under the condition of 15% CaO composition

composition of calcium oxide. Five different silica components located in the three different initial temperature of liquid phase formation have been selected to study the relationship between high-temperature wettability of bonding phase and sinter quality at a determined calcium oxide composition.

Figure 6 shows the liquid phase generation capability and liquid quantity of Fe_2O_3–CaO–SiO_2 ternary system under the condition of 15% CaO composition. The thermodynamic calculation results show that the Fe_2O_3–CaO–SiO_2 ternary system have three different initial temperature of liquid phase formation with the increase of SiO_2 composition, which is consistent with the phase diagram calculation. The final temperature of liquid phase formation decreased gradually at the initial stage and then increased with the increasing of the SiO_2 composition. When the initial temperature of liquid phase formation achieved, a certain amount of liquid phase is appeared and increased to 100% slowly. A significant turn appears on the curve of the liquid phase quantity and the first stage grows faster than the second, when the SiO_2 component is 0.5. A point between the second and third temperature range for liquid phase formation has been selected to calculate the properties of the liquid phase, which the SiO_2 component will increase with the increase of CaO component. The quantity of liquid phase less than 1% before the temperature achieved 1300 °C. When the temperature reaches 1300 °C, the quantity of liquid phase has an extremely rapid improvement, and then increased to 100% slowly and gradually.

Wetting Behaviour

The wetting behaviour of each melt is illustrated in Fig. 7. The wetting behaviour is started to appear with the increase of furnace temperature. The contact angle of

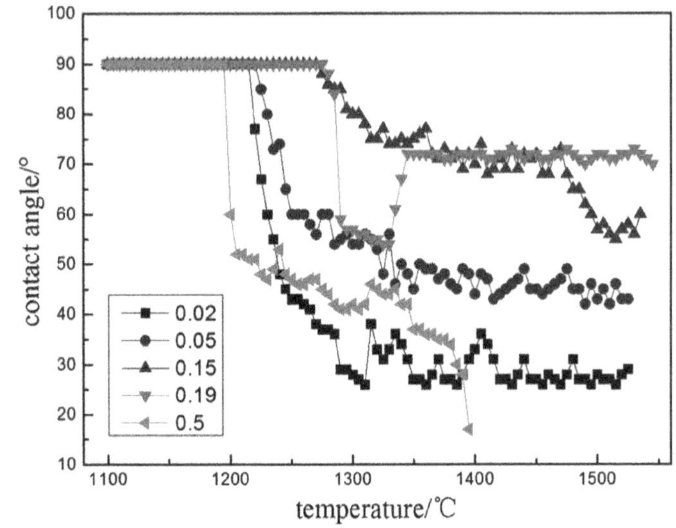

Fig. 7 Temperature change of apparent contact angle of different samples on Fe_2O_3 substrate under the condition of 15% CaO composition

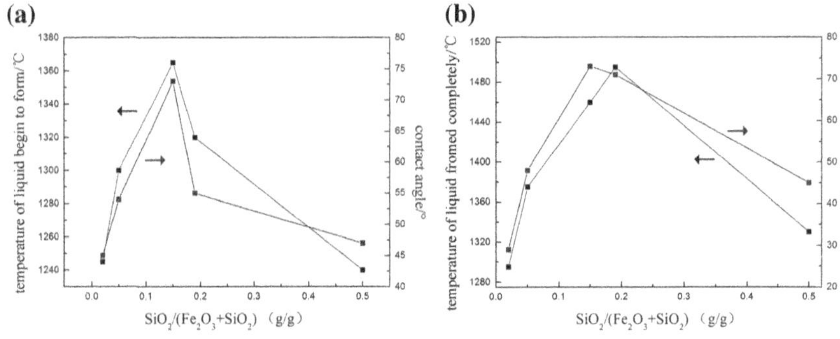

Fig. 8 Comparison of the contact angle and liquid phase generation ability

different samples on Fe_2O_3 substrate has the same change rule that the apparent contact angle gradually decreases with the increase of furnace temperature and then reaches to a stable state. When the composition of calcium oxide is 15 wt%, the wetting behaviours with different SiO_2 composition are not the same along with the temperature increase.

Figure 8 shows the comparison of the contact angle and liquid phase generation ability. The comparison of contact angle and the temperature of liquid begin to form is illustrated in Fig. 7a, and the comparison of contact angle and the temperature of liquid formed completely is illustrated in Fig. 7b. Both Figs. 7a, b have the same rule, which is the variation of contact angle is consistent with the variation trend of liquid phase generation ability. The researchers found a positive correlation between the contact angle and liquid phase generation ability.

Conclusion

A new method has been designed to research the high-temperature wettability of sinter bonding phase. The thermodynamic analysis of Fe_2O_3–CaO–SiO_2 ternary system has been calculated by FactSage 7.1. The phase diagram, Liquid phase generation capability and quantity of the liquid phase has been calculated and analysed, respectively. Two cryogenic liquid generated areas of andradite and calcium ferrite formation appeared in the phase diagram of Fe_2O_3–CaO–SiO_2 ternary system by thermodynamic analysis. With the increase of SiO_2 composition, the initial temperature of liquid phase formation of Fe_2O_3–CaO–SiO_2 ternary system was divided into three levels and five points were selected in the three levels to study the high-temperature wettability. The contact angle decreased gradually with the increase of furnace temperature until a constant value was reached. A positive correlation has been found between contact angle and liquid phase generation ability. The result of this experiment shows that this method is meaningful for studying the high-temperature wettability of bonding phase.

References

1. Li HG, Zhang JL, Pei YD (2011) Melting characteristics of iron ore fine during sintering process. J Iron Steel Res Int 18(5):11–15
2. Jiang T (ed) (2016) Principle and technology of agglomeration of iron ores. Central south University Press, Changsha
3. Li GS (2008) The mechanism and reasonable composition of bond relative sinter strength. Ph.D. thesis, Northeastern University
4. Li GS, Dou LW, Shang C (2009) Study on self-strength of calcium ferrite binding phase in Sinter. J Northeastern Univ (Nat Sci) 30(2):225–228
5. Zhang GL, Wu SL, Su B (2015) Influencing factor of sinter body strength and its effects on iron ore sintering indexes. Int J Miner Metall Mater 22(6):553–561
6. Yu B, Lv X, Xiang S, Bai C, Yin J (2015) Wetting behavior of Al_2O_3 substrate by calcium ferrite series melts. Trans Iron Steel Inst Japan 55(3):483–490
7. Nakashima K, Saito N, Shinozaki N (2007) Wetting and penetration behavior of calcium ferrite melts to sintered hematite. ISIJ Int 44:2052–2056
8. Yu B, Lv XW, Xiang SL, Xu J (2015) Dissolution kinetics of Al_2O_3, SiO_2 into calcium ferrite based slag. In: Proceedings of the 10th CSM Steel Congress & the 6th Baosteel Biennial Academic Conference, Shang Hai, 1–9 Oct 2015
9. Zhu CE, Wu K, Huang DJ (2017) Research on characteristic number about bonding strength of binding phase in sinter. Energy Metall Ind 36(3):23–26

Part IV
High-Temperature Processing

Research on Dezincification of Ironmaking and Steelmaking Ashes

Yan Zhang, Xiaojiang Wu, Meng Xu, Gele Qing, Haoyu Cai, Wenbin Huang, Yunqing Tian and Wenwang Liu

Abstract In this paper, research on dezincification in ironmaking and steelmaking processes was carried out. Ashes of ironmaking and steelmaking, amylum and bentonite were used as the experimental materials. Bentonite has no significant effect on quality of green pellet by pelletizing process; pressing pressure should be controlled at 10 MPa while using press machine to produce cold-bonded pellet; falling strength of cold-bonded pellet would have a sharp increase while adding a little sodium hydroxide in amylum; ZnO will be removed at a high temperature with carbon contained in experimental materials, but removal rate decreases with the increase of $CaCl_2$; less than 30% of ZnO would be removed by adding calcium chloride without carbon.

Keywords Dezincification · Ashes · Pelletizing · Cold-bonded pellet

Introduction

Iron-containing dust and sludge are generated during ironmaking and steelmaking processes [1 3]. However, many of the dust and sludge contain zinc which is harmful to blast furnace[4, 5]. Therefore, dezincification should be carried out before the dust and sludge can be utilized.

Herck et al. reported a hydrometallurgical process, whereby the sludge was leached under both acid (HCl) and oxidizing conditions [6]. Vereš et al. reported a method of zinc removal by using microwave [7].

In this paper, experiments of pelletizing green pellet with discs and producing cold-bonded pellet with press machine have been carried out to verify whether

Y. Zhang · M. Xu · G. Qing (✉) · H. Cai · Y. Tian
Shougang Research Institute of Technology, Beijing, China
e-mail: qinggele_68@163.com

X. Wu · W. Huang · W. Liu
Shougang Jingtang United Iron and Steel Co. Ltd., Tangshan, Hebei, China

© The Minerals, Metals & Materials Society 2019

T. Jiang et al. (eds.), *10th International Symposium on High-Temperature Metallurgical Processing*, The Minerals, Metals & Materials Series,
https://doi.org/10.1007/978-3-030-05955-2_15

ashes of ironmaking and steelmaking can be used; zinc removal experiment has been conducted in an electric tube furnace as well.

Materials and Experimental Methods

Ashes of ironmaking and steelmaking, amylum and bentonite were used in the experiments. Chemical composition and particle size of the ashes used in experiment are listed in Table 1.

ZnO content of ironmaking ash and steelmaking ash is 0.86 and 6.69%, respectively. Dezincification must be carried out before the ash has been cyclically utilized especially for steelmaking ash. Carbon content of ironmaking ash is 17.27%. Ironmaking and steelmaking ashes all have coarse particle sizes; percentages of −200 meshes are 56.3 and 48.1, respectively.

Pelletizing Experiments

Green pellets were prepared for 10 min in pelletizing disc with diameter of 800 mm; pellets with size between 10 and 12.5 mm were screened out.

The pelletizing schemes are shown in Table 2. Five pelletizing schemes were conducted.

Table 1 Chemical composition and particle size of materials (mass%)

Name	TFe	FeO	SiO_2	Al_2O_3	CaO	MgO	ZnO	C	<200 meshes
Ash of ironmaking	43.87	24.38	7.11	4.06	4.64	0.71	0.86	17.27	56.3
Ash of steelmaking	56.46	32.92	1.50	0.47	5.12	1.25	6.69	0.75	48.1

Table 2 Pelletizing schemes

No.	Ash of ironmaking (%)	Ash of steelmaking (%)	Bentonite (%)	Falling strength (0.5 m/No.)	Compression strength (N/P)	Water content (%)
1	10	90	2.0	4.2	8.98	18.5
2	10	90	2.5	4.0	10.18	19.0
3	20	80	3.0	2.6	7.44	19.3
4	10	90	8.0	4.6	11.64	22.1
5	10	90	10.0	4.8	12.15	23.4

The results show that: with the increase in bentonite content from 2 to 10%, falling strength of green pellet remains stable and compression strength improves slightly. But the quality of green pellet decreased sharply with the increase in ironmaking ash from 10 to 20%.

Cold-Bonded Pellet Experiments

Press machine was used to produce cold-bonded pellet with ironmaking and steelmaking ashes.

The cold-bonded pellet experiments with different pressing pressures are shown in Table 3. The results show that: falling strength is increasing from 3 to 5 with the increase in pressing pressure from 5 to 10 MPa, while falling strength decreases from 5 to 1 when the pressing pressure further increased up to 40 MPa.

It is suggested that pressing pressure should be controlled at 10 MPa while using press machine to produce cold-bonded pellet.

The cold-bonded pellet experiment with amylum is shown in Table 4. Falling strength remains 1 with the increase in amylum from 2 to 3% despite using

Table 3 Cold-bonded pellet experiment schemes with different pressing pressure

No.	Ash of ironmaking (%)	Ash of steelmaking (%)	Bentonite (%)	Pressing pressure (MPa)	Falling strength (1 m/No.)
1	10	90	10.0	5	3
2	10	90	10.0	10	5
3	10	90	10.0	20	4
4	10	90	10.0	30	1
5	10	90	10.0	40	1

Table 4 Cold-bonded pellet experiment with amylum

No.	Ash of ironmaking (%)	Ash of steelmaking (%)	Amylum (%)	Sodium hydroxide (%)	Pressing pressure (MPa)	Falling strength (1 m/No.)
1		100	2		10	1
2		100	3		10	1
3		100	2	0.5	10	3
4		100	3	0.5	10	5
5	100		2		10	1
6	100		3		10	1
7	100		2	0.5	10	3
8	100		3	0.5	10	3

Table 5 Cold-bonded pellet experiment with cement

No.	Ash of ironmaking (%)	Ash of steelmaking (%)	Cement (%)	Curing time (h)	Pressing pressure (MPa)	Falling strength (1 m/No.)
1		100	5	0	10	6
2		100	5	24	10	7
3	100		5	0	10	2
4	100		5	24	10	6

ironmaking ash or steelmaking ash. However, falling strength of cold-bonded pellet would have a sharp increase while adding a little sodium hydroxide in amylum. Because amylum will react with sodium hydroxide and hydrophilicity of amylum will have a significant increase.

The cold-bonded pellet experiment with cement is shown in Table 5. Falling strength will reach 6 when only steelmaking ash is used to produce cold-bonded pellet. Falling strength is 2 when ironmaking ash is used, but it will have a sharp rise after a curing of 24 h.

Dezincification Experiment

Due to the high melting point and boiling point of ZnO, 1973 and 2360 °C, respectively, ZnO can be hardly removed at oxidizing atmosphere. However, the melting point and boiling point of Zn are 419 and 911 °C. Dezincification can be realized easily if ZnO is reduced to Zn.

Melting point and boiling point of $ZnCl_2$ are only 283 and 732 °C. If the reaction below occurred, ZnO can be removed as well.

$$ZnO + CaCl_2 \rightarrow ZnCl_2 + CaO$$

Two kinds of ironmaking ashes and one kind of steeling making ash were used in the experiment. Chemical compositions are listed in Table 6.

Dezincification experiment schemes are listed in Table 7. Firing experiments were conducted in an electric tube furnace, and the firing temperature was 1100 °C under air condition. ZnO removal rates of different experiments were calculated.

Table 6 Chemical compositions of ashes

Content (%)	K_2O	Na_2O	ZnO	C
Ironmaking ash 1	0.61	0.40	0.79	17.34
Ironmaking ash 2	3.58	1.47	3.55	20.22
Steelmaking ash	1.54	0.76	6.69	0.75

Table 7 Dezincification experiment schemes by mixing ironmaking and steelmaking ashes

No.	Ironmaking ash 1 (%)	Ironmaking ash 2 (%)	Steelmaking ash (%)	$CaCl_2$ (%)	Bentonite (%)	Carbon content (%)	ZnO removal rate (%)
1	50	50			3	18.23	93.0
2	50	50		1	3	18.06	97.8
3	50	50		3	3	17.72	92.8
4	50	50		5	3	17.39	73.9
5			100	3	3	0.71	25.2
6			100	5	3	0.69	27.4
7			100	10	3	0.66	27.0
8			100	12	3	0.65	27.6

Removal rate of ZnO is defined as follows: where M_1 is the initial ZnO content of mixing materials and M_2 is the ZnO content of mixing material after heat treatment.

$$\text{ZnO removal rate }(\%) = \frac{M_1 - M_2}{M_1}$$

As shown in Fig. 1, ZnO could react with carbon at high temperature, and removal rate of ZnO would be more than 90% with about 18% carbon contained in the mixing materials. However, removal rate of ZnO in ironmaking ashes slightly decreases with the increase of $CaCl_2$.

Figure 2 shows that removal rate of ZnO had a slight increase with the increase of $CaCl_2$ with no more than one per cent carbon contained in the mixing matcrials, but the removal rate was no more than 30%.

Fig. 1 Removal rate of ZnO with carbon-contained ironmaking ashes

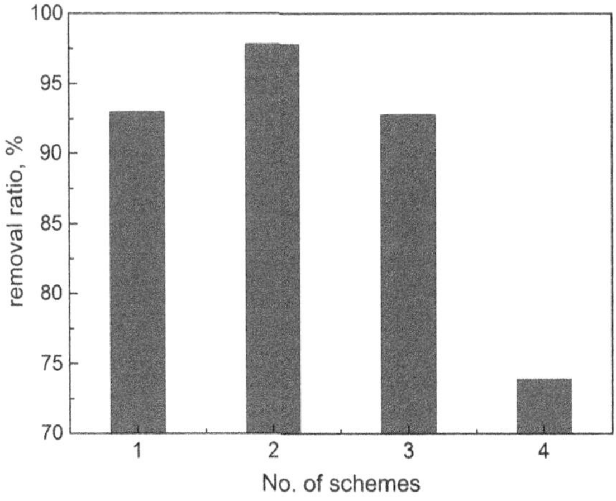

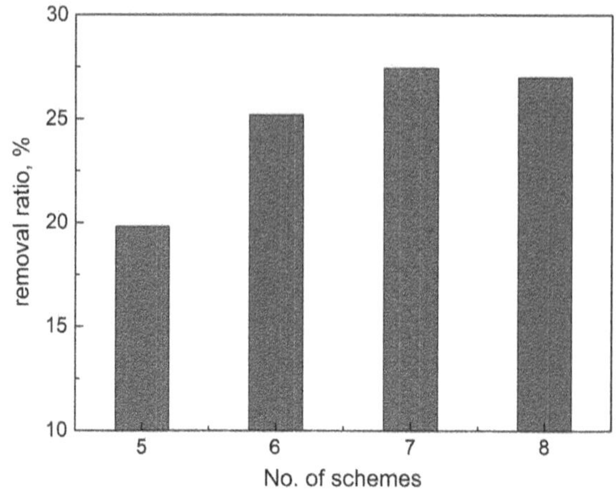

Fig. 2 Removal rate of ZnO with $CaCl_2$ added in materials

Conclusions

Research on dezincification in ironmaking and steelmaking processes was carried out. Ashes of ironmaking and steelmaking, amylum and bentonite were used as the experimental materials. The following conclusions can be drawn from this work:

1. Bentonite has no significant effect on quality of green pellet by pelletizing process;
2. Pressing pressure should be controlled at 10 MPa while using press machine to produce cold-bonded pellet;
3. Falling strength of cold-bonded pellet would have a sharp increase while adding a little sodium hydroxide in amylum;
4. Falling strength will have a sharp rise after a curing of 24 h when ironmaking ash is used;
5. ZnO will be removed at a high temperature with carbon contained in experimental materials, but removal rate decreases with the increase of $CaCl_2$;
6. Less than 30% of ZnO will be removed by adding calcium chloride without carbon.

References

1. Kelebek S, Yörük S, Davis B (2004) Characterization of basic oxygen furnace dust and zinc removal by acid leaching. Miner Eng 17(2):285–291
2. Su F, Lampinen H-O, Robinson R (2004) Recycling of sludge and dust to the BOF converter by cold bonded pelletizing. ISIJ Int 44(4):770–776
3. Mikhail SA, Turcotte A-M (1998) Thermal reduction of steel-making secondary materials: I. Basic-oxygen-furnace dust. Thermochimica Acta 311(1):113–119

4. Li Z (2002) Damage of element of zinc to blast furnace and its inhibition. Bao Steel Technol 6 (18–20):24
5. Esezobor DE, Balogun SA (2013) Zinc accumulation during recycling of iron oxide wastes in the blast furnace. Ironmaking Steelmaking 33(5):419–425
6. Van Herck P, Vandecasteele C, Swennen R, Mortier R (2000) Zinc and lead removal from blast furnace sludge with a hydrometallurgical process. Environ Sci Technol 34(17): 3802–3808
7. Vereš J, Lovás M, Jakabský Š, Šepelák V, Hredzák S (2012) Characterization of blast furnace sludge and removal of zinc by microwave assisted extraction. Hydrometallurgy 129–130:67–73

Effects of Temperature and Alkali Carbonates on Graphitization and Metallurgical Properties of Coke

Rongjin Zhu, Shengfu Zhang, Guangsheng Suo, Yue Wu, Xiaohu Zhou and Shuxing Qiu

Abstract In this paper, metallurgical cokes containing different proportion of alkali carbonate were subjected to treatment with annealing under N_2 temperature from 1100 to 1600 °C to investigate graphitization and metallurgical properties of coke. The annealed coke powder was characterized using weight method, I-drum tumbling, X-ray diffraction and optical microscope. The results clearly showed that the weight loss increased with the increasing of temperature. The Na_2CO_3 (or K_2CO_3) also effectively accelerated the weight loss. Combined with the crystallite structure, it could be shown that the L_a and L_c of coke increased with the increasing of temperature, while the d_{002} revealed the opposite trend. It could also be shown that alkali carbonates destroyed crystallite structure. From optical micrographs, it could be seen that the porosity of coke increased with the increasing of temperature. The large pores decreased while the micropores increased gradually which revealed that alkali carbonates destroyed pore structure. The destruction of potassium carbonate was more obvious.

Keywords Coke · Alkali carbonate · Crystalline structure · Porosity

Introduction

Coke is an indispensable material in iron-making, playing a crucial role in the process [1–4] and acting as:

(1) A fuel, to provide 70–80% heat for the blast furnace operation;
(2) A reductant (the fixed carbon in coke and carbon monoxide are all reductant for iron ores);

R. Zhu · S. Zhang (✉) · G. Suo · Y. Wu · X. Zhou · S. Qiu
College of Materials Science and Engineering, Chongqing University, Chongqing 400044, China
e-mail: zhangsf@cqu.edu.cn

© The Minerals, Metals & Materials Society 2019

T. Jiang et al. (eds.), *10th International Symposium on High-Temperature Metallurgical Processing*, The Minerals, Metals & Materials Series,
https://doi.org/10.1007/978-3-030-05955-2_16

(3) Carburizing agent for molten iron;
(4) Supporting spine of burden to maintain the permeability of the blast furnace.

In recent years, with the development of large-scale blast furnace, increase of pulverized coal injection rate, the role of coke as supporting the furnace stack is becoming increasingly prominent and the framework function of coke can't be undertook [5]. However, the factors of coke degradation in the blast furnace come from various aspects. The coke reactivity in the blast furnace is increased due to the addition of catalysis, both the strength and size of coke are decreased because of the mechanical damage, thermal stress, background reaction and slag corrosion [2, 6]. Thus, mechanical and thermal strength of metallurgical coke is much higher than of the coke load in blast furnace. In recent years, the degradation behaviour of coke in blast furnace was deeply studied by domestic and foreign scholars [2, 7, 8]. The results showed that mechanical damage, thermal stress and background reaction were the dominant reasons leading to crushing and powdering of coke. It was also found that the circulation and accumulation of harmful elements generated by high furnace pressures, increased the solution loss reaction and accelerated the degradation of coke in lower zone of the blast furnace. Wang et al. [5] have studied the influence of zinc on the solution loss reaction of coke using impregnation method of zinc acetate solution. Potassium and sodium are present in different zones of the blast furnace [9]. Both theoretical calculation and investigation of coke samples from dissected blast furnace have demonstrated the presence of potassium and sodium mainly as carbonates in the coke at 900–1200 °C [10]. Kaczorowski and Lindstad [11] have investigated the effect of potassium as a catalyst for the solution loss reaction using a gaseous impregnation technique. Mei et al. [12] studied the mineral transformation behaviours during sodium carbonate-catalysed CO_2 gasification of high-alumina coal, whereas Xing et al. [13] have evaluated the effect of annealing temperature with gasification on coke weight loss and graphitization.

In this paper, the effects of catalysis of Na_2CO_3 (K_2CO_3) and annealing temperature on the coke powdering, coke pore structure, carbon structure were systemically studied under simulated temperature in the blast furnace by using X-ray diffraction (XRD) and optical microscope.

Experimental and Methods

Materials

Four different types of coal (Coking coal (ZJ), 1/3 coking coal (1/3ZJ), fat coal (FM), lean coal (SM)), from Anshan Steel, were chosen as coal samples. The proximate and ultimate analyses of these coals are presented in Table 1, respectively. These raw coals were crushed and sieved to particles from 0.5 to 3.0 mm, dried in an oven at 105($\pm$5) °C until the mass was constant.

Table 1 Proximate and ultimate analyses of ZJ, 1/3ZJ, FM and SM coal

Coal	Proximate analysis/%				Ultimate analysis/%				
	M_{ad}	A_d	V_d	FC_d	C_d	H_d	N_d	O_d	S_d
1/3ZJ	2.34	7.12	31.95	60.94	78.30	5.25	0.94	8.24	0.15
FM	1.74	9.61	28.67	61.72	77.27	4.61	1.36	5.68	1.47
ZJ	2.30	7.52	20.57	71.90	81.35	4.58	1.45	4.73	0.37
SM	1.70	13.92	15.78	70.30	76.44	3.89	1.25	4.50	1.06

ad: air-dry basis; d: dry basis

Carbonization Process

In the experiment, a total of 2 kg dry blending coal with 20wt% of fat coal, 30wt% of coking coal, 40wt% of 1/3 coking coal, and 10wt% of lean coal was mixed uniformly. The blending coal was placed in the reactor (vertical tube type coke oven) with inner diameter of 100 mm and height of 500 mm, whose stack density was 0.9 t/m^3. The coke oven was firstly heated from room temperature to 800 °C (10 °C/min), then to 1100 °C (5 °C/min) and finally kept at the constant temperature for 6 h.

Preparation of Coke Samples

Solution impregnation method was used to prepare alkali carbonate-enriched coke in this study. The experiment scheme of Na_2CO_3 (K_2CO_3) loading is shown in Table 2. The procedure was as follows:

(1) A 20 g coke sample with particle size from 10 to 15 mm;
(2) The sample was placed into an aqueous solution of 10mass% Na_2CO_3 (K_2CO_3) and the boiled at 70 °C at different time for loading various concentration of Na_2CO_3 (K_2CO_3);
(3) Setting mixture aside at room temperature for 24 h;
(4) Addition of alkali carbonate-enriched coke into oven at 105(±5) °C for 48 h.

Table 2 Experiment scheme of Na_2CO_3 (K_2CO_3) loading

Label	Concentration of Na_2CO_3	Load of Na_2CO_3 (%)	Concentration of K_2CO_3	Load of K_2CO_3 (%)
1	10% + boiling 50 min	3.25	–	–
2	–	–	10% + boiling 60 min	4.59

Annealing

A 20(±0.5) g of the above prepared coke samples were charged into corundum crucible placed in high-temperature silicon-molybdenum furnace in N_2 atmosphere (99.99%) at the rate of 5 L/min all the time. The coke samples in the furnace were firstly heated with the heating rate of 10 °C/min to 1100 °C. Then the heating rate was changed to 5 °C/min. The samples were kept for additional 2 h at target temperatures of 1100°, 1200°, 1300°, 1400°, 1500°, 1600 °C, and then cooled in air to the room temperature. The original coke sample was labelled as raw coke; while samples without loading Na_2CO_3, after annealing, were labelled sequentially as C-1100 to C-1600. Samples loaded with Na_2CO_3 (or K_2CO_3), after annealing, were named as CN-1100 to CN-1600 (or CK-1100 to CK-1600).

The coke samples heated at different temperatures were taken out for I-type drum and rotated (20 rotations/min for 100 rotations). Then, the annealed samples were sieved by different screens (15, 10, 5, 1, 0.5, 0.074 mm) to obtain the particle size distribution.

The weighing method was used to measure the weight loss of samples before and after annealing, and then evaluated by Eq. (1):

$$\text{Weight loss} = \frac{m_b}{m_b - m_a} \tag{1}$$

where m_b and m_a are the mass of samples before and after annealing, respectively.

X-Ray Diffraction

Carbon structure has an essential influence on the property of coke. X-ray diffraction technique is one of the most common methods to evaluate carbon structure parameters. XRD specimens were prepared by using fine powder about 200 mesh in size. The Escalab 250Xi powder diffractometer was used to acquire XRD spectrum over the angular range from 10 to 90° using Cu Kα radiation, at the step size of 0.02. The crystallites sizes of coke were evaluated by analysing the width of crystallite layer in the horizontal direction(L_a), the average stack height (L_c) and the interlayer spacing (d_{002}) by using Eqs. (2–4):

$$L_a = \frac{1.84\lambda}{\beta_{100}\cos(\theta_{100})} \tag{2}$$

$$L_c = \frac{0.89\lambda}{\beta_{002}\cos(\theta_{002})} \tag{3}$$

$$d_{002} = \frac{\lambda}{2\sin(\theta_{002})} \tag{4}$$

where λ is the wavelength of X-ray, θ_{100} and θ_{002} are the corresponding scattering angle (100 and 002), respectively. The β_{100} and β_{002} are the full with at half maximum of the diffraction peak (100 and 002).

Analysis of Pore Structure

A representative analysis of each sample was based on 10 images taken from 10 different lumps of sample. Microscopical analysis was carried out by a Zeiss microscope using Nikon camera (magnification ×5). The pores and walls of cokes were distinguished by black and white areas by using software Imagine Pro Plus 6.0. The porosity was also calculated by using the same software [14].

Results and Discussion

Weight Loss of Cokes During Annealing

A series of chemical and physical reactions occurred under high-temperature heat treatment resulting in weight loss of cokes (see Fig. 1). Results show clearly that with the increasing of annealing temperature, weight loss of cokes increased. K_2CO_3-absorbed coke has approximately 2–5 times than higher weight loss at lower temperature, than the Na_2CO_3-absorbed coke and raw coke, respectively. Nevertheless, as the temperature increases to 1600 °C, the weight loss of K_2CO_3-loaded coke is nearly 1.5–2 times lower in comparison with the Na_2CO_3-absorbed coke and raw coke. The highest weight loss after annealing showed K_2CO_3 sample, and the lowest the raw coke.

Coke Particle Size Distribution After Annealing

Due to different temperature in different parts of the blast furnace, the particle size distribution of cokes clearly changed as a function of carbonization. After annealing in the I-type drum, three coke samples were screened to obtain particle size distribution, as shown in Fig. 2. The results showed that the particle size of three cokes exhibited the same trend, e.g. the percentage of coke with size within 0.074–0.5, 0.5–1, 1–5 mm increased with the increase of annealing temperature. On the contrary, the content of coke within 10–15 mm decreased with the increasing of

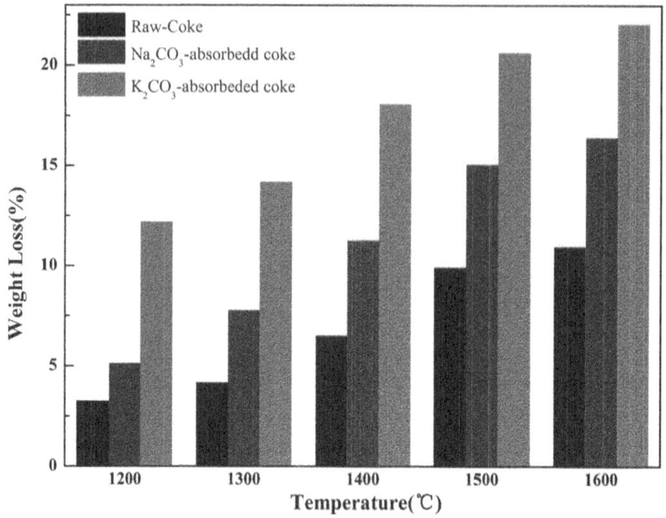

Fig. 1 Weight loss of cokes after annealed

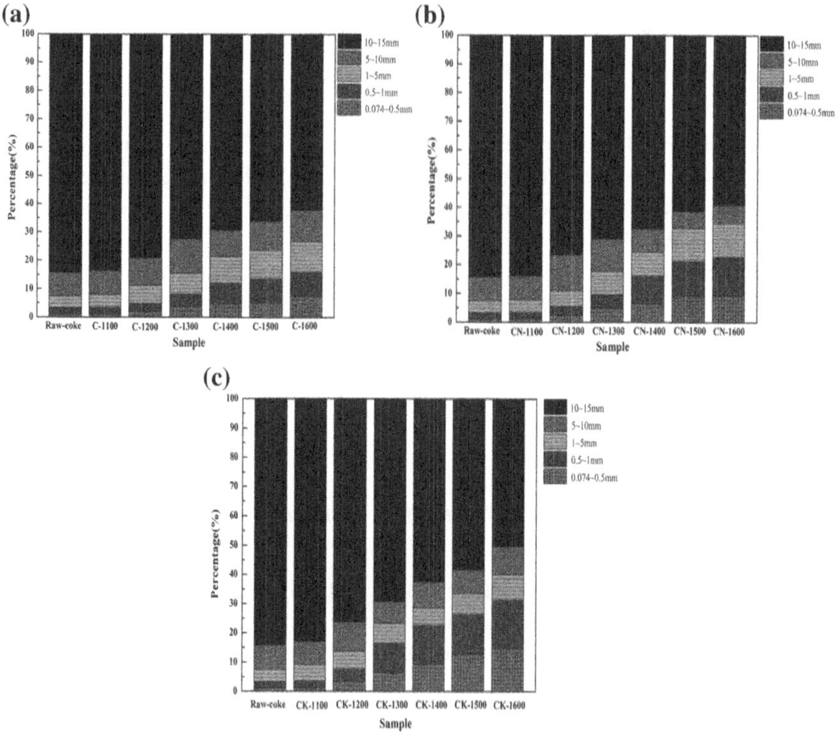

Fig. 2 Particle size distribution of raw coke, Na_2CO_3-absorbed coke and K_2CO_3-absorbed coke after different annealing temperatures

annealing temperature. No obvious changes were observed for the samples with particle size between 5 and 10 mm. In comparison with the raw coke, it was also found that with the addition of K_2CO_3 (Na_2CO_3), the percentage of coke powder (particle size of 0.074–0.5, and 0.5–1 mm) clearly increased. The reason for this phenomenon is probably due to reaction of sodium carbonate, K_2CO_3 and Na_2CO_3 with coke ash to albite and nepheline. Particularly, the nepheline formation leads generally to volume expansion. At the same time, K_2CO_3 may react with coke matrix to form C_6K (interlayer compound) which destroys coke matrix and accelerates coke pulverization.

Evolution of Carbon Crystallite Structure

The change of carbon crystallite parameters (L_c, L_a, d_{002}) under different annealing temperatures are shown in Fig. 3. In the temperature range of 1100–1600 °C, the crystallite size of all samples increased with increasing annealing temperatures. The results clearly indicate the further increase in the ordering of the carbon structure. In contrast, it could be found that the d_{002} values of samples obtained with increasing

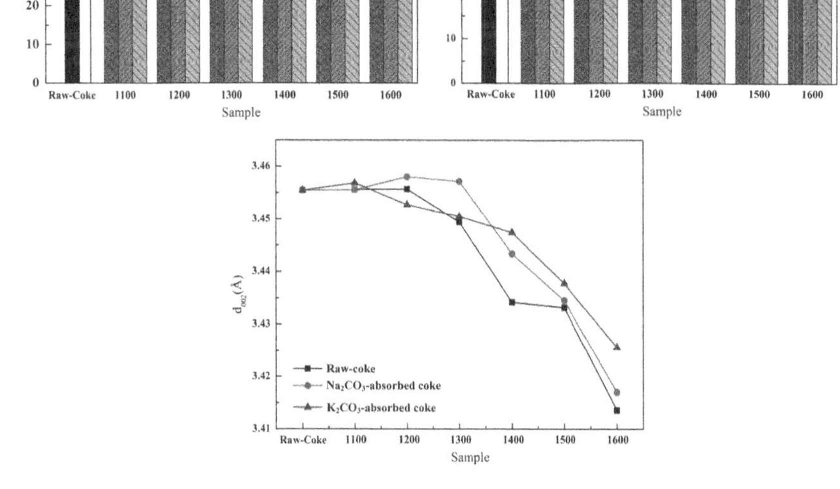

Fig. 3 Crystallite size of raw coke, Na_2CO_3-absorbed coke and K_2CO_3-absorbed coke after different annealing temperatures

temperatures decreased gradually varying from 3.456 to 3.413 Å for raw coke, from 3.456 to 3.417 Å for Na_2CO_3-absorbed coke and from 3.457 to 3.426 Å for K_2CO_3-absorbed coke, respectively. The results indicate that the interlayer spacing is increased gradually with loading Na_2CO_3 and K_2CO_3 on cokes which illustrates the carbon crystallite is destroyed by Na_2CO_3 and K_2CO_3.

Morphological Analysis

The porosity of all samples is presented in Fig. 4. The porosity of the raw coke is 41.2%, the increase of sample porosity was more noticeable after annealing, especially with the increased annealing temperature. When the temperature increased to 1600 °C, the porosity of sample reached 56.5%. Thus, the heat treatment of sample at 1600 °C caused 15% porosity increase. The highest porosity was observed for K_2CO_3-absorbed sample and the smallest for raw coke.

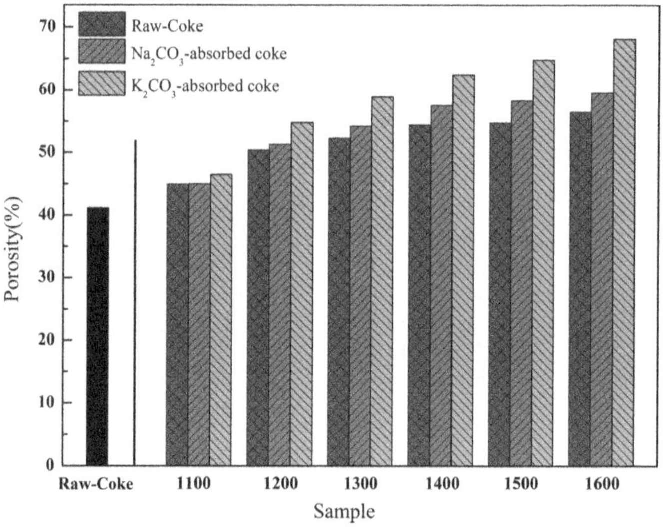

Fig. 4 Porosity of raw coke, Na_2CO_3-absorbed coke and K_2CO_3-absorbed coke after different annealing temperatures

Conclusions

The effect of annealing and alkali carbonate on graphitization and metallurgical properties of coke were studied by weight loss method, X-ray diffraction and image analysis. The major conclusions are summarized below:

1. Annealing temperatures and alkali carbonate increase the powdering of cokes.
2. The crystallite size of all samples increased with increasing annealing temperatures.
3. The carbon crystallite is destroyed by Na_2CO_3 and K_2CO_3.
4. Heating treatment increases the porosity of all cokes.

Acknowledgements This work was supported by the National Natural Science Foundation of China (Grant No. 51474042 & 51774061) and the Fundamental Research Funds for Central Universities (Grant No. 106112017CDJQJ138801). The authors also acknowledge the support provided by the Fund of Chongqing Science and Technology (Project No. cstc2018jscx-msyb0988).

References

1. Li K, Khanna R, Zhang J (2014) The evolution of structural order, microstructure and mineral matter of metallurgical coke in a blast furnace: a review. Fuel 133(5):194–215
2. Hilding T, Gupta S, Sahajwalla V (2005) Degradation behaviour of a high CSR coke in an experimental blast furnace: effect of carbon structure and alkali reactions. ISIJ Int 45(7):1041–1050
3. Watakabe S, Takeda K, Nishimura H (2006) Development of high ratio coke mixed charging technique to the blast furnace. ISIJ Int 46(4):513–522
4. Peacey JG (1979) The iron blast furnace. theory and practice. Pergamon Press, New York
5. Wang W, Wang J, Xu R (2017) Influence mechanism of zinc on the solution loss reaction of coke used in blast furnace. Fuel Process Technol 159:118–127
6. Nomura S, Terashima H, Sato E, Naito M (2007) Some fundamental aspects of highly reactive iron coke production. ISIJ Int 47(6):823–830
7. Hilding T, Kazuberns K, Gupta S (2005) Effect of temperature on coke properties and CO_2 reactivity under laboratory conditions and in an experimental blast furnace. In: Iron & steel technology conference, Vietnam, 9–12 May 2005
8. Xing X, Rogers H, Zhang G (2016) Coke degradation under simulated blast furnace conditions. ISIJ Int 56(5):786–793
9. Kazuberns K, Gupta S, Grigore M (2008) Coke mineral transformations in the experimental blast furnace. Energy Fuels 22(5):514–522
10. Li K, Zhang J, Barati M (2015) Influence of alkaline (Na, K) vapors on carbon and mineral behavior in blast furnace cokes. Fuel 145(37):202–213
11. Kaczorowski J, Lindstad T (2007) The effect of potassium impregnation on the boundouard reactivity of selected single-source and commercial cokes. Ifapa 12(5):584–593

12. Mei Y, Wang Z, Fang H (2017) Na-containing mineral transformation behaviors during Na_2CO_3-catalyzed CO_2 gasification of high-alumina coal. Energy Fuels 31(2):1235–1242
13. Xing X, Rogers H, Zhang G (2016) Coke degradation under simulated blast furnace conditions. ISIJ Int 56(5):786–793
14. Xing X, Ostrovski O (2013) Effect of annealing on properties of carbonaceous materials. Part II: porosity and pore geometry. Metall Mater Trans B Process Metall Mater Process Sci 44 (4):862–869

Field-Assisted Sintering of Nickel-Based Superalloy Powder for High-Temperature Hybrid Turbine Disc Applications

Charis I. Lin, Sebastian Niuman, Namiko Yamamoto, Anil Kulkarni and Jogender Singh

Abstract Turbine discs are currently made of nickel-based superalloys, known for their high strength and creep resistance at high temperatures. Hybrid or dual-microstructure turbine discs allow for significant weight savings, but current methods of joining dissimilar nickel-based superalloys such as friction welding exhibit a heat-affected zone and localized melting at the interface, leading to weak bonding. Here, we show that field-assisted sintering technology may be used to sinter the nickel-based superalloy powder CM247LC to high relative density, and the effect of sintering temperature and time on porosity, grain size, and mechanical properties of CM247LC is investigated. We also show that the same technology may be used to form hybrid discs with a solid Inconel 718 core and a powder-sintered CM247LC rim without the formation of a heat-affected zone at the interface. Two different joining angles between the two materials in the hybrid discs are explored, and preliminary results suggest that the joining angle does not affect the tensile properties of the material interface.

Keywords FAST · Hybridization · Ni-based superalloys · Sintering · Turbine disc

C. I. Lin (✉) · S. Niuman · J. Singh
Penn State Applied Research Laboratory, State College, PA 16804, USA
e-mail: cwl5451@psu.edu

N. Yamamoto
Department of Aerospace Engineering, The Pennsylvania State University, University Park, PA 16802, USA

A. Kulkarni
Department of Mechanical Engineering, The Pennsylvania State University, University Park, PA 16802, USA

© The Minerals, Metals & Materials Society 2019

T. Jiang et al. (eds.), *10th International Symposium on High-Temperature Metallurgical Processing*, The Minerals, Metals & Materials Series,
https://doi.org/10.1007/978-3-030-05955-2_17

Introduction

Effective hybridization of the turbine disc improves life cycle costs by allowing for enhanced thrust-to-weight ratios and thermal stabilities in the high-temperature and high-stress environment. Currently, a heavy mechanical joint connects single crystal turbine blades to a polycrystalline disc. Hybridization, or joining components directly without a mechanical joint, results in a significant weight reduction, and the component materials can be tailored to meet the different mechanical and thermal requirements for the different portions of the assembly, as seen in Fig. 1. One approach to hybridization is joining dissimilar nickel (Ni)-based superalloys in the turbine disc using solid-state joining, but current joining methods result in weak bonding with localized melting at the interface. In the current study, field-assisted sintering technology (FAST) is used to manufacture hybrid discs via solid-state joining because of FAST's versatility, short sintering times, and joining capabilities.

To achieve high thermal and mechanical properties, polycrystalline turbine discs are made from Ni-based superalloys via powder metallurgy followed by forging. Ni-based superalloys display high strength and creep resistance at high temperatures, with current rim temperature capabilities above 700 °C and are one of the only materials that can withstand the high-temperature turbine environment [1]. Although ceramics exhibit high strength and thermal properties, their brittleness makes them unfit for use in turbine disc applications. The current difficulty in forming hybrid or dual-microstructure discs is joining dissimilar Ni-based superalloys without a weak interface while maximizing weight savings, so a novel joining method is needed.

Previous efforts in forming hybrid or dual-microstructure turbine discs include inhomogeneous heat treatments and friction welding. Dual microstructure was achieved by heating the rim of the disc over the γ' solvus temperature while keeping the core under the solvus temperature using a variety of heating methods [2–5]. The result was a gradient variation in grain size, with the rim of the discs having a larger grain size than the core. The limitation of forming dual-structured turbine discs using heat treatments is that the same base material is used throughout the disc, and the capabilities of the base material, therefore, limit the changes in microstructure.

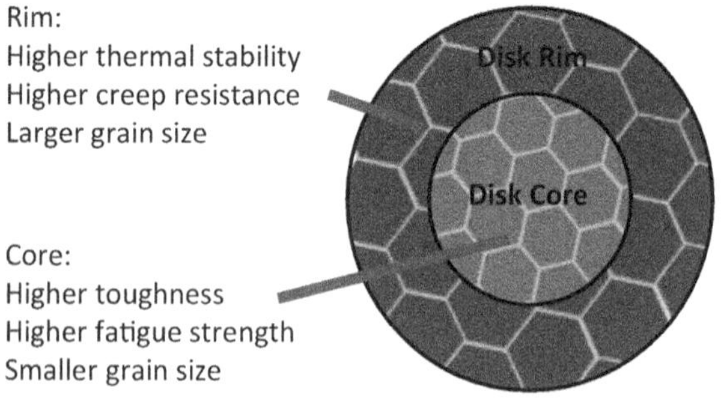

Fig. 1 Targeted properties of a hybrid turbine disc. Relative grain size (not to scale) is illustrated

Thus, joining dissimilar alloys to form one disc would lead to greater potential for optimization of the disc materials. Currently, the most successful joining process for disc materials is friction welding, a solid-state joining method where localized friction leads to a plasticized interface and localized melting. After the relative movement stops, a forging force consolidates the weld [6]. One series of studies joined polycrystalline MAR-M247 and single crystal low solvus high refractory (LSHR) Ni-based superalloys using inertia friction welding, but the weld led to inconsistencies at the weld interface, including a 4–5 mm heat-affected zone, localized melting, and residual stresses [7, 8]. An improved method to join dissimilar Ni-based superalloys without such deformations is desired.

Our proposed method of manufacturing hybrid turbine discs is to perform solid-state joining of dissimilar Ni-based superalloys using field-assisted sintering, also known as spark plasma sintering (SPS). FAST combines direct-current-driven temperature with pressure to sinter materials, as seen in Fig. 2. A wide variety of materials, including metals, refractory metals, intermetallics, and ultra-high-temperature ceramics, can be sintered using FAST [9]. Meanwhile, few studies exist regarding the sintering of Ni-based superalloys via FAST. FAST has also been used to join dissimilar materials through functional grading or solid-state joining, resulting in interfaces free of localized melting [9–11]. Because of its versatility and joining capabilities, FAST may be a candidate for manufacturing hybrid discs.

Before taking advantage of this technology for hybrid disc applications, more research is needed to understand the processing–structure–property relationships during the joining of dissimilar Ni-based superalloys using FAST. In this work, we examine the effect of FAST parameters on the microstructure and mechanical properties of sintered CM247LC powder and then study the resulting characteristics of hybrid discs joined using FAST. Microscopy, microindentation, and tensile testing are used as the main analysis tools.

Materials and Experimental Procedures

Both single-material discs and hybrid discs were formed in this study. CM247LC powder was chosen as the material for the single-material discs and the hybrid disc rim. CM247LC is a Ni-based superalloy specifically designed for directionally solidified turbine blade applications. The material was derived from the superalloy MAR-M247 and exhibits an increased ductility and carbide stability [12]. The CM247LC powder used was spherical with a diameter of $\sim$105–149 μm. The density of the sintered samples was measured using Archimedes' method. The measured density of sintered samples exhibiting zero porosity was 8.62 g/cc, slightly greater than the density of 8.52 g/cc listed in previous literature [12]. The difference in density possibly results from impurities in the powder or a slight deviation of the powder material from the nominal composition. The hybrid disc core material was Inconel 718, a Ni-based superalloy used for high-temperature applications.

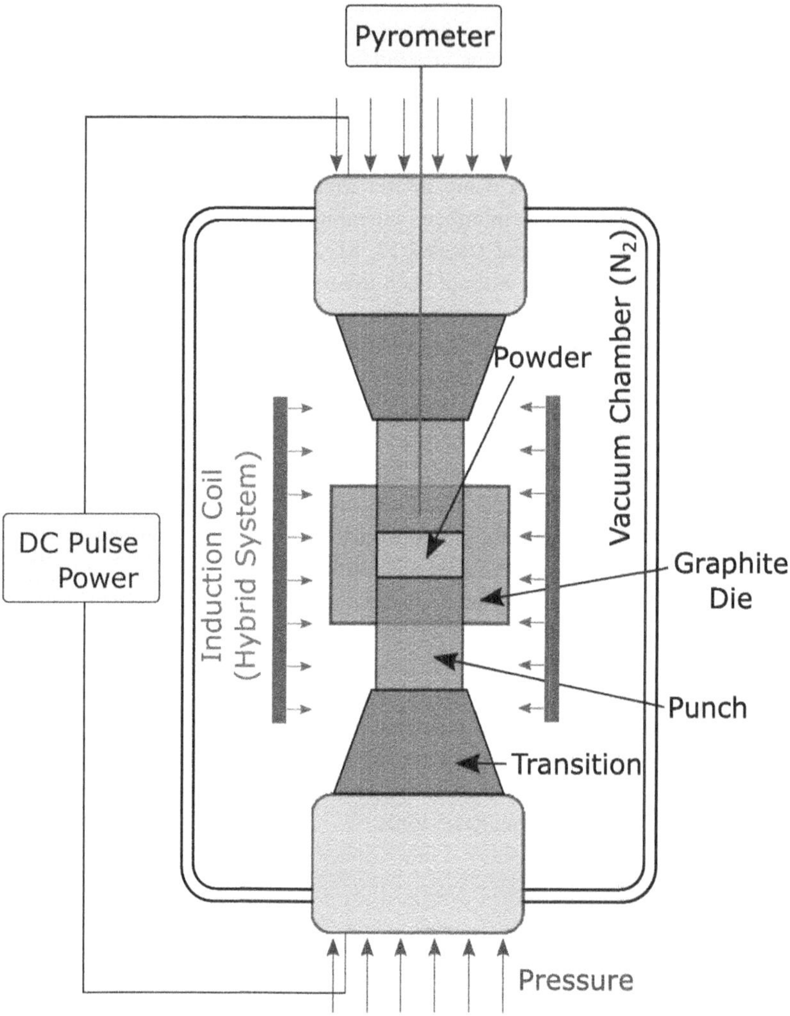

Fig. 2 FAST sintering mechanism

In order to understand the material structures and behaviours of FAST-sintered CM247LC samples, single-material CM247LC discs were sintered before fabricating hybrid discs. Sintered single-material discs underwent a predetermined sintering process, with sintering temperature, time, and pressure as variables, as seen in Fig. 3. First, the rams steadily applied 5 kN of force to initially compact the powder in the graphite die. Second, the material temperature increased to the specified temperature at a rate of 50 °C per minute. The pressure simultaneously increased, initially at a lower rate ($\sim$7 kN/min) and then at a higher rate ($\sim$28 kN/min) so the

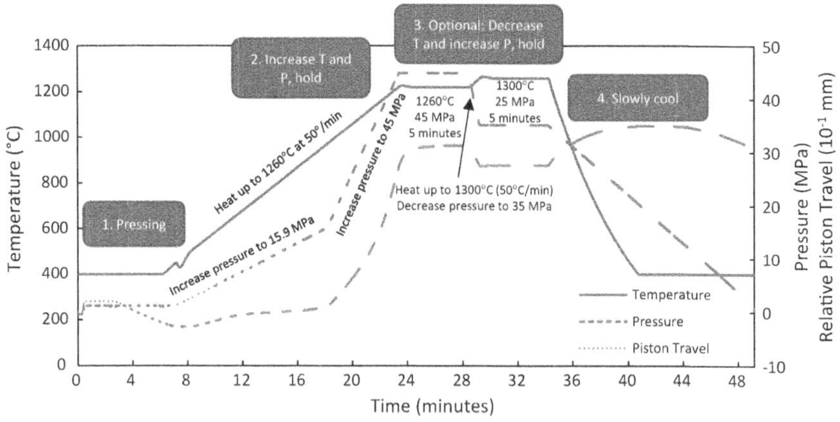

Fig. 3 Sintering parameters for an 80 mm diameter CM247LC disc. The disc was sintered first at 45 MPa and 1260 °C for 5 min, then at 25 MPa and 1300 °C for 5 min before cooling

desired sintering pressure was reached at the same time as the desired temperature. The material was held at this pressure and temperature for a given time, usually 15 min. Third, an optional hold step with an increased temperature and decreased pressure immediately followed; increasing temperature for the second hold step allowed the material to reach a higher temperature and exhibit further grain growth while decreasing pressure minimized flashing. Finally, after the hold step(s), water-cooled rams quickly cooled the die assembly while pressure was gradually released.

The sintered samples' size and sintering parameters were constrained by the FAST unit capabilities, die material strength, powder temperature capabilities, and dimensions needed for tensile testing. The thickness of the 40 and 80 mm diameter CM247LC discs varied between 3 and 4.5 mm, the hold step pressure varied between 35 and 45 MPa, the hold step temperature varied between 1150 and 1260 °C, and the hold step time varied between 15 and 25 min. The discs' microstructures and mechanical properties were measured, as summarized in the following section.

Based on our understanding of FAST-sintered CM247LC discs, we formed hybrid discs of 70 mm diameter by sintering CM247LC powder for the rim material and simultaneously bonding the powder to a solid Inconel 718 disc. Producing hybrid discs using FAST required a more unconventional die assembly. The die assembly in Fig. 4 allowed for the inner Inconel 718 disc to be held in place while the powder was compacted and sintered around the disc. Using this die assembly, several discs with a 45° angle between the two materials were manufactured, as seen in Fig. 4, as well as discs with a joining surface parallel to the thickness of the disc (0°). Powder height was adjusted so the thickness of the rim powder matched the powder height above the solid inner disc, allowing the pressure on the disc to be maintained while the powder compacted. To simplify the manufacturing process, the optional second hold step was not used. The list of hybrid discs manufactured is shown in Table 1.

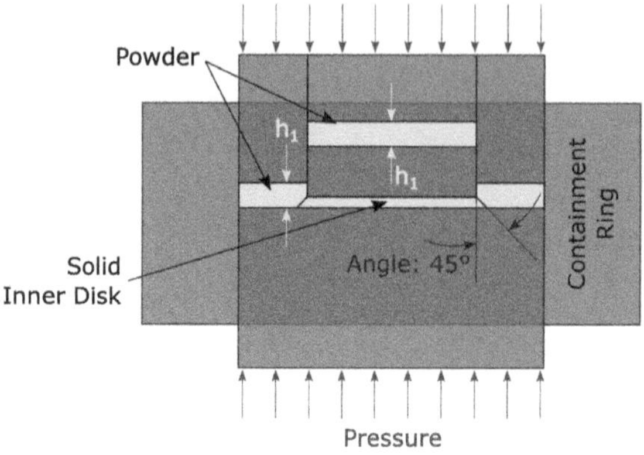

Fig. 4 Assembly of graphite or carbon fibre reinforced carbon die, machined inner disc, and powder rim for manufacturing of hybrid disc

Table 1 Processing parameters (joining angle, thickness *th.*, pressure *P*, temperature *T*, and time *t*) for hybrid discs and resulting densities ρ

Sample #	Joining Angle (°)	th. (mm)	P (MPa)	T (°C)	t (min.)	ρ (g/cc)	ρ (%)
1	45	3	45	1225	15	8.4039	97.5
2	45	3	45	1200	30	8.4720	98.3
3	45	3	45	1200	25	8.4518	98.0
4	45	2.5	45	1200	25	8.4494	98.0
5	45	2.5	45	1225	15	8.5547	99.2
6	45	2.5	45	1225	15	8.5256	98.9
7	0	3.2	45	1200	25	8.3465	96.8
8	0	3.2	45	1225	20	8.5156	98.8
9	0	3.2	45	1225	20	8.5050	98.7
10	0	3.2	45	1225	20	8.5492	99.2
11	45	3.2	45	1225	15	8.4565	98.1
12	45	2.5	45	1225	15	8.5082	98.7

Results and Discussion

Effect of Sintering Parameters on Microstructure

Sintered CM247LC discs of 80 mm diameter showed decreasing porosity with increasing sintering temperature, with greater porosity near the outer edges of the discs. The gradient of porosity throughout the sample may result from a temperature gradient through the disc; the outer edges of the disc see more heat loss than the centre of the disc, allowing greater diffusion between powder particles near the centre of the disc. The porosity gradient may also be caused by a difference in

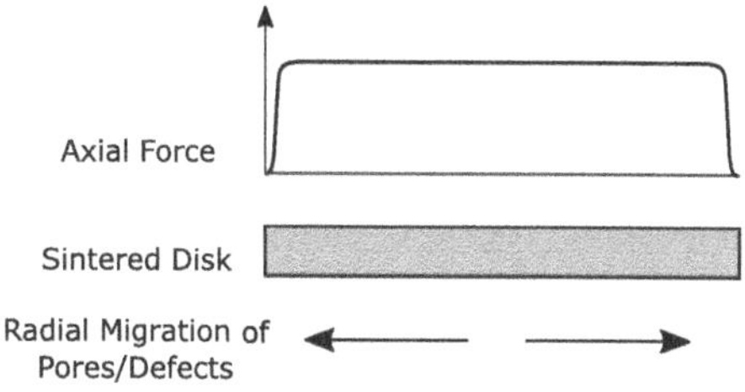

Fig. 5 Possible variation of axial force with radius; hypothesized migration of pores and defects with increasing sintering temperature is shown

pressure throughout the sample. Even if the powder initially had a uniform thickness before sintering, there may be a slightly lower pressure where the powder contacts the ductile graphite foil lining the inside edge of the die, causing the porosity to exit the powder from the centre of the sample outwards, as hypothesized in Fig. 5. An increase in temperature from 1225 to 1260 °C reduced porosity in the disc centre region, but a small amount of porosity remained at the disc edges. Sintering temperatures higher than 1260 °C may further decrease porosity. However, with an increase in sintering temperature, flashing also increased at the edges of the sintered discs; an increase in material leakage between mould surfaces resulted from the increased ductility of the samples at high temperatures.

The introduction of a second sintering step of increased temperature and decreased pressure was observed to effectively decrease porosity, even at the regions close to the edges of the discs, while minimizing flashing. Although porosity was significantly reduced with the two-part hold step process, especially when the temperature of the second hold step was increased further, slight porosity still remained at the edge of the samples, as seen in Fig. 6. The reduction in porosity with increasing temperature demonstrates that volumetric sintering dominates at the centre of the disc, and porosity and defects migrate to the edge of the disc during sintering. Further optimization of sintering time, pressure, and temperature is needed to achieve minimum porosity and defects at the edge of the discs.

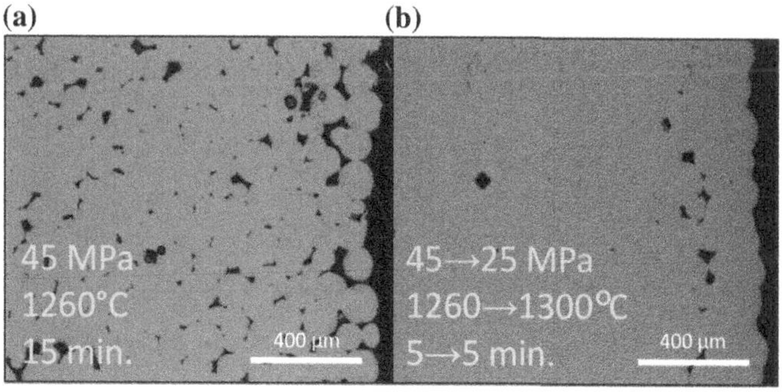

Fig. 6 Optical micrographs of the edge of sintered CM247LC samples. **a** Single hold step, **b** two-part hold step

Alternatively, if porosity only remains at the edges of the sintered samples, as seen with the samples sintered with a two-part sintering step, machining the edges off the sintered samples could eliminate porosity.

Next, the effect of sintering parameters on grain size and shape was evaluated. The centre region of sintered CM247LC discs was inspected using optical microscopy. As shown in Fig. 7, increased sintering temperature led to an increase in grain size, reduction in porosity and reduction in the number of retained spherical structures that remained from the spherical superalloy particles. The grain size of the powder sintered at 1225 °C was approximately 16 μm, which is quite small. Because the CM247LC would be used as the rim material, a larger grain size is desired to improve creep resistance at high temperatures. An increase in sintering temperature to 1260 °C increased the grain size to 31 μm, and the number of spherical structures also decreased, with the sample sintered at 1260 °C generally showing equiaxed grains. Introduction of the second sintering step with increased temperature and decreased pressure was observed to further increase the grain size to 40 μm, as seen in Fig. 7d.

Fig. 7 Optical micrographs of sintered CM247LC samples of 80 mm diameter. Waterless Kalling's Etch. The samples were sintered at **a** 1225 °C, **b** 1250 °C, **c** 1260 °C, **d** 1260 °C followed by 1300 °C

Effect of Sintering Parameters on Mechanical Properties of FAST-Sintered Single-Material CM247LC Discs

Sintering temperature did not significantly affect the elastic moduli of FAST-sintered CM247LC discs. Elastic modulus was estimated by measuring ultrasonic sound velocity through the sintered samples, with resulting values seen in Table 2. The elastic moduli for all the samples are comparable, with most around 230–240 GPa. The sample sintered at 35 MPa exhibited a lower elastic modulus than the samples sintered at higher pressures. However, porosity seen in the samples sintered at 35 MPa may have skewed the sound velocity measurement.

Meanwhile, the FAST-sintered single-material CM247LC discs exhibited slightly higher hardness values when sintered at higher temperatures (>1260 °C), despite the larger grain size. The hardness of the sintered samples was measured using a 1 kg load and a standard Vicker's indenter, with values shown in Table 2. The higher hardness may be due to precipitation of secondary phases with higher sintering temperatures.

Mechanical Properties of Hybrid Discs

After exploring the effect of sintering parameters on the properties of single-material CM247LC discs, the mechanical properties of hybrid discs formed using FAST were also explored. The rim–core interface was inspected at the cross section, as shown in Fig. 8. Some oxide formation was observed that may be associated with an oxide layer at the surface of the powder particles. Some porosity was still seen within the sintered CM247LC material, so processing parameters need to be further optimized to reduce porosity.

First, the mechanical properties of the rim and core materials were measured separately. The elastic modulus of both the sintered CM247LC rim and the Inconel 718 core was comparable to the parent materials, as seen in Table 3. The hardness across the interface showed a sharp transition between the two materials. When

Table 2 Hardness and elastic modulus of sintered CM247LC discs of 80 mm diameter

Pressure (MPa)	Temperature (°C)	Modulus (GPa)	Hardness (HV)
35	1225	215	401 ± 23
45	1240	238	411 ± 5
45	1250	236	394 ± 15
45	1260	235	405 ± 7
45–25	1260–1275	238	418 ± 5
45–25	1260–1300	240	417 ± 2
45–35	1260–1300	241	415 ± 4

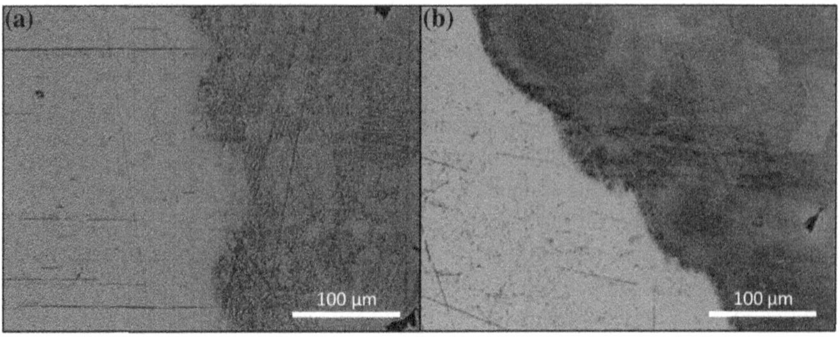

Fig. 8 Optical micrographs of joined subscale hybrid discs. **a** Disc with 0° interface, joined at 45 MPa and 1225 °C for 20 min, **b** disc with 45° interface, joined at 45 MPa and 1225 °C for 15 min

Table 3 Elastic modulus of hybrid discs sintered at 45 MPa and 1225°C

Joining angle	Sintering time (min.)	Modulus of CM247LC rim (GPa)	Modulus of IN718 core (GPa)
0°	15	204	202
45°	15	226	225

comparing discs with a 0° joining angle to discs with a 45° joining angle, elastic modulus as measured via the acoustic sound velocity method did not differ significantly between the two joining angles. However, the 45° hybrid disc measured showed a significantly higher hardness (428 and 232 VHN) than the 0° hybrid disc (400 and 199 VHN) in both rim and core materials, respectively. This may be because of the thickness of the samples; due to the die mould set-up, the 45° hybrid discs were less thick than the 0° discs, which may result in a higher temperature gradient during the sintering and joining process. The difference may also be attributed to the sintering time: the 0° hybrid disc used to determine hardness experienced a hold time five minutes longer than the 45° hybrid disc's 15-minute hold time because nominal density was not reached after 15 min.

Second, subscale room temperature tensile testing was performed to determine the tensile properties of parent materials and hybrid discs, with results seen in Table 4. Tensile behaviours of reference samples should first be noted. It was observed that the sintered CM247LC disc had a significantly higher tensile strength than the as-received solid commercial plate. However, at room temperature, the CM247LC disc showed a more brittle failure than the Inconel 718, with significantly less necking.

The tensile properties of hybrid discs were then compared to the parent materials. The hybrid discs with a CM247LC rim and an Inconel 718 core showed a

Table 4 Tensile properties of hybrid discs and parent materials

Material	Samples	YS (MPa)	UTS (MPa)	Elongation (%)
CM247LC	8	926 ± 5	1347 ± 35	14.8 ± 2.0
IN718 (as-received)	6	467 ± 2	868 ± 4	49.7 ± 1.9
Hybrid Disc (45° join)	5	348 ± 6	615 ± 41	15.7 ± 4.2
Hybrid Disc (0° join)	5	343 ± 6	557 ± 67	13.4 ± 4.8

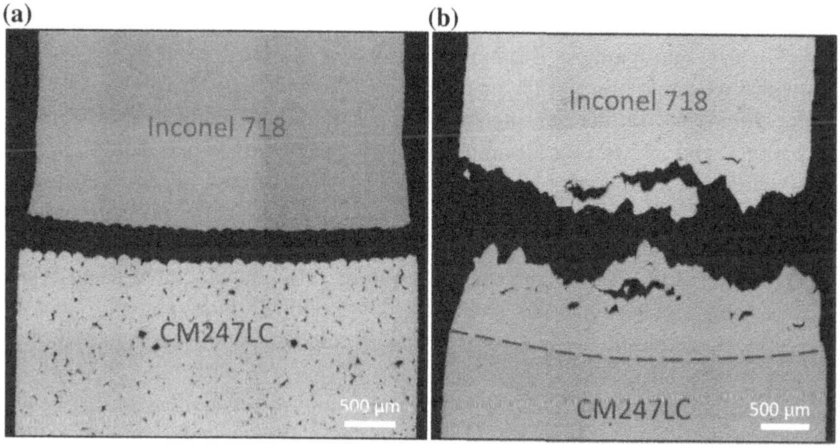

Fig. 9 Optical micrographs of broken tensile samples cut from hybrid discs. Discs were sintered at 45 MPa and 1225 °C for 15 min. **a** 0° joining angle, **b** 45° joining angle

combination of the parent materials' tensile properties, with tensile strengths slightly lower than the as-received Inconel 718 plate and a room temperature elongation similar to the more brittle CM247LC material. Both hybrid discs showed similar tensile properties, suggesting that the joining angle of the hybrid discs does not significantly affect yield strength, ultimate tensile strength, and elongation. The decrease in the hybrid discs' tensile strengths compared to the as-received solid Inconel 718 material likely results from decomposition of the strengthening γ'' phase to the stable γ phase at temperatures higher than 650 °C [13].

Although the hybrid discs had similar numerical tensile properties to the parent materials, the location of the fracture was slightly different for the two joining angles, as seen in Fig. 9. With the 45° joining angle, the fracture occurred mostly in the Inconel 718 material. In contrast, although necking occurred mostly in the Inconel 718 material, the 0° hybrid disc broke cleanly at the interface, with the fracture surface clearly showing the spherical CM247LC grains. However, for both joining angles, fracture occurred at or near the interface of the two materials; therefore, interface effects caused tensile failure.

Conclusions

This work studied the effect of FAST processing parameters on the Ni-based superalloy CM247LC, and discs of high relative density were formed. Hybrid discs with a sintered CM247LC rim and a solid Inconel 718 core were then manufactured using FAST, and the resulting discs exhibited little porosity between the two materials. When studying microstructure and mechanical properties of the hybrid discs and parent materials, it was found that:

- The sintering temperature of the CM247LC discs affected material porosity, grain size, and hardness. Nominal density was achieved in 80 mm discs via a two-step heating process: the material was first sintered at 1260 °C and 45 MPa for 5 min, and further densification and grain growth followed with heating at 1300 °C and 25 MPa for 5 min. An increase in sintering temperature from 1225 to 1300 °C increased grain size as well as hardness, likely because of microstructural changes. Elastic modulus was not affected by sintering temperature.
- Hybrid discs of 70 mm diameter were successfully manufactured using FAST. The interface between the two materials exhibited little porosity, and the hardness and elastic modulus of the rim and core were comparable to that of the parent materials.
- Joining angle between the two materials did not affect the hybrid disc tensile properties, although fracture location differed for the two joining angles. Fracture with the 0° interface occurred at the material interface while fracture with the 45° interface occurred near the interface but mostly in the Inconel 718 material.

In conclusion, this work showed that it is possible to use FAST to join dissimilar superalloys via solid-state joining. No previous method successfully joined dissimilar superalloys without localized melting and high residual stresses at the interface. More research is needed to determine the processing parameters of hybrid discs for optimal bond strength, and high-temperature mechanical properties of hybrid discs manufactured using FAST need to be understood.

References

1. Gabb TP, MacKay RA, Draper SL, Sudbrack CK, Nathal MV (2013) The mechanical properties of candidate superalloys for a hybrid turbine disk. NASA Technical Reports. https://ntrs.nasa.gov/archive/nasa/casi.ntrs.nasa.gov/20140000732.pdf. Accessed 16 July 2018
2. Mathey GF (1994) Method of making superalloy turbine disks having graded coarse and fine grains. US Patent 5,312,497. 17 May 1994
3. Gayda J, Gabb TP, Kantzos PT (2004) The effect of dual microstructure heat treatment on an advanced nickel-base disk alloy. Superalloys 2004:323–329

4. Mourer DP, Williams JL (2004) Dual heat treat process development for advanced disk applications. Superalloys 2004:401–407
5. Lemsky J, Gayda J (2005) Assessment of NASA dual microstructure heat treatment method utilizing Ladis SuperCooler(trademark) cooling technology. NASA Technical Reports. https://ntrs.nasa.gov/archive/nasa/casi.ntrs.nasa.gov/20050161950.pdf. Accessed 2 Sept 2018
6. Vairis A, Frost M (1998) High frequency linear friction welding of a titanium alloy. Wear 217 (1):117–131
7. Senkov ON, Mahaffey DW, Semiatin SL (2014) Inertia friction welding of the dissimilar superalloys Mar-M247 and LSHR. Metall Mater Trans A 45(12):5545–5561
8. Mahaffey DW, Senkov ON, Shivpuri R, Semiatin SL (2016) Effect of process variables on the inertia friction welding of superalloys LSHR and Mar-M247. Metall Mater Trans A 47 (8):3981–4000
9. Guillon O, Gonzalez-Julian J, Dargatz B, Kessel T, Schierning G, Räthel J, Merrmann M (2014) Field-assisted sintering technology/spark plasma sintering: mechanisms, materials, and technology developments. Adv Eng Mater 16(7):830–849
10. Belmonte M, González-Julián J, Miranzo P, Osendi MI (2010) Spark plasma sintering: a powerful tool to develop new silicon nitride-based materials. J Eur Ceram Soc 30:2937–2946
11. Chanthapan S, Rape A, Gephart S, Kulkarni AK, Singh J (2011) Industrial scale field assisted sintering an emerging disruptive manufacturing technology: applications. Mater Sci Forum 169:25–28
12. Harris K, Erickson GL, Schwer RE (1984) MAR M 247 derivations—CM 247 LC DS alloy, CMSX® single crystal alloys, properties and performance. Superalloys 1984:221–230
13. Sundararaman M, Mukhopadhyay P, Banerjee S (1988) Precipitation of the δ-Ni 3 Nb phase in two nickel base superalloys. Metall Trans A 19(3):453–465

Sintering Test Research of High Proportion Limonite

Qiang Zhao and Jilai Xue

Abstract The limonite was from Australia, the Fe grade was 56.09%, the grade was up to about 61.88% after ignition, the SiO_2 content was 5.62%, Al_2O_3 content was 3.87%, the gangue content was a little high, the harmful element S,P and alkali metal was low, the main metallic minerals are haematite and limonite, pyrolusite and pyrite is little; for gangue minerals, quartz is in majority, followed by kaolinite. The proportion of limonite was high up to 30% for sintering, through rational combination with other iron ore and process parameter optimization, the sintering was qualified for blast furnace.

Keywords Limonite · Sintering · Metallurgical performance

Introduction

The research and development of high proportion limonite was mainly in Australia, Japan and China. The most prominent was Hide Y [1] from Japan. After Years of research, the sintering technology about lenticular limonite own densification and high melting point liquid (SHS) [2] was developed by them. Currently, imported ore fines were mainly haematite in China, because of years mining, the haematite reserves decreased year by year [3, 4]. It was predicted that the Chinese haematite mining area would be closed one after the other, in contrast, the limonite reserves in Australia was huge. For making rational use of resources and adapting to changes in the market, increasing the limonite proportion in sintering was imperative. In this paper, the limonite was from Australia, the different sintering conditions tests were taken in laboratory to search the appropriate sintering conditions for limonite.

Q. Zhao · J. Xue (✉)
School of Metallurgical and Ecological Engineering, University of Science and Technology Beijing, Beijing 100083, China
e-mail: jx@ustb.edu.cn

Q. Zhao
e-mail: 13319532520@163.com

© The Minerals, Metals & Materials Society 2019

T. Jiang et al. (eds.), *10th International Symposium on High-Temperature Metallurgical Processing*, The Minerals, Metals & Materials Series,
https://doi.org/10.1007/978-3-030-05955-2_18

Raw Material Characteristics

Characteristics of Raw Material

Iron Ore

The main chemical compositions and iron phase analysis of limonite were shown in Tables 1 and 2. It is shown that limonite fines was characterized by medium iron grade, assaying 56.09% total iron and upgraded to about 61.88% after ignition, SiO_2 content is 5.62%, Al_2O_3 content is 3.87%, the contents of detrimental elements S, P and alkali metals are low. It is known from Table 2 that the major composition was limonite. The main problem is the high content of aluminium (the appropriate content is 1.5–2.0%), the increase in aluminium content may prevent the transformation of calcium orthosilicate crystals, it is propitious to prevent sintering mineralization and increase strength. However, raw material is easy to appear during sintering and the melting point increases during smelting, which is not conducive to slag formation (Table 3).

Table 1 Chemistry of limonite/wt%

TFe	FeO	Fe_2O_3	SiO_2	Al_2O_3	CaO	MgO	MnO
56.09	0.35	79.80	5.62	3.87	0.017	0.095	0.11

K_2O	Na_2O	P	S	Cu	Pb	Zn	As	LOI
0.023	0.011	0.076	0.026	0.0006	0.0042	0.0082	0.0085	9.35

Table 2 Iron phase of limonite /wt%

Iron phase	Iron carbonate	Iron sulfide	Magnetite	Martite	Limonite	Ferrosilite	TFe
Content	0.22	0.02	0.09	2.43	53.02	0.31	56.09
Ratio	0.39	0.04	0.16	4.33	94.53	0.55	100

Table 3 Chemistry of other raw materials/wt%

Element	TFe	FeO	SiO_2	Al_2O_3	CaO	MgO	P	S	LOI
Domestic concentrate	54.22	22.80	6.55	1.80	6.05	2.17	0.12	1.42	2.38
Blast furnace returns	51.76	6.50	5.64	2.46	11.55	2.45	0.071	0.078	2.00
Brazil ore	64.06	0.66	2.36	1.01	0.23	0.062	0.023	0.028	2.13
South Africa concentrate	65.27	27.74	0.95	1.33	1.14	2.66	0.084	0.05	1.32

Fluxes and Coke

The fluxes are used to adjust basicity in the sintering test. The chemistry of fluxes (limestone, burntlime and dolomite) is present in Table 4, it is a qualified flux. The aim of adding flux in sintering material is to obtain a certain basicity sinter, make the blast furnace smelting with no or less flux, and improve the blast furnace smelting intensity, reduce coke rate. The coke breeze was supplied by XISC. Analysis items including industrial analysis and chemical analysis of coke ash are tabulated are tabulated in Tables 5 and 6, respectively. As shown in Table 5, the C_{ad} of coke breeze is so high that it is a good fuel for sintering. In the sintering process, coke has good granulation performance, and coke particles larger than 0.25 mm can be grown as nuclear particle adhesive powder. The mixing grain size and fixed carbon segregation cannot be synchronized, and the control of most coke size after granulation to the grain size of 1–2 mm is beneficial to achieve synchronous and reasonable segregation of mixture grain size and fixed carbon. Good granulation performance of coke will have an important impact on its combustion behaviour.

Table 4 Chemical compositions of the fluxes /wt%

Flux	TFe	SiO_2	Al_2O_3	CaO	MgO
Burntlime	0.059	0.91	0.065	65.86	0.58
Limestone	0.18	6.47	0.070	49.78	0.36
Dolomite	–	0.71	0.66	32.64	19.83

Table 5 Industry analysis of coke breeze /wt%

Reducing agent	C_{ad}	M_{ad}	A_{ad}	V_{ad}
Cokebreeze	81.70	3.90	16.02	2.28

Note C_{ad} fixed carbon content on an air dried basis, V_{ad} volatile compound content on an air-dried basis; M_{ad} moisture content on an air-dried basis; A_{ad} ash content on an air-dried basis

Table 6 Chemical analysis of coke ash /wt%

Items	TFe	SiO_2	Al_2O_3	CaO	MgO
Coke ash	14.27	39.42	23.33	11.36	4.82

The Process Mineralogy of Limonite

Thermo-Analysis of Limonite (DTA and TGA)

In order to find out the behaviour of limonite in heating process, thermo-analysis was carried out by thermal analyser. The result is shown in Fig. 1. From it, we can know that the beginning disintegrating temperature of crystal water is 260.4 °C and the content of crystal water is about 8.8%.

From reaction formula:

$$mFe_2O_3 \cdot nH_2O \rightarrow mFe_2O_3 + nH_2O + 4932\,kJ/kgH_2O$$

The limonite ore, containing high combined water, which require more coke to supply heat for water removal. The decomposition heat of crystal water is 434,016 kJ/t, so we need 14.83 kg standard coal for it. This is why the solid fuel consumption of limonite is higher than that of magnetite and haematite.

Mineral Composition Analysis of Limonite

The ore is yellow-brown or red-brown under naked eye.

The results of microscopic analysis, X-ray diffraction and scanning electron microscopy show that species of mineral in the sample is relatively simple, the main metallic minerals are haematite and limonite, pyrolusite and pyrite is little; for gangue minerals, quartz is in majority, followed by kaolinite.

Overall, crystallization degree of essential mineral in the sample is lower, because the background value of X-ray diffraction analysis is higher and diffraction peak is very dispersive (Fig. 2). The weight contents of the essential minerals in the sample are listed in Table 7.

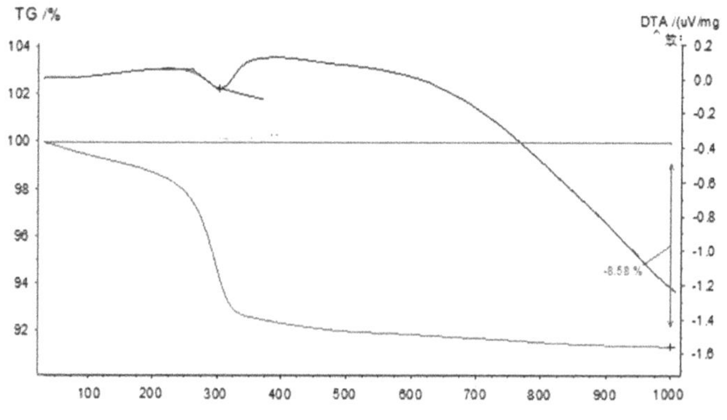

Fig. 1 Thermal analysis of limonite

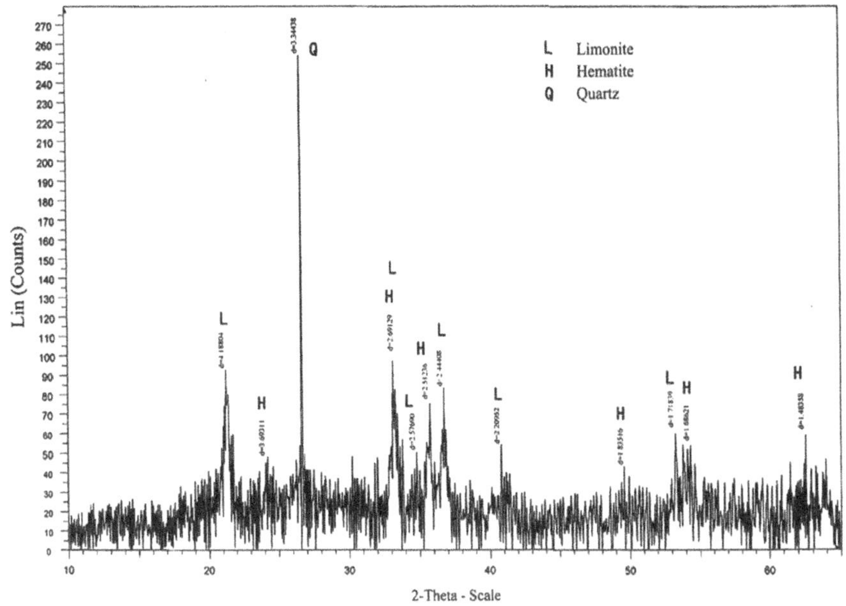

Fig. 2 X-ray diffraction spectrum of samples

Table 7 Content of essential mineral in sample /wt%

Mineral	Haematite	Limonite	Pyrolusite	Pyrite	Quartz	Kaolinite	Others
Content	29.3	63.8	<0.1	Trace	3.7	2.9	0.2

Research Method

According to the testing program, the raw material moisture was detected, then they were weighed on the basis of ingredients structure. Mixing included first mixing and second mixing. The objective of first mixing was blending and adding moderate water for wetting. The objective of second mixing was granulating. The diameter of drum mixer was 0.6 m and its length was 1 m. The granulating parameters of drum mixer were as follows: rotate speed was 17 r/min, granulating time was 4 min. Then the mixed sintering blend was put into sinter pot for sintering. Before loading the sintering mixture, 1 kg grate-layer was loaded in the pot. The main sintering parameters were as follows: layer height was 700 mm, ignition temperature was 1100 °C, ignition time was 1.5 min, ignition negative pressure was 5000 Pa, sintering negative pressure was 10,000 Pa. The sintering was dropped down three times from 2 m high first and then screened, screen size was 40, 25, 16, 10, 5 mm respectively, after screening, the 7.5 kg sintering was taken in proportion for tumble inspection. Finally, the vertical sintering speed, yield, productivity and tumble were calculated, respectively.

Test Results and Analysis

Small Pot Sintering Test Results and Analysis

Effect of Moisture on Sintering Performance

Tables 8 and 9 showed the effect of moisture on the sintering. In general, the sinter yield, unit productivity and tumble index increased with an increase in the moisture of sinter mixture from 6.4 to 7.2%. However, they began to decline when the moisture was higher than 7.2%. Therefore, the optimum moisture was recommended at 7.2%.

Effect of Coke Breeze on Sintering Performance

The effect of coke levels on sintering was tabulated in Tables 10 and 11. With an increase in the coke, the sinter yield and tumble index were improved. The optimum coke was recommended at 4.5%, on this condition, the sinter yield, unit productivity and tumble were achieved as follows: 81.06%, 1.661 t/m^2 h and 62.5%, respectively.

Table 8 Effect of moisture on sintering performance

Index no.	Moisture/%	Permeability index	Size distribution of granules/%					
			Size/mm					
			−0.5	0.5–1	1–3	3–5	5–8	+8
1	6.4	0.09	14.6	10.4	31.9	19.7	12.3	11.1
2	6.8	0.17	4.0	6.3	35.1	28.5	15.5	10.6
3	7.2	0.15	8.9	6.7	41.7	21.2	12.1	9.5
4	7.6	0.19	4.6	5.5	37.7	28.7	13.5	10.0

Table 9 Effect of moisture on sintering performance

Index no.	Flame speed/ mm/min	Yield/%	Productivity /t/m^2 h	Solid fuel/kg/t	Tumble/%
1	13.96	81.99	1.119	62.77	62.20
2	20.52	82.70	1.590	64.85	60.73
3	19.51	82.90	1.487	65.27	62.40
4	21.85	81.99	1.751	61.82	61.73

Table 10 Effect of coke breeze on sintering performance

Index no.	Moisture /%	m	Size distribution of granules/%					
			Size/mm					
			−0.5	0.5–1	1–3	3–5	5–8	+8
3	7.2	0.15	8.9	6.7	41.7	21.2	12.1	9.5
5	7.2	0.18	4.7	3.0	32.0	30.1	17.8	12.4
6	7.1	0.17	6.6	4.0	41.4	25.7	14.3	8.0

Table 11 Effect of coke breeze on sintering performance

Index no.	Flame speed /mm/min	Yield/%	Productivity /t/m^2 h	Solid fuel/kg/t	Tumble /%
3(cokebreeze 5.0%)	19.51	82.90	1.487	65.27	63.1
5(cokebreeze 4.5%)	21.69	81.06	1.661	59.49	62.5
6(cokebreeze 4.0%)	18.89	76.83	1.385	55.60	62.2

Large Pot Sintering Test

The parameters of Φ150 mm sintering test were demonstrated by Φ250 mm sintering test. The test conditions of large pot were the same as those of small pot test of 5. The results were shown in Table 12. The results were similar to those of small pot sintering test.

Chemical and Physical Properties of Sintering

Size Distribution

The size distribution was tabulated in Table 13.

Table 12 Results of large pot sintering test

Permeability index	Flame speed /mm/min	Yield/%	Productivity /t/m^2 h	Solid fuel/kg/t	Tumble /%
0.15	16.92	82.68	1.433	58.48	62.2

Table 13 Size distribution of sintering/wt%

Size/mm					
>40	40–25	25–16	16–10	10–5	<5
13.09	17.82	19.75	15.87	15.51	17.95

Chemistry of Sinter

The chemistry of sinter was shown in Table 14. The chemical composition was wonderful, FeO 7.94%, MgO 2.44% and basicity2.21 were similar to the mill's requirements, but Zn was a little higher, in a word, it was a good raw material for blast furnace.

Mineralogy of Sinter

The mineral compositions of sinter, with and without blending WPF iron ore fines into sinter mixtures, were determined by using Japanese microscope equipped with reflection and transmission. All the sinter samples consist of the minerals as follows: calcium ferrite ($CaO \cdot 2Fe_2O_3$, $CaO \cdot Fe_2O_3$, $2CaO \cdot Fe_2O_3$), haematite (Fe_2O_3), magnetite (Fe_3O_4), calcium and iron containing fayalite ($CaO \cdot FeO \cdot SiO_2$, $2FeO \cdot SiO_2$), calcium silicate ($CaO \cdot SiO_2$), magnesium ferrite ($MgO \cdot Fe_2O_3$), iron pyroxene ($2FeO \cdot 2SiO_2$) and glass phase. Identified under the microscope, the essential iron minerals of sample were magnetite, haematite and calcium ferrite, followed by magnesium ferrite. The content of gangue minerals including vitreous silica, calcium silicate, olivine and quartz was relatively low. The approximate weight contents of the essential minerals in the sample were listed in Table 15 (Photos 1 and 2).

Metallurgical Performance

The measured metallurgical performance included reducibility index (RI), low temperature reduction degradation index (RDI) and softening and melting performance, which would be used to predict the behaviours of sinter in blast furnace (Tables 16 and 17).

Table 14 Chemistry of sinter of sintering/%

TFe	FeO	SiO_2	Al_2O_3	CaO	MgO
55.36	7.94	4.66	2.10	10.30	2.44

K_2O	Na_2O	P	S	Cu	Pb	Zn	As
0.073	0.048	0.055	0.020	0.026	0.043	0.15	0.029

Table 15 Content of essential mineral in sample/wt%

Minerals	Magnetite	Haematite	Calcium ferrite	Magnesium Ferrite	Calcium silicate	Vitreous silica	Quartz
Content	32	18	43	3	1	2	<1

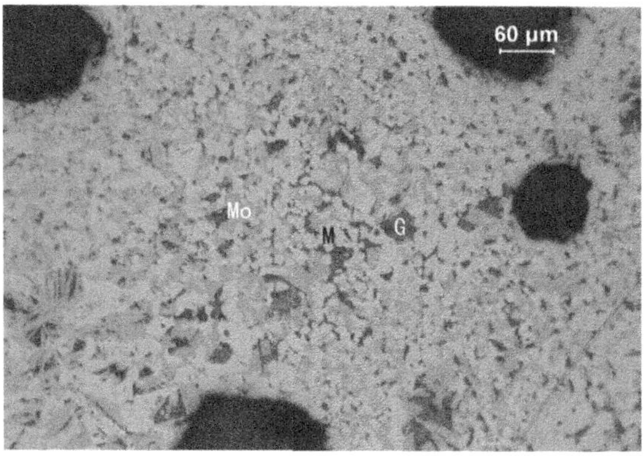

Photo 1 Superfine granular and vermiform magnitite (M) and bladed calcium ferrite (Mo) mixing and intergrowing, G–vitreous silica, Black-pore reflecting light

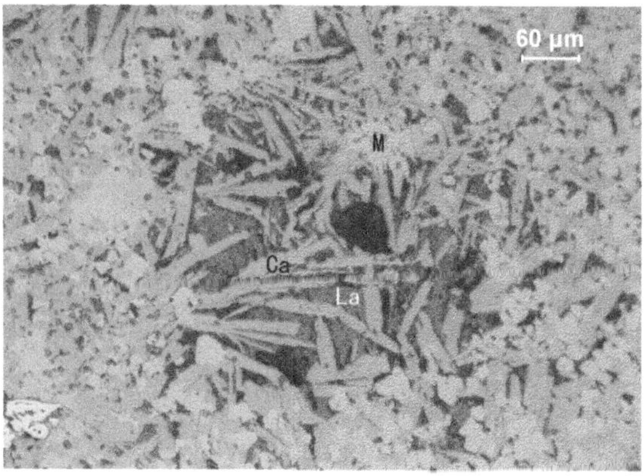

Photo 2 Fine acicular bladed calcium ferrite (Mo) polymerize into irregular agglomerate aggregates and replace with calcium silicate (La), M–magnetite, Black-pore reflecting light

Table 16 RDI and RI index/wt%

RDI			RI
$RDI_{+6.3\ mm}$	$RDI_{+3.15\ mm}$	$RDI_{-0.5\ mm}$	
50.60	79.23	6.03	86.01

Table 17 Softening and melting performance/°C

Softening performance			Melting performance		
T_a	T_s	ΔT_{sa}	T_p	T_m	ΔT_{mp}
1181	1267	86	1319	13,510	191

Conclusion

(1) The limonite was characterized by medium iron grade, assaying 56.09% total iron and upgraded to about 61.88% after ignition, in which SiO_2 content was 5.62%, Al_2O_3 content was 3.87%, the contents of detrimental elements S,P and alkali metals were low. The metallic minerals of WPF mainly were haematite

and limonite, pyrolusite and pyrite was little; for gangue minerals, quartz is in majority, followed by kaolinite.

(2) According to the results of sintering tests, limonite had good granulation behaviours and sintering performance. Through the optimization of ore proportioning and sintering parameters, the sintering with high tumble index, high productivity, low solid fuel consumption, good metallurgical behaviours and good mineralogy was obtained. It is qualified for blast furnace. The sintering results could be used to guide the best possible use of WPF fines for potential customers in China.

References

1. Hide Y (1993) Pisolitic limonite self-densification and high melting point liquid phase sinter process. In: The 6th international agglomeration conference proceeding, pp 246–251
2. Chen W, Xing JM, Wang K (2007) Experimental study of sinter about Australian FMG fine ore. Metal World 2:26
3. Fu JY, Jiang T, Zhu DQ (2006) Sintering and Pelletizing. Central South University Press, p 68
4. Li HC, Hou YH (1994) The effect of MgO content on limonite sintering. Sintering and Pelletizing 6:30–32

A Machine Learning Method for State Identification of Superheat Degree with Flame Interference

Shiwei Zhao, Yongfang Xie, Weichao Yue and Xiaofang Chen

Abstract The superheat degree in the process of aluminium electrolysis is an important indicator for judging the condition of the electrolysis cell. In the actual production process, the artificial observation of the fire hole is usually used for judgment and decision of cell condition. However, the decreasing number and frequent flow of experienced technicians make it difficult to guarantee the accuracy of this complex work. Although there exist some methods for state identification of superheat degree, they do not consider flame interference, resulting in decreasing of accuracy. In view of this fact, a method for state identification of superheat degree with flame interference is proposed, and the proposed method is compared with the existing method on 17 aluminium electrolysis cells. The vilification result shows that the proposed method has a better performance than the existing methods. Moreover, it reveals that the proposed method is feasible for identification with flame interference. In addition, it can provide suggestions for the technicians to judge the state of superheat degree.

Keywords Superheat degree · Electrolysis cell · Fire hole · Flame interference
Features extraction

Introduction

Aluminium reduction process is characteristic with high temperature, corrosion, variables coupling. Because of complexity of process, it is difficult to get a precise control decision [1]. Superheat degree (SD), which is comprehensive reflection of aluminium reduction cell condition, is the difference value between electrolyte temperature and liquidus temperature, and the state of superheat is divided into three classes, including high, normal and low. If SD is high, the increasing of

---

S. Zhao · Y. Xie · W. Yue (✉) · X. Chen
School of Information Science and Engineering, Central South University,
Changsha 410083, China
e-mail: yue_weichao@163.com

© The Minerals, Metals & Materials Society 2019

T. Jiang et al. (eds.), *10th International Symposium on High-Temperature Metallurgical Processing*, The Minerals, Metals & Materials Series,
https://doi.org/10.1007/978-3-030-05955-2_19

energy consumption will result in waste of resources. If SD is low, the Al_2O_3 don't have a thoroughly dissolution, and the sediment may be produced on the cathode. In practice, the SD adjustment is usually used to get a high current efficiency. In aluminium reduction process, the estimation of SD state always relies on artificial experience, which has some shortcomings, including: (1) There exists subjectivity of technicians; (2) It creates a lot of human, material, financial waste; (3) It is difficult to maintain consistent stability of reduction cell. Consequently, scientific method is expected to identify the state of SD. According to the definition of SD, the electrolyte temperature can be detected by infrared thermometer. However, the different composition of electrolyte, liquidus temperature shows the difference. The detection of liquidus temperature is a complex work, which is unable to meet the real-time requirements [2].

The problem has attracted many attention of researchers. Chen Z, Li Y, et al. proposed an intuitive fuzzy hyper-graph semantic network model to determine the SD state. However, a large amount of fire hole videos with labels were necessary to detect the parameters of model. In practice, these videos with labels are difficult to obtain [2]. Zeng S, et al. proposed the model of evaporation rate of aluminium fluoride with the parameters temperature and AlF_3 feeding amount. Based on the analysis and computation of electrolyte composition, the soft-measuring model was proposed. However, because of complicacy of reduction process, this model may not capture the features of SD calculation [3]. Yang J. et al. also proposed a soft-measuring model of SD, and a novel attenuation of data weight method and definition of rule confidence were given. First, the static rules were extracted from historical data based on map reduce of rough set theory. And then the rule incremental update method is constructed based on the newly defined rule tree. However, because of the limitation of stochastic disturbance and complexity, the calculation of SD based on the aforementioned method is easily affected [4]. Random forest method for relative density noise filtering was proposed by Liu Y, et al. to predict SD, and then to control SD. However, because of sever environment, the sensors may be affected, resulting in accuracy decreasing of measured data. Therefore, the trained model based on these historical data may be difficult to apply in the aluminium reduction process [5].

The experiential technicians are necessary for the artificial judgment of SD state. However, because of the frequent flow and dwindling in numbers of experiential technicians, artificial judgment of SD state has been unable to meet the needs of collectivization of aluminium electrolysis industry and large size of reduction cell. Therefore, a method that can simulate the artificial judgment is necessary in the future, resulting in inheriting the experiential knowledge [1]. Chen X, et al. [6] had proposed the method that computer vision and expert rules were used for the first time to identify SD state of aluminium reduction cell. The proposed method combining with artificial experience and fire hole information meets the requirements of real-time detection to some extent. However, the high temperature environment of aluminium reduction cell results in producing flame in the fire hole, as shown in Fig. 1. Flame in the hole has a great influence on features extraction, and it may even cause the wrong features. However, the aforementioned method which

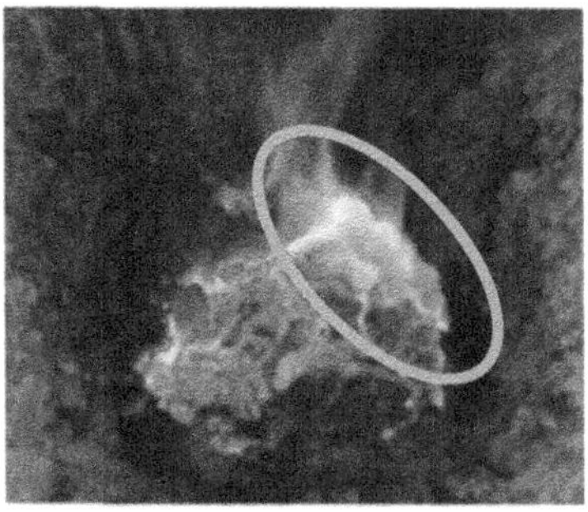

Fig. 1 Picture of fire hole in aluminum electrolysis cell with flame

may have a poor performance on the flame interference does not take into account the effect of the flame [6]. In view of this fact, a method is proposed to eliminate the flame interference and to extract features of fire hole, and then these features are used for the state identification of SD.

The outline of this study is as follows. In Sect. 2, the feature extraction method for fire hole is proposed, and the features include amplitude and colour of electrolyte. Moreover, the *k* nearest neighbour algorithm is introduced to classify. In Sect. 3, the proposed method is verified on the historical data, and the discussion is performed. In Sect. 4, the conclusion is given.

The Features Extraction Method of Fire Hole with Flame Interference

SD plays an important role in the aluminium reduction production process [7]. Because of high measurement cost and measured online with difficulty, the online detection of SD has become a challenge. The major component of electrolyte is Na_2AlF_6–Al_2O_3, and the chemical reaction in the cell is illustrated by Eq. (1) with no consideration of other side reactions.

$$2Al_2O_3 + 3C = 4Al + 3CO_2 \uparrow \tag{1}$$

Based on analyzing the aluminium reduction mechanism, we can see that between amplitude and colour changing of electrolyte have the following relationship. If the state of SD is high, electrolyte viscosity and surface tension would be small, the production of reaction include CO_2 which may result in small amplitude of electrolyte. In addition, the longer it takes for the electrolyte to go from red to black. If the state of SD is low, electrolyte viscosity and surface tension would be large, the production of reaction includes CO_2 which may result in large amplitude of electrolyte. In addition, the shorter it takes for the electrolyte to go from red to black. Therefore, the amplitude and rate of colour changing can be used to reflect the state of SD. Based on the above analysis, the amplitude of electrolyte is affected by electrolyte surface viscosity, and the flame has little influence on amplitude. Consequently, suppose the fire had no influence on the amplitude in this paper.

The Features Extraction of Amplitude

Because electrolyte movement belongs to fluid movement, the traditional moving target tracking algorithm, such as inter-frame difference method, may have poor performance. Corner points are a characteristic point in the neighbourhood having two principal directions. The region is usually a stable, information-rich region in the image. In addition, these regions are characteristic with rotation invariance, scale invariance and affine invariance. Therefore, corner points are widely used for target tracking, 3D reconstruction and motion estimation. Commonly used angle detection methods included Harris algorithm [8], Shi-Tomasi algorithm [9] and so on. Shi-Tomasi algorithm was considered the extended Harris algorithm, and it has shown a better performance than Harris algorithm in many researches. Accordingly, Shi-Tomasi algorithm was used to detect the corner points, and also pyramid LK optical flow method was used to trace the corners to extract the features of amplitude [9, 10]. Since the amplitude feature extraction is basically free from fire, the feature extraction method of this part is the same under the interference of fire or not. Therefore, the method in [6] is introduced to extract amplitude.

Colour Change Rate Feature Extraction with Flame Interference

Besides the amplitude feature, the colour change feature also has the very important meaning to the superheat judgment. The details are below. If the state of SD is low, the Al_2O_3 concentration is high, resulting in carbon residue precipitates from the electrolyte, and the colour changes from red to black faster. On the contrary, if the state of SD is high, and the Al_2O_3 concentration is low, and the colour changes from red to black shorter. Therefore, colour change rate can be used to reflect the state of SD.

Colour is usually described by three independent attributes. These three independent variables have comprehensive utilization, which naturally constitutes a spatial coordinate, namely the colour space. However, the described colour object is objective. Different colour spaces only measure the same object from different perspectives. According to the basic mechanism, colour spaces can be divided into two categories: primary colour space and bright separated colour space. The former is typically RGB, while the latter includes YUV and HSV. RGB is most important and most common colour model to process image. RGB is established in the Cartesian coordinate system, and red, green and blue three are basic colours. A rich and extensive colour is produced based on the superposition of different degrees.

HSV is a colour space created by A. R. Smith based on the intuitive nature of colour, also known as the hexagonal vertebral body model [11]. HSV colour model refers to a subset of visible light in the three-dimensional colour space of H, S and V, which contains all colours in a certain colour field. Compared with HSV colour space, RGB colour space is intuitive and easy to understand. In addition, the three components of R, G and B are highly correlated, that is, if a certain component of a colour changes to some extent, then the colour will probably change, which can well represent the colour change characteristics. Therefore, RGB colour space is adopted in this paper.

As we all know, the flame don't always appear in the same position. It often flutters back and forth in a range. The flame appears in a location in a moment, but it may disappear in the next moment. Because the computer is to convert video to continuous processing images, which can show the aforementioned phenomenon well, as shown in Fig. 2. Because the grey image is vulnerable to external interference, and also it has a poor performance on expressing the colour change rate. In this paper, by extracting a maximum of three colour channels for each pixel corresponding pixel values. As shown in Fig. 3, there existing flame in the fire hole bring a lot of interference to colour features extraction, and even lead to the result

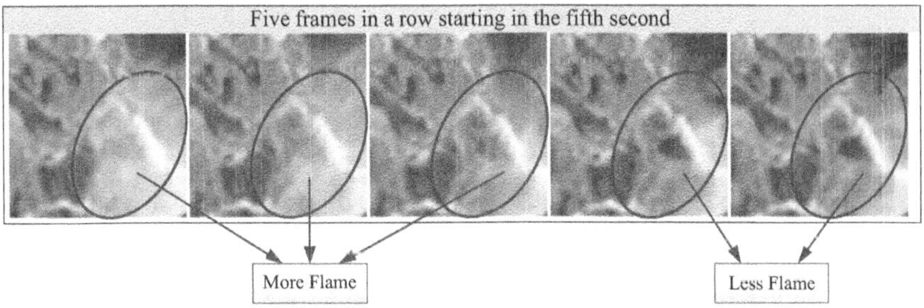

Fig. 2 Flame appears randomly in the same position

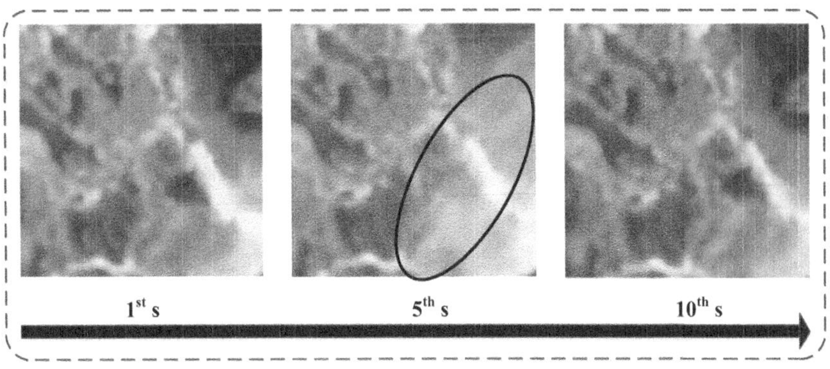

Fig. 3 Pictures of fire hole in 1st, 5th, 10th second

error. In view of this fact, a method is proposed based on region partition to eliminate fire interference in this paper.

Step 1: Start with the first second in video, take five consecutive frames of images every five seconds and take the first ten seconds to get the picture set, as shown by Eq. (2)

$$\{I_t^N(x,y), N = 1,\ldots,5, t = 1,5,10\} \tag{2}$$

Where t is the time to start the collection of pictures; N represents the Nth frame image index after collection from start; $I(x, y)$ represents the image.

Step 2: The collected images are segmented according to the segmentation method of the amplitude extraction process. In order to reduce the calculation amount, the area with the centre of fire hole is considered as the centre of image. The side length is the half of the maximum height of fire hole, which is considered as the approximation of the fire hole area, and the image is cropped to obtain the following picture set, as shown by Eq. (3).

$$\{I_t^{\prime N}(x,y),\ N = 1,2,\ldots,5,\ t = 1,5,10\} \tag{3}$$

where $I_t^{\prime N}(x,y)$ is the Nth image from tth second.

Step 3: The first frame image starting from 1st, 5th, 10th seconds (as shown in Fig. 2) is divided into regions for filling. The details are below. Take the fifth second as an example, each frame image in the set $\{I_5^{\prime N}(x,y),\ N = 1,2,\ldots,5\}$ is uniformly divided into 16 rectangular and equal areas. The average value of the colour channel corresponding to the maximum value of the corresponding position area is compared. Select the region corresponding to the image with the smallest average value to fill in the first frame image, as shown in Fig. 4. We will get the first frame image of the fifth second after removing the fire interference.

Step 4: The sum of the maximum value of the colour channel corresponding to each second is obtained, as shown by Eq. (4). Then the difference is divided by the time interval, namely, the average colour change rate. In order to reduce the error, the average colour change rate of the first ten seconds and the average colour change rate of the first five seconds are summed up in this paper, as shown by Eq. (5). Take the average and get the average colour change rate.

$$\max\_channel\_img = \left(\sum_{i=0}^{n}\sum_{j=0}^{m} \max(I_{kt}^{\prime 1}\ k = 0,1,2),\ t = 1,5,10\right) \tag{4}$$

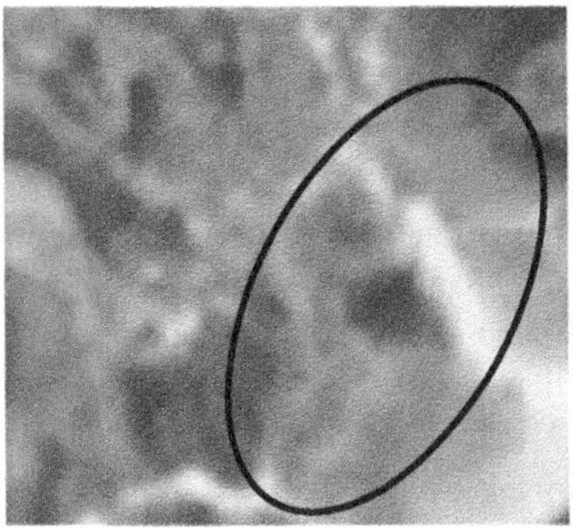

Fig. 4 Picture of the 5th second after flame interference is removed

$$ave\_color\_change = \frac{1}{2} * \left(\frac{1}{5} * (\max\_channel\_img_{10} - \max\_channel\_img_{5}) + \frac{1}{10}(\max\_channel\_img_{10} - \max\_channel\_img_{1})\right) \quad (5)$$

k-*Nears Neighbour Algorithm*

Because k-Nears Neighbour (kNN) algorithm is easy to understand, no parameter estimation, and other advantages, it is widely used for multi-classification [12–14]. The details of kNN are below.

Input: Training dataset

$$T = \{(x_1, y_1), (x_2, y_2), \ldots, (x_N, y_N)\} \quad (6)$$

where $x_i \in R^n$, n is the number of sample features. $y_i \in \{c_1, c_2, \ldots, c_K\}$, k is the number of sample categories.

Output: The category y to which the instance x belongs.

(1) Data normalization;
(2) According to the given distance metric, the k nearest neighbour points are found in the training set. Neighbourhood of x which covers the k points is recorded as $N_k(x)$;
(3) The category y of x is determined by classification decision rules, such as majority vote, in $N_k(x)$, as illustrated by Eq. (7).

$$y = \arg\max_{c_j} \sum_{x_i \in N_k(x)} I(y_i = c_j), \ \ i = 1, 2, \ldots, N; \ j = 1, 2, \ldots, K \quad (7)$$

Result and Discussion

In order to verify the effectiveness of the proposed method, experiments are performed with 17 groups fire hole video, which are from the same aluminium reduction workshop, compared with the results of reference [6]. Firstly, the cause of the change of electrolyte amplitude and colour is analysed, and the result that the flame affects colour change is obtained. Then according to the random flutter of a flame, a method based on region partition is proposed to eliminate fire interference. The features are extracted from these videos, and then these features are used for identification of SD state based on *k*NN to void subjectivity of technicians. The details are as follows. Take one sample in order as the test sample, and others are considered as the training sample. This process is repeated 17 times, and we will get the category of each cell. In the *k*NN algorithm, let $k = 4$. Experimental results are gathered in Table 1.

As shown in Table 1, the accuracy of method proposed in [6] is 64.7% with flame interference. In contrast, the accuracy of method proposed in this paper is 76.47%, which reveals that the proposed method in this paper has a better performance. That is to say, the accuracy of SD classification has been improved to some extent. At the same time, due to the introducing of *k*NN, the subjectivity could be eliminated. In summary, the proposed method in this paper is feasible.

Table 1 Feature extraction and superheat state prediction results of this study and Chen [6] method

ID	Label	AveAmp	Ave_colour_change (Chen)	Ave_colour_change (this study)	Predict (Chen)	Predict (this study)
1	Normal	12.85115	−0.63601	−0.03934	Normal	Normal
2	Normal	16.48215	2.120958	0.920962	Cold	Normal
3	Normal	8.19725	0.301496	0.463617	Hot	Normal
4	Normal	9.340452	0.072359	0.931329	Normal	Normal
5	Hot	5.758534	1.020779	1.591017	Cold	Cold
6	Cold	5.390055	1.789803	2.865898	Cold	Cold
7	Cold	3.538292	4.835451	4.765882	Cold	Cold
8	Cold	2.150642	4.995921	3.806577	Cold	Cold
9	Normal	36.97835	0.579563	3.096458	Normal	Normal
10	Normal	7.046891	2.428155	0.301592	Cold	Hot
11	Hot	2.421264	0.043427	0.626652	Cold	Cold
12	Hot	2.617564	−1.44359	−1.08244	Hot	Hot
13	Cold	3.810315	3.032787	3.661108	Cold	Cold
14	Hot	6.595656	0.478175	1.947057	Cold	Cold
15	Cold	7.640349	0.979662	1.357588	Cold	Cold
16	Cold	3.874096	1.858326	2.957373	Cold	Cold
17	Cold	4.87692	1.120229	1.844163	Cold	Cold

Conclusion

Due to difficulty in measurement and the subjectivity for SD state judgment, machine learning methods had been proposed in existing researches for SD classification. However, external interference didn't be considered in these methods. In view of this fact, a method was proposed in this paper for SD classification with flame interference. Firstly, the production environment of aluminium electrolysis is characterized by high temperature, and the combustion conditions will be achieved in most cases. Therefore, fire always exists in the fire hole, and the causes of amplitude, colour change and the relationship with superheat are analysed. Secondly, Shi-tomasi corner point detection algorithm is adopted and pyramid LK optical flow method is adopted to track the corner point to get the amplitude feature. For the fire hole image under the flame interference, the regional filling method is adopted to eliminate the flame interference. Finally, the extracted features are applied for SD classification based on *k*NN algorithm. Experimental results show that the proposed algorithm is effective to classify the SD state.

Acknowledgements Supported by National Natural Science Foundation of China (61773405, 61533020, 61751312, 61725306, 61621062), the teachers-students co-innovation project of Central South University (502390003).

References

1. Yue W, Chen X, Gui W et al (2017) A knowledge reasoning fuzzy-bayesian network for root cause analysis of abnormal aluminum electrolysis cell condition. Front Chem Sci Eng 11 (3):414–428
2. Chen Z, Li Y, Chen X et al (2017) Semantic network based on intuitionistic fuzzy directed hyper-graphs and application to aluminum electrolysis cell condition identification. IEEE Access 5:20145–20156
3. Shuiping Z, Fuwei C (2010) Dynamic decision model for amount of AlF3 addition in industrial aluminum electrolysis. In: Proceedings of the 8th world congress on intelligent control and automation, Jinan, China, 7–9, July, 2010
4. Yang J-S, Yu H, Chen X-F, Zou Z (2017) Soft measuring model of superheat degree in the aluminum electrolysis production. In: Chinese automation congress (CAC), Jinan, China, 20–22, Oct 2017
5. Liu Y, Xia S, Yu H, Wang G (2017) Prediction of aluminum electrolysis superheat based on relative density noise filtering random forest. In: Chinese automation congress (CAC), Jinan, China, 20–22, Oct 2017
6. Xiaofang C, Xiaowei Y, Keke H (2017) Identification of superheat of aluminum electrolytic cell based on computer vision and expert rule. In: Chinese automation congress (CAC), Jinan, China, 20–22, Oct 2017
7. Gui W, Yue W, Chen X et al (2018) Process industry knowledge automation and applications in aluminum reduction production process. Control Theory Control Appl 35(07):887–899
8. Huang, H, Ma, G, Zhuang Y (2008) Vehicle license plate location based on Harris corner detection. In: 2008 IEEE international joint conference on neural networks (IEEE world congress on computational intelligence), Beijing, China, Sept 28–Oct 4, 2008

9. Kenney CS, Zuliani, M, Manjunath BS (2005) An axiomatic approach to corner detection. In: IEEE computer society conference on computer vision and pattern recognition, San Diego, California, 20–26, June, 2005
10. Xia T, Jing XS, Zou WJ, SO Automation (2018) A moving object detection method based on pyramid LK optical flow under dynamic background. Navigation & Control
11. Smith AR (1978) Color Gamut Transform Paris. In: ACM-SIGGAPH'78 conference proceedings, pp 12–18
12. G Guo, H Wang, D Bell, Y Bi, K Greer.(2003) KNN model-based approach in classification. Lecture Notes in Computer Science
13. Zhang S, Li X, Zong M et al (2017) Learning k for kNN classification. Acm Trans Intell Syst Technol 8(3):43
14. Guo H, Li Y, Li Y, et al (2016) BPSO-Adaboost-KNN ensemble learning algorithm for multi-class imbalanced data classification. Eng Appli Artif Intell 49(C):176–193

Part V
Extraction and Recovery of Metals

Study on Volatizing Tin from Tin-Bearing Middling by Carbothermic Reduction in Rotary Kiln

Jianfa Jing, Yufeng Guo, Feng Chen, Fuqiang Zheng and Lingzhi Yang

Abstract The tin-bearing middling obtained by flotation in Huanggang, Inner Mongolia, is with SnO_2 content of 3–5 wt%, which was stacked and cannot be utilized, leading to the tin resource wasting and environmental pollution. In order to recover this tin resource, the tin volatilization behaviour in the tin-bearing middling during carbothermic reduction in rotary kiln process was studied in this paper. The results show that the volatilization rate can reach 87.4% and the SnO_2 content in the residue decreased to 0.57 wt% under the conditions of Kiln tail: Kiln middle: Kiln head (coal addition ratio) with 1:2:7, the proportion of ball and coal was 20 kg: 20 kg and the pellets size was 5–10 mm at 1000 °C for 2 h. The research could provide a technical reference for the utilization of the tin-bearing middling.

Keywords Tin-bearing middling · Carbothermic reduction · Volatilization rate

Introduction

Recently, the tin oxides are an important resource which were widely used in many fields, such as gas sensors, electrode conductors, liquid crystal displays, packaging containers, electronics industry, transportation, construction and so on [1–3]. The tin resource was reduced with the rapid development of the tin metallurgy. The tin-bearing middling obtained by flotation in Huanggang, Inner Mongolia, is with SnO_2 content of 3–5 wt%, which was stacked and cannot be utilized, leading to the tin resource wasting and environmental pollution. Therefore, the tin-bearing middling was considered a popular feedstock for the production of tin.

J. Jing · Y. Guo (✉) · F. Chen · F. Zheng · L. Yang
School of Minerals Processing and Bioengineering, Central South University, Changsha 410083, China
e-mail: yfguo@csu.edu.cn

© The Minerals, Metals & Materials Society 2019

T. Jiang et al. (eds.), *10th International Symposium on High-Temperature Metallurgical Processing*, The Minerals, Metals & Materials Series,
https://doi.org/10.1007/978-3-030-05955-2_20

Many methods were proposed to recovery tin which mainly includes reduction roasting, sulfuration and chlorination. In the method of chlorination, tin phases are transformed to chlorides and evaporated making tin removal rate up to 94% [4, 5], but the shortcoming is the corrosion to the facilities, enhancing the production cost and polluting the environment. In the process of sulfuration, the tin removal rate can reach 93% [6, 7], but the added sulphurizing reagent have a severe corrosion for the equipment and the extra addition have a high economic losses. The reduction roasting for recovering tin is a simple technological process and has a desirable result.

During the reduction roasting, the reduction reagent was also the main factor for tin recovery. Zhang [4, 8], respectively, studied the behaviour of tin volatilization by carbothermic reduction roasting which considered the graphite, coke, petroleum coke, semi-coke, anthracite, charcoal, lignite etc. as the reduction reagent. The result showed that the tin removal rate can reach 80%. Su [9–11] proposed reducing SnO_2 by a gas agent (H_2,CO–CO_2) and the results indicated that the volatilization rate reached 91% at the strict reduction atmosphere. However, the reduction atmosphere is difficult to control which lead to these methods not industrialize.

In this study, the coke powder as the reduction reagent for reducing the tin in kiln process. The roasting time, temperature, coal addition rates, and the size of pellets were system studied. These results provided a technology reference for tin recovery in industrially.

Experimental

Raw Material

The tin-bearing middling used in this study was collected from Inner Mongolia, in Huanggang. Chemical composition (wt%) of it is depicted in Table 1. From Table 1, we can see that the tin-bearing middling contains only 3.73% Sn and the main oxides are CaO and SiO_2. The XRD (XRD, D/max 2550PC, Japan Rigaku Co., Ltd.) patterns of the raw ore and iron concentrate are presented in Fig. 1. Figure 1 shows that the major mineral phases in the tin-bearing middling are fluorite, cassiterite, and magnesium aluminosilicate, while the main mineral phases in the concentrates are fluorite and a small amount of cassiterite. Figure 2 is the EDS analysis of tin-bearing middling, it can be seen that the main phases were fluorite and silicate which was corresponded with the XRD phase.

Table 1 Chemical composition of the tin middling (wt%)

Sn	TFe	Ca	SiO_2	MgO	Al_2O_3	S	F	P	Zn	As	Burning loss
3.73	15.32	25.54	20.60	1.54	3.94	1.84	4.89	0.18	1.33	2.26	7.28

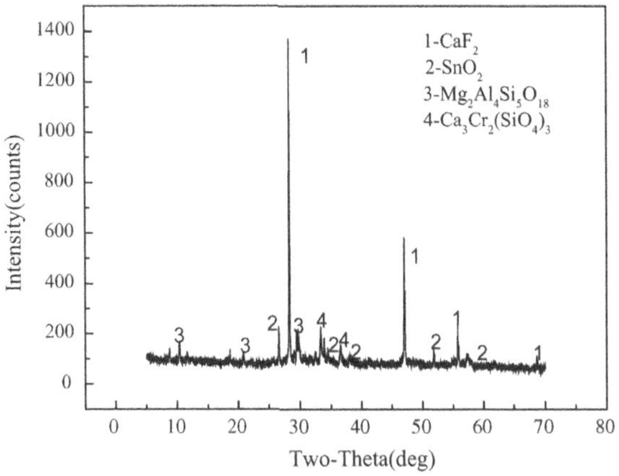

Fig. 1 XRD pattern of raw material

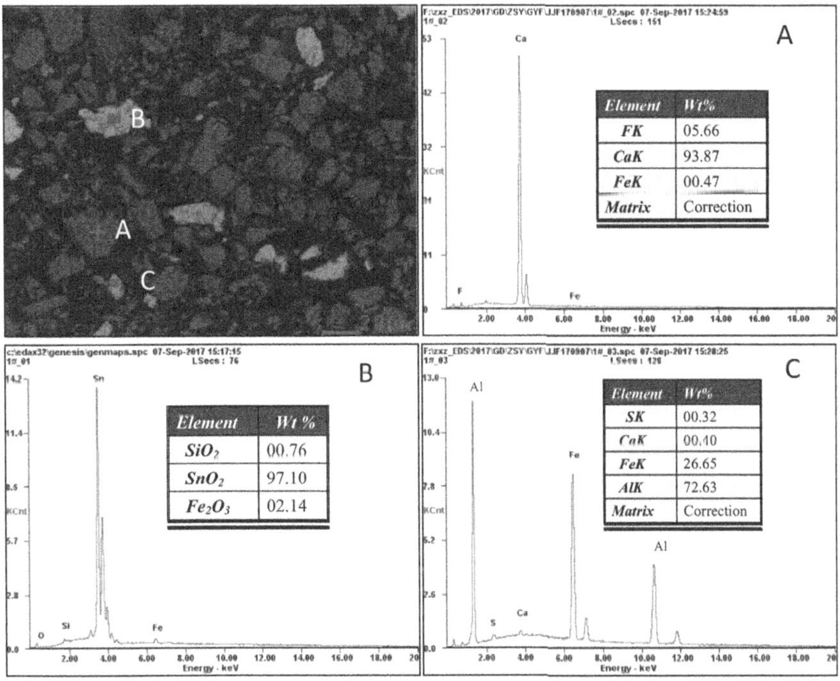

Fig. 2 The EDS analysis of the tin-bearing middling

Methods

Figure 3 shows the flow sheet of a typical tin extractive metallurgy in kiln process from tin-bearing middling. Firstly, the samples of tin-bearing middling were dried in an oven at 120 °C for 4 h; the dried samples were crushed to less than 1 mm. Secondly, weight the crushed samples mixed with 8% water and agglomerate into pellets. Thirdly, 20 kg of pellets (5–10 mm) were mixed with 8 kg coal which was crushed to 25 mm and blew to the reduction roasting kiln when the temperature reached the designed temperature. During the roasting process, the coal was added into kiln from the door of coal addition with 2 kg/5 min in 30 min; keep the aim temperature for an hour and a half. Finally, the roasted pellets were weighed and the residual tin content was analyzed by chemical titration when the temperature reached room temperature.

Characterization

The content of tin in the residue was analyzed by chemical titration (GB/T212-2008). The mineral constituents of the roasted samples were identified by X-ray diffraction (XRD, D/max 2550PC, Japan Rigaku Co., Ltd.) with the step of 0.005° at 10° min^{-1} in ranging from 10° to 80°. The microstructure and elemental composition of the reduced pellet were analyzed by using environmental scanning electron microscopy (ESEM; FEI QUANTA 200; FEI, Eindhoven, The Netherlands) equipped with an EDAX energy dispersive X-ray spectroscopy (EDS) detector (EDAX Inc., Mahwah, NJ). ESEM images were recorded by backscattering in the low-pressure environment (0.5 Torr and 20 keV).

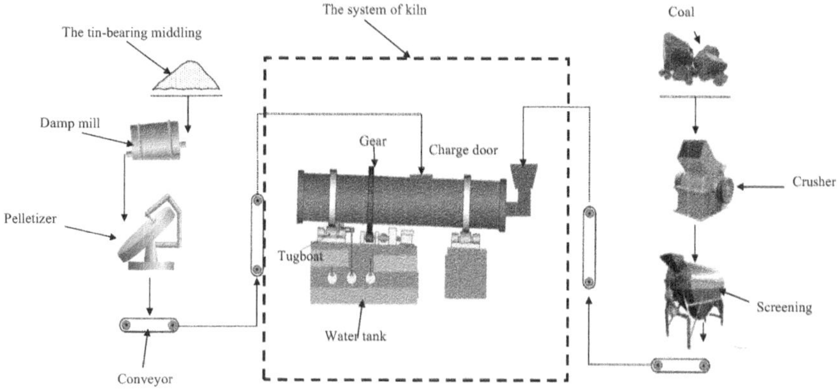

Fig. 3 The flow sheet of a typical tin extractive metallurgy in kiln process

Calculation

The volatilization ratio of Sn was calculated according to the following equation:

$$X = (\beta_1 m_1 - \beta_2 m_2)/\beta_1 m_1 \times 100 \quad (1)$$

where X is the volatilization ratio of Sn, %; m_1 is the mass of the sample before reduction roasting, g; m_2 is the mass of the sample after reduction roasting, g; β_1 is the content of tin in the raw material, (by GB/T15924-2010) wt%; β_2 is the content of the mineral after reduction roasting, (by GB/T15924-2010) wt%.

Discussion and Results

Effect of Roasting Temperature

Figure 4 shows that the heating system in kiln process. It can be seen that the heating system can reach the aim temperature in 1 h when the pellets was added, and then the heating system was homoeothermic. Due to the slow reaction rate between the pellets and coals, the reduction of SnO_2 is the reaction between SnO_2 and CO produced by the Boudouard reaction of carbon instead of coal. The tin was volatilized as SnO according to the reaction when CO % (provided by the coke powder burning) is in the range of 9–15% [4]. Figure 5 presents the effect of roasting temperature on the volatilization ratio of tin from tin-bearing middling. The results indicated that the roasting temperature has an obvious effect on the tin

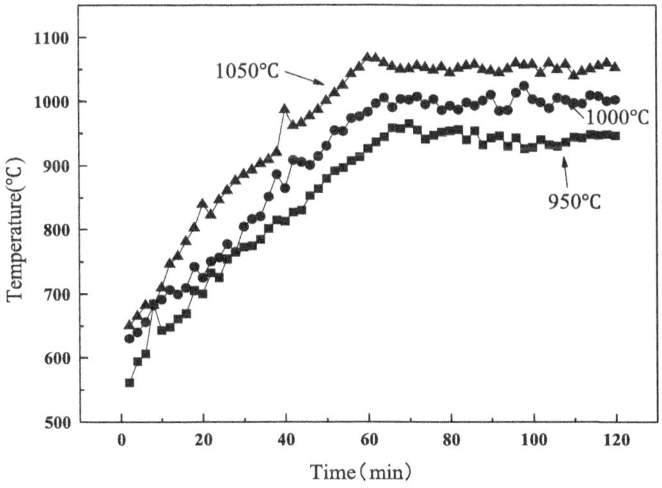

Fig. 4 The heating system of kiln

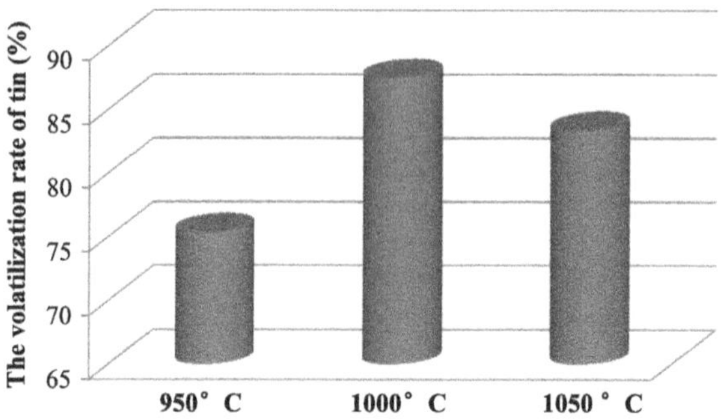

Fig. 5 The effect of roasting temperature on the volatilization of tin

volatilization. The Sn volatilization ratio firstly increased and then decreased slightly with the temperature increased from 950 to 1050 °C. The maximum Sn volatilization ratio of 87.4% is obtained at 1000 °C. The reason for why the Sn volatilization ratio decrease is that the SnO_2 was reduced to Sn according to the reaction (3). To decrease the process energy consumption and improve tin removal rate, residence temperature is fixed at 1000 °C.

$$SnO_2 + CO = SnO + CO_2 \tag{2}$$

$$SnO + CO = Sn + CO_2 \tag{3}$$

Effect of the Size of Pellets and the Roasting Time

SnO_2 was first reduced to SnO and then to metallic Sn with a low concentration of CO in the gaseous phase; the transition to metallic Sn is facilitated by an elevated CO partial pressure. The CO was produced by carbon (Boudouard reaction). The mixing uniformity of coal and the pellets is the main factor for providing suitable reduction conditions. Therefore, it is necessary for studying the effect of the size of the pellets on Sn volatilization ratio. Table 2 shows that the effect of the size of pellets on Sn volatilization. It can be seen that there is a desirable volatilization when the size of pellets is between 5 and 10 mm. The reason for why the size of pellets effecting tin volatilization is that the pellets are easy to segregate with coal. The model of the segregation between coal and pellets is shown in Fig. 6. From the Figure (a), we can see that the coal was segregated in sides and the pellets in centre;

Table 2 The effect of the size of pellets on volatilization

Size of pellets (mm)	The contents of tin in residue (%)	The recovery rate of tin (%)
−5	1.24	77.2
+10	3.59	49.1
5–10	0.88	87.4

Fig. 6 The model of the segregation between coal and pellets

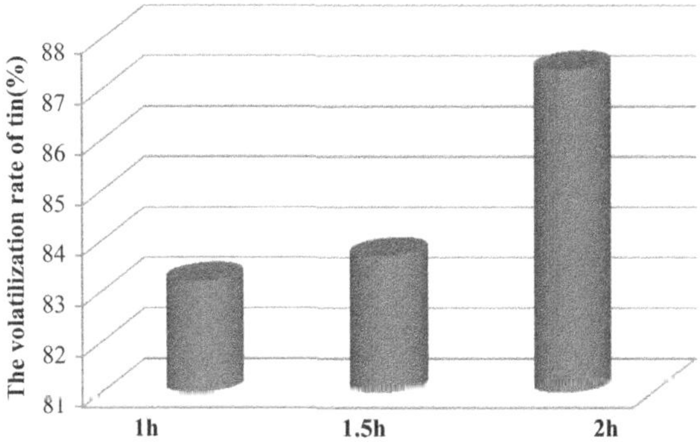

Fig. 7 The effect of reduction time on the Sn volatilization ratio

the Figure (b) shows that the coal was segregated on the surface and the pellets in substrate; the Figure (c) shows that the coal and the pellets have been fully mixed. The Sn volatilization ratio can reach 87.4% when chosen the pellets whose sizes were 5–10 mm.

The effect of reduction time on the Sn volatilization ratio is shown in Fig. 7. The results indicate that the Sn volatilization ratio increases sharply with the time increasing. The tin volatilization ratio can reach 87.4% and the tin contents of residue are 0.88% when the roasting time was 2 h.

Effect of the Ratio of Coal Addition

In the roasting process, the coke powder was selected to reduce the SnO_2 from tin-bearing middling. The ratio of coal addition determines the strength of the reducing atmosphere in the rotary kiln, and the strength of the reducing atmosphere affects the volatilization effect of tin. Table 3 shows the effect of coal addition on tin volatilization. The recovery rate of tin can reach 87.4% when the Kiln tail: Kiln middle: Kiln head (coal addition ratio) was 1:2:7. At the beginning, the reaction was very severe which indicated that the little coke powder was enough to provide a

Table 3 The effect of coal addition on volatilization

Kiln tail: kiln middle: kiln head (coal addition ratio)	The contents of tin in residue (%)	The recovery rate of tin (%)
2:3:5	0.784	80.6
1:2:7	0.628	87.4

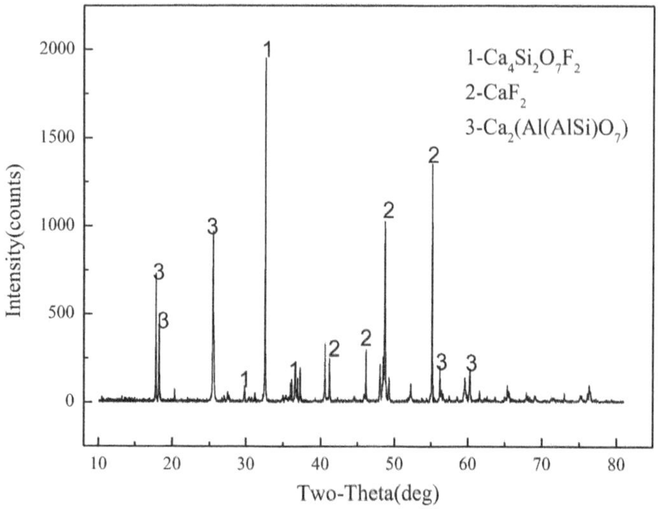

Fig. 8 XRD pattern of residue after reduction roasting in kiln

weak reduction atmosphere. The reaction tends to slow down with the reaction going on and it needs much more coke powder to bring the reduction atmosphere. Therefore, study the coal addition can ensure the reducing atmosphere of the layer in the constant temperature reduction stage. The result could provide a technical reference for the industrial utilization of the tin-bearing middling. Figure 8 was the phase after reduction roasting in kiln. Compared with the raw material, the tin phase after reduction roasting in kiln was vanished, and the CaF_2 still was the main phase. The tin is effectively recovery during reduction roasting in kiln which provided a reference for industrial.

Conclusion

In the reduction process, the coke powder as reductants which ensure the CO% at a lower level due to its weak reactivity with CO_2. The result was showed that the volatilization rate can reach 87.4% and the SnO_2 content in the residue decreased to 0.57 wt%. The desirable conditions, respectively, were coal addition ratio in Kiln tail: Kiln middle: Kiln head with 1:2:7, the proportion of ball and coal was 20 kg: 20 kg and the pellets size was 5–10 mm at 1000 °C for 2 h. These finding provided a technology reference for recovery tin in industrial utilize.

References

1. Kumar Sonker R, Sharma A, Shahabuddin M, Tomar M, Gupta V (2013) Low temperature sensing of NO_2 gas using SnO_2–ZnO nanocomposite sensor. Adv Mater Lett 4:196–201
2. Nanto H, Morita T, Habara H, Kondo K, Douguchi Y, Minamia T (1996) Doping effect of SnO_2 on gas sensing characteristics of sputtered ZnO thin film chemical sensor. Sens Actuators
3. Platteeuw JC, Meyer G (1956) The system tin-oxygen. Trans Faraday Soc 52:1066–1073
4. Zhang Y, Liu B, Su Z, Chen J, Li G, Jiang T (2015) Volatilization behavior of SnO_2 reduced under different CO–CO_2 atmospheres at 975–1100 °C. Int J Miner Process 144:33–39
5. Li G, You Z, Zhang Y, Rao M, Wen P, Guo Y, Jiang T (2014) Synchronous volatilization of Sn, Zn, and As, and preparation of direct reduction iron (DRI) from a complex iron concentrate via CO reduction. JOM 66:1701–1710
6. Li L, Yu Y, Sang X (2015) Removing tin from tin-bearing iron concentrates with sulfidation roasting using high sulfur coal. ISIJ Int
7. Li L, Yu Y, Zhang R (2015) Reduction roasting of tin-bearing iron concentrates using pyrite. ISIJ Int
8. Zhang YB, Li GH, Jiang T, Guo YF, Huang ZC (2012) Reduction behavior of tin-bearing iron concentrate pellets using diverse coals as reducers. Int J Miner Process 110–111:109–116
9. Kim B-S, Lee J, Yoon H-S, Kim S-K (2011) Reduction of SnO_2 with hydrogen. Mater Trans 52:1814–1817
10. Zhang YB, Su ZJ, Zhou YL, Li GH, Jiang T (2013) Reduction kinetics of SnO_2 and ZnO in the tin, zinc-bearing iron ore pellet under a 20%CO–80%CO_2 atmosphere. Int J Miner Process 124:15–19
11. Su Z, Zhang Y, Liu B, Chen Y, Li G, Jiang T (2017) Effect of CaF_2 on the reduction volatilization of tin oxide under CO–CO_2 atmosphere. Miner Process Extr Metall Rev 38:207–213

Isothermal Sulphation Roasting of Nickel Sulphide Minerals in a Static Air Atmosphere

Lizhen Wei, Caixiang Yu, Guangshi Li, Xiaolu Xiong, Hongwei Cheng, Qian Xu and Xionggang Lu

Abstract Sulphation roasting offers the potential to simultaneously extract valuable metals from multi-metal sulphide minerals. In this study, the influences of roasting temperature and sodium sulphate addition on the sulphation of valuable metals were investigated under isothermal condition in a static air atmosphere. Additionally, the reaction mechanism during sulphation roasting process was characterized by thermogravimetry and differential scanning calorimetry (TG-DSC), X-ray diffraction (XRD) and scanning electronic microscopy (SEM). The results show that the optimum sulphation roasting temperature range is 600–700 °C. The leach extraction efficiencies of Ni, Co and Cu exceeded 60% when about 10 wt% sodium sulphate was added during roasting. The role of sodium sulphate as a sulphation promoter is attributed to an intermediate phase that acts as a reservoir of SO_2.

Keywords Nickel · Sulphide minerals · Sulphation roasting · Sodium sulphate · Leaching

Introduction

Nickel, copper and cobalt are regarded as important metal materials in modern industry and widely used in electrical engineering, mechanical manufacturing, construction, transportation and other fields. Currently, most of these non-ferrous

L. Wei · C. Yu · G. Li (✉) · X. Xiong · H. Cheng · Q. Xu · X. Lu (✉)
School of Materials Science and Engineering, Shanghai University, Shanghai 200444, People's Republic of China
e-mail: lgs@shu.edu.cn

X. Lu
e-mail: luxg@shu.edu.cn

G. Li · H. Cheng · X. Lu
State Key Laboratory of Advanced Special Steel & Shanghai Key Laboratory of Advanced Ferrometallurgy, Shanghai University, Shanghai 200444, People's Republic of China

© The Minerals, Metals & Materials Society 2019

T. Jiang et al. (eds.), *10th International Symposium on High-Temperature Metallurgical Processing*, The Minerals, Metals & Materials Series,
https://doi.org/10.1007/978-3-030-05955-2_21

metals are pyrometallurgically produced by smelting sulphide minerals [1–4]. With the depletion of high-grade sulphide resources, low-grade polymetallic mineral resources are being developed, and the traditional pyrometallurgical process is not suited for the treatment of these complex materials. Previous researchers have described various combinations of roasting and leaching processes [5–10]. Roasting mainly includes oxidation roasting, sulphation roasting and chlorination roasting. Sulphation roasting makes use of the contained sulphur to convert valuable non-ferrous metals into water-soluble sulphates [11]. Sodium sulphate has been used to enhance the sulphation of non-ferrous metals during the roasting process [12]. In this study, nickel concentrate was roasted with the addition of sodium sulphate under isothermal condition in a static air atmosphere.

Experimental Methods

Raw Materials

The raw material used in this study was nickel sulphide flotation concentrate from Jinchuan Group Ltd. (Gansu, China). The main chemical components of the sample are shown in Table 1. Figure 1 is the X-Ray diffraction (XRD) pattern of the nickel concentrate sample. Phase identification analysis shows that the main valuable metal phases of the concentrate sample are pentlandite ($(Ni,Fe)_9S_8$), chalcopyrite ($CuFeS_2$), pyrrhotite (Fe_7S_8), pyrite (FeS_2) and small amount of cubanite ($CuFe_2S_3$). The gangue minerals mainly include serpentine ($Mg_3Si_2O_5(OH)_4$), ferromagnesite ($Mg_xFe_{3-x}O_4$), quartz (SiO_2) and talc ($Mg_3(OH)_2Si_4O_{10}$). The chemical reagent used in the test was commercially pure sodium sulphate.

Characterization Methods

The composition of the sulphide concentrate sample was determined by XRF (XRF-1800, SHIMADZU LIMITED) with full spectrum scanning at 8°/min. The crystal phase composition of the raw materials and leaching residue was characterized by XRD (D8 ADVANCE, Bruker). The XRD data were collected at 50 kV/50 mA and with a 2-theta range from 10 to 90° (0.02°/step, 0.3 s/step). Thermoanalysis of roasting process was performed using thermogravimetric and

Table 1 Chemical composition of the nickel sulphide flotation concentrate sample (mass fraction, %)

Fe	O	S	Ni	Mg	Cu	Si	Al	Co	K
28.25	22.43	19.61	8.73	7.25	6.51	6.46	0.38	0.24	0.12

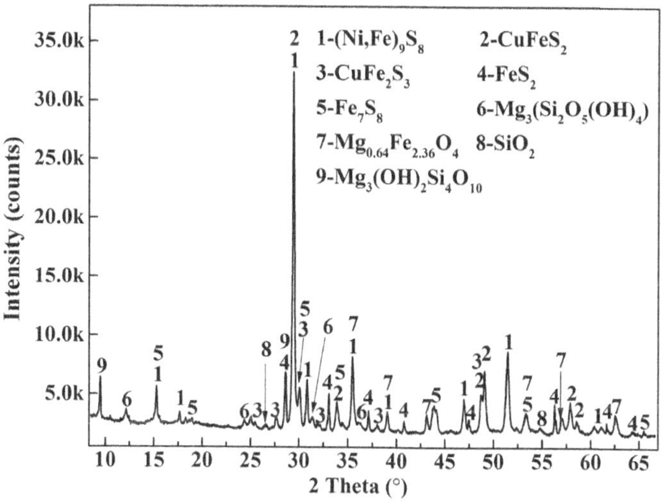

Fig. 1 XRD pattern of the flotation nickel sulphide concentrate sample

differential scanning calorimetry (TG-DSC, STA 449 F3, NETZSCH). The thermoanalysis was conducted under air atmosphere with a flow rate of 40 mL/min and heated to 1000 °C with a heating rate of 10 °C/min. The concentration of metal ions in the leachate was analysed by using an inductively coupled plasma optical emission spectrometer (ICP-OES, Optima 7300DV, PerkinElmer). The microstructure of the calcine samples was characterized by a tungsten filament scanning electron microscope (SEM, SU-1500, HITACHI).

Experiment Procedure

In preparation for the roasting experiments, the nickel sulphide concentrate sample was dried at 90 °C for 24 h. Dried material that passed through a 200-mesh sieve was designated as A#. A fraction of A# was separated for XRF and XRD tests. One hundred (100) g of A# and 10 g of sodium sulphate was ground and mixed to uniformity with an agate mortar; the resulting mixture was designated as B#. Twenty (20) mg of B# was separated for TG-DSC analysis.

Table 2 presents the experimental scheme of roasting and leaching. In each experiment, a 5-g sample (A# for I, II, III and IV tests; B# for VI test) was poured into corundum crucible and placed in a muffle furnace. The calcined product was placed in a 500 mL Erlenmeyer flask and immersed in a solution of 10:1 in a liquid–solid ratio. The leach extraction efficiencies were calculated on the basis of the Fe, Co, Ni, Cu and Mg concentrations in the leachate

Table 2 Roasting/L leach experiment parameters

Oxidation roasting			Sulphation roasting			Water leaching		
Category	Temperature (°C)	Time (h)	Dosage of Na_2SO_4 (%)	Temperature (°C)	Time (h)	Dosage of Na_2SO_4 (%)	Temperature (°C)	Time (h)
I	600	2	–	–	–	10	60	2
II	600	2	–	–	–	–	60	2
III	700	2	–	–	–	10	60	2
IV	700	2	–	–	–	–	60	2
V	600	2	10	700	2	–	60	2
VI	–	–	10	700	2	–	60	2

Results and Discussion

TG-DSC Analysis

In order to investigate the reaction mechanism of roasting process, the TG-DSC analyses were carried out on the A# and B# materials. The TG-DSC curves are presented in Fig. 2. The exotherm peaks of A# at 420 and 490 °C correspond to the oxidation reactions of sulphides. The TG curve of A# indicates weight gain after 420 °C, followed by rapid weight loss above 570 °C. This outcome indicates that the initial stage of oxidation is a weight-increasing reaction for the formation of sulphates [13, 14]. However, only one oxidation exothermal peak is evident with B#, which is obviously shifted to high temperature at about 455 °C. Moreover, the temperature range of weight gain reactions of B# is significantly wider than that of A#, especially in the temperature range from 570 to 700 °C (the shaded area in Fig. 2); the weight

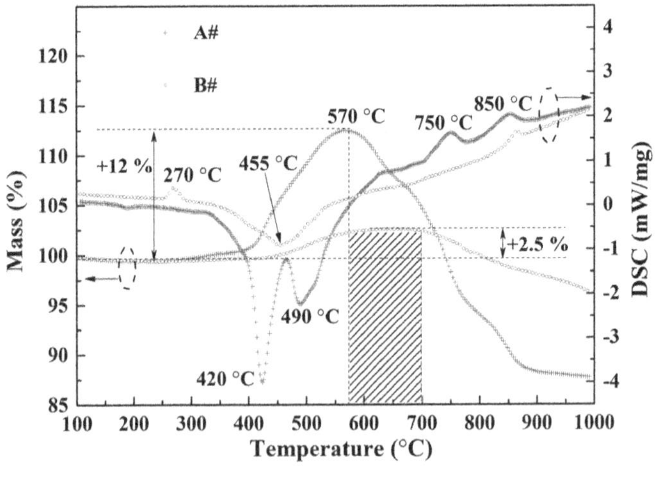

Fig. 2 TG–DSC curves of A# and B# under 40 mL/min of air and a 10 °C/min heating rate

gain of the TG curve of B# remains basically unchanged, and the weight loss of the A sample begins. The data show that sodium sulphate promotes non-ferrous metal sulphation by broadening the reaction temperature range.

Sulphation Roast-Water Leach Process

According to the experiment scheme defined in Table 2, the leach extraction efficiencies for the metals of interest are summarized in Fig. 3. Comparison of tests I and II, or tests III and IV, revealed that sodium sulphate addition during roasting had no effect on the water leach extraction efficiencies. The Ni and Mg extractions in test II were significantly lower than that in test IV, and the leaching yields of Fe, Co and Cu in test II were slightly higher than that in test IV. These findings indicate that the sulphating reaction temperatures of Ni and Mg were higher than that of Fe, Co and Cu [12]. The leaching rates of Ni and Mg in test VI were notably higher than seen in test IV, suggesting that sodium sulphate plays a greater role in promoting Ni and Mg sulphation. In particular, the highest leaching yields were realized in test V. Oxidation–sulphation two-stage roasting process can greatly promote the sulphating reactions of non-ferrous metals [15].

The Characterization of Calcine and Leaching Residue

The SEM topographies of the calcined product of tests I, III, V and VI are shown in Fig. 4.a, b, c and d, respectively. Figures 4a and b reveal a typical loose porous morphology that may attributed to the outward escape of SO_2 gas from inner

Fig. 3 Comparison of the leach extraction efficiencies for the tests outlined in Table 2

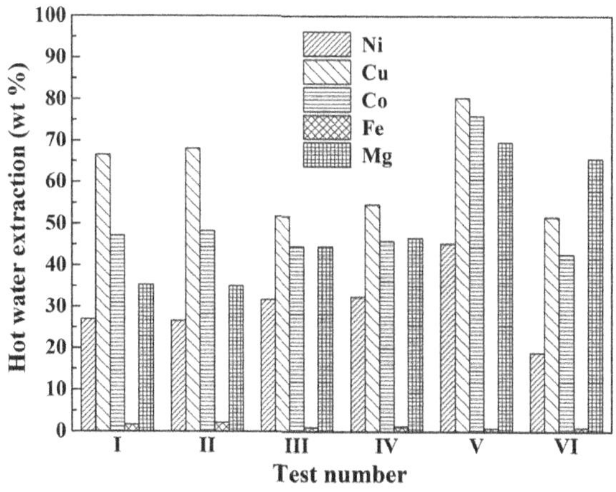

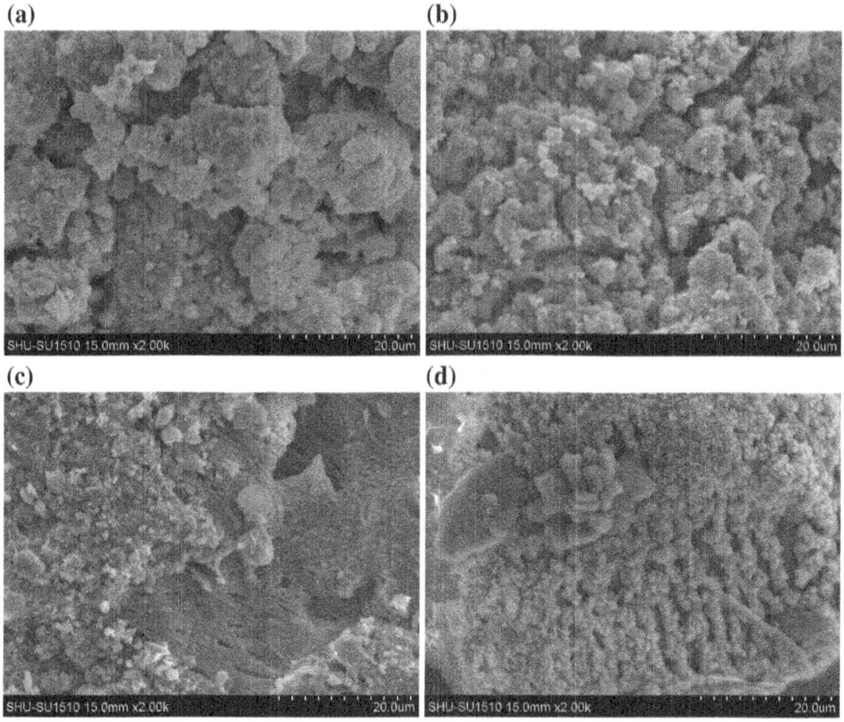

Fig. 4 SEM images of calcine samples produced under different roasting conditions: figures a, b, c and d show the products of test I, III, V and VI, respectively

reaction interface during the oxidation of sulphides. However, Figs. 4c and d show significant liquid phase sintering morphology. This morphology may be due to the fact that after the addition of sodium sulphate, an intermediate liquid phase was formed during the roasting process. The composite mineral's sulphate liquid phase can be used as a buffer for SO_2 to stabilize the partial pressure of SO_2 and then to achieve high-efficiency sulphation of valuable metals [6, 12, 16].

As shown in Fig. 5, the leaching residues are mainly composed of Fe_2O_3 and (Ni, Cu, $Fe)_xFe_{3-x}O_4$. Since the roasting temperature of test I was lower than test III, unoxidized NiS phase was also detected in the leaching residue of test I. Comparing the XRD patterns of the leach residue samples from tests V and VI, it can be seen that the two-stage roasting with adding sodium sulphate can greatly inhibit the formation of (Ni, Cu, $Fe)_xFe_{3-x}O_4$, which is consistent with the higher leaching rate of the valuable metals in test V (Fig. 3).

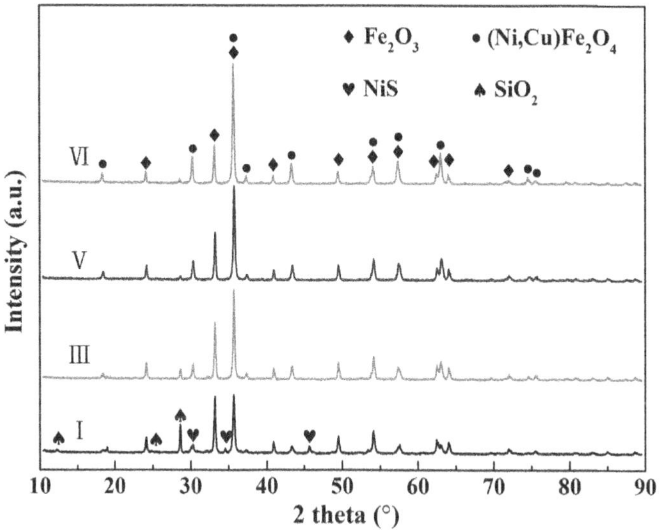

Fig. 5 XRD pattern of the calcine samples produced under different roasting conditions

The Effect of Addition of Sodium Sulphate on the Sulphation of Metals

Roasting-water leaching experiments were carried out to investigate the influence of the dosage of sodium sulphate on the sulphation roasting of nickel sulphide concentrates. Twenty-five (25) g of sample A# was divided into five equal portions and oxidized at 600 °C for 2 h in air. The five calcine samples were mixed to uniformity with different dosages of sodium sulphate. The mixtures were roasted at 700 °C for 2 h in air; finally, the calcine samples were leached with a water–solid ratio of 10:1 at 60 °C for 2 h; the metal ion concentration of the leachate was analysed to calculate the leach extraction efficiencies reported in and the leaching rate of metals were calculated as shown in Fig. 6. It is obvious that the leaching rate of Ni, Co and Cu gradually increases with the increasing of sodium sulphate addition and remains unchanged after the dosage exceeds 10 wt%. The results show that the addition of appropriate amount of sodium sulphate into the pre-oxidized sulphide nickel concentrate can significantly increase the formation of water-soluble phases of non-ferrous metals (Ni, Co and Cu), and the optimum dosage in this work is about 10 wt% [16].

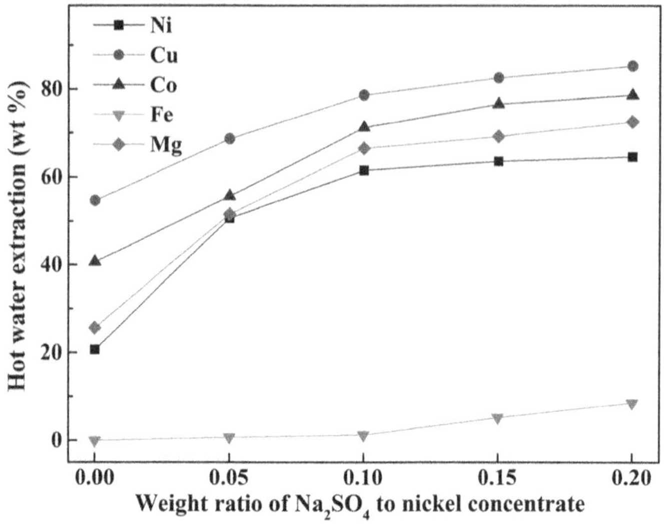

Fig. 6 Effect of the dosage of sodium sulphate on the formation of water-soluble species

Conclusions

The process of sulphation roasting-water leaching of nickel sulphide concentrate under air atmosphere was investigated. The nickel sulphide concentrate was oxidized and sulphated in two stages, and about 10 wt% sodium sulphate was added during the sulphation roasting process. After leaching, the leaching rates of Ni, Co and Cu could reach 60%, and the leaching rate of Fe was controlled below 1%. This process combines the advantages of pyrometallurgy and hydrometallurgy to achieve green, efficient and simultaneous extraction of non-ferrous metals.

Acknowledgements This work is financially supported by Steel Joint Research Foundation of National Natural Science Foundation of China–China Baowu Iron and Steel Group Co. Ltd. (Grant No. U1860203), National Natural Science Foundation of China (Grant No. 51576164) and the CAS Interdisciplinary Innovation Team.

References

1. Shuai G (1999) Development of the nickel metallurgy in Jinchuan. China Nonferrous Metall (S1):10–11, 45–122
2. Tang B (1988) Energy-saving in the copper and nickel metallugy. Energy Metall Ind 04:18–21
3. Wenshi Xie, Zhongsheng Li, Wendong Yang, Yunzhi Zhao, Jingang Wang (2007) Discussion of prsvent situations and development strategy of copper metallurgy technology. Non-Ferrous Min Metall 06:68–71

4. Cui H, Zhou M (1995) Development of metallurgical technology of copper in China. Min Metall 03:120–126
5. Song N, Liu C (2005) A new technique to make copper removing iron from chalcopyrite. Nonferrous Metals Eng 02:84–87
6. Ying B, Zheng X (1991) The effects of added sodium sulfate on the sulfation roasting of refined multi-metal sulfide ore. J Zhejiang Sci-Tech Univ (Natural Sciences Edition)(1):55–59
7. Weijie Z, Yuyie Z, Shuxia C, Yun Z, Lin Z (1994) Experimental studies on roasting of copper-cobalt complex sulphides. Chin J Process Eng 1:57–60
8. Yu Y, Tian Y, Zhu C, Sun C, Xu Q (2016) Reaction behaviour of CuS during ammonium roasting. Chin J Nonferrous Metals 26(3):681–688
9. Liu X, Hu W, Li H, Feng Y (2014) Leaching process and mechanism of high magnesium low-grade nickel sulfide ore. Chin J Nonferrous Metals (06):1599–1606
10. Liu X, Feng Y, Li H, Cai Z, Yang Z (2012) Leaching of valuable metals from nickel pyrrhotite by ammonium sulfate roasting method. Chin J Nonferrous Metals (05):1520–1526
11. Yu D, Utigard TA (2012) TG/DTA study on the oxidation of nickel concentrate. Thermochim Acta 533:56–65
12. Yu D, Utigard TA, Barati M (2013) Fluidized bed selective oxidation-sulfation roasting of nickel sulfide concentrate: part II. Sulfation roasting. Metall Mater Trans B 45(2):662–674
13. Dunn J, Kelly C (1980) A TG/MS and DTA study of the oxidation of pentlandite. J Therm Anal Calorim 18(1):147–154
14. Li G, Cheng H, Xiong X, Lu X, Xingli Z, Qian X (2017) In-situ XRD and EDS method study on the oxidation behaviour of Ni-Cu sulphide ore[J]. Sci Rep 7(1):3212
15. Yu D, Utigard TA, Barati M (2013) Fluidized bed selective oxidation-sulfation roasting of nickel sulfide concentrate: part I. oxidation roasting. Metall Mater Trans B 45(2):653–661
16. Li G, Cheng H, Chen S, Lu X, Xu Q, Lu C (2018) Mechanism of Na_2SO_4 promoting nickel extraction from sulfide concentrates by sulphation roasting–water leaching. Metall Mater Trans B 49(3):1–13

Thermal Transformations of Main Components in Molybdenite Concentrates Under SO_2-Containing Atmosphere

Hu Sun, Guanghui Li, Junjie Yu, Jun Luo, Mingjun Rao and Tao Jiang

Abstract Oxidative roasting-ammonia leaching is the most popular method to extract Mo from molybdenite concentrates. However, the pyrometallurgical process has long suffered from the exhaust gas treatment for acid-making using low-concentration SO_2. Technically, gas circulation can work as an effective means to enrich SO_2 in exhaust gas and meantime to reduce overall emissions. However, with SO_2-enriched atmosphere, roasting behaviours of the components in concentrates would be changed, thereby influencing the recovery of Mo. In this study, both thermodynamic and experimental studies were carried out to explore the thermal transformations of main components in molybdenite concentrates under SO_2-containing atmosphere. Thermal behaviours including oxidation of sulphides and formation of molybdates and sulphates were discussed. It was found that SO_2 has a minor effect on sulphides oxidation, and raising the temperature and proportion of SO_2 and O_2 are effective means to extract the associated Mo from molybdates, which demonstrate the potential of applying exhaust gas circulation (EGC) technology in oxidative roasting of molybdenite concentrates.

Keywords Molybdenite concentrate · Oxidative roasting · Exhaust gas circulation SO_2 treatment

Introduction

In global molybdenum market, China has long been the leading producer to supply molybdenum raw materials for making ferromolybdenum, Mo metal powder, and various chemical products such as pure MoO_3 and molybdates of ammonia and sodium. [1] The Mo content in molybdenite concentrates varies from below 30% to

H. Sun · G. Li (✉) · J. Yu · J. Luo · M. Rao · T. Jiang
School of Minerals Processing & Bioengineering, Central South University, Changsha 410083, Hunan, China
e-mail: liguangh@csu.edu.cn

© The Minerals, Metals & Materials Society 2019

T. Jiang et al. (eds.), *10th International Symposium on High-Temperature Metallurgical Processing*, The Minerals, Metals & Materials Series,
https://doi.org/10.1007/978-3-030-05955-2_22

over 57%, and common impurities include silicon, calcium, iron, lead and copper. As of now, most of molybdenite concentrates are initially processed into technical-grade MoO_3 through oxidative roasting [2], which has advantages of simple technology and equipment, high oxidation efficiency and production capacity. However, the pyrometallurgical process for oxidizing MoS_2 into MoO_3 at 500–700 °C inevitably produces large amounts of SO_2-containing gas. To avoid air pollution, emissions are usually be made into sulphuric acid, but low-concentration SO_2 exhaust generally produced at mid-late stage of oxidative roasting brings tough problems to both technics and costs of acid-making [2, 3].

In steel-making industry, EGC technology has been intensively explored to reduce environmental costs and meantime to promote the heat utilization efficiency [4]. Referring to this, we think it is also feasible to recycle the exhausts generated from oxidative roasting of molybdenite concentrates. From perspectives of environmental protection and energy saving, low-concentration SO_2 exhaust could be easily enriched during circular use and turns out to be ideal supply for making sulphuric acid. The pollution source of SO_2-containing gas could be converted to value-added product, and the heat that ought to be taken away by hot current is also largely saved due to the decrease of total emissions. However, SO_2 enrichment must come with oxygen consumption, so gas compositions would be changed, which might exert significant influence on thermal behaviours of molybdenite concentrates.

Transformations of molybdenite concentrates during oxidative roasting have been extensively reported. During thermal treatment, MoS_2 is oxidized into ammonia-soluble MoO_3, and impurities such as Ca, Fe, Cu and Pb are turned into corresponding molybdates [5, 6]. However, most laboratory experiments in terms of oxidative roasting are performed under air atmosphere and the generated SO_2 could not stay among the concentrates but be immediately swept out by air flow. Actually, the practical gas phase during oxidative roasting in multi-hearth furnace, rotary kiln or fluidized bed comprises SO_2 with concentration varying from below 0.5–5%, which is distinguished from the gas compositions in existing work. When EGC technology is adopted, SO_2 in exhaust would be further concentrated, so key issues in terms of SO_2 influence on transformation behaviours of components in raw ore are urgently to be figured out.

Herein, both thermodynamic and experimental studies were carried out to explore the thermal transformations of main components in molybdenite concentrates under SO_2-containing atmosphere. Roasting behaviours including the oxidation of Mo, Fe, Pb, Cu sulphides and the formation of Ca, Fe, Pb, Cu molybdates, sulphates were exhaustively discussed. The total work shows encouraging results to support the application of EGC technology in oxidative roasting of molybdenite concentrates and also provide new insights for optimizing the existing Mo extraction process.

Materials and Methods

Thermodynamic studies were performed with software of HSC 6.0 using "Reaction Equations" and "TPP-Diagrams" modules. Gibbs free energy changes varying with temperatures in standard state and predominance diagrams were calculated to evaluate the occurrence tendency of reactions.

Pure reagents including MoO_3, $CaCO_3$, Fe_2O_3, PbO, CuO (Aladdin, 99.5%) and gases including N_2, O_2, SO_2 with purity over 99.99% were used for experimental work. Given that MoO_3 is volatile during roasting, each time, MoO_3 with 10 mol% excess (referring to the normal molybdate) was mixed with another metal oxide or carbonate, and ground in an agate mortar for 10 min. The dispersed mixtures were moistened with 3.0 wt% ethanol, pressed into cylindrical pellets with diameter 6 mm and height 10 mm, and dried for roasting experiments.

Pellet roasting tests were performed in a shaft furnace assembled with gas supply system, as seen in Fig. 1. During each test, a quartz crucible with perforated bottom lining loading the pellets was charged into a quartz tube and then directly put into the high-temperature furnace under the protection of N_2. When the temperature stabilized, gas flows consisting of different proportions of N_2–O_2–SO_2 with total flux 4000 sccm were inducted in. After roasting for 90 min, nitrogen was resupplied and immediately the quartz tube was pulled out along with the roasted pellets. To ascertain the phase compositions of products, the roasted pellets were cooled and grinded into powdery samples.

The roasted samples were identified by X-Ray Powder Diffraction (XRD, BRUKER, Advance D8) with Cu–Kα radiation (λ = 1.54 Å) performed at a scanning rate of 8° 2θ min^{-1}. Sulphur contents of samples were measured using Infrared Carbon-sulfur Analyzer (Leco Corporation, CS844).

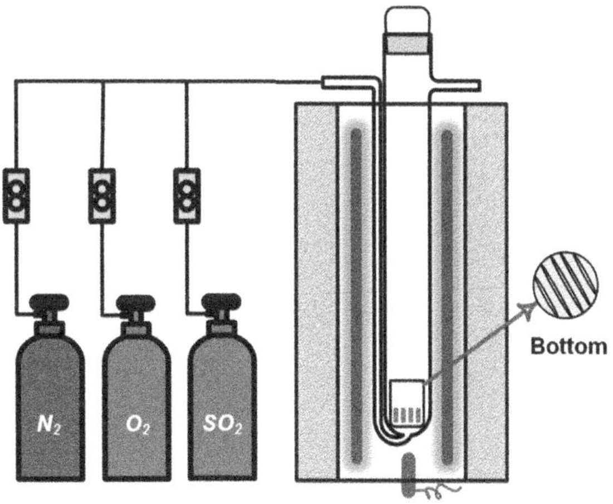

Fig. 1 Set-up for pellet roasting

Results and Discussion

Oxidation of Sulphides

In molybdenite concentrates, molybdenum and other impurity elements usually exist in form of sulphide minerals including molybdenite, galena, pyrite, chalcopyrite and chalcocite. During roasting at high temperatures with air supply, metal sulphides are oxidized as oxides with SO_2 being released. Standard Gibbs free energy changes of sulphides oxidation as function of temperature are shown in Fig. 2. It is seen that oxidation reactions of PbS and Cu_2S (reactions 1 and 2) have the lowest occurring tendency, but still the corresponding standard Gibbs free energy changes reach below −300 kJ mol^{-1}. Standard Gibbs free energy changes of molybdenite oxidation (reaction 3) at 500–700 °C also reach below −800 kJ mol^{-1}, indicating that the reaction is easy to happen. Generally, oxidation of sulphides varies in small range of reacting tendency with temperature change and exhibits absolute feasibility in thermodynamics.

Predominance diagram of Mo-O-S system determined by partial pressures of SO_2 and O_2 at 600 °C is depicted in Fig. 3a. It is seen that the oxidation of MoS_2 into MoO_3 seems to be accompanied by molybdenum oxides with intermediate valence including MoO_2, Mo_4O_{11}, Mo_8O_{23} and Mo_9O_{26}. Existing work reported that Mo_4O_{11}, Mo_8O_{23}, Mo_9O_{26} can only be formed from MoO_3 reduction with H_2 [7], but MoO_2 could be produced from oxidation of MoS_2 or combination between MoS_2 and MoO_3 during oxidative roasting [8]. According to the predominance diagram, transformations of MoS_2 into MoO_2 and Mo_9O_{26} into MoO_3 only require very low oxygen proportion over 10^{-15} and $10^{-9.}$ respectively, while SO_2 proportion has no effect on oxidation products when oxygen proportion reach over 10^{-15}, which means the oxidation of molybdenite under high-concentration SO_2 atmosphere is entirely feasible.

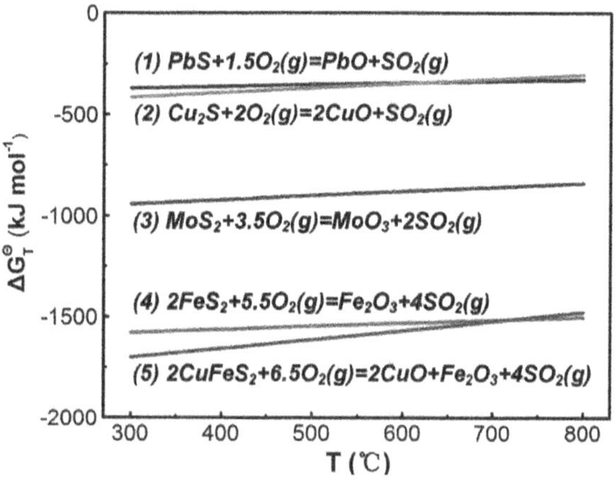

Fig. 2 Standard Gibbs free energy changes of sulphides oxidation as function of temperature

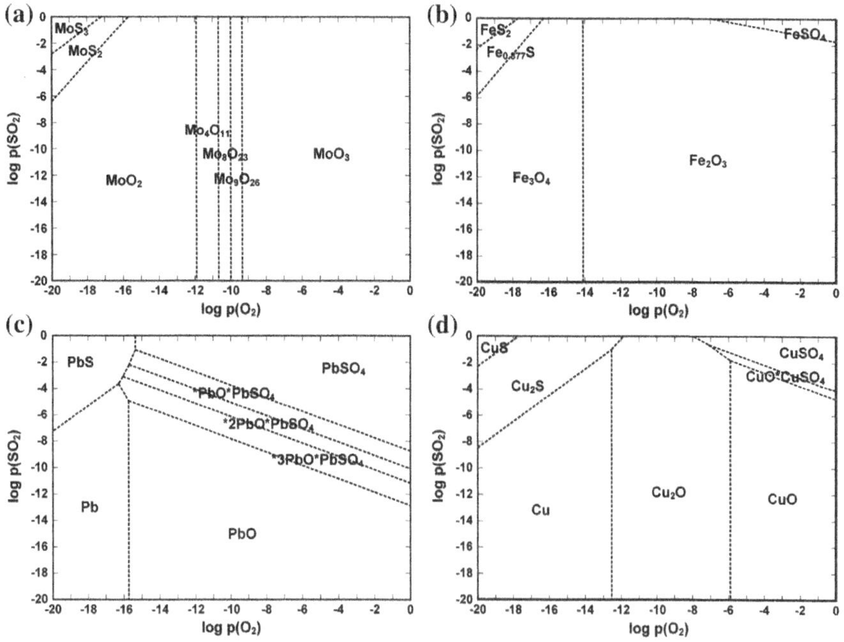

Fig. 3 Predominance diagrams of **a** Mo-O-S system, **b** Fe–O–S system, **c** Pb–O–S system, **d** Cu–O–S system at 600 °C

Predominance diagrams of Fe–O–S system, Pb–O–S system and Cu–O–S system at 600 °C are also displayed in Fig. 3b, c, d. Similarly, the transformations of FeS_2, PbS, Cu_2S into Fe_2O_3, PbO, CuO only call for oxygen proportion over 10^{-14}, 10^{-15}, 10^{-6}, respectively, and partial pressure of SO_2 also exerts less influence on the oxidation of FeS_2, PbS, Cu_2S. However, unlike the Mo-S-O system, there exists possibility that SO_2 might combine with high valence state oxides of Fe, Pb, Cu to form corresponding sulphates under high partial pressure of SO_2 and O_2. According to the predominance area of each sulphate, it could be confirmed that $PbSO_4$ is more stable than $FeSO_4$ and $CuSO_4$ at 600 °C and may be formed during roasting under high-concentration SO_2 atmosphere.

Based on above thermodynamic analysis, it can be concluded that sulphides of Mo, Fe, Pb, Cu are easy to be oxidized even under high-concentration SO_2 atmosphere, which provides a theoretical support for the application of EGC technology in oxidative roasting of molybdenite concentrates. The original exhaust gas with SO_2 content below 5% could be reused for several times since oxygen proportion is high enough for sulphides oxidation.

Formation of Molybdates

After being oxidized, sulphides of Mo, Fe, Pb, Cu are transformed into corresponding oxides first, and then combination reactions between MoO_3 and other metal oxides may happen. In addition to Fe, Pb and Cu, Ca existing in form of calcite is also a major impurity in all types of molybdenite concentrates. During roasting, $CaCO_3$ may decompose into CaO first or stay in original form. Thus, either $CaCO_3$ or CaO may combine with MoO_3 to produce $CaMoO_4$. Standard Gibbs free energy changes of $CaCO_3$ decomposition and molybdates formation as function of temperature are shown in Fig. 4. The formation of iron molybdate is not included for the vacancy of relevant thermodynamic data. It is seen that standard Gibbs free energy changes of $CaCO_3$ decomposition (reaction 6) at 300–800 °C appear as total positive numbers, which means calcite could not decompose into CaO and CO_2 during oxidative roasting. However, in the presence of MoO_3, $CaCO_3$ could be directly transformed into $CaMoO_4$ (reaction 8) and meantime release CO_2. Similarly, PbO and CuO are also able to combine with MoO_3 to form $CuMoO_4$ and $PbMoO_4$ (reactions 7 and 9).

Formations of molybdates from solid–solid reactions between MoO_3 and $CaCO_3$, Fe_2O_3, PbO, CuO were also investigated using pure reagents. XRD patterns of the roasted mixtures under conditions of 600 °C, 80% N_2-20% O_2 atmosphere are shown in Fig. 5 along with the standard PDF cards of $CaMoO_4$, $Fe_2Mo_3O_{12}$, $PbMoO_4$, $CuMoO_4$. As expected, XRD peaks of the four products match well with standard cards of corresponding molybdates and no peaks relating to $CaCO_3$, Fe_2O_3, PbO, CuO are found, indicating that they have been totally transformed into molybdates within 90 min. In practical system, the formation of lead and copper molybdates would lower the melting point of local positions where the impurities are occasionally concentrated, and produce compact sinters [5], which is adverse to the inner diffusion of oxygen. Not only that, the associated molybdenum in molybdates of calcium, lead, iron is hard to be leached out by

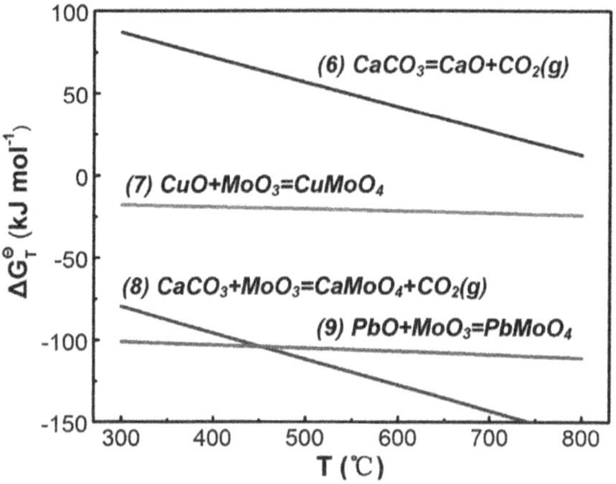

Fig. 4 Standard Gibbs free energy changes of $CaCO_3$ decomposition and molybdates formation as function of temperature

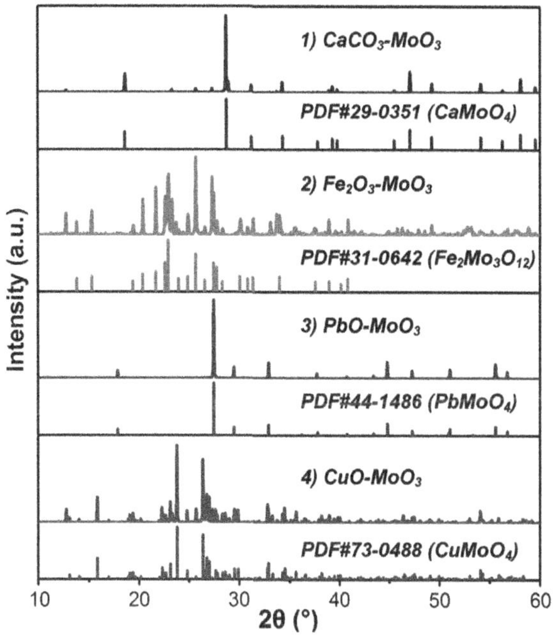

Fig. 5 XRD patterns of roasted mixtures of MoO_3 with $CaCO_3$ (1#), Fe_2O_3 (2#), PbO (3#) and CuO (4#) under conditions of 600 °C, 80% N_2-20% O_2 atmosphere, and standard PDF cards of $CaMoO_4$, $Fe_2Mo_3O_{12}$, $PbMoO_4$, $CuMoO_4$

ammonia water, and thus requires extra acid pickling process to transform it into molybdic acid, which is also a corrosive process for reactors due to use of strong acids such as HCl or HNO_3 [3].

Formation of Sulphates

Above research has proven that molybdates of $CaMoO_4$, $Fe_2Mo_3O_{12}$, $PbMoO_4$, $CuMoO_4$ are easy to be formed from solid–solid reactions, but sulphates may also be produced since the gas phase comprises both SO_2 and O_2. Herein, standard Gibbs free energy changes of sulphates formation from molybdates as function of temperature are shown in Fig. 6. Relevant reactions are listed as below:

$$2CaMoO_4 + 2SO_2(g) + O_2(g) = 2CaSO_4 + 2MoO_3 \tag{10}$$

$$2PbMoO_4 + 2SO_2(g) + O_2(g) = 2PbSO_4 + 2MoO_3 \tag{11}$$

$$2CuMoO_4 + 2SO_2(g) + O_2(g) = 2CuSO_4 + 2MoO_3 \tag{12}$$

The reaction involving iron molybdate and sulphate is still excluded for the lack of data. It is seen that standard Gibbs free energy changes of transformations of Ca, Pb, Cu molybdates into corresponding sulphates (reactions 10, 11 and 12) appear as

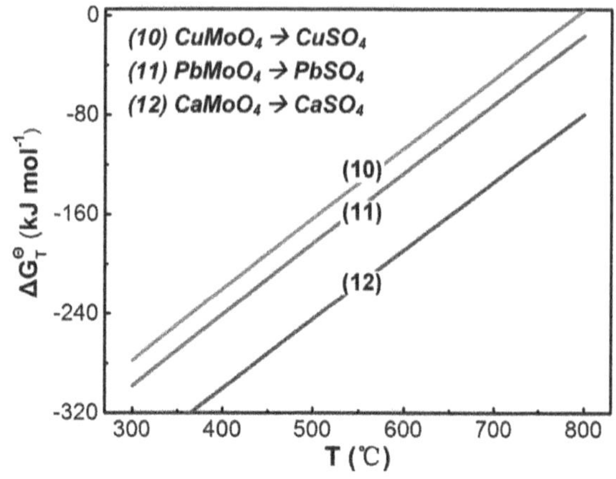

Fig. 6 Standard Gibbs free energy changes of sulphates formation from molybdates as function of temperature

total negative numbers at 500–700 °C, suggesting that all the reactions are able to happen in theory. Comparatively, calcium and lead sulphates are easier to be produced than copper molybdate, which is consistent with the results in Fig. 3c, d. With temperature rising, standard Gibbs free energy changes also increase, indicating a drop of reaction tendency. Nevertheless, from kinetics' perspective, the thermal motion of microscopic particles is also activated, which contributes to faster reaction rate.

Experiments in terms of effects of temperature and gas proportion on sulphates formation were conducted. XRD patterns of roasted mixtures of MoO_3-$CaCO_3$ and standard PDF cards of $CaSO_4$, MoO_3, $CaMoO_4$, $CaCO_3$ are shown in Fig. 7. The positions of major peaks in standard cards are marked with imaginary line for identifications. Products 5#, 6#, 7# were obtained under 80%N_2-10% O_2-10% SO_2 atmosphere at varying temperatures. It is observed that product 5# roasted at 500 °C comprises $CaMoO_4$ and much unreacted $CaCO_3$, whereas products 6# and 7# roasted at higher temperatures consist of both $CaSO_4$ and $CaMoO_4$, suggesting that the formation of calcium molybdate and sulphate can be activated by raising temperature. Products 8#, 9#, 10# were obtained at 600 °C under varying gas proportions of N_2–O_2–SO_2. It is seen that peaks relating to $CaSO_4$ are only not detected in product 8# with low SO_2 proportion of 1.5%. Except this, the intensity of peaks relating to $CaSO_4$ and MoO_3 in products 6#, 9#, 10# increases with the rise of SO_2 and O_2 proportion over the peaks matching $CaMoO_4$. Obviously, raising the proportions of SO_2 and O_2 is also an effective means to promote the transformation of $CaMoO_4$ into $CaSO_4$, whereby associated Mo is released as MoO_3 from the molybdate. Besides, it is noticed that the formation of $CaSO_4$ seems slower than $CaMoO_4$. One possible reason is that roasting experiments were conducted using pelletized raw materials, which could help the occurrence of solid–solid reactions but is adverse to inner diffusion of SO_2 and O_2.

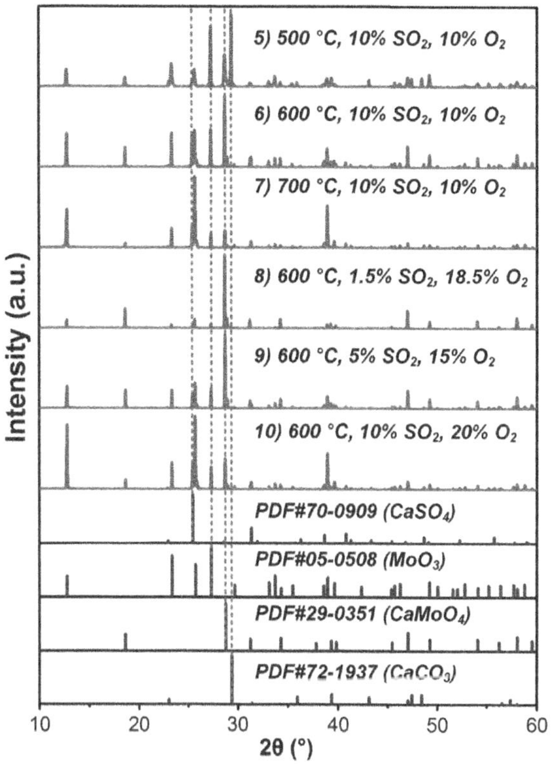

Fig. 7 XRD patterns of roasted mixtures of MoO_3–$CaCO_3$ (5-10#) under varying temperatures and gas proportions of N_2–O_2–SO_2, and standard PDF cards of $CaSO_4$, MoO_3, $CaMoO_4$, $CaCO_3$

Mixtures of MoO_3 with Fe_2O_3, PbO, CuO roasted at 600 °C under 80% N_2-10% O_2-10% SO_2 atmosphere were also analysed by XRD as shown in Fig. 8. It is seen that roasted mixtures of Fe_2O_3–MoO_3 (product #11) are composed of $Fe_2Mo_3O_{12}$ and surplus MoO_3, while no ferric or ferrous sulphate is detected. Roasted mixtures of PbO–MoO_3 (product #12) are composed of both $PbSO_4$ and $PbMoO_4$, indicating the conversion of partial molybdate to sulphate. As for product 13#, much peaks relating to $CuMoO_4$ and MoO_3 are detected, but peaks relating to $CuSO_4$ are not found. Notwithstanding, some peaks in 2θ of 9.5°, 17.8°, 20.7° and 25.0° cannot be identified, so it is unable to determine that $CuMoO_4$ is stable under SO_2-containing atmosphere. Additionally, sulphur contents in products 6# and 12# are measured to be 8.27 and 4.70 wt%, respectively, which means $CaSO_4$ and $PbSO_4$ account for 35.1 and 44.5 wt% in each product. Sulphur content in product 11# is measured to be 0.18 wt%, which is consistent with the result in Fig. 8 that less ferric or ferrous sulphate is generated, while sulphur content in product 13# turns out to be 2.88 wt%, suggesting that the unidentified peaks relate to sulphur-containing compound, that is, partial $CuMoO_4$ has been transformed.

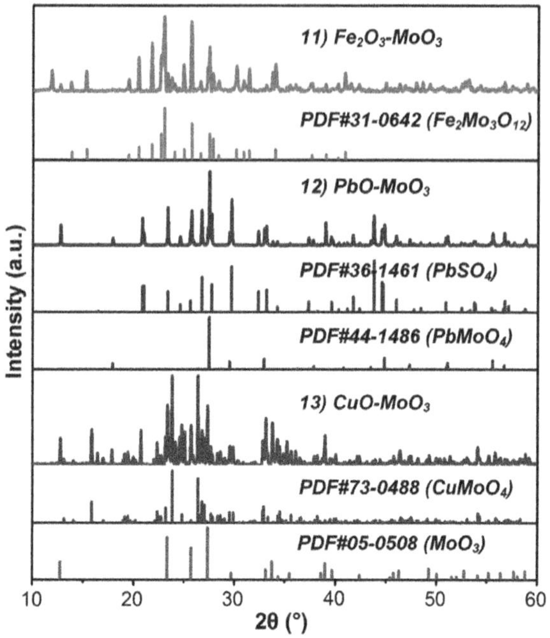

Fig. 8 XRD patterns of roasted mixtures of MoO_3 with Fe_2O_3 (11#), PbO (12#), CuO (13#) at 600 °C under 80% N_2-10% O_2-10% SO_2 atmosphere, and standard PDF cards of $Fe_2Mo_3O_{12}$, $PbSO_4$, $PbMoO_4$, $PbSO_4$, $CuMoO_4$, MoO_3

Conclusions

Thermodynamic studies indicate that sulphides of Mo, Fe, Pb, Cu are easy to be oxidized under SO_2-containing atmosphere and molybdates of Ca, Pb and Cu formed from solid–solid reactions could be transformed into corresponding sulphates, whereby associated Mo is released in form of MoO_3. Roasting experiments further demonstrate that molybdates of Ca, Pb and Cu could also be transformed during roasting under SO_2-containing atmosphere, but iron molybdate would remain unchanged. Besides, both raising the temperature and proportions of SO_2 and O_2 are effective means to promote the transformation of $CaMoO_4$ into $CaSO_4$. Overall, the results provide good support for the application of EGC technology in oxidative roasting of molybdenite concentrates, and the new finding of transformations of molybdates into sulphates might help improve the subsequent Mo extraction process.

Acknowledgements This work was supported by the National Natural Science Foundation of China (No. 51874355) and the Co-Innovation Center for Clean and Efficient Utilization of Strategic Metal Mineral Resources.

References

1. Braithwaite ER, Haber J (1994) Molybdenum: an outline of its chemistry and uses. Elsevier, New York
2. Xiang TG (2009) Molybdenum metallurgy. Central South University Press, Changsha (in Chinese)
3. Chen X, Liu X, Zhao Z, Hao M (2015) Research of the dissolving capacity of molybdenite in the white matte. Int. J. Refract. Met. Hard Mater. 52:1–5
4. Jiang T et al (2011) A simulation study on flue gas circulating sintering (FGCS) for iron ores. Paper presented at the 2nd international symposium on high-temperature metallurgical processing, San Diego, California, February 27–March 3, 2011
5. Li GH, You ZX, Sun H, Sun R, Peng ZW, Zhang YB, Jiang T (2016) Separation of rhenium from lead-rich molybdenite concentrate via hydrochloric acid leaching followed by oxidative roasting. Metals 6(11):282
6. Sun H, Li GH, Yu JJ, Luo J, Rao MJ, Peng ZW, Zhang YB, Jiang T (2018) Preparation of high purity MoO_3 through volatilization of technical-grade Mo calcine in water vapor atmosphere. Int J Refract Met Hard Mater 77:1–7
7. Dang J, Zhang GH, Chou KC (2014) Phase transitions and morphology evolutions during hydrogen reduction of MoO_3 to MoO_2. High Temp Mat Pr-isr 33:305–312
8. Utigard T (2009) Oxidation mechanism of molybdenite concentrate. Metall Mater Trans B 40:490–496 (2009)

A Novel Method of Recovering Rare Earths from Bayan Obo Rare-Earth Concentrate Under Super-Gravity Field

Xi Lan, Jintao Gao and Zhancheng Guo

Abstract A novel method of effectively recovering rare earths (REEs) from rare-earth concentrate under super-gravity was proposed in this paper. The reconstructions of rare-earth phases in normal gravity and the separation behaviours of REEs under super-gravity were investigated. It was indicated by the results of the mineral phase reconstructions that the rare-earth phases were transformed from original minerals of bastnaesite and monazite into britholite, rare-earth oxide fluoride, rare-earth ferrate and monazite particles at 1423–1523 K, and there was the only solid phase—rare-earth oxide fluoride phase in the sample at 1773 K. In a super-gravity field, all the rare-earth oxide fluoride crystals were overall intercepted by the filter along the super-gravity direction and effectively separated from the slag melts. The mass fractions of $\sum$ReO and Ce_2O_3 in the separated rare-earth oxide fluoride phase increased to 88.31 and 60.76 wt%, the recovery of $\sum$ReO and Ce_2O_3 were 90.22 and 95.41%, respectively.

Keywords Mineral reconstructions · Recovery · Rare earths · Rare-earth concentrate · Super-gravity

Introduction

Rare-earth elements (REEs) play a key role in the development of future sustainable technologies, some of which are necessary to achieve a greener global energy profile. Although not exactly rare, these elements are hard to mine and extract due to the fact that they rarely form continuous ore bodies and they share similar chemical properties [1]. The REE-Nb-Fe (Rare earth-Niobium-Iron) deposit at Bayan Obo, Inner Mongolia, China, currently represents the largest REEs resource worldwide [2]. However, it has been primarily adopted by the mineral process

---

X. Lan · J. Gao (✉) · Z. Guo
State Key Laboratory of Advanced Metallurgy, University of Science and Technology Beijing, Beijing 100083, People's Republic of China
e-mail: jintaogao@ustb.edu.cn

© The Minerals, Metals & Materials Society 2019

T. Jiang et al. (eds.), *10th International Symposium on High-Temperature Metallurgical Processing*, The Minerals, Metals & Materials Series,
https://doi.org/10.1007/978-3-030-05955-2_23

involved in blast furnaces to refine iron. As a result of this process, large quantities of REEs were discarded [3]. Recently, the Bayan Obo ore has been primarily performed by multistage magnetic flotation method to transform the REEs into the rare-earth concentrate [4]. Yet it was difficult to extract REEs from the rare-earth concentrate through the conventional beneficiation or hydrometallurgy methods resulting from the complexity and variety of primary minerals [5, 6]. Therefore, the escalating demand of REEs in various applications and the low utilization rate of rare-earth concentrate have laid emphasis to recovery REEs from rare-earth concentrate by developing energy-efficient and eco-friendly processes [7].

If REEs could be enriched into a specific mineral on the basis of their similar ionic radii [8], it would be conducive to recover the REEs from rare-earth concentrate. As for the precipitation and crystallization behaviours of rare-earth minerals, most investigations were focused on the rare-earth bearing blast furnace slag [9] and the direct-reduced rare-earth ore [10, 11]. However, few investigations on the precipitation of rare-earth concentrate could be referenced due to its complex mineral composition and the lack of relevant thermodynamic data of REEs. Therefore, the reconstructions of various RE-rich phases could be first observed in this work.

Moreover, if the RE-rich phases could be separated directly from other minerals at a specific high temperature, at which the REEs were enriched into a solid phase while other minerals formed molten slag, it would be beneficial to effectively separate the two different phases, yet it was infeasible to accomplish this task under the conventional conditions. Some investigations on morphology of rare-earth bearing slags revealed that the precipitation phases were embedded intimately in the slag [12], which were difficult to be separated effectively from the slag by adopting the conventional beneficiation methods; because the free sedimentation driven by the difference in density between the two phases was insufficient to accomplish solid–liquid separation due to the high viscosity of the solid–liquid mixture. As one of the cutting-edge process intensification technologies, super-gravity has received considerable attention and extensively explored in many fields, such as absorptions, stripping, distillation, heat transfer, adsorption, extraction, nanoparticles preparation and so on. Inspired by Ramshaw's exploitation of centrifugal apparatus and its use [13], the mass migration of different phases was improved in a super-gravity field. Zhao [14] reported that the refined grains of pure aluminum could be effectively removed in a super-gravity field, and the grain size of aluminum decreased rapidly with gravity coefficient increasing. Previous investigations on separating various compounds of perovskite [15], anosovite [16], olivine [17] and calcium phosphate [18] from slag melts indicated that the super-gravity field significantly enhanced the migration and concentration of the different phases. Consequently, the solid–liquid separation between the RE-rich phase and slag phase was carried out further in a super-gravity field in current study.

The aim of this paper was improving our understanding on the reconstructions of rare-earth phases in the rare-earth concentrate, and initially proposed a novel method to effectively recover REEs from rare-earth concentrate by super-gravity.

Experimental

Materials

Bayan Obo rare-earth concentrate was employed as the raw material in this study; its chemical composition measured by the XRF incorporated with ICP methods as listed in Table 1, the mass fractions of Ce_2O_3, La_2O_3, Pr_6O_{11} and Nd_2O_3 are 26.42, 11.84, 2.99 and 2.84 wt%, respectively. Combined with the XRD pattern of the raw material as shown in Fig. 1, bastnaesite, monazite and fluorite are the primary minerals in the Bayan Obo rare-earth concentrate.

Apparatus

A centrifugal apparatus incorporated with heating function is developed, as illustrated in Fig. 2, for achieving the recovery of the REEs from Bayan Obo rare-earth concentrate at their respective precipitation temperatures under super-gravity. A heating furnace and a counterweight are fixed symmetrically onto a centrifugal rotor, which rotate from a vertical to horizontal orientation when the centrifuge commenced operation. The isothermal zone of the heating furnace is 80 mm long, and the temperature is controlled by a program controller with a B-type thermocouple within the observed precision range of ± 3 K.

Mineral Reconstructions of Rare-Earth Concentrate

The mineral reconstructions of rare-earth concentrate with temperature rising were investigated through conducting hot-quenching method at different temperature ranges combined with offline analysis methods. 200 g rare-earth concentrate powder was evenly filled into eight magnesia crucibles with an inner diameter of 18 mm and a height of 60 mm. The eight samples were sequentially heated to 1423, 1473, 1523, 1573, 1623, 1673, 1723 and 1773 K, respectively, under argon atmosphere in a muffle furnace at a heating rate of 1 K/min. After holding the samples at each targeted temperature for 30 min, the magnesia crucibles were taken out and water-quenched immediately. Thereafter, the samples were measured by the X-ray diffraction method (XRD, monochromatized CuK_α radiation ($\lambda = 1.5418$Å) from 15 to 65°) to determine the variations in mineral compositions of various rare-earth phases at different temperature ranges.

Table 1 Chemical compositions (wt%) of Bayan Obo rare-earth concentrate

Composition	CaO	P_2O_5	Fe_2O_3	SO_3	F	SiO_2	MgO	BaO	MnO	Ce_2O_3	La_2O_3	Pr_6O_{11}	Nd_2O_3
Content	14.34	9.63	8.09	4.93	8.41	1.09	3.46	1.59	0.37	26.42	11.84	2.99	2.84

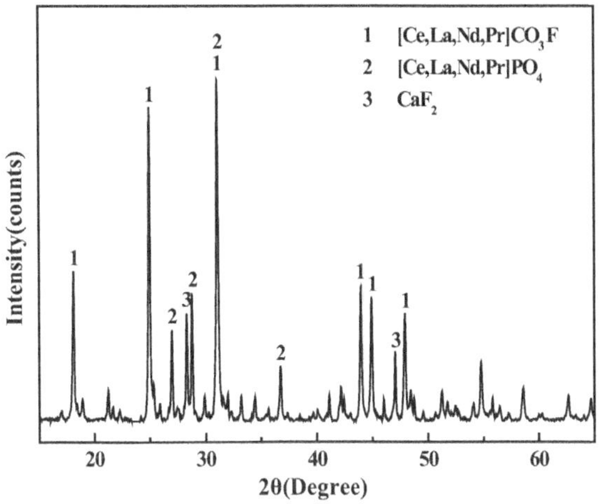

Fig. 1 XRD pattern of Bayan Obo rare-earth concentrate

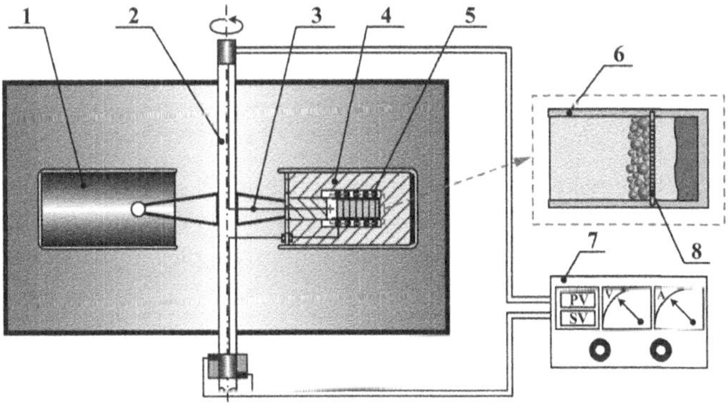

Fig. 2 Sketch of the centrifugal apparatus: (1) counterweight, (2) centrifugal axis, (3) thermocouple, (4) insulating layer, (5) resistance wire, (6) magnesia crucible, (7) temperature controller and (8) filter plate

Recovery of Rare Earths Under Super-Gravity Field

Based on the mineral reconstructions of the rare-earth concentrate with temperature rising, various RE-rich phases were precipitated at different temperatures. The recovery experiment of the rare earths was carried out further in a super-gravity field generated by the centrifugal apparatus. Twenty grams rare-earth concentrate powder was placed into a magnesia crucible with 0.01 mm pore size embedded in the filter plate and heated to 1773 K in the heating furnace of the centrifugal apparatus for 10 min to make other minerals forming molten slag while keeping the specific

RE-rich phase in a solid state, and then the centrifugal apparatus commenced operation and was adjusted to 1892 rpm angular velocity to achieve gravity coefficient of $G = 1000$ as calculated via Eq. (1). After centrifugal rotating at the constant temperature for 10 min, the apparatus was shut off and the crucible was water-quenched.

$$G = \frac{\sqrt{g^2 + (\omega^2 x)^2}}{g} = \frac{\sqrt{g^2 + \left(\frac{N^2 \pi^2 x}{900}\right)^2}}{g} \tag{1}$$

where G is gravity coefficient, g is the normal gravitational acceleration ($g = 9.8\ m/s^2$), ω is the angular velocity (rad/s), x is the distance between the centrifugal axis and sample centre (m) and N is the rotating speed (r/min).

Subsequently, the sample was sectioned longitudinally along the centre axis to form a macrograph view, and the separated samples were analysed via the SEM-EDS and XRD methods to determine the variations in microstructures and mineral compositions of the separated RE-rich phases attained by super-gravity. In addition, the mass fractions of $\sum$ReO in the separated samples were measured further by XRF incorporated with ICP methods, and the recovery ratios of REEs in the separated samples were calculated via Eq. (2), respectively.

$$R_{\mathrm{Re}-u} = \frac{m_u \times \omega_{\mathrm{Re}-u}}{m_u \times \omega_{\mathrm{Re}-u} + m_b \times \omega_{\mathrm{Re}-b}} \tag{2}$$

where, R_{Re-u} is the recovery ratio of REEs in the upper sample; m_u and m_b are the mass of the upper and bottom samples, respectively (kg); ω_{Re-u} and ω_{Re-b} are the mass fractions of $\sum$ReO in the upper and bottom samples, respectively (%).

Results and Discussion

Mineral Reconstructions of Rare-Earth Concentrate

The variations in the mineral compositions of various rare-earth phases in Bayan Obo rare-earth concentrate with temperature rising are shown in Fig. 3. It was obvious that the primary rare-earth mineral—bastnaesite ([Ce,La,Nb,Pr]CO_3F) decomposed with temperature increasing, and the rare-earth phases transformed from the original bastnaesite and monazite ([Ce,La,Nd,Pr]PO_4) minerals into the britholite (Ca_3[Ce,La,Nd,Pr]$_2$[(Si,P)O_4]$_3$F), the rare-earth oxide fluoride ([Ce,La,Pr, Nd]$_3O_4F_3$), the rare-earth ferrate ([Ce,La,Pr,Nd]FeO_3) and the monazite particles at a low-temperature range of 1423–1523 K as shown in Fig. 3.

With temperature increasing to a higher temperature range of 1573–1723 K as shown in Fig. 3, the rare-earth phases were consisted of britholite, rare-earth oxide fluoride and rare-earth ferrate. When temperature increasing further to 1723 K, the

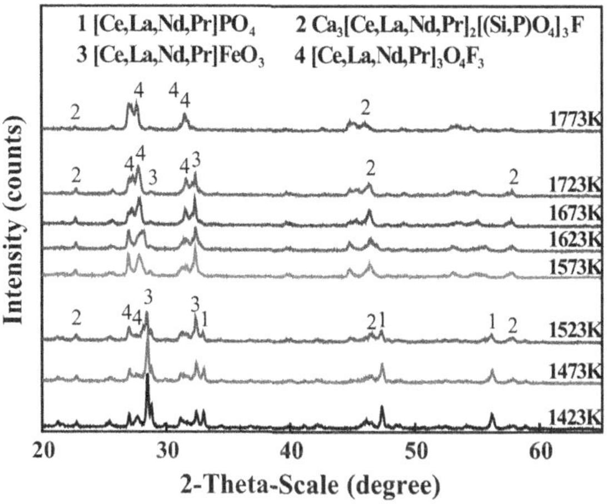

Fig. 3 Variations in XRD patterns of the rare-earth concentrate with temperature rising

monazite and rare-earth ferrate disappeared completely, and the britholite decreased significantly instead of forming the slag melt, whereas the rare-earth oxide fluoride phase was the only solid phase.

Recovery of Rare Earths Under Super-Gravity Field

To further recover REEs from rare-earth concentrate, isothermal separation experiment was carried out at the high-temperature range of 1773 K in a super-gravity field. As the macrograph obtained by super-gravity with gravity coefficient of G = 1000 and G = 1 at 1773 K shown in Fig. 4a and b, apparently, the entire sample was blocked by the filter, and a uniform structure was taken on by the sample attained under a normal gravity field (G = 1), as presented in Fig. 4a. In contrast, the sample was separated into two parts by the filter in a super-gravity field (G = 1000).

A series of SEM-EDS images and XRD patterns of the separated samples attained by super-gravity are shown in Figs. 5 and 6, respectively. As accordingly verified, the single rare-earth oxide fluoride crystals were the only solid phase in the molten rare-earth concentrate at the temperature range of 1773 K, while the other minerals formed slag melts. The slag melts evidently flowed through a filter driven by super-gravity, and the britholite crystals with a hexagonal prism structure were precipitated further from the separated slag melts after the water-quenching process. Conversely, all of the equiaxed rare-earth oxide fluoride crystals were intercepted by the filter along the super-gravity direction, and effectively separated from the slag melts.

Furthermore, the chemical compositions of the separated samples obtained by super-gravity are given in Table 2. After separating in a super-gravity field with gravity coefficient of G = 1000 at 1773 K for 10 min, the mass fraction of $\sum$ReO

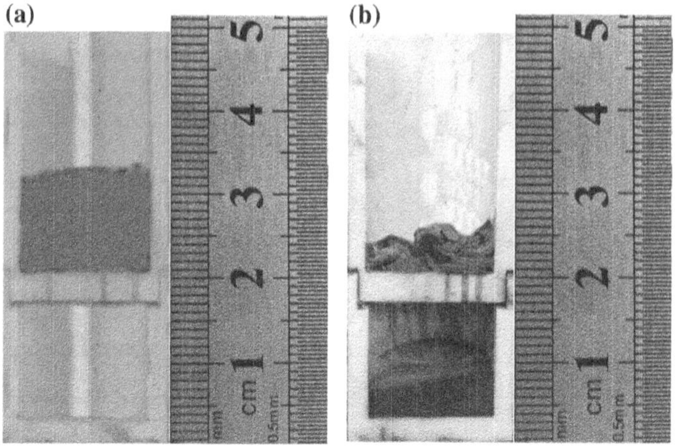

Fig. 4 Vertical sections of samples attained by super-gravity compared with parallel samples: **a** $G = 1$, T = 1773 K, **b** $G = 1000$, T = 1773 K

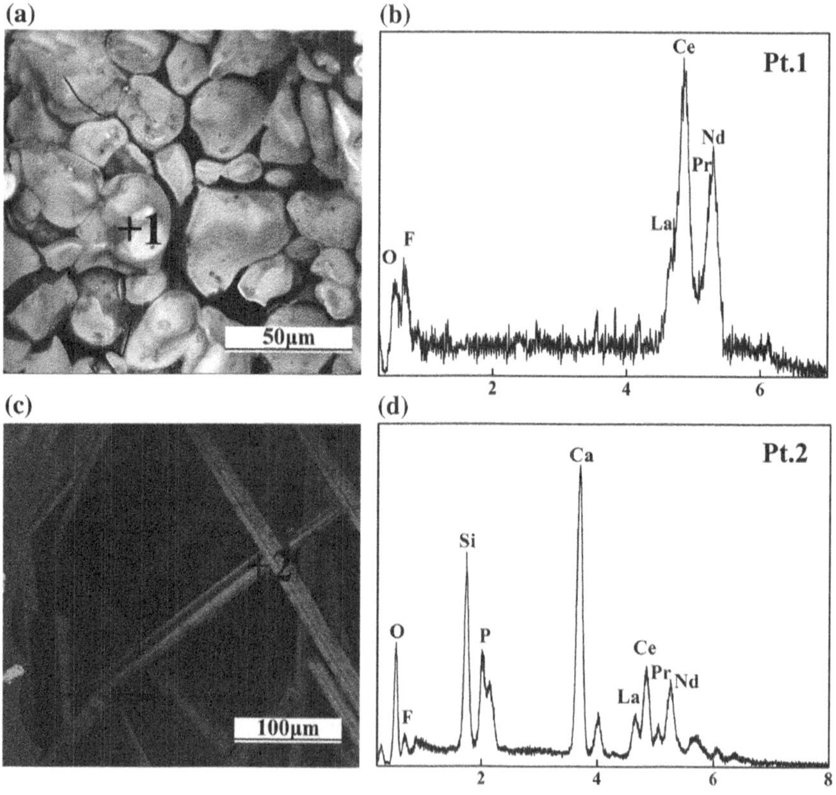

Fig. 5 Variations in SEM-EDS of the separated samples attained by super-gravity with $G = 1000$ and T = 1773 K: **a**, **b** SEM and EDS-upper sample, **c**, **d** SEM and EDS-bottom sample

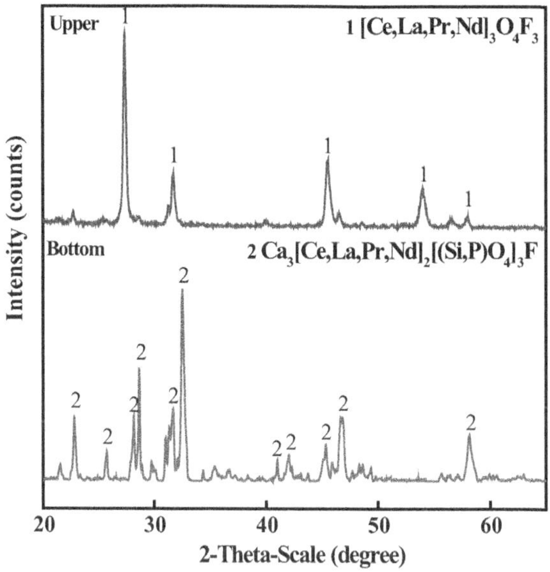

Fig. 6 XRD patterns of the separated samples attained by super-gravity with $G = 600$ and T = 1773 K

Table 2 Chemical compositions of the samples after super-gravity separation (wt%)

Samples	Ce_2O_3	La_2O_3	Pr_6O_{11}	Nd_2O_3	$\sum$ReO	Recovery ratio of $\sum$ReO	Recovery ratio of Ce_2O_3
Upper sample	60.76	20.18	3.78	3.59	88.31	90.22	95.41
Bottom sample	2.08	3.67	1.35	1.06	8.16	–	–

in the rare-earth oxide fluoride phase was up to 88.31 wt%, and that of Ce_2O_3 was up to 60.76 wt%. The recovery of $\sum$ReO and Ce_2O_3 were 90.22 and 95.41%, respectively. In contrast, the mass fractions of $\sum$ReO and Ce_2O_3 in the britholite containing slag phase were decreased to 8.16 and 2.07 wt%, respectively.

Discussion

With respect to the Stokes' law via Eq. (3), the effect of super-gravity field on the separation behaviours of rare-earth oxide fluoride crystals in the slag melt can be explained as follows.

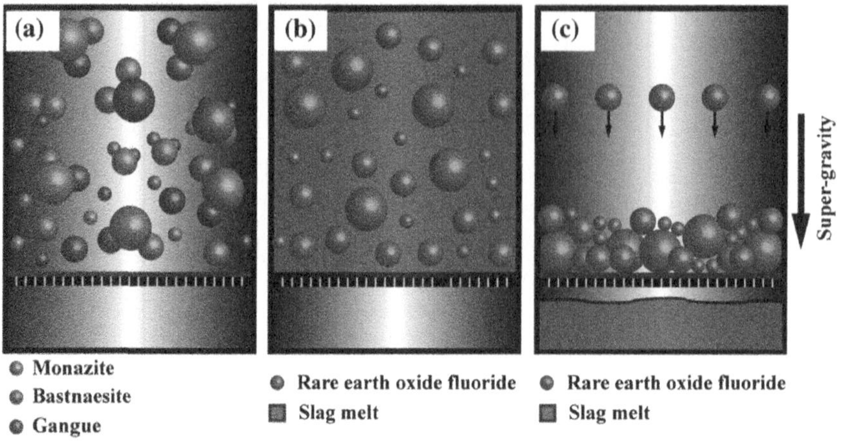

Fig. 7 Systematic diagram of recovering REEs from rare-earth concentrate under super-gravity: **a** room temperature, **b** 1773 K, $G = 1$, **c** 1773 K, $G = 1000$

$$V_r = \frac{dr}{dt} = \frac{d^2 \Delta\rho}{18\eta} w^2 R \tag{3}$$

where V_r is the migration velocity of rare-earth oxide fluoride crystals (m/s), d is the diameter of rare-earth oxide fluoride crystals (m), $\Delta\rho$ is the density difference between rare-earth oxide fluoride crystals and slag melt (kg/m^3) and η is the viscosity of slag melt (N s/m^2).

The reconstructions of RE-rich phases and the recovery of REEs from rare-earth concentrate under super-gravity can be intuitively described by Fig. 7 based on above results. The rare-earth oxide fluoride phase was the only solid phase while the other minerals formed slag melts at the high temperature of 1773 K. However, the viscosity of slag melts increased as the existence of rare-earth oxide fluoride crystals, so the crystals were difficult to be separated from the slag melts in the normal gravity field. In case of a super-gravity field, the slag melts evidently followed through a filter driven by super-gravity. Conversely, all of the equiaxed rare-earth oxide fluoride crystals were intercepted by the filter along the super-gravity direction, and effectively separated from the slag melts. Moreover, this work could potentially develop a novel approach for recovering various elements from some other complex slag melts in a super-gravity field.

Conclusion

In this paper, a novel method of recovering REEs from rare-earth concentrate by super-gravity was proposed. The Bayan Obo rare-earth concentrate was employed as the raw material, the reconstructions of RE-rich phases with temperature rising, and recovery of REEs in a super-gravity field were investigated. The following conclusions can be drawn:

(1) It was indicated by the reconstructions results that the rare-earth phases transformed from original bastnaesite and monazite minerals into britholite, rare-earth oxide fluoride, rare-earth ferrate and monazite particles at 1423–1523 K, which transformed further into the only solid phase—rare-earth oxide fluoride at 1773 K.
(2) In a super-gravity field, all of the rare-earth oxide fluoride crystals were intercepted by the filter along the super-gravity direction, and effectively separated from the slag melts, and the mass fractions of $\sum$ReO and Ce_2O_3 in the separated rare-earth oxide fluoride phase were up to 88.31 and 60.76 wt%, the recovery of $\sum$ReO and Ce_2O_3 were 90.22 and 95.41%, respectively.

Acknowledgements This study is supported by the National Natural Science Foundations of China (No. 51774037 and No. 51404025).

References

1. Cristian T, Martina P, Christian E, Teodora R (2016) A hydrometallurgical process for the recovery of rare earth elements from fluorescent lamp waste fractions. Sep Purif Technol 161:172–186
2. Yang KF, Fan HR, Santosh M, Hu FF, Wang KY (2011) Mesoproterozoic carbonatitic magmatism in the Bayan Obo deposit, Inner Mongolia, North China: Constraints for the mechanism of super accumulation of rare earth elements. Ore Geol Rev 40:122–131
3. Zhang B, Liu CJ, Li CL, Jiang MF (2014) A novel approach for recovery of rare earths and niobium from Bayan Obo tailings. Miner Eng 65:17–23
4. Liu YZ, Zhou X, Chen Y, Zhang Z, Zhou LL (2015) Clostridium butyricum can be used as a potential alternative for the antibiotic in Cherry Valley ducks. J Anim Plant Sci 25:99–106
5. Huang XW, Long ZQ, Wang LS, Feng ZY (2015) Technology development for rare earth cleaner hydrometallurgy in China. Rare Met 34:215–222
6. Vijayan S, Melnyk AJ, Singh RD, Nuttall K (1989) Rare earths: their mining, processing, and growing industrial usage. Miner Eng 41:13–18
7. Archana K, Rekha P, Manis KJ, Kumar JR, Lee JY (2015) Process development to recover rare earth metals from monazite mineral: a review. Miner Eng 79:102–115
8. Krishnamurthy N, Gupta CK (2004) Extractive metallurgy of rare-earths. CRC Press, Florida, p 195
9. Li DG (2005) Selective precipitation and separation of valuable constituent in blast furnace slags. N. U. 25–32
10. Ding YG, Wang JS, Wang G, Xue QG (2012) Innovative methodology for separating of rare earth and iron from Bayan Obo complex iron ore. ISIJ Int 52:1772–1777
11. Wang FQ, Gao JT, Lan X, Guo ZC (2017) Direct concentration of iron, slag and britholite-(Ce, La, Pr, Nd) at 1473 K in a super gravitational field. ISIJ Int 57:200–202
12. Li JC, Guo ZC (2014) Innovative methodology to enrich britholite (Ca3Ce2[(Si, P)O4]3F) phase from rare earth-rich slag by super gravity. Metall Mater Trans B 45:1272–1280
13. Ramshaw C, Mallinson RH (1984) Mass transfer apparatus and its use. Europe Patent 0002568

14. Zhao L, Guo ZC, Wang Z, Wang M (2010) Influences of super-gravity field on aluminum grain refining. Metall Mater Trans A 41:670–675
15. Gao JT, Zhong YW, Guo ZC (2016) Separation of iron phase and P-bearing slag phase from gaseous-reduced, high-phosphorous oolitic iron ore at 1473 K (1200 °C) by super gravity. Metall Mater Trans B 47:2459–2467
16. Lu Y, Gao JT, Wang FQ, Guo ZC (2017) Separation of anosovite from modified titanium-bearing slag melt in a reducing atmosphere by supergravity. Metall Mater Trans B 48:749–753
17. Gao JT, Li Y, Xu GL, Wang FQ, Lu Y, Guo ZC (2017) Separation of olivine crystals and borate containing slag from $CaO–SiO_2–B_2O_3–MgO–Al_2O_3$ system by utilizing super-gravity. ISIJ Int 57:587–589
18. Li C, Gao JT, Wang Z, Ren HR, Guo ZC (2017) Separation of Fe-bearing and P-bearing phase from the steelmaking slag by super gravity. ISIJ Int 57:767–769

Part VI
Ironmaking and Steelmaking

Sintering Characteristic and Consolidation Behaviour of Chromite Fines

Xiaohui Fan, Guojing Wong, Min Gan, Xuling Chen, Zhiyun Ji, Xunwei Zhou and Wei Lv

Abstract A study on sintering of chromite fines was conducted through sintering pot test. The effect of sintering parameters including mix moisture content, coke breeze rate, bentonite dosage and sintering suction was investigated. It was found that conventional sintering process can be applied to the agglomeration of chromite fines as well. Appropriate parameters for chromite fines sintering were 8.0% moisture, 7.0% coke breeze, 2.0% bentonite and 14 kPa sintering suction. Furthermore, the consolidation characteristic of chromite sinter was observed by using a microscope, scanning electron microscope (SEM) and energy dispersive spectrometer (EDS). It can be found that the sinter mainly consisted of chromohercynite, forsterite and monticellite. Besides, the element distribution of Fe and Cr in chromohercynite was quite different. Fe mainly distributed on the outer layer while Cr distributed in the inner layer. The chromohercynite formed from liquid phase has higher Fe and Mg content.

Keywords Sintering · Chromite fines · Consolidation characteristic

Introduction

As an essential element in modern industry, chromium is widely used in steel-making process, which improves the property of steel [1]. Therefore, special steel can be produced with the application of chromium. Besides, chromium can also be used for the smelting of special alloy with other elements, such as nickel, cobalt, and tungsten. Both the special steel and the special alloy are indispensable to spacecraft, automobiles and other manufacturing [2–4]. Hence, as a typical chromium contained resource, chromite is reserved by governments as a strategic resource [5].

X. Fan · G. Wong (✉) · M. Gan · X. Chen · Z. Ji · X. Zhou · W. Lv
School of Minerals Processing & Bioengineering, Central South University,
No. 932, South Lushan Road, Changsha, Hunan 410083, People's Republic of China
e-mail: wgjcsu@126.com

© The Minerals, Metals & Materials Society 2019

T. Jiang et al. (eds.), *10th International Symposium on High-Temperature Metallurgical Processing*, The Minerals, Metals & Materials Series,
https://doi.org/10.1007/978-3-030-05955-2_24

Nowadays, nearly 80% chromite belongs to ore fines. Even though the fines have a higher grade and lower price [6], it will also cause the deterioration of permeability and furnace condition during the smelting process and finally worsens the technical and economical index. Therefore, realizing the agglomeration of chromite ore fines is a functional way to reduce the cost of ferrochrome production [7]. The methods of agglomeration include briquetting, pelleting and sintering [8–10]. According to the research of Deqing Zhu [9] and Bikash Nandy [11], sintering is much more suitable for the agglomeration of chromite due to the better sinter strength, reasonable particle size distribution and good reducibility. Generally, chromite ((Fe, Mg) (Cr, Al)$_2$O$_4$) has a high melting point so that some fluxes, like chlorite or ophiolite, should always be added to help generate low melting point materials which can produce more liquid phase and improve the sintering indexes [12]. Owing to the better capacity of mineralization, ophiolite is always chosen [13].

China has been the largest manufacturer and consumer of stainless steel as well as the consumer of chromite. However, China is short of chromium resource, which accounts for only 0.15% of the global reserves [14]. Therefore, how to utilize the chromite ore fines is of significance for China. In this paper, the sintering test of chromite fines with ophiolite has been operated. The sintering characteristic and consolidation behaviour have also been researched.

Materials and Methods

Properties of Raw Materials

In this test, chromite, ophiolite, coke breeze and bentonite were used to produce sinter. Chromite came from South Africa and Turkey. Based on the characteristic of particle size distribution, the chromite from Turkey was divided into the coarse one and the fine one. And the chromite fines were abbreviated to SFC, TCC and TFC, respectively. The chemical compositions of these materials were given in Table 1. Obviously, the chromium content of SFC and TCC was much higher than that of

Table 1 Chemical compositions of materials/wt%

Material types	TFe	FeO	Cr_2O_3	SiO_2	Al_2O_3	CaO	MgO	P	S	LOI
SFC	18.91	18.49	40.60	6.85	12.29	1.29	10.46	0.01	0.01	1.04
TCC	9.78	10.93	42.44	11.22	6.53	1.03	19.85	0.01	0.01	5.30
TFC	10.89	11.00	33.95	10.49	7.67	1.35	18.56	0.01	0.10	11.01
Ophiolite	5.93	1.89	0.068	36.77	0.94	2.83	36.18	0.01	0.02	14.02
Coke breeze	2.21	0.08	1.82	9.01	3.24	0.64	1.62	0.03	0.09	81.26
Bentonite	3.10	–	–	39.13	6.87	11.62	6.09	–	–	21.51

Table 2 Size distributions of materials

Material types	Size distributions (mm)/%							Average diameter/ mm
	+8	5–8	3–5	1–3	0.5–1	0.25–0.5	−0.25	
SFC	2.58	14.73	8.03	3.93	21.77	17.02	31.94	1.92
TCC	0.00	0.55	8.90	31.08	19.02	4.85	35.60	1.23
TFC	0.00	0.35	2.61	11.01	16.45	6.97	62.62	0.60
Coke breeze	0.58	1.61	7.50	17.39	19.33	5.76	47.34	1.07
Ophiolite	0.00	0.17	9.04	37.76	25.56	9.82	17.65	1.38
Return fines	0.00	0.00	13.57	38.36	22.60	7.93	17.54	1.54

TFC. Besides, TCC and TFC have a higher content of MgO than SFC, while SFC has a higher content of Al_2O_3. The size distributions of raw materials were shown in Table 2. The contents of fine particles under −0.5 mm in SFC, TCC and TFC were 48.96, 40.45 and 69.59%, respectively. The size of ophiolite was mainly -3 mm, which accounted for nearly 90%. Hu [15] pointed out that the particle size of ophiolite should be less than 3 mm when ophiolite was used for sintering. Hence, the ophiolite utilized in this test met the requirement.

Methods

The sintering tests were carried out through a sintering pot (as shown in Fig. 1), of which the size is Φ180 mm × 700 mm. The sintering process mainly included blending, mixing, granulation, feeding, ignition, sintering and cooling. The granulation was practiced in a drum with a diameter of 600 and 800 mm length at 15 rpm for 4 min. Before feeding, 1 kg hearth of chromite sinter with size between 10 and 16 mm should be placed in the sintering pot. The ignition was operated with natural gas at 1100 ± 50 °C for 2 min, and the suction was controlled at 7.5 kPa. Sintering suction was set to 14 kPa when the ignition was done. Then, the sintering finished until the flue gas temperature reached the peak value. The sinter cooled for 6 min at 7.5 kPa suction. Then, the tumble index, yield and productivity were calculated as follows to help evaluate the quality and yield of sintered ore.

$$TI = \frac{m_1}{m_0} \times 100\% \tag{1}$$

$$\eta_{yield} = \frac{S_0 - R}{S_0} \times 100\% \tag{2}$$

$$\gamma = \mathrm{k}\frac{M}{D^2 t} \tag{3}$$

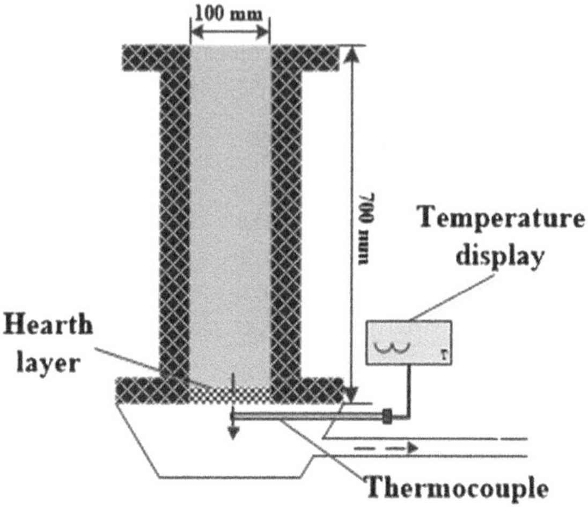

Fig. 1 Effects of moisture on sintering indexes

where *TI* is the Tumbler index, %; m_1 is the mass of sample (+6.3 mm) after tumble, kg; m_0 is the mass of sample for tumble, kg; η_{yield} is the yield, %; S_O is the mass of sinter without hearth, kg; R is the mass of return fine (−5 mm), kg; γ is the productivity, t m^{-2} h^{-1}; k is a conversion factor, 7.65 * 10^4; M is the mass of finished sinter, kg; D is the diameter of sintering pot, mm; t is the sintering time, min.

After the finished sinter was cooled down, their mineralogical characteristics were observed through XRD (D8 ADVANCE, Bruker), optical microscope (DMRXP, Lecia) and SEM-EDS ((Nova NanoSEM230, FEI Electron Optics B.V).

Results and Discussion

Effects of Sintering Conditions on Sintering Indexes

The sintering test was carried out with a sintering pot under the conditions (shown in Table 3). The effects of moisture content, coke breeze rate, bentonite dosage and sintering suction on sintering indexes were researched.

Table 3 Experimental conditions of sintering test

SFC: TCC: TFC	MgO/ Al_2O_3	Moisture (%)	Coke breeze rate (%)	Dosage of bentonite (%)	Sintering suction kPa
6:2:2	1.5	7.0 → 8.5	7	2	14
6:2:2	1.5	8	4 → 8	2	14
6:2:2	1.5	8	7	1.5 → 3	14
6:2:2	1.5	8	7	2	12 → 15

Effects of Moisture Content on Sintering Indexes

The moisture of mixture played an important role during the sintering process, especially for the granulation and combustion and heat transfer. The effects of moisture on sintering indexes were given in Fig. 2. As the moisture increased, the granulation was improved so that the permeability of sinter layers was also improved and vertical sintering velocity was accelerated. Faster vertical sintering velocity meant rapid combustion of fuel, which affected the metallogenetic process and caused the reduction of yield. Excess moisture needed more heat to be evaporated. Hence, insufficient heat and the thickness of over-wet layer would lower the TI. In this test, it would be better to control the moisture at 8.0%.

Effects of Coke Breeze Rate on Sintering Indexes

The mineral composition, microscopic structure and sintering indexes were influenced by the coke breeze rate. The effects of coke breeze rate on sintering indexes were shown in Fig. 3. As the rate of coke breeze increased, the yield, TI, productivity and solid fuel consumption went up. However, the granulation process would be deteriorated, and the productivity would be reduced when the rate was more than 7.0%. Therefore, the coke breeze rate should be controlled at 7.0%.

Effects of Bentonite Dosage on Sintering Indexes

Bentonite was widely used for improving the granulation and the permeability of sinter layers because of its better water absorption capability. Besides, during the sintering process, the bentonite can also react with the silicate minerals under high temperature. The reaction can promote the formation of liquid phase, which improves the sinter strength. The effects of bentonite dosage on sintering indexes

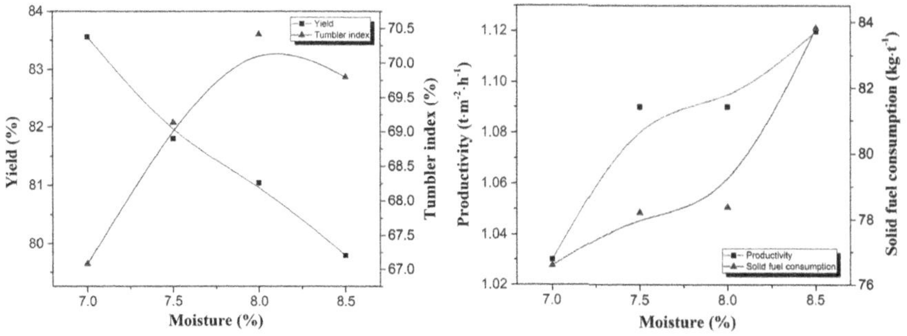

Fig. 2 Effects of moisture on sintering indexes

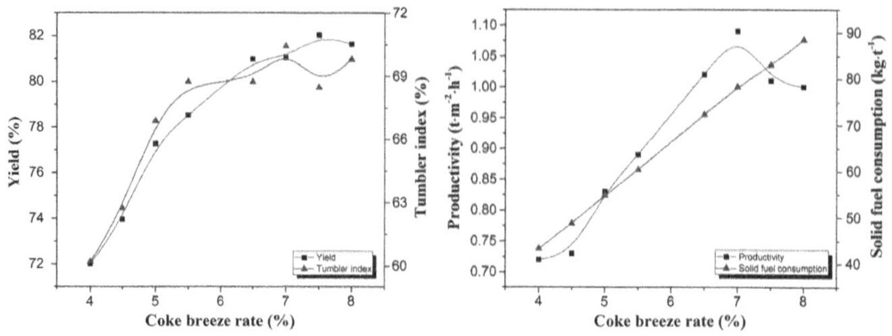

Fig. 3 Effects of coke breeze rate on sintering indexes

were shown in Fig. 4. To avoid the deterioration caused by the increase of bentonite, the suitable dosage of bentonite should be controlled at 2.0%.

Effects of Sintering Suction on Sintering Indexes

The sintering suction was researched when the moisture, the rate of coke and the dosage of bentonite were 8.0, 7.0 and 2.0%, respectively. The results were shown in Fig. 5. When the sintering suction was relatively lower, the blast capacity passed through the layers per unit time was less so that the combustion of the fuel was incomplete and the sintering indexes deteriorated. While the sintering suction increased, the sintering indexes were improved. However, higher sintering suction meant higher consumption. Therefore, it would be better to control the sintering suction at 14 kPa.

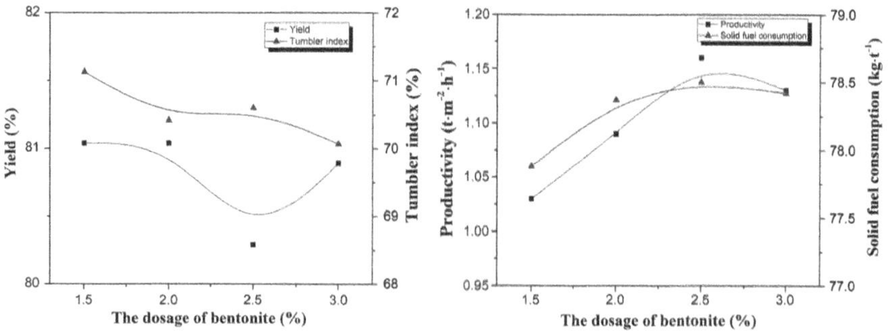

Fig. 4 Effects of bentonite dosage on sintering indexes

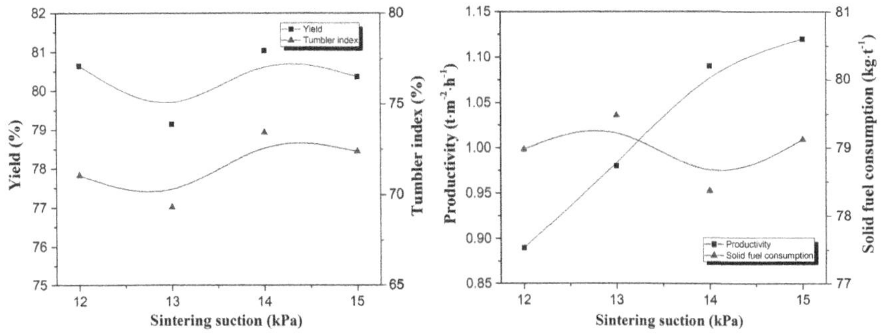

Fig. 5 Effects of sintering suction on sintering indexes

Consolidation Behaviour

The sintering test was operated when the moisture, the rate of coke breeze, the dosage of bentonite and the sintering suction were 8.0%, 7.0%, 2.0% and 14 kPa, respectively. The consolidation behaviour of finished sinter was researched.

The chemical compositions of the finished sinter were given in Table 4. As shown in Fig. 6, the finished sinter mainly consisted of chromohercynite ((Fe, Mg)(Cr, Fe, Al)$_2$O$_4$)

Table 4 Chemical compositions of finished sinter/wt%

Material	TFe	FeO	Cr_2O_3	SiO_2	Al_2O_3	CaO	MgO
Finished sinter	13.92	14.70	34.58	10.66	9.33	1.52	13.99

Fig. 6 XRD pattern of finished sinter

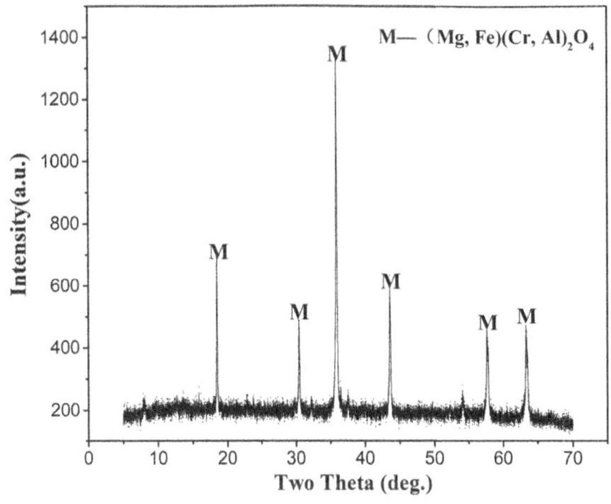

The microstructure of finished sinter was given in Fig. 7. The structure of this finished sinter was compact. There were mainly four kinds of structures, the chromohercynite, the spinel with higher content of Fe and lower content of Cr, the kirschsteinite and the forsterite. The chromohercynite and kirschsteinite held higher proportion, while the spinel with higher content of Fe and lower content of Cr existed between the chromohercynite and kirschsteinite like a boundary. And the forsterite mainly came from the decomposing reaction of ophiolite.

The characteristic of mineral composition in the finished sinter were analysed by SEM-EDS, and the results were shown in Fig. 8 and Table 5.

It can be found that the element distribution of Fe and Cr in chromohercynite was quite different, of which Fe mainly focused on the outer layer (Region 2 shown in Fig. 8), while Cr distributed in the inner layer (Region 1). The chromohercynite came from liquid phase has a small size and higher content of Fe and Mg (Region 3).

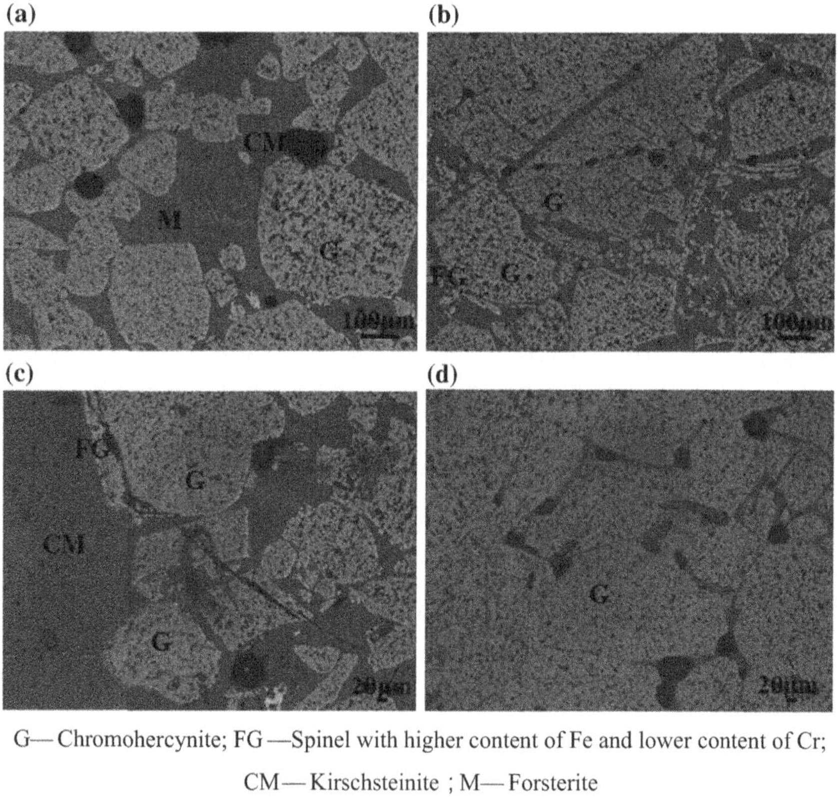

Fig. 7 Microstructure of finished sinter. G—Chromohercynite; FG—Spinel with higher content of Fe and lower content of Cr; CM—Kirschsteinite; M—Forsterite

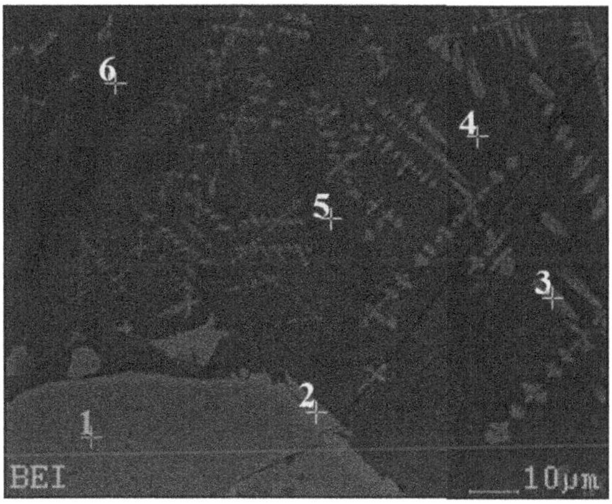

Fig. 8 SEM pattern of finished sinter

Table 5 Chemical compositions of different regions in finished sinter

Composition	Cr_2O_3		Fe_2O_3		MgO		Al_2O_3		SiO_2		CaO	
	Wt/ %	Mol/ %	Wt/ %	Mol/ %	Wt/ %	Mol/ %	Wt/ %	Mol/ %	Wt/ %	Mol/ %	Wt/ %	Mol/ %
Region 1	51.4	35.4	19.4	12.7	13.9	36.0	14.3	14.7	–	–	0.2	0.3
Region 2	1.6	1.2	78.9	56.8	9.9	28.1	5.5	6.2	3.0	5.7	0.4	0.8
Region 3	0.3	0.2	71.3	45.8	12.2	30.9	6.0	6.0	8.1	13.8	0.8	1.5
Region 4	0.2	0.1	8.7	3.5	9.3	14.7	11.2	7.0	56.9	60.2	11.3	12.8
Region 5	–	–	5.7	1.8	44.5	56.4	1.6	0.8	46.8	39.9	0.8	0.8
Region 6	–		5.4	1.7	48.1	59.7	–	–	46.2	38.5	–	–

Note: "-" means not detected

Besides, the forsterite, monticellite and other silicates also existed in the finished sinter, and the chemical composition of them was quite different (Region 4 to 6).

As shown in Fig. 9, the white parts were chromohercynite, of which the brighter one contained more Fe. The forsterite was rod-like, and parts of them were monticellite. Fe mainly distributed in the outer layer of chromohercynite. The content of Mg distributed in hortonolite was more than that in the chromohercynite. Al mostly distributed in the monticellite and chromohercynite. The content of Ca and Si were relatively lower, and they mainly distributed in the silicate.

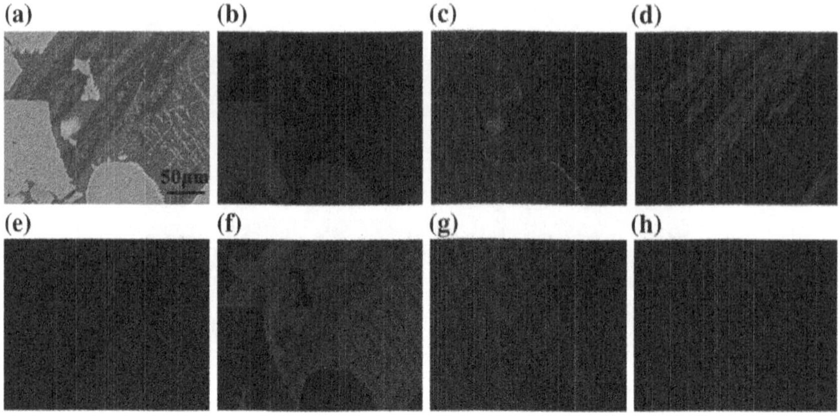

Fig. 9 Distribution of different elements in the finished sinter. **a** Figure of SEM, **b** distribution of Cr, **c** distribution of Fe, **d** distribution of Mg, **e** distribution of Al, **f** distribution of Si, **g** distribution of Ca, **h** distribution of O

Conclusions

The sintering characteristic of chromite fines was investigated through sintering pot test. Appropriate parameters were confirmed: moisture 8.0%, rate of coke breeze 7.0%, dosage of bentonite 2.0%, sintering suction 14 kPa. Good sintering indexes have been obtained, which meant that sintering process was also applicative for chromite fines. The microstructure of finished sinter was impact. The finished sinter consisted of chromohercynite, forsterite and monticellite instead of calcium ferrite and fayalite. The distribution of Fe and Cr in chromohercynite was quite different, of which Fe mainly focused on the outer layer while Cr distributed in the inner layer. The chromohercynite came from liquid phase has a small size and higher content of Fe and Mg. In this test, ophiolite was selected as flux. Therefore, the contents of Ca and Si were relatively lower, and they mainly distributed in the silicate.

Acknowledgements The research was financially supported by the National Natural Science Foundation of China (No. U1760107, U1660206), Hunan Provincial Co-Innovation Center for Clean and Efficient Utilization of Strategic Metal Mineral Resources and Innovation Driven Plan of Central South University (No.2015CX005)

References

1. Murthy YR, Tripathy SK, Kumar CR (2011) Chrome ore beneficiation challenges & opportunities—a review. Miner Eng 24(5):375–380
2. Zhao NC, Zhang QX (1998) Practical technical manual for ferro-alloy production. Metallurgical Industry Press, Beijing

3. Cui HJ, Chen J, Feng XM, Ling N, Liu JY (2007) Numerical modeling of temperature rising characteristic curve for carbon containing chromite fines in microwave field. China Metall 17(1):30–37
4. Yan JF, Chen JX, Hu L (2007) Chromium metallurgy. Metallurgical Industry Press, Beijing
5. Gu F, Wills BA (1988) Chromite-mineralogy and processing. Miner Eng 1(3):235–240
6. Liu PX, Chu SJ (2011) Research on cold briquet with chromite fines. Res Iron Steel 39(3): 12–15
7. Zhang MJ, Liu FQ (1997) The study on the chromite sintering. Ferro-Alloys 3(1):7–10
8. Zhu DQ, Yang CC, Pan J, Lu LM, Liu XQ (2018) An integrated approach for production of stainless steel master alloy from a low grade chromite concentrate. Powder Technol 335 (C):103–113
9. Zhu DQ, Lee J, Fan XH, Pan J (2004) Literature review of the agglomeration of chromite fines. Sinter Pelletizing 29(2):27–29
10. Qiu WJ (2003) Balling-up mechanism and consolidation of cold briquetting from fine ore. Ferro-Alloys 3(1):7–9
11. Nandy B, Chaudhury MK, Paul J (2009) Sintering characteristics of Indian chrome ore fines. Metall Mater Trans B 40(5):662–675
12. Fukagawa S, Shimoda T (1987) Smelting reduction mechanism of chromium ore sinter by solid carbon. Trans Iron Steel Inst Jpn 27(8):609–617
13. Wu SL, Han HL, Jiang WZ, Zhu JM, Feng GS, Zhang ZC (2009) MgO interaction mechanism in sinter. J Univ Sci Technol Beijing 31(4):428–432
14. He ZF, Jiang RB, Liu SC (2016) Analysis of chromite ore resources security in China. China Ming Mag 25(6):7–11
15. Xiong FW, Hu BS, Gui YL (2011) Influence of the sintering process of serpentine in TangSteel. J Hebei Politechnic Univ (Natural Science Edition) 33(1):32–36

Construction and Practice on Energy Flow Network of New Generation Recyclable Iron and Steel Manufacturing Process

Fuming Zhang

Abstract Shougang Jingtang iron and steel plant is a new generation recyclable iron and steel plant designed according to the concept and principle of circular economy. The steel plant is provided with the comprehensive functions of high quality steel product manufacture, high-efficiency energy conversion and waste disposal. In order to realize the cooperation and high efficiency of iron and steel manufacturing process, a full process energy flow network with carbon flow as the core is designed and constructed to realize energy high efficiency conversion and low carbon green manufacturing. Since Jingtang steel plant was put into production, the efficiency of energy conversion has been continuously improved, and remarkable results have been obtained in high-efficiency energy utilization and high value conversion. The emission of CO_2 and pollutants has been greatly reduced, and cleaning production and low-carbon metallurgy have been realized.

Keywords Iron and steel · Energy saving · Energy flow network · Emission control · Circular economy

Introduction

In order to solve the non-capital functions of Beijing at the beginning of the twenty-first century, Shougang decided to move out of Beijing region and build a new steel base, Jingtang Iron and Steel Plant, in Caofeidian industrial district, Tangshan city, Hebei province. The project is designed and built independently according to the concept of a new generation of recyclable iron and steel processes and has a large steel project with advanced international technical merit in the twenty-first century. It is also a large-scale modern steel plant which produces thin strip and sheet. Shougang Jingtang Iron and Steel Plant has "three functions" [1], it not only pays attention to high quality and high-efficiency iron and steel products

---

F. Zhang (✉)
General Engineer Office of Shougang Group Co Ltd., Beijing, China
e-mail: zhangfuming2018@163.com

© The Minerals, Metals & Materials Society 2019

T. Jiang et al. (eds.), *10th International Symposium on High-Temperature Metallurgical Processing*, The Minerals, Metals & Materials Series,
https://doi.org/10.1007/978-3-030-05955-2_25

manufacturing function, but also pays attention to high efficiency, clean energy conversion function, handles large amount of social waste and realizes the function of reusing resources.

The design annual output of Shougang Jingtang Iron and Steel Plant is 8.7–9.2 million ton. The engineering erection began on 12 March 2007, No. 1 blast furnace blew in on 21 May 2009, No. 2 blast furnace was put into production on June 26 June 2010 and the complete iron and steel manufacturing process put into production smoothly. The high efficiency and clean steel production process with 2 blast furnaces, 1 steelmaking plant, 2 hot rolling lines and 4 cold rolling lines are integrated and constructed. 5500 m^3 blast furnace, high efficiency and low cost clean steel production platform, 50,000 tons per day seawater desalination facility and other representative advanced technological equipment have been designed and constructed. It plays an important role in guiding and demonstrating the design and construction of a new generation of recyclable steel plant in the twenty-first century.

Design Philosophy and Main Objectives

Engineering Concept

Shougang Jingtang Iron and Steel Plant is a modern large-scale iron and steel project with international advanced level in the twenty-first century, the first real coastal port construction and the whole production of thin strip and sheet, according to the concept of the new generation of recyclable steel process and the independent design and construction of China. The engineering design concept and goal are: guided by the theory of metallurgical process engineering [2], taking the construction of the new generation recyclable steel plant "Three Functions" as the core design philosophy, according to the "high starting point, high standard and high requirement" put forward by the central government, achieve the "First Class" project construction objectives of product, management, environment and benefit. In engineering design, independent integrated innovation has been carried forward and Shougang Jingtang has become a clean steel manufacturing base, a circular economy demonstration base and an independent innovation demonstration base for the production of thin strip with high quality and high efficiency.

In engineering design, on the basis of deep understanding of the physical essence of the dynamic operation of steel plant and the development trend of iron and steel industry in the new century, the concept of a new generation of recyclable iron and steel technological process is followed. Based on the theory and method of modern iron and steel metallurgical engineering design, the conceptual design, top-level design, dynamic precision design and three-dimensional simulation design of Shougang Jingtang Iron and Steel Plant project are innovated and practised, a new generation of recyclable steel manufacturing process is designed and constructed, and become a new generation of recyclable steel manufacturing process demonstration project [3].

Engineering Objectives

The main goals of Shougang Jingtang project are according to the construction concept of the new generation recyclable iron and steel manufacturing process, research and development of a new producing process of clean steel with high efficiency and low cost, metallurgical gas dry dust removal and the high temperature with high pressure dry coke quenching power generation. In order to realize the effective utilization of low quality waste heat and social waste resources, the integrated technology is applied to Shougang Jingtang as a systematic technological integration and strives to build the twenty-first-century world-class advanced iron and steel plant.

The new generation recyclable iron and steel manufacturing process is not a representation or transformation of existing processes or facilities, but a new iron and steel manufacturing process based on the dynamic integration of substance flow, energy flow and information flow around the core concepts of high efficiency, low cost and recyclable process. To realize the three functions of product manufacture, energy conversion and disposal of social waste, and integrate construction in Shougang Jingtang.

- High efficiency and low cost clean steel manufacturing platform.
- High-efficiency energy conversion and energy saving steel plant.
- Clean steel plant with recycling resources and energy.

The dynamic and precise concept of engineering design is established in the top-level design of Shougang Jingtang. Around the dynamic running rule of "Flow" in the whole iron and steel manufacture process, the basic concept of process optimization is dynamic-orderly and synergetic-continuous. In the top-level design, the principles of overall, hierarchical, dynamic, relevance and environmental adaptability are emphasized, including process or facility element selection, process structure optimization, function extension, as well as efficiency excellence.

(1) Factor selection

In the capital, land, resource, market, environment, labour force and technology and so on, the reasonable matching of the basic elements of industry and economy is realized, and the technology integration system is built, and the development mode of interaction, cooperation, integration and evolution is realized.

(2) Structure optimization

Advanced and large-scale technical equipment are adopted and configured in Shougang Jingtang to promote the operational efficiency, product quality and economic benefit. Two blast furnaces, one steelmaking plant, two hot rolling plants and the corresponding four cold rolling and coating production line are constructed by adopting the most advanced technological process in the world and the "2-1-2-4" type manufacturing process structure. The annual production capacity of

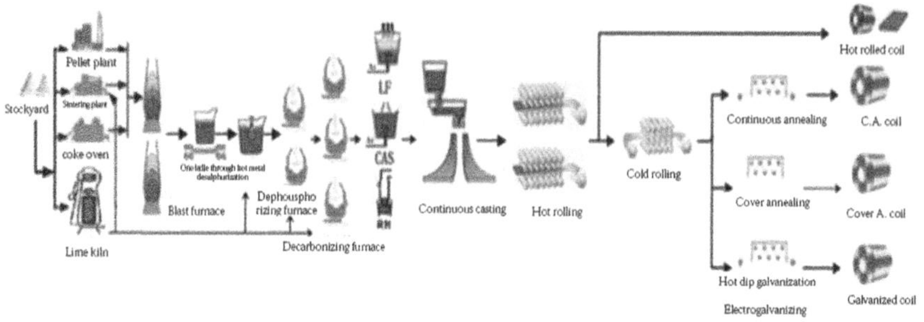

Fig. 1 Iron and steel manufacturing process flow of Shougang Jingtang steel plant

crude steel is between 8.7 and 9.2 million ton. The iron and steel manufacturing process flow of Shougang Jingtang Iron and Steel Plant is shown in Fig. 1.

(3) Function extension

The "Three Functions" concept of new generation steel plant was carried out during the engineering design. The multi-functions of modern steel plant are explored from single function of steel products manufacturing to three functions and circular economic zone.

(4) Efficiency excellence

The coordination relationship between time and space is established, the construction of process network and the relationship between processes are emphasized and the new interface technologies between different operational units are developed and applied. The integration and innovation of top-level design are emphasized, pay attention to the stability, reliability and efficiency of the whole process.

Construction of Full Process Energy Flow Network

Energy Flow Network Design

In the design of energy flow network structure, according to the behaviour and conversion characteristics of energy flow and different energy medium operating process, a perfect energy supply system and energy conversion network system are designed and built based on real-time monitoring and online scheduling. Process control, centrally managed by Energy Control Center [4]. The efficient conversion of energy and the optimal allocation of energy structure are studied in depth and the system is optimized. The secondary energy of the steel manufacturing process is fully recovered, and the waste heat of the steel plant is fully utilized to generate electricity. The power generation rate of iron and steel plant is more than 94%, and all kinds of associated gas in iron and steel metallurgical process are "zero emission".

All kinds of energy media are connected with each manufacturing process through pipeline, and each production process acts as the node on the energy network to become the pivot of energy medium conversion, transmission and storage, so as to realize the energy network management of the whole plant.

In the engineering design of circular economy, green manufacturing, energy saving and emission reducing, lots of advanced large-scale technology and equipment are adopted to improve production efficiency and energy use efficiency. The two sets of 260 t/h high temperature and high pressure dry quenching (CDQ) device are configured to four series 70 batteries 7.63 m coke oven and the generating electricity up to 105 kWh/t. The low calorific value blast furnace gas (BFG) is applied for 5500 m^3 blast furnace hot blast stove combustion, The high efficiency preheating technology of gas and combustion air is applied, the high air temperature dome combustion hot blast stoves are configured, so that the blast temperature reaches above 1250 °C, and the highest monthly average blast temperature reaches 1300 °C [5]. The blast furnace gas cleaning system is innovated by full dry de-dusting technology and high-efficiency top gas recovery turbine (TRT) coupling technology [6]. The top gas residual pressure turbine generator set with installed capacity of 36.5 MW, and the power generation is up to 50 kWh/thm, the dust content in purified gas is 2–4 mg/m^3. The dry dust removal technology is adopted in BOF gas de-dusting system [7], the BOF gas recovery reaches 96 m^3/t and steam recovery reaches 115 kg/t, dust content in purified gas is less than 10 mg/m^3. All kinds of the associated gas in the metallurgical process of the iron and steel plant are recovered completely, used for the combustion and heating of the iron and steel manufacturing process, the surplus gas is used for generating electricity, and the "spent steam" after generating electricity as the heat source for the 50,000 t/d low temperature and multi-efficiency seawater desalination facility, is used for desalination of seawater.

Figure 2 shows the network flow diagram of energy flow operation in Shougang Jingtang Steel Plant.

Technical Feature

- Energy management networking. Through the GIS (Geography Information System) simulation system to realize the accurate space and time positioning in the control of various energy media, and the network management information.
- Real-time online monitoring, scheduling and management are carried out for any energy medium through space and time systems, and operational decision making is modelled.
- The control and forecast model of energy network operation is developed to realize the intelligent management of energy network. Includes process model, cost model and decision model.

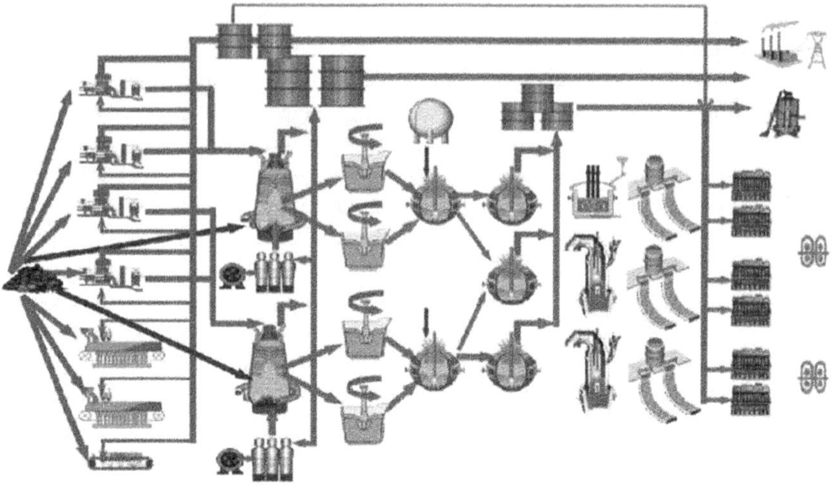

Fig. 2 Operation network of energy flow in Shougang Jingtang plant

Operating Practice

- The systematic energy saving is achieved. The energy saving of the whole process system can save 91,600 ton standard coal per year.
- The gas release rate is reduced. BOF gas is "zero release", coke oven gas emission rate is 0.043% and blast furnace gas is less than 1.96%.
- The pressure fluctuation of gas pipeline network is reduced. The range of pressure fluctuation of blast furnace gas is reduced to ±0.3 kPa.
- The steam pipe network runs stable. The fluctuation range of steam pressure is less than 0.1 MPa, and the steam "zero release" is realized.

Energy Saving and Energy Recycling Practice

Technical Concept

In the design and production, the energy supply structure is optimized, the high-price energy is replaced by the low-price energy. The blast oxygen enrichment and pulverized coal injection technology are adopted to reduce the coke consumption, and the secondary energy (such as coke oven gas, blast furnace gas and converter gas) in the iron and steel production process could be fully recovered to reduce the costs of energy purchase and outsourced energy. All kinds of residual heat generated by various production processes (including sintering, hot blast stove, converter, heating furnace, etc.) are recovered and reutilized. The clean production is implemented, reduce primary energy consumption from the source and reduce emissions of various pollutants. It has been proved that energy saving and consumption reducing are the most effective and economic measures for environmental protection because of the reduction of all kinds of emissions from the iron and steel production source.

Technical Route and Basic Principles

(1) According to the 3R principle of circular economy of "Reduce, Reuse and Recycle", the emphasis of energy saving is on optimizing production technology. Through the active use of energy saving technology and technology in the design, the production process and product structure optimization.

(2) The new energy saving technologies are applied and popularized, such as dry coke quenching and metallurgical gas dry dust removal, to promote the reuse and recycling of resources and energy.

As a result of ironmaking system accounts for about 70% of the total energy consumption of iron and steel production, ironmaking process is the focus of energy saving. The overall implementation of the concentrate policy in the design mainly includes controlling the fluctuation range of grade of ferrous content of raw materials within ±0.3% and ore grade reaching more than 60%. Increasing coke strength, reducing coke ash and sulfur content. Giving play to the advantages of comprehensive synthesis raw material yard, stabilizing blast furnace raw material composition. Adopting the high blast oxygen enrichment, high blast temperature and high pulverized coal injection technologies to reduce the coke consumption. The TRT device power generation for recovery of top gas residual pressure and the waste heat of hot blast stove flue gas are configured to realize the high efficiency energy recovery and conversion. The reasonable burden composition of high basicity sinter ore with acid pellet (or lump ore) is adopted to ensure the blast furnace stable and smooth operation, to realize the long campaign life and to promote the hot metal quality.

(3) The secondary energy and residual pressure, waste heat resources are fully recycled, and the energy structure of enterprises is optimized. The recovery and utilization of blast furnace gas, coke oven gas and converter gas are improved. The comprehensive utilization of all kinds of energy medium on the basis of energy balance is emphasized.

High Efficiency Energy Conversion and Highlight Energy Saving Technology

(1) Advanced large-scale modern technologies and equipment for energy saving and emission reducing are applied.
The main manufacturing process is configured advanced and large-sized facilities, such as 5500 m^3 blast furnace, 500 m^2 sintering machine, 504 m^2 travelling grate induration machine, 7.63 m coke oven, 300 t converter, secondary refining, high efficiency slab continuous casting machine, 2250 mm hot continuous rolling, 1580 mm hot continuous rolling, 2230 mm cold continuous rolling, 1700 mm cold continuous rolling, 1420 mm cold continuous rolling, hot galvanizing and electro tinning, coating production line, etc. These equipment are modern large-scale production facilities with high production efficiency and low energy consumption.

- The large-scale and high-efficiency equipment is integrated and applied
- The 5500 m^3 blast furnace is adopted, compare with the 4000 m^3 blast furnace, the capacity is increased 32% and the fuel ratio is reduced approximate 4.8%.
- The 500 m^2 sintering machine compare with 360 m^2 sintering machine, the production capacity is increased by 62%, the energy consumption is reduced by 8%, the consumption of fuel and electricity is reduced by 10%.
- The 7.63 m coke oven configured with 260 t/h dry coke quenching facility, compare with the 6.0 m coke oven configured with 140 t/h CDQ, the production efficiency is improved 50%, energy consumption is reduced 6.0%, and the CDQ power generation efficiency is improved 12%.

(2) Advanced energy saving and emission reducing technologies are adopted and innovated.

- The two sets of 260 t/h CDQ facility are equipped with two sets 30 MW high temperature and high pressure steam generator. The power output reached 447 million kWh and per ton coke was 113.5 kWh in 2016.
- Dry gas dust removal technology is developed in 5500 m^3 blast furnace. The dust content of purified gas reaches 2–3.5 mg/m^3. Compared with wet dust removal technology, the water saving is 4000 t/d and the electricity saving is 36000 kWh/d. The power generation is increased by more than 45%. The energy saving 7.24 kg standard coal per ton hot metal, and annual emission reduction of CO_2 approximate 169,100 ton.
- 300t converter gas dry dust removal technology, the clean gas dust content is less than 10 mg/m^3, compared with converter gas wet dust removal, save electricity and water about 1/3, respectively, reducing construction land about 1/2, energy saving reach 4.5 kg standard coal per ton steel, and reduce annual CO_2 emissions approximate 107,600 tons.

Practice and Effect of Energy-Saving Technology Innovation

(1) 5500 m^3 super large blast furnace technology is adopted.
The dome combustion hot blast furnace technology is developed and adopted in 5500 m^3 super large blast furnace in the world for the first time. The dome combustion hot blast stove can promote the heat transfer efficiency and reduce the fuel consumption. The hot blast temperature can achieve more than 1250 °C under the condition of only burning blast furnace gas. The comprehensive utilization ratio of energy is increased by 20% compared with another large blast furnace.
The 11,000 m^3/min blast volume (Standard Temperature and Pressure) axial intake blower is adopted, which is the largest in the world and energy saving and high efficiency at present, and a large capacity dehumidification system is provided. Compared with the traditional blast furnace blower, the power consumption is reduced greatly, the power can be saved by 8 million kWh per year and it is beneficial to the stability of the blower operation in summer and to increase the stability of blast furnace smelting.

(2) Clean steel manufacturing platform is constructed.
The hot metal pretreatment of desulphurization, dephosphorization, desilication is developed and adopted firstly in China, and the 4 sets KR device for desulphurization, 2 sets of 300t converter for dephosphorization and desilica tion 3 sets of converter for steelmaking. The process route of "full three removal" is configured for clean steel manufacturing in Shougang Jingtang. A small amount slag steelmaking and total molten steel refining are adopted to create a clean steel production platform. The lime consumption is reduced to 30 kg/t steel when the decarburized slag is returned to the dephosphorization converter.
In order to realize slag splashing and converter protection under the condition of less slag, the technology of triple slag circulation has created a solid foundation for the reduction and reuse of steel slag. By using ladle capping technology, the tapping temperature of converter can be reduced by 9 °C, the consumption of coke oven gas is reduced by 6.5 m^3/t, and the stable operating ratio is more than 99.97%. The total contents of five harmful elements (sulphur, phosphorus, nitrogen, hydrogen, oxygen) in steelmaking can reach the international advanced level of less than 45 ppm.

(3) BF-BOF interface technology is developed and applied.
The hot metal transportation direct from blast furnace to steelmaking plant by open ladle technology is adopted. The temperature of hot metal to the desulphurizing station is 1390 °C, compared with the torpedo tank, the temperature drop of hot metal transportation is reduced by 50 °C, and the dust produced in the process of pouring the ladle is reduced by 4700 ton per year; the lifting height of the crane is reduced, and the electricity consumption is saved by 11.4 million kWh/year. After put into production, the hot metal ladle cover measure is applied, and the drop of hot metal temperature is further reduced.

(4) Large-scale oxygen production technology is configured.
    The VAROX process is adopted in 75,000 m^3/h oxygen generator, it can adjust the load in extremely large range variable, the oxygen output can be adjusted in 36,250–98,750 m^3/h in working condition to require the iron and steel enterprise discontinuous oxygen demand, achieves the oxygen zero emission. It can save the electricity 20 million kWh and save energy 2458 ton standard coal per year.
(5) Energy management center is constructed to strengthen energy management.
    The whole plant energy management center system (EMS) has been built into an intelligent control system which integrates process monitoring, energy management and energy scheduling. The system plays a significant role in reducing gas emission, improving environmental quality, reducing energy consumption, improving labour productivity and energy management level.

Conclusions

Shougang Jingtang Iron and Steel Plant design and construction are based on a new generation of recyclable iron and steel manufacturing process. It has three functions: product manufacturing, energy conversion and waste disposal. Through the optimization of the design, the large-scale technical equipment and energy-saving technology are adopted to effectively construct the energy flow network, to realize the high efficiency energy conversion and utilization. The energy consumption is greatly reduced, and the effect of energy saving and emission reducing is remarkable. The significant achievement and performance have been achieved.

References

1. Yin R (2016) Theory and method of metallurgical process flow integration. Metallurgical Industry Press, Beijing, pp 284–287
2. Yin R (2016) Metallurgical process engineering. Metallurgical Industry Press, Beijing, pp 379–386
3. Zhang F, Xie F (2017) Green iron and steel manufacturing process of Shougang Jingtang plant. In: TMS energy materials 2017 conference proceedings, Mar 2017, pp 17–29
4. Gu L (Mar 2011) Theory and practice of the control and management system of energy in Shougang Jingtang United Iron & Steel Co., Ltd. Metall Autom 35(3):24–28
5. Zhang F, Mei C, Yin G et al (Mar 2012) Design and research of BSK dome combustion hot blast stove of Shougang Jingtang 5500 m^3 blast furnace. China Metall 22(3):27–32 (in Chinese)
6. Zhang F (Jan 2011) Study on large scale blast furnace gas dry bag dedusting technology. Ironmaking 30(1):1–5 (in Chinese)
7. Zhang F, Zhang D, Zhang L et al (Feb 2013) Research and application on large BOF gas dry dedusting technology. Iron Steel 48(2):1–9 (in Chinese)

Dependency of Microstructure and Inclusions on the Different Growth Rate for Directionally Solidified Non-quenched and Tempered Steel

Hui Liu, Jianbo Xie, Honggang Zhong, Qijie Zhai and Jianxun Fu

Abstract Using a Bridgman directional solidification furnace, the solidification experiments of a non-quenched and tempered steel were directionally conducted at a temperature ingredient with various growth rates (v = 10–300 μm/s). The values of the dendrite arm spacing and mean diameter of MnS (d_{MnS}) inclusions in the steady uni-solidification zone were measured. Based on these results, d_{MnS} is more directly related to the secondary dendrite arm spacing (λ_2) than the primary one. Besides, the values of d_{MnS} and λ_2 decrease with increasing growth rate. The relationship between λ_2 and d_{MnS} with growth rate was determined by linear regression analysis. Therefore, $\lambda_2 = 472.06\ v^{-0.375}$, $d_{MnS} = 4.45\ v^{-0.125}$, and the linear fitting exponent values obtained in this work were nearly same with the previous data in similar systems.

Keywords Directional solidification · Dendrite · MnS · Growth rate

Introduction

Quenching and tempering heat treatment cost can reduced during the production process of non-quenched and tempered steel. And the sulphur-containing Non-quenched and tempered steel are mainly used in the automobile industry, engineering machinery, etc. The sulphur existing in the steel is mainly in form of manganese sulphide (MnS) inclusions. As the key factors to the performance of steel, the morphology and distribution of MnS are extensively influenced by the precipitation behaviour of MnS [1–3]. And MnS has a positive effect in the cutting

H. Liu · J. Xie · H. Zhong · Q. Zhai · J. Fu (✉)
School of Materials Science and Engineering, Shanghai University,
Shanghai 200072, People's Republic of China
e-mail: fujianxun@shu.edu.cn

H. Zhong · Q. Zhai · J. Fu
State Key Laboratory of Advanced Special Steel, Shanghai University,
Shanghai 200072, People's Republic of China

© The Minerals, Metals & Materials Society 2019

T. Jiang et al. (eds.), *10th International Symposium on High-Temperature Metallurgical Processing*, The Minerals, Metals & Materials Series,
https://doi.org/10.1007/978-3-030-05955-2_26

performance and can promote the formation of intragranular ferrite nucleation [4, 5]. Further effort is required to understand and better control the parameters dominating the formation and variation of the MnS inclusions during the solidification process.

The foundation of modern directional solidification and single crystal growth techniques was established and firstly presented by Bridgman et al. [6]. Subsequently, directional solidification was widely used to prepare various structural materials [7–9]. The process of the directional solidification will begin after the crucible was placed on the withdrawal rod, which is consist with the process of that in a continuous casting mould. The mode of primary dendrites in columnar zone was regarded as a step-wise succession of localized steady state directional ones [10]. Hence, particular emphasis is placed on the study in the area of columnar zone. It has been previously concluded that the primary dendrite arm spacing does not depend on the composition of steel based on the research on low carbon steels [11] and medium carbon steels [12]. And the spacing is mainly influenced by the temperature gradient and the solid-liquid interface growth rate. In the current study, the microstructure and precipitated MnS under various cooling rate were investigated in medium carbon, non-quenched and tempered steel containing sulphur. And the formation of MnS during the solidification process was calculated using the FactSage software.

Experimental Methods

Directional solidification experiments were carried out in a Bridgman furnace, as shown in Fig. 1. The samples were machined to rods (8 mm in diameter and 100 mm in length) and cleaned in anhydrous ethanol for 15 min to ensure cleanliness, and then the specimen was placed into the 99.99 pct pure alumina crucible.

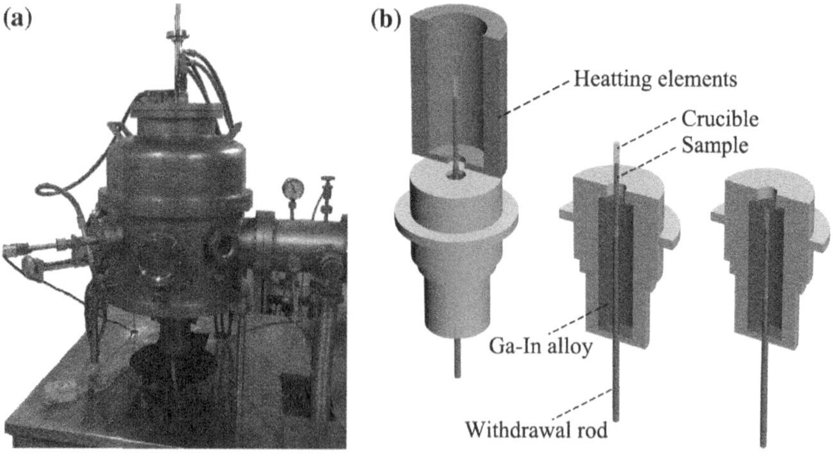

Fig. 1 **a** The directional solidification furnace and **b** schematic of the experimental process

Table 1 Chemical compositions of specimens used in this work (wt%)

steel	C	Si	Mn	P	S	Al	O	N	Cr	Fe
49 MnVS	0.48	0.35	0.9	0.013	0.047	0.012	0.0015	0.007	0.2	Balance

The chemical compositions of samples used in present work were shown in Table 1. The vacuum room was filled with high-pure argon to prevent the sample from oxidation after the chamber of the equipment was vacuumized to the level of 1.0×10^{-4} Pa. The sample was transformed into molten steel entirely after heat preservation 30 min at 1550 °C and then move downward at a constant velocity to form directional solidification microstructure. The sample quickly moves into the Ga-in alloy to preserve the in situ microstructure after a movement of 60 mm. The growth rate of the solid–liquid interface was controlled by the withdrawal rod and set as 10, 50, 100, 200 and 300 μm/s, respectively.

In order to obtain the three dimensional morphology of the concerned MnS inclusions, the most suitable extraction solution, non-aqueous electrolyte (tetramethylammonium chloride–methanol), was adopted in the current study. This solution can avoid a serious dissolution of MnS according to the comparison of different extraction methods conducted by Janis et al. [13]. The solidification structures of the steel samples were observed by optical microscopy (Avio Image. A2 m, Zeiss). The processed samples were characterized by a scanning electron microscope (SEM) combined with Energy Dispersive Spectrometer (EDS) in order to confirm the inclusions precipitated during the solidification process. SEM images were performed on a VEGA3 TESCAN SEM at 20 kV.

Results and Discussion

Effect of growth rate on microstructure

The morphologies of microstructures near the solid–liquid interface at different growth rate were shown in Fig. 2. The solid–liquid interface can be clearly observed at the end of the dendrite tips. The values of dendrite arm spacing are mainly determined by the heat conduction at the solid–liquid interface in directional solidification processing [14] and decrease with the increasing capability of heat conduction. The primary dendrite arm spacing (λ_1) and secondary dendrite arm spacing (λ_2) decreased with the increase of growth rate (V), as shown in Fig. 2f. The microstructure evolution of steels using directional solidification was studied in previous literatures [11, 15, 16].

A minimum square method was used to regression analysis the relationship between secondary dendrite arm spacing (λ_2) and growth rate (V) through the power function ($y = m \times x^n$, m and n are constants) and the experimental values based on the previous researches [12, 17–19]. The fitting result was expressed as Eq. (1), and the regression coefficient of this fit is $R^2 = 0.989$. With a view to

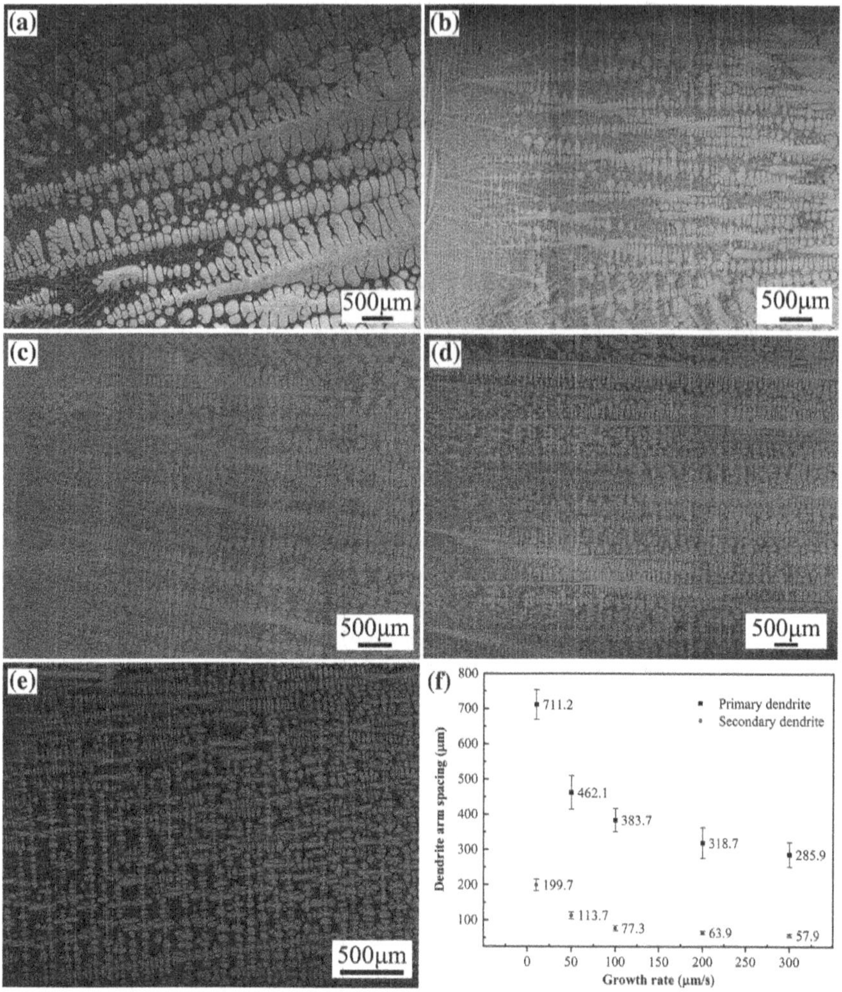

Fig. 2 Dendrite morphologies of samples in longitudinal sections with various growth rate: **a** 10 μm/s; **b** 50 μm/s; **c** 100 μm/s; **d** 200 μm/s; **e** 300 μm/s; and **f** dendritic arm spacing

determining the correlation between the secondary dendritic arm spacing (λ_2) and the growth rate (V), the fitted curves with experimental data were plotted in Fig. 3. The power function Eq. (1) seems to be adequate to characterize the variations of dendritic arm spacing with the growth rate.

$$\lambda_2 = 472.063 \times V^{-0.376} \quad (1)$$

Fig. 3 The relationship between the secondary dendrite arm spacing and the growth rate

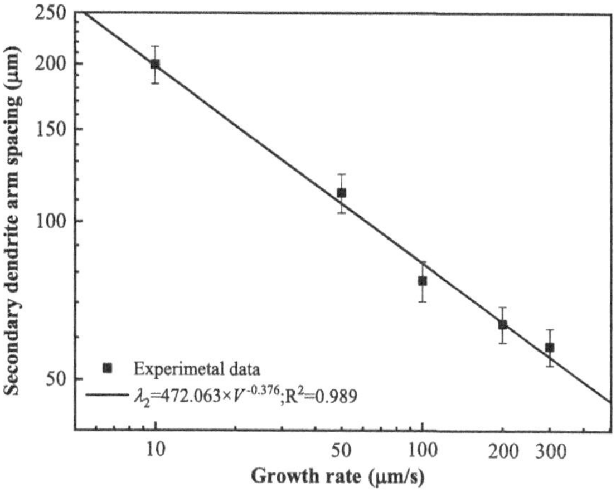

Effect of growth rate on inclusions

Figure 4 shows the metallographic observation of the directional solidification area under different growth rate. The three dimensional morphology of the concerned inclusions after electrolysis processing was shown in Fig. 5, and the EDS analysis indicated that the inclusions were mainly MnS particles. As presented in Fig. 4, the mean diameter of the MnS inclusions (d_{MnS}) were decreased with the increasing of growth rate.

The relationship between temperature and concentration of MnS in molten steel was calculated by the equilibrium solidification model using FactSage 7.0 (Thermfact Ltd., Montreal, Canada and GTT-Technologies, Aachen, Germany) based on the FactPS, Ftoxid and FSstel databases [20–22], as shown in Fig. 6. Figure 6 shows that the MnS phase precipitates rapidly in the mushy zone during the later solidification state. Thus, the formation of the MnS inclusions was closely related to the microstructure during the directional solidification area, especially have a direct correlation with secondary dendrite arm spacing. Similarly, the relationship between the mean diameter of MnS inclusions (d_{MnS}) and the growth rate (V) was obtained by using linear regression analysis, and the results were given as:

$$d_{MnS} = 4.560 \times V^{-0.125} \quad (2)$$

The regression coefficient of this fit is R^2 = 0.934. Figure 7 shows the variations of the mean diameter of MnS (d_{MnS}) with the growth rate (V), and the fitting results verified the strongly relationship between d_{MnS} and V.

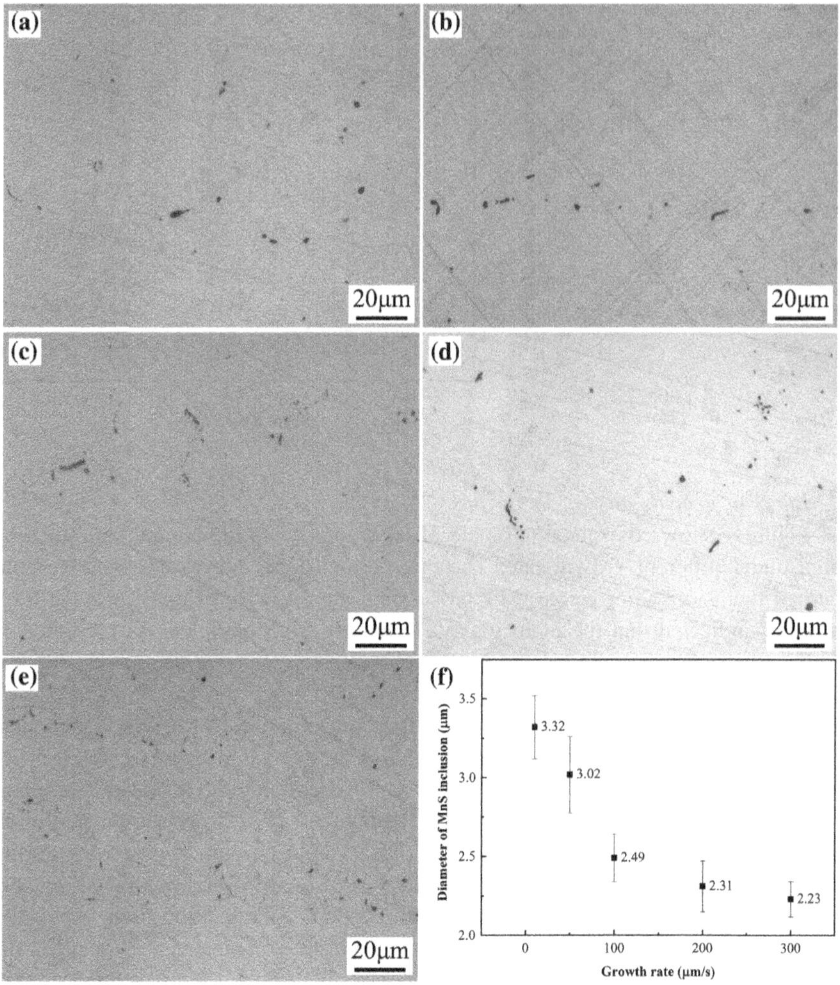

Fig. 4 Metallographic observation of the directional solidification area under different growth rate: **a** 10 μm/s; **b** 50 μm/s; **c** 100 μm/s; **d** 200 μm/s; **e** 300 μm/s; and **f** mean diameter of MnS inclusions

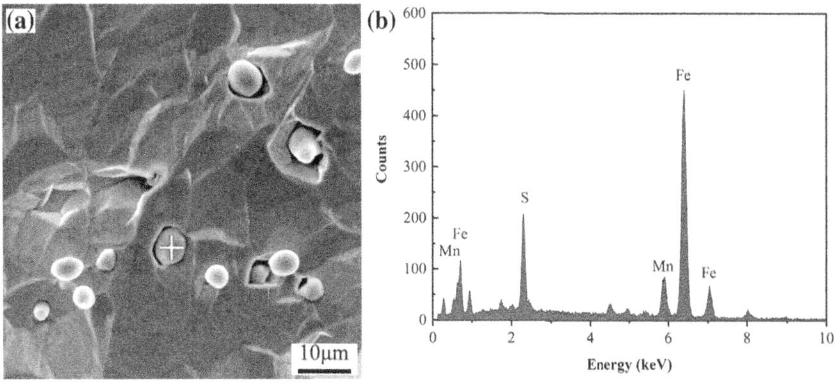

Fig. 5 Morphologies of typical inclusions with EDS analysis

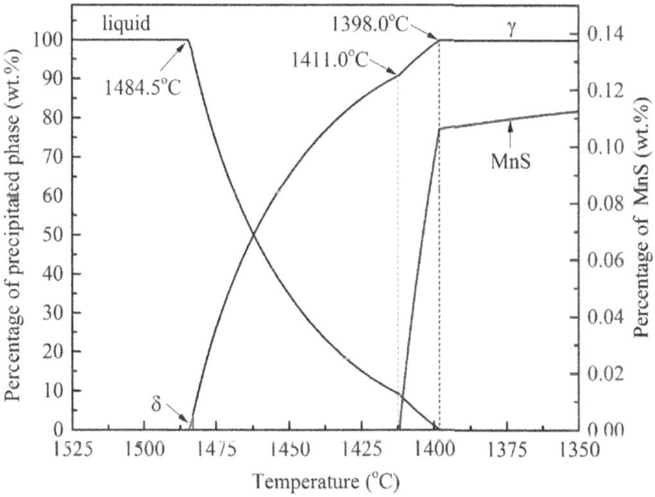

Fig. 6 Phase transformation and MnS precipitation of 49 MnVS steel calculated by FactSage 7.0

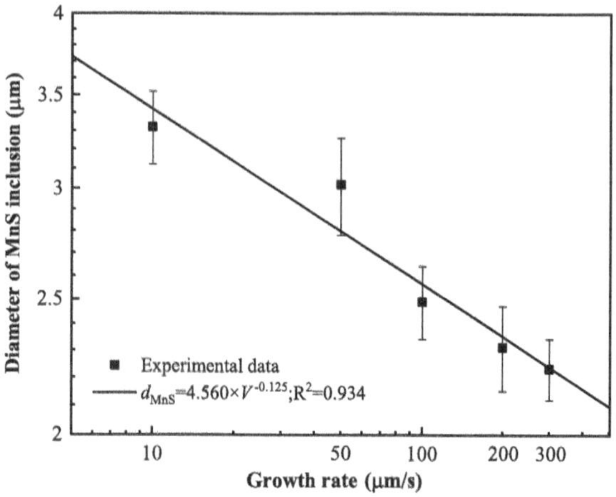

Fig. 7 The relationship between the diameter of MnS inclusions and the growth rate

Conclusions

The dependency of dendrite arm spacing (λ) and mean diameter of precipitated MnS (d_{MnS}) on the interface growth rate (V) of directionally solidified 49MnVS steel was investigated in current work. The dendrites were obtained by using the directional solidification experimental. The MnS particles were precipitated gradually in the residual liquid phase during the later state of the 49MnVS solidification. The values of primary (λ_1) and secondary (λ_2) dendrite arm spacing and mean diameter of precipitated MnS (d_{MnS}) were decreased with the increase of growth rate (V). The relationships between λ_2 with d_{MnS} and V were given by the mathematical expressions as: $\lambda_2 = 472.063 \times V^{-0.376}$; $d_{MnS} = 4.560 \times V^{-0.125}$.

Acknowledgements This work is supported by National Key Research and Development Program of China (2018YFB0704400), and the National Natural Science Foundation of China (Grant numbers: 51671124, 51474142)

References

1. Oikawa K, Ohtani H, Ishida K, Nishizawa T (1995) The control of the morphology of MnS inclusions in steel during solidification. ISIJ Int 35(4):402–408
2. Kim HS, Lee HG, Oh KS (2002) Evolution of size, composition, and morphology of primary and secondary inclusions in Si/Mn and Si/Mn/Ti deoxidized steels. ISIJ Int 42(12):1404–1411
3. Liu JH, Su XF, Liu HB, Han ZB, He Y, Qiu ST (2017) Solidification interface transformation and solute redistribution of FeCrAl stainless stee. Metall Res Technol 114(4):409
4. Madariaga I, Gutiérrez I (1999) Role of the particle–matrix interface on the nucleation of acicular ferrite in a medium carbon microalloyed steel. Acta Mater 47(3):951–960
5. Xiao GH, Dong H, Wang MQ, Hui WJ (2011) Effect of sulfur content and sulfide shape on fracture ductility in case hardening steel. J Iron Steel Res Int 18(8):58–64

6. Liu L, Huang TW, Qu M, Liu G, Zhang J, Fu HZ (2010) High thermal gradient directional solidification and its application in the processing of nickel-based superalloys. J Mater Process Technol 210(1):159–165
7. Zhong HG, Cao X, Chen XR, Zhang JY, Zhai QJ (2013) Numerical and experimental investigation of solidification structure in horizontal directional solidification process of Al-Cu alloy. Chin J Nonferrous Met 10:2792–2799 (in Chinese)
8. Bai L, Wang B, Zhong HG, Ni J, Zhai QJ, Zhang JY (2016) Experimental and numerical simulations of the solidification process in continuous casting of Sla. Met Open Access Metall J 6(3):53
9. Zhong HG, Chen XR, Ma YZ, Ou L, Li RX, Zhai QJ (2013) A method for simulated horizontal growth process of solidification microstructure.
10. Imagumbai M, Takeda T (1994) Influence of calcium-treatment on sulfide—and oxide-inclusions in continuous-cast slab of clean steel-dendrite structure and inclusions. ISIJ Int 34(7):574–583
11. Taha MA (1986) Influence of solidification parameters on dendrite arm spacings in low carbon steels. J Mater Sci Lett 5(3):307–310
12. Taha MA, Jacobi H, Imagumbai M, Schwerdtfeger K (1982) Dendrite morphology of several steady state unidirectionally solidified iron base alloys. Metall Trans A 13(12):2131–2141
13. Janis D, Inoue R, Karasev A, Jönsson P (2014) Application of different extraction methods for investigation of nonmetallic inclusions and clusters in steels and alloys. Adv Mater Sci Eng 2014(2):1–7
14. Kurz W, Fisher DJ (1989) Fundamentals of solidification. Trans Tech Publications, p 71–92
15. Xia YJ, Wang FM, Li CR, Wang JL, Wu ZY (2012) Study on the formation behavior of sulfides in free-cutting steel by unidirectional solidification. J Univ Sci Technol Beijing 34 (2):118–124 (in Chinese)
16. Li X, Gagnoud A, Fautrelle Y, Ren ZM, Moreau R, Zhang YD (2012) Dendrite fragmentation and columnar-to-equiaxed transition during directional solidification at lower growth speed under a strong magnetic field. Acta Mater 60(8):3321–3332
17. Young KP, Kerkwood DH (1975) The dendrite arm spacings of aluminum-copper alloys solidified under steady-state conditions. Metall Trans A 6(1):197–205
18. Jacobi H, Schwerdtfeger K (1976) Dendrite morphology of steady state unidirectionally solidified steel. Metall Trans A 7(6):811–820
19. Bertelli F, Brito C, Ferreira IL, Reinhart G, Nguyen-Thi H, Mangelinck N, Cheung N, Garcia A (2015) Cooling thermal parameters, microstructure, segregation and hardness in directionally solidified Al–Sn-(Si;Cu) alloys. Mater Des 72:31–42
20. Bale CW, Chartrand P, Degterov SA, Eriksson G, Hack K, Ben Mahfoud R (2002) FactSage thermochemical software and databases. Calphad 26(2):189–228
21. Bale CW, Bélisle E, Chartrand P, Decterov SA, Eriksson G, Hack K (2009) FactSage thermochemical software and databases—recent developments. Calphad 33(2):295–311
22. Bale CW, Bélisle E, Chartrand P, Decterov SA, Eriksson G, Gheribi AE (2016) Reprint of: factsage thermochemical software and databases, 2010-2016. Calpha 55:1–19

Development of Offshore Steel for High Heat Input Welding

Xiaodong Ma, Peng Zhang, Tingliang Dong, Feng Wang and Baojun Zhao

Abstract To reduce the construction cost and period of steel structure, high-efficiency welding methods applying high heat input have been widely employed. However, deterioration of the strength and toughness, especially low-temperature impact toughness, due to the coarse grains in the heat-affected zone (HAZ) of weld steels often comes along with high heat input welding. Thick offshore steel has been developed by Ti deoxidization for the use of high heat input welding. The inclusions of the steel were analysed by EPMA and SEM, and the major inclusions were identified to be Al_2O_3-(MgO) and Al_2O_3-CaO-(TiO_2) types. The steel plates with different thickness over 40 mm were welded with heat input over 100 kJ/cm, and Charpy impact energies of the welding joints at −40 °C were over 75 J. The nano-sized TiN particles in HAZ were observed and were considered to retard the grain coarsening in the HAZ during high heat input welding.

Keywords Offshore steel · High heat input welding · Heat-affected zone · Inclusions

Introduction

The demands of thick steel plates are increasingly used in shipbuilding, architectural construction, offshore structure and pipelines. To reduce the construction cost and period of steel structure, high-efficiency welding techniques with high heat input normally over 50 kJ/cm have been widely employed. However, deterioration of the strength and toughness, especially low-temperature impact toughness, due to the coarse grains in the heat-affected zone (HAZ) of the weld steels often occurs with high heat input welding. Those offshore engineering structures, ships and pipelines

X. Ma · B. Zhao (✉)
The University of Queensland, Brisbane, Australia
e-mail: baojun@uq.edu.au

P. Zhang · T. Dong · F. Wang
Hebei Iron and Steel Group Co., Ltd., Shijiazhuang, China

© The Minerals, Metals & Materials Society 2019

T. Jiang et al. (eds.), *10th International Symposium on High-Temperature Metallurgical Processing*, The Minerals, Metals & Materials Series,
https://doi.org/10.1007/978-3-030-05955-2_27

have high requirements for the weldability and mechanical properties of the weld joints. To improve the HAZ toughness of high heat input welding joints, oxide metallurgy technology has been developed by employing the fine inclusion particles to control the microstructure of HAZ. [1, 2] The concept of oxide metallurgy not only included the utilization of oxide inclusions, but also extended to use nitride, sulphide and carbide particles. Basically, the inclusions or particles took effect in either pinning at the austenite grain boundary to suppress the grain coarsening at elevated temperatures or being a nucleate site of acicular ferrite growth during austenite-ferrite transformation. [3–5] This study will focus on the development of offshore steel by Ti deoxidation for high heat input welding, as well as illustrating the mechanism of improving HAZ toughness after high heat input welding.

Experimental

The samples employed in this study were as-received industrial EH40 thick plates. The manufacturing process of thick plates was featured by "BF → Convertor → LF → VD → CC → Hot Rolling → TMCP → Plates". The chemical composition of the EH40 steel is shown in Table 1. The yield strength of EH40 steel is over 450 MPa, which is usually used for the construction of a ship hull. High heat input welding was applied to weld the EH40 thick plates with a thickness of 40, 50, 80 and 100 mm, respectively. The 40 mm plate was welded by electro-gas welding with a heat input of 300 kJ/cm. The plates with a thickness of 50, 80 and 100 mm were welded by flux copper backing submerged arc welding with a heat input of 100 kJ/cm. The Charpy impact energy of welding joints was above 75 J at −40 °C impact test. The microstructures of welding joints were examined by OM and SEM after grinding, polishing and etching of 2% Nital solution. The inclusions in the steel samples were measured by EPMA. The TEM specimens were prepared by FIB-SEM. The second phase particles of welding joints were analysed by TEM.

Results and Discussion

1. Inclusions in the EH40 Steel

The inclusions in the EH40 steel by Ti deoxidation were analysed by EPMA. The inclusions were classified by their major oxide components. The types of

Table 1 The chemical composition of EH40 steel (wt%)

C	Si	Mn	P	S	V	Ni	N	Ti	Cr	Mo	Cu	C_{eq}
0.06	0.14	1.39	0.006	0.005	0.002	0.11	0.0041	0.016	0.04	0.07	0.01	0.32

$C_{eq} = C + Mn/6 + (Cr + Mo + V)/5 + (Cu + Ni)/15$

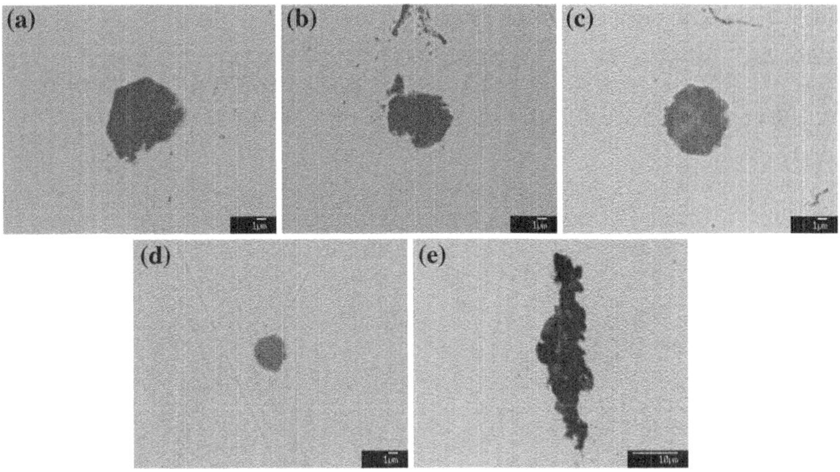

Fig. 1 Different types of inclusions in the steels. **a** Al_2O_3, **b** Al_2O_3-MgO, **c** Al_2O_3-CaO-"TiO_2", **d** Al_2O_3-CaO, **e** Al_2O_3-CaO-SiO_2-"TiO_2"

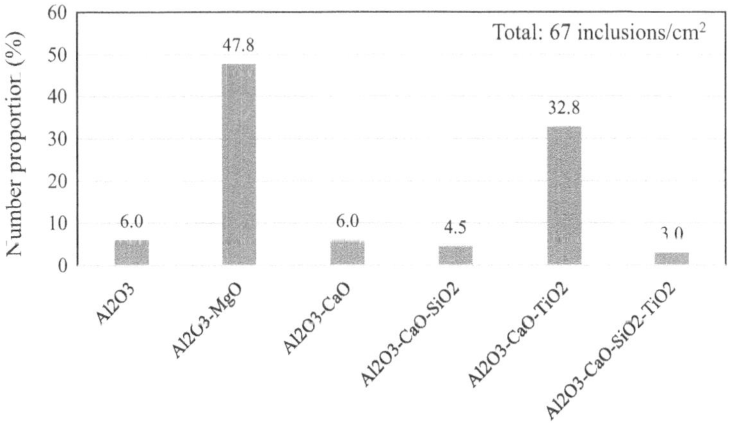

Fig. 2 Number proportion of different type inclusions

inclusions are shown in Fig. 1. The number density of total inclusions larger than 1 μm is around 67 inclusions/cm^2. The number proportion and size of each type of inclusion are shown in Fig. 2 and Table 2. The major inclusions are Al_2O_3-MgO and Al_2O_3-CaO-TiO_2 inclusions, which account for 80% of total inclusions. The compositions of Al_2O_3-MgO-(CaO) and Al_2O_3-CaO-TiO_2 are normalized and plotted in Fig. 3. The formation of major inclusions can be explained by the following reactions taking place in molten steel during Ti deoxidation.

Table 2 The size range of different inclusions

Inclusion type	Al_2O_3	Al_2O_3-MgO	Al_2O_3-CaO	Al_2O_3-CaO-SiO_2	Al_2O_3-CaO-TiO_2	Al_2O_3-CaO-SiO_2-TiO_2
Size (μm)	3.7–8.2	2.8-11.4	3.6–7.2	4.3–4.5	3.0–8.1	4.2–41.6

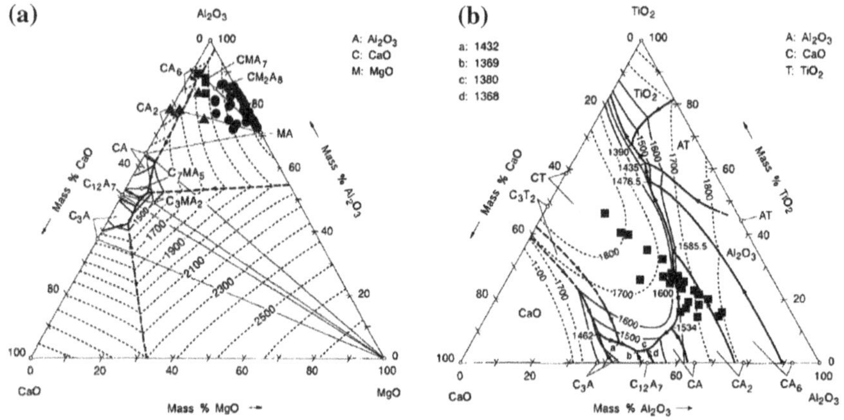

Fig. 3 Composition distribution of **a** Al_2O_3-MgO-(CaO) and **b** Al_2O_3-CaO-TiO_2 inclusions

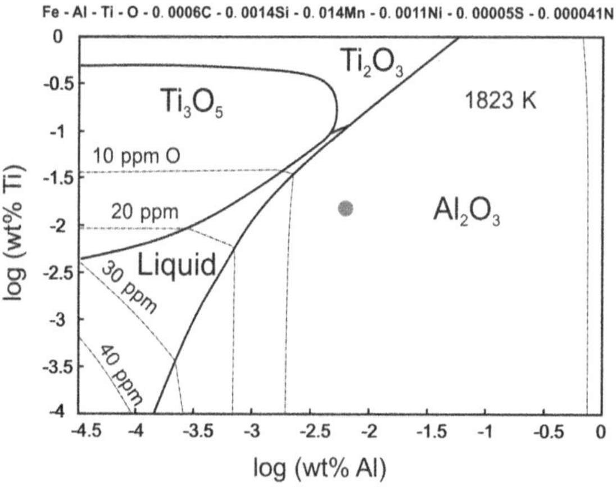

Fig. 4 Stability diagram of inclusions at 1600 °C

$$[Al] + [O] \rightarrow Al_2O_3 + MgO \rightarrow Al_2O_3 - MgO \quad (1)$$

$$[Al] + SiO_{2\,\text{in inclusion}} \rightarrow [Si] + Al_2O_{3\,\text{in inclusion}} \quad (2)$$

$$[Ti] + SiO_{2\,\text{in inclusion}} \rightarrow [Si] + TiO_{2\,\text{in inclusion}} \quad (3)$$

FactSage software was used to calculate the inclusion stability with the thermodynamic databases to steelmaking processes [6]. The inclusion diagram of EH40 steel at 1600 °C is calculated and shown in Fig. 4. It can be seen from Fig. 4 that the Al_2O_3 inclusion is stable at a very low level of dissolved Al in molten steel. Al is usually introduced into the steel through ferroalloys. TiO_x inclusion is not thermodynamically stable when a trace amount of dissolved Al is present.

2. Microstructures of Welding Joints after High Heat Input Welding

The microstructures of FCB welded joint were observed by OM and shown in Fig. 5. The typical structure of weld joint consists of a base metal (BM), heat-affected zone (HAZ) and weld metal (WM). The HAZ mainly consists of acicular ferrites (AF), bainites (B) and grain boundary ferrites (GBF). The average grain size from base metal to weld metal of weld joints is shown in Fig. 6. The grain sizes were coarsened in HAZ, and they increased with increasing heat input energy of welding. For the same heat input of welding, the average grain size of welding joints was similar among different thickness plates.

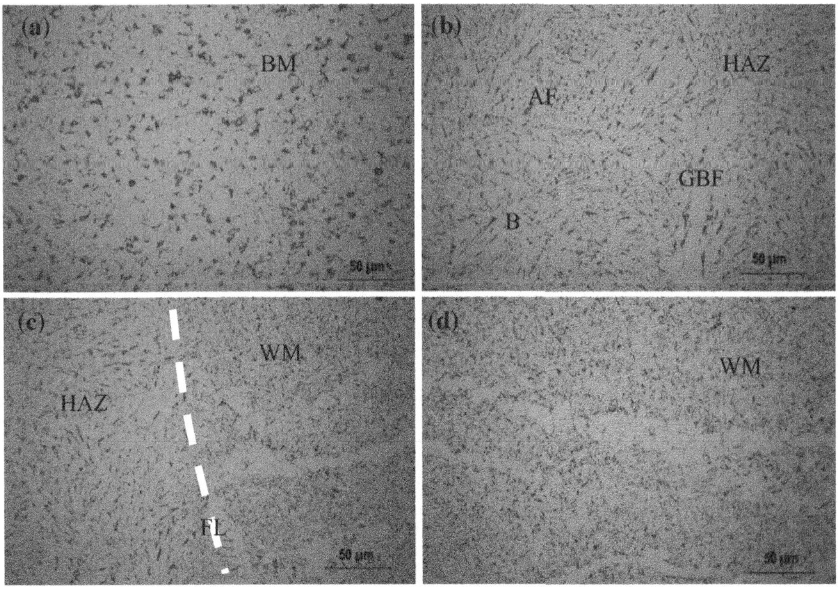

Fig. 5 Microstructures of FCB welded joint of 50 mm thickness and 100 kJ/cm heat input **a** base metal, **b** HAZ, **c** fusion line (FL), **d** weld metal

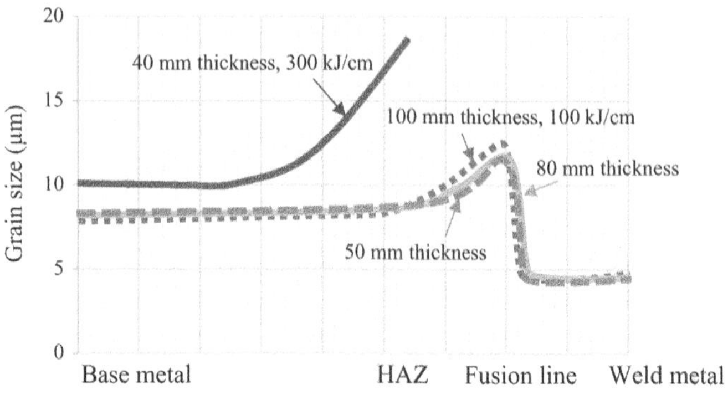

Fig. 6 The average grain size of different area of weld joints

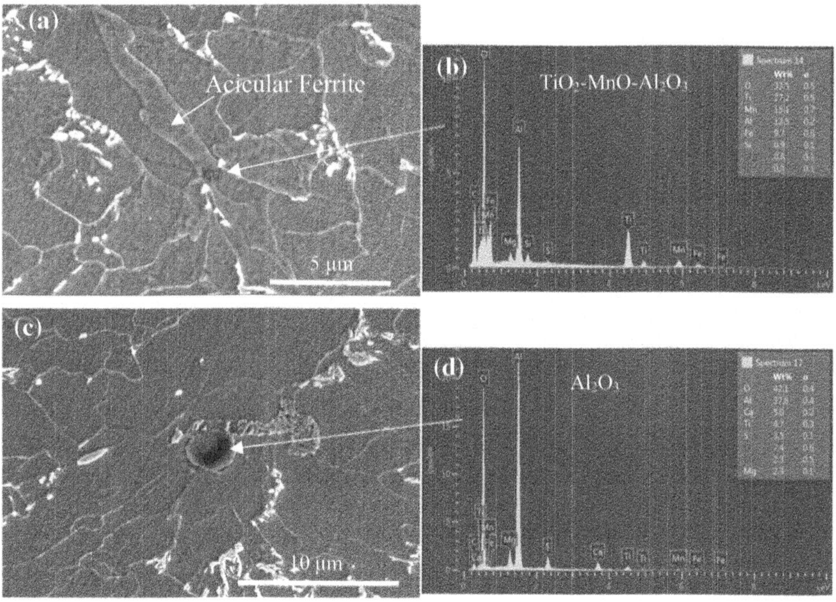

Fig. 7 Microstructures of HAZ (**a**), (**c**) and EDS of inclusions (**b**), (**d**)

3. Second Phase Particles of Welding Joints

The effect of inclusions on the transformation of acicular ferrite was observed. Figure 7 shows the ferrite morphologies on the inclusions. In Fig. 7a, the TiO_2-MnO-Al_2O_3 inclusion worked as an active nucleate site of acicular ferrite, while the Al_2O_3 was an inert inclusion for the nucleate site of acicular ferrite in Fig. 7c. However, it was noted that the number density of active inclusions for acicular ferrite growth is very low, so the contribution of inclusions to the improvement of HAZ toughness was limited.

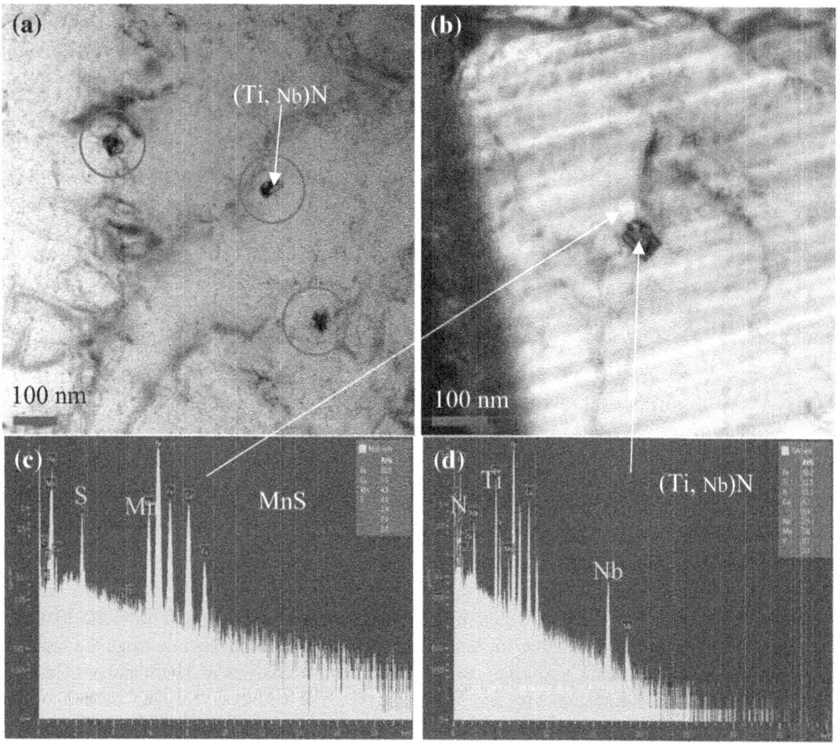

Fig. 8 Nano-sized particles in HAZ (**a**), (**b**) and EDS of the particles (**c**), (**d**)

To clarify the mechanism of HAZ toughness improvement after high heat input welding, the fine second phase particles of HAZ were analysed by TEM. Figure 8 shows the distribution of fine (Ti, Nb)N particles. In order to minimize the coarse

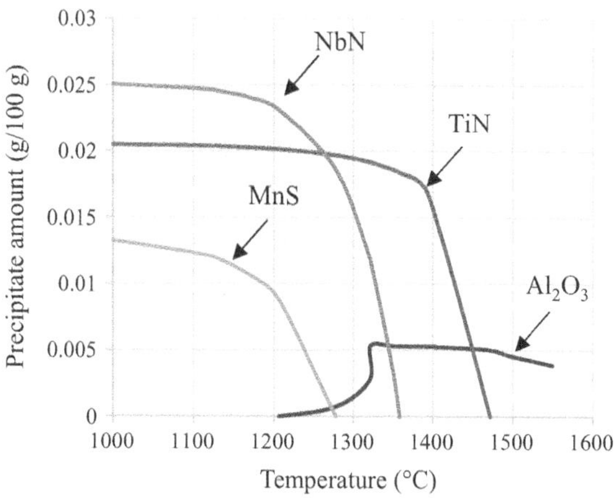

Fig. 9 Precipitation of second phase particles during equilibrium cooling by FactSage calculation

grain region after high heat input welding, it was necessary to control austenite grain growth at high temperature. Nitrides, which were stable at elevated temperatures, were generally used to control the austenite grain growth. FactSage calculation shows the precipitation of TiN above 1400 °C in the Fig. 9, which was closed to the peak temperature of HAZ during the high heat input welding. Therefore, the nano-sized TiN particles in HAZ are considered to retard the grain coarsening in the HAZ during high heat input welding.

Conclusions

The EH40 offshore steels have been developed for the application of high heat input welding, with the excellent low-temperature impact toughness. The inclusions present in the steels were mainly Al_2O_3-MgO and Al_2O_3-CaO-TiO_2. Ti element was found to be present as nitride particles and Al_2O_3-CaO-TiO_2 inclusions. The nano-sized TiN particles in HAZ were observed and were considered to retard the grain coarsening in the HAZ during high heat input welding.

Acknowledgements The authors would like to thank the financial support from HBIS Group through HBIS-UQ Innovation Centre for Sustainable Steel. The authors acknowledge the facilities, and the scientific and technical assistance, of the Australian Microscopy & Microanalysis Research Facility at the Centre for Microscopy and Microanalysis, The University of Queensland.

References

1. Ogibayashi S (1994) Advances in technology of oxide metallurgy. Nippon Steel Techn Rep 61:70–76
2. Mu W, Jonsson PG, Nakajima K (2017) Recent aspects on the effect of inclusion characteristics on the intragranular ferrite formation in low alloy steels: a review. High Temp Mater Processes 36:309–325
3. Sarma DS, Karasev AV, Jonsson PG (2009) On the role of non-metallic inclusions in the nucleation of acicular ferrite in steels. ISIJ Int 49:1063–1074
4. Kojima A, Kiyose A, Uemori R, Minagawa M, Hoshino M, Nakashima T, Ishida K, Yasui H (2004) Super high HAZ toughness technology with fine microstructure imparted by fine particles. Nippon Steel Tech Rep 90:2–6
5. Tomita Y (1994) Improvement in HAZ toughness of steel by TiN-MnS addition. ISIJ Int 34:829–835
6. Jung IH (2010) Overview of the applications of thermodynamic databases to steelmaking processes. CALPHAD: Comput Coupling of Phase Diagrams and Thermochemistry 34: 332–362

Slag Basicity: What Does It Mean?

G. A. Brooks, M. M. Hasan and M. A. Rhamdhani

Abstract The concept of "basicity" has been central to our understanding of slag chemistry for many decades. Traditionally, basicity has been connected to the level of networking present in a slag, acid slags having a great deal of networking and basic slag having less networking. Qualitatively, this approach provides quite reasonable explanations to the viscosity of slags, their general interaction with refractories and their ability to dissolve oxidised elements from the metal/matte phase. Various quantitative measures of basicity have been proposed, such as optical basicity and the NBO/T ratio, and there have some successful attempts to link these measurements to important properties. However, in general, these measures have not been successfully linked to common thermodynamic relationships (other than empirically) or direct measurements from the plant. In industry, weight ratio's and empirical relationships dominate the control logic of the metallurgical processes. In this paper, we will review what is the current state of knowledge of "basicity", recent research into connecting NBO/T ratios to quantitative measurements and thermodynamic quantities and propose a way forward towards a more rigorous and useful understanding of "basicity".

Keywords Basicity · Slag chemistry · Slag structure · Slag properties

Introduction

It is well understood in pyrometallurgy that tight control of slag chemistry is critical to controlling metal chemistry, providing process stability, increasing refractory life and often, improving productivity of the process. One of the most famous examples of the importance of controlling slag chemistry was the development of the use of lime-based refractories in the Bessemer process by Gilchrist and Thomas in the

G. A. Brooks (✉) · M. M. Hasan · M. A. Rhamdhani
Swinburne University of Technology, Melbourne, Australia
e-mail: gbrooks@swin.edu.au

© The Minerals, Metals & Materials Society 2019

T. Jiang et al. (eds.), *10th International Symposium on High-Temperature Metallurgical Processing*, The Minerals, Metals & Materials Series,
https://doi.org/10.1007/978-3-030-05955-2_28

early 1880s [1]. The underlying principle of that development was the idea that by adding a "basic" oxide (CaO) into the system would stabilise an "acid" oxide impurity (P_2O_5) in the slag phase and limit the presence of the impurity in the metal. This idea is extended to the choice of refractories for a process; a "basic" refractory will have limited reaction with a "basic" slag but will dissolve readily into an "acid" slag.

However, in slag chemistry literatures, whilst the concept of "basicity" is much referred to, the definitions provided are often quite qualitative and difficult to directly relate to measurements and thermodynamic calculations. Some quantitative definitions have put forward, such as "V" ratios and optical basicity, but these quantities have obvious weaknesses and can only be connected to thermodynamics via empirical relationships [2–6]. Rosenqvist was pessimistic when he stated "Thus, an absolute measure of slag basicity does not exist and we cannot compare the basicity of an oxide with that of another on an absolute scale" [5]. In this paper, some of the central concepts of "slag basicity" will be examined, particularly, recent attempts to link ratios of bonding types (NBO/T) to quantitative measurements (FTIR and Raman spectroscopy) to properties and thermodynamic descriptions of slags.

Concepts of Acid and Base

The idea that some fluids are acid and others are base is an ancient idea relating to observations about how different fluids react, smell and taste [7]. For much of human history, chemistry developed as a practical art and the theoretical basis was tied up with philosophical and metaphysical descriptions of nature [8]. Arrhenius provided the first rigorous definition of acid and basic solution in 1884. In his definition, the addition of an acid compound to water would increase the concentration of H^+ ions in the water, whereas a basic compound would increase the concentration of OH^- ions in the solutions [7]. For example, when HCl and NaOH are dissolved into water, the following reactions occur:

$$HCl_{(aq)} = H^+_{(aq)} + Cl^-_{(aq)} \tag{1}$$

$$NaOH_{(aq)} = Na^+_{(aq)} + OH^-_{(aq)} \tag{2}$$

These two can react to neutralise each other and, in an aqueous environment, produce water and a salt (NaCl). We now understand that acidity in aqueous solutions is connected to the concentration of H_3O^+ ions, as protons are not stable in solution, but it is still common to refer to "H^+" concentration, such as in the definition of pH, in describing acids. Arrhenius's concept was expanded to a more general definition by Bronsted and Lowry in 1923, where they defined an acid as a proton donor and base compound as a proton acceptor. For example, the reaction between hydrochloric acid and ammonia to form a salt can take place in an aqueous environment (as per Arrhenius's definition) but also in other solvents (e.g. benzene) [7]:

$$H_3O^+_{(aq)} + Cl^-_{(aq)} + NH_3 = Cl^-_{(aq)} + NH^{4+}_{(aq)} + H_2O \quad (3)$$

$$HCl_{(benzene)} + NH_{3(benzene)} = NH_4Cl_{(s)} \quad (4)$$

Lewis provided an alternative definition of an acid in 1932, as an electron-pair acceptor. In this definition, reaction (3) involves the oxygen atom in H_3O^+ accepting an electron pair in reaction with a base compound (electron donor). Bronsted and Lowry's definition is what is generally used in chemistry today, but the term "Lewis acid" is reserved for compounds that can accept a pair of electrons but do not contain ionisable hydrogen.

Acidic and Basic Oxides

Solid oxides are well known to react with solutions and their behaviour can be used classified using the concepts of acids and bases developed from aqueous chemistry. For example, phosphorous oxides react readily with water to form an acid:

$$P_4O_6 + 6\ H_2O = 4\ H_3PO_4 \quad (5)$$

In this situation, the acid oxide is behaving like an Arrhenius acid (providing H^+ ions from the dissolution of acid product in water). Basic oxides form hydroxide compounds (e.g. sodium oxide reacts with water to form sodium hydroxide). Acid oxides are formed from non-metals reacting with oxygen, and basic oxides are formed from the oxidation of metals, with oxides formed from alkali metals and alkali earth metals being the most basic. [9] Some elements, such as aluminium, form an oxide that can behave like acidic or basic in solution, and this behaviour is termed "amphoteric" in oxide chemistry, a term also used to slag chemistry. This concept of acid and basic solid oxides is loosely connected to the concept of acidity and basicity used in slag chemistry. The solid-state version of acid/base relationship refers to a reaction between a solid and an aqueous environment, where the molten oxide concept relates to bonding and structural changes in the molten state.

Basicity in Slag Chemistry

In 1947, Flood and Forland related this concept of acid and basic oxides to slag behaviour [10]. They built on earlier work by Lux [11] in defining basic and acid oxides in terms of exchanging the O^{2-} ion, expressing the key reactions as:

$$\text{Base} = \text{Acid} + O^{2-} \tag{6}$$

This definition cannot be directly related to Bronsted and Lowry's definitions of acid and base, and Flood and Forland described the connection as being by "analogy". Providing the following examples:

$$SiO_4^{4-} = SiO_3^{2-} + O^{2-} \tag{7}$$

$$SiO_3^{2-} = SiO_2 + O^{2-} \tag{8}$$

$$CaO = Ca^{2+} + O^{2-} \tag{9}$$

Flood and Forland than argued that adding an oxygen ion (donated from a basic oxide) cause bridged two silicon ions (an "acid") to break down into its corresponding base, expressed as:

$$\vdots Si - O^{2-} - Si\vdots + O^{2-} \rightarrow 2\left[\vdots Si - O^{2-}\right] \tag{10}$$

In our modern understanding of this reaction, the introduction of a base has broken down the network of the acid silica structure, but modern slag chemists would not refer to the smaller silica network structure as "basic". This concept of basicity is linked to the ionic model of slags, which was developed later into a general treatment of slag metal equilibria by Flood and Grojtherm [12]. In this model of a molten slag, "O^{2-}" ions have an activity within the melt, which is a troubling idea, as no such entity exists within molten oxides [13].

The concept of acid oxides as "network formers" and basic oxides as "network breakers", formulated by Richardson [14] around the same period of Flood and Forland's paper, had the following central tenements:

- Acid oxides in slags behave as network formers and formed from oxide species where covalent bonding dominates over ionic bonding (i.e. the elements bonding with oxygen have a high Pauling electronegativity).
- Basic oxides in slags behave as network breakers and formed from oxide species where ionic bonding dominates over covalent bonding (i.e. the elements bonding with oxygen have a low Pauling electronegativity).
- As a result, acid oxides tend to form highly viscous oxide melts and the addition of basic oxides tends to lower the viscosity of an acid melt, through breakdown of the network structure, as shown schematically in Fig. 1.

This approach to slag basicity successfully connects chemical behaviour (CaO stabilising P in a slag) with bonding at the molecular level and physical properties. The key elements of the theory are broadly consistent with observed data, namely (i) the addition of basic components to an acid slag will lower the viscosity and (ii) basic oxides will readily dissolve into acid oxides and vice versa. The

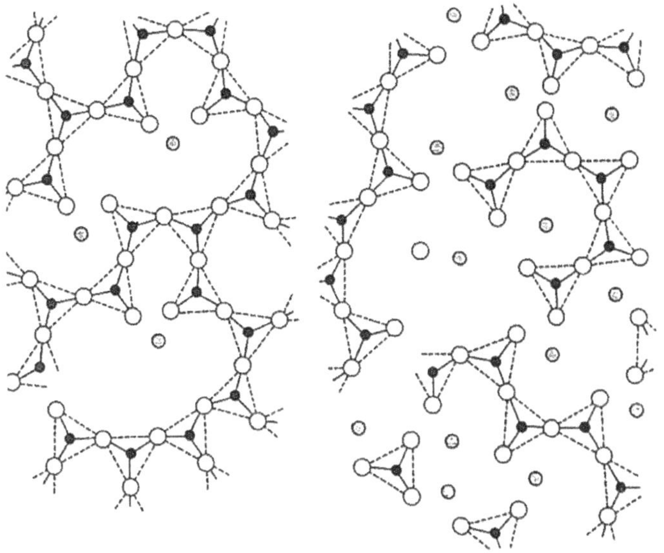

Fig. 1 Schematic representation of the change in a slag structure when the concentration of network breakers (basic oxides species) is added to a silica-rich melt and the network structure is broken down [6]

underlying principle is that acid oxides are predominately covalently bonded and basic oxides are dominated by ionic bonding. This concept is simple to understand and conceptually elegant but leaves some critical questions, namely:

- What is the correct quantitative measure of basicity using these concepts? One candidate is molar ratios, but this would need some modification to reflect different "strengths" of the acid and basic components. Another candidate would be a measure of networking in the melt. How could this be measured and what scale should be used?
- The degree of polymerisation of a slag is also a function of temperature and concentration of species. Can the effects of temperature and concentration on networking in the molten slag be quantitatively linked to a measure of basicity?
- How is this definition of basicity linked to chemical thermodynamics? Can the basicity of a slag be mathematically related to thermodynamic quantities such as Gibbs free energy and activity?
- Can this approach describe quantitatively the amphoteric nature of some oxides? There is experimental evidence that some oxides (notably alumina) behave as an acid in certain slags but can behave as a basic oxide. In this theoretical structure, there is no apparent quantitative approach to predicting these changes in chemical behaviour.

There have been numerous studies to address these issues and some progress has been made but a widely accepted quantitative measurement of basicity based on the "network former/network breaker" theoretical approach has not emerged. In industry, weight ratios of basic components over acid components are extensively expressed as basicity and much of the slag literature listing properties (viscosity, activity of species, partitioning of particular element in equilibrium between a metal and slag) is listed against weight ratios. Some attempts have been made to include factors to compensate for the different "strengths" of oxide components. For example, Turkdogan [6] provides five different definitions of basicity for steel-making, all based on weight ratios. None of these definitions can deal with the issues identified above and are in essence empirical definitions of basicity that provide some guidance to operators.

There have been also various attempts to relate basicity with other characteristics of the oxide that should relate to the level of networking associated with a particular oxide. [3, 13] Some of these values are provided in Table 1 for comparison. The quantities compared are: (a) ionic bond fraction—a measure of how much of the bond between the cation and oxygen ion is ionic based on calculation using the Pauling Electronegativity of the two ions [3]; (b) ion-oxygen attraction (z/a^2)—a numerical calculation to identify the electrostatic bond strength and to relate basicity with it. Here, 'z' is the cationic charge and 'a' is the sum of the ionic radii of the ions [13]; and (c) optical basicity—which is based on physical measurements involving absorption of light to determine the average negative charge borne by the oxygen atoms (charges varies due to electronegativity differences and delocalization effects) [1].

These approaches are more sophisticated than weight rations, and in the case of optical basicity, a body of literature have been generated that relates optical basicity

Table 1 Properties associated common slag components and various measures of basicity [3, 13]

Oxide	Ionic bond fraction [3]	Ion-oxygen attraction, Z/a^2 [3]	Optical basicity (Pauling Electronegativity) [13]	Optical basicity (electron density) [13]	Character of oxide [3]
Na_2O	0.65	0.18	1.15	1.1	Basic
BaO	0.65	0.27	1.15	1.08	
SrO	0.61	0.32	1.07	1.04	
CaO	0.61	0.35	1	1	
MnO	0.54	0.42	0.59	0.95	Intermediate
FeO	0.47	0.44	0.51	0.93	
MgO	0.44	0.69	0.78	0.92	
Cr_2O_3	0.41	0.75	0.55	0.69	
Fe_2O_3	0.41	0.75	0.48	0.69	
Al_2O_3	0.38	0.83	0.605	0.68	
TiO_2	0.36	0.93	0.61	0.64	
SiO_2	0.36	1.22	0.48	0.47	Acidic
P_2O_5	0.28	1.66	0.4	0.38	

(normally calculated from electronegativity values) to important slag properties and chemical behaviours [15–18]. However, whilst optical basicity can provide useful empirical correlations (e.g. sulphide capacity of ladle slags [13]), these values have not been successfully linked to fundamental thermodynamics or measurements in the molten state. Wagner also proposed measuring basicity as a ratio of carbonate capacity over a reference carbonate capacity [19]. This approach has the challenge of connecting basicity to thermodynamics but includes quantities not easy to measure or link to structural units.

Structural Studies of Silicates

In progress to the search of basicity in a quantitative fashion, it was not possible until the discovery of Raman spectroscopy in 1928. In the early stages of the development of the technique, the field was dominated by glass researchers studying the different Raman signals of simple alkali silicate systems. In the 1980s, geologists progressed with the measurement of the vibrational characteristics of molten alkaline earth-silicate systems [20]. They assumed that silicate melt to consist of bulk NBO/Si (non-bridging oxygen per silicon atom) ratio for the melt [NBO/Si = (2 × O − 4 × Si)/Si, between 4 (orthosilicate) and 0 (tectosilicate)]. It was clear that the abundance ratio of low NBO/Si units (2 to 1) decreases with increasing ionization potential of the metal ions. They also identified the NBO/Si ratio as a measure of the degree of polymerisation of the melt, which was later widely expressed as NBO/T (non-bridging oxygen per SiO_4 tetrahedra). In 1982, Mysen's group deconvoluted Raman spectra to quantify the different structural units of silicate melts [20]. They successfully characterised the vibrational modes of the different silicate units with respect to NBO/T, as listed in Table 2.

Mysen et al. [20] also expressed the equilibrium of the anionic structural units with a number of disproportion reactions at different range of NBO/Si ratio (Eq. 11–13).

For melts with NBO/Si $\geq$ 2:

$$2Si_2O_7^{6-} \leftrightarrow Si_2O_8^{8-} + Si_2O_6^{4-} \tag{11}$$

Table 2 Raman frequency characterisation related to specific silicon anionic structural units [20]

Structural unit	NBO/T	Frequency (cm^{-1})	Characteristics of the vibrational mode
$Si_2O_8{}^{8-}$	4	850–870	s, p; symmetrical stretch
$Si_2O_7{}^{6-}$	3	3905–920	s, p; symmetrical stretch
$Si_2O_6{}^{4-}$	2	2970–950	s, p; symmetrical stretch
$Si_2O_5{}^{2-}$	1	1050–1100	s, p; symmetrical stretch
$Si_2O_4^0$	0	1065, 1200	vw, dp; antisymmetrical stretch

For melts with 2 > NBO/Si $\geq$ 1:

$$3Si_2O_6^{4-} \leftrightarrow Si_2O_8^{8-} + 2Si_2O_5^{2-} \quad (12)$$

For melts with NBO/Si < 1:

$$2Si_2O_5^{2-} \leftrightarrow Si_2O_6^{4-} + Si_2O_4^{0} \quad (13)$$

Jong and Schramm [21] used magic angle spinning nuclear magnetic resonance (MAS-NMR) to study the local structure of Li_2O-SiO_2 glass system. They successfully expressed the abundance of different anionic structural units in the system and the evolution of the structure with adding Li_2O. They suggested the Q_m distribution theory, the representation of the structure in terms of the number of bridging oxygen present per tetrahedral unit (later this was expressed as Q^n, and here, n is the number of bridging oxygen, ranging from 0 to 4). Some researchers expressed the structure of silicates in terms of the number of O^0 (bridging oxygen), O^- (non-bridging oxygen) and O^{2-} (free oxygen) present in the melt. The reaction of metal oxide with silicates using this nomenclature is shown in Eq. 14.

$$\left(Si - O^0 - Si\right) + M^{2+} + O^{2-} = (Si - O^-)\ldots M^{2+}\ldots(Si - O^-) \quad (14)$$

It is possible to calculate the bulk bridging oxygen (BO) or NBO/T from the different Q units using Eqs. 15 and 16.

$$NBO = 4 - BO \quad (15)$$

$$BO = \left(0 \times Q^0\right) + \left(1 \times Q^1\right) + \left(2 \times Q^2\right) + \left(3 \times Q^3\right) + \left(4 \times Q^4\right) \quad (16)$$

Along with diffraction (electron, x-ray and neutron) technique, Raman, FTIR and NMR spectroscopes, X-ray photoluminescence spectroscope (XPS), X-ray absorption fine structure (XAFS), X-ray absorption near edge structure (XANES) and Mossbauer spectra have also been used by different researchers to study the structure of silicates [22–25]. The use of different techniques sometime gives complimentary information that can be used to understand the structure in a better way. For instance, Mossbauer and XPS spectroscopy can provide information about valence state and the coordination environment of different iron species in the silicates. FTIR spectroscopy is sensitive to hetero-nuclear functional group (asymmetric bonds) vibrations and polar bonds, and Raman is sensitive to homo-nuclear molecular bonds (symmetric bonds). Both Raman and FTIR techniques have been used in a complimentarily way to study bond length, bond angle and ratio of different structural units of different bonds.

Mysen and co-workers calculated the abundance pattern of different anionic structural units depending on the bulk NBO/T for different systems [26]. Mysen and co-workers determined the relative abundance of different structural units as a

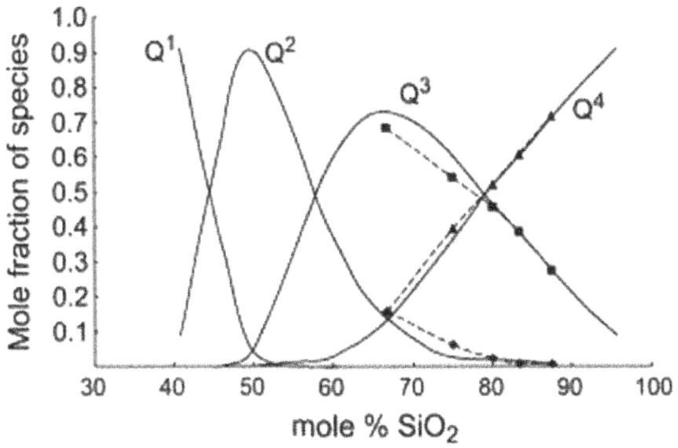

Fig. 2 Abundance of different structural units as a function of percentage silica in Na_2O-SiO_2 melt. Solid lines are the thermodynamic properties for melt species obtained by fitting phase equilibria and dotted lines are the experimentally measured values [27]

function of percentage silica in melt, as shown in Fig. 2 [27]. Essentially, the demonstrated how networking changes with composition in an idealised acid basic slag.

Slag Structure Relationship to Physical Properties/ Thermodynamics

A number of researchers have related Raman and FTIR studies of slag structures with physical properties [28–32]. Min et al. [28] established a relation between viscosity with the NBO/T of different silicate slag systems, as shown in Fig. 3. The strength of this structural model is the ability to correlate NBO/T with viscosity for

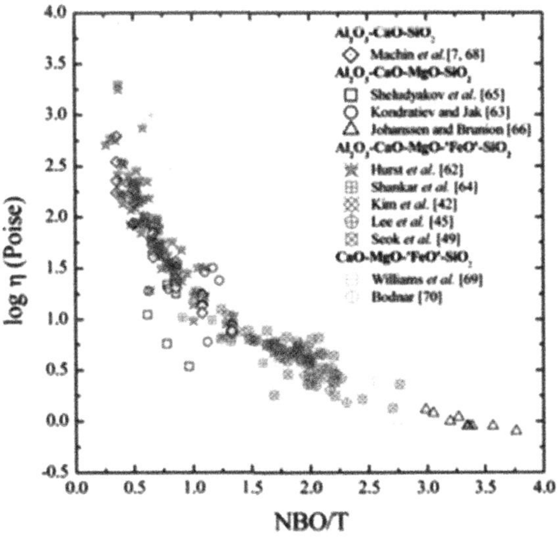

Fig. 3 Effect of NBO/T on the viscosity of metallurgical slag systems [28]

the silicate melts containing amphoteric oxides. For instance, the addition of Al_2O_3 (up to 10 wt%) in CaO-SiO_2 slag was found to increase the melt's viscosity [29] but further addition decreased the viscosity. These measurements could be explained in terms of structural view of silicate slags. Park and Min [30] successfully related the foaming index of the silicate slags with NBO/T in an empirical equation. Maroufi et al. [31] also studied the diffusivity of SiO_2 in silicomanganese slag and CaO in aluminosilicate slag as functions of the NBO/T ratio.

All these researches found relationships between physical properties of the silicate systems and the structure defined in terms of NBO/T. According to Mills [32], the introduction of various network breaking oxides causes to the formation of smaller units resulting in depolymerisation of the silicate structure. For a specific mole fraction of metal oxide, the overall degree of polymerisation may remain constant; however, the proportion of various units may vary. This suggests that describing the structure of silicates in terms of different fractions of Q^n rather than NBO/T could be more useful in understanding the properties and thermodynamics of the slag.

Geologists have correlated the partitioning of different elements in melt and minerals with NBO/T measurements [26]. Recently, Shuva et al. [33, 34] attempted to correlate the equilibrium distribution ratio of germanium and palladium in slag and metal with optical basicity. Using linear regression analysis, semi-empirical equations were developed to relate process parameters (temperature and partial pressure of oxygen) and the equilibrium distribution ratio with structure of slag (degree of polymerisation, expressed as Q^3/Q^2) [35]. The next step would be to form thermodynamic relationships for each reaction involving different Q species and the cation being partitioned between the metal and the slag. This would result in three to four equilibrium equations to describe partitioning in a specific system. Apart from theoretical complications with developing these relationships (e.g. what activity would each complex have in the melt?), whether such calculations could be successfully used in practice by operators remains to be established. In principle, FTIR measurements from the plant could be directly related to a thermodynamic calculation using such an approach and this would provide a more rigorous understanding of basicity.

Conclusions

Historically, the definition of basicity of oxides has evolved with time. The networking forming/breaking model developed in the 1950s provides a good qualitative explanation for the behaviour of slags but is difficult to relate in a quantitative way to measurements and thermodynamic relationships. In more recent times, Raman and FTIR measurements of slag structures and determination of structural parameter, such as NBO/T and the degree of polymerisation (Q^3/Q^2), suggest that more rigorous definitions of basicity that link structure, properties and thermodynamics are emerging. Further work is required to quantify the thermodynamic relationships using this approach and demonstrate the practical usefulness of these techniques in industry.

Acknowledgements This paper is dedicated to the memory of the late Dr Andrew Ducret who patiently explained the fundamentals of slag chemistry to the first author of this paper between doing experiments, playing cards in the student common room and enjoying each other's company at the University of Melbourne in 1990. We are also grateful to conversations with Mr Lesley Beyers from KU Leuven on these topics.

References

1. Thomas and Gilchrist (1929) Blockow and Vaughan 1879-1928. Vaughan & Co., Ltd., Middlesbrough, Blockow
2. Bodsworth C, Bell HB (1972) Physical chemistry of iron and steel manufacture. Longman, London, pp 66–89
3. Gilchrist JD (1989) Extraction metallurgy, 3rd edn. Pergamon Press, London, pp 198–200
4. Oeters F (1989) Metallurgy of steelmaking. Berlin, Stahl & Eisen, p 3–14
5. Rosenqvist T (2004) Principles of extractive metallurgy. Tapir Academic Press, Trondheim, pp 295–320
6. Turkdogan E (1996) Fundamentals of steelmaking. Institute of Materials, London, pp 138–179
7. Laird BB, Chang R (2009) University Chemistry. McGraw-Hill, Boston, pp 556–610
8. Sherwood Taylor F (1976) The alchemists. Palladin, London
9. Kurushkin M, Kurushkin D (2018) Acid-base behavior of 100 element oxides: visual and mathematical representations. J of Chem Edu 95(4):678–681
10. Flood H, Förland T (1947) The acidic and basic properties of oxides. Acta Chem Scand 1:592–604
11. Lux H (1939) Acids and bases in a fused salt bath: the determination of oxygen-ion concentration. Z Elektrochem Soc 45:303–310
12. Flood H, Grojtherm K (1952) J Iron Steel Inst 171:64
13. Sommerville I, Sosinsky D (1994) Solubility, capacity and stability of species in metallurgical slags and glasses. Pyrometallurgy Complex Mater Wastes: p 73–91
14. Richardson F (1948) Slags and refining processes. The constitution and thermodynamics of liquid slags. Discussions of the Faraday Society **4**:244–257
15. Duffy J, Ingram M (1976) An interpretation of glass chemistry in terms of the optical basicity concept. J Non-Cryst Solids 21(3):373–410
16. Sommerville I, Sosinsky D. The application of the optical basicity concept to metallurgical slags. In: Proceedings in 2nd international symposium on metallurgical slags and fluxes. metallurgical society of the AIME, pp 1015–1026
17. Nakamura T, Ueda Y, Toguri J (1986) A new development of the optical basicity. J Jpn Inst Met 50(5):456–461
18. Sommerville I, Masson C (1992) Group optical basicities of polymerized anions in slags. Metall Trans B 23(2):227–229
19. Wagner C (1975) The concept of the basicity of slags. Metall Trans B 6(3):405
20. Virgo D, Mysen B, Kushiro I (1980) Anionic constitution of 1-atmosphere silicate melts: implications for the structure of igneous melts. Science 208(4450):1371–1373
21. De Jong B, Schramm CM, Parziale VE (1984) Silicon-29 magic angle spinning NMR study on local silicon environments in amorphous and crystalline lithium silicates. J Am Chem Soc 106(16):4396–4402

22. Farges F, Brown Jr GE (1997) Coordination chemistry of titanium (IV) in silicate glasses and melts: IV. XANES studies of synthetic and natural volcanic glasses and tektites at ambient temperature and pressure. Geochimica et Cosmochimica Acta, 61(9):1863–1870
23. Farges F, Brown Jr GE, Rehr J.J (1996) Coordination chemistry of Ti (IV) in silicate glasses and melts: I. XAFS study of titanium coordination in oxide model compounds. Geochimica et Cosmochimica acta, 60(16):3023–3038
24. Sohn I et al (2012) Influence of TiO2 on the viscous behavior of calcium silicate melts containing 17 mass% Al2O3 and 10 mass% MgO. ISIJ Int 52(1):158–160
25. Chaskar V, Richards G, McCammon C (1993) A mössbauer study of the behavior of iron cations in iron oxide-containing melts at 1400 °C. Metall Trans B 24(1):101–111
26. Mysen BO, Virgo D, Seifert FA (1982) The structure of silicate melts: implications for chemical and physical properties of natural magma. Rev Geophys 20(3):353–383
27. Halter WE, Mysen BO (2004) Melt speciation in the system Na2O–SiO2. Chem Geol 213(1–3):115–123
28. Min DJ, Tsukihashi F (2017) Recent advances in understanding physical properties of metallurgical slags. Met Mater Int 23(1):1–19
29. Park JH, Min DJ, Song HS (2004) Amphoteric behavior of alumina in viscous flow and structure of CaO-SiO 2 (-MgO)-Al 2 O 3 slags. Metall Mater Trans B 35(2):269–275
30. Park Y, Min DJ (2017) A Structural Study on the Foaming Behavior of CaO-SiO 2-MO (MO = MgO, FeO, or Al 2 O 3) Ternary Slag System. Metall Mater Trans B 48(6):3038–3046
31. Maroufi S et al (2016) Diffusion coefficients and structural parameters of molten slags. In: Advances in molten slags, fluxes, and salts: proceedings of the 10th international conference on molten slags, fluxes and salts 2016, Springer
32. Mills KC (1993) The influence of structure on the physico-chemical properties of slags. ISIJ Int 33(1):148–155
33. Shuva MAH, Rhamdhani MA, Brooks GA, Masood S, Reuter MA (2016) thermodynamics behavior of germanium during equilibrium reactions between FeOx-CaO-SiO2-MgO slag and molten copper. Metall Mater Trans B 47(5):2889–2903
34. Shuva MAH, Rhamdhani MA, Brooks GA, Masood S, Reuter MA (2017) Thermodynamics of palladium (Pd) and tantalum (Ta) relevant to secondary copper smelting. Metall Mater Trans B 48(1):317–327
35. Shuva MAH (2017) Analysis of thermodynamics behaviour of valuable elements and slag structure during e-waste processing through copper smelting. Ph.D. thesis, Swinburne University of Technology

Investigation on Clogging of Submerged Entry Nozzles for GCr15 Bearing Steels

Gong Cheng, Lifeng Zhang, Wenbo Wang, Qiangqiang Wang and Piotr Roman Scheller

Abstract Clogging of submerged entry nozzles for GCr15 bearing steels has been investigated by industrial experiments coupled with mathematical simulations. It was found that clogs on the submerged entry nozzle with four ports were mainly $CaO\text{-}MgO\text{-}Al_2O_3$ inclusions with the size of 2–6 μm, which cannot be avoided by removal of inclusions from the liquid steel. Cathodoluminescence microscopy was also applied for the characterisation of submerged entry nozzle clogging. Mathematical simulations were carried out to predict the main clogging location at the submerged entry nozzle with different types. The results showed that the percentage of inclusions entrapped on the cylindrical submerged entry nozzle and four-port one was 39.7 and 46.5%, respectively. It indicated that the cylindrical submerged entry nozzle helped reduce the probability of submerged entry nozzle clogging and was beneficial for the improvement of the fatigue property of bearing steels.

Keywords Inclusion · Clogging · Mathematical simulation

Introduction

Since the demand of high-quality steel increases year by year, enhancing cleanliness is always one of the main tasks of steelmakers [1–4]. Aluminum as the deoxidizer is widely used to improve cleanliness of the molten steel. The molten

G. Cheng · L. Zhang (✉) · W. Wang · P. R. Scheller
School of Metallurgical and Ecological Engineering, University of Science and Technology Beijing (USTB), Beijing 100083, China
e-mail: zhanglifeng@ustb.edu.cn

G. Cheng · L. Zhang · W. Wang
Beijing Key Laboratory of Green Recycling and Extraction of Metal, University of Science &Technology Beijing, Beijing 100083, China

Q. Wang
College of Materials Science and Engineering, Chongqing University, Chongqing 400044, China

© The Minerals, Metals & Materials Society 2019

T. Jiang et al. (eds.), *10th International Symposium on High-Temperature Metallurgical Processing*, The Minerals, Metals & Materials Series,
https://doi.org/10.1007/978-3-030-05955-2_29

steel flows into the mould through the submerged nozzle entry (SEN) during continuous casting, and high melting point inclusions in the molten steel, such as Al_2O_3 and spinel [3, 5–7], tend to accumulate on the inner wall of the nozzle, causing partial or complete blockage of SEN. SEN clogging would change the flow pattern of the molten steel at the outlet of the nozzle, resulting in the partial flow of the molten steel flowing out of SEN, which could induce the defect of the surface quality of the slab and cause the leakage of steel. The drop of blockage during casting is also one of main sources of large inclusions. The origin of clogs on the SEN has not yet been determined, and there are some main standpoints [8–10]. The viewpoint that Al_2O_3 inclusions and other deoxidation products are the main origin of clogs has been gained more and more consensus. For some steel grades, calcium treatment is extensively applied to modify Al_2O_3 inclusions [11–15]. Calcium adding in the molten steel could modify oxide inclusions, desulfurize and control shape of sulfide inclusions. But calcium treatment is prohibited for bearing steels because large-size globular calcium aluminate inclusions are most detrimental to bearing steels due to their size. Control of inclusions in bearing steels affecting rolling contact fatigue life has attracted much attention. But less attention has been paid to clogging of SEN for GCr15 bearing steels.

In the current work, an experimental investigation on clogging of SEN was carried out. And mathematical simulations about the effect of SEN type on clogging were also performed, in order to reduce the blockage of SEN.

Experimental Procedure

The production process of GCr15 bearing steels is EAF, LF refining, vacuum degassing, CC. Five samples were taken from different locations of the four-port type SEN, including near the slag line (sample 1), 40 mm below the slag line (sample 2), 40 mm above the outlet (sample 3), near the outlet (sample 4), and at the bottom of SEN (sample 5), as shown in Fig. 1. These samples were detected by the field emission electron microscope to observe the morphology, size and chemical compositions of clogs on the SEN. Another sample below the slag line was mounted in epoxy resin, ground with 320- and 800-grit SiC paper, the section perpendicular to SEN inner of which was analysed by cathodoluminescence microscopy (CLM) coupled with ASPEX Explorer to investigate clogging layers.

Besides, mathematical simulations about different type SEN were executed by Fluent 14.0 to predict the main clogging location at SEN. Parameters for the mathematical simulation of four-port SEN are shown in Table 1. Main parameters for the mathematical simulation of cylindrical SEN are the same as that of four-port SEN.

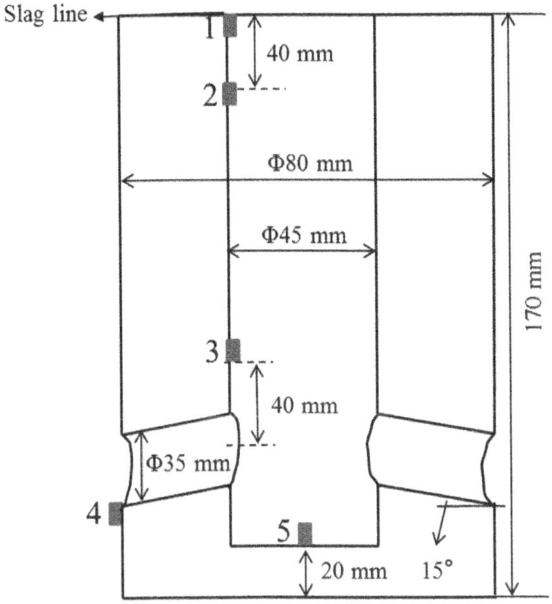

Fig. 1 Sample schematic

Table 1 Main parameters for the mathematical simulation

Parameter	Value	Parameter	Value
SEN type	Four ports	Inlet velocity	0.8209 m/s
SEN bore diameter	45 mm	Inlet turbulent kinetic energy	0.0038 m^2/s^2
SEN outer diameter	80 mm	Inlet turbulent energy dissipation rate	0.0536 m^2/s^3
Total nozzle length	900 mm	Density of molten steel	7020 kg/m^3
Port angle	15° down	Viscosity of molten steel	0.0067 kg/(m s)
Port width × height	35 mm × 30 mm	Gravitational acceleration	9.81 m/s^2
Bloom cross section	480 mm × 320 mm	Casting speed	0.51 m/min

Results and Discussion

High-resolution field emission electron microscope was utilised to observe three-dimensional morphologies of clogs on the SEN, and photographs of the clog morphologies at five locations are presented in Fig. 2. In general, the clogs were coral-like and consisted of many polygonal particles that bonded to form clusters. Especially, the clogs on the outer wall of the outlet were denser and comprised of rod-like particles. Twenty particles at each of the five locations (100 particles in total) were randomly selected. The particles were mainly 2–6 μm inclusions of

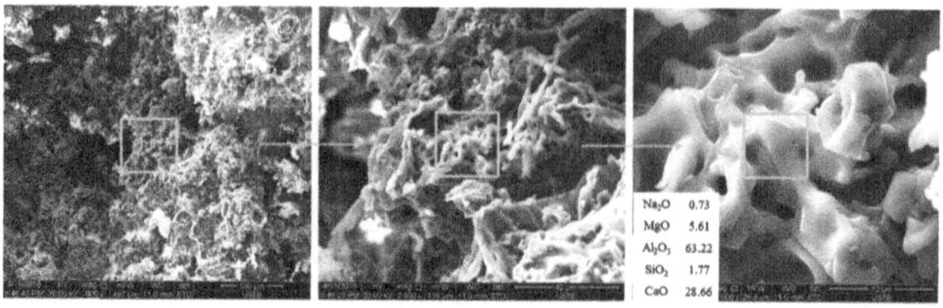

(a) Near the slag line

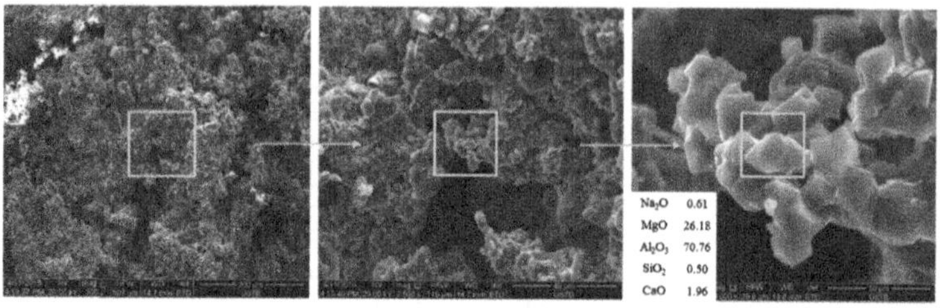

(b) 40 mm below the slag line

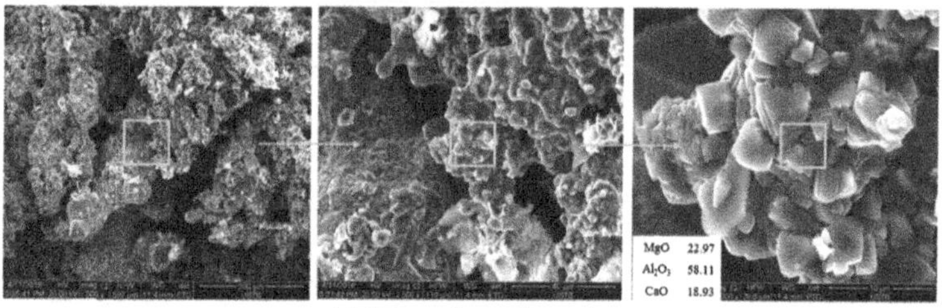

(c) 40 mm above the outlet

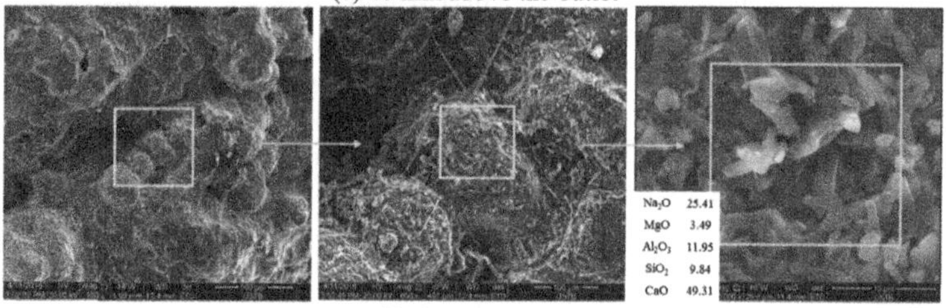

(d) Near the outlet

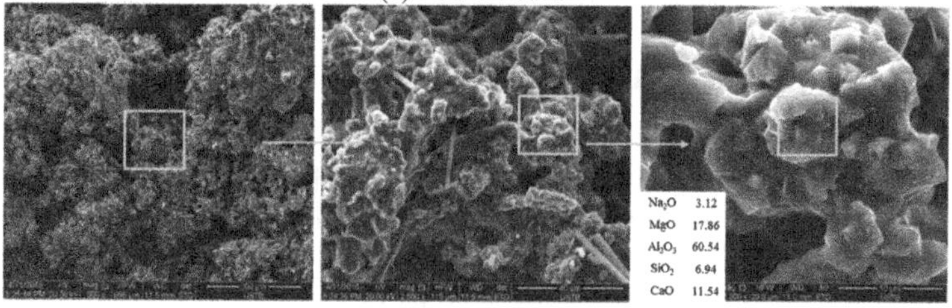

(e) At the bottom of the SEN

Fig. 2 Morphology of clogs on the SEN at five locations

$CaO-MgO-Al_2O_3$. The compositions of clogging materials and inclusions in the molten steel in the tundish are shown in Fig. 3. Clogs at the slag line were $CaO-Al_2O_3-MgO$ inclusions, containing about 7 mass% SiO_2 and 8 mass% Na_2O. Forty millimetre below the slag line, $MgO{\cdot}Al_2O_3$-based inclusions were mainly observed, as shown in Fig. 2b. Clogs above the outlet were $CaO-Al_2O_3-MgO$ inclusions, containing about 9 mass% SiO_2. On the outer wall of the outlet, clogs were $CaO-Al_2O_3-MgO-SiO_2$ inclusions with about 5 mass% Na_2O, which originated from the mould flux. High Na_2O inclusions were also detected on the outer wall of the outlet, as shown in Fig. 2d. As shown in Fig. 3b, inclusions in the molten steel in the tundish were $CaO-Al_2O_3-MgO$ and $MgO{\cdot}Al_2O_3$, which were mainly smaller than 6 μm. The composition of clogging materials was also similar to inclusions in the molten steel, indicating that clogging materials were from the molten steel in the tundish.

Inclusion composition governs the mode of removal. The contact angle of Al_2O_3 and $MgO{\cdot}Al_2O_3$ inclusions is far larger than that of $CaO-Al_2O_3-(MgO)$ inclusions [16–18]; thus, it is relatively difficult to remove $CaO-Al_2O_3-(MgO)$ inclusions. In particular, it is relatively easy for solid $CaO-Al_2O_3-(MgO)$ inclusions to cause SEN clogging. Large-size globular calcium aluminate inclusions are detrimental to rolling contact fatigue life. Therefore, the key point for GCr15 bearing steels is to reduce CaO content in inclusions and restrain the transformation of spinel or alumina inclusions into calcium aluminate.

Cathodoluminescence (CL) is the phenomenon by which non-metallic solids emit cold light when bombarded by an electron beam under vacuum. Cathodoluminescence microscopy (CLM) was applied in combination with ASPEX Explorer. Three macro-layers from the original side to the molten steel side were visually observed, including a decarburization layer (black), a dense inclusion layer (grey), a loose inclusion layer (yellow), as shown in Fig. 4. Phase distribution of the clogging SEN is shown in Fig. 5. The SEN matrix was mainly composed of Al_2O_3 and carbon. Under the observation of CLM, as shown in Fig. 5a, the red area

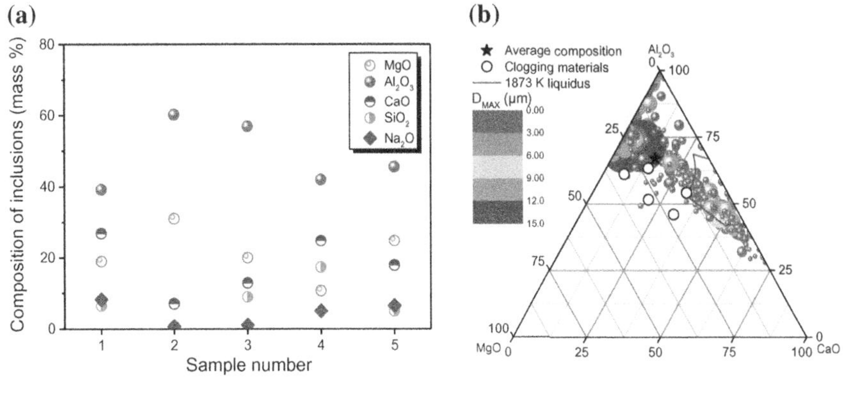

Fig. 3 Composition of **a** clogging materials and **b** inclusions in the molten steel in the tundish

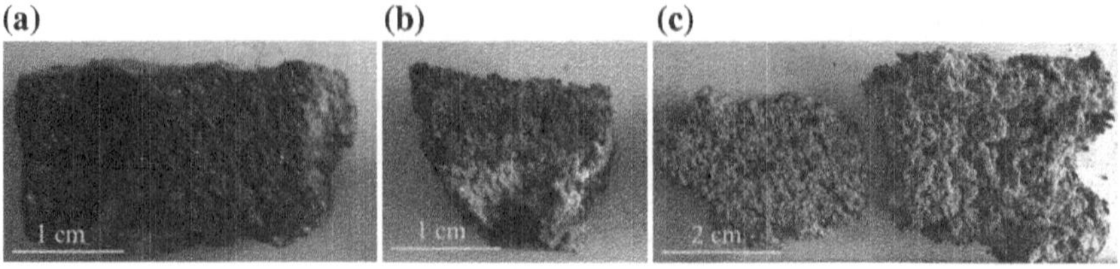

Fig. 4 Three layers for the clogging materials: **a** decarburization layer, **b** dense inclusion layer, **c** loose inclusion layer

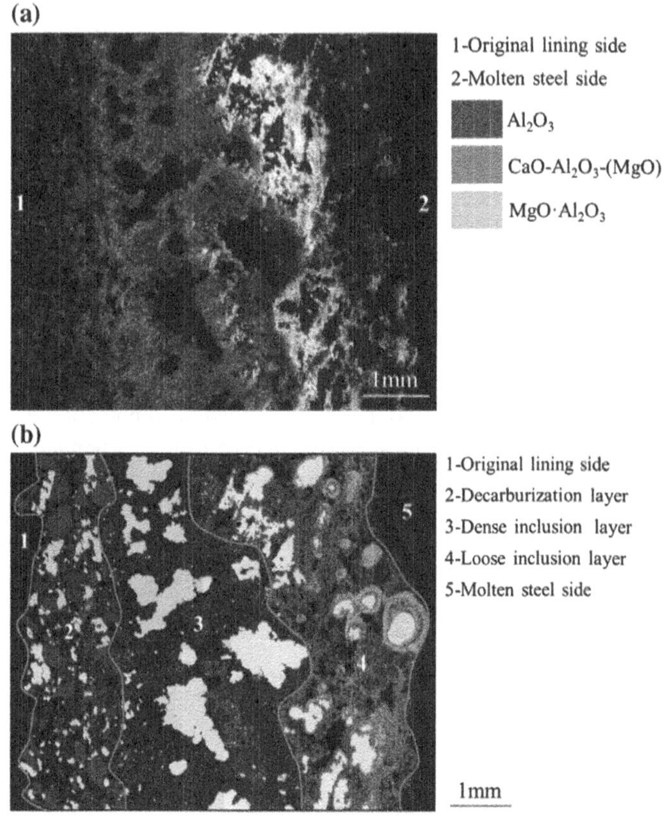

Fig. 5 Phase distribution of the clogs under the observation of **a** CLM and **b** ASPEX Explorer

denotes Al_2O_3 phase, which is the SEN matrix. The green area represents calcium aluminate with low MgO content, and the yellow colour represents spinel phase. According to the area of each phase, it could be concluded that the clogs were mainly calcium aluminate with low MgO content, which was in agreement with the observation of the field emission electron microscope. Besides, there was some solidified steel in the layer of inclusions, which appeared black in Fig. 5a. And under the observation of ASPEX Explorer, as shown in Fig. 5b, three macro-layers were easily divided, except original lining side and molten steel side.

The decarburization layer was adjacent to the original SEN lining refractory. Some holes were filled with CaO-Al_2O_3-(MgO) inclusions, and lots of Fe-enriched droplet supposed to be from the frozen steel, which were black and white under the observation of CLM and ASPEX Explorer, respectively. The original SEN lining refractory was made up of Al_2O_3 and carbon with about 7 mass% SiO_2. SiO and CO gas generated due to electrochemical reaction between the molten steel and the refractory [19], resulting in the decrease of density of the refractory. After preheating SEN before casting, carbon in the refractory was burnt; thus, the inner surface of SEN was not flat, which may promote the formation of decarburization layer. Due to low superheat or high rate of heat loss, large-size Fe-enriched droplets were frozen and attached to the lining wall, and then inclusions easily precipitated or were captured on the surface of these droplets, resulting in the formation of dense inclusion layer. According to the boundary layer theory, the speed of the molten steel near the inner surface of SEN was close to zero. Inclusions in the molten steel had a chance to collide with the inner wall due to the effect of turbulence and thus be adsorbed. And the loose inclusion layer formed under the influence of the occurrence of air absorption, capture of bubbles and inclusion attachment to bubbles.

Mathematical simulations were carried out to predict the main clogging location at SEN. Twenty thousand one hundred and sixty inclusions with diameter of 5 μm were injected into SEN in the simulation. For the four-port type SEN, the molten steel lashes against the mould after flowing out the nozzle, and then flows upward, which can activate the meniscus and is beneficial for slagging. However, the molten steel flows downward after flowing out the cylindrical type nozzle, and the temperature at the meniscus is relatively low. Final entrapment positions of inclusions on the SEN are presented in Fig. 6.

For the four-port type SEN, 46.5% of the total number of inclusions were entrapped. Thirty-five percent were attached to the inner wall and 6.5% were on the inner wall of ports. However, for the cylindrical SEN, 39.7% of the total number of inclusions were entrapped on the SEN. Approximately 35.7% were attached to the inner wall and 3.7% were on the outer wall of the SEN. It indicated that the cylindrical SEN helped reduce the probability of SEN clogging and was beneficial for the improvement of the fatigue property of bearing steels.

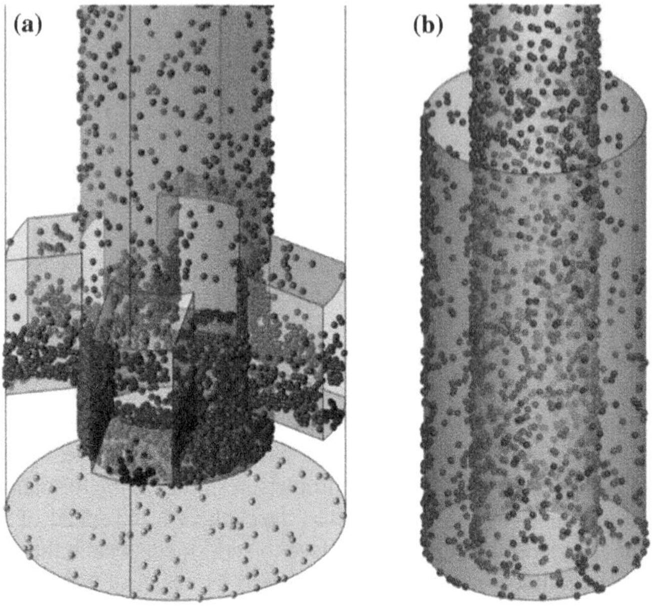

Fig. 6 Final entrapment positions of inclusions on the **a** four-port type and **b** cylindrical type SEN

Conclusions

Investigation on clogging of submerged entry nozzles for GCr15 bearing steels was performed by experiments conducted at the industrial scale coupled with mathematical simulations. The following conclusions were obtained.

(1) Clogging materials were mainly $CaO\text{-}MgO\text{-}Al_2O_3$ inclusions with the size of 2–6 μm. Three macro-layers were observed: a decarburization layer, a dense inclusion layer and a loose inclusion layer.
(2) The percentage of inclusions entrapped on the cylindrical submerged entry nozzle and four-port one was 39.7 and 46.5%, respectively. The cylindrical submerged entry nozzle helped reduce the probability of submerged entry nozzle clogging.
(3) The key point for GCr15 bearing steels is to reduce CaO content in inclusions and restrain the transformation of spinel or alumina inclusions into calcium aluminate.

Acknowledgements The authors are grateful for support from the National Science Foundation China (Grant No. U1860206), the Fundamental Research Funds for the Central Universities (Grant No. FRF-TP-15-001C2), Beijing Key Laboratory of Green Recycling and Extraction of Metals (GREM) and the High-quality Steel Consortium (HQSC) at the School of Metallurgical and Ecological Engineering at University of Science and Technology Beijing (USTB), China.

References

1. Zhang L, Thomas B (2003) State of the art in evaluation and control of steel cleanliness. Trans Iron Steel Inst Jpn 43(3):271–291
2. Zhang L (2006) Indirect methods of detecting and evaluating inclusions in steel—a review. J Iron Steel Res (Int) 13(4):1–8
3. Park J, Todoroki H (2010) Control of MgO·Al_2O_3 spinel inclusions in stainless steels. ISIJ Int 50(10):1333–1346
4. Zhang L (2013) Nucleation, growth, transport, and entrapment of inclusions during steel casting. JOM 65(9):1138–1144
5. Zhang L, Thomas B (2006) State of the art in the control of inclusions during steel ingot casting. Metall Mater Trans B 37(5):733–761
6. Zhang L, Wang Y, Zuo X (2008) Flow transport and inclusion motion in steel continuous-casting mold under submerged entry nozzle clogging condition. Metall Mater Trans B 39(4):534–550
7. Xie W, Bao Y, Wang M, Zhang X (2015) Relationship between high frequency defect detection and inclusions in GCr15 Bearing Steel. Iron Steel 50(3):44–48
8. Andersson M, Appelberg J, Tilliander A, Nakajima K, Shibata H, Kitamura S, Jonsson L (2006) Some aspects on grain refining additions with focus on clogging during casting. ISIJ Int 46(6):814–823
9. Pielet H, Bhattacharya D (1984) Thermodynamics of nozzle blockage in continuous casting of calcium-containing steels. Metall Trans B 15(4):743
10. Szekely J, Dinovo S (1974) Thermal criteria for tundish nozzle or taphole blockage. Metall Trans 5(3):747–754
11. Verma N, Pistorius P, Fruehan R, Potter M, Lind M, Story S (2011) Transient inclusion evolution during modification of alumina inclusions by calcium in liquid steel: part I. Background, experimental techniques and analysis methods. Metall Mater Trans B 42 (4):711–719
12. Yang S, Wang Q, Zhang L, Li J, Peaslee K (2012) Formation and modification of MgO·Al_2O_3—based inclusions in alloy steels. Metall Mater Trans B 43(4):731–750
13. Ren Y, Zhang L, Li S (2015) Transient evolution of inclusions during calcium modification in linepipe steels. ISIJ Int 54(12):2772–2779
14. Yang G, Wang X, Huang F, Yang D, Wei P, Hao X (2015) Influence of calcium addition on inclusions in LCAK steel with ultralow sulfur content. Metall Mater Trans B 46(1):145–154
15. Zhao D, Li H, Bao C, Yang J (2015) Inclusion evolution during modification of alumina inclusions by calcium in liquid steel and deformation during hot rolling process. Trans Iron Steel Inst Jpn 55(10):2115–2124
16. Ogino K, Nogi K, Yamase O (1983) Effects of selenium and tellurium on the surface tension of molten iron and the wettability of alumina by molten iron. ISIJ Int 23(3):234–239
17. Cramb A, Jimbo I (1989) Calculation of the interfacial properties of liquid steel-slag systems. Steel Res 60(3):157–165
18. Zouvelou N, Mantzouris X, Nikolopoulos P (2007) Interfacial energies in oxide/liquid metal systems with limited solubility. Int J Adhes Adhes 27(5):380–386
19. Ruan G, Li N (2005) Local corrosion mechanism of refractory at slag-iron(steel)interface. Mater Rev 19(2):47–49

Part VII
Preparation of Alloys and Materials I

Effects of Electrolytic Parameters on the Preparation of Al–Sc Master Alloy in Na_3AlF_6-K_3AlF_6-AlF_3 Melt

Kai Yang, Zhongliang Tian, Xun Hu, Yanqing Lai and Jie Li

Abstract The effects of electrolytic parameters on the content of Sc in Al–Sc master alloy prepared by electrolysis in Na_3AlF_6-K_3AlF_6-AlF_3 melt were studied. It is observed that the content of Sc in Al–Sc master alloy increases and then decreases with increasing the electrolysis temperature and the cathodic current density. Under the electrolysis condition (temperature 950 °C, cathodic current density 1.50 A cm^{-2} and concentration of Al_2O_3 1.0 wt%), the content of Sc in the production is increased from 0.81 to 3.87 wt% as the value of the concentration ratio of Sc_2O_3 to Al_2O_3 ($[Sc_2O_3]/[Al_2O_3]$) increased from 2 to 6. The analysis of SEM indicates that the production of Al–Sc master alloy is composed of the aluminum and the Al_3Sc. The distribution of Sc in Al–Sc master alloy is uniform.

Keywords Al–Sc master alloy · Electrolytic parameters Na_3AlF_6-K_3AlF_6-AlF_3 melt · Electrolysis

Introduction

Scandium is the best modifier for casting grain structure and the strongest recrystallization inhibitor for semi-finished processing of aluminum alloys. If a small amount of Sc (0.1–0.4 wt%) is introduced into the traditional aluminum alloys, the comprehensive properties (strength, toughness, stress corrosion resistance and thermal stability) of the traditional aluminum alloys can be greatly improved [1]. Hence, the aluminum alloy-added scandium is widely used in the aerospace, military, civilian industry and other fields [2–4]. However, because of the large melting point difference between Sc (1541.0 °C) and Al (660.3 °C) and the high chemical activity of the metal Sc, scandium must be added in the form of

K. Yang · Z. Tian (✉) · X. Hu · Y. Lai · J. Li
School of Metallurgy and Environment, Central South University, 932 Lushan South Road, Changsha 410083, Hunan, China
e-mail: tianzhongliang@csu.edu.cn

© The Minerals, Metals & Materials Society 2019

T. Jiang et al. (eds.), *10th International Symposium on High-Temperature Metallurgical Processing*, The Minerals, Metals & Materials Series,
https://doi.org/10.1007/978-3-030-05955-2_30

Al–(2.0 wt%)Sc master alloy.. Therefore, the Al–(2.0 wt%)Sc master alloy becomes the key raw material for producing Al alloys containing Sc [5].

There are some methods to prepare the Al–Sc master alloy, including the mixing method, the metallothermics and the molten salt electrolysis. Compared with the other methods, the molten salt electrolysis method has obvious advantages in preparing the Al–Sc master alloy [6–9]. Firstly, the cost of production can be reduced because the low purity scandium oxide can be used as the cheap raw material. Secondly, the segregation phenomenon of the Al–Sc master alloy can be effectively avoided. In recent years, the electrolysis preparation of Al–Sc master alloy by using the fluoride melt has attracted extensive attention [10–16]. Liu et al. [11] prepared the Al–Sc master alloy with the Sc content of 0.44 and 0.73 wt% in the KF-AlF_3-Sc_2O_3 melt at the current densities of 0.50 and 1.00 A cm^{-2}, respectively. Yang [12] adopted the Na_3AlF_6-MgF_2-CaF_2 melt with the mole concentration ratio of Sc_2O_3 to Al_2O_3 above 2:1 and prepared the Al–Sc master alloy with the content of Sc more than 2.0%. However, the content of Sc in Al–Sc master alloy prepared by the above-mentioned methods is low, because the solubility of Sc-containing raw materials in the melt is limited [13, 14]. In comparison with the other electrolyte melt, the potassium cryolite melt has the lower primary crystallization temperature and the higher oxide solubility [6].

On the basis of the previous studies, the solubility and thermodynamic analyses of Sc_2O_3 in the Na_3AlF_6-K_3AlF_6-AlF_3 melt show that the Al–Sc master alloy can be prepared by the electrolysis [15, 16]. In this work, the technological parameters of Al–Sc master alloy prepared by electrolysis in the Na_3AlF_6-K_3AlF_6-AlF_3 melt were studied. The purpose is to provide the selection basis of the technological parameters for preparing the Al–Sc master alloy with high Sc content and uniform Sc distribution.

Experimental

Chemicals and Materials

The electrolyte was made of Na_3AlF_6, K_3AlF_6, AlF_3, Al_2O_3 and Sc_2O_3. The Na_3AlF_6, K_3AlF_6 and Al_2O_3 were used as reagent grades. The AlF_3 was purified by sublimation (99.5 wt%.). The purity of Sc_2O_3 was 99.99 wt%. All components were dried at 200 °C for at least 48 h in the vacuum oven to remove the water.

Cell Design and Electrolysis Procedure

The schematic diagram of the experimental set-up is shown in Fig. 1. The graphite crucible with corundum lining was filled with 500 g electrolyte. It was necessary to provide the extra heat by placing the experimental cell in a vertical laboratory

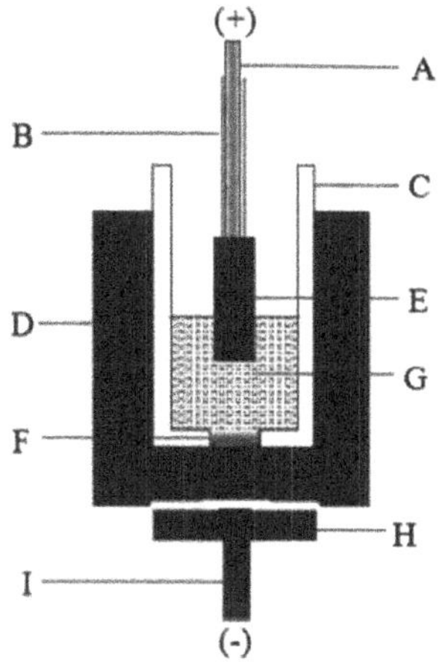

Fig. 1 Schematic drawing of the electrolysis setup: A—Stainless steel rod, B—Corundum sleeve, C—Corundum crucible, D—Graphite crucible, E—Graphite anode, F—Al–Sc master alloy, G—Electrolyte, H—Iron plate, I-Cathode rod

furnace which was then heated to the target temperature. The temperature of the melt was measured with Pt/Pt-10% Rh thermocouple; the uncertainty of temperature measurement was ±1 °C. After the melting of electrolyte, the anode was immersed into the molten salt. Then, the graphite crucible containing the melt was quickly removed out of the furnace, and the melt with the Al–Sc master alloy liquid at the bottom of the crucible was poured into the self-made cooling tank.

Material Characterization

The content of Sc in the master alloy was analysed by inductively coupled plasma-atomic emission spectroscopy (ICP-AES). The microstructure of the master alloy sample was characterized by scanning electron microscopy (SEM, JSM-5600LV), and the sample composition was analysed by energy disperse spectroscopy (EDS, GENESIS).

Results and Discussion

Electrolysis Temperature

As shown in Fig. 2, the content of Sc in the alloy obtained at the cathode increased and then decreased with the electrolysis temperature increasing from 940 to 970 °C. At the electrolysis temperature of 940 °C, the content of Sc in the cathodic production was 0.46 wt%. The Al–Sc master alloy with the content of Sc 1.11 wt% could be obtained at electrolysis temperature of 950 °C. With increased electrolysis temperature from 940 to 950 °C, the content of Sc in the Al–Sc master alloy was

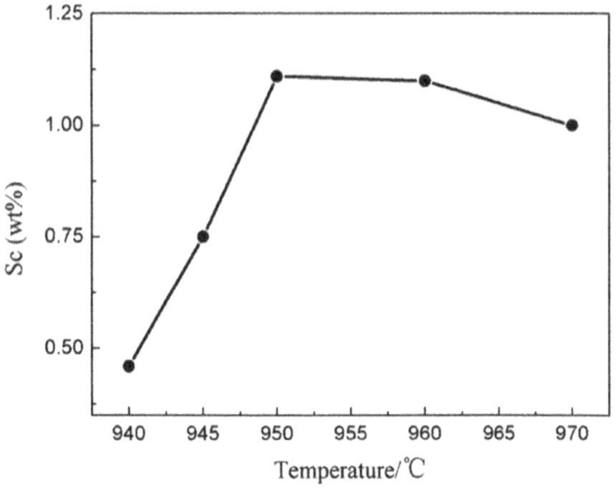

Fig. 2 Effect of temperature on the content of Sc in the Al–Sc master alloy (cathodic current density: 1.50 A cm^{-2}, concentration of Al_2O_3:1.0 wt %, concentration of Sc_2O_3:3.0 wt%, t: 90 min)

between 0.29 and 0.36 wt%. However, the content of Sc in the production was reduced to 1.00 wt% as the electrolysis temperature increased to 970 °C. At 950 °C, the content of Sc in the Al–Sc master alloy slowly decreased by 0.01 and 0.1 wt% for every 10 °C temperature increase.

Cathodic Current Density

The effect of cathodic current density on the Sc content is shown in Fig. 3. The content of Sc in Al–Sc master alloy increased and then decreased with increasing cathodic current density from 1.00 to 2.00 A cm^{-2}. At the cathodic current density of 1.00 A cm^{-2}, the content of Sc in the alloy was 0.44 wt%. The content of Sc in the cathodic production only increased by 0.04 wt% at higher cathodic current

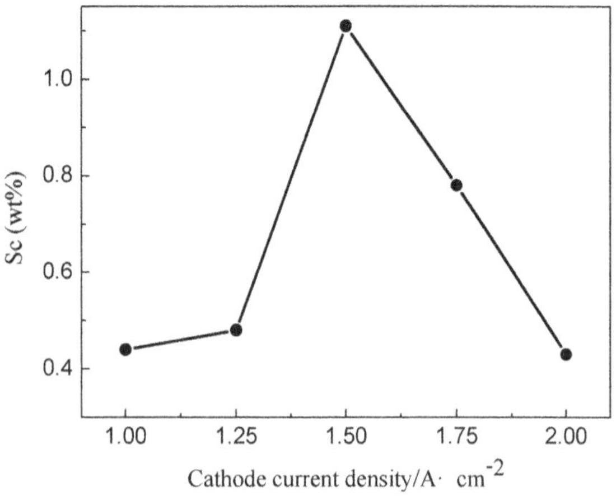

Fig. 3 Effect of cathodic current density on Sc content in Al–Sc master alloy (T: 950 °C, concentration of Al_2O_3: 1.0 wt%, concentration of Sc_2O_3: 3.0 wt%, t: 90 min)

density of 1.25 A cm^{-2}. Further, the cathodic production with the content of Sc 1.11 wt% could be obtained if the cathodic current density was increased to 1.50 A cm^{-2}. However, the content of Sc in Al–Sc master alloy was reduced to 0.43 wt% when the cathodic current density was increased to 2.00 A cm^{-2}. At the higher cathodic current density (over 1.50 A cm^{-2}), the content of Sc in the production decreased by 0.33 and 0.35 wt% for each 0.25 A cm^{-2} increase in the cathodic current density.

Concentration Ratio of Sc_2O_3 to Al_2O_3

Figure 4 shows the effect of concentration ratio of $[Sc_2O_3]/[Al_2O_3]$ on the content of Sc in Al–Sc master alloy. The content of Sc in Al–Sc master alloy increased from 0.81 to 3.87 wt% with the increasing concentration ratio of $[Sc_2O_3]/[Al_2O_3]$ from 2 to 6. At the concentration ratio of $[Sc_2O_3]/[Al_2O_3]$ of 2, the content of Sc in the alloy was 0.81 wt%. With increased concentration ratio of $[Sc_2O_3]/[Al_2O_3]$ from 2 to 5, the content of Sc in the production increased by 0.77, 0.49 and 0.36 wt%, respectively. The highest Sc content in Al–Sc master alloy with 3.87 wt% was observed at the concentration ratio of 6.

As shown in Fig. 5, the Al–Sc master alloy prepared under different concentration ratios is composed of aluminum matrix (grey colour) and numerous secondary phases (white colour) with different shapes. The EDS analysis confirmed that the Al–Sc master alloy mainly consists of aluminum matrix and Al_3Sc intermetallic compounds. The distribution of Al_3Sc in the intermetallic compounds is higher with increased concentration ratios of $[Sc_2O_3]/[Al_2O_3]$ (Fig. 5). The latter can be explained due to increased Sc content in Al–Sc master alloy which accelerates the reaction between the aluminum and scandium in the melt. In addition, the Al_3Sc particles in all samples are evenly distributed throughout the Al–Sc master

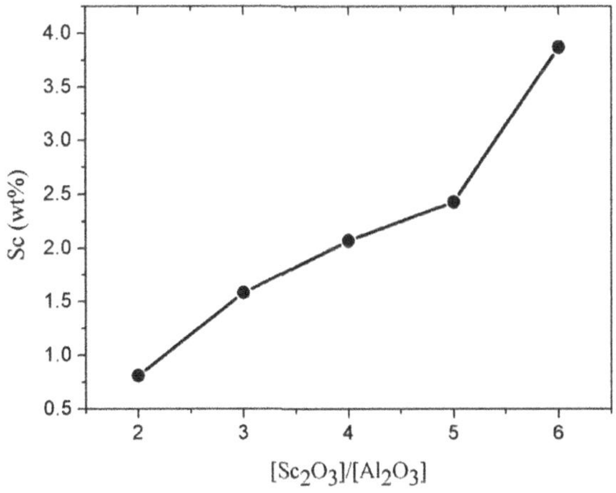

Fig. 4 Effect of concentration ratio of $[Sc_2O_3]/[Al_2O_3]$ on Sc content in Al–Sc master alloy

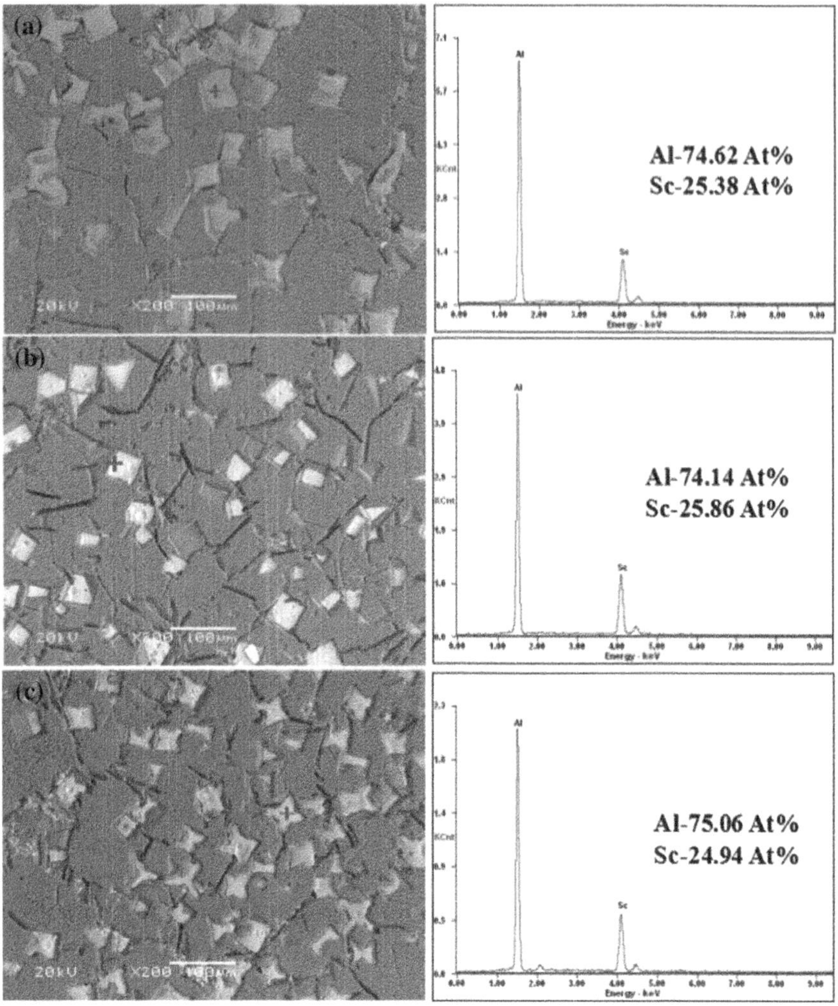

Fig. 5 SEM-EDS of the Al–Sc master alloy at concentration of Al_2O_3 1.0 wt%

alloy (Fig. 5). This indicates that concentration ratio of $[Sc_2O_3]/[Al_2O_3]$ cannot affect the uniform distribution of Sc in Al–Sc master alloy.

Conclusion

The effects of the electrolysis temperature, the cathodic current density and concentration ratio of $[Sc_2O_3]/[Al_2O_3]$ on the content of Sc in Al–Sc master alloy prepared by electrolysis in Na_3AlF_6-K_3AlF_6-AlF_3 melt were studied. The results showed that the content of Sc in Al–Sc master alloy increased and then decreased with increasing electrolysis temperature and cathodic current density. Under the electrolysis temperature of 950 °C, cathodic current density 1.50 A cm^{-2} and concentration of Al_2O_3•1.0 and Sc_2O_3 3.0 wt%, the content of Sc in the production reached 1.11 wt%. Additionally, the content of Sc in Al–Sc master alloy increased with increasing concentration ratio of $[Sc_2O_3]/[Al_2O_3]$. The analysis of SEM-EDS indicated that the Al–Sc master alloy mainly consists of the aluminum and the Al_3Sc.

Acknowledgements The authors gratefully acknowledge the financial support from the State Natural Science Fund (No. 51674301).

References

1. Rã,Yset J, Ryum N (2005) Scandium in aluminium alloys. Metall Rev 50(1):19–44
2. Davydov VG, Rostova TD, Zakharov VV et al (2000) Scientific principles of making an alloying addition of scandium to aluminium alloys. Mater Sci Eng A 280(1):30–36
3. Asta M, Ozoliņš V (2001) Structural, vibrational, and thermodynamic properties of Al-Sc alloys and intermetallic compounds. Phys Rev B 64(9):094104
4. Venkateswarlu K, Pathak LC, Ray AK et al (2004) Microstructure, tensile strength and wear behaviour of Al–Sc alloy. Mater Sci Eng A 383(2):374–380
5. Teng GC, Chai YC, Yang Y et al (2010) Preparation of Al-Sc master alloy. World Nonferrous Met 1:70–71 (in Chinese)
6. Liu QC, Xue JL, Zhu J et al (2012) Processing Al-Sc alloys at liquid aluminum cathode in KF-AlF_3 molten salt. ECS Trans 50(11):483–489
7. Guan CY, Xue JL, Zhu J, et al (2012) Preparing Aluminum-Scandium alloys using direct hall reduction process. In: 3rd international symposium on high-temperature metallurgical processing. Springer, Cham, pp 243–250
8. Harata M, Yasuda K, Yakushiji H et al (2009) Electrochemical production of Al-Sc alloy in $CaCl_2$-Sc_2O_3 molten salt. J Alloy Compd 474(1–2):124–130
9. Cheng T, Lv ZJ, Zhai XJ, Zhang MJ, Tu GF (2010) Preparation of Al-Sc alloys investigated in the systems of $CaCl_2$-LiF. J Mater Metall 9(1):39–42 (in Chinese)
10. Sun BL, Zhai YC, Tian YW (1998) Preparation of Al-Sc alloy in fluoride molten salts with molten salt electrolysis. Chin J Rare Met 22(3):191–193 (in Chinese)
11. Liu QC, Xue JL, Zhu J, et al (2012) Preparing aluminum-scandium inter-alloy during reduction process in KF-AlF_3-Sc_2O_3 melts. Light Met 685–688
12. Yang S (2003) The research on direct electrolytic Al-Sc alloys in molten salt. Zhengzhou University (in Chinese)
13. Lu GM, Liu XS (1999) Dissolution of Sc_2O_3 in fluoride molten salt. Chin J Nonferrous Met 9 (3):624–626 (in Chinese)
14. Yang S, Li Q, Gu QS (2003) Solubility of Sc_2O_3 in *n* NaF•AlF_3-Al_2O_3 melts. Chin J Rare Met (4):418–420 (in Chinese)
15. Tian Z, Hu X, Lai Y, et al (2015) Solubility of Sc_2O_3 in Na_3AlF_6-K_3AlF_6-AlF_3 melts. In: 6th international symposium on high-temperature metallurgical processing. Springer, Cham, pp 105–112
16. Tian Z, Lai Y, Zhang K, et al (2016) Preliminary study on preparation of Al-Sc master alloy in Na_3AlF_6-K_3AlF_6-AlF_3 melt. In: 7th international symposium on high-temperature metallurgical processing. Springer, Cham, pp 157–163

Numerical Simulation Study on the Position Layout of the Permeable Brick at the Bottom of 300t Reblown Converter

Yun Huang, Chengbo Wu, Yong Zhong, Haitao Zhang and Gaopeng Zhang

Abstract Based on 300 t top-bottom combined blowing converter bottom air brick location layout problems, ANSYS Fluent software is adopted to establish the three-dimensional model, which simulated the change after the location of the four air brick at the bottom of the flow of molten steel in the converter. The results show that when the bottom blow flow is constant, the four bottom permeable bricks should adopt asymmetric staggered layout, which is beneficial to uniform converter flow field. At the same time, it was found that the angle of the bottom-blown gas supply element is different and the influence on the wall scour is different. When the position of the bottom blown air brick is determined, the inclination angle of the air supply element is 10°, the molten steel has relatively less erosion on the furnace wall of the converter, the flow field in the converter is more reasonable, and the cost is saved for the converter smelting.

Keywords 300t converter · Bottom blown breathable brick · Gun position angle

Introduction

The top-bottom blow-up of oxygen converter was a new technology developed in the late 1970s [1] and also served as the main method for steelmaking. However, how to improve the production efficiency and the quality of molten steel, as well as the cost of production, is the most important research direction of converter steelmaking.

At present, converter steelmaking usually adopts a composite blowing method, that is, top blowing uses a supersonic oxygen jet, and bottom blowing uses nitrogen or argon for intensive stirring. The method is simple in equipment and low in investment cost, so that the original top-blown converter can be easily converted

---

Y. Huang · C. Wu (✉) · Y. Zhong · H. Zhang · G. Zhang
College of Materials Science and Engineering, Chongqing University,
Chongqing 400044, China
e-mail: 824734696@qq.com

© The Minerals, Metals & Materials Society 2019

T. Jiang et al. (eds.), *10th International Symposium on High-Temperature Metallurgical Processing*, The Minerals, Metals & Materials Series,
https://doi.org/10.1007/978-3-030-05955-2_31

into a top-bottom composite blowing equipment, and the economic cost can be quickly recovered. For this reason, the converter top-bottom composite blowing method has been widely used and developed in various countries around the world. So far, dozens of compound blowing processes have been developed [2].

At present, in order to pursue a faster smelting rhythm, steelmaking enterprises often adopt a method of increasing the oxygen supply intensity. However, due to the problem of the arrangement of the bottom-blown permeable bricks, the flow field distribution during the re-blowing process is uneven, and the molten steel mixing time is too long. According to the literature [3–5], in the case that the bottom blowing flow rate is constant, as the number of bottom blown ventilating bricks increases, the bottom blowing mixing time also increases, and the mixing effect is not good, when the bottom blowing permeable brick. When the bottom ventilation brick is arranged in four pieces, the bottom blowing mixing effect is best. The literature [6–8] pointed out that the bottom-blown permeable bricks have an influence on the mixing effect of molten steel by the asymmetric arrangement and the symmetrical arrangement. In general, the asymmetric arrangement has a better mixing effect on the molten steel.

In this paper, the numerical simulation of the converter was carried out by changing the blowing angle of the bottom blown permeable brick and the symmetrical and asymmetric arrangement of the permeable brick. By comparing the distribution of the molten steel flow field in the converter, the optimal bottom blow ventilation brick arrangement is obtained.

Model Establishment

Geometric Model

The focus of this study is the influence of the position distribution of the bottom blown gas permeable brick of the 300t double-blowing oxygen converter on the flow field and mixing time of the molten steel, so the oxygen lance part is not added. The geometric model of the bottom blown ventilation brick arrangement of the 300t double-blown converter is shown in Fig. 1. The physical properties of the molten steel and argon used in the calculation are shown in Table 1.

Mathematical Model

Basic assumptions

(1) Ignore all chemical reactions of molten steel during steelmaking;
(2) The argon bubble size is uniform;

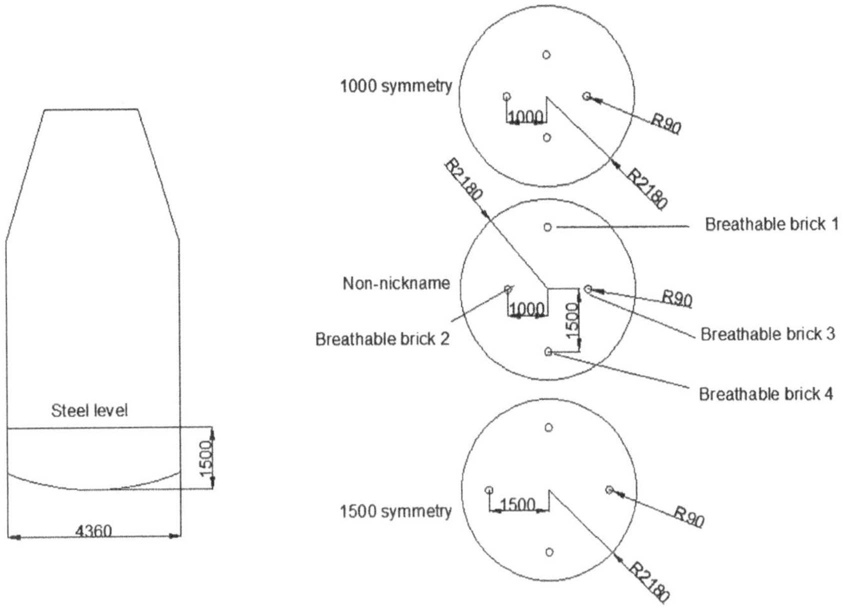

Fig. 1 Physical model of the bottom blown ventilation brick arrangement of 300t double-blown converter

Table 1 Physical properties of molten steel and air

Parameter	Steel water density	Molten steel viscosity	Argon density	Argon viscosity
	$Kg\ m^{-3}$	$Kg\ m^{-1}\ s^{-1}$	$Kg\ m^{-3}$	$Kg\ m^{-1}\ s^{-1}$
numerical value	7000	0.0067	1.6228	2.125×10^{-5}

(3) The bottom blowing element is equivalent to a circular hole having a certain section;
(4) The flow of molten steel in the converter is regarded as an incompressible flow.

Control equation

Blowing at the bottom of the converter is a complicated gas–liquid two-phase flow process. The molten steel is subjected to a circulating two-phase flow under the agitation of the bottom blown permeable bricks, thereby driving the molten steel to flow and functioning to stir the molten steel. In this paper, the Euler multiphase flow model and the turbulent $k - \varepsilon$ model in ANSYS Fluent 14.0 software are used to calculate the molten steel flow in the converter. The governing equations used in the specific calculation model are as follows:

The k equation describing the kinetic energy [9]:

$$\rho\left(\frac{\partial k}{\partial \tau}+U_j\frac{\partial k}{\partial_{x_j}}\right)=\frac{\partial}{\partial_{x_j}}\left(\frac{\mu_{eff}}{\sigma_k}\frac{\partial k}{\partial_{x_j}}\right)+G-\rho\varepsilon \tag{1}$$

$$G=\frac{C_\mu \rho K^2}{\varepsilon}\left(\frac{\partial \mu_i}{\partial_{x_j}}+\frac{\partial \mu_j}{\partial_{x_i}}\right) \tag{2}$$

where ρ is density; k is turbulent flow energy; τ is time; x_j is displacement in j direction; μ_{eff} is turbulent viscosity; $\sigma_k = 1.0$; ε is turbulent energy dissipation rate; $C_\mu = 0.09$; u_i is i direction turbulent viscosity; u_j is the turbulent viscosity in the j direction; x_i is the displacement in the i direction.

Describe the ε equation for turbulent energy dissipation [10]:

$$\rho\left(\frac{\partial \varepsilon}{\partial \tau}+U_j\frac{\partial \varepsilon}{\partial_{x_j}}\right)=\frac{\partial}{\partial_{x_j}}\left(\frac{\mu_{eff}}{\sigma_k}\frac{\partial \varepsilon}{\partial_{x_j}}+\frac{\varepsilon}{k}\right)(C_{\delta 1}G-C_{\delta 2}\rho\varepsilon) \tag{3}$$

where U_j is the fluid velocity in the j direction; $C_{\delta 1} = 1.44$; $C_{\delta 2} = 1.92$; $\sigma_\varepsilon = 1.3$.

Boundary Conditions

(1) Argon gas velocity inlet condition is adopted at the bottom of the converter, argon gas enters the molten steel from the gas permeable brick, and the volume fraction of argon gas is 1.
(2) The converter outlet is a pressure outlet. It is assumed in the calculation that only gas will produce reflux.
(3) The wall surface of the converter adopts no-slip boundary condition, and the near wall surface is treated by standard wall function.

Numerical Solution

The test was performed using ANSYS Fluent version 14.0. Using the PISO solver, the flow field time step was calculated as 0.01 s and the residual was set to 10-3. The calculation results were processed using Tecplot software to obtain the flow field of the converter.

Calculation Results and Discussion

According to the characteristics of a steel enterprise converter and the demand of the production process, the variation in the flow field of the molten steel under the argon blowing mode at the bottom of the converter is calculated. The arrangement of the bottom blown ventilated brick to the centre of the converter is 1000 mm (1000 symmetry for short); the arrangement of the bottom blown ventilated brick to the centre of the converter is 1500 mm (referred to as 1500 symmetry); the distance between the two permeable bricks to the center of the converter is 1000 mm, and the other two bottom blown permeable bricks are arranged at a distance of 1500 mm from the bottom of the converter (abbreviated as asymmetry). The degree to which the bottom blowing air supply element is outwardly offset along the centreline is a low blowing angle. See Table 2 for the air blowing method of the brick bottom blow blasting brick under different working conditions (Figs. 2, 3 and 4).

Bottom blowing adopts 1000 symmetric arrangement for smooth calculation
Case 1 is respectively a bottom speed cloud diagram and a bottom blowing speed line diagram of the bottom blowing using a 1000 symmetrical 0° arrangement. As can be seen from the figure, since the bottom blowing air supply elements are closely spaced and arranged symmetrically, the bottom blowing is sufficient for the centre position of the converter. However, it can be seen from the velocity cloud map and the speed line graph that the gas supply elements have no inclination angle

Table 2 Blowing method of blowing brick at the bottom of converter under a different case

No	Arrangement	Blowing volume ($m^3\ h^{-1}$)	Bottom blowing angle (°)	Remarks
1	1000 symmetry	275	0	4 breathable bricks blown
2	1000 symmetry	275	10	4 breathable bricks blown
3	1000 symmetry	275	20	4 breathable bricks blown
4	1500 symmetry	275	0	4 breathable bricks blown
5	1500 symmetry	275	10	4 breathable bricks blown
6	1500 symmetry	275	20	4 breathable bricks blown
7	asymmetric	275	0	4 breathable bricks blown
8	asymmetric	275	10	4 breathable bricks blown
9	asymmetric	275	20	4 breathable bricks blown

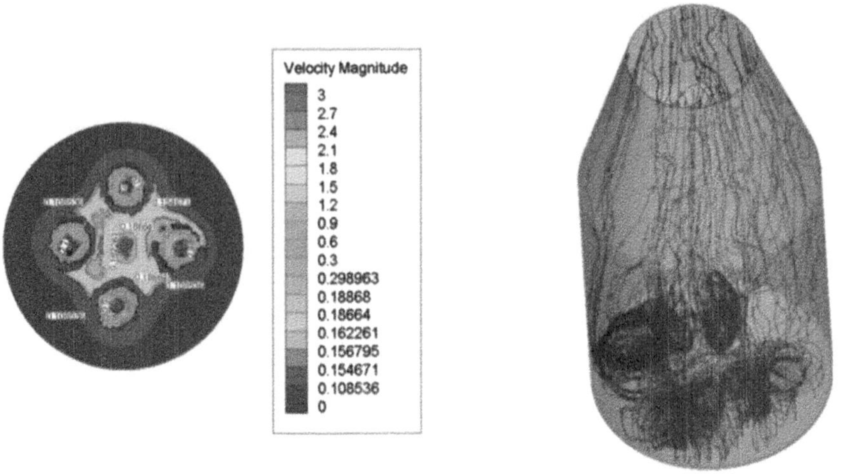

Fig. 2 1000 symmetrical 0° bottom speed map and bottom blow speed line diagram

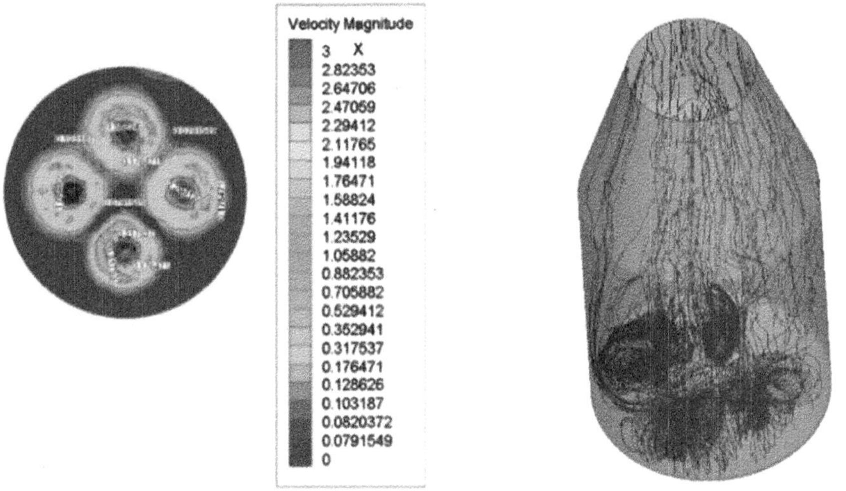

Fig. 3 1000 symmetrical 10° bottom speed map and bottom blow speed line diagram

because they are close to each other. Therefore, there is almost no agitation on the position on the side of the converter.

Case 2 respectively shows the bottom speed waveform and the bottom blowing speed line diagram of the bottom blowing using a 1000 symmetrical 10° arrangement. As can be seen from the figure, since the gas supply element is inclined by 10°, the range of the unheated furnace near the furnace wall is reduced by a part,

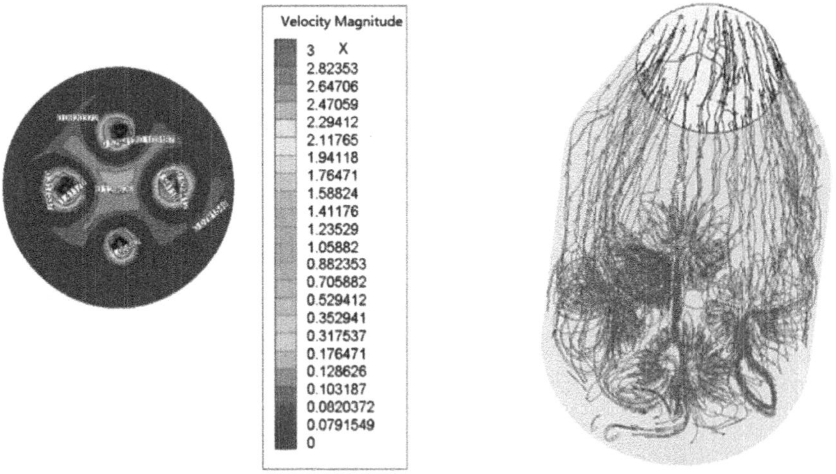

Fig. 4 1000 symmetrical 20° bottom speed map and bottom blow speed line diagram

and the velocity flow field is more uniform. Therefore, after changing the inclination angle of the air supply element, the flow field has a positive influence.

Case 3 respectively shows the bottom speed waveform and the bottom blowing speed line diagram of the bottom blowing using a 1000 symmetrical 20° arrangement. It can be seen from the figure that the flow field of the centre of the converter is affected by the inclination of the gas supply element. By the influence, the stirring effect is reduced, so the inclination angle of the gas supply component should not be too large (Figs. 5, 6 and 7).

Bottom blowing adopts 1500 symmetric arrangement for smooth calculation Case 4 is respectively a bottom speed cloud diagram and a bottom blowing speed line diagram of a bottom blowing using a 1500 symmetrical 0° arrangement. As can

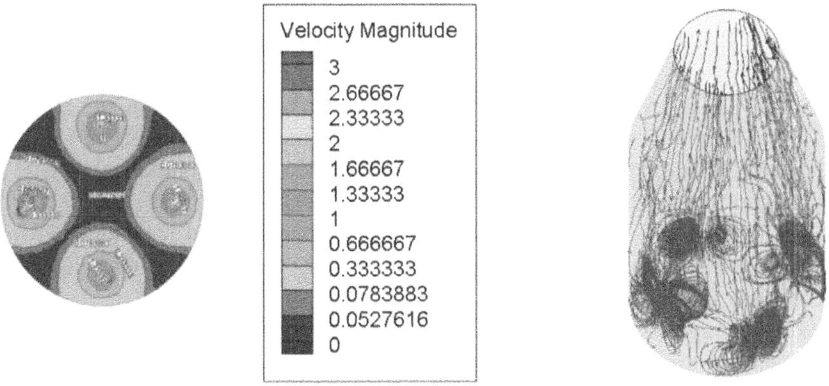

Fig. 5 1500 symmetrical 0° bottom speed curve and bottom blowing speed line diagram

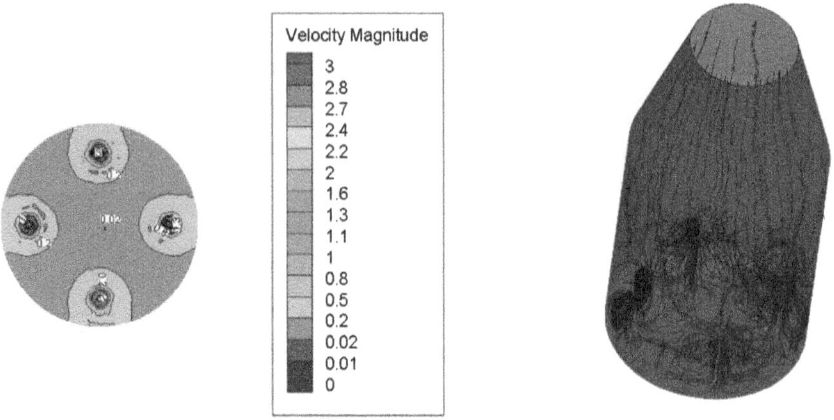

Fig. 6 1500 symmetrical 10° bottom speed cloud diagram and bottom blowing speed line diagram

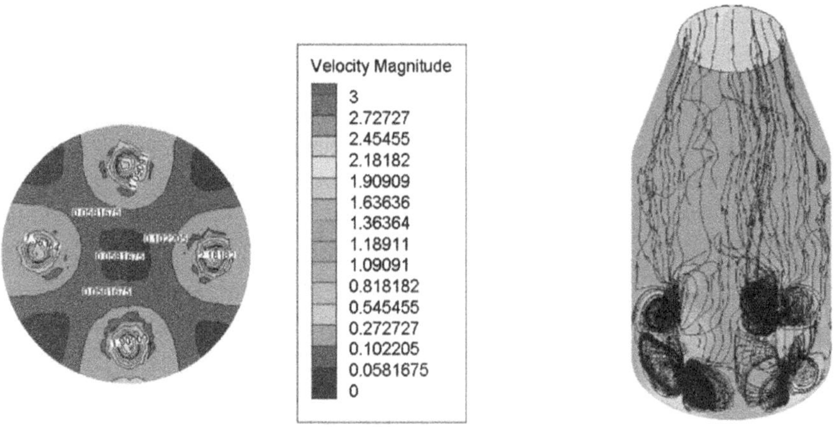

Fig. 7 1500 symmetrical 20° bottom speed map and bottom blow speed diagram

be seen from the figure, as the distance from the gas supply element to the center of the converter increases, the entire converter flow field becomes relatively active. However, because the four gas supply components are too far apart, the flow field in the center of the converter is not active.

Case 5 is respectively a bottom speed cloud diagram and a bottom blowing speed line diagram of a bottom blowing using a 1500 symmetrical 10° arrangement. According to the figure, after the air supply element is added with an inclination angle, although the flow field movement near the furnace wall is strengthened, the molten steel in the centre of the converter hardly flows. Therefore, in the case of increasing the distance of the gas supply element, increasing the inclination angle is not conducive to the activity of the entire converter flow field.

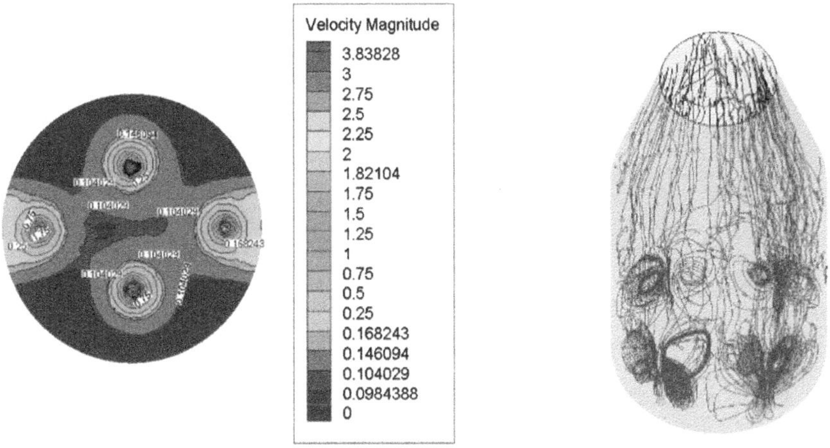

Fig. 8 Asymmetric 0° bottom speed cloud diagram and bottom blowing speed line diagram

Case 6 is respectively a bottom speed cloud diagram and a bottom blowing speed line diagram of a bottom blowing using a 1500 symmetrical 20° arrangement. According to the cloud diagram and the line graph, it can be seen that as the angle increases, the range of the flow field in the centre of the converter does not increase, and the active area of the flow field is concentrated in the furnace wall area (Figs. 8, 9 and 10).

Bottom blowing adopts asymmetric arrangement and smooth calculation

Case 7 is respectively a bottom speed cloud diagram and a bottom blowing speed line diagram of the bottom blowing using an asymmetric 0° arrangement. According to the figure, when the gas supply element is arranged asymmetrically,

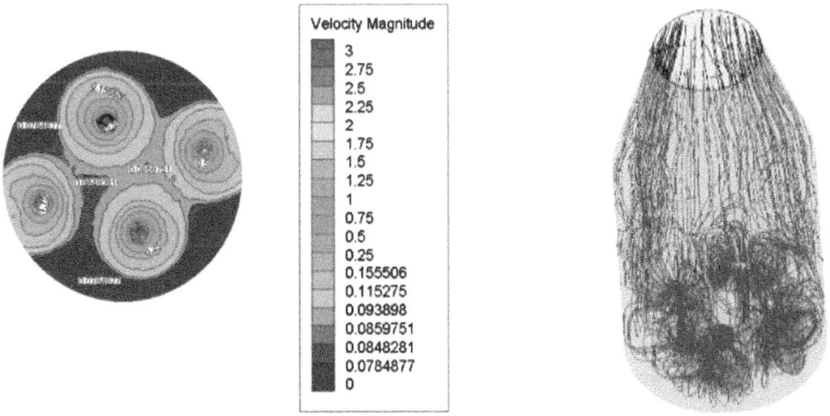

Fig. 9 Asymmetric 10° bottom speed map and bottom blow line diagram

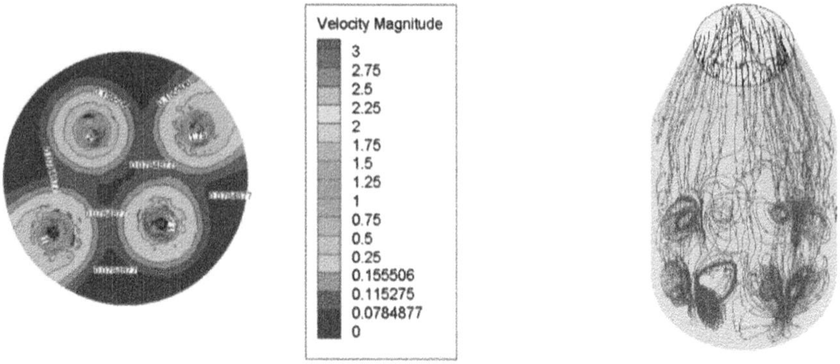

Fig. 10 Asymmetric 20° bottom speed map and bottom blow line diagram

the active area of the converter flow field is obviously increased, and the flow field of both the centre part and the edge of the furnace wall is active.

Case 8 is respectively a bottom speed cloud diagram and a bottom blowing speed line diagram of the bottom blowing using an asymmetric 10° arrangement. It can be concluded from the figure that when the permeable bricks are arranged in an asymmetric manner, the active area of the flow field can be greatly improved, and when the gas supply element is inclined by 10°, the active area of the flow field is maximized and active. The centre of the converter and the edge of the furnace wall reached the expected results of this simulation.

Case 9 is respectively a bottom speed velocity diagram and a bottom blowing speed diagram of the bottom blowing using an asymmetric 20° arrangement. It can be seen from the figure that the whole flow field distribution of the converter is also reasonable, but because the inclination angle is too large, the two gas supply elements with the distance of 1500 mm from the centre cause a flushing effect on the furnace wall, which is not conducive to extend the life of the converter. At the same time, because the angle is too large, the flow field in the central area is not active.

Conclusions

When the bottom blown ventilating brick is arranged in a non-nickname, the flow field distribution of the converter is relatively uniform, and the entire converter flow field is more active. At the same time, if the bottom blowing air supply element has no inclination angle, the molten steel is not active in the furnace wall area, so there should be a certain inclination angle, but the inclination angle should not be too large, otherwise the molten steel will be excessively eroded to the converter furnace wall. In summary, the case 8 (non-nickname 10°) ventilation brick layout is the most reasonable.

References

1. Xiaofang Jiang (2002) Current status and development of Baotou steel's 300t converter combined blowing. Baosteel Technol 3:1–4
2. Pei Pei, Fuhai Liu, Rong Zhu (2017) Research on optimization of bottom blowing arrangement of 300t converter. Ind Heating 46(3):20–22
3. Mori A et al (1989) Optimum arrangement of botom tuyeres in 160t top and bottom blowing BOF. In: Steelmaking conference proceedings, 1989, pp 443–446
4. Kyoji Nakanishi et al (1980) Cold model study on the mixing rates of slag and metal bath in Q-BOP. J Iron Steel Inst Jpn 66(9):1307–1316
5. Paul S, Ghosh. DN (1986) Model study of mixing and mass transfer rates of slag-metal in top and bottom blown converters. Metall Trans 17B(9):461–469
6. Roth Christian, Peter Michael et al (1995) Cold model investigations into the effects of bottom blowing in metallurgical reactors. Steel Res 66(8):325–330
7. Tatjana S et al (2002) Bottom blowing investigations on a cold model reactor to optimize mixing behavior in metallurgical process. Steel Res 73(9):373–377
8. Rongsheng Zhang, Longgen Chen, Yisheng Zhou (1983) Hydraulic research on stirring of top bottom blowing and smelting pool. J Iron Steel Res Inst 3(1):14–22
9. Wenjin Wang (2013) Three-dimensional compressible fluid simulation of porous jet top-blown converter. Mod Metall 2(1):32–40
10. Yan W, Li Y, Zhu M (2011) Numerical simulation of gas-liquid two-phase prevalence in top-bottom combined blown converter. J Process Eng 2011, 11

Optimization of Zn-Al-Fe Alloy Vacuum Distillation Experiments by Response Surface Methodology

Zhenghao Pu, Yifu Li, Bin Yang and Huan Zhang

Abstract Vacuum distillation of Zn-Al-Fe alloy was discussed based on the experimental investigations of the distillation temperature, the holding time and the pressure of pressing block by response surface methodology. The experimental results showed that zinc can be satisfactorily obtained from Zn-Al-Fe alloy with suitable distillation temperature, holding time and the pressure of pressing block. The zinc in Zn-Al-Fe alloy was effectively recovered at 1073–1123 K for 60–75 min, and pressure of pressing block was below 50 MPa. When the temperature was 1073 K, the holding time was 75 min and the pressure of pressing block was 40 MPa, the zinc content in the volatiles was about 98.35% and the aluminium content in the residue was about 92.22%.

Keywords Zn-Al-Fe alloy · Vacuum distillation · Response surface

Introduction

With the rapid development of national economy and the continuous adjustment of steel structure, the demand for galvanized steel in China is increasing dramatically. Hot galvanizing slag (Zn-Al-Fe alloy) is a by-product of hot dip galvanizing. It will

Z. Pu · Y. Li (✉) · B. Yang · H. Zhang
National Engineering Laboratory for Vacuum Metallurgy, Kunming University of Science and Technology, Kunming 650093, People's Republic of China
e-mail: 326686058@qq.com

Z. Pu · Y. Li · B. Yang · H. Zhang
Faculty of Metallurgical and Energy Engineering, Kunming University of Science and Technology, Kunming 650093, People's Republic of China

Z. Pu · Y. Li · B. Yang · H. Zhang
The State Key Laboratory of Complex Nonferrous Metal Resources Clean Utilization, Kunming University of Science and Technology, Kunming 650093, People's Republic of China

© The Minerals, Metals & Materials Society 2019

T. Jiang et al. (eds.), *10th International Symposium on High-Temperature Metallurgical Processing*, The Minerals, Metals & Materials Series,
https://doi.org/10.1007/978-3-030-05955-2_32

produce large quantities of Zn-Al-Fe alloy every year for such a huge yield [1]. Meanwhile, with the rapid development of the economy and industry, the consumption of zinc and aluminium will increase. Extraction of zinc and aluminium from primary mineral resources has been difficult to satisfy the requirements of industry. Hot dip galvanizing slag and scrap aluminium alloy contain a large amount of zinc and aluminium [2]. It will result in huge waste of resources and environmental pollution if it abandons the hot dip galvanizing slag and scrap aluminium alloy directly [3]. Accordingly, it is necessary to recycle these metals.

Vacuum distillation is a new kind of metallurgy method, which carries out in a closed container with the characteristics of high efficiency and clean [4]. The problems of long technological process, high energy consumption, low yield and poor economic benefit of traditional metallurgical method can be solved effectively. Vacuum distillation has been widely used in the separation of alloys and the refining of crude metals [5–7]. Pu [8] studied the separation of hot dip galvanizing by response surface methodology under vacuum condition. The results showed that the zinc in Zn-Al-Fe alloy was effectively recovered at 1073–1123 K for 60–75 min and the thickness of the raw materials showed little influence in vacuum distillation. However, the further work and more factors should be considered.

Theoretical Basis

Separation of zinc from Zn-Al-Fe alloy was the core of vacuum distillation experiment, whether it can be separated by the vacuum distillation depending on the difference in the saturated vapour pressure of the pure metal. Generally speaking, the greater the difference in saturated vapour pressure, the better the effect of alloy separation [9]. And it will be existed in liquid phase if the saturated vapour pressure was small.

The saturated vapour pressures can be calculated by the formula:

$$\log P = A + BT^{-1} + C \log T + DT \tag{1}$$

A, B, C and D indicated the volatilization constants respectively. The relation curves of saturated vapour pressure and temperature of zinc, aluminium and iron are shown in Fig. 1.

As can be seen in Fig. 1, the saturated vapour pressure of the metal increased with the increase in temperature. The saturated vapour pressure of zinc was far greater than aluminium and iron, which means zinc can evaporate easily into the gas phase, so that zinc can be separated from aluminium and iron.

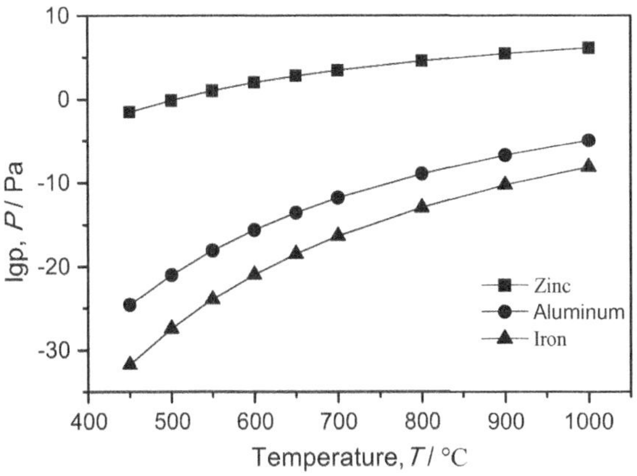

Fig. 1 Change curve of saturated vapour pressure and temperature of zinc, aluminium and iron

Experimental Procedure

Sample of the Zn-Al-Fe alloy for experiment is obtained from a factory in China. The information about materials and samples is concluded in Table 1. The vacuum furnace is shown in Fig. 2. The working pressure for the vacuum furnace was about 1 Pa, and the working temperature can be set from room temperature to about 2000 K. The vacuum furnace used Pt-Rh thermocouple and McLeod vacuum gauge. The experimental samples must be dried and placed in a graphite crucible at the constant temperature zone of the vacuum furnace. Experiments were carried out under preset experimental parameters. When the furnace temperature dropped to room temperature, the condenser and evaporator should be taken out for analysis.

Hot galvanizing slag cannot be used in the experiment directly. The author pressed the block after grinding the sample for the experiment. This research adopted the experimental design software Design-Expert 8.0.6 experimental design (BBD), in order to study the influence factors of the interaction in the distillation process of distillation temperature and time, and the pressure of pressing block on the refining of Zn-Al-Fe alloy. The vacuum distillation experiments by the response surface method were carried out with the distillation temperature from 973 to 1173 K, holding time from 15 to 75 min and pressure of pressing block from 40 to 60 MPa.

The direct yield was calculated by the following formulas:

$$\text{The direct yield} = \left(\frac{x_i' \times m'}{x_i \times m}\right) \times 100 \tag{2}$$

where x_i is the content of i in raw material, x_i' is the content of i in the volatiles, m is the weight of the feeding materials, m' is the weight of the volatiles.

Table 1 Chemical composition of raw materials

Sample	x_{Zn}	x_{Al}	x_{Fe}
Raw material	0.3652	0.6023	0.0198

Fig. 2 Schematic diagram of the internal structure of the vertical vacuum furnace: 1—furnace cover; 2—vacuum furnace body; 3—furnace chassis; 4—electrode; 5—condensation plate; 6—observation hole; 7—thermal insulation sleeve; 8—heater; 9—crucible

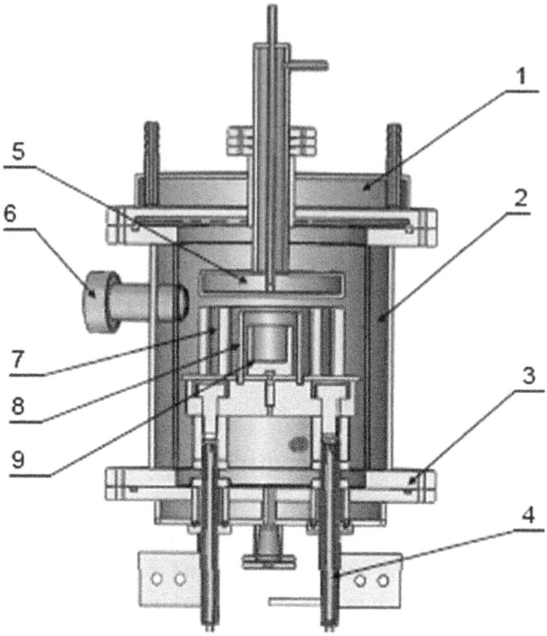

This paper studied the repeatability and random error and provides additional freedom for error estimation. The central point in the experimental design matrix was repeated 5 times.

The relationship between the factors and the results is shown in Eq. (3):

$$Y = \beta_0 + \sum_{i=1}^{k} \beta_i X_i + \sum_{i=1}^{k} \beta_{ii} X_i^2 + \sum_{i}^{i<j} \sum_{j} \beta_{ij} X_i X_j \tag{3}$$

where Y is the response quantity; K is the number of factors; β_0 is a constant term; β_i is a linear coefficient; β_{ii} is the two-order coefficient; β_{ij} is the coefficient of interaction; X_1, X_2, X_3 represent the experimental factors. The equation was based on the least square method to estimate the experimental parameters. The experimental design is shown in Table 2.

Table 2 Factors and levels of experimental design

Factor	Code	Coding level		
		−1	0	1
Distillation temperature/K	X_1	973	1073	1173
Pressure/Mpa	X_2	40	50	60
Holding time/min	X_3	15	45	75

Results and Discussion

The experimental results of the vacuum distillation with contents of zinc, aluminium and iron in residue (liquid phase) and volatiles (vapour phase) are listed in Table 3. The effect of factors is shown in Fig. 3.

The Effect of Factors on the Zinc Content in Gas Phase

As shown in Fig. 3, Zn-Al-Fe alloy was under the distillation temperature from 973 to 1173 K, holding time from 15 to 75 min and pressure of pressing block from 40 to 60 MPa, with pressure 5 Pa. The effect of these factors on the direct yield of zinc was examined.

As can be seen from Fig. 3a, with the increase in temperature and holding time, the content of zinc increased gradually in the volatiles. The main reason was that the saturated vapour pressure of zinc increased gradually with the increase in temperature. With the increase in temperature, the direct yield of zinc increased sharply at the beginning. When the temperature was 973 K and the holding time was 15 min, the direct yield of zinc was approximately 78.29% and the content of zinc was 88.24% in the volatiles. When the temperature was 1073 K and the

Table 3 Elements distribution of distillation product

Sample	T/K	Time/min	Pressure/MPa	x_{Zn}	x_{Al}	x_{Fe}	y_{Zn}	y_{Al}	y_{Fe}	The direct yield of Zinc/%
1#	1173	45	40	0.0002	0.9289	0.0632	0.9821	0.0003	<0.0001	97.6
2#	1173	45	60	0.0001	0.9126	0.0693	0.9772	0.0017	0.0002	97.9
3#	1073	15	60	0.0010	0.9267	0.0500	0.9209	0.0007	0.0001	86.7
4#	973	45	40	0.0152	0.8941	0.0758	0.9087	0.0001	<0.0001	75.8
5#	1173	15	50	0.0005	0.9278	0.0579	0.9719	0.0004	<0.0001	96.1
6#	1073	45	50	0.0023	0.9256	0.0618	0.9828	0.0022	0.0006	97.1
7#	1073	15	40	0.0046	0.9065	0.0664	0.9673	0.0006	0.0003	87.6
8#	973	45	60	0.0086	0.8923	0.0733	0.9328	0.0022	0.0006	90.4
9#	1073	45	50	0.0004	0.9280	0.0521	0.9837	0.0006	0.0002	93.2
10#	1073	45	50	0.0029	0.9232	0.0637	0.9833	0.0003	0.0001	94.5
11#	1073	45	50	0.0003	0.9276	0.0623	0.9788	0.0004	<0.0001	95.9
12#	973	75	50	0.0096	0.9013	0.0884	0.9554	0.0003	<0.0001	88.6
13#	1073	75	40	0.0003	0.9323	0.0554	0.9835	0.0003	<0.0001	97.2
14#	973	15	50	0.0091	0.8716	0.0636	0.8824	0.0002	<0.0001	78.3
15#	1173	75	50	0.0002	0.9408	0.0344	0.9836	0.0005	<0.0001	98.4
16#	1073	75	60	0.0003	0.9026	0.0861	0.9807	0.0002	<0.0001	93.6
17#	1073	45	50	0.0029	0.9267	0.0611	0.9798	0.0003	0.0001	94.1

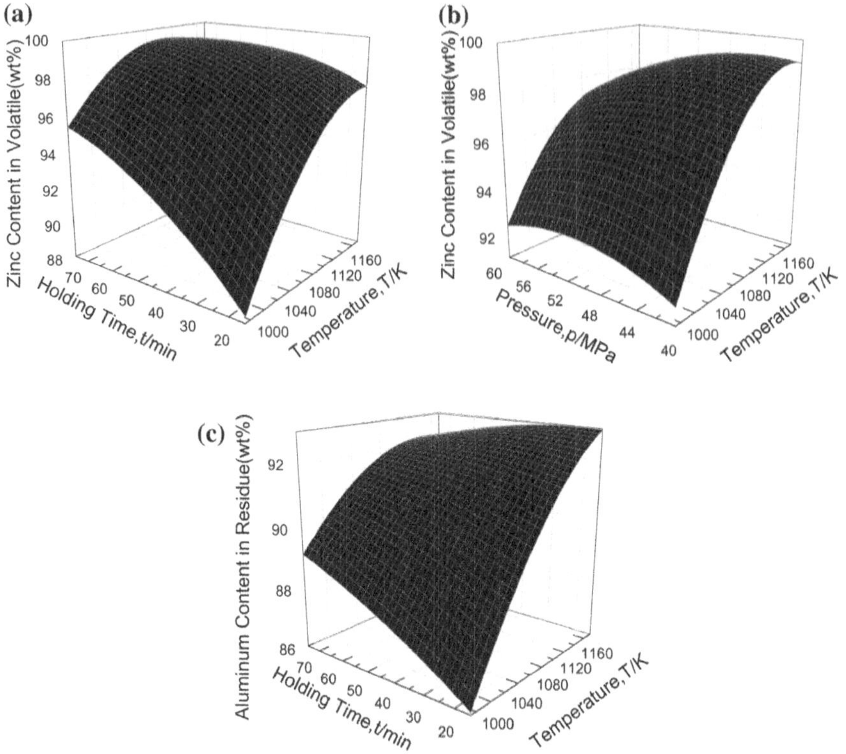

Fig. 3 Effect of factors **a** The effect of factors on the content of zinc in gas phase (holding time), **b** the effect of factors on the content of zinc in gas phase (pressure), **c** the effect of factors on the content of aluminium in liquid phase

holding time was 45 min, the direct yield of zinc was approximately 97.12% and the content of zinc was 98.28%. At this time, the direct yield and the content of zinc changed little with the increase in temperature and holding time. When the distillation temperature was above 1073 K and holding time was above 60 min, the zinc content in the volatiles was no longer rising, even a little reduce. That was because long cooling time process led the zinc to reflux. When the temperature was 1073 K and the holding time was 75 min, the zinc content in the volatiles was about 98.35%.

As can be seen from Fig. 3b, with the increase in pressure of pressing block, the content of zinc increased firstly and then decreased. The main reason was that when the pressure of pressing block increased, zinc particles became more close in the raw material and it was not conducive to evaporation after reaching a certain pressure. When the temperature was 973 K and the pressure of pressing block was 60 MPa, the direct yield of zinc was approximately 75.81% and the content of zinc was 90.87%. When the temperature was 1073 K and the pressure of pressing block was 40 MPa, the direct yield of zinc was approximately 97.19% and the content of zinc was 98.35%. The zinc content in the volatiles was no longer rising, as the distillation temperature was above 1073 K and pressure of pressing block was below 50 MPa.

The Effect of Factors on the Aluminium Content in Liquid Phase

As, the can be seen from Fig. 3c, with the increase in temperature and holding time, the content of aluminium increased gradually in the residue. The aluminium content had a great relationship with the zinc content. This was because with the increase in temperature and the extension of time, the zinc in the Zn-Al-Fe alloy evaporates basically, the aluminium and iron were left, and so the content of aluminium increased. When the temperature was 973 K and the holding time was 15 min, the content of aluminium was 87.16%. When the temperature was 1073 K and the holding time was 45 min, the content of aluminium was 91.80% in the residue. When the distillation temperature and holding time were above 1073 K and 45 min, respectively, the aluminium content in the residue was no longer rising. When the temperature was 1073 K and the holding time was 45 min, the aluminium content in the residue was about 92.03%. For this residue, aluminium can be recycled in other ways.

The iron content in the residue change was not very obvious due to the less content in raw materials.

Conclusion

The experimental results showed that vacuum distillation was an advanced technology of clean metallurgy. For the processing of Zn-Al-Fe alloy, zinc can be recovered effectively.

The zinc in Zn-Al-Fe alloy was effectively recovered at 1073–1123 K for 60–75 min and pressure of pressing block was below 50 MPa. When the temperature was 1073 K, the holding time was 75 min and the pressure of pressing block was 40 MPa, the zinc content in the volatiles was about 98.35% and the aluminium content in the residue was about 92.22%. And the experiment provided an efficient and convenient idea to guide the process of vacuum metallurgy.

Acknowledgements This work has been funded by the Fund of Yunnan science and technology plan project under Grant No. 2013FZ012, the National Natural Science Foundation of China Youth Foud of Study on mechanism of vapour condensation from complex gas of As-Pb No. 51504115, the Cultivating Plan Program for the Leader in Science and Technology of Yunnan Province under Grant No. 2014HA003.

References

1. Ren X, Wei Q, Hu S et al (2010) The recovery of zinc from hot galvanizing slag in an anion-exchange membrane electrolysis reactor. J Hazard Mater 181(1):908–915
2. Yuan X, He M, Wang S, Zhao X, Zhao X (2007) Formation cause and recovering technology of hot Galvanizing Dross. Yunnan Metall 36(1):32–35
3. BaoQiang XU, Yang B, Liu DC et al (2007) The study on recovering metal zinc from the hot Galvanizing Slag by vacuum distillation. Non-Ferrous Mining Metall
4. Li Dongsheng, Dai Yongnian, Yang Bin et al (2013) Purification of indium by vacuum distillation and its analysis. J Central South Univ 20(2):337–341
5. Pu Z, Han J, Li Y et al (2018) Removal of arsenic from crude tin by vacuum distillation. Mater Trans 59(2):311–315
6. Kong LX, Yang B, Xu BQ et al (2012) Application of MIVM for Pb–Sn–Sb ternary system in vacuum distillation. Metall Mater Trans B 43(6):1649–1656
7. Yang HW, Zhang C, Yang B et al (2015) Vapor–liquid phase diagrams of Pb–Sn and Pb–Ag alloys in vacuum distillation. Vacuum 119:179–184
8. Pu Z, Han J, Li Y, et al (2018) Application of the Wilson equation for Zn-Al-Fe alloy in vacuum distillation. Mater Trans 59(3):443–449
9. Dai YN, Yang B (2000) Vacuum metallurgy of nonferrous metals. Metallurgical Industry Press

Review on Preparation of Medium- and Low-carbon Ferrochrome Alloys

Ting Hu, Hua Liu, Bingguo Liu, Linqing Dai, Libo Zhang and Shenghui Guo

Abstract As one of the important strategic materials, chromium has been widely used in metallurgical, refractory, and chemical industry. Ferrochrome is an important additive in the production of alloy steel. According to the carbon content, it is classified as high-carbon ferrochrome (carbon 4–8%), medium-carbon ferrochrome (0.5–4%), low-carbon ferrochrome (0.15–0.50%), micro-carbon ferrochrome (carbon 0.06%), and ultra-micro-carbon ferrochrome (less than 0.03%). But the most widely used one in metallurgical is the first three. This article mainly summarizes the current development of domestic and international medium-low-carbon ferrochrome and summarizes the preparation methods and the influence factors in the smelting process including reduction temperature, reduction time, the reduction agent, and alkalinity.

Keywords Chromite · Smelting method · Medium- and low-carbon ferrochrome · Influence factors

---

T. Hu · H. Liu · B. Liu · L. Dai (✉) · L. Zhang · S. Guo
State Key Laboratory of Complex Nonferrous Metal Resources Clean Utilization, Kunming University of Science and Technology, Kunming 650093, China
e-mail: linqingdai@163.com

T. Hu
e-mail: ht_tranquillity@163.com

T. Hu · H. Liu · B. Liu · L. Dai · L. Zhang · S. Guo
Faculty of Metallurgical and Energy Engineering, Kunming University of Science and Technology, Kunming 50093, China

T. Hu · H. Liu · B. Liu · L. Dai · L. Zhang · S. Guo
Key Laboratory of Unconventional Metallurgy, Ministry of Education, Kunming 650093, China

T. Hu · H. Liu · B. Liu · L. Dai · L. Zhang · S. Guo
National Local Joint Laboratory of Engineering Application of Microwave Energy and Equipment Technology, Kunming 650093, China

© The Minerals, Metals & Materials Society 2019
T. Jiang et al. (eds.), *10th International Symposium on High-Temperature Metallurgical Processing*, The Minerals, Metals & Materials Series,
https://doi.org/10.1007/978-3-030-05955-2_33

Introduction

Chromium is an important metal, which contents only 0.01% in the crust and ranking 17th. In nature, iron and chromium symbiotic formation of ferrochrome spinel ($FeO \cdot Cr_2O_3$), whose mineralogical name is chromite [1]. In the world, the chromium-containing ore is mainly chromite. About 60% of chromite is used in the metallurgical industry to produce ferrochrome (manufacturing alloy steel and stainless steel), 20% for refractory materials, 12% for the chemical industry, and the remaining 8% for the foundry industry [2]. According to the compositions and uses, chromite ore is classified as the metallurgical grade, refractory grade, and chemical grade and used for casting stone. Ferrochrome alloy is one of the important alloy materials used in the steel industry. Besides containing the main components of chromium and iron, it also contains other impurity elements such as carbon, silicon, sulphur, and phosphorus. Ferrochrome alloy is used to produce a variety of special steels with corrosion resistance, high-temperature resistance, high strength, and oxidation resistance. Special steel is an essential material for the production of artillery, rockets and missiles, ships, etc. in the aerospace, automotive, shipbuilding, and defence industries. As the steelmaking technology advance, the requirements of quality and cleanliness of special steels are becoming increasingly high. Meanwhile, with the refinement of steel grades, the demand for medium-low-carbon ferrochrome alloy is also increasing [3, 4].

The global reserves of chromite are about 5 billion tons, which is not particularly rich. Compared with other countries, about 62.4% of chromite reserves are produced in South Africa, ranked first in the world. And Zimbabwe about 32.7%, the former Soviet about 1.2%. More than 96% of the chromite resources are concentrated in these three countries, followed by the Philippines (about 0.3%), Turkey (about 0.1%), and other countries (about 3.3%) [2, 5–7]. According to the US Geological Survey, exceed 12 billion tons of chromium are available, enough to sustain consumption for several centuries. Chromite resource distribution is more concentrated, about 95% of chromite mainly in Kazakhstan and South Africa. The natural reserves of chromite in the crust exceed 1.8 billion tons, and the recoverable reserves exceed 810 million tons. Proved China's reserves of chromite increased slightly in recent years, remained stable at 12 million tons. As the end of 2016, the country's proven reserves 12,331,900 tons, down 1%.

Compared with other national reserves, China's chromite resources are few, and its chromite deposits are mostly small and medium-sized deposits. Our country is the main consumer of chromite reserves less than 0.15% of world reserves. Meanwhile, our ferrochrome resources are faced with many problems such as small-scale mineral deposits, scattered distribution, poor development and utilization conditions, more lean ore, and less open mining [8–12]. This article discusses the current domestic and international forms from the preparation methods of medium- and low-carbon ferrochrome alloys and the influencing factors in the smelting process.

Medium-Low-Carbon Ferrochrome Alloys Preparation Methods

Medium-low-carbon ferrochrome alloys are used to produce medium-low-carbon structural steel, chrome steel, and alloy structural steel. Chrome steel is commonly used in the manufacture of gears and gear shafts. Currently, the smelting methods mainly include an electron-silicothermic method and oxygen blowing method [4, 13].

Electron-Silicothermic

At the beginning of the twentieth century, French metallurgists studied the related aspects of medium-low-carbon ferrochrome production through electron-silicothermic, which was the earliest application of the electron-silicothermic.

The silicon thermal method is to add the raw material chromite, silicon chrome alloy, and lime into the electric furnace in the test furnace under the condition of making the alkali slag in the electric furnace to melt the material so that the silicon in the silicon-chromium alloy reduces the chromium and iron in the chromite.

The basic principles of electrothermal smelting are as follows:

$$2Cr_2O_3(S) + 3[Si] = 4[Cr] + 3(SiO_2) \tag{1}$$

$$2(FeO) + [Si] = 2[Fe] + (SiO_2) \tag{2}$$

Tang et al. [14] prepared low-carbon ferrochrome by electric silicon thermal method. The results show that the low-carbon and micro-carbon ferrochrome in the production of silicon-chromium alloy can meet the requirements of users with $C \leq 0.010\%$ or less. The system has a high pass rate; the process of producing low-carbon ferrochrome by electro-silicon thermal method was studied by Tianhu et al. [15], and the main factors affecting production indicators and mechanisms were found. By adjusting the production process and strengthening technical management, there is a certain effect; Wen et al. [16] discussed the process of electrothermal production.

The chromium-iron alloy prepared by the electrothermal method has a carbon content of less than 0.03% and belongs to micro-carbon ferrochrome [15]. However, these methods require three electric furnaces to complete the smelting process, one of which is for the production of carbon ferrochrome. One for the production of silicon-chromium alloys, the other for the refining of chromium and silicon-chromium alloys. Since three electric furnaces are required, the main disadvantage of this method is high power consumption (about 1800 kWh/t). The carbon electrode in the smelting is in contact with the liquid alloy so that the ferrochrome alloy is easy to add carbon and cannot meet the quality requirements of the medium- and low-carbon ferrochrome alloy. The process is long, and the intermediate product needs to be repeatedly cast, crushed, finished, and stored, thus greatly reducing the recovery rate of chromium [17].

Oxygen Blowing

The low-carbon ferrochrome used in the oxygen blowing process is a converter, so it is called the converter method. According to different oxygen supply methods, oxygen blowing can be divided into four types: side blowing, top blowing, bottom blowing, and top and bottom blowing. China adopts the top blowing converter method.

The oxygen blowing method is to produce oxygen low-carbon ferrochrome by directly blowing oxygen into liquid high-carbon ferroniobium to decarburize it. The working principle is that the temperature required for the high-carbon ferrochrome solution fed into the converter is usually between 1723 and 1873 K. Due to the oxidation of chromium, iron, and silicon, the temperature in the molten pool will rise rapidly, the decarburization reaction begins, and the temperature is higher. High, the better the decarburization reaction. Under normal pressure, a product containing 2% carbon can be obtained at a temperature of 2038 K, a product containing 1% carbon can be obtained at a temperature of 2217 K, and a product having a carbon content of 0.5% can be obtained at a temperature of 2432 K.

As early as around 1970, Shanghai Ferroalloy Plant and Beijing Iron and Steel Research Institute [18] jointly studied the low-carbon ferrochrome in converter oxygen blowing, mainly studied the blowing process, from raw materials, oxygen supply, temperature, slag, smoke. Detailed studies have been carried out on gas, furnace gas, and its molten iron casting. The research results show that "high temperature, fast" is the basic principle of converter smelting ferrochrome. Hu [19] proposed the process of extending the slag method of the converter age for some problems such as short furnace age and high refractory consumption in the production of medium- and low-carbon ferrochrome in pure oxygen top-blown converter. The theoretical basis of the process operation and the influence of the slag composition on the slag and the specific operation of the slag are discussed. According to production practice, the use of reasonable slagging system and correct slagging method has an obvious effect on the age of the furnace; Narita et al. [20] deeply discussed the effect of lime-based flux on dephosphorization of molten iron, taking oxygen injection. The top blowing method has been studied. The research shows that the dephosphorization rate is better in this way, the high phosphorus distribution ratio (% P_2O_5)/[%P] = 500–1500, when the CaO/SiO_2 in the Cao residue exceeds 3, the flux will be When mixed with soda ash, the dephosphorization rate can be maintained at a high level, and the desulphurization rate is remarkably improved.

Masuda et al. [21] discussed the development of stainless steel refining oxygen top-blowing argon blowing process; Yu et al. [22] discussed the fuzzy PID oxygen-rich bottom blowing bath temperature control method. The research results show that the high temperature (1800–2000 °C) is a demand of smelting by oxygen blowing method, so the damage to the furnace lining is serious, so the liquid high-carbon ferrochrome must be strictly required to enter the furnace, the temperature is 1450–1600 °C, the molten iron contains silicon $\leq$ 1.5%, molten iron containing sulphur $\leq$ 0.036%, while the recovery rate of chromium is relatively low, only about 81% [23].

Another Method

For the problem of traditional medium-low-carbon ferrochrome production, it is necessary to research and develop a new technology and process with short process, cost-effective and simple operation to adapt to the advancement of steelmaking technology.

In recent years, microwave heating has been applied increasingly in the metallurgical field. At the same time, many good research results have been achieved, and a new metallurgical technology has been formed. Different from the conventional heating, microwave heating is the use of heat generated by the energy consumed by the electromagnetic field of the dielectric material to generate heat. Compared with the traditional heating method, the characteristics of microwave metallurgy are as follows: (1) heating uniformity and penetrating properties of materials; (2) volumetric heating mode and rapid heating of materials; (3) selective heating and dielectric properties of the material; (4) it does not produce any gas itself, and it is one of the effective ways to achieve clean metallurgical industry production. With these advantages, new metallurgical techniques and processes that cannot be achieved under conventional heating conditions can be developed. Realize the transformation of some traditional metallurgical processes and technologies, improve the processing level of metallurgical products, improve product structure, metallurgical processes achieve high efficiency, energy saving, and environmental protection. Currently, microwave heating technology has been used for the ore crushing [24, 25], pretreatment of refractory gold ore [26–28], recovery of gold from low-grade ore and tailings, extraction of rare metals and heavy metals from ore [29], etc. In the future, as the metallurgical technology advance, microwave metallurgy will play an indispensable role in the industry.

Chen et al. [30] studied the temperature rise characteristics of carbon-containing chromite ore powder in the microwave field. The results showed that chromite ore fines containing coal had a better temperature-rising characteristic in the microwave field of 2.45 GHz in frequency. Under the condition of the microwave power of 10 kW and the ore fines' mass of 1 kg, chromite ore fines containing coal reached 1100 °C in 7 min with a temperature-rising rate of 157.1 $°C \cdot min^{-1}$, which stated chromite has good absorbing properties. It provides a new method for the smelting of low-carbon ferrochrome, which has certain theoretical and practical value. However, carbon is used as a reducing agent, and carbon is not easily removed completely so that it is difficult to obtain a medium-low-carbon ferrochrome. The best way is to use silicon as a reducing agent to achieve short process smelting. And in the reduction process, a flux can be added to achieve slag and metal separation. Wu et al. [31, 32] studied the microstructure of voluminal reduction on chromite ore fines containing silicon (basicity: 1.60–1.90) heated by microwave. It has been revealed that at the temperature of 1100, 1200, and 1300 °C (at atmospheric pressure without protective gas), the microstructure of voluminal reduction on chromite ore fines containing silicon heated by microwave consisted mainly of the thermic fractured reduction structure from the edge to inside at the first. The fines thermic fractured reduction structures were dominated by the increase of temperature. Finally, at the temperature of 1300 °C, complete separation occurred, and Fe, Cr content in metallic Cr-Fe phase increased with temperature. And analyse the chemical composition of reduction products Cr-Fe alloy, the results show that the reduction product Cr-Fe alloy conglomerates to block and separates from slags preferably at smelting temperature of 1 300 °C, holding time of 40 min and end-point temperature of 1 450 °C. The composition of reduction products Cr-Fe alloy meets requirements of (GB5683-87)FeCr55C25II. Liu [33] studied the thermal reduction process of chromite ore in microwave field. The result clearly revealed that the optimum process parameters are as follows: microwave power: 1500 W, reduction time: 60 min, raw material size: −200 mesh, reducing agent dosage: 40%; 1300 °C at temperature, heat preservation at 60 min, and 1450 °C for bulk reduction. The reduction product has been made the SEM analysis and phase analysis, low-carbon ferrochrome prepared chromite reduction method that microwave thermal field of silicon, the main chemical components to meet the requirements of $FeCr_{55}C_{25}$ II (GB5683-87).

Influence Factors

Reduction Temperature

Xu et al. [34] studied the spheronization and reduction catalysis of carbon-containing chrome ore and studied the effect of reduction temperature on the metallization rate. Zambrano et al. [35] analysed the effects of reduction temperature and excess reducing agent on ferrosilicon using raw materials (chromite, petroleum coke, Portland cement, lime, and silica); Liu [33] also conducted related research. The results show that the metallization rate of chromium increases with the increase of the reduction temperature. These have achieved good results. It can be seen that compared with the conventional method, recent research has greatly reduced the reduction temperature, thus reducing the loss. These studies have laid the foundation for future research work.

Reduction Time

It is known from the literature [34, 36] that the reduction time has a great influence on the metal conversion rate in the preparation of medium- and low-carbon ferrochrome. Li [37] also conducted related research, controlling the reduction temperature to 1400 °C, and used bituminous coal as a reducing agent to study the effect of reduction time on the pre-reduction results. The results show that the reduction time is more conducive to the reduction of chromium. The reduction time increased from 2.0 to 4.5 h, and the metallization rate of iron was relatively small, indicating that the reduction rate of iron was faster, and the metallization rate was higher after 2 h. When the reduction time is 2 h, the chromium metallization rate is only 13.11%. After the reduction time is extended to 4 h, the chromium metallization rate reaches 47.01%, but the reduction time is prolonged and the reduction effect is not obvious. Therefore, it is not the longer the reduction time, the better the reduction effect. Liu [33] also did this research. She studied the effect of reduction time on the heating performance and chromium metallization rate of chromite in the microwave field. It's shown that the longer the reduction time, the higher the metallization rate of chromium. When the metallization rate reaches a certain value, the reduction time has no effect on the metallization rate of chromium.

Reduction Agent

At present, many studies have shown that reducing agents have a great influence on the preparation of low-carbon ferrochrome. Li [37] also studied the effect of coal ash on the reduction process. The results show that the content of ash and volatiles in bituminous coal has a great influence on the reduction process, the lower ash content, the higher volatile content, the better reduction effect. Zhou et al. [38] studied the effect of different reducing agents on the metallization rate; Liu [33] studied the effect of reducing agent dosage on the heating performance and chromium metallization rate of chromite in the microwave field. Studies have shown that as the amount of reducing agent increases, the metallization rate of chromium increases in turn. When the amount of the reducing agent was controlled at 40%, the metallization ratio of chromium was 95.47%. The amount of reducing agent continues to increase, and the metallization rate of chromium is not obvious and is basically stable.

Alkalinity

Control the slag basicity is important in the preparation of medium-carbon ferrochrome. It has a lot of related research has been done at home and abroad. Fu [39] studied smelting medium- and low-carbon ferrochrome by argon-oxygen refining (AOD), it was claimed that the alkalinity of slag is an important factor affecting product quality and energy consumption. Ren et al. [40] analysed the influencing factors of Cr_2O_3 content in slag during the EAF smelting process of trace carbon chrome-iron slag and introduced some specific measures to improve the utilization of chromium. Studies have shown that the performance of raw materials has the greatest impact on the utilization of chromium. The alkalinity (CaO/SiO_2) in the melting period is controlled at about 1.4–1.6, the refining period is controlled at about 1.7–1.8, and the [Si] is controlled at about 1–1.5 and final slag (MgO) is advantageous for increasing the recovery of chromium. Wu [6] used FactSage software to simulate SiO_2-CaO-Cr_2O_3 ternary slag and SiO_2-CaO binary slag, also to correction the slag basicity depend on the slag system is complicated, from the data can be concluded that out-feeding temperature should be controlled at 1450 °C and basicity should be controlled at 1.6 to 1.8, and the final slag contains lower Cr_2O_3 and higher metal chromium recovery.

Granularity

The finer the mineral powder particles, the larger the contact area, it's helpful for the chemical reaction proceeds. Many studies have shown that during the thermal reduction of chromite ore powder, the rate of chemical reaction increases with the decrease of the particle size of solid particles [41].

Liu [33] had done this research. The results of the study indicate that The smaller the particle size, the faster the heating rate, and the higher the metallization rate of chromium. As the particles are reduced, the heating rate is gradually increased, and the time for heating up to 1200 °C is gradually reduced. When the particle size is less than 100 mesh and 150 mesh, there is no significant difference in the rate of temperature increase. However, when the particle size is less than 300 mesh, the heating rate is fast, especially when the heating time is about 15 min, and the temperature is rapidly raised to 1200 °C. When the particle diameter is less than 300 mesh and 200 mesh, the time required for heating to 1200 °C is 3–4 min. The particle size is less than 300 mesh, the chromium metallization rate is the highest, and the chromium metallization rate is 72.13%.

Conclusion

In summary, now, the ferrochrome alloys' smelting methods are extensive. But in order to meet the national requirements of energy conservation and green metallurgy, we should improve traditional production processes. And because of the production of medium-low-carbon ferrochrome, the grade and melting properties of chrome ore directly affect the smelting process and various technical and economic indicators. Therefore, in the smelting process, it is necessary to know not only the grade of chrome ore but also the melting property of different chrome ore, which contributes to the rational use of resources [42].

In recent years, microwave technology has attracted much attention because microwave energy is a new type of heating energy that is clean and efficient [43–46]. The biggest advantage is that the powder material can be heated selectively [47–50]. It can simple and efficient processing and utilization of ferrochrome powder.

Acknowledgements This research was financially supported by the National Natural Science Foundation of China (No.51504114) and Kunming University of Science and Technology Foundation for analysis and testing (No.2018M20172202034). This support is gratefully acknowledged. The authors are grateful to the reviewers for the discerning comments on this paper.

References

1. Dai W, Li S (1999) Iron alloy metallurgical engineering, vol 92. Metallurgical Industry Press, Beijing, pp 121–122
2. X Wei (1992) The present situation of chromium industry at home and abroad. Yunnan Metall 2(1–5):19
3. Chakraborty D, Ranganathan S, Sinha SN (2005) Investigations on the carbothermic reduction of chromite ores. Metall Mater Trans B 36(4):437–444
4. Yu H, Wang H, Chu S, et al (2014) Introduction of CO_2 branch storage and production of oxygen blowing method for low carbon ferrochrome role. In: The iron alloy symposium, 2014
5. Weber P, Eric RH (2006) The reduction of chromite in the presence of silica flux. Miner Eng 19(3):318–324
6. Wu K (2015) Smelting low-carbon ferrochromium by microwave reduction of silicon. Kunming University of Science and Technology
7. Jia Z (1989) The technology of pretreatment of chromite. Iron alloy 3:39–40
8. Shi F, Zhu K, Qian Y (2000) The new progress of chromite. China Chromium Iron Ore Mining Ind 15 (zk): 4–10
9. Fang S (1998) The real situation and party gram of foreign chromite mine. Geol prospect 34 (2):16
10. Liu J (2005) Beneficiation of low-grade chromite. Sci Technol Jiuquan Iron steel 2:4–9
11. Chen X, Xie Q (2006) Present situation and the prospect of China Chrome iron ore. Ind Technol 35(5):38
12. Hu D (2004) Analysis of the supply strategy of chromium iron ore resources in China. Economic Manag Geol Technol 26(2):22–25
13. Yan J, Chen J, Hu L (2007) Chromium metallurgy. Metallurgical Industry Press, Beijing
14. Tang X, Li X, Bo S (2002) Methods of producing low carbon ferrochrome. Iron Alloy 33 (5):7–9
15. Xiang T, Zuo B (2007) The electrosilicothermic process of producing low carbon ferrochrome and practice. Iron alloy 38(6):6–11
16. Wen X, Co J (2015) Discussion on medium-low carbon ferromang anese raw material selection in electrosilicothermic process production. Ferro-Alloys
17. Guo J (2012) Study on low carbon ferrochrome in the production of argon-oxygen refining. Iron alloy 43(5):1–6

18. Shanghai iron alloy, Beijing Iron and Steel Institute. Low carbon ferrochrome smelting converter blowing
19. Hu L (1982) Production of low carbon ferrochrome prolongs the life of converter oxygen blowing method for the practice of. Iron alloy 2:16–21
20. Narita K, Makino T, Matsumoto H et al (1983) Dephosphorization and desulfurization of hot metal by lime based flux injection-oxygen top blowing method(progress in pre-treatment of hot metals). Tetsu- to- Hagane 69(15):1825–1831
21. Masuda S, Taga M, Nakajima H et al (2009) Development of oxygen top blowing and argon bottom blowing method in refining of stainless steel. Tetsu- to- Hagane 72:1301–1308
22. Yu H, Sun P (2013) Oxygen enriched bottom blowing bath smelting temperature control method based on variable universe fuzzy-PID. In: International Conference on Intelligent Networks and Intelligent Systems. IEEE, 2013, pp 127–130
23. Li Q (1982) Production of medium and low carbon ferrochrome by combined blowing and blowing process of oxygen converter. Iron alloy 3:18–19
24. Ding YL, Warner NA (1997) Kinetics and mechanism of reduction of carbon-chromite composite pellets. Ironmak Steelmak 24:224–229
25. Walkiewicz JW, Kazonich G, McGill SL (1988) Microwave heating characteristics of selected minerals and compounds. Miner Metall Process 5(1):39–42
26. Kingman SW, Rowson NA (2000) The effect of microwave radiation on the magnetic properties of minerals. J Microw Power Electromagn Energy A Publ Int Microw Power Inst 35(3):144–150
27. Kingman SW, Vorster W, Rowson NA (2000) The influence of mineralogy on microwave-assisted grinding. Miner Eng 13(3):313–327
28. Amankwah RK, Khan AU, Pickles CA et al (2005) Improved grindability and gold liberation by microwave pretreatment of a free-milling gold ore. Miner Process Extr Metall Imm Trans 114(1):30–36
29. Amankwah RK, Pickles CA, Yen WT (2005) Gold recovery by microwave augmented ashing of waste activated carbon. Miner Eng 18(5):517–526
30. Chen J, Li N, Wang S et al (2007) Temperature rising characteristics of chromite ore fines in the microwave field. J Proc Chinese Soc Sci Eng 29(9):000880–906
31. Wu K, Zhu H, Peng J et al (2014) Analysis of the microscopic structure of silicon chromite powder by microwave heating reduction. J Electron Microsc 6:526–530
32. Kui-lin Wu, Li L, Hong-bo Zhu et al (2015) Study on the process of microwave reduction of chromite powder containing Si. Powder Metall Ind 25(5):22–25
33. Liu H (2018) Study on thermal reduction process of chromite silicon in microwave field. Kunming University of Science and Technology
34. Xu R, Ni R, Zhang S et al (1995) Pelletization of carbon-containing chromite and catalytic reduction of. Sintering and pelletizing, 1995(2)
35. Zambrano B, Pillihuaman A (1969) Study of reduction in the self-reducing pellet of chromites. Revue Danthropologie Des Connaissances 5(3):473–491
36. Wang H, Dong Y, Wang S (2000) The thermodynamic analysis on smelting medium carbon and low carbon ferrochrome with converter practice. Ferro-Alloys
37. Jianchen Li (2017) Study on the pre-reduction characteristics of chromite pellets. Iron Alloy 48(2):19–22
38. Zhou X, Hao X, Ma Q (2017) Effects of compound chemical activators on the hydration of low-carbon ferrochrome slag-based composite cement. J Environ Manage 191:58–65
39. Fu Y-Y (2013) AOD furnace smelting low carbon ferrochrome slag basicity optimization and prediction technology [D]. Changchun University of Technology
40. Ren Z, Zhou S, Yang Z et al (2007) Increasing the utilization of EAF in the production of Cr micro carbon ferrochrome. Mater Rev 21 (z2)
41. Qing Zhao, Chengjun Liu, Maofa Jiang et al (2013) Influence factors of chromite solid carbon thermal reduction process. Sci Technol Herald 31(5):40–43
42. Guo W (1979) The melting of chrome ore in smelting low carbon ferrochrome. Iron Alloy 1979(4):24–30

43. Mou Q, Li X (2004) Application and research progress of microwave heating technology. Physics 33(6):438–442
44. Waiquan Cai, Huiquan Li, Yi Zhang (2005) The application of microwave technology in metallurgical. J Process Eng 5(2):228–232
45. Nengsheng Liu, Jin-hui Peng, Li-bo Zhang et al (2009) Advances in the application of microwave technology in metallurgy of rare and precious metals. Noble Metals 30(4):48–51
46. Lin Jin, Su Jie, Peng Jinhui, et al. (2016) Microwave technology in metallurgical smelting technology application status and prospects of. Vac Electr Technol (6):36–42
47. Al-Buraik KA (2016) Utilization of microwave technology in enhanced oil recovery process for deep and shallow applications: US 20140027109 A1 [P]
48. Company S A O. (2016) Utilization of microwave technology in enhanced oil recovery process for deep and shallow applications
49. Zhou Y, Peng KY, Wang SJ (2006) Characteristics of temperature Increasing of metallurgical Zinc-bearing dust and sludge in the microwave field. Research on Iron & Steel
50. Nowak D, Granat K, Opyd B (2014) Measurement of electrical properties as effectiveness appraisal of microwave absorption by moulding and core sands subject to microwave utilization. Arch Metall Mater 59(2):713–716

Study on Separation of Sn-Sb Alloy by Vacuum Distillation

Yanjun You, Zhenghao Pu, Yifu Li, Bin Yang and Junjie Xu

Abstract Sn-Sb alloy is the secondary resources of tin and antimony in metal recovery process and massively exist in the smelting and refining of crude tin. Vacuum distillation has many advantages such as high recovery efficiency, less pollution and easy operation in disposing of the Sn-Sb alloys. In this study, important factors of Sn-Sb alloy separation by vacuum distillation were researched, which includes distillation temperature, soaking time and feeding materials (thickness). The optimal experimental conditions of Sn-Sb alloy for vacuum distillation were determined. Under these conditions, temperature was 1474 K, soaking time was 45 min and feeding materials (thickness) were 125 g(0.8 cm), the content of Sn in liquid phase was 98.77%, and the content of Sb in gas phase was 98.77% and the direct yield of Sn was 96.01%. It was indicated that vacuum distillation is an effective method for Sn-Sb alloy separation and tin and antimony collection.

Keywords Sn-Sb alloy · Vacuum distillation · Separation

Y. You · Z. Pu · Y. Li · B. Yang (✉) · J. Xu
National Engineering Laboratory for Vacuum Metallurgy, Kunming University of Science and Technology, No. 253, Xuefu Road, Kunming 650093, Yunnan Province, People's Republic of China
e-mail: kgyb2005@126.com

Y. You
e-mail: 874074548@qq.com

Y. You · Z. Pu · Y. Li · B. Yang · J. Xu
Key Laboratory of Vacuum Metallurgy for Nonferrous Metal of Yunnan Province, Kunming 650093, People's Republic of China

Y. You · Z. Pu · Y. Li · B. Yang · J. Xu
The State Key Laboratory of Complex Nonferrous Metal Resources Clean Utilization in Yunnan Province, Kunming 650093, People's Republic of China

© The Minerals, Metals & Materials Society 2019

T. Jiang et al. (eds.), *10th International Symposium on High-Temperature Metallurgical Processing*, The Minerals, Metals & Materials Series,
https://doi.org/10.1007/978-3-030-05955-2_34

Introduction

Tin and antimony are widely used in the fields of information technology industry, biological industry, new material industry and so on with excellent physical and chemical properties [1–4]. However, tin and antimony resources storage guarantee period worldwide is shorter, while the consumption of tin and antimony persistently high, and simply relying on ore mining has been impossible. Vigorously developing renewable resources is one of the solutions of the contradiction between the shortage of mineral resources and consumption growth.

Sn-Sb alloy is the secondary resources of tin and antimony in metal recovery process, also existing in the smelting and refining of crude tin [5]. Vacuum distillation is a new kind of field metallurgy method, which is carried out in a closed container with the characteristics of high efficiency and cleanliness [6]. It can effectively solve the problems such as long process, high energy consumption, low yield and poor economic benefit in traditional metallurgical method of separation of alloy. Vacuum distillation has been widely used in the separation of alloys and the refining of crude metals [7]. Kunming University of Science and Technology has been engaged in small scale, expanded and industrialized experiments of alloy separation for many years and achieved good results in those experiments [8, 9]. However, there is less research work on theoretical basis and further work is needed.

Theoretical Basis

The principle of vacuum distillation to separate two components of the alloy is the difference of saturated vapour pressure. The greater saturated vapour pressure, the easier the metal will be volatile. At the same time, the one having lower saturated vapour pressure will exist in liquid phase. Thus, it can achieve the purpose of separation of two components and the greater the difference, the greater the separation effectiveness [10]. The relationships between saturated vapour pressure and temperature for tin and antimony are shown in Fig. 1.

From Fig. 1, it can be easily seen that the saturated vapour pressure of each component increases with the increase of temperature. The saturated vapour pressure of antimony is greater than tin. It means that under the suitable conditions, antimony can be volatilized into the gas phase, so that tin can be separated simply.

Methods

Experimental Procedure

A sample of the Sn-Sb alloy system for experiments was obtained from a factory in China. The chemical composition of the sample is shown in Table 1. The

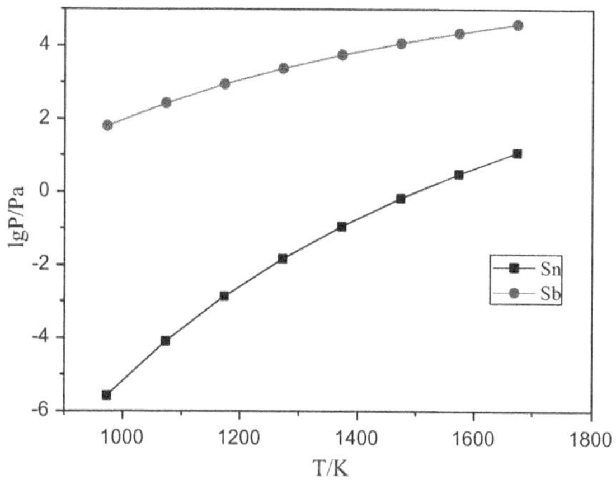

Fig. 1 The relationships between saturated vapour pressure and temperature for tin and antimony

Table 1 Chemical composition of raw materials

Sample	X_{Sn}	X_{Sb}
Raw material	0.513	0.487

experiments were carried out in the vacuum furnace shown in Fig. 2. The range of working pressure for the vacuum furnace is 1 Pa to atmospheric pressure, and the working temperature is from normal temperature to about 2000 K. The vacuum furnace adopts silicon-controlled transformer to controlling temperature and measures the residual gas pressure by stand McLeod vacuum gauge. Temperature detection system was comprised by Pt-100 probes which connects to a digital temperature metre (ANTHONE LU-900M) and predicts to 0.01 K. The residual gas pressure can be measured. The sample was placed in the graphite crucible at the constant temperature zone of the vacuum furnace after accurate weighing and drying treatment. When the vacuum degree met the requirement, the silicon-controlled transformer was regulated. The vacuum furnace should do heat preservation processing when it was heated to the preset temperature. Switch the power off when the experiment was over. Intake valve was opened to make sure the pressure in the furnace reach to the atmosphere when the temperature was below 373 K. Then, the furnace was opened to collect residues and volatiles, respectively.

The recovery can be calculated by the formula as follow:

$$\text{Recovery} = \left(\frac{x_i' \times m}{x_i \times m}\right) \times 100 \quad (1)$$

where x_i is the content of i in raw material, x_i' is the content of i residue, m is the weight of the feeding materials, m' is the weight of the residue.

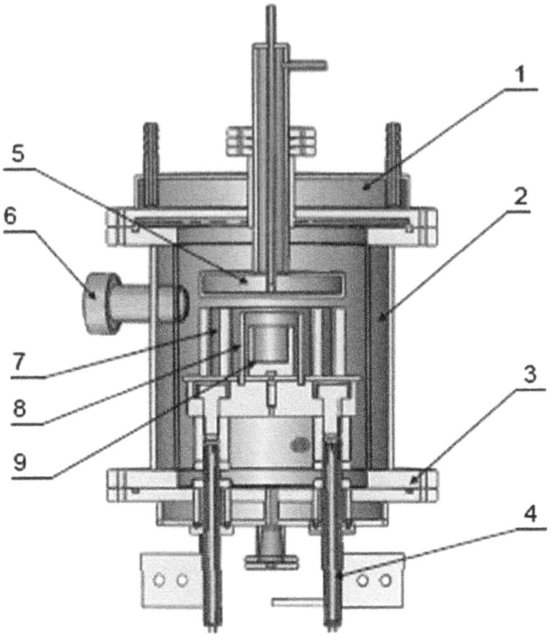

Fig. 2 Schematic diagram of the internal structure of the vertical vacuum furnace: 1—furnace cover; 2—vacuum furnace body; 3—furnace chassis; 4—electrode; 5—condensation plate; 6—observation hole; 7—thermal insulation sleeve; 8—heater; and 9—crucible

Results and Discussion

The Effect of Distillation Temperature

As shown in Table 2, Sn-Sb alloy was subjected to the distillation temperatures in the range of 1173–1673 K, while soaking time was 35 min, weight of feeding materials was 125 g(0.8 cm) and pressure being 5 Pa. The effect of distillation temperature on the content in liquid phase and gas phase was analyzed.

Figure 3 indicates the volatilization rate of Sn and Sb is small, and the volatilization rate of Sb is much larger than that of Sn at 1173 K, the content of Sb in liquid phase still 23.05%. When the distillation temperature rises to 1473 K, the volatilization of Sn cannot be ignored. When the temperature continues to rise, it is not conducive to recycling Sn because it began to evaporate into the gas phase largely.

The influence of the distillation temperature on the Sn content in the liquid phase and the direct yield of Sn is shown in Fig. 4. When the distillation temperature is less than 1473 K, the liquid phase content of Sn increased rapidly from 76.89% (1173 K) to 98.56% (1473 K) with distillation temperature, the growth rate is 28.18%; when the distillation temperature is higher than 1473 K, the liquid phase content of Sn increases rapidly from by 98.56% (1473 K) to 99.52 (1673 K), the growth rate of only 1%, but the direct yield of Sn decreases rapidly with the distillation temperature increasing. Integrating of all factors, the optimum distillation temperature was determined as 1473 K.

Table 2 The experimental results for the influence of distillation temperatures

Distillation temperature/K	Quantity/g	Liquid phase			Gas phase			Direct yield of Sn/%
		Sn/%	Sb/%	Quantity/g	Sn/%	Sb/%	Quantity/g	
1173	125.34	76.89	23.05	83.05	1.05	98.95	42.29	99.31
1273	124.79	85.57	14.41	73.32	2.48	97.52	51.47	98.00
1373	125.04	92.34	7.59	66.52	4.65	95.35	58.52	95.76
1473	124.87	98.56	1.42	61.09	6.03	93.97	63.78	93.99
1573	124.91	99.07	0.88	57.81	10.14	89.86	67.1	89.38
1673	125.05	99.52	0.47	54.68	14.03	85.97	70.37	84.61

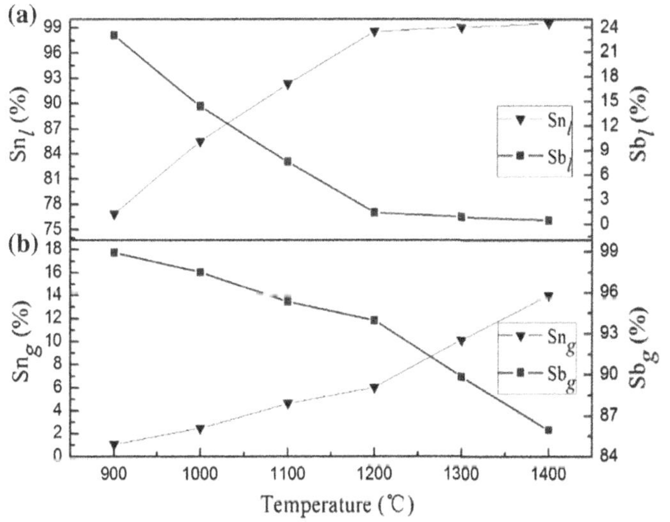

Fig. 3 The relationship between the composition of **a** liquid phase and **b** gas phase and distillation temperature

The Effect of Soaking Time

As shown in Table 3, Sn-Sb alloy was subjected to the distillation temperature 1473 K, while soaking time was in the range of 15–65 min, weight of feeding materials was 125 g(0.8 cm) and pressure being 5 Pa. The effect of soaking time on the content in liquid phase and gas phase was analyzed.

Table 3 indicates that with the increasing of soaking time, the content of Sn in liquid phase is increasing, but the direct yield of Sn is declining. Figure 5 shows the effect of soaking time on the content of Sn in liquid phase and gas phase. The content of Sn in liquid phase is 97.13%, Sb is 2.85%, meanwhile, the content of Sn in gas phase is 5.29%, Sb is 94.71% when soaking time is 15 min. Extending the soaking time to 45 min, the content of Sn in liquid phase increases to 98.79%.

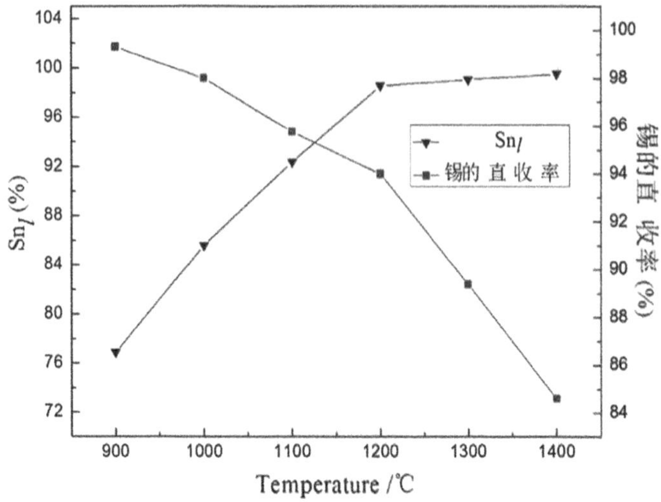

Fig. 4 Effect of temperature on the content of Sn in liquid phase and the direct yield of Sn

Table 3 The experimental results for the influence of soaking time

Soaking time/ min	Quantity/ g	Liquid phase			Gas phase			Direct yield of Sn/%
		Sn/%	Sb/ %	Quantity/ g	Sn/ %	Sb/%	Quantity/ g	
15	125.25	97.13	2.85	62.74	5.29	94.71	62.5	94.86
25	125.11	97.69	2.27	61.83	5.96	94.04	63.27	94.13
35	124.89	98.44	1.53	61.13	6.10	93.90	63.76	93.92
45	124.96	98.79	1.15	60.88	6.18	93.82	64.08	93.82
55	124.88	98.35	1.64	60.83	6.62	93.38	64.05	93.39
65	125.04	98.83	1.15	60.54	6.69	93.31	64.5	93.27

However, when the soaking time is 55 min, the content of Sn in liquid phase decreases, which may due to the error of the experimental operation or test. The growth rate of Sn in liquid phase is slowed down when the soaking time is more than 45 min, so it will not significantly increase the content of Sn in the liquid phase.

The influence of the soaking time on the content of Sn in the liquid phase and the direct yield of Sn is shown in Fig. 6. With the increasing of soaking time, the content of Sn in liquid phase increased, while direct yield of Sn decreased. When the soaking time is more than 45 min, the growth of direct yield of Sn decreased, while the content of Sn in liquid phase tends to be unchanged. Therefore, considering the effect of soaking time on the content of Sn in the liquid phase and the direct yield of Sn, the optimal soaking time is 45 min.

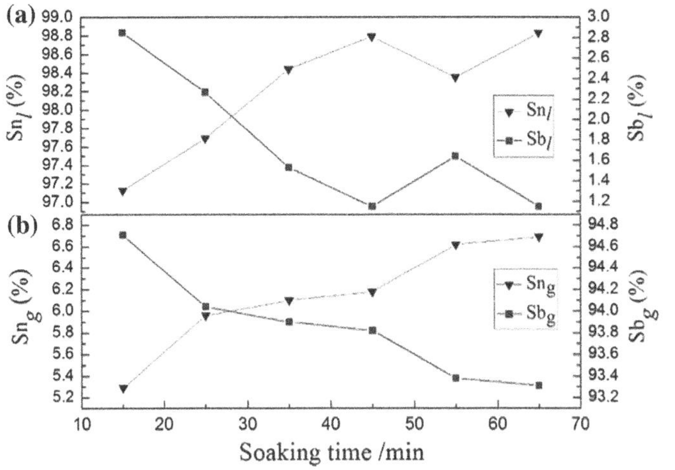

Fig. 5 The relationship between the composition of **a** liquid phase and **b** gas phase and soaking time

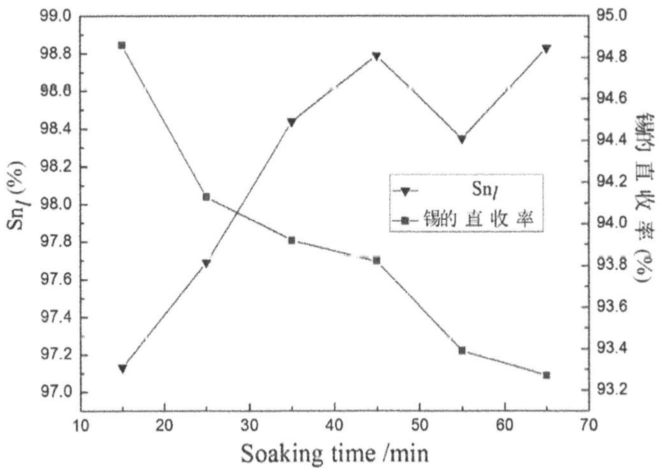

Fig. 6 Effect of soaking time on the content of Sn in liquid and the direct yield of Sn

The Effect of Feeding Materials (Thickness)

As shown in Table 4, Sn-Sb alloy was subjected to the distillation temperature 1473 K, while soaking time was 45 min, weight of feeding materials (thickness) was in the range of 85 g(0.4 cm) to 185 g(1.4 cm) and pressure being 5 Pa. It is the effect of feeding materials (thickness) on the content in liquid phase and gas phase.

Table 4 The experimental results for the influence of feeding materials (thickness)

Quantity/ g	Thickness/ cm	Liquid phase			Gas phase			Direct yield of Sn/ %
		Sn/%	Sb/ %	Quantity/ g	Sn/ %	Sb/%	Quantity/ g	
85.13	0.4	98.98	0.87	41.21	6.54	93.46	43.91	93.42
104.87	0.6	98.83	1.04	51.52	5.40	94.60	53.35	94.64
125.05	0.8	98.77	1.17	62.36	4.08	95.92	62.69	96.01
145.00	1	98.51	1.28	72.38	4.25	95.75	72.76	95.85
164.79	1.2	98.25	1.54	82.74	3.96	96.04	82.05	96.16
185.04	1.4	97.99	1.94	93.22	3.90	96.10	91.82	96.23

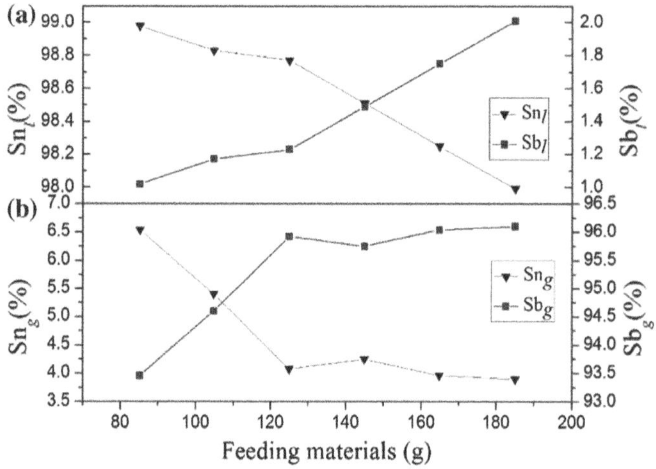

Fig. 7 The relationship between the composition of **a** liquid phase and **b** gas phase and feeding materials

Figure 7 shows the effect of feeding materials on the content of Sn in liquid phase and gas phase. When the weight of feeding materials (thickness) is 85.13 g (0.4 cm), the content of Sn in liquid phase is 98.98%, Sb is 0.87%, meanwhile, the content of Sn in gas phase is 6.54%, Sb is 93.46%, the content of Sn in liquid and gas phase is higher. This is because the thickness of the material layer is small, the mass transfer is faster in Sn-Sb alloy, and the volatilization is large of Sn and Sb. The content of Sn in liquid phase is 98.77%, the content of Sb in gas phase is 98.77%, when the weight of feeding materials (thickness) is 125.05 g(0.8 cm).

The influence of the weight of feeding materials (thickness) on the content of Sn in the liquid phase and the direct yield of Sn is shown in Fig. 8. With the increasing of quantity of raw material, the content of Sn in liquid phase decreased, while direct yield of Sn increased. When the weight of feeding materials (thickness) is more than 125.05 g(0.8 cm), the content of Sn decreases rapidly in liquid phase. Therefore,

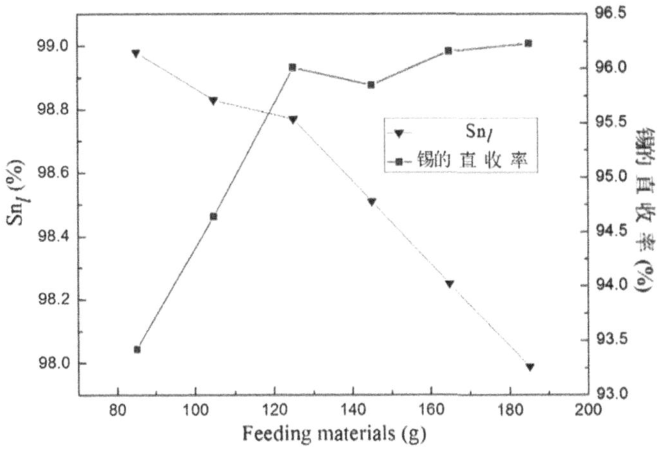

Fig. 8 Effect of feeding materials on the content of Sn in liquid phase and the direct yield of Sn

comprehensive consideration all the effects of feeding materials (thickness), the optimal feeding materials (thickness) is 125 g(0.8 cm).

Conclusion

The experimental results indicate that Sn can be recovered from Sn-Sb alloy at suitable distillation temperature, soaking time and feeding materials (thickness). The optimum experimental conditions of Sn-Sb alloy for vacuum distillation are distillation temperature of 1474 K, soaking time of 45 min and weight of feeding materials (thickness) being 125 g(0.8 cm). Under this condition, the content of Sn in liquid phase is 98.77%, the content of Sb in gas phase is 98.77% and the direct yield of Sn is 96.01%. It provides an efficient and convenient method for Sn-Sb alloy separation and tin and antimony collection.

Acknowledgements This work was supported by the National Key Research and Development Program of China (2016YFC0400404), Joint Foundation of the NSFC-Yunnan Province (U1502271), the Cultivating Plan Program for the Leader in Science and Technology of the Yunnan Province (2014HA003), the Program for Nonferrous Metals Vacuum Metallurgy Innovation Team of Ministry of Science and Technology (2014RA4018), the Program for Innovative Research Team in University of Ministry of Education of China (IRT_17R48).

References

1. Corma A, Nemeth I, Renz M et al (2001) Sn-Zeolite beta as a heterogeneous chemoselective catalyst for Baeyer-Villiger oxidations. Nature 412(6845):423–425

2. Grigoriants I, Sominski L, Li H et al (2005) The use of tin-decorated mesoporous carbon as an anode material for rechargeable lithium batteries. Chem Commun 7(7):921
3. Gonzalez R, Salem P Jr, Syamal BP et al (2011) Use of antimony in the treatment of leishmaniasis: current status and future directions. Mol Biol Int 2011(2011):571242
4. Arias AC, Roman LS, Kugler T et al (2000) The use of tin oxide thin films as a transparent electrode in PPV based light-emitting diodes. Thin Solid Films 371(1–2):201–206
5. Xiong Y, Yang B, Xiong H et al (2012) Separation of Sn-Fe alloys by vacuum distillation. Chin J Vac Sci Technol 32(9):820–824
6. Li D, Dai Y, Yang B et al (2013) Purification of indium by vacuum distillation and its analysis. J Central South Univ 20(2):337–341
7. Wan W, Yang B, Liu DC et al (2006) Study on obtaining pure selenium by vacuum distillation. J Kunming Univ Sci Technol
8. Kong LX, Yang B, Xu BQ et al (2012) Application of MIVM for Pb–Sn–Sb ternary system in vacuum distillation. Metall Mater Trans B 43(6):1649–1656
9. Yang HW, Zhang C, Yang B et al (2015) Vapor–liquid phase diagrams of Pb–Sn and Pb–Ag alloys in vacuum distillation. Vacuum 119:179–184
10. Dai YN, Yang B (2000) Vacuum metallurgy of nonferrous metals. Metallurgical Industry Press

Statistical Optimization of Tungsten Carbide Synthesis Parameters

Grant C. Wallace, Jerome P. Downey, Jannette Chorney, Katie Schumacher and Trenin Bayless

Abstract Commercial methods for synthesizing tungsten carbide require large energy inputs due to the high temperatures and grinding processes associated with production. In this study, tungsten carbide was synthesized by adsorbing tungstate anions from aqueous solutions onto an activated carbon matrix. The tungsten-loaded precursor was carburized under mixtures of hydrogen, methane, and carbon monoxide to produce tungsten carbide at temperatures significantly lower than most commercial operations. Tungsten carbide was synthesized at temperatures below 1000 °C with conversion to tungsten carbide exceeding 90%. The adsorption and carburization processes were modelled and optimized using experimental design techniques. The effects of time, temperature, initial tungsten concentration, and pH were considered for modelling adsorption behaviour, while carburization behaviour was modelled on the effects of time, temperature, carbon content, and reducing gas composition. Carburization products were characterized using X-ray diffraction and scanning election microscopy. The adsorption of tungstate anions was measured using inductively coupled plasma optical emission spectroscopy.

Keywords Tungsten carbide · Synthesis · Statistical optimization

Introduction

The group of materials known as ceramic carbides includes a wide range of materials that vary greatly in terms of structure and properties. Metal carbides are some of the hardest materials in existence. Known as hardmetals, metal carbides such as tungsten carbide (WC) possess hardness values approaching that of

G. C. Wallace (✉) · J. P. Downey · J. Chorney · K. Schumacher · T. Bayless
Montana Technological University, 1300 W. Park St., Butte, MT 59701, USA
e-mail: gwallace@mtech.edu

J. P. Downey
e-mail: jdowney@mtech.edu

© The Minerals, Metals & Materials Society 2019
T. Jiang et al. (eds.), *10th International Symposium on High-Temperature Metallurgical Processing*, The Minerals, Metals & Materials Series,
https://doi.org/10.1007/978-3-030-05955-2_35

diamond as well as being resistant to wear and chemical attack at relatively high temperatures. Because of their ability to resist corrosion and abrasion, carbides have been applied to a number of industrial applications where other traditional metals do not perform as well. Such applications include high-temperature cutting tools, drill bits, surgical implements, and alloying agents [1]. Research has also shown that some hard metals possess catalytic properties. Tungsten carbide has been shown to have catalytic abilities that are comparable to precious metal catalysts such as platinum and palladium, [2] and molybdenum carbide (Mo_2C) has been shown to facilitate the production of synthesis gas (syngas) from methane gas [3].

Current methods of processing ceramic carbides are energy intensive. Most carbides are produced at relatively high temperatures (>1500 °C) by reacting metals, or oxide powders, with a source of carbon. Extensive milling operations are then required to reduce the carbides into a fine powder for further use [1]. To produce tungsten carbide, for example, tungsten oxide powders are first reduced under a hydrogen (H_2) atmosphere at temperatures ranging from 700 to 900 °C to produce elemental tungsten powder. This tungsten powder is then mixed with carbon and heated under H_2 again at significantly higher temperatures (1400–1600 °C) in order to convert the tungsten powder into tungsten carbide. Finally, the tungsten carbide powder is crushed, ground, and sieved to produce commercial tungsten carbide powder products [4].

The use of reducing gas atmospheres has reduced the operating temperatures required to synthesize tungsten carbide. Tungsten carbide was synthesized under mixtures of hydrogen and carbon-containing gases, such as CH_4, at temperatures lower than 1000 °C [5]. The reaction rate of the tungsten carburization reaction in the presence of reducing gases (H_2 and CH_4) is dependent on a number of factors including the partial pressures of the reducing gases and the size of the tungsten particles. Research has indicated that the initial reaction between tungsten particles and carbon-containing gases is controlled at the surface of the tungsten particles by the CH_4-W reaction. As the surface of the tungsten particles is carburized, the reaction mechanism is believed to change to one controlled by diffusion of CH_4 through the carburized surface layer [5].

Activated carbon is used in a number of industries to remove metals and metal complexes from aqueous solutions. The use of aqueous solutions in the production of carbide ceramics has shown potential in reducing the energy requirements associated with carbide production. Tungsten and molybdenum ions have been shown to readily adsorb from solution onto activated carbon in concentrations that promote its use in producing tungsten-carbon precursors for carburization [6]. Tungsten and molybdenum adsorption onto activated carbon are improved by low-pH solutions and the addition of NaCl to the solution [7]. The use of aqueous solutions in the carburization process also has the potential to produce submicron carbide particles. The mechanical properties of sintered carbide materials, particularly tungsten carbide, have also been shown to improve as the size of the tungsten carbide crystals decreases in size [8].

Experimental Procedure

Carburization Precursor Production

Sodium tungstate dihydrate and sodium chloride were added to 2 L of distilled water to produce solutions with concentrations of 18,000 ppm tungsten and 5844 ppm sodium chloride. The pH of the solutions was lowered to approximately 2.0 using concentrated hydrochloric acid. A 50 g sample of activated carbon (Sigma-Aldrich, −100/+400 Mesh) was added to each solution and the mixture was agitated to allow the activated carbon to remain suspended in solution. After two hours, the activated carbon was removed from the solution using vacuum filtration, and the activated carbon was placed in a drying oven. The parameters for the tungstate solutions were established through a design of experiments in order to maximize tungsten loading onto the activated carbon matrix [9].

Carbide Synthesis

Carburization samples were prepared in 25 g batches by blending W-loaded precursor with virgin activated carbon in a ceramic ball mill for five minutes so that each carburization sample was composed of 20–50% virgin activated carbon. This preparation step was added because previous research indicated that additions of virgin activated carbon increased the production of WC on the W-loaded precursor [10]. The blended samples were placed inside a quartz kiln and heated inside an ATS 3210 rotary furnace to a temperature of 850–950 °C under argon gas at a rate of 5.3 °C/min. Once the kiln interior reached the desired final temperature, a reducing gas mixture of methane, hydrogen, and carbon monoxide was introduced to the furnace with the flow of each gas regulated by Omega flowmeters. The ratio of methane to hydrogen gas was set at 2:1 or 8:1 for each carburization experiment, and the total gas flow through the furnace was held at approximately 0.5 L/min. Samples were held at the desired temperature for 6 or 8 h. Following carburization under the reducing gas atmosphere, the samples were cooled back to room temperature under an argon purge gas and removed from the furnace for analysis. The compositions of the carburized samples were determined with X-ray diffraction using a Rigaku Ultima IV diffractometer with a Cu-kα X-ray source. Qualitative analyses of all samples were done using the whole pattern powder fitting method (WPPF).

Carburization Modelling

A preliminary response surface model of carburization behaviour was produced using the statistical analysis software, DesignExpert 9 (StatEase Inc.). Eleven carburization experiments were used to model the effects of temperature, time, gas composition, and activated carbon content on the conversion o6f the W-loaded precursor to tungsten carbide. The WPPF data for the relative weight per cent of tungsten carbide within each carburized sample was used to measure tungsten carbide conversion. The WPPF data was fitted to an inverse square root transform and a modified quadratic relationship. The preliminary model was verified using a series of statistical diagnostics including analysis of variance (ANOVA) and Cook's distance.

Results and Discussion

The parameters and resulting relative weight percent values of tungsten carbide for each of the eleven experiments used to model carburization of the W-loaded precursor are provided in Table 1.

From the data, it is evident that Experiment 6 produced the highest amount of tungsten carbide with a composition containing 93.3% WC. This value for tungsten carbide conversion was achieved at a carburization temperature of 950 °C, a reaction time of 8 h, a gas composition containing a 8:1 ratio of methane to hydrogen, and an activated carbon content of 20% of the total sample mass prior to carburization.

A three-dimensional response surface model showing the effects of the four parameters of interest on tungsten carbide production is presented in Figs. 1, 2 and 3. In each figure, tungsten carbide content is plotted on the z-axis, temperature is plotted on the x-axis, and the amount of virgin activated carbon within each sample prior to carburization is plotted on the y-axis. For each response surface presented, the CH_4: H_2 ratio was set as 8:1 and carburization time was varied at 6, 7, and 8 h so that the effects of temperature and activated carbon content could be observed

Table 1 Carburization results and experimental parameters

Experiment no.	Time (h)	Temp (°C)	% Activated carbon	CH_4:H_2	% WC
1	6	850	20	2	72.1
2	8	850	20	2	73.0
3	6	950	20	2	81.9
4	6	850	20	8	66.4
5	8	850	20	8	79.1
6	8	950	20	8	93.3
7	8	950	20	8	87.0
8	6	850	50	8	74.9
9	8	850	50	8	81.8
10	6	950	50	8	65.0
11	8	950	50	8	79.9

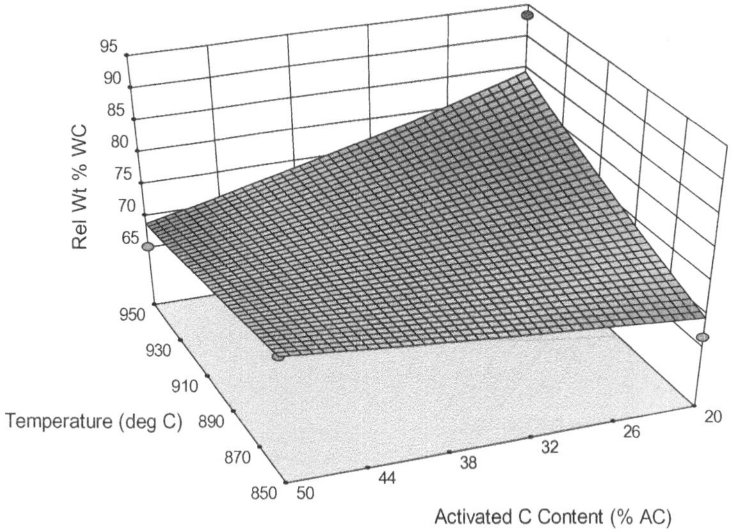

Fig. 1 Response surface model of WC conversion (time = 6 h)

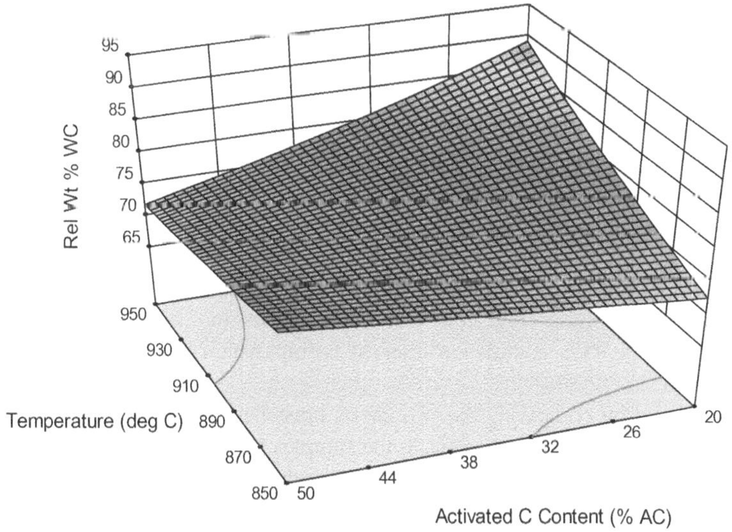

Fig. 2 Response surface model of WC conversion (time = 7 h)

without interference. Altering gas composition produced no significant effect on the amount of tungsten carbide produced during carburization.

From the response surface diagrams, it is evident that tungsten carbide conversion is significantly improved by increasing temperatures to 950 °C provided

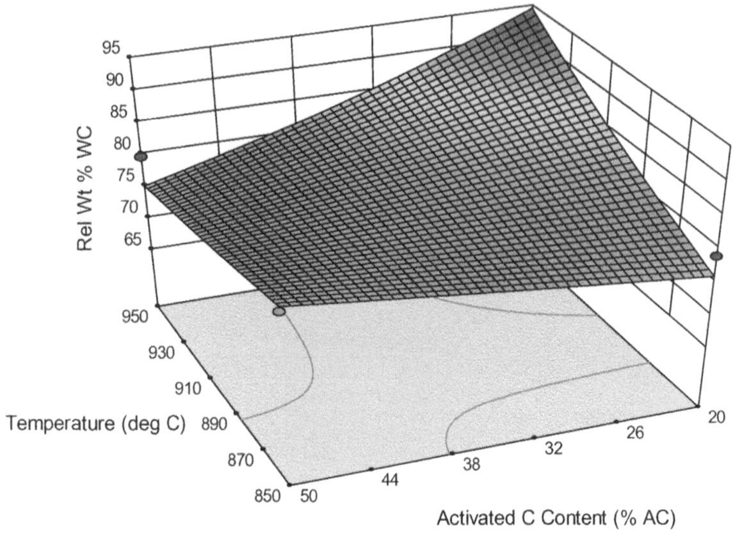

Fig. 3 Response surface model of WC conversion (time = 8 h)

that the activated carbon content added to the W-loaded precursor prior to carburization remains low (20%). By comparing the variations on the carburization model presented in Figs. 1–3, conversion to WC is shown to increase with increasing carburization time.

The relationship between carburization temperature and the effect of activated carbon content on tungsten carbide production is demonstrated in Fig. 4.

The relationship between activated carbon content and carburization temperature and their effect on carburization was further investigated. The one-factor analysis presented in Fig. 4 shows carburization as a factor of carburization temperature when the activated carbon content is 20% of total sample mass (Fig. 4a) and 50% of the total sample mass (Fig. 4b). The dashed lines in each chart represent the boundaries of a 95% confidence interval within which the experimental results should exist.

It is evident by comparing the two charts presented in Fig. 4 that increasing the amount of activated carbon present in the samples beyond 20% of the total sample mass induced a negative effect on tungsten carbide synthesis. An over-abundance of activated carbon may act as a kinetic inhibitor, limiting the carburization reaction by restricting contact between the adsorbed tungsten and the carbonaceous gas mixture.

The negative effect of higher additions of activated carbon can also be seen in the one-factor analyses presented in Fig. 5. Again, Fig. 5a represents the effect of reaction time on tungsten carbide synthesis from samples containing 20% additional activated carbon, and Fig. 5b represents samples containing 50% additional activated carbon.

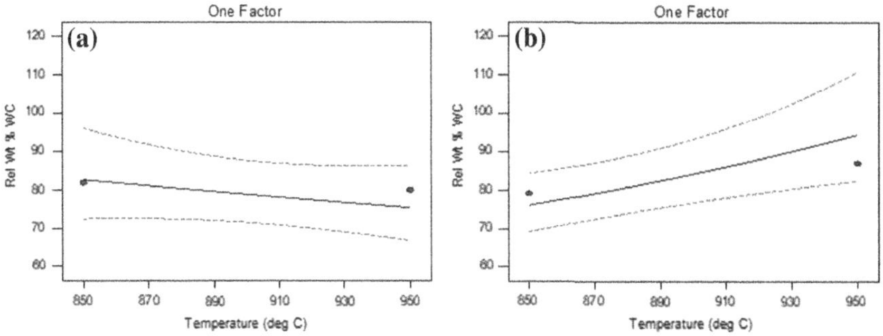

Fig. 4 Effect of added activated carbon on carburization temperature-dependence

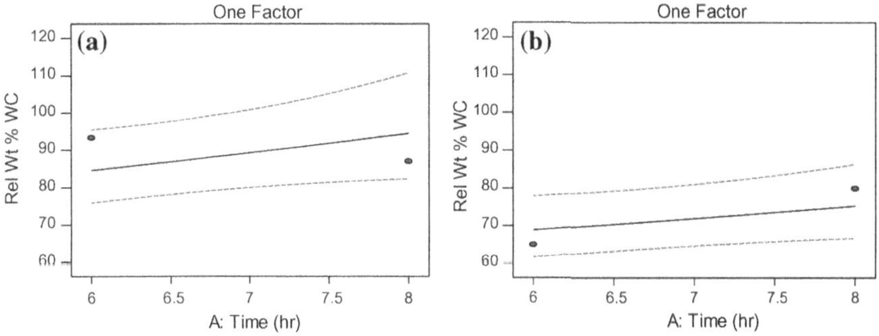

Fig. 5 Effect of added activated carbon on carburization time-dependence

The additions of activated carbon greater than 20% are shown to significantly reduce the total amount of tungsten carbide present within the carburized products. The results presented in Fig. 5 indicate that additions of activated carbon should be limited to 20% of the total mass of the precursor sample prior to carburization.

Conclusions

Tungsten carbide was synthesized at 950 °C under a reducing gas mixture of methane, hydrogen, and carbon monoxide from a precursor material consisting of tungstate anions adsorbed onto an activated carbon matrix. Conversion to the desired tungsten carbide species, WC, was achieved with values exceeding 90%.

Experimental results were used to produce a preliminary model of carburization behaviour for the W-loaded precursor. From this model, it was determined that maximum conversion to tungsten carbide can be achieved through the use of temperatures approaching 950 °C, a reaction time of 8 h, and an activated carbon content that does not exceed 20% of the total mass of the sample prior to carburization. Modelling carburization behaviour indicated that activated carbon additions greater than 20% of the total sample mass reduced the conversion of the W-loaded precursor to tungsten carbide and may kinetically inhibit this process.

Future Work

Further refinement of the response surface model through additional experimentation is needed in order to improve the fit of the model to the experimental data. Additional work will also focus on applying these same modelling principles to understand the carburization behaviour of precursors containing molybdate anions on an activated carbon matrix.

Acknowledgements Research was sponsored by the Army Research Laboratory and was accomplished under Cooperative Agreement Number W911NF-15-2-0020. The views and conclusions contained in this document are those of the authors and should not be interpreted as representing the official policies, either expressed or implied, of the Army Research Laboratory or the US Government. The US Government is authorized to reproduce and distribute reprints for Government purposes notwithstanding any copyright notation herein.

References

1. Total Materia, Metal Carbides, November 2006. [Online]. Available: http://www.totalmateria.com/page.aspx?ID=CheckArticle&site=ktn&NM=158. Accessed 16 April 2016
2. Zaoxue Y, Mei C, Pei KS (2013) Nanosized tungsten carbide synthesized by a novel route at low temperature for high performance electrocatalysis. Sci Rep 3(1646):1–7
3. Brungs AJ, York AP, Claridge JB, Marquez-Alvarez C, Green ML (2000) Dry reforming of methane to synthesis gas over supported molybdenum carbide catalysts. Catal Lett 70(3–4):17–122
4. Zawrah MF (2007) Synthesis and characterization of WC-Co nanocomposites by novel chemical method. Ceram Int 33:155–161
5. Davidson CF, Alexander GB, Wadworth ME (1978) Initial kinetics of tungsten carburization by methane. Metall Trans B 9B:553–557
6. Ohashi K, Murakami K, Yamamoto K (1983) Adsorption behavior of tungsten (VI) onto activated carbon. The Jpn Soc Anal Chem 32:313–319
7. Cruywagen JJ, Pienaar AT (1989) The adsorption of tungsten(VI) on activated carbon from 1.0 M Na(H)Cl solution. Polyhedron 8(1):71–76
8. Seegopaul P, McCandish LE, Shinneman FM (1997) Production capability and powder processing methods for nanostructured WC-Co powder. Int J Refract Metals Hard Mater 15(1–3):133–138
9. Wallace GC, Downey JP, Chorney J, Mallard A, Hutchins D (2017) Synthesis of carbide ceramics via reduction of adsorbed anions on an activated carbon matrix. In: 8th international symposium on high-temperature metallurgical processing, San Diego, CA, 2017
10. Wallace G, Downey J, Chorney J, Schumacher K, Mallard A (2018) Synthesis of nanocrystalline carbide ceramics via reduction of anion-loaded activated carbon precursors. In: 9th international symposium on high-temperature metallurgical processing, Phoenix, AZ, 2018

Part VIII
Preparation of Alloys and Materials II

Effects of CeO_2 on Melting Temperature, Viscosity and Structure of Mold Fluxes for Casting Rare Earths Alloyed Steels

Zeyun Cai, Bo Song, Longfei Li and Zhen Liu

Abstract Effects of CeO_2 on melting temperature, viscosity and structure of mold fluxes for the continuous casting of rare earths alloyed heavy railway steels were studied. Results showed that the melting temperature of mold fluxes with 0, 4, 8 and 12 wt% CeO_2 contents were 1066, 1088, 1090 and 1102 °C, respectively. Moreover, the viscosity values at 1300 °C were 0.40, 0.38 and 0.34 Pa s, but the viscosity of slags with 12 wt% CeO_2 could not be measured because slags were solid. Therefore, the viscosity at 1300 °C decreased with the increasing of CeO_2 contents, although the melting temperature increased. Raman spectroscopy and X-ray photoelectron spectroscopy (XPS) confirmed that CeO_2 enhanced the de-polymerization of mold fluxes, and the relative percentage content of O^{2-} in melts increased, leading to the viscosity decrease at 1300 °C.

Keywords CeO_2 · Melting temperature · Viscosity · Structure
RE steel

Introduction

Mold fluxes play the key role in the smooth process and producing high-quality steels [1–4]. Mold fluxes mainly protect the steel from oxidation and insulation. Besides, mold fluxes with reasonable viscosity properties and melting performance

Z. Cai · B. Song (✉) · L. Li · Z. Liu
School of Metallurgical and Ecological Engineering, University of Science and Technology Beijing, Beijing 100083, China
e-mail: songbo@metall.ustb.edu.cn

Z. Cai
e-mail: zexiaoyun@126.com

L. Li
e-mail: aifei_0105@126.com

Z. Liu
e-mail: liuzhen_ustb@163.com

© The Minerals, Metals & Materials Society 2019

T. Jiang et al. (eds.), *10th International Symposium on High-Temperature Metallurgical Processing*, The Minerals, Metals & Materials Series,
https://doi.org/10.1007/978-3-030-05955-2_36

are beneficial in providing the lubrication and controlling the horizontal heat transfer between the mold and the shell [5–7].

Rare earth has been used in steels to improve the tensile strength, wear resistance and corrosion resistance. However, because of the manufacturing process of RE alloy heavy rail steels, rare earth oxides enter the mold flux and the amount of rare earth oxide is even up to 17.88% in previous literatures. Previous researches showed that the viscosity decreased with the increasing content of any rare earth additions [8, 9]. In addition, studies of the effect of rare earth oxides on the CaO–Al_2O_3-based mold flux were reported, which presented that amounts of Li_2O were needed to balance the effect [10]. However, effects of CeO_2 on CaO–SiO_2-based mold flux for casing the rare earth alloy heavy rail steel and the relationship of viscosity and structure were not revealed.

In this work, effects of CeO_2 on the mold flux for casing the rare earth alloy heavy rail steel were studied and the relationship of viscosity and structure was also discussed.

Materials and Methods

Experimental Slags

Reagent grade powders of CaO, SiO_2, Al_2O_3, Na_2CO_3, CaF_2 and CeO_2 were used as raw materials. CaO were calcined at 1000 °C in a muffle furnace for 10 h and others were heated at 500 °C for 5 h to remove impurities and moisture. Firstly, all of them were well mixed according to the required proportion of industrial mold fluxes for continuous casting rare earths alloyed heavy railway steels to obtain the original mold flux. Then 0, 4, 8 and 12 wt% CeO_2 were added into the original mold flux to get four experimental slags. Slags were grounded in the agate mortar with ethanol as the mixing media. Finally, slags were put into the graphite crucible and melt in a silicon molybdenum furnace at approximately 1400 °C for 30 min in Ar atmosphere followed by water quenching to form glassy mold fluxes. Then part of dried glass slags was crushed, grounded and screened through a 200-mesh sieve, and compositions were analysed using X-ray fluorescence (XRF) as shown in Table 1.

Melting Temperature Tests

The melting temperature of experimental slags was determined by the hemisphere method as shown in Fig. 1 using RDS-04 automatic melting point tester. The pre-melting mold fluxes were placed and uniformly mixed in an evaporating dish with a little dextrin solution and made into cylindrical samples of $\varphi 3$ mm × 3 mm using a sample preparation device. Cylindrical samples were compacted and pushed out with a spring-loaded rod. All prepared samples were naturally dried for 24 h

Table 1 Chemical composition of experimental mold fluxes

Sample number	Composition (wt%)					
	CaO	SiO_2	Al_2O_3	Na_2O	CaF_2	CeO_2
1	27.35	39.79	6.46	11.95	14.45	0
2	25.46	38.14	6.27	11.44	14.64	4.05
3	24.94	37.01	6.11	10.68	13.33	7.93
4	22.94	35.56	5.83	10.08	13.87	11.72
5	23.04	34.81	5.90	11.19	11.17	13.89
6	22.50	33.99	5.77	10.93	10.89	15.92
7	21.94	33.14	5.16	10.66	10.62	18.03

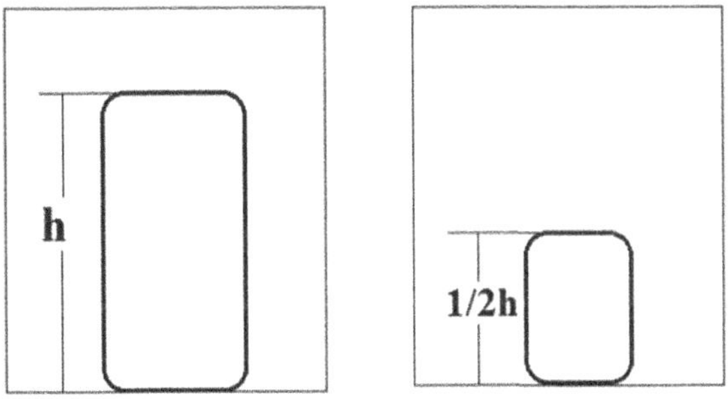

Fig. 1 Height change of samples during the melting point test

before measurement. Then cylindrical samples were placed on the gasket and heated at rate of 5–10 °C/min during the test. The melting point was obtained when the height of the sample became half of the original height during the heating process. Experiments were performed at least three times.

Viscosity and Structure Tests

The rotating cylinder viscometer (Brookfield DV2T) was employed, and its schematic diagram is shown in Fig. 2. The $MoSi_2$ heating electric resistant furnace was used for melting slags. A molybdenum crucible and a molybdenum bob with Pt-30Rh thermal couple were employed to the rotating system. Before the test, the viscometer should be calibrated at room temperature (25 °C) using standard oil with known viscosity of 0.495 Pa s. The viscometer inaccuracy was less than 2%, and viscosity values can be calculated using the equation as below:

$$\eta = \text{TK} \times \text{SMC} \times \frac{10}{\text{RPM}} \times \text{Torque} \quad (1)$$

where η is the measured viscosity, Pa s; TK is DV2T the torque constant of 0.09373; SMC is the spindle multiplier constant of 41.75; torque is the measured value between 10 and 100%.

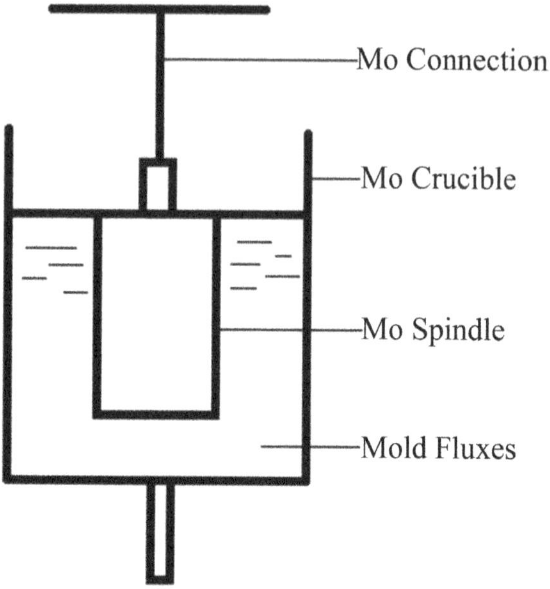

Fig. 2 Schematic diagram of inner cylinder rotating viscometer

Then, 150 g pre-melting mold fluxes were placed into a molybdenum crucible with the dimension of 38 mm and the height of 100 mm. Then samples were heated at the rate of 10 °C/min from 25 to 1400 °C and held at 1400 °C for 20 min with protective gas (90% Ar and 10% H_2). Viscosity values were measured during the cooling process. The Mo spindle was used to stir melts about 15 min before the measurement to keep it stable.

Structure of glassy samples was analysed using Raman spectrometer (LabRAM HR Evolution) and X-ray photoelectron spectroscopy (AXIS Ultra^{DLD}). Four kinds of water-quenched samples were subjected to the room temperature Raman spectroscopic analysis. In this work, Raman spectra were recorded in the frequency range (600–2000 cm^{-1}), and measurement was performed using excitation wavelength (532 nm). O_{1s} XPS spectra were analysed by using a spectrometer. The main O_{1s} peak due to partial surface charging was calibrated by C_{1s}. Peaks were fitted a mixed Gauss–Lorentz function after subtraction of the background.

Results and Discussion

Effects of CeO_2 on Melting Temperature

Specific melting temperatures of experimental mold fluxes are presented in Table 2, and the function relationship of melting temperature with the CeO_2 content was shown in Fig. 3. The melting temperature increased with the addition of CeO_2. The melting temperature of mold fluxes without CeO_2 was only 1066 °C. However,

Table 2 Melting temperature of experimental mold fluxes

Sample number	Melting temperature (°C)			Mean value (°C)
	1	2	3	
1	1067	1066	1066	1066
2	1089	1088	1088	1088
3	1090	1090	1090	1090
4	1104	1100	1102	1102
5	1162	1152	1167	1160
6	1186	1191	1176	1183
7	1206	1208	1217	1210

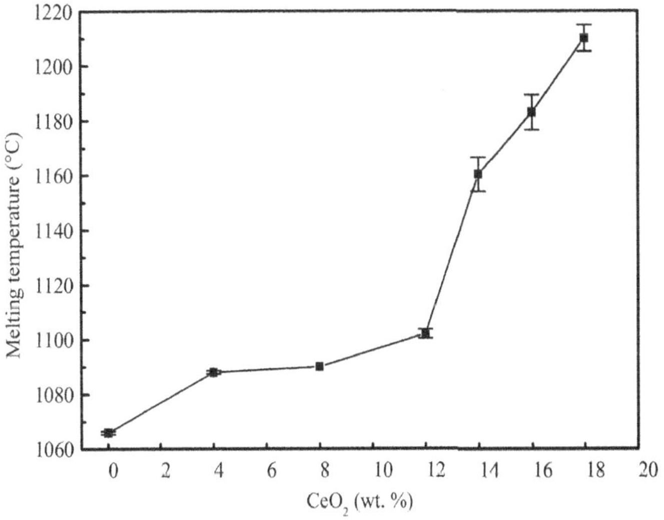

Fig. 3 Relationship of melting temperature with the CeO_2 content

when the CeO_2 content was 18 wt%, the melting temperature of sample has increased by nearly 150 °C compared to the slag without the addition of CeO_2. The increase of the melting temperature was ascribed to CeO_2 with high melting temperature of 2400 °C. The content of CeO_2 in experimental mold fluxed should not exceed 12 wt% since the melting temperature of the mold flux for casting rare earths alloyed heavy railway steels is between 1050 and 1110 °C.

Effects of CeO_2 on Viscosity

Effects of CeO_2 on the viscosity of experimental mold fluxes are shown in Fig. 4. For every curve, the viscosity value increased with the decreasing temperature, and there was a breaking point during the cooling measurement. The breaking

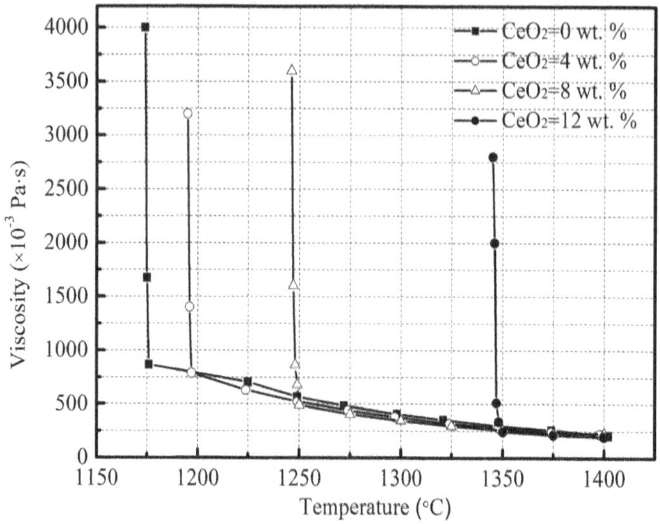

Fig. 4 Effects of CeO_2 content on the viscosity of experimental mold fluxes

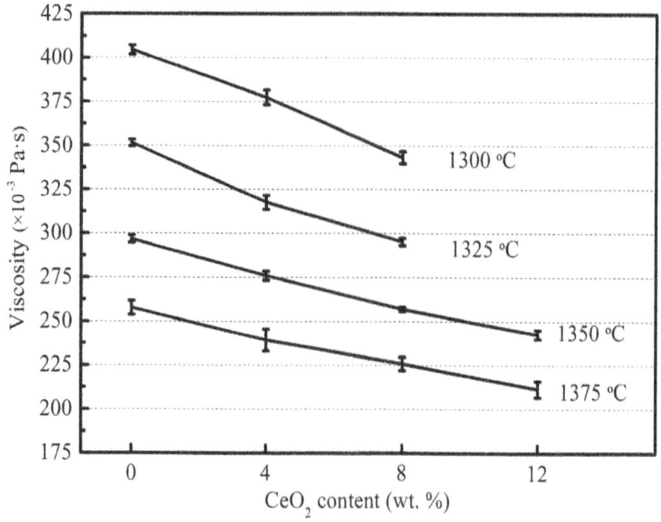

Fig. 5 Effects of CeO_2 content on the viscosity values from 1375 to 1300 °C

temperature increased with the addition of CeO_2. The breaking temperature of mold fluxes without CeO_2 was 1175 °C. When the CeO_2 addition was up to 12 wt%, the breaking temperature has been up to 1350 °C, which indicates that CeO_2 enhanced the crystallization of experimental mold fluxes.

Effects of CeO_2 on viscosity values at some certain temperatures between 1375 and 1300 °C were presented in Fig. 5. The viscosity at same temperature decreased with the increasing of CeO_2. The viscosity values at 1300 °C of samples were 0.40, 0.38 and 0.34 Pa s, respectively. Viscosity values at 1300 and 1325 °C of mold fluxes with 12 wt% CeO_2 could not be obtained because the experimental mold flux has almost become solid, which is beyond the measurable range of the viscosity. Thus, the addition of CeO_2 should be lower than 12 wt%, which indicates that the mold flux was not applicable to the continuous casting process of rare earths alloyed heavy railway steels. In conclusion, CeO_2 increased viscosity values and the breaking temperature, which is equivalent to enhancing the crystallization ability.

Effects of CeO_2 on Structure

Raman Spectra Analysis

Raman spectra signals of experimental mold fluxes with 0, 4, 8 and 12 wt% CeO_2 at room temperature were shown in Fig. 6. The Raman peak of every curve mainly appeared in two regions, the medium frequency region of 600 cm^{-1}–800 cm^{-1} due to the stretching vibration of bridging oxygen and the high-frequency region of 800 cm^{-1}–1100 cm^{-1} corresponding to the bending vibration of bridging oxygen [11, 12]. The Raman shift of the band at about 650 cm^{-1} was almost unchanged, but the Raman peak of the band at about 970 cm^{-1} shifted to the left region. Previous literatures reported that different alkaline cations affected the Raman peak position [13]. Specific analyses can be seen as below.

Deconvolved results of Raman signals in Fig. 6 using Gaussian deconvolution methods similar to the one proposed by Mysen et al. are presented in Fig. 7 after the background subtraction. All Raman peaks in high-frequency region were fitted, and there were four peaks. Actually, five kinds of Si–O tetrahedron exist in glassy silicate. Q_0 is $(SiO_4)^{4-}$ with zero bridging oxygen in a monomer structure unit, Q_1 is $(Si_2O_7)^{6-}$ with one bridging oxygen in dimmer structure unit, Q_2 is $(Si_2O_6)^{4-}$ with two bridging oxygen in chain structure unit, Q_3 is $(Si_4O_{11})^{6-}$ with three bridging oxygen in sheet structure unit, and Q_4 is $(SiO_2)^0$ with four bridging oxygen in network structure unit [11, 12]. Q_0, Q_1, Q_2, Q_3 and Q_4 Raman bands are presented at 850–880, 900–930, 950–980, 1040–1060 and 1200 cm^{-1}, respectively.

For finding out the change of the structural network, area ratios of Q_i Raman peaks were introduced to represent the amount of Q_i. The area ratio is shown in Fig. 8. With the continuous increasing CeO_2, area ratios of Q_0 and Q_1 increased, Q_2 increased firstly and then decreased, and Q_3 decreased all the time. It indicated that CeO_2 enhanced the de-polymerization of the structural network and played the role of network modifier.

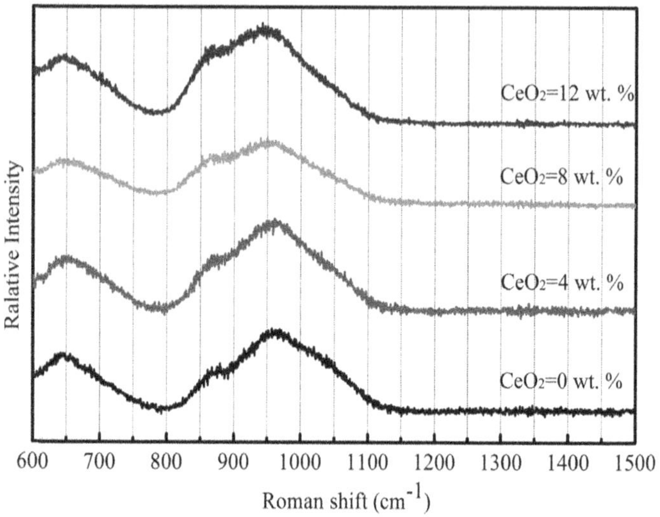

Fig. 6 Raman spectra signals of studied samples with different contents of CeO_2

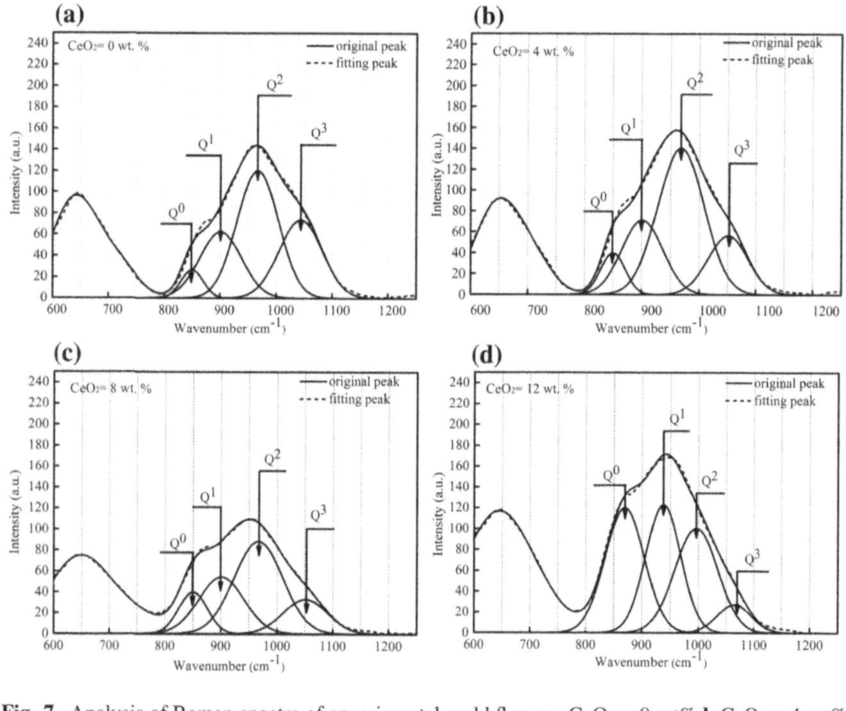

Fig. 7 Analysis of Raman spectra of experimental mold fluxes **a** $CeO_2 = 0$ wt% **b** $CeO_2 = 4$ wt% **c** $CeO_2 = 8$ wt% **d** $CeO_2 = 12$ wt%

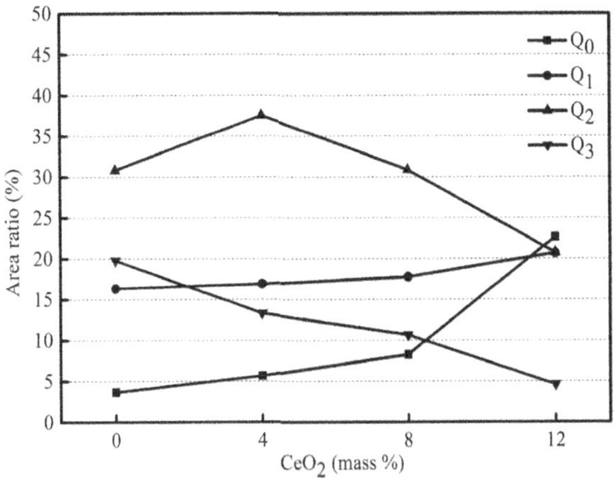

Fig. 8 Area ratios of different bands in four experimental mold fluxes

XPS Spectra Analysis

Three different kinds of oxygen units exist in glassy silicate, such as O^0 bonded to two silicon cations, O^- connected with a single silicon cation, and O^{2-} associated with no silicon cation. In Fig. 9, effects of CeO_2 content on structure of mold fluxes are shown, which revealed deconvoluted XPS results and peaks of different oxygen forms. Statistical results of the mole fraction of O^0 and O^{2-} are presented in Fig. 10. The amount of O^0 decreased with the increase of CeO_2, but the mole fraction of O^{2-} increased. Newly introduced O^{2-} supplied by CeO_2 could decrease the viscosity value by de-polymerizing the structural network and reducing the amount of strongly bonded O^0 and increasing the fraction of O^{2-}.

Conclusions

Effects of CeO_2 content on melting temperature, viscosity and structure of mold fluxes for casting rare earths alloyed heavy railway steels have been studied, and the relationship of viscosity and structure has been discussed. The conclusions are as follows:

(1) CeO_2 increased the melting temperature, and the content of CeO_2 in experimental mold fluxes should not exceed 12 wt%. CeO_2 decreased the viscosity and the breaking temperature, and the reasonable fraction should be lower than 12 wt%.
(2) CeO_2 decreased the viscosity by playing the role of network modifier in structural network of mold fluxes and introducing the increased amount of O^{2-}.

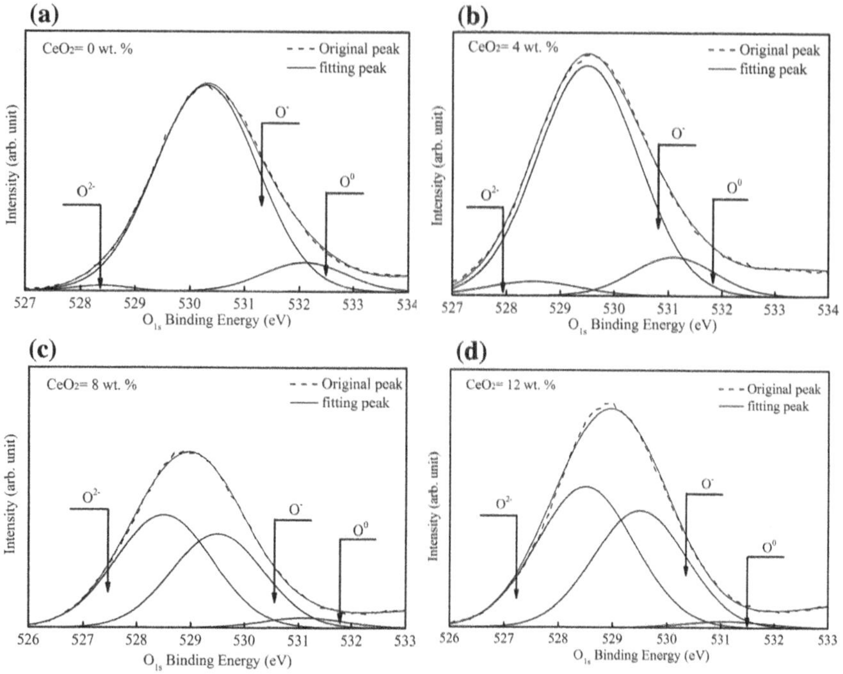

Fig. 9 Deconvoluted O_{1s} binding energy of experimental mold fluxes **a** CeO_2 = 0 wt% **b** CeO_2 = 4 wt% **c** CeO_2 = 8 wt% **d** CeO_2 = 12 wt%

Fig. 10 XPS results of mold fluxes with varying CeO_2 contents

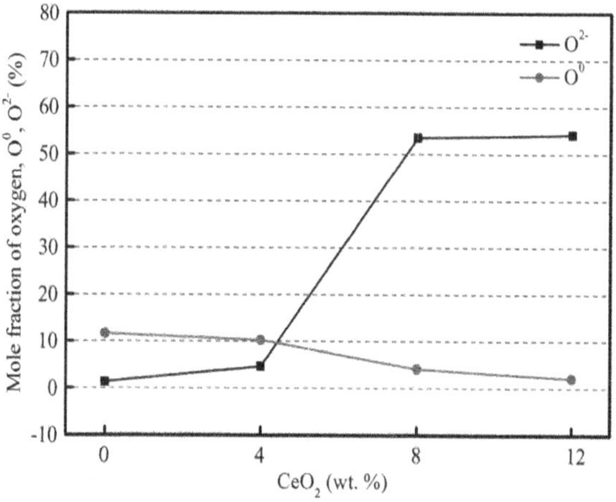

Acknowledgements This work was supported by National Natural Science Foundation of China (Grant No. 51774024).

References

1. Yamauchi A, Sorimachi K, Sakuraya T, Fujii T (1993) Heat transfer between mold and strand through mold flux film in continuous casting of steel. ISIJ Int 33(1):140–147
2. Cho J, Shibata H, Emi T, Suzuki M (1998) Thermal resistance at the interface between mold flux film and mold for continuous casting of steels. ISIJ Int 38(5):440–446
3. Mahapatra RB, Brimacombe JK, Samarasekera IV (1991) Mold behavior and its influence on quality in the continuous casting of steel slabs: part II. Mold heat transfer, mold flux behavior, formation of oscillation marks, longitudinal off-corner depressions, and subsurface cracks. Metall Mater Trans B 22(6):875–888
4. Narita K (1975) Physical chemistry of the groups IVa (Ti, Zr), Va (V, Nb, Ta) and the rare earth elements in steel. Trans ISIJ 15(3):145–152
5. Fu H, Xiao Q, Kuang J, Jiang Z, Xing J (2007) Effect of rare earth and titanium additions on the microstructures and properties of low carbon Fe–B cast steel. Mater Sci Eng A 466(1–2):160–165
6. Waudby PE (1978) Rare earth additions to steel. Int Met Rev 23(1):74–98
7. Wang M, Mu S, Sun F, Wang Y (2007) Influence of rare earth elements on microstructure and mechanical properties of cast high-speed steel rolls. J Rare Earth 25(4):490–494
8. Shimizu F, Tokunaga H, Saito N, Nakashima K (2006) Viscosity and surface tension measurements of RE_2O_3–MgO–SiO_2 (RE = Y, Gd, Nd and La) melts. ISIJ Int 46(3):388–393
9. Wang D, Jiang M, Liu C, Shi P, Yao Y, Wang H (2005) Effects of rare earth oxide on viscosity of mold fluxes for continuous casting. J Rare Earth 23(1):68–72
10. Qi J, Liu C, Zhang C, Jiang M (2017) Effect of Ce_2O_3 on structure, viscosity, and crystalline phase of CaO–Al_2O_3–Li_2O–Ce_2O_3 slags. Metall Mater Trans B 48(1):11–16
11. Li J, Shu Q, Chou K (2014) Structural study of glassy CaO–SiO_2–CaF_2–TiO_2 slags by Raman spectroscopy and MAS-NMR. ISIJ Int 54(4):721–727
12. Chromčíková M, Liška M, Gavenda T, Macháček J (2014) Structure of Na_2O–MgO–CaO–SiO_2 glasses by combined Raman spectroscopy and thermodynamic modeling approach. J Therm Anal Calorim 118(2):835–840
13. Hill RG, Da Costa R, Law RV (2005) Characterization of a mould flux glass. J Non-Cryst Solids 351(1):69–74

Experimental Study on the Mechanism of Lead Vapor Condensation Under Vacuum

Huan Zhang, Zhenghao Pu, Yifu Li, Junjie Xu, Baoqiang Xu and Bin Yang

Abstract Under vacuum conditions, ranging 5 to 80 Pa, orthogonal experiments were conducted by controlling the distillation temperature and holding time to study the lead vapor condensation conditions and nucleation growth mechanism. The results showed that the condensing temperature range of lead vapor concentrates in the region between 510 and 720 °C. Among them, the condensation effect was best in the region from 550 to 660 °C, and more than 80% of the lead condensed in this region. When the ambient pressure changed from 5 to 80 Pa, the temperature of the lead vapor condensate would change around 50 °C. By scanning electron microscopy, the condensation of lead vapor was found to be droplet condensation. The presence of temperature, pressure and gas all affected the nucleation and growth of lead. As the degree of undercooling increases, the condensed particles of lead gradually decreased.

Keywords Lead vapor · Condensation · Nucleation · Vacuum

H. Zhang · Z. Pu · Y. Li · J. Xu · B. Xu · B. Yang
National Engineering Laboratory for Vacuum Metallurgy, Kunming University of Science and Technology, Kunming 650093, People's Republic of China

H. Zhang · Z. Pu · Y. Li · J. Xu · B. Xu · B. Yang
Faculty of Metallurgical and Energy Engineering, Kunming University of Science and Technology, Kunming 650093, People's Republic of China

H. Zhang · Z. Pu · Y. Li (✉) · J. Xu · B. Xu · B. Yang
The State Key Laboratory of Complex Nonferrous Metal Resources Clean Utilization, Kunming University of Science and Technology, Kunming 650093, People's Republic of China
e-mail: 29376154@qq.com

© The Minerals, Metals & Materials Society 2019

T. Jiang et al. (eds.), *10th International Symposium on High-Temperature Metallurgical Processing*, The Minerals, Metals & Materials Series,
https://doi.org/10.1007/978-3-030-05955-2_37

Introduction

Lead has made a significant contribution to the development of human society and the progress of civilization. Lead and lead compounds could be used in the battery, construction, transportation, ceramics, glass, rubber, petrochemical and metallurgical industries [1–4]. Among them, lead storage batteries were the most widely used. Lead was the world's highest metal recycling rate. Common lead wastes included lead-acid batteries and lead-smelting slag. This type of lead-containing waste contained certain toxicity. China's renewable lead resources were very large and increased year by year. In recent years, lead-acid batteries in China have consumed more than 3 million tons of lead per year. The scrapped lead-acid battery was also increased accordingly. At the same time, with the decline in the quality of raw ore in recent years, slag produced in large quantities. Generally, for every 1 ton of crude lead produced, 1–2 tons of slag would be produced. The resulting slag volume is 3–6 times that of crude lead. Such wastes were piled up for a long time, which polluted the environment and wasted resources. Therefore, it is necessary to recycle lead-containing waste.

The main components of lead storage batteries were lead, lead paste, electrolytes and other materials. The recycling methods of waste lead were mainly divided into three types, namely fire method, wet method, fire method and wet method combination. The fire process was often smelted using a conventional reverberatory furnace. However, reverberatory furnace smelting led to a large lead loss due to its low productivity. The most advanced method for treating waste lead storage batteries was osmite furnace smelting. This method had high productivity and could produce soft lead and lead-bismuth alloys. However, the furnace spray gun was easy to be ground. Due to the many problems of the fire treatment of waste lead storage batteries, it was difficult to meet environmental protection requirements. Therefore, extensive research has been carried out at home and abroad in recent years, such as RSR process, Placid process, USBM process and solid phase electrolytic reduction method. [5–9]. The main mechanism of these methods was the desulfurization of the lead paste. It was then dissolved with a certain reagent to form an electrolyte. Finally, lead was obtained by electrolysis. The main methods for treating lead-containing slag were rotary kiln method, electric furnace method and smouldering method [10–12]. Among them, the tobaccoification method was a widely used method in the industry.

Vacuum technology had the advantages of high resource utilization, low pollution, low labour intensity, etc. It was widely used in the process of removal of volatile impurities in alloy separation and metal refining. Therefore, in this paper, lead was used as raw material, and the condensation experiment was carried out under vacuum conditions. The effect of vacuum and residence time on the condensation temperature region was investigated.

Table 1 Boiling temperature of lead at different pressures

Pressure/Pa	1.33×10^{-4}	1.33×10^{-3}	1.33×10^{-2}	1.33×10^{-1}	1.33	13.3	1.33×10^{2}	1.33×10^{3}
Pb/°C	435	489	554	633	722	837	977	1171

Theoretical Calculation

Pure Metal Boiling Temperature

Through calculation, the boiling temperature of lead at different pressures was obtained. It could be seen from Table 1 that at 13 Pa, the boiling temperature of lead was 837 °C. In the experiment, since the vacuum degree of the furnace can only reach about 30 Pa, the initial heating temperature was set to 900 °C.

Pure Metal Saturated Vapor Pressure

The saturated vapor pressure of pure metals was the basis for vacuum distillation of alloys. There are many relationships between saturated vapor pressure and temperature. But the calculation results for each formula were similar. Equation 1 was the relationship between saturated vapor pressure and temperature. Table 2 showed the saturated vapor pressure constants of lead.

$$\lg p = AT^{-1} + B\lg T + CT + D \tag{1}$$

According to the evaporation constant of lead in Table 2 and Eq. 1, a schematic diagram of the relationship between lead saturated vapor pressure and temperature was plotted, as shown in Fig. 1. It can be seen from Fig. 1 that the lead saturated vapor pressure increases at a power level between 300 and 1800 °C. It shows that,

Table 2 Vapourization constant of saturated vapor pressure of As and Pb

Element	A	B	C	D
Pb	−10130	−0.985	0	13.28

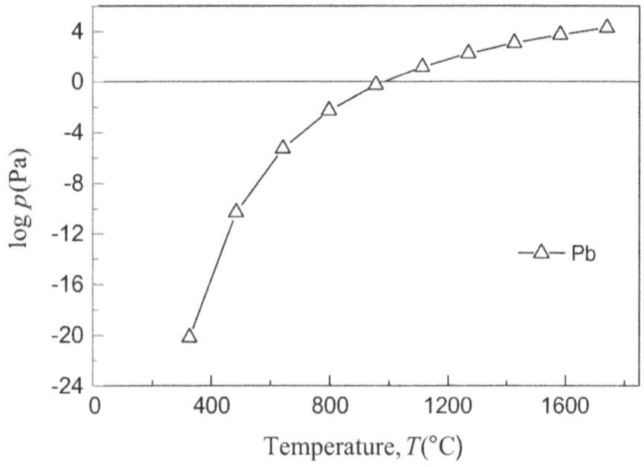

Fig. 1 Diagram of saturated vapor pressure versus temperature for Pb

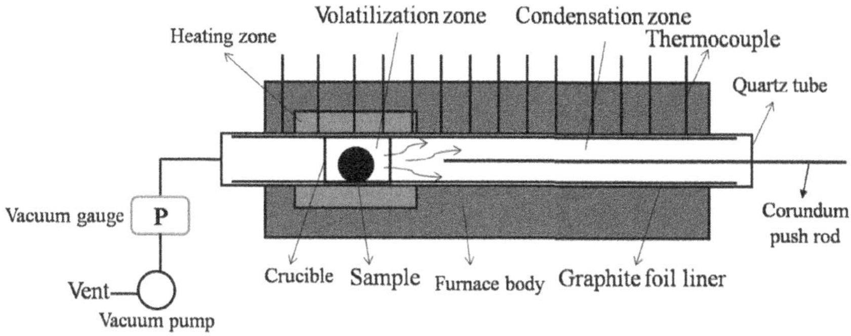

Fig. 2 Schematic of the vacuum tube furnace of fractional condensation

when the temperature raised, the evaporation rate of the metal is increased, and the holding time was shortened.

Experimental

Raw Materials and Equipment

The experimental material was an alloy of 6 N lead (99.9999%) produced by a Chinese company. The experimental equipment adopted the vacuum tube furnace of fractional condensation. As shown in Fig. 2, the setup consists of one quartz tube forming two distinct zones for: (1) volatilization zone (2) condensation zone. A total of 14 thermocouples were arranged from left to right. The distance between the first 5 thermocouples was indefinite. The distance between NO.5 and NO.6 thermocouple was 5.4 cm. And the distance between the other thermocouples was 2.7 cm. The temperature difference between the inside and outside of the pipe wall was 30 °C by the temperature calibration test.

Method and Procedure

The experimental variables were the two factors of holding time and distillation temperature. The experimental conditions were as shown in Table 3. Metal recovery in the experiment can be expressed as formula (2).

$$\omega = \frac{m_1}{m_2} \times 100\% \quad (2)$$

Table 3 Experimental design of vapor condensation of arsenic

Holding time/min	60	80	100
Distillation temperature/°C	900		
	950		
	1000		
	1050		

where ω is the metal yield, m_1 is the mass of metal collected after condensation and m_2 is the mass of the material.

The system pressure was 30 Pa. When the set temperature was reached, lead was pushed to the heating zone in a horizontal tubular. The low temperature part of the tube was used as condenser to condensed lead vapor. Measurements of the axial temperature profile, calculation of volatilization rates and analysis of condensation products within the condenser allowed for identification of condensation temperatures.

Results and Discussion

The evaporation temperature was set at 900, 950, 1000 and 1050 °C. The holding time was 60 min, 80 min and 100 min, respectively. The experiment was carried out at a pressure of 1–50 Pa. Figure 3 was a graph showing the results of the experimental lead vapor condensation experiment of 5#. It could be found from the figure that the lead was condensed on the graphite foil and had a grey-black drop structure. The purity of the condensate Pb was detected to be above 99.99%.

Tables 4 and 5 showed the temperature distribution and condensate. Among them, the thermocouples were recorded in the order of the tube furnace from left to right (serial number) (the thermocouple temperature was the data when the furnace temperature was relatively stable). The condensing mass was the mass of lead condensate between two adjacent thermocouples.

It could be seen from the table that as the temperature of the heating zone increased, the temperature gradient in the tubular vacuum furnace increased. When

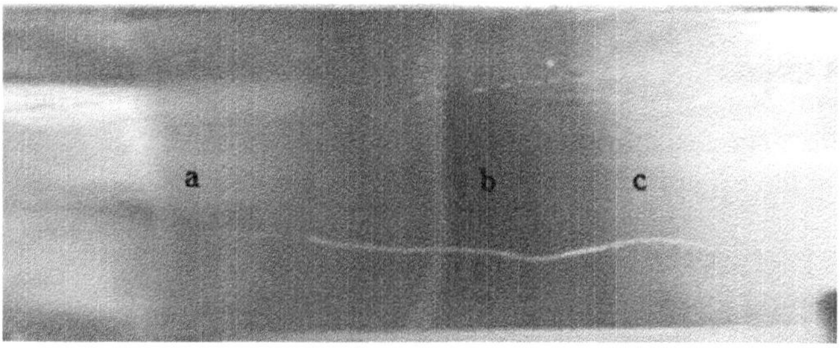

Fig. 3 Pb vapor condensation experimental results

Table 4 Pb vapor condensation experiment thermocouple data (°C)

Thermocouples	900 °C			950 °C			1000 °C			1050 °C		
	60 min	80 min	100 min	60 min	80 min	100 min	60 min	80 min	100 min	60 min	80 min	100 min
1	838.4	836.6	837.0	865.3	872.8	861.2	924.9	939.1	938.8	980.2	986.9	993.4
2	889.1	894.2	891.6	921.5	933.7	919.4	987.0	988.9	990.1	1026.0	1038.0	1036.0
3	900.0	901.0	902.0	949.0	950.0	948.0	1000.0	999.0	1001.0	1048.0	1048.0	1049.0
4	842.4	859.4	861.3	860.3	896.2	884.4	948.1	954.7	957.7	992.1	1002.4	1010.4
5	798.3	808.4	815.2	800.3	851.6	830.2	899.3	910.2	911.7	930.2	957.5	968.4
6	604.2	622.1	640.3	651.4	670.2	685.9	707.1	726.5	743.2	762.7	782.8	797.5
7	537.3	555.3	564.6	593.0	593.3	622.0	623.1	643.8	662.6	682.3	705.8	720.3
8	472.3	483.5	503.2	529.4	541.9	561.3	561.9	575.1	606.1	629.1	643.7	656.3
9	429.0	448.0	465.0	490.0	506.0	523.0	520.0	531.0	561.0	567.0	592.0	609.0
10	382.4	397.6	416.3	440.1	460.4	472.2	457.7	474.8	509.8	527.3	545.2	561.2
11	342.9	358.8	377.7	398.2	417.1	430.4	411.4	428.4	458.4	460.1	485.9	500.1
12	313.5	321.3	339.5	360.1	372.7	386.8	364.7	379.0	410.4	420.3	436.7	450.2
13	272.3	290.2	309.4	312.5	337.4	352.6	329.3	341.5	372.5	373.6	397.4	405.8
14	240.2	252.4	273.2	279.4	297.6	311.7	285.1	291.8	325.7	335.2	350.2	363.1

Table 5 Pb vapor condensation experiment condensate mass table (g)

Experiment number	4–5	5–6	6–7	7–8	8–9	9–10	10–11	Not evaporated	Raw material	Recovery rate
1# (900 °C, 60 min, 10 Pa)		3.21	4.12	2.20				0.49	10.08	99.40%
2#(900 °C, 80 min, 10 Pa)		1.54	4.28	4.17					10.02	99.70%
3# (900 °C, 100 min, 10 Pa)		5.18	3.76	1.06					10.05	99.50%
4# (950 °C, 60 min, 10 Pa)		5.67	3.43	0.98					10.11	99.70%
5#(950 °C, 80 min, 20 Pa)			2.91	5.38	1.65				10.05	98.91%
6# (950 °C, 100 min, 20 Pa)		5.19	4.11	0.75					10.10	99.50%
7# (1000 °C, 60 min, 20 Pa)		0.42	4.87	4.02	0.72				10.09	99.41%
8# (1000 °C, 80 min,20 Pa)			3.46	4.28	2.25				10.04	98.50%
9# (1000 °C, 100 min, 20 Pa)			0.61	4.48	3.96	0.85			9.97	99.30%
10#(1050 °C, 60 min, 40 Pa)				5.74	3.38	0.85			10.07	99.00%
11# (1050 °C, 80 min, 40 Pa)			0.49	4.92	3.41	1.18			10.08	99.21%
12# (1050 °C, 100 min, 40 Pa)				2.10	4.87	2.67	0.31		10.02	99.30%

the heating time was 60 min, the four groups of 1#, 4#, 7# and 10# experiments, the temperature of the heating zone was sequentially increased by 50 °C (thermocouple No. 3). The temperature difference between the 2–3 thermocouples on the right side of the heating zone was still maintained at around 40 °C. The temperature difference of the rightmost thermocouple had been reduced to about 20 °C. It was the right end close to room temperature, although the temperature of the heating zone was different. But as the distance from the heating zone increased and approaches room temperature, the temperature gradient reduced. When the heating temperature was 950 °C, the temperature gradients of the 4#, 5# and 6# experiments were similar, and the temperature difference between them was only about 20 °C. Cross-comparison 3# and 4# experiments, although the heating zone temperature and holding time of the two groups were different. However, the experimental temperature gradients were similar in both groups, and the lead condensation positions were similar. Ten sets of experimental metals had higher recovery rate.

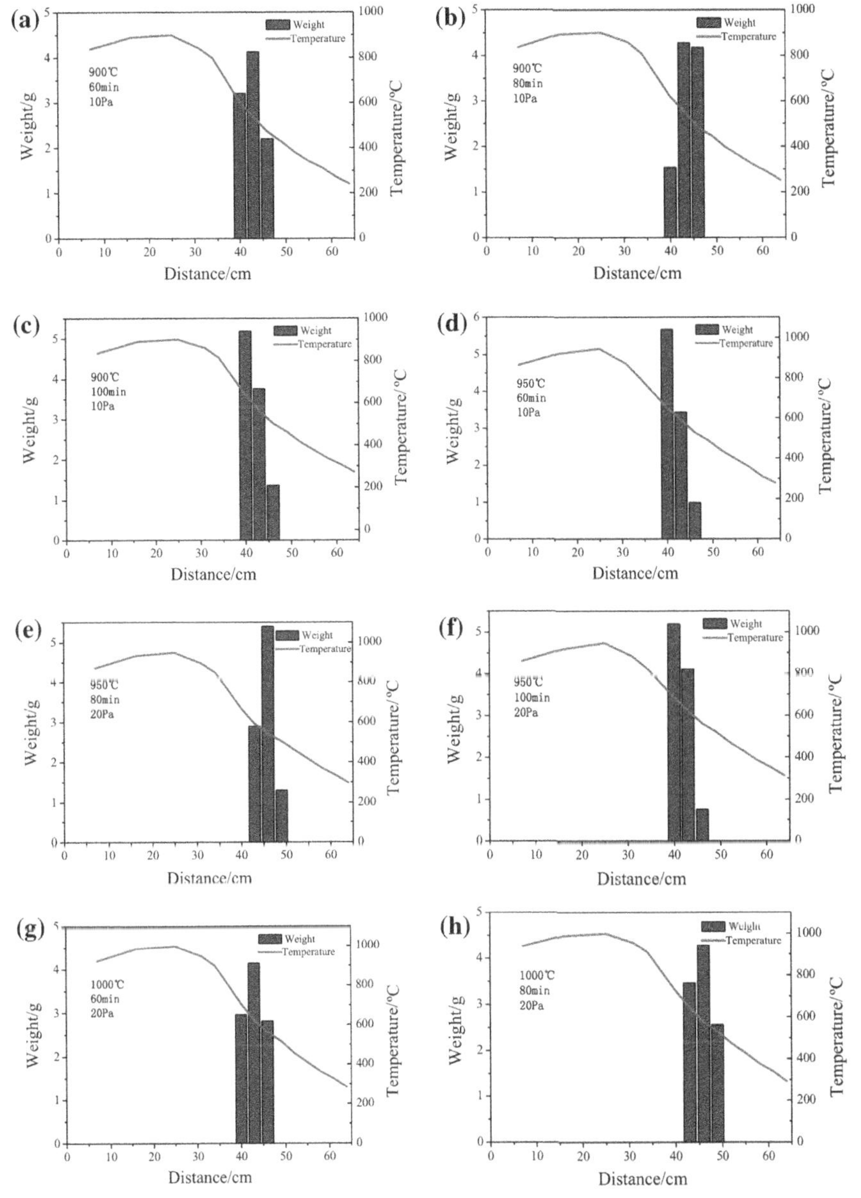

Fig. 4 Pb vapor condensation experiment results

Based on the data listed in Tables 4 and 5, the results of the twelve sets of experimental results were shown in Fig. 4. In the figure, the abscissa was the length from the left to the right of the tube furnace, the red curve indicated the temperature

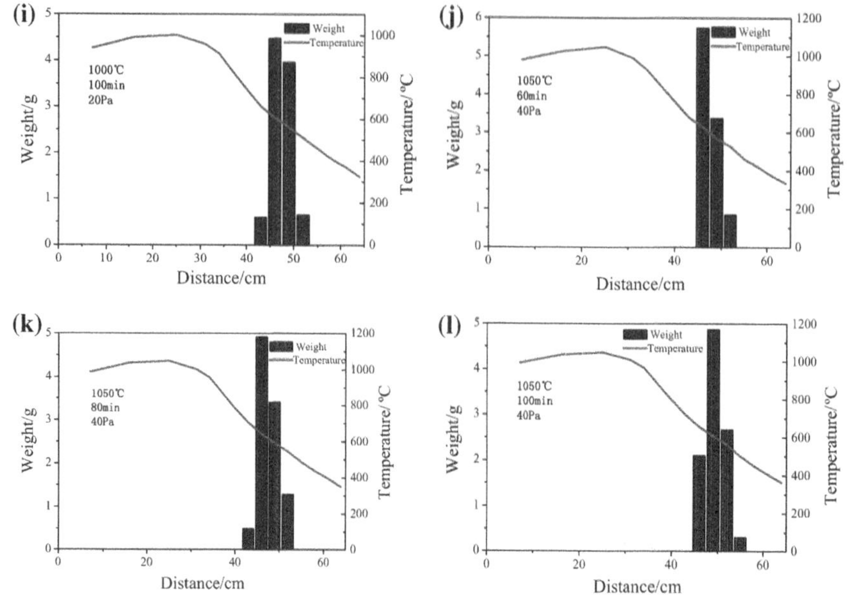

Fig. 4 (continued)

gradient of the tube furnace, and the blue bar graph indicated the mass of the condensate lead.

Known by (a), lead was condensed in the middle of the tube furnace. There was a mass of 3.21 g of condensate between the temperatures of 604.2 and 798.3 °C. There was 4.12 g of condensate between 537.3 and 604.2 °C. There was 2.20 g of condensate between 472.3 and 537.3 °C. Due to the large distance between the thermocouple NO.5 ~ NO.6 in the experimental equipment, although there was partial condensation in the experiment 1# between the thermocouple NO.5 ~ NO.6, that is, between 604.2 and 798.3 °C. However, almost all of them were 1 cm near the thermocouple NO.6, and the distance from the thermocouple NO.5 was large (4 cm). Therefore, when the distillation temperature was 900 °C and the holding time was 60 min, the lead condensation was concentrated in the region between 540 and 700 °C.

From (b) to (i), the experiment 2# ~ 9# situation was similar to the 1# conclusion. As can be seen in Figure (j), the temperature gradient in the tube furnace was higher than the previous groups due to the higher temperature in the heating zone. The lead condensation area moved to the right but was the same as the previous sets of experiments. When the distillation temperature was 1050 °C and the holding time was 60 min, the lead condensation was concentrated in the region between 530 and 680 °C. The rules in (k) and (l) were similar to those in figure (j). However, the temperature gradient in the tube furnace was higher and the condensation area was again shifted to the right.

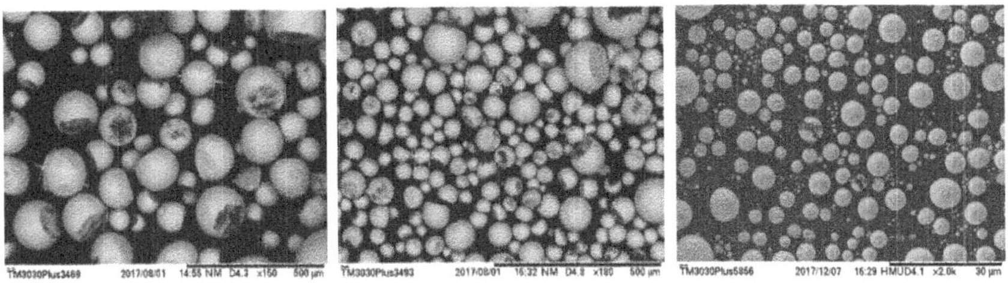

Fig. 5 Scanning electron micrograph of Pb condensate

Compared the three groups of experiments #1, 2# and 3#, the heating zone temperature was 900 °C, but the holding time was different. Most of the lead was condensed between 520 and 710 °C. Cross-contrast 3#, 4# and 9#, 10# experiments, the heating temperature and holding time of the four sets of experiments were different. However, the temperature gradients were similar. Lead condensed at 520–690 °C.

Scanning electron microscopy analysis was performed on the three condensation zones a, b and c in Fig. 3. The results are shown in Fig. 5. By comparing Fig. 4e and Fig. 5, it was found that the temperature at the position of a was about 660 °C. The lead condensation particles were about 150 μm. As the temperature decreases, the temperature at the position of b dropped to about 560 °C, and the condensation particles became about 60 μm. As the temperature continued to decrease to 510 °C, the degree of supercooling increased, and lead condensate particles of about 5 μm can be observed at the position of c.

Conclusion

When the ambient pressure was 5–10 Pa, the lead condensation concentrated in the region between 510 and 720 °C. Wherein the 550–660 °C region had the best condensation effect, that is, more than 80% of the lead was condensed in this region. The condensation of lead in the graphite foil was in the form of drops. The presence of temperature, pressure and gas could affect the nucleation and growth of lead. And as the degree of supercooling increases, the lead condensation particles became smaller.

Acknowledgements The authors were grateful for financial support from the National Nature Science Fund of China (grant NO. 51,504,115), the NSCF-Yunnan Joint Fund (grant NO. U1502271) and the National Key Research and Development Program of China (grant NO. 2016YFC0400400).

References

1. Irfandi AF (2015) Analisis kandungan kadmium (cd) dan timbal (pb) pada air sumur gali penduduk di sekitar industri daur ulang aki, dan gangguan, kesehatan pada masyarakat, desa bandar khalipah kabupaten, deli serdang, tahun 2013, Lingkungan Dan Kesehatan Kerja, 2
2. Deng YH, Cao XM, Zhang QG (2010) Study of solar energy battery PbSO4 stored system. Electron Qual 10:66–68
3. Zhou JT, Xiao QH (2002) The Intelligent Management of Plumbum Acid Storage Battery. Packag Eng 23:88–89
4. Zhou Y, Lu DH, Zhang M (2017) Performance test of lead steel pipe damper. China Civil Eng J 1:46–52
5. Wang ZP, Wen ZG (2018) Life cycle assessment of recycled lead alloy from waste lead-acid battery. Acta Sci Circum 3:1245–1255
6. Zhang ML (2016) Study on recycling and harmless treatment process of lead-rich slag, pp 11–12
7. Ogundiran MB, Buluku TG, Babayemi JO (2015) Waste rechargeable electric lamps: characterisation and recovery of lead from their lead-acid batteries. J Mater Cycles Waste Manage 19:1–9
8. Tong YS, Zhang K, Cui H (2011) Production practice of lead recovery from waste lead-acid batteries and lead wastes. Chin Lab Man 4:165–167
9. Lin DQ (2011) Study on new process of recycling and utilization of waste lead storage battery and lead anode slime, Central south university
10. Yuan PX, Li CL (2006) Production practice of reducing lead in blast furnace slag by SKS lead smelting process. China Nonferrous Met 6:10–12
11. Zhu WY (2004) Ways to reduce lead in blast furnace slag. Hunan Nonferrous Met 5:26–28
12. Li SD, Shan TY (2011) Discussion on process of extraction indium from slag with plumbum. Hunan Nonferrous Met 4:22–24

Effect of Al on the Formation of IAF in Al–Ti–Mg Deoxidized and RE-Treated Steel

Xiaokang Cui, Bo Song, Zhen Liu and Longfei Li

Abstract To clarify the effect of aluminium on the formation of intragranular acicular ferrite (IAF) in Al–Ti–Mg deoxidized and rare earth-treated steels with different S content, the thermodynamic calculation and scanning electron microscope (SEM) with energy dispersive spectrometer (EDS) were used to study the distribution and composition of inclusions. The results showed that the addition of Al led to the increase of mean size of effective inclusions. With the content of Al increasing from 0.0027 to 0.0086 wt%, the main inclusions in the low S steels changed from La_2O_3 to $LaAlO_3$ and the proportion of IAF decreased rapidly. However, the inclusions in the high S samples were mainly composed of La_2O_2S or $LaAlO_3$ and MgO with MnS precipitating on the surface of them. As the Al content increased from 0.0030 to 0.0097 wt%, thc main composition of core evolved from La_2O_2S to $LaAlO_3$, but the proportion of IAF slightly changed. As the Al content approached 0.03 wt%, the proportion of IAF decreased rapidly and a plenty of upper bainite or widmanstatten were discovered.

Keywords Intragranular acicular ferrite · Aluminum · Inclusions

X. Cui · B. Song (✉) · Z. Liu · L. Li
School of Metallurgical and Ecological Engineer, University of Science and Technology Beijing, Beijing 100083, China
e-mail: songbo@metall.ustb.edu.cn

X. Cui
e-mail: cuixiaokang126@126.com

Z. Liu
e-mail: liuzhen_ustb@163.com

L. Li
e-mail: aifei_0105@126.com

© The Minerals, Metals & Materials Society 2019
T. Jiang et al. (eds.), *10th International Symposium on High-Temperature Metallurgical Processing*, The Minerals, Metals & Materials Series,
https://doi.org/10.1007/978-3-030-05955-2_38

Introduction

With the service condition of steels becoming more and more complex, the requirement for steels is more and more comprehensive, for example, the offshore platforms, long-distance oil and gas pipelines require steel materials not only have higher strength but also have better low-temperature toughness and better weldability [1, 2]. Acicular ferrite (IAF) is a kind of interlocking structure with large angle grain boundaries and high dislocation density. It cannot only refine grain effectively, but also restrain crack extension due to the interlocking effect of grain. Therefore, the strength and toughness of steel that contain a lot of IAF are improved simultaneously.

A large number of studies [3–5] show that difference of deoxidizer has great influence on the type, size, and quantity of inclusions and then affects the formation of acicular ferrite. As an important deoxidizer in steel, Al has an important influence on inclusions. Song et al. [6] reported that increasing the content of Al in RE-treated steel was not conducive to the formation of acicular ferrite, and Zheng et al. [7] reached the similar conclusion. However, Zhang and Farrar [8] reported that the formation of Al_2O_3 contributes to the formation of IAF. Although a number of mechanisms have been proposed for the nucleation of IAF, there are two most widely accepted mechanisms, the inclusions having high coherency with ferrite [9] and Mn-depleted zone (MDZ) forming around the inclusions [10–12]. With regard to the origin of the local MDZ around inclusions, most researchers reported that the MDZ was produced by MnS precipitation on the inclusions during the cooling process.

The effect of Al on the formation of IAF in Al–Ti–Mg deoxidized and RE-treated steel has not been studied. The aim of this work is to study the effect of Al on the compositions, size distribution of inclusions, and the microstructures evolution and discuss the nucleation theory of IAF in the steel with different S content.

Experimental

In the experiment, molybdenum wire furnace was used to smelt electrolytic iron of about 300 g in argon atmosphere. After the iron was melted in a corundum crucible at 1600 °C, alloys were added to adjust the compositions of steel, as shown in Fig. 1. After the smelting process, the experimental steels cooled to 1100 °C and then crucible was taken out quickly and quenched into water. The steel composition of each sample is shown in Table 1.

The samples were sectioned under the same process in the radius of 1/2 of experimental steels. They were mechanically ground and polished to prepare metallographic samples and then were etched for 4–5 s in 4 vol% nital. Optical microscope (OM) was used to observe microstructure. SEM +EDS were used to

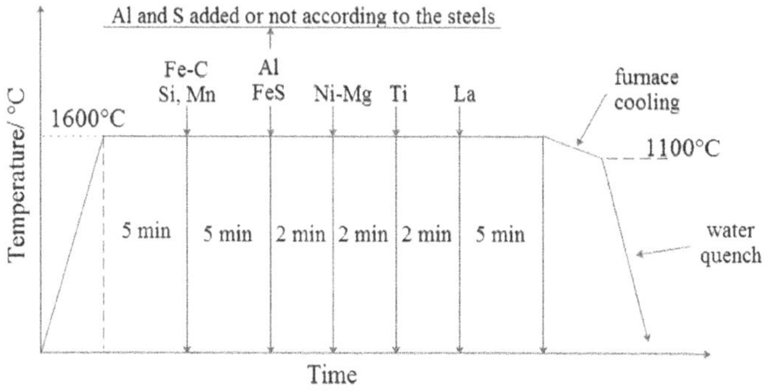

Fig. 1 Experimental procedure diagram

Table 1 Composition of each tested steels/wt%

Sample	C	Si	Mn	S	Al	Mg	La	Ti	O	Fe
AS1	0.11	0.23	1.61	0.0025	0.0029	0.0042	0.017	0.020	0.0059	Bal.
AS2	0.10	0.24	1.63	0.0026	0.0085	0.0043	0.016	0.019	0.0061	Bal.
AS3	0.10	0.25	1.59	0.0030	0.0220	0.0044	0.015	0.017	0.0054	Bal.
AS4	0.09	0.26	1.62	0.0300	0.0027	0.0044	0.016	0.018	0.0057	Bal.
AS5	0.10	0.25	1.61	0.0310	0.0097	0.0045	0.018	0.016	0.0046	Bal.
AS6	0.11	0.23	1.60	0.0330	0.0290	0.0040	0.017	0.018	0.0060	Bal.

observe the morphology of inclusions and analyse the composition of inclusions. The size distribution of effective inclusions (inclusions that promote the acicular ferrite formation) was analysed statistically.

Results and Discussion

Effect of Al on the Inclusion

The Effect of Al on the Composition of Inclusion

FactSage was used to calculate the equilibrium weight per cent of inclusions in steels with S content of 0.0025 and 0.030 wt% with different Al content at 1600 °C. As shown in Fig. 2a, in low S samples, when the Al content increases from 0 to 0.005 wt%, La_2O_3 inclusion disappear, $LaAlO_3$ and La_2O_2S content increase significantly. The content of various inclusions slightly change, when Al content exceeds 0.005 wt%. As shown in Fig. 2b, in high S samples, when the Al content increases from 0 to 0.005 wt%, the content of La_2O_2S inclusions decreases rapidly and the content of $LaAlO_3$ inclusions increases significantly. The content of

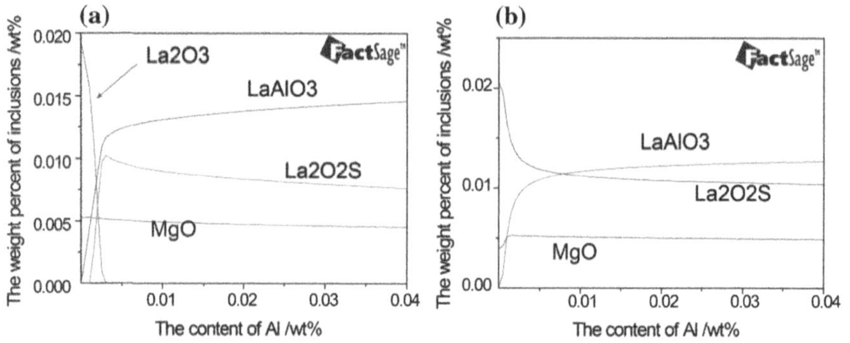

Fig. 2 Calculated equilibrium weight per cent of inclusions in steels at 1600 °C **a** low S content (0.0025 wt%) sample; **b** high S content (0.030 wt%) sample

inclusions slightly changes with further increase of Al content. As the Al content increase, the content of MgO basically has no change in two kinds of steels.

When the S content was low, the inclusions in sample AS1 were mainly MgO and La_2O_3, and IAF was formed around the inclusions, as shown in Fig. 3a, b. In sample AS2, the main inclusions were MgO and $AlLaO_3$, and IAF was formed around MgO, but not around $LaAlO_3$, as shown in Fig. 3c, d. In sample AS3, the main inclusions were MgO and $LaAlO_3$, but there were not IAF formed around inclusions, as shown in Fig. 3e. Moreover, there were not MnS being found on the surface of inclusions in low S steels.

When the S content was high, the main inclusions were obviously spherical composite with a two-layer complex structure. The main inclusions in AS4 steel were MgO and La_2O_2S with MnS precipitating on the surface of them, as shown in Fig. 4a, and IAF formed around the inclusions. The inclusions in AS5 were mainly $LaAlO_3$ and MgO with MnS on the surface of them, and the inclusions in AS6 were the same with AS5. However, there were IAF nucleation around the inclusions in sample AS5, but there was not IAF appeared around the inclusions in sample AS6, as shown in Fig. 4b, c. The composition of the inclusions is basically in agreement with the calculation results.

Effect of Al on the Size of Effective Inclusions

The size distributions of effective inclusions in AS1, AS4, and AS5 steels were statistically analysed, as shown in Fig. 5. The size of the effective inclusions was mostly at the range of 1–5 μm. The main size of effective inclusions of sample AS1, AS4, and AS5 was 1–4 μm, 2–5 μm, and 2–5 μm, respectively. By comparing AS1 and AS4 samples, when the Al content was about 0.0030 wt%, it was found that the average size of effective inclusion in the high S samples was 3.47 μm, and that in the low S samples was 2.47 μm. The rise of sulphur content resulted in the increase of the mean size of the effective inclusions. The formation of MnS on the oxide

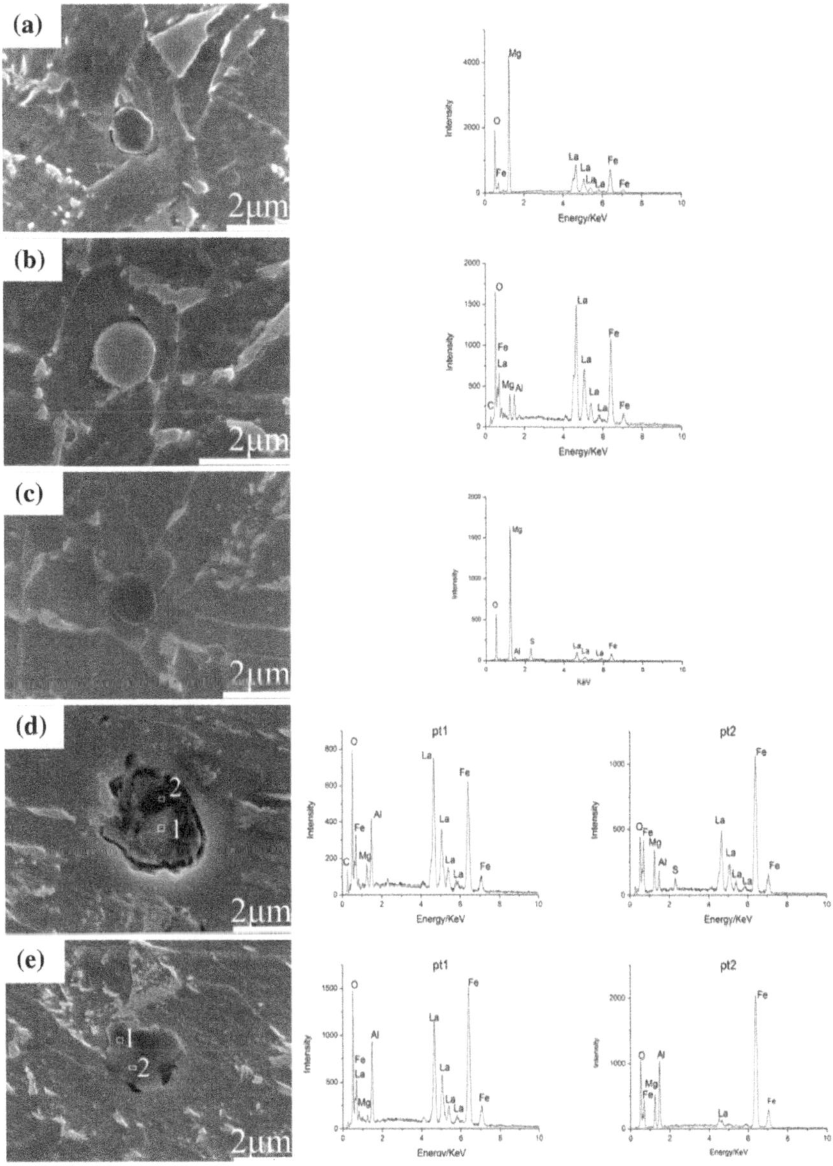

Fig. 3 Micrographs and compositions of typical nonmetallic inclusions in the low S samples **a** and **b** AS1; **c** and **d** AS2; **e** AS3

surface caused the increase of mean size of inclusions. By Comparing AS4 and AS5, when the S content was about 0.03 wt%, the average size of inclusions increased from 3.47 μm in AS4 steels to 3.72 μm in AS5 steels. The addition of Al could lead to the increase of average size of effective inclusions.

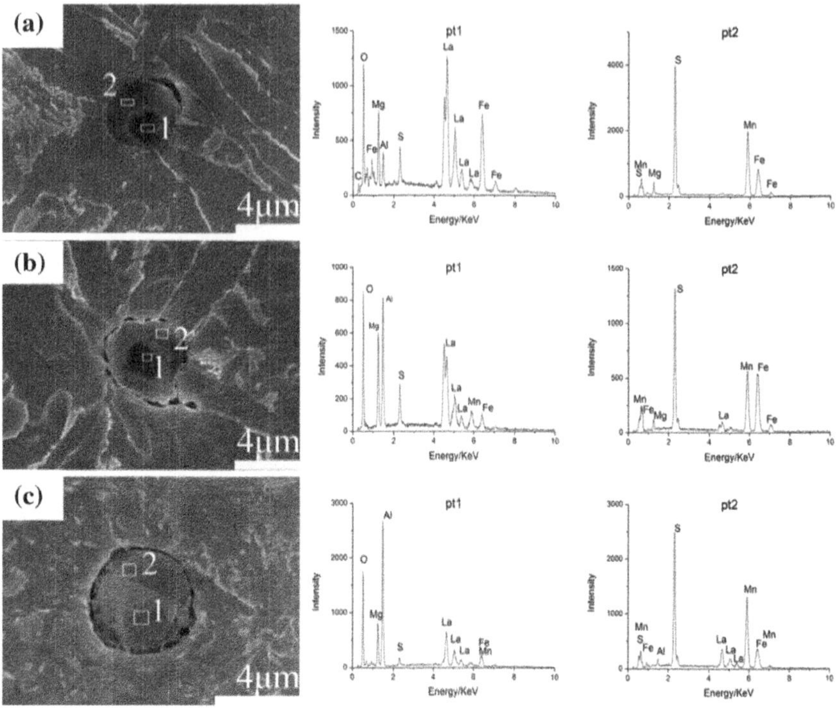

Fig. 4 Micrographs and compositions of typical nonmetallic inclusions in the high S samples **a** AS4; **b** AS5; **c** AS6

Fig. 5 Size distribution of effective inclusions

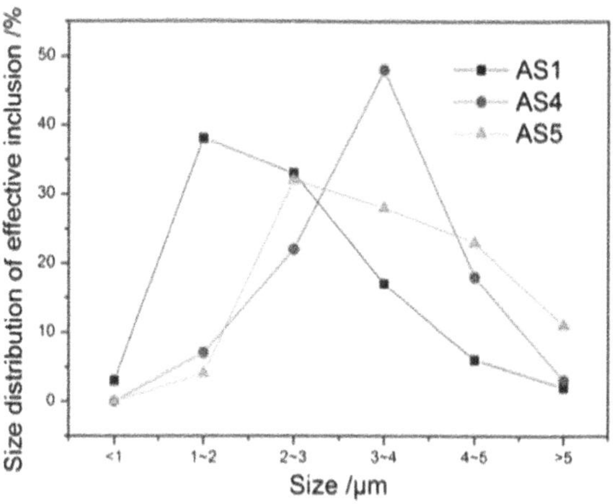

The Effect of Al on the Microstructure

Optical micrographs in Fig. 6 showed the microstructures of steels AS1, AS2, and AS3. The microstructures of AS1 steel consisted of polygonal ferrite (PF), ferrite side plates (FSP), and a lot of IAF which mainly nucleated on inclusions. In AS2 steel, only a small amount of IAF was found and upper bainite appeared. In the low S steels, there was not MnS on the surface of inclusions, the main inclusions of AS1 were La_2O_3 and MgO, and that of AS2 were $LaAlO_3$ and MgO. According to the two-dimensional misfit (2-D misfit) theory, the small lattice misfit between inclusion and α-Fe could facilitate IAF nucleation. Table 2 shows the lattice misfit between inclusions and α-Fe. The misfit between $LaAlO_3$ and α-Fe are bigger than the misfit between La_2O_3, MgO, and α-Fe. That led to the content of IAF decreased in AS2 steel comparing to AS1. As the content of Al increased further, the upper bainite and/or widmanstatten ferrite side plate structure were dominant, as shown in Fig. 6c.

Optical micrographs in Fig. 7 showed the microstructure of steels AS4, AS5, and AS6. The microstructure of AS4 and AS5 consisted of a lot of IAF together with a small amount of PF, when the Al ≤ 0.097 wt%. In high S samples, MnS precipitated on the surface of La-containing inclusions. Yamamoto et al. [13] found that the formation of MnS could cause the formation of MDZ. Although there was no direct measurement of MDZ in this work, we could speculate that the MDZ was formed around the inclusions and played an important role in inducing the nucleation of acicular ferrite. When the content of Al reached 0.033 wt%, the upper bainite and/or widmanstatten ferrite side plate structure were the dominant structure just like AS3 steel.

The increase of Al content in both low S and high S specimens was not conducive to the formation of acicular ferrite. Shim et al. [3] pointed out that Al could change the microstructure mainly because Al had a greater oxygen affinity than Ti, and Al_2O_3 was likely to replace Ti_2O_3 at high Al contents. Song et al. [6] had studied the effect of Al content on microstructure of rare earth-treated C–Mn steel, and he considered that higher Al content was detrimental to IAF formation. Moreover, Al was ferrite stabilized element, and it could increase the temperature of Ar3, promote the formation of grain boundary ferrite and bainite/widmanstatten and

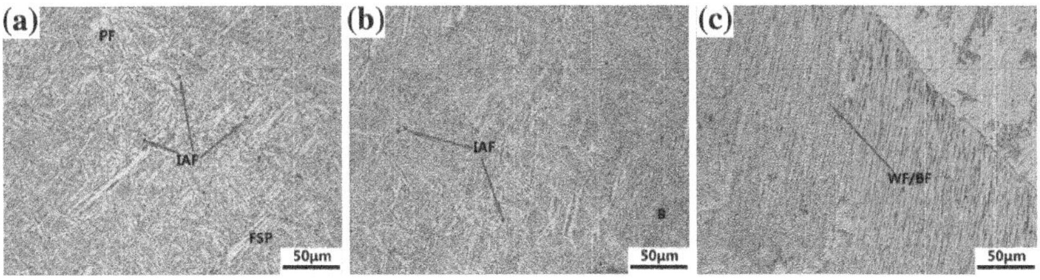

Fig. 6 Microstructure with different content of Al in low S samples **a** AS1; **b** AS2; **c** AS3

Table 2 Lattice mismatch between inclusion and the α-Fe

Inclusion	Lattice structure	Lattice parameters *a*/nm	$a/a_{\alpha Fe}$	2-D misfit/%
La_2O_3	Cubic	$a = 0.3943$	3.971	2.8 [6]
La_2O_2S	Hcp	$a = 0.4062$, $c = 0.6943$	1.417	0.2 [6]
$LaAlO_3$	Cubic	$a = 0.3807$	1.328	6.1 [6]
MgO	Cubic	$a = 0.4306$	1.502	2.8 [4]

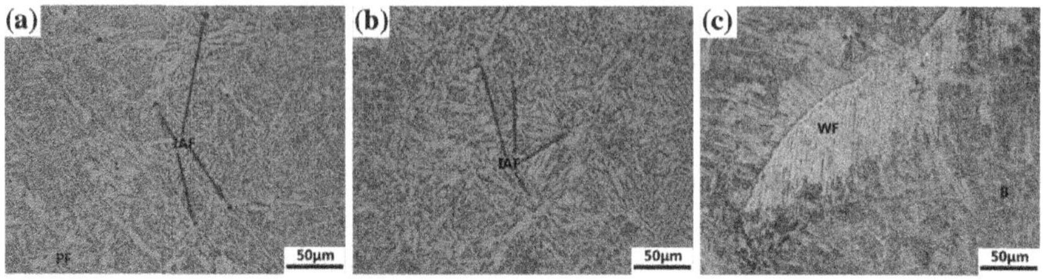

Fig. 7 Microstructure with different contents of Al in high S samples **a** AS4; **b** AS5; **c** AS6

then inhibit the nucleation of acicular ferrite. Therefore, the content of Al ought to be controlled in a lower range in these experimental steels.

Conclusion

In the low S steels, as the Al content increased from 0.0029 to 0.0085 wt%, the main inclusions changed from La_2O_3, La_2O_2S, and MgO to $LaAlO_3$, La_2O_2S, and MgO. The evolution of inclusion caused the rise of the lattice misfit between α-Fe and inclusions. Thus, the addition of Al inhibited the formation of IAF.

In the high S steels, as the Al content increased from 0.0027 to 0.0097 wt%, the main inclusions changed from La_2O_2S + MgO + MnS to $LaAlO_3$ + MgO + MnS. IAF was formed around the inclusions. Mn-depleted zone played an important role in promoting the nucleation of IAF. Therefore, the addition of Al hardly affected the formation of IAF, but it resulted in the increase of the mean size of effective inclusions.

Acknowledgements This work is financially supported by National Nature Science Foundation of China under grant No. 51774024

References

1. Shin SY, Oh K, Kang KB (2010) Improvement of Charpy impact properties in heat affected zones of API X80 pipeline steels containing complex oxides. Mater Sci Technol 26(9):1049–1058
2. He K, Edmonds DV (2002) Formation of acicular ferrite and influence of vanadium alloying. Mater Sci Technol 18(3):289–296
3. Shim JH, Cho YW, Shim JD (2001) Effects of Si and Al on acicular ferrite formation in C–Mn steel. Metall Mater Trans A 32(1):75–83
4. Wu Z, Zheng W, Li G (2015) Effect of inclusions' behavior on the microstructure in Al–Ti deoxidized and magnesium-treated steel with different aluminum contents. Metall Mater Trans B 46(3):1226–1241
5. Ma W, Bao Y, Wang M (2014) Effect of Mg and Ca treatment on behavior and particle size of inclusions in bearing steels. ISIJ Int 54(3):536–542
6. Song M, Song B, Yang Z (2017) Effects of Mn and Al on the intragranular acicular ferrite formation in rare earth treated C–Mn steel. High Temp Mater Process Lond 36(7):683–691
7. Zheng W, Wu Z, Li G (2014) Effect of Al content on the characteristics of inclusions in Al–Ti complex deoxidized steel with calcium treatment. ISIJ Int 54(8):1755–1764
8. Zhang Z, Farrar RA (1996) Role of non-metallic inclusions in formation of acicular ferrite in low alloy weld metals. Mater Sci Technol 12(3):237–260
9. Bramfitt BL (1970) The effect of carbide and nitride additions on the heterogeneous nucleation behavior of liquid iron. Metall Trans 1(7):1987–1995
10. Shigesato G, Sugiyama M (2002) Development of in situ observation technique using scanning ion microscopy and demonstration of Mn-depletion effect on intragranular ferrite transformation in low-alloy steel. J Electron Microsc 51(6):359–367
11. Kang Y, Han K, Park JH (2014) Mn-depleted zone formation in rapidly cooled high-strength low-alloy steel welds. Metall Mater Trans A 45(11):4753–4757
12. Byun JS, Shim JH, Cho YW (2003) Non-metallic inclusion and intragranular nucleation of ferrite in Ti-killed C–Mn steel. Acta Mater 51(6):1593–1606
13. Yamamoto K, Hasegawa T, Takamura J (1996) Effect of boron on intra-granular ferrite formation in Ti-oxide bearing steels. ISIJ Int 36(1):80–86

Effect of Ce Treatment on the Composition of Nucleation Inclusion in Ti–Mg Complex Deoxidized C–Mn Steel

Zhen Liu, Bo Song, Longfei Li, Zeyun Cai and Xiaokang Cui

Abstract FactSage software and SEM + EDS were introduced to determine the composition of nucleation inclusion in Ce-treated Ti–Mg deoxidized C–Mn steel with Ce content ranging from 0 to 0.036 wt%. The nucleation mechanism of inclusion in tested steels was investigated simultaneously. The experimental results showed that the composition of inclusion in the Ti–Mg complex deoxidized steel changed from $MgO + MgAl_2O_4/MgTi_2O_4 + MnS$ to $MgO + MgAl_2O_4/MgTi_2O_4 + CeAlO_3/Ce_2O_2S + MnS$ after Ce treatment, while there was no significant change in the content of Ti(C, N). The mass fraction of Ce_2O_2S was markedly increased with the increase of Ce. The addition of Ce enhanced the probability of nucleation of acicular ferrite on inclusion. The nucleation mechanism of acicular ferrite induced by inclusion containing Ce was manganese-depleted zone (MDZ) based on the line scanning results and the low lattice mismatch between Ce_2O_2S and α-Fe.

Keywords Ce treatment · Composition · Nucleation inclusion Ti–Mg complex deoxidized

Z. Liu · B. Song (✉) · L. Li · Z. Cai · X. Cui
School of Metallurgical and Ecological Engineering, University of Science and Technology Beijing, Beijing 100083, China
e-mail: songbo@metall.ustb.edu.cn

Z. Liu
e-mail: liuzhen_ustb@163.com

L. Li
e-mail: aifei_0105@126.com

Z. Cai
e-mail: zexiaoyun@126.com

X. Cui
e-mail: cuixiaokang126@126.com

© The Minerals, Metals & Materials Society 2019

T. Jiang et al. (eds.), *10th International Symposium on High-Temperature Metallurgical Processing*, The Minerals, Metals & Materials Series,
https://doi.org/10.1007/978-3-030-05955-2_39

Introduction

Inclusions or complex precipitates had been usually considered as deleterious for toughness of steel until a phenomenon, intragranular ferrite(IGF) nucleated on these particles, being demonstrated. Followed that, Takemura and Mizoguchi [1] and Mizoguchi and Takamura [2] presented "oxide metallurgy" attracting great interests in the metallurgy field with the concept of adopting the second-phase particles to microstructure controlling.

So far, numbers of cases applying "oxide metallurgy" to the heat-affected zone (HAZ) in welding process proved to be successful since oxide inclusions instead of precipitates which resolved in high temperature such as Ti(C, N) nucleated plenty of IGF plates refining the microstructure as well as improving the toughness of welding joints [3–6]. Meanwhile, choosing modification reactants, mainly deoxidization elements or alloys, was also in hotspot for their direct interaction with the activity of nucleation inclusions. Generally speaking, it experienced three development phases from typical deoxidizing and alloying element Ti to strong deoxidizers represented by Zr, Mg and Ce to complex deoxidizers employing both the weak and strong deoxidization elements, which was proved to be more practical in controlling the level of oxygen in steel [7, 8].

When it comes to complex deoxidization, however, systematic investigation concerning Ce is rarely seen in the field, though Ce has good performances in many fields as follows:

(a) strong deoxidizer and desulfurizer [9];
(b) modifying the inclusions, especially MnS [10];
(c) reducing the harm of residual elements such as Pb, As [9, 11];
(d) pinning the growth of austenite grain and refining the microstructure so as to improve the mechanical properties of steel [12–14].

There is no doubt that Ce has many advantages over other elements, but it also does have disadvantages like a coin having two sides. For instance, Ce treatment may coarsen the inclusion enabling the size of them increasing which is not good for the performances of materials. Nevertheless, it was reported that Mg addition refined the size of inclusion [15–17], which may in turn make up for the shortcomings of Ce. Thus, study on the complex deoxidization concerning Ce and Mg is of great significance for the industrial production.

According to the principle of complex deoxidization, adding the deoxidizers in the order of weak to strong, the present paper studied the evolution of the inclusion, especially the composition of those nucleation ones in Ti–Mg complex deoxidized and Ce-treated C–Mn steel. The nucleation mechanism was discussed in terms of the observed results by FE-SEM equipped with EDS as well as calculation results.

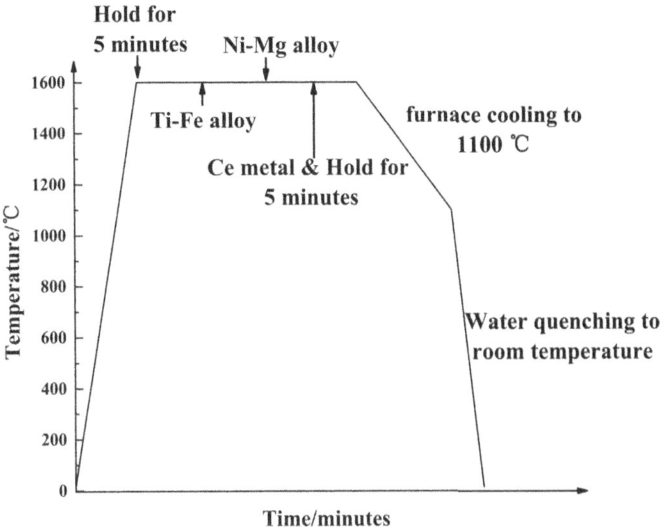

Fig. 1 Process of alloys addition

Experimental

Molybdenum furnace with accuracy of ±2 °C was used to melt the samples. During the whole of process of melting, argon gas was introduced to protect steel from oxidation. The alloys of Ti–Fe, Ni–Mg and Ce metal were inserted into the melt after the temperature rose to 1600 °C. Figure 1 shows the process of alloys addition. The interval time was 5 min between alloys so that the alloy elements could become homogenization. The corundum crucible as well as sample was quenched in water after cooled to 1100 °C in furnace. Samples were cut from a half of the radius of the tested steels to observe inclusions and microstructure and analysed chemical composition of them. The chemical composition of tested steels was listed in Table 1.

Table 1 Chemical composition of the experimental steels/wt%

Heat	C[a]	Si[b]	Mn[b]	P[b]	S[a]	N[d]	Al[b]	O[a]	Ti[b]	Mg[b]	Ce[c]
TM-B	0.17	0.37	1.35	0.03	0.031	0.0027	0.007	0.0052	0.014	0.0035	0
C-1	0.17	0.37	1.35	0.03	0.031	0.0027	0.0066	0.0049	0.014	0.0036	0.0059
C-2	0.17	0.37	1.35	0.03	0.031	0.0027	0.007	0.0057	0.014	0.0034	0.023
C-3	0.17	0.37	1.35	0.03	0.031	0.0027	0.007	0.0048	0.014	0.0034	0.036

[a]Infrared spectroscope (IR)
[b]Inductively couple plasma atomic emission spectrometric (ICP-AES)
[c]Inductively couple plasma mass spectrometry (ICP-MS)
[d]Inert gas fusion thermal conductivity method (IGF-TC)

The mass fraction of potential products was calculated by means of FactSage software based on the analysis results in Table 1. And the Equilib-Scheil cooling model as well as the FToxide and FSstel databases were used for the calculation. FE-SEM and EDS were used to test the elements distribution and observe the morphology of nucleation inclusions as well as the microstructure of samples. Finally, line scanning analysis and lattice mismatch calculation were carried out to illustrate the mechanism of nucleation.

Results and Discussion

The Formation Thermodynamics of Inclusions in Ce-Treated and Ti–Mg Complex Deoxidized C–Mn Steel

The main constitution of inclusion in Ti–Mg deoxidized steel was MnS, $MgAl_2O_4$, MgO, TiC, TiN and $MgTi_2O_4$, which was consistent with previous literature [17–19]. $AlCeO_3$ and Ce_2O_2S appeared after Ce-treated, while both formation temperatures and mass fraction of $MgAl_2O_4$ and $MgTi_2O_4$ decreased sharply with the increase of Ce. MnS was formed at the end of solidification of liquid iron though Mn and S elements existed in various phases or inclusions at high temperatures. The results suggested that it was when the Ce content increased to 0.023% that Ce played a role on desulphurization for MnS decreased significantly and Ce_2O_2S became the dominant inclusion containing Ce. On the contrary, Ti(C, N) changed little with the content of Ce (Fig. 2).

The Composition of the Nucleation Inclusions in Tested Steels

As illustrated in Fig. 3, the morphology of nucleation inclusion in tested steels was spherical with multilayers. Meanwhile, the results of spectrum suggested that elements distribution varied both in a single inclusion and among the inclusions. The domain composition of inclusion in Ti–Mg deoxidized sample was Mg–Al–Ti–O complex inclusions with MnS coating the surface. Once Ce was added into the steel with the range of 0.0059–0.023%, it reacted with O and S immediately, and thus, the type of inclusion after Ce treatment became Mg–Al–Ti–Ce–O–S complex inclusion on the condition of the low content of Ce and Mg–Ce–O–S complex inclusion in high Ce content, respectively. It was the same case with the sample with Ce content as high as 0.036% except that Mg content in the inclusions decreased sharply. Moreover, it indicated that Ce segregated in certain part of inclusions where the corresponding grey value got maximum just as Fig. 3c, d showed, which was due to a large number of electrons existing in Ce element gotten

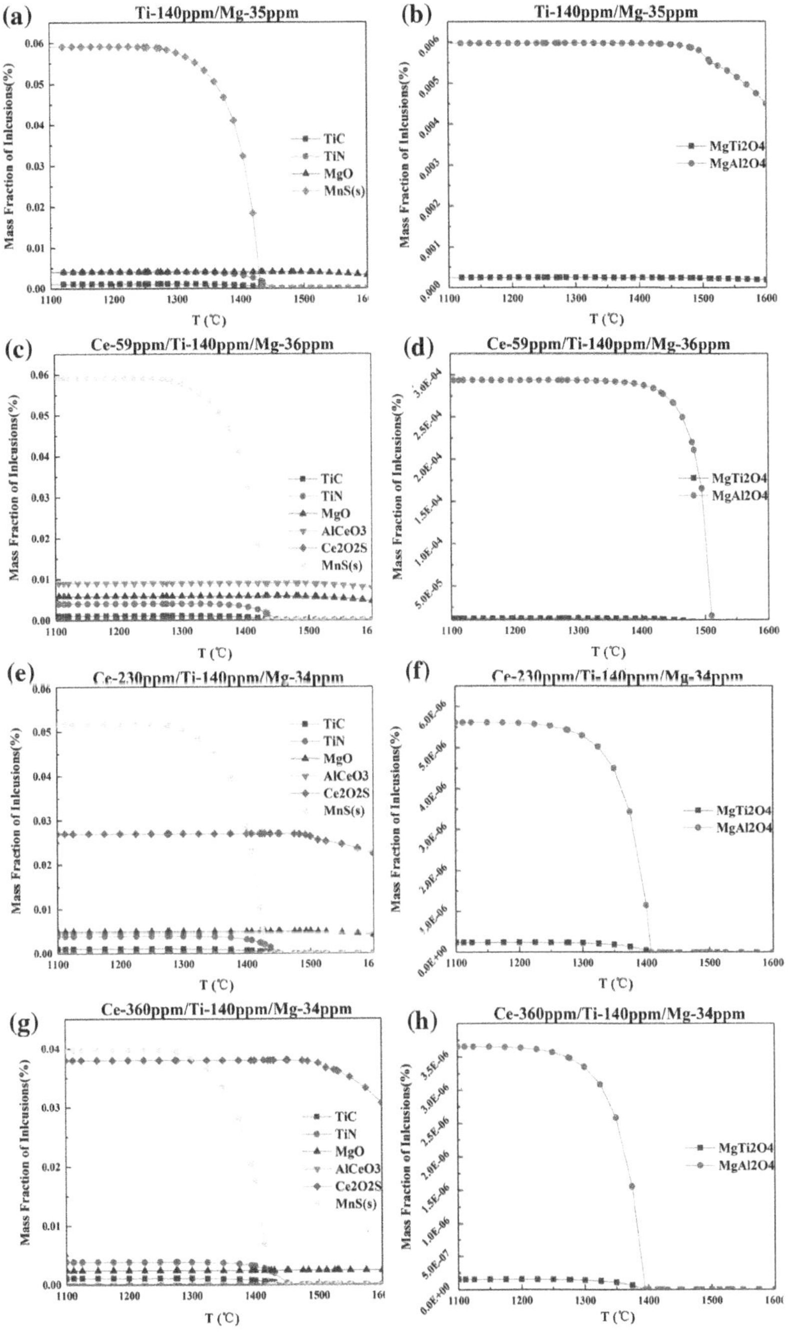

Fig. 2 Calculation results of inclusions in Ce-treated Ti–Mg complex deoxidized C–Mn steel by FactSage, **a** and **b** TM-B, **c** and **d** C-1, **e** and **f** C-2, **g** and **h** C-3

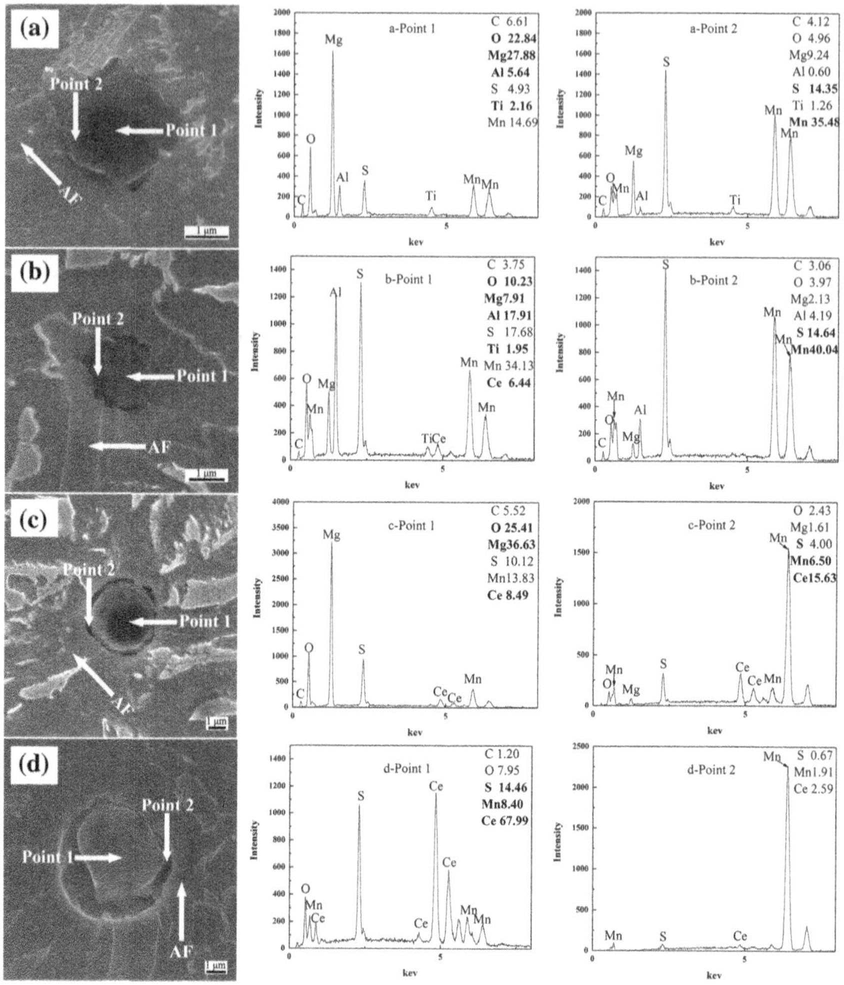

Fig. 3 Morphologies and energy spectrums of nucleation inclusions in the tested steels **a** TM-B, **b** C-1, **c** C-2, **d** C-3

provoked. The phenomenon indicated that the Ce element was unevenly distributed inside the inclusions.

Figure 4 shows the changes of Ti, Mg, Ce contents in inclusions of different samples, based on the data of EDS. More than 40 inclusions were characterized. According to the distribution of data, it could be indicated that more Mg and Ce instead of Ti were composed of the inclusion. This was connected with the relative activity of three elements reacting with oxygen. At the same time, it could also be seen that the more Ce content is in the tested steels, the more Ce content is in the inclusions. Combined with the results of calculation mentioned above, the transformation of the composition for inclusion was in such manner:

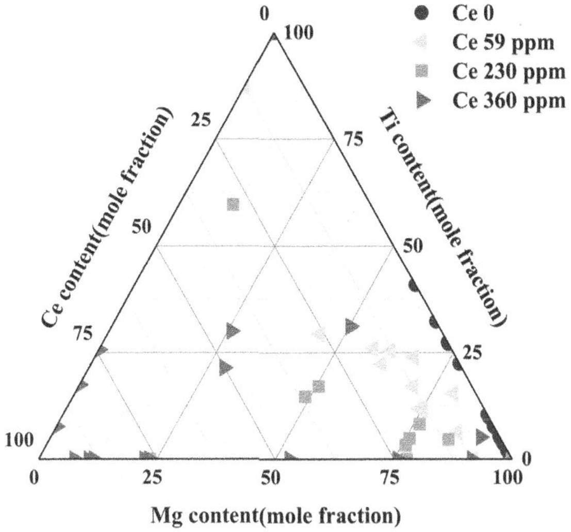

Fig. 4 Ternary phase diagram of the composition of inclusions

$MgO + MgAl_2O_4 + MgTi_2O_4$ to $MgO + MgAl_2O_4 + MgTi_2O_4 + AlCeO_3$ to $MgO + Ce_2O_2S$ finally to Ce_2O_2S with MnS adhering to their surface all the time.

The Nucleation Mechanism of Inclusion in Experimental Steels

According to the concept of oxide metallurgy, inclusions could nucleate acicular ferrite as heterogeneous nucleation sites during the process of solidification and the microstructure would be controlled eventually. In this way, there is hope that microstructure could be optimized by optimizing the characteristics of inclusions. As mentioned above, Ce modified the composition of inclusions greatly. Figure 5 shows the nucleation phenomenon of inclusions modified by Ce. According to the figure, inclusions in both the TM-B and C-1 to C-3 played a part on nucleating acicular ferrite during solidification. However, there were some differences between them. On the one hand, the size of nucleation inclusions in TM-B was smaller than other samples with Ce addition, the similar conclusion having been drawn by Zhang et al. [20]. On the other hand, the average number of acicular ferrite laths nucleated on a single inclusion was least in TM-B with a value of approximately 2, while the number increased significantly after Ce treatment as shown in Fig. 6. It could be concluded that the optimum of Ce content was about 0.023% in terms of the nucleation activity of inclusions.

Detailed information accounting for AF nucleation on the inclusions was obtained by literature and experimental results in terms of two popular nucleation mechanisms: lattice mismatch mechanism and MDZ mechanism. Table 2 is the

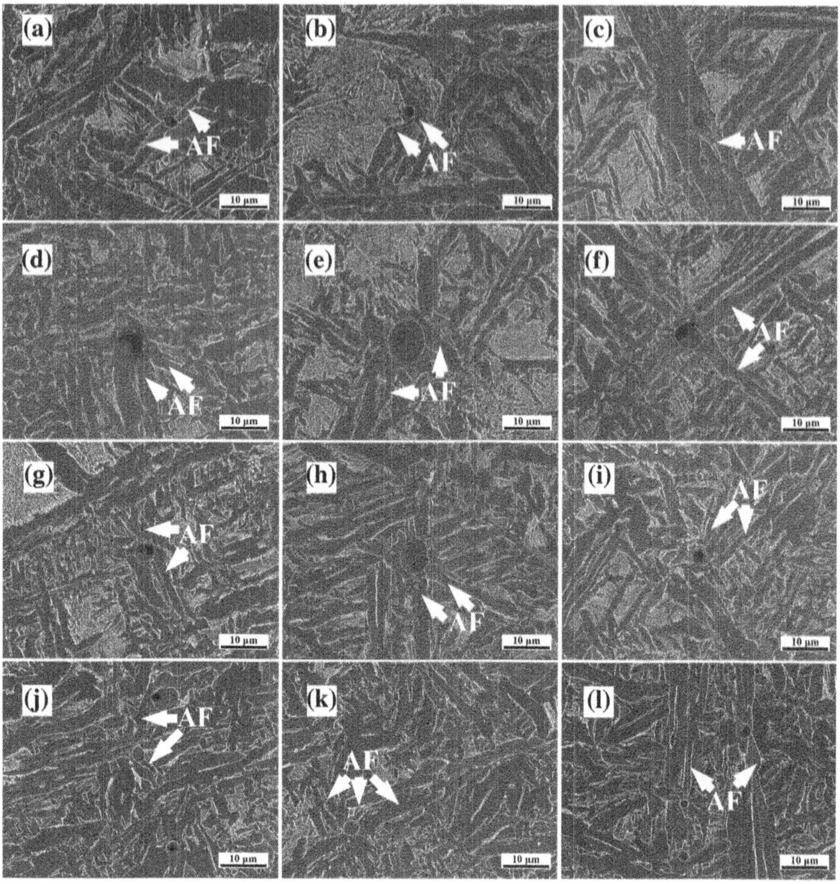

Fig. 5 Morphology of acicular ferrite(AF) nucleated on inclusions in steels **a**, **b** and **c** TM-B; **d**, **e** and **f** C-1; **g**, **h** and **i** C-2; **j**, **k** and **i** C-3

lattice mismatch between the inclusion and α-Fe. Generally speaking, the smaller the value is, the higher the nucleation activity of the inclusion is, and the critical value of lattice mismatch is 6 [21]. Thus, it could be seen that $MgAl_2O_4$, $CeAlO_3$, Ce_2O_2S could be regarded as potential nucleation sites. When it came to MnS, its formation consumed Mn in the matrix, enabling MDZ formed due to low transformation coefficient of Mn in steel matrix. Manganese was known as austenite stable element, and then, the transformation from austenite to ferrite surrounding the inclusions would take place in higher temperature. The MDZ, therefore, contributed to the nucleation of acicular ferrite. Figure 7 shows line scanning results of inclusions in Ce-treated and Ti–Mg deoxidized steels. It suggested that the concentration of Mn fluctuated more drastically around inclusion of Ce-treated samples compared with those in Ti–Mg deoxidized one. The fluctuation of Mn

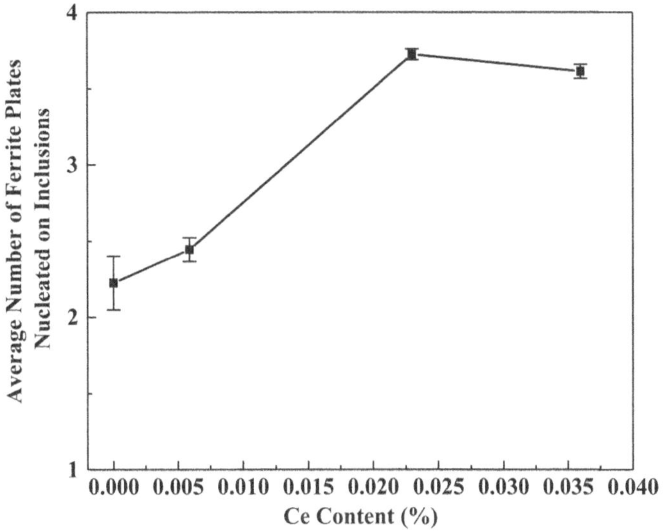

Fig. 6 Average number of ferrite plates nucleated on active inclusions with the increase of Ce

Table 2 Lattice mismatch between the inclusion and α-Fe

Inclusion type	Lattice mismatch	References
$MgAl_2O_4$	0.6	[22]
$CeAlO_3$	7.0	[23]
Ce_2O_2S	1.2	[23]
MnS	8.8	[24]

concentration enables inclusions containing Ce more active in nucleating acicular ferrite. Under the same conditions, tested steel named C-3 was, exceptionally, unable to detect the MDZ around the inclusion, but its main inclusions Ce_2O_2S had

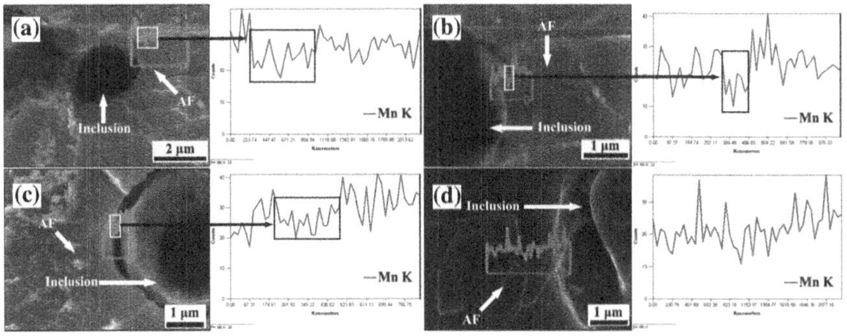

Fig. 7 Line scanning results of nucleation inclusion in the experimental steels **a** TM-B, **b** C-1, **c** C-2, **d** C-3

low lattice mismatch with α-Fe, 1.2, which indicated that lattice mismatch dominated in the condition of high Ce content when it was as high as 0.036%.

Conclusion

1. FactSage software calculation results suggested that Ce treatment modified $MgAl_2O_4$ and $MgTi_2O_4$ significantly and played a part on desulphurization with its content reaching 0.023 wt%. Accordingly, inclusions containing Ce changed from $CeAlO_3$ to Ce_2O_2S.
2. The observation results of SEM + EDS showed that Ce segregated into the centre of inclusions, while Ti and Al decreased sharply on the same condition with the increase of Ce content. MnS occupied the surface of them despite its content decrease responding to Ce treatment. The observation results were consistent with the calculated ones.
3. MDZ accounted for nucleation of acicular ferrite on the active inclusions under low Ce condition, while it was mainly lattice mismatch that dominated this activity of inclusions when Ce content was as high as 0.036 wt%.

Acknowledgements The authors would like to gratefully acknowledge the financial support from the National Nature Science Foundation of China (No. 51774024).

References

1. Takemura JI, Mizoguchi S (1990) Roles of oxides in steel performance. Paper presented at the proceeding of the 6th international iron and steel congress. ISIJ, Nagoya, 20–22 Mar 1990
2. Mizoguchi S, Takamura J (1990) Control of oxides as inoculants. Paper presented at the proceeding of the 6th international iron and steel congress. ISIJ, Nagoya, 20–22 Mar 1990
3. Huang Q, Wang XH, Jiang M, Hu ZY, Yang CW (2016) Effects of Ti–Al complex deoxidization inclusions on nucleation of intragranular acicular ferrite in C–Mn steel. Steel Res Int 87(4):445–455
4. Luo X, Niu YW, Chen XH, Tang H, Wang ZD (2017) High performance in base metal and CGHAZ for ferrite-pearlite steels. J Mater Process Technol 242:101–109
5. Nako H, Okazaki Y, Speer JG (2015) Acicular ferrite formation on Ti-rare earth metal-Zr complex oxides. ISIJ Int 55(1):250–256
6. Lee J, Pan Y (1992) Effect of silicon content on microstructure and toughness of simulated heat affected zone in titanium killed steels. Mater Sci Technol 8:236–244
7. Zhang TS, Li YD, Mu HY, Liu CJ, Jiang MF (2017) Influence of Mg/Ti complex addition on evolution and thermodynamics of the inclusion in Al-killed steel. Metall Res Technol 114(3):1–8
8. Song MM, Song B, Hu CL, Xin WB, Song GY (2015) Formation of acicular ferrite in Mg treated Ti-bearing C–Mn steel. ISIJ Int 55(7):1468–1473
9. Waudby PE (1978) Rare earth additions to steel. Int Mater Rev 23(1):74–98

10. Torkamani H, Raygan S, Garcia Mateo C, Rassizadehghani J, Palizdar Y, San-Martin D (2018) Contributions of rare earth element (La, Ce) addition to the impact toughness of low carbon cast niobium microalloyed steels. Met Mater Int 24(4):773–788
11. Wang HP, Xiong L, Zhang L, Wang Y, Shu YY, Zhou YH (2017) Investigation of RE–O–S–As inclusions in high carbon steels. Metall Mater Trans B 48(6):2849–2858
12. Yan N, Yu S, Chen Y (2017) In situ observation of austenite grain growth and transformation temperature in coarse grain heat affected zone of Ce-alloyed weld metal. J Rare Earths 35 (2):203–210
13. Huang YR, Jin X, Cai GJ (2017) Evolution of microstructure and mechanical properties of a new high strength steel containing Ce element. J Mater Res 32(20):3894–3903
14. Zhang S, Yu Y, Wang S, Li H (2017) Effects of cerium addition on solidification structure and mechanical properties of 434 ferritic stainless steel. J Rare Earths 35(5):518–524
15. Wang Y, Han YH, Zhu LG, Zhang QJ, Wang SM, Zhang CJ (2017) Effect of microalloy elements V and Mg on organization at high heat input welding shipbuilding structure steel. Paper presented at the 8th international symposium on high-temperature metallurgical processing. San Diego, California, 26 Feb–3 Mar 2017
16. Wang LZ, Yang SF, Li JS, Zhang S, Ju JT (2017) Effect of Mg addition on the refinement and homogenized distribution of inclusions in steel with different Al contents. Metall Mater Trans B 48(2):805–818
17. Kong H, Lin H, Yue Q, Cai ZY (2017) Effects of Ti–Mg complex inclusions on acicular ferrite nucleation. High Temp Mater Process Lond 36(5):459–465
18. Zheng W, Fu XH, Qu Y, Xiong S, Li GQ, Qin QW (2017) Inclusion-induced pitting corrosion behaviors in Al–Mg/Ti–Mg deoxidized steel. J Wuhan Univ Sci Technol 40(2):81–87
19. Ono H, Nakajima K, Maruo R, Agawa S, Usui T (2009) Formation conditions of Mg_2TiO_4 and $MgAl_2O_4$ in Ti–Mg–Al complex deoxidation of molten iron. ISIJ Int 49(7):957–964
20. Zhang TS, Wang DY, Jiang MF (2014) Effect of magnesium on evolution of oxide and sulfide in liquid iron at 1873 K. J Iron Steel Res Int 21(12):1073–1080
21. Bramfitt LB (1970) The effect of carbide and nitride additions on the heterogeneous nucleation behavior of liquid iron. Metall Trans 1(7):1987–1995
22. Wen B, Song B, Pan N, Hu QY, Mao JH (2011) Effect of SiMg alloy on inclusions and microstructures of 16Mn steel. Ironmaking Steelmaking 38(8):577–583
23. Wen B (2011) Fundamental research on application of cerium and magnesium in oxides metallurgy. Ph.D. thesis. University of Science and Technology, Beijing
24. Mills AR, Thewlis G, Whiteman JA (1987) Nature of inclusions in steel weld metals and their influence on formation of acicular ferrite. Mater Sci Technol U K 3(12):1051–1061

Effect of La Addition on Inclusions, Microstructures, and High-Temperature Mechanical Properties of As-Cast FeCrAl Alloys

Yang He, Jianhua Liu, Yindong Yang and Alex McLean

Abstract The effects of La addition on the inclusions, microstructures, and high-temperature mechanical properties of as-cast FeCrAl alloys were investigated by using scanning electron microscope (SEM), energy disperse spectroscopy (EDS), and hot tensile tests. The characteristics of inclusions were examined, and further discussions focused on the analysis of thermodynamic calculations. The results showed that after La addition, AlN inclusions were likely to have cores of La-containing particles. Under the conditions of this study, La_2O_2S and La_2O_3 inclusions were the main La-containing precipitates and could act as the nucleus for the heterogeneous nucleation of δ-Fe to refine grains. The microstructure comparison of the two studied alloy ingots indicated that both the columnar and the equiaxed grains were refined after La addition. A series of tensile tests were carried out at temperatures from 973 to 1673 K under a constant strain rate of 1×10^{-3} s^{-1} to investigate the high-temperature mechanical properties. Both the tensile stress and the hot ductility were improved at the studied temperatures after La addition.

Keywords FeCrAl alloys · La addition · Inclusions · Microstructures Mechanical properties

Introduction

FeCrAl alloys are used as the material for the metal substrates of automobile exhaust gas purifying systems due to their high-temperature oxidation and corrosion resistance [1]. The thickness of alloys for this application is approximately

Y. He (✉) · J. Liu
Institute of Engineering Technology, University of Science and Technology Beijing, Beijing 100083, People's Republic of China
e-mail: ustbheyang@163.com

Y. He · Y. Yang · A. McLean
Department of Materials Science and Engineering, University of Toronto, Toronto M5S 3E4, Canada

© The Minerals, Metals & Materials Society 2019

T. Jiang et al. (eds.), *10th International Symposium on High-Temperature Metallurgical Processing*, The Minerals, Metals & Materials Series,
https://doi.org/10.1007/978-3-030-05955-2_40

50 μm, which demands superior microstructures and mechanical properties [2]. However, the solidified structures of these alloys have well-developed columnar grain regions and coarse ferritic grains because these alloys have narrow freezing ranges and lack transitions from ferrite to austenite during solidification [3]. Our previous study reported that the ferrite grain coarsening is more prone to occur at high temperatures and seriously decreases the hot ductility of these alloys [4]. The occurrence of dynamic recovery and dynamic recrystallization in the FeCrAl alloys, with high stacking fault energy, was beneficial for the grain refinement and increased the hot ductility during hot deformation. Since this phenomenon only occurs under certain conditions, such as at certain temperatures and at relatively low strain rates, it is of great value to develop the optimization of as-cast microstructure for better mechanical properties. The alloying of lanthanum (La) has recently been applied in steel and alloy production [5]. Through its grain refinement and purification effect, La can enhance the strength, plasticity, and toughness of steels. Some researchers have studied the effect of La modification on the stability of coating alumina layer on FeCrAl alloys [6, 7]. However, the knowledge of La addition on the solidification characteristics and properties of these alloys is still needed. The present work aims to study the effect of La addition on the inclusions, microstructures, and high-temperature mechanical properties in as-cast FeCrAl alloys and provides knowledge about the mechanism of developing the microstructure optimization of FeCrAl alloys by La addition for better mechanical properties.

Materials and Methods

Two experimental FeCrAl alloy ingots were produced by a 25 kg capacity vacuum induction melting furnace using commercial purity materials. Table 1 shows the chemical compositions of the two studies alloy ingots. After the alloying process, the longitudinal and the transverse sections of the ingots were polished and etched with a solution consisting of 20 g $FeCl_3$, 30 mL HCl, and 100 mL distilled water to investigate the solidified microstructures. Scanning electron microscopy (SEM) and energy dispersive spectrometer (EDS) were utilized to determine the inclusions on the polished cube specimens that were cut from the inner and the outer of the two ingots and with the side length of 10 mm.

Hot tensile tests were carried out using a Gleeble-3500 facility to evaluate the high-temperature mechanical properties. The cylindrical tensile specimens were cut

Table 1 Chemical compositions of the studied FeCrAl alloy ingots (in wt%)

	C	Cr	Al	Si	Mn	S	N	O	La	Fe
Alloy1	0.028	19.96	5.08	0.11	0.13	0.0046	0.0062	0.0022	–	Bal.
Alloy2	0.025	19.92	5.07	0.13	0.16	0.0041	0.0054	0.0018	0.052	Bal.

from the columnar grain regions of the two ingots by the electric spark method. The specimens were isothermally tested at a temperature interval of 50 K in the range between 973 and 1673 K, at a constant true strain rate of 1×10^{-3} s^{-1}, which is typical of unbending operations during continuous casting. All the tests were performed under an argon atmosphere to avoid specimen oxidation. Specimens were first heated to 1573 K at a rate of 10 K/s and held for 180 s to homogenize the microstructure. Specimens were then heated or cooled to the test temperature at a rate of 5 K/s and held again for 120 s. Finally, the specimens were strained until fracture and cooled down immediately after rupture under a stream of argon. The reductions in area were measured as a function of deformation temperatures to evaluate the hot ductility of the studied alloys.

Results and Discussion

Inclusions

Inclusions significantly affect the microstructure and mechanical properties of FeCrAl alloys. According to SEM observation and EDS spectrum, AlN inclusions are the dominant precipitates in Alloy1 without La addition. After La addition, AlN inclusions still exist in Alloy2, but they are likely to have cores of La-containing particles. Figure 1 shows the SEM micrographs of typical duplex inclusions in Alloy2. These La-containing particles seem to act as the nucleus for the heterogeneous nucleation of AlN inclusions, which implies that La addition is beneficial to AlN formation. However, the number density of AlN inclusions is basically similar before and after La addition. Yue and Du [8] have reported that rare earth could increase the solubility of N in liquid iron. Thus, La addition should have complicated effects on AlN formation. AlN inclusions are not effective nucleus for heterogeneous nucleation of δ-Fe and may initiate cracks during mechanical processing to deteriorate mechanical properties. Nevertheless, the size of the AlN inclusions in Alloy2 is generally smaller than those in Alloy1. The smaller AlN inclusions after La addition can reduce the risk of the deterioration of mechanical properties.

La is easy to form La–S–O precipitates in melts due to its strong affinities with O and S. LaS, La_2S_3, La_3S_4, La_2O_2S, and La_2O_3 have been reported as possible precipitates in molten steel [9]. According to SEM observation and EDS spectrum, in Alloy2, La_2O_2S and La_2O_3 inclusions are the dominant precipitates after La addition; however, no La–S inclusions are found. The predominant formations of La–S–O precipitates are strongly affected by the activity of O and S [10, 11]. The standard free energies of these La–O–S precipitates that are calculated for varying activities of S and O allows prediction of the predominant formations of these precipitates in molten steel [9, 11]. Based on the inclusion observation and thermodynamic calculation for Alloy2, the possible precipitation path of La–S–O

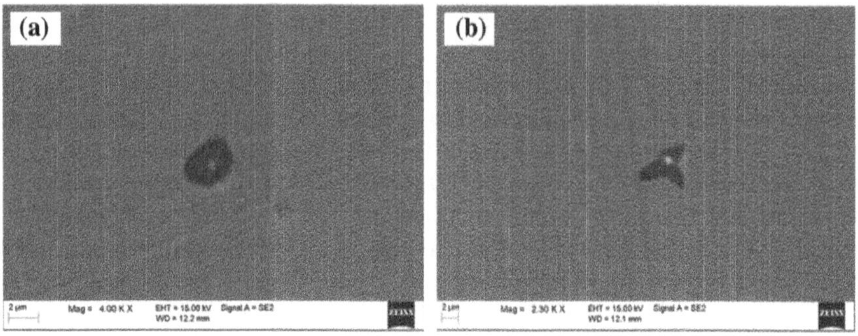

Fig. 1 SEM micrographs of typical complex inclusions with La-containing cores covered by AlN in Alloy2

precipitates can be predicted in the FeCrAl molten alloy. Figure 2 shows the predominant stable phase fields and the possible projected precipitation path for La–S–O reaction products at 1873 K. At the first period of La alloying, the O activity in the molten alloy was in the predominant stable phase field of La_2O_3. During the La_2O_3 formation, the O activity continuously decreases and then reaches the predominant stable phase field of La_2O_2S. Both the activities of O and S decrease during the La_2O_2S formation. Considering that no La sulphides were observed in Alloy2, the experimental melting condition of this study might fail to satisfy the formation condition of La sulphides, which requires extremely low O activity and relatively high S activity.

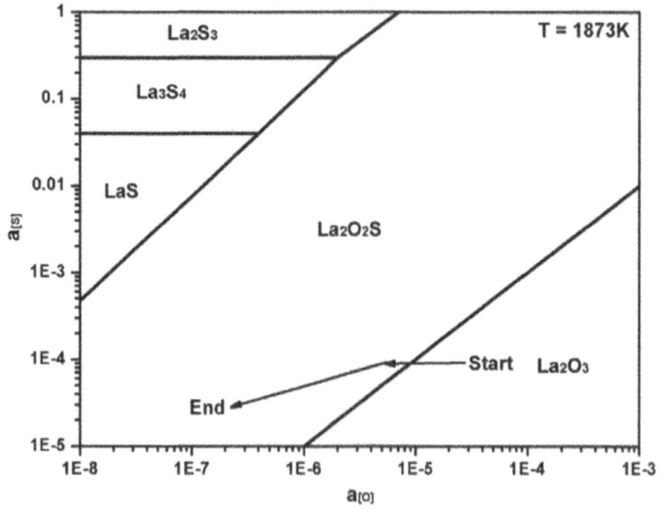

Fig. 2 Stable phase fields and the projected precipitation path of La–S–O precipitates at 1873 K

During the melting process, La_2O_2S and La_2O_3 primarily precipitate in the molten alloy due to their high melting points and La's strong affinities with O and S. In general, precipitates are more likely to be the nucleus for substrates when the planar disregistries between the precipitates and the substrates are small [12]. Both La_2O_2S and La_2O_3 have small disregistries with δ-Fe and can act as the nucleus for heterogeneous of δ-Fe to refine grains [13]. In this way, La addition is expected to optimize the as-cast microstructure of FeCrAl alloys for better mechanical properties.

Solidified Structures

Figure 3 shows the solidified microstructures of the longitudinal and the transverse sections of the two alloy ingots. It is obvious that both the columnar and the equiaxed grains are refined in Alloy2 after La addition, which agrees reasonably with the increase of the active nucleation site rate in FeCrAl alloys after La addition. Moreover, the structures of the grain boundaries in the two alloys are different: straight grain boundaries in Alloy1 but zigzag grain boundaries in Alloy2. According to the elemental mapping of the grain boundaries in Alloy2 by SEM and EDS, as shown in Fig. 4, it is observed that the La enriches at the grain boundaries. The rare earth metals are soluble with molten steel, but they have extremely low distribution coefficients in iron-base solid solution [14]. Thus, during the solidification process of molten steel, the rare earth metals are very likely to concentrate at the solidifying fronts, which not only induces the constitutional supercooling to promote nucleation ratio and grain refinement, but also purifies the impurity elements, such as C and N, to reduce the formation of carbides and nitrides. The zigzag grain boundaries are mostly due to the pinning effect by the La enrichment at grain boundaries. The pinning effect can hinder the grain growth to avoid the grain coarsening at high temperatures.

Mechanical Properties

The microstructures will largely affect the high-temperature mechanical properties of the alloys. Figure 5a shows the tensile stress of the two alloys with varying temperature. The tensile stress decreases continuously with increasing temperature; however, the tensile stress of Alloy2 is larger than that of Alloy1, especially below 1273 K. Based on the study of the microstructures of the two studied alloys, the improvement of tensile stress by La addition is mostly due to the grain refinement and the grain boundary strengthening. However, the solid solution strengthen of La is not considered as one of the main reasons for the improvement of the tensile stress because La has an extremely low distribution coefficient in iron-base solid solution.

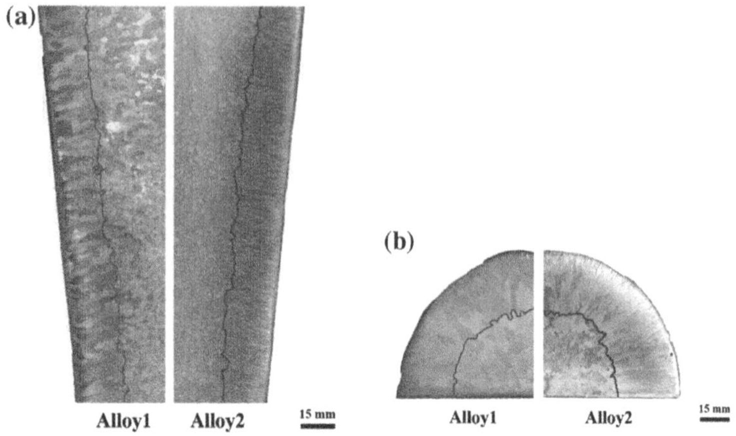

Fig. 3 Solidified microstructure of different ingots: **a** longitudinal sections; **b** transverse sections

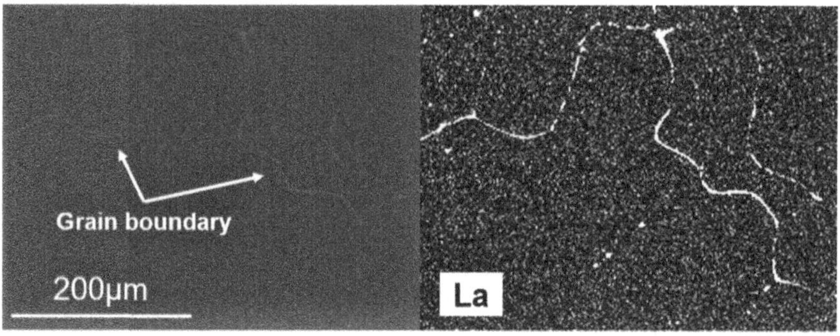

Fig. 4 SEM image and EDS mapping diagram of grain boundaries in Alloy2

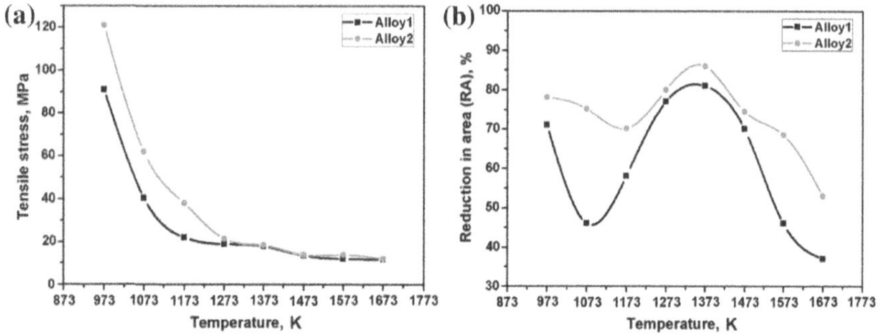

Fig. 5 Peak stress (**a**) and hot ductility (**b**) as a function of temperature for the different alloys

Fig. 6 Morphologies of fractures for the two alloys after tensile test

Figure 6 shows the morphologies of fractures for the two alloys after tensile test. The reductions in area were measured as a function of deformation temperatures to evaluate the hot ductility. Figure 5b shows the variations of the area reduction versus temperature for the two studied alloys. The hot ductility of the alloys mostly depends on both the temperatures and the La addition. For Alloy1, the reductions in area decrease from 973 K to a minimum at around 1073–1173 K, then sharply increase to a maximum at around 1373 K before decreasing again. Temperature ranges of 1073–1173 K and above 1473 K clearly represent ranges of low ductility. Our previous study [4] indicated that the ranges of low ductility resulted from the precipitation of carbides and the coarsening of the ferrite grains. However, for Alloy2, the hot ductility between 1073 and 1173 K is obviously improved. Since the precipitation of carbides has been considered as one of the main causes of the low ductility between 1073 and 1173 K, the purification effect of La is beneficial to improve the hot ductility by hindering the formation of carbides. The hot ductility above 1473 K is also improved, although it still drops at this temperature range. Due to no ferrite–austenite transition existing in FeCrAl alloys, the ferrite grain coarsening is more prone to occur at high temperatures to reduce hot ductility. However, La addition can hinder the grain growth at high temperatures by pinning effect at grain boundaries, which can be reflected by the much slower drop of the hot ductility above 1473 K. Overall, La addition has an advantage in improving the high-temperature mechanical properties in FeCrAl alloys.

Conclusions

(1) After La addition, AlN inclusions were likely to have cores of La-containing particles. La_2O_2S and La_2O_3 inclusions are the dominant precipitates under the conditions of this study. They have small lattice mismatch values with δ-Fe and can act as the nucleus for heterogeneous nucleation of δ-Fe to refine grains.
(2) The solidified structure of the FeCrAl alloy was obviously refined after La addition. The zigzag grain boundaries appeared after La addition and were mostly due to the pinning effect by the La enrichment at grain boundaries.

(3) The improvement of tensile stress by La addition is mostly due to the grain refinement and the grain boundary strengthening. The purification effect of La is beneficial to improve the hot ductility between 1073 and 1173 K by hindering the formation of carbides. La addition can hinder the grain growth at high temperatures by pinning effect at grain boundaries.

Acknowledgements The authors gratefully acknowledge the financial support of this research by the National Science Foundation of China (Grant No 51374023).

References

1. He Y, Liu JH, Qiu ST, Deng ZQ, Yang YD, McLean A (2018) Interfacial morphology evolution in directional solidified FeCrAl alloys. Mater Charact 139:303–310
2. Kim DH, Yu BY, Cha PR, Yoon WY, Byun JY, Kim SH (2012) A study on FeCrAl foam as effective catalyst support under thermal and mechanical stresses. Surf Coat Technol 209:169–176
3. Han Z, Liu J, He Y, Li K, Ji Y, Liu J (2015) Determination of the liquidus and solidus temperatures of FeCrAl stainless steel. Int J Miner Metall Master 22(11):1141–1148
4. He Y, Liu JH, Qiu ST, Deng ZQ, Yang YD, McLean A (2018) Microstructure and high temperature mechanical properties of as-cast FeCrAl alloys. Mater Sci Eng A Struct 726:56–63
5. Smirnov LA, Rovnushkin VA, Oryshchenko AS, Kalinin GY, Milyuts VG (2016) Modification of steel and alloys with rare-earth elements. Part 2. Metallurgist 60(1–2):38–46
6. Ozawa M, Araki KI (2015) Effect of La modification on the stability of coating alumina layer on FeCrAl alloy substrate. Surf Coat Tech 271:80–86
7. Buscail H, Nguyen CT, Cueff R, Issartel C, Riffard F, Perrier S (2009) Effect of the La_2O_3 sol-gel coating on the alumina scale adherence on a model Fe-20Cr-5Al alloy at 1100 °C. J Mater Sci 44(15):3968–3974
8. Yue K, Du T (1989) Thermodynamical study on Fe–Ce–N–O, Fe–Y–N–S solutions. J Chin Rare Earth Soc 1:12–15
9. Vahed A, Kay DAR (1976) Thermodynamics of rare earths in steelmaking. Metall Trans B 78:375–383
10. Wilson WG, Kay DAR, Vahed A (1974) The use of thermodynamics and phase equilibria to predict the behavior of the rare earth elements in steel. JOM 26(5):14–23
11. Wang LM, Du T (1986) [La]–[S]–[O] equilibrium in liquid iron. Chin Rare Earth 2:27–34
12. Bramfitt BL (1970) The effect of carbide and nitride additions on the heterogeneous nucleation behavior of liquid iron. Metall Trans 1(7):1987–1995
13. Wang AQ, Li M, Ma DQ, Wu QJ, Xie JP (2016) Effect of lanthanum on microstructures and properties of ASTM A216 steel. Kem Ind 65(1–2):11–16
14. Wang LM, Ting D, Lu XL, Li ZB, Gai YC (2003) Thermodynamics and application of rare earth elements in steel. J Chin Rare Earth Soc 21(3):251–254

Removal of Copper from Fe–Cu Alloy by Using Iodine

Yuichi Takamatsu and Takashi Nagai

Abstract Impurities such as copper, tin, zinc are included in a steel scrap. It is needed to remove these impurities when the steel scraps are recycled, since the elements are harmful for mechanical characteristic and workability of some steels. However, copper cannot be removed by a traditional smelting by oxidation. In this study, it is investigated to remove copper from steel by using iodine. Iodine was added to molten Fe–Cu alloy at high temperature, Copper iodide (CuI_2) gas was formed, and the copper content in the alloy decreased. By this method, the copper in the alloy can be removed to gaseous phase easily. The relationships between removal rate of copper and experimental conditions, such as temperature, mass of iodine, and so on, were investigated.

Keywords Steel · Recycling · Copper · Iodine

Introduction

Steel is most important and imminent metal. The demand of steel in the world is increasing. Amount of wasted steel is also increasing, while it is produced by iron ore. Then, it is important to recycle of steel scraps. However, there is a problem of tramp elements in steel scrap recycling. The tramp elements, such as copper, tin, zinc, and so on, reduce the performance of steel, dramatically. Therefore, the elements should be removed in the recycling process.

As copper removal method from iron, evaporation refinement and flux refinement have been reported. In the evaporation refinement, copper is removed by difference of vapour pressure of copper and iron. Since vapour pressure of copper is

---

Y. Takamatsu · T. Nagai (✉)
Department of Mechanical Science and Engineering, The University of Chiba Institute of Technology, 2-17-1 Tsudanuma, Narashino, Chiba 275-0016, Japan
e-mail: takashi.nagai@it-chiba.ac.jp

© The Minerals, Metals & Materials Society 2019
T. Jiang et al. (eds.), *10th International Symposium on High-Temperature Metallurgical Processing*, The Minerals, Metals & Materials Series,
https://doi.org/10.1007/978-3-030-05955-2_41

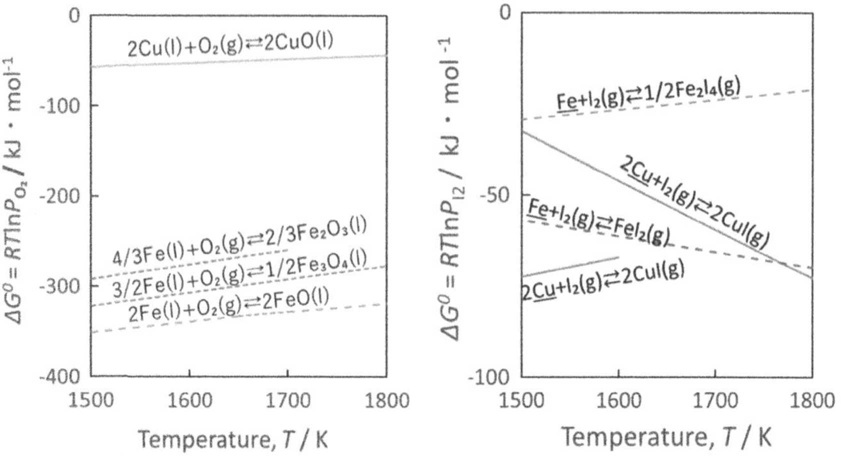

Fig. 1 Gibbs energy changes of the oxidization and iodination reaction

higher than that of iron, copper is removed to gas phase. However, high vacuum degree is needed to remove copper effectively [1–3].

In case of flux refinement, a flux containing FeS is used to remove copper. Copper is removed to flux as CuS or Cu_2S, but this process needs a large quantity of flux [4–12].

Gibbs energy changes of the oxidization and iodination reaction of iron and copper are shown in Fig. 1 [13]. It is difficult to remove the copper from Fe–Cu alloy by oxidization, since iron is oxidized before copper. However, there is possibility of removal of copper form Fe–Cu alloy, based on following reaction.

$$2\mathrm{Cu(in\ Iron)} + \mathrm{I_2(g)} = 2\mathrm{CuI(g)}$$

In this study, the removal of copper from Fe–Cu alloy by iodination was investigated.

Experimental

Preparation of Fe–Cu Alloy

Fe–Cu alloy was perpetrated by melting Iron (electrolytic iron), Copper (regent grade 99.9%), and Carbon powder (regent grade) in electro resistance furnace (Fig. 2) at 1673 K under Ar atmosphere. The copper content of the alloy was adjusted to 2 wt%.

The carbon-saturated iron with 2 wt% copper was perpetrated.

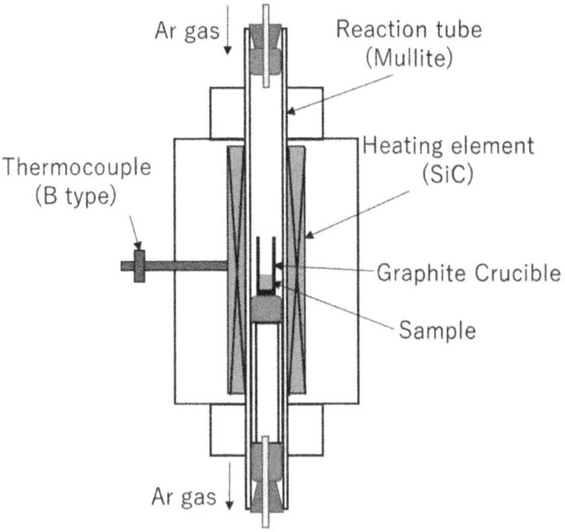

Fig. 2 Furnace used in experiment

Removal of Copper Form Fe–Cu Alloy

The sample (carbon-saturated iron with 2 wt% copper) of ca. 0.1 g set in small graphite crucible was melted in electron resistance furnace with injecting iodine system (Fig. 3) at 1473, 1523, or 1573 K in Ar atmosphere. After the inside of

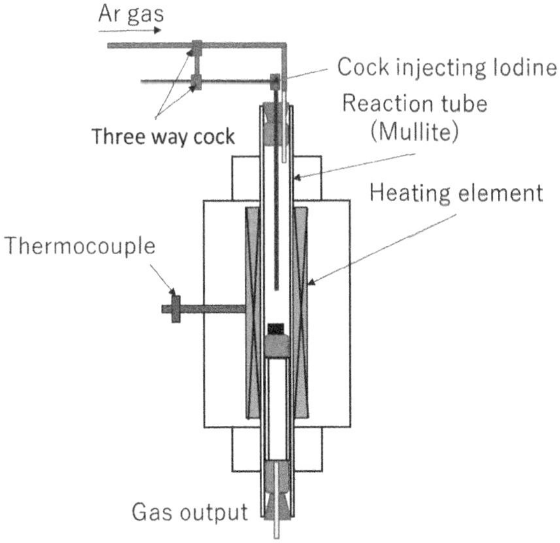

Fig. 3 Furnace with injecting iodine system

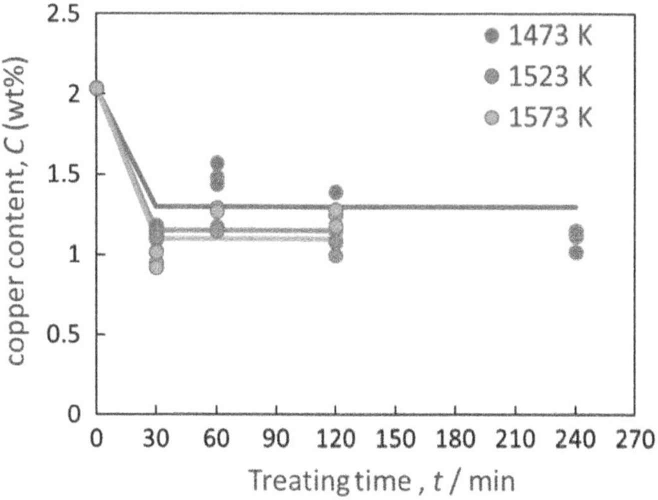

Fig. 4 Copper content in alloys after experiments

furnace achieved to experimental temperature, iodine powder was injected from upper side of the furnace, and heated for 30–240 min. About 0.3 g iodine powder was injected in each experiment.

After heating treatment, the sample was dissolved to nitric acid. The copper concatenation is measured by ICP-OES. Then copper content in the alloys after experiments was estimated.

Results and Discussion

The copper contents in sample after experiments and removal rate of copper by iodination are shown in Figs. 4 and 5, respectively. The copper content decreased to around 1 wt% from 2 wt% during 30 min treatment. The experiments for over 60 min, the copper content did not decrease from 1 wt%. It is thought the reaction of iodination of copper was finished within 30 min.

The remarkable difference by temperature between 1473 and 1573 K was not able to observe.

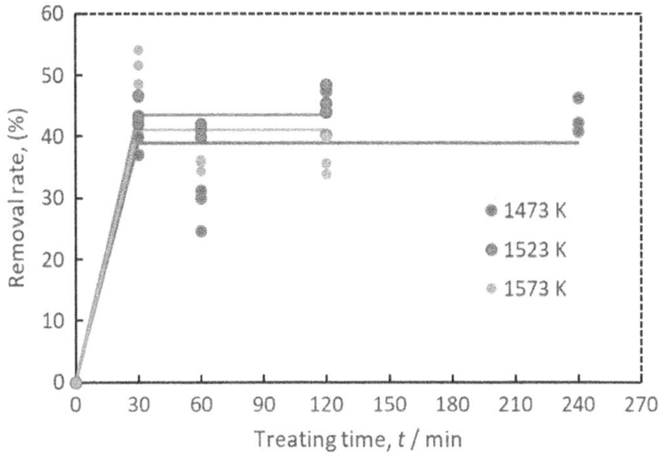

Fig. 5 Removal rate of copper

Conclusion

Iodination of copper is useful to removal of copper from Fe–Cu alloy.

The reaction of copper iodination is finished within 30 min from injection iodine to the furnace.

It is needed to achieve high removal rate.

References

1. Ward RG (1963) Evaporation of Mn and Cu from liquid iron. J Iron Steel Inst 201:11
2. Ohno R, Ishida T (1968) Rate of evaporation of manganese, copper, tin, chromium and sulphur from molten iron under vacuum. J Iron Steel Inst 206:904
3. Fischer WA, Janke D, Stahlschmidt K (1974) Vaporization of copper, manganese, and chromium from melts of the steel X 5 CrNi 18 9 under reduced pressure. Arch Eisen 45:361
4. Langenberg FC, Lindsay RW, Robertson DP (1955) Removal of copper from iron-copper-carbon melts by use of sodium sulfide slags. Blast Furnace Steel Plant 43:1142
5. Imai T, Sano N (1988) The copper partition between Na_2S bearing fluxes and carbon-saturated iron melts. Trans Iron Steel Inst Jpn 28:999
6. Wang C, Nagasaka T, Hino M, Ban-ya S (1991) Copper distribution between molten FeS-NaS0. 5 flux and carbon saturated iron melt. ISIJ Int 31:1300
7. Wang C, Nagasaka T, Hino M, Ban-ya S (1991) Copper distribution between FeS-Alkaline or-Alkaline earth metal sulfide fluxes and carbon saturated iron melt. ISIJ Int 31:1309
8. Liu X, Jeffes JHE (1986) The removal of tramp elements in steelmaking. In: 5th international iron and steel congress—process technology proceedings, vol 6. Washington, DC, p 759
9. Chen X, Nakashima K, Mori K (1996) Decopperizing rate from high carbon iron melt by Na_2S-FeS flux. In: CAMP-ISIJ, vol 9, p 684
10. Uchida Y, Matsui A, Kishimoto Y, Miki Y (2014) Removal of copper from molten iron with Na_2CO_3-FeS fluxes. Tetsu-to-Hagane 100:55

11. Takahashi Y, Nakase K, Kikuchi N, Uchida Y, Miki Y (2012) Development of a copper removal technique from molten iron with mechanical stirring. In: CAMP-ISIJ, vol 25, p 971
12. Hui K, Jianjun W, Shangxing G, Li Z, Jie L (2009) Copper removal from molten steel with FeS-Na_2S Slag. High Temp. Mater Process 28:67
13. Barin I (ed) (1989) Thermochemical data of pure substance. VCH Verlagsgesellschaft mbH, Weinheim

Part IX
Treatment and Recycling of Wastes

Effect of Chemical Composition on the Crystallization Behaviour of Rare Earth Phase in Slag

Tengfei Ma, Fu Feng, Xuefeng She, Jingsong Wang and Qingguo Xue

Abstract In order to figure out the effect of chemical composition on the crystallization behaviour of rare earth phase in rare earth slag, the quaternary system SiO_2–CaO–CaF_2–Ce_2O_3 slags were made by pure chemical reagent at 1673 K in muffle furnace. The mineral phases and micromorphology of slags were detected by the X-ray diffraction and scanning electron microscope (SEM), respectively. The results show that grains of RE-containing mineral grow very well and the equivalent diameter of grains is about 100 μm when the basicity is about 1. The effect of CaF_2 on the crystallization of the rare earth phase is complex. With the increase of basicity, the crystallization of the rare earth phase is increasing and the RE-concentrating phase exists in a rod-like structure. But when the alkalinity is up to 2.0, the distribution of RE-concentrating phase becomes dispersive and the crystals of rare earth phase have no definite shape. The major minerals are fluorite, cuspidine, RE-concentrating phase and substrate phase.

Keywords Rare earth slag · SiO_2–CaO–CaF_2–Ce_2O_3 · Crystallization behaviour

Introduction

The Bayan Obo Mine in the Inner Mongolia region of North China is the largest iron-REE-niobium ore deposit of hydrothermal origin known in the world. It accounts for 35% rare earth resource of world's proven reserves and nearly 80% in China [1–3]. Due to the special properties, rare earths are essential for modern technology [4–6]. Unfortunately, since it is mined mainly as an iron ore, the recovery rate of rare earth elements is only about 10%. As the fact that rare earth element can't be reduced by carbon at blast furnace condition, nearly all the rare earths are enriched into the blast furnace slag. Unfortunately, the RE-containing

T. Ma · F. Feng · X. She (✉) · J. Wang · Q. Xue
State Key Laboratory of Advanced Metallurgy, University of Science and Technology Beijing, Beijing 100083, China
e-mail: shexuefeng@ustb.edu.cn

© The Minerals, Metals & Materials Society 2019

T. Jiang et al. (eds.), *10th International Symposium on High-Temperature Metallurgical Processing*, The Minerals, Metals & Materials Series,
https://doi.org/10.1007/978-3-030-05955-2_42

slag with 14pct RE_2O_3 was finally stacked in tailing dams which makes it very difficult for recycle. Therefore, the comprehensive utilization of rare earth in slag has become an urgent task for us. There are many ways to enrich and separate rare earth elements in slag, such as, mineral dressing based on gravity separation and magnetic separation, super-gravity separation and hydrometallurgy [7–11]. Although the latter two methods have a high recovery rate, there are still problems. At first, it is difficult for the super-gravity separation method to apply to industry; besides, there are serious environmental problems in hydrometallurgy procedure.

Ore dressing is widely used for mineral separation. Usually, the blast furnace slag will be slowly cooled for mineral recrystallization. Then, it is separated by gravity separation, magnetic separation and flotation and so on. Actually, the slow cooling recrystallization behaviour of rare earth-enriched slag is concerned with temperature, chemical composition, pressures and cooling rate. Although the cooling rate has been studied by many researchers, the effect of chemical composition on the crystallization behaviour of rare earth phase in slag is not clear. By researching the effect of chemical composition, it not only help us to find out the most suitable composition for the crystallization of rare earth phase, but also contribute to the construction of RE-containing slag phase diagram. As the lack of thermodynamic data of rare earth oxides, there is little information about RE-containing slag can be obtained in thermodynamic software. So it is necessary to obtain the thermodynamic data of rare earth in slag by the means of experiment.

In this study, the mineral phases and micromorphology of slags were detected by the X-ray diffraction and scanning electron microscope (SEM). The volume fraction and average grain size of RE-concentrating phase at different compositions were obtained by the image analyzer software. This study aims to providing reliable data support for the industrial application and theoretical study of rare earth slag.

Experimental

The Selection of Slag Composition

Many people have researched the crystallization behaviour of rare earth slag, and the chemical composition of rare earth slag is given in Table 1 [10–13]. As we can see, the main compositions are CaF_2, RE_2O_3, CaO and SiO_2, the sum of which has exceeded 80%. To simplify experiment process, the CaF_2–Re_2O_3–CaO–SiO_2 slag was chosen for the experiment. As listed in Table 2, the RE_2O_3 content in the slag is fixed as 15%, while the basicity changes between 0.5 and 2.0 and the mass percentage of CaF_2 covers from 15 to 35%.

Table 1 Chemical composition of rare earth-rich slag of predecessors

Researchers	CaF_2	RE_2O_3	CaO	SiO_2	Fe	Al_2O_3	MgO	P_2O_5	R
Xu X.Y	30.38	11.58	18.49	17.12	0.48	5.60	1.07	–	1.08
Ding Y.G	36.82	14.19	18.19	18.01	1.61	1.72	1.42	2.75	1.01
Li D.G	18.67	10.70	32.20	18.37	–	5.18	3.68	–	1.75

Table 2 Chemical composition of rare earth-rich slag

No.	CaF_2 (%)	CaO (%)	SiO_2 (%)	RE_2O $(\%)_3$	R
1	15	23.3	46.6	15	0.5
2	20	21.7	43.4	15	0.5
3	25	20	40	15	0.5
4	30	18.3	36.6	15	0.5
5	35	16.7	33.3	15	0.5
6	15	35	35	15	1
7	20	32.5	32.5	15	1
8	25	30	30	15	1
9	30	27.5	27.5	15	1
10	35	25	25	15	1
11	15	42	28	15	1.5
12	20	39	26	15	1.5
13	25	36	24	15	1.5
14	30	33	22	15	1.5
15	35	30	20	15	1.5
16	15	46.7	23.3	15	2.0
17	20	43.3	21.7	15	2.0
18	25	40	20	15	2.0
19	30	36.7	18.3	15	2.0
20	35	33.3	16.7	15	2.0

Slag Preparation

Reagent grade oxides powders of CaF_2, Re_2O_3, CaO and SiO_2 were employed to synthesize the slags, which were calcined at 1273 K for 4 h to evaporate the moisture and impurities, respectively. The chemical powders were carefully weighted and fully mixed. Then, the mixtures were placed into graphite crucible and pre-melted using $MoSi_2$ furnace. When muffle furnace was heated to 1673 K at 5 k/min, the graphite crucibles were placed on a graphite plate which was placed inside the hot zone of the furnace at 1673 K for 4 h to completely homogenize the slags. After that, turn off the muffle furnace and make the sample cool with the furnace. The cooling rate is about 10 degrees per minute when the furnace temperature is above 1373 k.

The mineral composition of sample was detected by the X-ray diffraction with a Rigaku diffractometer (DMAX-RB 12 kW, Rigaku Corporation, Tokyo, Japan) using CuKa radiation. The mineralogy of slag was observed by scanning electron microscope (SEM; JEOLJSM-6480A, JEOL Ltd., Toyo, Japan) with an energy-dispersive spectroscope.

Experimental Results

SEM-EDS Analysis of Rare Earth Slag

The furnace-cooling slags were mounted in epoxy resin, then sectioned and polished for cross section observation by SEM-EDS. The micro structures of the samples are shown as Figs. 1, 2, 3 and 4. The photographs of the slag at basicity 0.5 were taken at 2000 magnifications because of fine grain, while the others were taken at 200 magnifications. In the microstructural diagrams, the white part is RE-concentrating phase, the black part is CaF_2 and the most widely distributed grey part is substrate phase. From Fig. 1, we can learn that fine dispersed RE-concentrating phases were distributed in all over of the substrate even at 2000 magnifications. In the slags of this study, the network units are [SiO_4]-tetrahedron units. When the basicity is low, the molten slag has a higher polymerization so that the process of mass transfer is suppressed. With the increase of basicity, the crystallization of the rare earth phase is increasing and the RE-concentrating phase exists in a rod-like structure. CaO plays as a network modifier in the slag. With the increase of CaO content, the [SiO4]-tetrahedron units are depolymerized and the viscosity of slag decreases. The mass transfer condition is improved so that the RE-concentrating phase is better crystallized. EDS analyses show that RE element coexists with Si, Ca, O and F. But when the alkalinity is up to 2.0, the distribution of RE-concentrating phase becomes dispersive and the crystals of rare earth phase have no definite shape. EDS analyses of the mineral were made so as to ascertain its chemical composition. The results showed that rare earth elements (RE) only coexist with oxygen (O), which indicates rare earth oxide is the main form of the rare earths. The reason for this may be higher content of CaO. Binding capacity of SiO_2 with CaO is stronger than that with RE_2O_3 so that RE_2O_3 exists in a free state. Under the same total alkalinity in this experiment, the microstructures of the slag show no obvious change with the increase of CaF_2 content.

Mineral Composition of Rare Earth Slag

The mineral composition is also an important parameter for mineral dressing efficiency. It is assumed that the less the number of minerals in the slag is and the

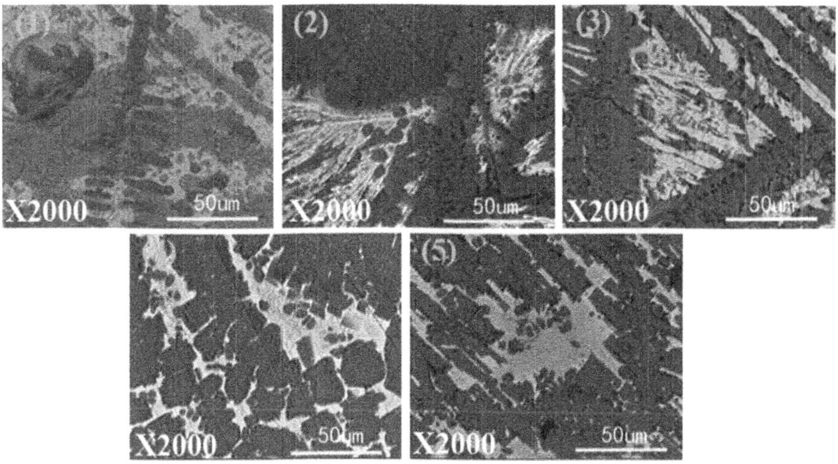

Fig. 1 Micro structure of rare earth slag at basicity 0.5

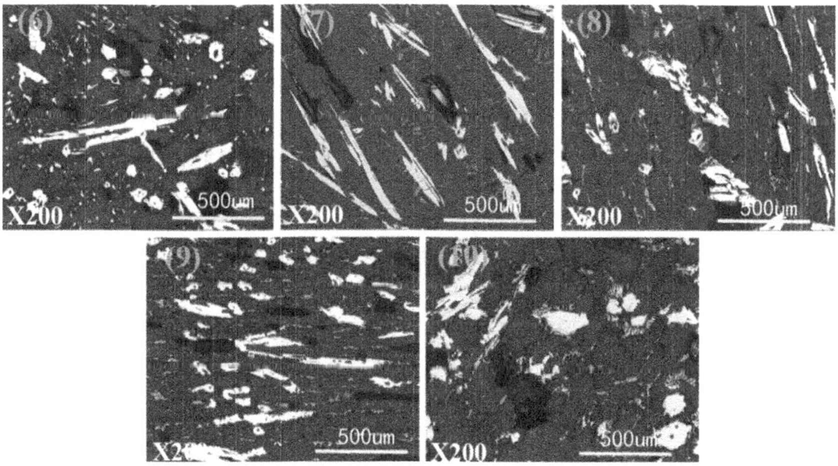

Fig. 2 Micro structure of rare earth slag at basicity 1.0

greater the difference in specific gravity between the minerals is, the better the ore dressing will be. Therefore, the mineral composition of the slag was detected and the XRD patterns are shown in Figs. 5, 6, 7 and 8. The major minerals are fluorite, cuspidine, RE-concentrating phase and substrate phase. When basicity is 0.5, there are other phases in the slag, such as SiO_2 and $CaSiO_3$ with fluorite content below 25%. The diffraction peaks of rare earth phase is similar to that of britholite [$Ca_{2.15}$ $Ce_{2.85}$ $(SiO_4)_3(F_{0.5}OH_{0.5})$]. Because of there is no OH- in the slag, it can be inferred that the rare earth phase of this experiment has a similar crystal structure as britholite. When the slags basicity is about 1.0, there are only three minerals.

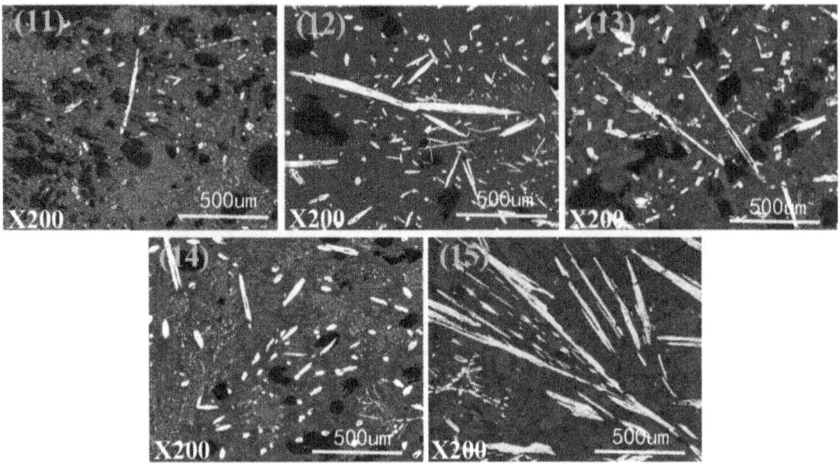

Fig. 3 Micro structure of rare earth slag at basicity1.5

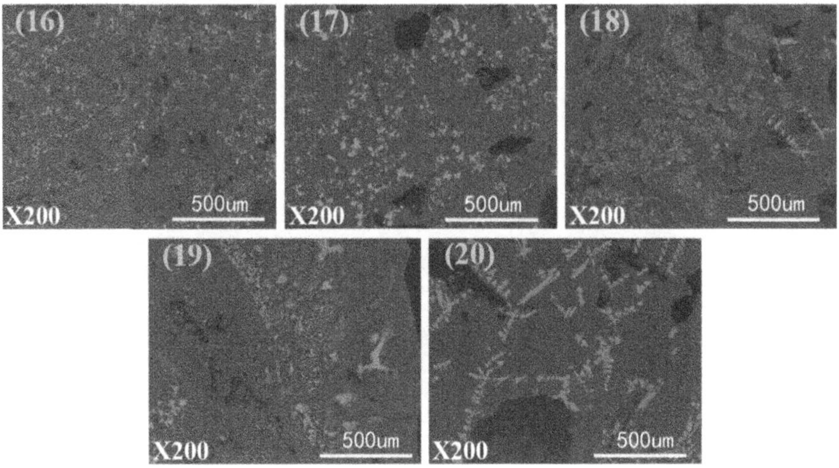

Fig. 4 Micro structure of rare earth slag at basicity 2.0

The diffraction peaks of britholite are shown at the bottom of Fig. 6, which is nearly consistent with that of the rare earth enriched phase in the slag. When the basicity is 1.5, there is Ca_2SiO_4, a new phase, appeared. With the increase of basicity, CaO binding capacity with SiO_2 is strengthened significantly. Ca_2SiO_4 is generated due to overdose of CaO. At the basicity of 2.0, rare earths exist in the form of RE_2O_3 on account of excessive basicity which makes rare earth oxides isolated with other acidic oxides. For the purpose of high-efficiency mineral separation, the basicity of the slag should be between 0.5 and 1.5.

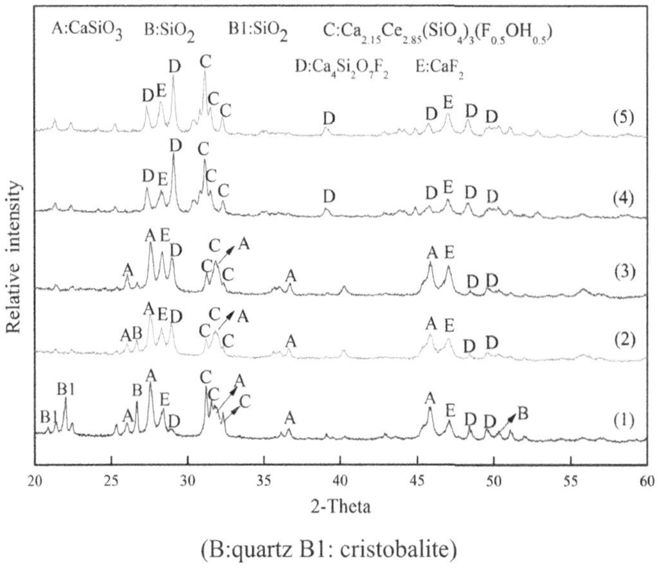

Fig. 5 XRD pattern for the sample at basicity 0.5

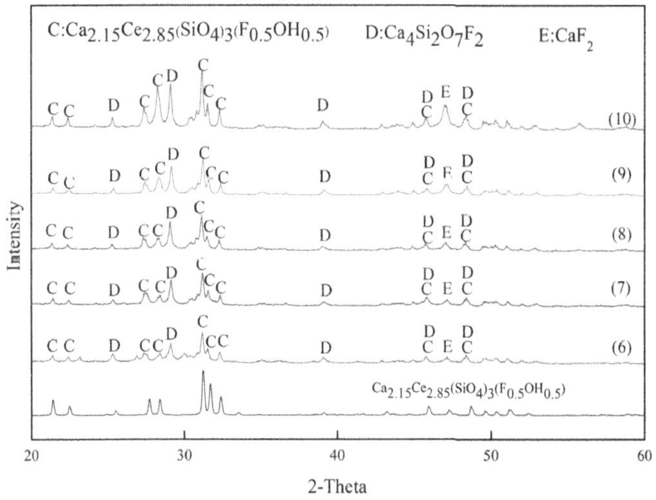

Fig. 6 XRD pattern for the sample at basicity 1.0

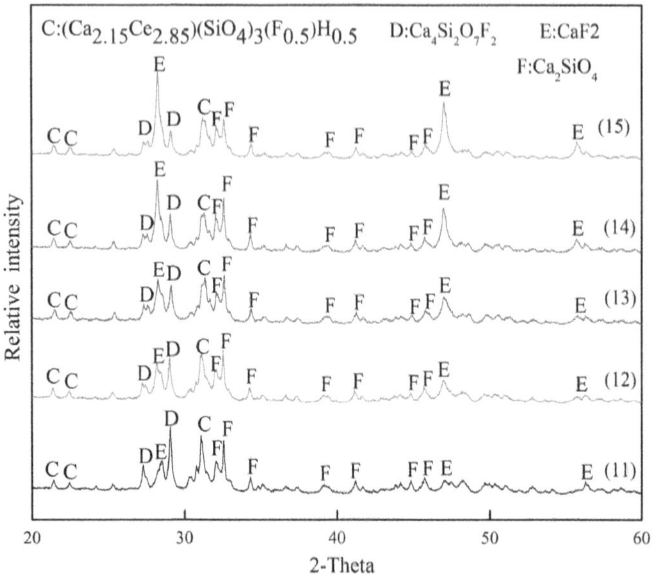

Fig. 7 XRD pattern for the sample at basicity 1.5

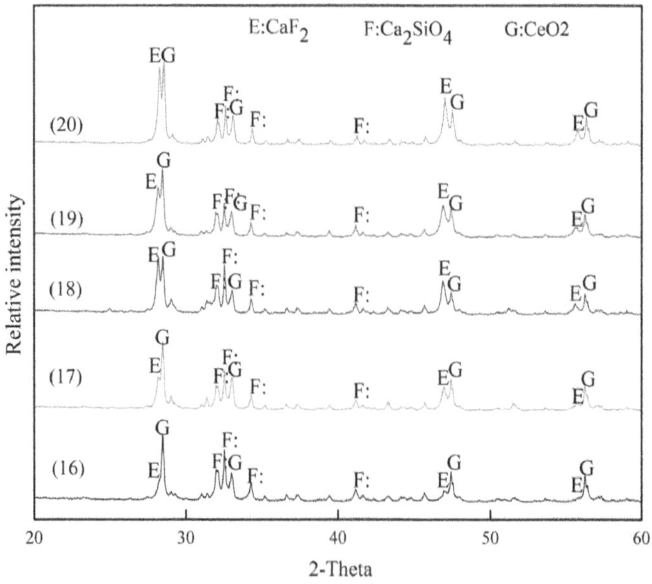

Fig. 8 XRD pattern for the sample at basicity 2.0

Conclusions

(1) For the purpose of high-efficiency mineral separation, the basicity of the slag should be 1.0. No matter the basicity is too low ($CaO/SiO_2 < 0.5$) or too high ($CaO/SiO_2 > 1.5$), rare earth phase in the slag can't meet the requirements of mineral processing.

(2) When the basicity of the slag is 1.0, RE-concentrating phase has a very good crystallization behaviour, which is perfect for the mineral processing. The equivalent diameter of RE-concentrating mineral grows to 100 μm. Meanwhile, there are only three minerals in the slag.

(3) According to this study, we can recovery rare earth efficiently from rare earth slag through mineral dressing, and it provides experimental support for the utilization of slag produced by Bayan Obo complex iron ore.

Acknowledgements The authors gratefully acknowledge the support of National Natural Science Foundation of China (No. 51874029) and the project of State Key Laboratory of Advanced Metallurgy (No. 41618018).

References

1. Cheng JZ, Che LP (2010) Current mining situation and potential development of rare earth in China. Chin Rare Earths 31(2):65–69
2. Yang X, Zheng Y, Yang X et al (2000) A Geochemical study of an REE-rich carbonatite dyke at Bayan Obo, Inner Mongolia, Northern China. Acta Geol Sin 74(3):605–612
3. Kanazawa Y, Kamitani M (2006) Rare earth minerals and resources in the world. J Alloy Compd 408(19):1339–1343
4. Ye-Sheng LI, Dong DQ, Zi-Ping WU (2006) Research status and application prospects of rare earth in magnesium alloys. Light Alloy Fabrication Technol 34(4):1–6
5. Li Wu, Hong Zhenhong (1996) Present Status and Future Prospect of Rare Earth Application in China. Chin Rare Earths 17(6):48–53
6. Zhongshan Wang, Qinggang Zhu et al (1999) Applications and prospects of rare-earth elements in national ecomomy. Sci Tech Inf Dev Econ 5:28–30
7. yongfu Yu (2001) Dress technology of REO ore and its development in china. J china univ min Technol 30(6):537–542
8. Zeqi Y, Qiang Z, Xiapeng Z (1992) Separation of cerite calcium and rare earth bearing barium fluorophlogopite. Nonferrous Met 44(1):46–53
9. Li J, Guo Z (2014) Innovative methodology to enrich britholite (Ca_3 Ce_2 [(Si, P) O_4] $_3$ F) phase from rare-earth-rich slag by super gravity. Metall Mater Trans B 45(4):1272–1280
10. Ding Y, Xue Q, Wang G et al (2013) Recovery behavior of rare earth from Bayan Obo complex iron ore. Metall Mater Trans B 44(1):28–36
11. Ding Y, Wang J, Wang G et al (2012) Innovative methodology for separating of rare earth and iron from Bayan Obo complex iron ore. ISIJ Int 52(10):1772–1777
12. Xu XY, Li MY, Hao MZ et al (1980) Crystallization and washability for blast furnace slags of Baotou rich rare earth iron mine. Chin Rare Earths 18(3):1–7
13. Li Bu, Qingcai Dagang et al (2005) Morphology of solidified slag for RE_2O_3–CaO–SiO_2–CaF_2–MgO–Al_2O_3System. J Iron Steel Res 16(1):30–33

Effects of Steel Scrap Oxidation on Scrap Preheating Process in an Electric Arc Furnace

Guangwu Tang, Yuchao Chen, Armin K. Silaen, Yury Krotov and Chenn Q. Zhou

Abstract The electric arc furnace (EAF) is used to produce steel, primarily from steel scraps, using electricity and chemical energy. Oxy-fuel burners firing natural gas (NG) are widely used in EAF operation during both the scrap preheating and melting stages. Excess oxygen is normally used to improve heating efficiency. Steel scrap oxidation increases the scrap temperature and reduces the scrap melting point. In this paper, a numerical study is conducted to simulate the steel scrap preheating process using oxy-fuel combustion. A three-dimensional computational fluid dynamics (CFD) model was developed with detailed consideration of gas flow, oxy-fuel combustion, scrap oxidation and multi-mode heat transfer between gas and solid scrap. The effects of scrap oxidation on the scrap preheating were also investigated. Model validation was conducted using experimental data obtained from a small-scale scrap preheating furnace. The model can be applied to industrial EAFs to investigate the effects of operating parameters on the scrap preheating process.

Keywords Electric arc furnace · CFD · Combustion · Heat transfer

Introduction

The electric arc furnace (EAF) is used to produce steel, primarily from steel scraps, using electricity and chemical energy. Worldwide, EAF steel scrap consumption has been increasing in the past 70 years [1]. The steel and foundry sectors consume approximately one-third of the world's steel scraps[2]. In the USA, the share of US steel production using EAF is approximately 67%[3]. Worldwide, crude steel production by the EAF process increased dramatically during the past 60 years, and

G. Tang · Y. Chen · A. K. Silaen · C. Q. Zhou (✉)
Center for Innovation Through Visualization and Simulation,
Purdue University Northwest, 2200 169th Street, Hammond, IN 46323, USA
e-mail: czhou@pnw.edu

Y. Krotov
Steel Dynamics Inc., 4500 County Road 59, Butler, IN 46721, USA

© The Minerals, Metals & Materials Society 2019

T. Jiang et al. (eds.), *10th International Symposium on High-Temperature Metallurgical Processing*, The Minerals, Metals & Materials Series,
https://doi.org/10.1007/978-3-030-05955-2_43

it is expected to increase in the next 30 years. The EAF steelmaking process is an energy-intensive process that consumes vast amounts of electrical and chemical energy. Raw materials and operating practices affect EAF efficiency and productivity. The energy efficiency of EAF steelmaking is only 55–65%. Raw materials charged into EAFs have changed since 1995, and worldwide, scrap accounts for around 75% of EAF materials input. The ability to meet the need for higher value-added steel grades will be difficult to achieve through scrap alone. Productivity and energy consumption are strongly dependent on alternative scrap materials [4] such as direct-reduced iron (DRI), hot briquetted iron (HBI), pig iron and hot metal. Worldwide, DRI and HBI account for around 15% of EAF materials input, with around 10% accounted for by pig iron and hot metal [5]. New research can help explore opportunities to further improve the productivity and efficiency for future EAF steelmaking.

Electricity and chemical energy typically account for around 65% and 35%, respectively, of energy supplied to EAF operation. New technologies intend to improve energy efficiency and reduce electrical power usage. Recently, oxy-fuel burners have been widely used in EAFs. Typically, 3–5 oxy-fuel burners are installed in the EAF to provide heat to melt the scrap located farthest from the electrodes and to provide coherent oxygen jets to stir the molten steel bath when a flat bath is achieved. Productivity increases by using oxy-fuel combustion burners in the EAF are in the range of 5–20% [6, 7]. In order to optimize the electricity and chemical energy input, gaining a complete understanding of the combustion, heat transfer and chemical reactions during the furnace operation is important. Computational fluid dynamics (CFD) simulation can reveal fluid flow, heat transfer and chemical reactions that occur in an EAF system. CFD modelling of the combustion process and electrode heating has been reported in the literature. [8–10] Odenthal et al. [11] reviewed EAF-related numerical modelling research and showed that very limited research was focused on oxy-fuel combustion and heat transfer to scrap piles. Mandal and Irons [6, 7] developed a CFD model that considered propane-oxygen combustion and heat transfer between combustion flue gas and scrap, which was treated as a porous medium. Model validation has been conducted by comparing simulation and experimental measurements of the temperature in experimental furnaces. The work done by Mandal and Irons [6, 7] provided a basic guideline on modelling approach and model parameters for scrap preheating simulation. In order to conduct numerical simulation of the industrial-scale EAF scrap preheating process, three basic points have to be addressed, based on the existing literature-reported models: (1) refinement of an oxy-fuel combustion model is needed to accurately predict the combustion process for oxy-fuel instead of traditional air-fuel combustion; (2) refinement of a radiation model is needed to account for the effects of high-CO-concentration gas atmosphere on the temperature field under oxy-fuel combustion conditions; and (3) chemical reactions such as scrap oxidation and other hydrocarbon combustion reactions in the scrap need to be considered to account for additional initial heat sources.

In this paper, comprehensive numerical modelling of the scrap preheating process has been conducted. A three-dimensional CFD model was developed with detailed consideration of gas flow, oxy-fuel combustion, scrap oxidation and heat transfer between gas and solid scraps. Model validation was conducted by comparing model predictions against measurement data which were obtained using a small-scale experimental furnace. The primary focus of this study is to understand the effects of scrap oxidation on the scrap preheating process when excess oxygen was supplied during the scrap preheating or initial melting stages in EAFs.

CFD Model

The scrap preheating CFD model was established based on the detailed physics and chemistry occurring during the EAF operation. The model considers gas flow, oxy-fuel combustion and multi-mode heat transfer including the heat exchange between gas and solid scrap. Governing equations were determined using the commercial software ANSYS Fluent [12, 13]. The scrap phase was treated as a porous medium, and its governing equations were input into ANSYS Fluent using the user-defined scalar (UDS). Gas momentum sink due to the porous medium is considered using a user-defined function (UDF). The energy source terms for both gas and solid phases are added using UDFs. The UDFs are solved together with other terms in the governing equations. The eddy dissipation concept (EDC) model was used to calculate the propane-oxygen combustion reactions with a four-step chemical reaction mechanism. [14] The discrete ordinates (DO) radiation model was used to model the dominant heat transfer flux, and a new weighted-sum-of-gray-gases model (WSGGM) for oxy-fuel combustion [15] was used to describe the radiation heat transfer in the combustion gases. The WSGGM for oxy-fuel combustion is implemented using a UDF. The governing equations were solved in transient conditions and are expressed as follows:

Gas Phase

The Mass Conservation Equation

$$\frac{\partial \rho}{\partial t} + \nabla \cdot (\rho \vec{v}) = 0 \quad (1)$$

The Momentum Conservation Equation

$$\frac{\partial}{\partial t}(\rho \vec{v}) + \nabla \cdot (\rho \vec{v}\vec{v}) = -\nabla p + \nabla \cdot (\bar{\bar{\tau}}) + \rho \vec{g} + \vec{F} + S_{gu} \quad (2)$$

where p is the static pressure, $\rho \vec{g}$ and $\vec{F}$ are the gravitational body force and external body force, respectively, and S_{gu} is a source term introduced by the scrap phase.

The momentum exchange between the gas phase and the solid scrap is described by an Ergun-type equation [16]. The source term for gas velocity in a radial direction is given as Eq. 3, and the source term for gas velocity in the vertical direction is similar.

$$S_{gu,i} = -\left(\frac{\mu}{\alpha}u_{g,i} + C_2\frac{\rho_g|u_{g,i}|}{2}u_{g,i}\right) \tag{3}$$

where

$$\alpha = \frac{(\psi_j d_j)^2 \emptyset^3}{150(1-\emptyset)^2} \tag{4}$$

$$C_2 = 3.5\frac{(1-\emptyset)}{(\psi_j d_j)\emptyset^3} \tag{5}$$

S_{gu} is the source term for the momentum equation of the gas phase (in units of N/m^3), u_g is gas superficial velocity (m/s), $\emptyset$ is scrap porosity, ψ_j is particle shape factor (1), d_j is particle diameter (m), ρ is the fluid density (kg/m^3), $\vec{v}$ is the fluid velocity having x, y and z components, and $\bar{\bar{\tau}}$ is the stress tensor which is given by

$$\bar{\bar{\tau}} = \mu\left[\left(\nabla\vec{v} + \nabla\vec{v}^T\right) - \frac{2}{3}\nabla\cdot\vec{v}I\right] \tag{6}$$

where is the molecular viscosity, I is the unit tensor, and the second term on the right-hand side is the effect of volume dilation. By the long-time averaged method, Eq. 2 can be converted to a Reynolds-averaged Navier-Stokes (RANS) equation as follows:

$$\frac{\partial}{\partial t}(\rho u_i) + \frac{\partial}{\partial x_j}(\rho u_i u_j) = -\frac{\partial p}{\partial x_i} + \frac{\partial}{\partial x_j}\left[\mu\left(\frac{\partial u_i}{\partial x_j} + \frac{\partial u_j}{\partial x_i} - \frac{2}{3}\delta_{ij}\frac{\partial u_l}{\partial x_l}\right)\right] + \frac{\partial}{\partial x_j}\left(-\rho\overline{u_i' u_j'}\right) \tag{7}$$

In Eq. 7, the Reynolds stresses may be related to the velocity gradient by the Boussinesq hypothesis as follows:

$$-\rho\overline{u_i' u_j'} = \mu_t\left(\frac{\partial u_i}{\partial x_j} + \frac{\partial u_j}{\partial x_i}\right) - \frac{2}{3}\left(\rho k + \mu_t\frac{\partial u_k}{\partial x_k}\right)\delta_{ij} \tag{8}$$

where δ_{ij} is the stress tensor due to molecular viscosity. The realizable k-ε turbulence model is used due to its ability to model flow with a strong streamline curvature, vortices and rotation.

The Energy Equation

$$\frac{\partial}{\partial t}(\rho E)+\nabla\cdot[\vec{v}(\rho E+p)]=\nabla\cdot\left[(k_{\mathrm{eff}}+k_r)\nabla T-\sum_j h_j\vec{J}_j+\left(\bar{\bar{\tau}}_{\mathrm{eff}}\cdot\vec{v}\right)\right]+S_h \quad (9)$$

where k_{eff} is the effective conductivity, and $\vec{J}_j$ is the diffusion flux of species j. The first three terms on the right-hand side of the equation represent energy transfer due to conduction, species diffusion and viscous dissipation, respectively. S_h includes the heat of chemical reaction, and any other volumetric heat sources. E is defined as:

$$E=h-\frac{p}{\rho}+\frac{v^2}{2} \quad (10)$$

where sensible enthalpy h is defined for incompressible flows as

$$h=\sum_j Y_j h_j+\frac{p}{\rho} \quad (11)$$

where Y_j is the mass fraction of species j, and h_j (the sensible enthalpy for species j) is defined as

$$h_j=\int_{298.15\ \mathrm{K}}^{T} C_{p,j}dT \quad (12)$$

where $C_{p,j}$ is the constant pressure specific heat of species j.

K_{r} is the radiative thermal conductivity, and can be expressed as:

$$K_{\mathrm{r}}=\frac{8\sigma T_j^3}{a+2s} \quad (13)$$

$$a=1.5\varepsilon(1-\emptyset)/d_j \quad (14)$$

$$s=1.5(2-\varepsilon)(1-\emptyset)/d_j \quad (15)$$

where a is the absorption coefficient, s is the scattering coefficient, T_j is the scrap temperature, $\emptyset$ is scrap porosity, d_j is the scrap size, σ is Stefan–Boltzmann constant, and ε is emissivity.

The heat transfer between the gas and solid scrap is described as Eq. 16, and the heat transfer coefficient proposed by Wakao et al. [17] and Mandal [18] is used to account for the porous medium effects on the heat transfer coefficient of a single particle. Re_j is the particle Reynolds number based on particle diameter and superficial gas velocity.

$$S_{gs} = \dot{n}_j h_j A_j (T_j - T_g) \tag{16}$$

$$\frac{h_j d_j}{k_g} = 2 + 1.1 Re_j^{0.6} Pr_g^{1/3} (T_j < 1373\,\text{K}) \tag{17}$$

$$h_j A_j = A * f(w) \frac{V_g^{0.9} T_j^{0.3}}{d_j^{0.75}} (T_j > 1373\,\text{K}) \tag{18}$$

where k_g is a constant, 0.0184, Pr_g is Prandtl number, 0.5, A is constant, 160.0, $f(w)$ is constant, 0.5, and V_g is gas velocity.

The species transport model was used to define the species and to calculate the species conservation and transportation. The local mass fraction of each species Y_i can be solved by the convection-diffusion equations. The conservation equation of each species can be expressed by

$$\frac{\partial}{\partial t}(\rho Y_i) + \nabla \cdot (\rho \vec{v} Y_i) = -\nabla \cdot \vec{J}_i + R_i \tag{19}$$

where R_i is the net rate of production of species i by chemical reactions. $\vec{J}_i$ is the diffusion flux term of species i, which arises due to gradients of concentration and temperature, under which the diffusion flux is given as follows:

$$\vec{J}_i = -\left(\rho D_{i,m} + \frac{\mu_t}{Sc_t}\right) \nabla Y_i - D_{T,i} \frac{\nabla T}{T} \tag{20}$$

where $D_{i,m}$ is the diffusivity for species i in the gas mixture, 2.88E-5 m^2/s, μ_t is the turbulent viscosity, and $D_{T,i}$ is the thermal diffusion coefficient, which is calculated from thermal conductivity divided by density and specific heat capacity at a constant pressure. Sc_t is the turbulent Schmidt number, which is 0.7. The oxy-propane combustion mechanisms and kinetics are listed in Table 1. [14].

Scrap oxidation reaction is also considered as the following mechanism:

$$\text{Fe(s)} + \frac{1}{2}\text{O}_2\text{(g)} = \text{FeO(s)}, \Delta H_f^{298} = -270.16\text{kJ/mol} \tag{21}$$

Based on the steel oxidation model developed for steel slab oxidation in reheating process, [19] this scrap oxidation rate for EAF scrap oxidation is calculated based on the rate constant and species concentrations defined as:

$$\left(\frac{W}{A}\right)^n = kt \cdot ([\text{Fe}][\text{O}_2])^n \tag{22}$$

Table 1 Reaction mechanism and kinetics (units in kg, m, s, kcal, mol and K)

Reaction	Reaction order	A	b	E
$C_3H_8 + 1.5O_2 \rightarrow 3CO + 4H_2$	$[C_3H_8]^{0.5}[O_2]^{1.25}$	4×10^{11}	0	30.0
$H_2 + 0.5O_2 \rightarrow H_2O$	$[H_2][O_2]^{0.5}$	1.8×10^{13}	0	35.1
$H_2O \rightarrow H_2 + 0.5O_2$	$[H_2O]$	5.337×10^{16}	−0.5	94.3
$CO + 0.5O_2 \rightarrow CO_2$	$[CO][O_2]^{0.5}[H_2O]^{0.5}$	1.3×10^{11}	0	30.0
$CO_2 \rightarrow CO + 0.5O_2$	$[CO_2][H_2O]^{0.5}$	1.213×10^{16}	−0.5	97.1

where $k = 4.00 \times 10^{-7} g^2/cm^4$ s, the term $\frac{W}{A}$ is weight gain per unit area, t is time, [Fe] is steel concentration in the local cell, and $[O_2]$ is oxygen concentration in the local cell.

Steel Scrap Phase

The general transport equation for the scrap phase is given as Eq. 23. The transient term for solid scrap phase is included to consider the change of scrap molar density, enthalpy and concentrations. In this study, since a simplified oxidation reaction is considered, only species Fe and FeO are used for scrap material. Due to the assumption that the scrap movement is not simulated, the flux term F_i is zero in Eq. 23. The diffusion term for the scrap molar density transportation equation is zero due to its solid form. [20] However, in the scrap enthalpy transportation equation, the thermal diffusivity is used.

$$\frac{\partial(\rho_s \emptyset_s)}{\partial t} + \frac{\partial}{\partial x_i}\left(F_i \emptyset_s - \Gamma_k \frac{\partial \emptyset_s}{\partial x_i}\right) = S_s \tag{23}$$

The general variables for the scrap phase are listed in Table 2.

The CFD simulations were conducted by using the commercial software package ANSYS Fluent 18.1. The finite volume method (FVM) was used to solve the partial differential equations by discretizing the equations using an upwind differencing scheme over the finite volumes. The SIMPLE algorithm was used to adjust the pressure and velocities after each iteration when solving the gas continuity and momentum equations.

Table 2 General variables for solid scrap phase

Scalar	$\emptyset_s$	ρ_s	F_i	Γ_k	S_s
Scrap molar density	1	1	0	0	0
Scrap enthalpy	H_s	ρ_s	0	Γ_k	S_{gH}
[Fe] concentration	Z_{Fe}	ρ_s	0	0	$-S_R$
[FeO] concentration	Z_{FeO}	ρ_s	0	0	S_R

Geometrical Model and Conditions

It is difficult to measure the scrap temperature inside an EAF during operation. In order to validate the CFD model, small-scale experimental results [18] were used to compare with the simulation. The experiment was conducted by Mandal [18] to measure the efficiency of scrap heating by a propane-oxygen burner. The volume of the experimental furnace is in the order of 1 m^3, as shown in Fig. 1a. The propane-oxygen burner was used in the lower part of the furnace, with burner power of 8.3 kW. The thermocouples are installed at different locations in the furnace without contacting the steel scrap. In the beginning, the furnace was purged with argon gas. During the experiment, steel scraps are placed in the furnace at room temperature. Then, the propane-air burner starts to heat the steel scrap, and the temperature measurements by the thermocouples were recorded. The scrap preheating model was applied to simulate the scrap preheating process by propane-oxygen combustion. The computational domain was developed based on the detailed dimensions of the experimental furnace as shown in Fig. 1b. The oxy-fuel burner power used is 8.3 kW with 10% excess oxygen. The scrap used for this simulation is the small shredded scrap. The mixed propane-oxygen was put into the furnace through a 4-mm-diameter lance. More details can be found in the literature [18] used for comparison.

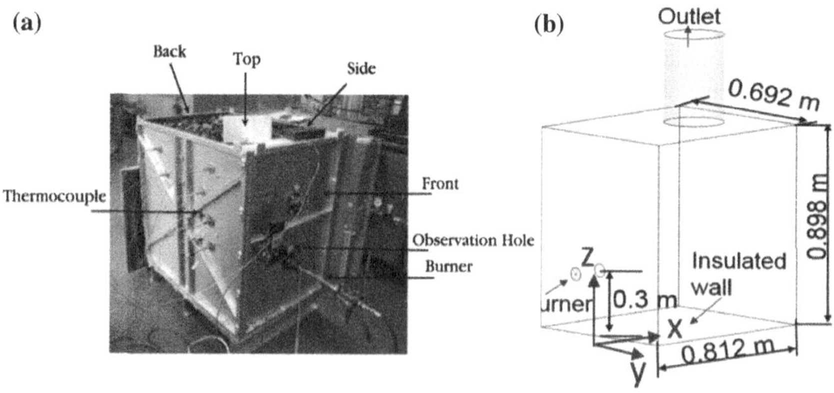

Fig. 1 a Experimental set-up, [18] b Computational domain

Results and Discussion

A numerical simulation was performed based on the real conditions used in the laboratory-scale experimental study. The simulation results shown in Fig. 2 demonstrate the process of scrap heating in the experimental furnace. The propane-oxygen gas mixture was ignited and put into the scrap piles. Due to the high temperature of oxy-fuel combustion, the density of the combustion flue gas is significantly reduced and flows upwards through the steel scraps to the top exit as indicated in Fig. 2. The combustion flame temperature of the propane-oxygen combustion can reach up to 3600 K as shown in Fig. 2. The corresponding highest gas velocity reaches 61 m/s. Due to the heat transfer from the hot combustion flue gas to the cold scrap, the scrap temperature increased most in the flame region as shown in Fig. 2. Quantitative comparisons are shown in Fig. 3. The measurement point (0.166 m, 0, 0.478 m) was chosen to compare.

Figure 3 shows comparisons of the gas temperature, scrap temperature and the thermocouple (TC) temperature between CFD simulation and experimental measurements. The TC measured temperature is neither the gas temperature nor the scrap temperature. Based on the assumption that the thermocouple temperature is changing slowly, a quasi-steady state thermocouple temperature can be calculated based on gas temperature, scrap temperature using the energy balance equation.

$$hA(T_{\mathrm{f}} - T_{\mathrm{th}}) = F_{\mathrm{th-sc}} A \varepsilon \sigma \left(T_{\mathrm{th}}^4 - T_{\mathrm{sc}}^4\right) \tag{24}$$

where h, A, T_{f}, T_{th}, $F_{\mathrm{th-sc}}$, ε, σ and T_{sc} are the heat transfer coefficient between gas and scrap, the area of the thermocouple, the fluid temperature, the thermocouple temperature, the shape factor, the thermocouple emissivity, the Stefan–Boltzmann constant and the scrap temperature, respectively.

The comparisons in Fig. 3 show three groups of data sets: the scrap temperature, gas temperature and TC temperature. Each group consists of CFD with oxidation,

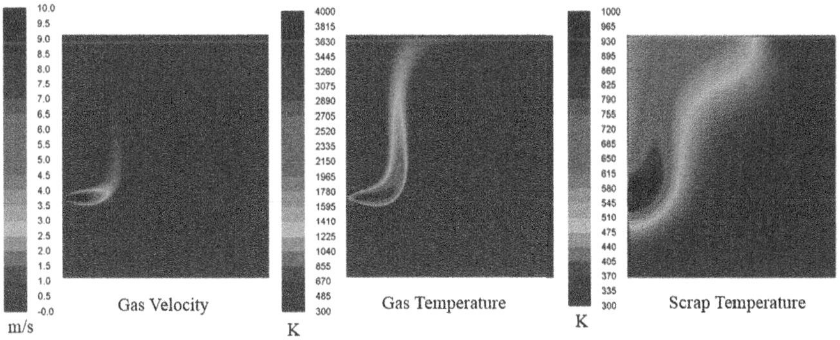

Fig. 2 Simulation results of **a** gas velocity, **b** gas temperature and **c** scrap temperature at t = 60 min

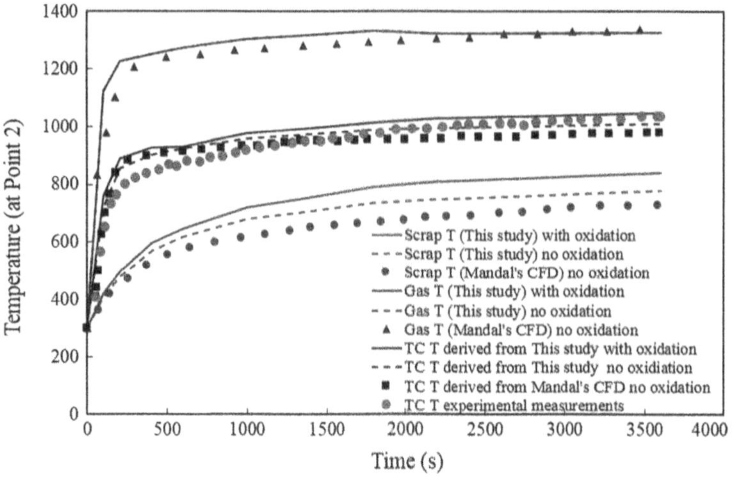

Fig. 3 Comparison of temperatures at a measured point between CFD and experiment

CFD without oxidation and Mandal's CFD. The TC temperature group also consists of results of experimental measurements. It can be seen that this study over-predicts the scrap temperature compared with Mandal's CFD. When the oxidation is considered, the scrap temperature is even higher due to the fact that the oxidation reaction is exothermic. In the numerical model, the scrap oxidation occurs in the steel scrap; therefore, 100% of the heat release is assumed to contribute to the scrap phase. When comparing the gas temperatures in Fig. 3, the effects of oxidation on the gas temperature change are minimal. Both CFD of this study and Mandal's CFD show good agreement on the trends and values. The TC temperature comparison indicates that both CFD simulations have a similar trend in scrap temperature change with respect to time. In general, the TC temperature will be higher when

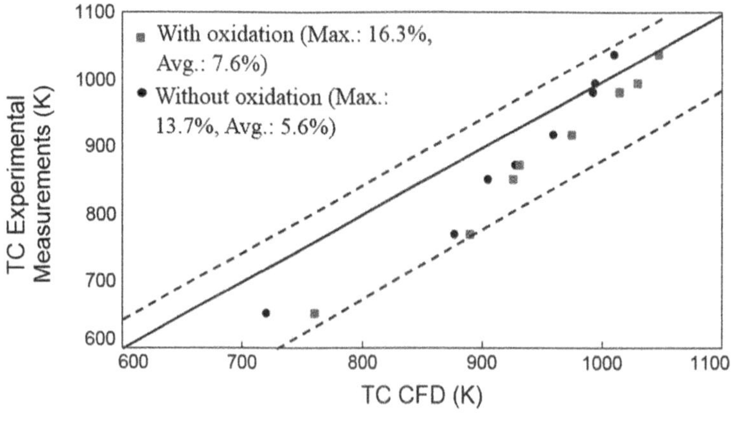

Fig. 4 Comparison of CFD and experiment on TC temperature

Fig. 5 Effects of oxidation on scrap temperature distribution and FeO concentration

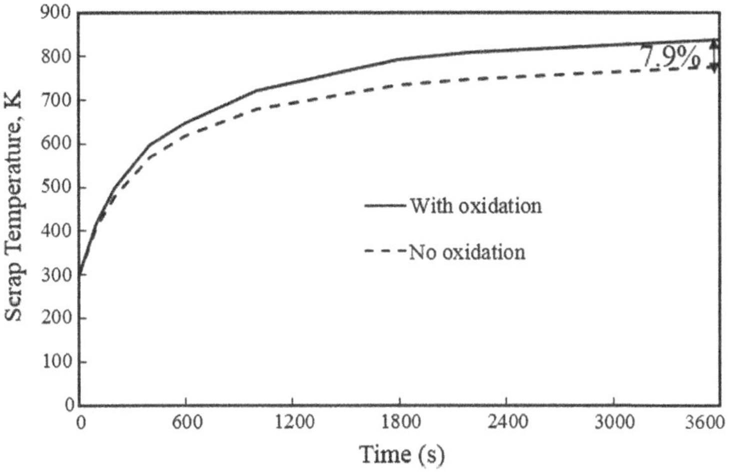

Fig. 6 Effects of oxidation on scrap temperature at the measurement point

oxidation is considered. The comparisons of the TC temperature predicted by this study with the measurements are shown in Fig. 4.

In experimental work, the small shredded scrap used may be subjected to be reused and cause oxidation. As shown in Fig. 4, the average percentage error of without oxidation is 5.6% which is lower than the 7.6% for with oxidation. It indicates that the scrap oxidation will increase the difference when compared to the experiment. Therefore, the oxidation during the experiment was not significant. However, in real EAF operation, the effects of oxidation might be more significant due to the fact that oxygen supply is over stoichiometric and furnace temperature is high. The contours of scrap temperature and oxidized scrap molar concentration are shown in Fig. 5. The scrap temperatures at the measurement point predicted by the CFD models are shown in Fig. 6.

Due to the multi-mode heat transfer, the temperature of the scrap near and above the oxy-fuel burner increases higher than the locations far away from the oxy-fuel burner as shown in Fig. 5. The scrap oxidation reaction increases the scrap temperature in general. In this simulation conditions, the oxidation reaction occurred mostly in front of the burner where the scrap temperature and oxygen concentrations are higher than any other places. Quantitatively, the oxidation would increase the scrap temperature by up to approximately 7.9% compared to no oxidation as shown in Fig. 6.

Conclusions

A three-dimensional CFD model for scrap preheating process by oxy-fuel combustion in EAFs has been developed. The model provides a detailed consideration of gas flow, oxy-fuel combustion, scrap oxidation and multi-mode heat transfer between gas and solid scrap. The CFD model has been validated against the measurement data obtained from a small-scale experimental furnace. The effects of scrap oxidation on the scrap preheating are investigated. The results indicate that scrap oxidation could increase scrap temperature by up to 7.9% in the experimental furnace. The model can be applied to industrial EAFs to optimize scrap preheating process.

Acknowledgements The authors would like to thank the Steel Manufacturing Simulation and Visualization Consortium (SMSVC) members for funding this project. The Center for Innovation through Visualization and Simulation (CIVS) at Purdue University Northwest is also gratefully acknowledged for providing all the resources required for this work. The authors also appreciate the help from Andrew Spencer (Steel Dynamics), Hamzah Alshawarghi (Praxair), Jianghua Li (AK Steel), Joe Maiolo (Praxair), John Lowry (Nucor), Michael Lowry (ArcelorMittal), Sunday Abraham (SSAB) and Yufeng Wang (SSAB). The authors also appreciate the help from Charlie Capek for proofreading.

References

1. Oda J, Akimoto K, Tomoda T (2013) Long-term Global Availability of Steel Scrap. Resour Conserv Recycl 81:81–91
2. Logar V, Dovz D, Škrjanc I (2011) Mathematical modeling and experimental validation of an electric arc furnace. ISIJ Int 51(3):382–391
3. Technology Roadmap Research Program for the Steel Industry, Final Report, December 31, 2010, American Iron and Steel Institute
4. Madias J (2014) Electric Furnace Steelmaking. Metallon, Buenos Aires, Argentina, pp 271–300
5. Steel Industry Technology Roadmap, Office of Energy Efficiency & Renewable energy, United States, Department of Energy, Dec 2001
6. Mandal K, Irons GA (2013) A study of scrap heating by Burners, Part I: Experiments. Metall Mater Trans B 44B:184–195
7. Mandal K, Irons GA (2013) A study of scrap heating by Burners, Part II: Numerical modeling. Metall Mater Trans B 44B:196–209
8. Li Y, Fruehan RJ (2003) Computational fluid dynamics simulation of post-combustion in the electric arc furnace. Metall Mater Trans B 34B:333–343

9. Cruber JC, Echterhof T, Pfeifer H (2016) Investigation on the influence of the arc region on heat and mass transfer in an EAF freeboard using numerical simulation. Steel Res Int 87:15–28
10. Yigit C, Coskun G, Buyukkaya E, Durmaz U, Guven HR (2015) CFD modeling of carbon combustion and electrode radiation in an electric arc furnace. Appl Therm Eng 90:831–837
11. Odenthal HJ, Kemminger A, Krause F, Sankowski L, Uebber N, Vogl N (2017) Review on modeling and simulation of the electric arc furnace (EAF). Steel Res Int :1700098
12. ANSYS, FLUENT 18.1, Theory Guide, FLUENT, Inc
13. ANSYS, FLUENT 18.1, User Guide, FLUENT, Inc
14. Hjartstam S, Normann F, Andersson K, Johnsson F (2012) Oxy-fuel combustion modeling: performance of global reaction mechanisms. Ind Eng Chem Res 51:10327–10337
15. Yin C, Johansen LCR, Rosendahl LA, Kar SK (2010) New weighted sum of gray gases model applicable to computational fluid dynamics (cfd) modeling of oxy-fuel combustion: derivation, validation, and implementation. Energy Fuels 24:6275–6282
16. Ergun S (1953) Pressure drop in blast furnace and in cupola. Ind Eng Chem 45(2):477–485
17. Wakao N, Kaguei S, Funazkri T (1979) Effect of fluid dispersion coefficients on particle-to-fluid heat transfer coefficients in packed beds: Correlation of nusselt numbers. Chem Eng Sci 34:325–336
18. Mandal K (2010) Modeling of scrap heating By Burners. Ph.D. Dissertation, McMaster University
19. Abuluwefa HT, Guthrie RIL, Ajersch F (1997) Oxidation of Low carbon steel in multicomponent gases: Part II. Reaction mechanisms during reheating. Metall Mater Trans A 28A:1643–1651
20. Fu D, Zhou CQ, Chen Y (2014) Numerical methods for simulating the reduction of iron ore in blast furnace shaft. J Therm Sci Eng Appl 6:021014

Enriching and Separating Iron Impurity from Galvanizing Dross by Super-Gravity Technology

Anjun Shi, Zhe Wang, Lei Guo, Ning Zhang and Zhancheng Guo

Abstract Hot-dip galvanizing dross is a valuable by-product and contains high levels of zinc and some iron impurities. In this work, the super-gravity technology was applied to removing iron impurities from an industrial galvanizing dross. Firstly, the enriching behaviour of iron phase was investigated in different super-gravity fields. The floatation movement of iron phase was greatly intensified by super gravity, leading to an iron-rich layer in the top area of the sample. Then, the separation of iron impurity from galvanizing dross was further carried out via filtration enhanced by super gravity. The majority of the iron impurities were retained in the upper residue by a filter felt, with the filtered zinc purified significantly. The efficiency of separating increased with increasing gravity coefficient (G). At $G = 600$, about 99.2 wt% iron was removed and the filtered zinc with a purity of about 99.9 wt% can be directly used as galvanizing zinc.

Keywords Galvanizing dross · Super gravity · Iron removal · Zinc recovery

Introduction

Hot-dip galvanizing is widely used in producing zinc coatings with the purpose of improving the corrosion resistance of steel [1]. This process is generally carried out by immersion of a steel article in a molten zinc bath at about 440–460 °C. Galvanizing dross is a combination of free zinc and Al–Fe–Zn intermetallic compounds formed by multiple reactions among zinc, aluminium and dissolved iron from the immersed steel, which can form swells on the galvanizing surface.

A. Shi · Z. Wang (✉) · L. Guo · N. Zhang · Z. Guo (✉)
State Key Laboratory of Advanced Metallurgy, University of Science and Technology Beijing, 30 Xueyuan Road, Beijing 100083, China
e-mail: zhewang@ustb.edu.cn

Z. Guo
e-mail: zcguo@ustb.edu.cn

© The Minerals, Metals & Materials Society 2019

T. Jiang et al. (eds.), *10th International Symposium on High-Temperature Metallurgical Processing*, The Minerals, Metals & Materials Series,
https://doi.org/10.1007/978-3-030-05955-2_44

Conventionally, the surface dross is skimmed off by labour or machine at intervals, while the bottom dross can be removed when the operation is shut down [2].

The galvanizing dross is a valuable by-product due to its high levels of zinc (over 80 wt%), motivating a search for technologies to allow the cost-effective recovery of the zinc content [3]. At present, there are many methods for the separation or recovery of zinc from galvanizing dross such as atmospheric or vacuum distillation, leaching and electrorefining [4]. However, these processes have limitations. The collected zinc products produced by distillation have very high purity [5], but high capital investment is required, and the productivity is relatively low [6]. Conventional leaching process is not appropriate for galvanizing dross as the large amount of metallic zinc in the dross consumes excessive lixiviating reagent [7]. The electrorefining process can produce high-purity zinc which is deposited on cathode with the zinc dross directly melted into anodes. But the electrolyte has to be purified frequently because almost all the iron impurity in the dross is transferred into the electrolyte, which correspondingly increases the cost of the electrorefining process [8].

In recent years, super-gravity technology has been successfully applied in removing impurities from metal melt on account of the difference of densities between impurities and melt. Matsubara et al. [9] studied the removal of Fe from liquid Al alloy under a super-gravity field during period of Al solidification, where the separation was attributed to macro-segregation caused by super gravity during solidification. Zhao et al. [10] studied the enrichment and remove behaviour of impurities in low concentrations from Al by solidifying it under a super-gravity field, which caused the impurity elements Fe and Si to be separated at the two ends of the sample along the direction of super gravity. The super-gravity technology has been proved to be a promising method for the large-scale purification of metals by these above-mentioned applications, and this technology would be an alternative method to effectively remove impurities and recover zinc from galvanizing dross.

In this study, the effect of gravity coefficient (G) on the efficiency of enriching and separating iron impurity from galvanizing dross has been investigated. Besides, the mechanism of super-gravity enrichment and separation of galvanizing dross was also examined.

Experimental

Raw Materials and Preparation

The galvanizing dross used in this study was supplied from Baotou Iron & Steel and contained 96.1 wt% Zn, 3.14 wt% Al, 0.75 wt% Fe and 0.01 wt% O. The zinc dross material was melted in an electric resistance furnace at 600 °C and cast into an ingot in a graphite crucible. The ingot was then machined through wire-electrode cutting into ~20 g cylinders of 20 mm diameter for the following super-gravity experiments.

Super-Gravity Experiments

The super-gravity field was generated by a centrifugal apparatus whose schematic was shown in Fig. 1. The facility mainly consists of a heating furnace and a counterweight which were installed symmetrically onto the centrifugal rotor. A program-controlled system controls temperature within the observed precision range of ±3 °C with an *R* type thermocouple. The super-gravity coefficient was calculated via Eq. (1).

$$G = \frac{\sqrt{g^2 + (w^2 g)^2}}{g} = \frac{\sqrt{g^2 + \left(\frac{N^2 p^2 g}{900}\right)^2}}{g} \tag{1}$$

where *G* is gravity coefficient, *g* is normal-gravitational acceleration ($g = 9.8\ \mathrm{m/s^2}$), ω is angular velocity ($\mathrm{rad/s^{-1}}$), *N* is rotating speed (r/min), and *R* is the distance between centrifugal axis and sample centre ($R = 0.25$ m).

In order to conduct the super-gravity enrichment experiment, a galvanizing dross sample was directly placed in a straight graphite crucible with the internal diameter of 21 mm. While carrying out super-gravity separation experiment, a galvanizing dross sample was first placed into a self-design graphite filter crucible consisting of two small crucibles with the same internal diameter of 21 mm. Twenty holes with a diameter of 1 mm were evenly distributed in the bottom of the upper crucible, and one layer of the carbon fibre felt (CFF) was laid at the bottom of the upper crucible

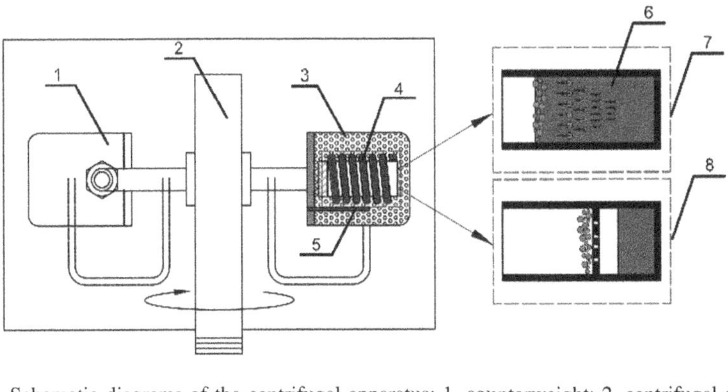

Fig. 1 Schematic diagrams of the centrifugal apparatus: 1. counterweight; 2. centrifugal axis; 3. miniature high-temperature furnace; 4. resistance wire; 5. thermocouple; 6. metal melt; 7. super-gravity enrichment; 8. super-gravity separation

as the filter medium. Both graphite crucibles were then heated to the target temperature for 15 min in the heating furnace. Afterwards, the centrifugal apparatus was started and adjusted to the desired rotational speed to conduct separation isothermally. The centrifugal apparatus was not shut off until the target separation time. After that, the graphite crucible was taken out from the furnace and quenched in water. The enrichment and separation zinc were collected, respectively, in preparation for further characterization.

Characterization

The original galvanizing dross, filtered zinc and residue were cut perpendicular to the top surface and then polished to a 1-mm finish for optical microscopy (Leica DM6000), scanning electron microscopy (SEM, MLA 250, FEI Quanta, USA), electron probe microanalyser (EPMA-1720, SHIMADZU, Japan) and X-ray diffraction (XRD, Smartlab, Rigaku, Japan) analysis. The zinc, aluminium and iron contents of original zinc dross, purified zinc and residue were analyzed by Inductively Coupled Plasma-Optical Emission Spectrometry (ICP-OES, Optima 7000DV, Perkin Elmer, USA), and the oxygen content was measured by the O, N and H analyzer (HORIBA, EMGR-830, Japan).

Results

Characterization of the Original Galvanizing Dross

Figure 2 shows the XRD pattern of the dross sample. It can be seen that the peaks for Al–Fe–Zn compounds ($Fe_2\ Al_5Zn_X$) and ZnO were very weak compared to those for metallic zinc, indicating the dominant phase in the galvanizing dross was metallic zinc.

Figure 3 shows an EPMA image and element distribution maps from a typical region of an original galvanizing dross sample. The Al–Fe–Zn dross particles, characterized by the dark grey colour and polygonal shape, were distributed in the continuous zinc matrix (light grey). Besides, a small amount of very fine and sinuous ZnO with the film shape was also found in the zinc matrix. Point analysis by EPMA of 20 random Al–Fe–Zn dross particles showed that these particles consisted of 64.5 at.% Al, 25.8 at.% Fe and 9.82 at.% Zn. This gave the Al–Fe–Zn dross particle a chemical formula of $Fe_2Al_5Zn_{0.8}$, which is in excellent agreement with the generally accepted chemical formula of $Fe_2Al_5Zn_X$ for the galvanizing dross particle in the previous XRD results.

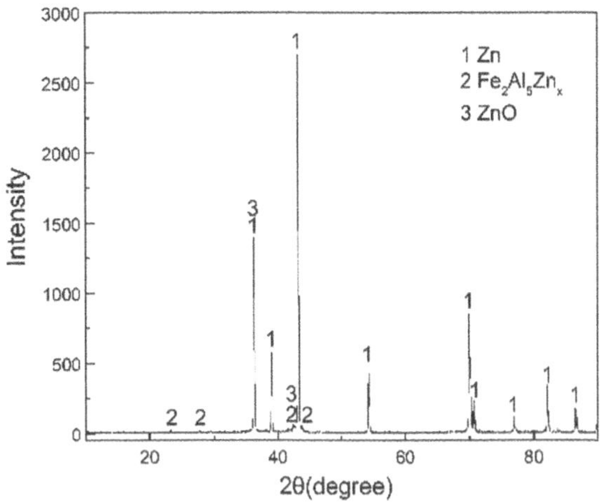

Fig. 2 XRD pattern of the original galvanizing dross sample

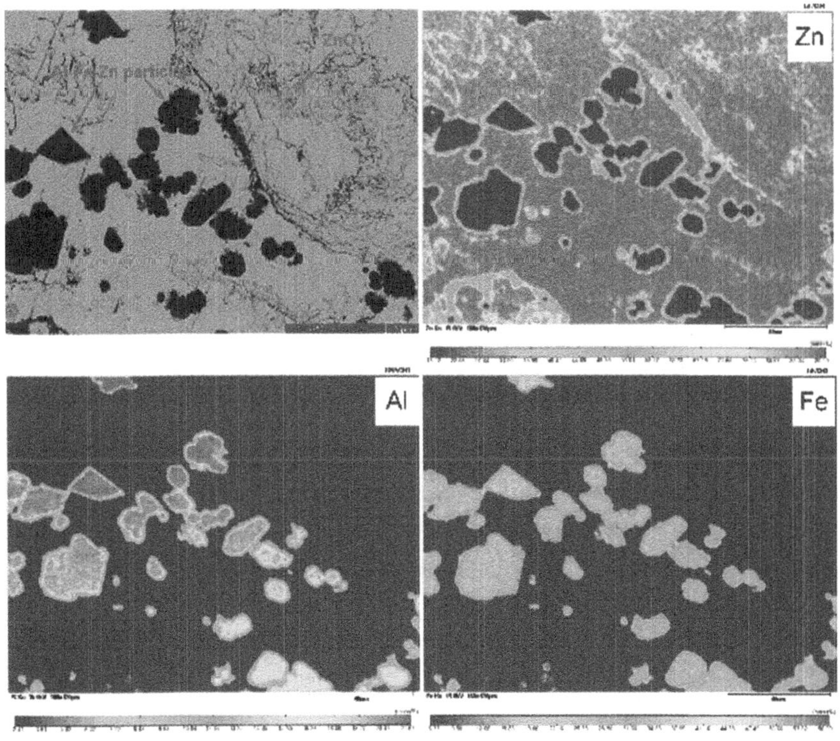

Fig. 3 EPMA mapping images of a typical region of the original galvanizing dross sample

Super-Gravity Enrichment of Galvanizing Dross

Figure 4 shows the cross section of the samples obtained by super gravity under the condition of $t = 180$ s, temperature $T = 510$ °C and gravity coefficient $G = 1$ (Fig. 4a) and 600 (Fig. 4b), respectively. It can be seen that a more cleared layer full of dross particles appeared at the top of the sample after super-gravity enrichment at $G = 600$ in comparison with the parallel sample ($G = 1$). To investigate the effect of the gravity coefficient on the distribution of dross particles, the different areas 'A'-'D' in Fig. 4 are observed by SEM, with the results shown in Fig. 5. The results show that the Al–Fe–Zn particles in the sample obtained by $G = 600$ presented gradient distribution along the centre axis, as indicated in Fig. 5b–d. A large number of precipitated phase aggregated to the upper part of the sample by super-gravity enrichment, while inclusions can hardly be found at the lower part of 10 mm (Fig. 5d).

Figure 6 shows the distribution of iron and zinc contents in the samples under different super-gravity fields. With the increase of distance from the bottom of the sample, the content of iron increased and the content of zinc decreased, which also indicated the enrichment of precipitated phase along the direction of super gravity.

Super-Gravity Separation of Galvanizing Dross

Figure 7 shows the images of the samples processed by super-gravity separation at various gravity coefficients, $t = 180$ s and $T = 510$ °C. It can be seen from Fig. 7a that the galvanizing dross sample was completely retained on the CFF in the upper crucible at the gravity coefficients of 1 due to the overlarge filtration resistance from the CFF. At $G = 600$, the filtration resistance was overcome and the filtered zinc was collected in the lower crucible, associated with the unfiltered residue intercepted in the upper crucible by CFF. The filtered zinc was quite dense, while the residue had a relatively porous morphology.

In order to further investigate the phase distribution in the residue, the microstructure of areas 'A and B' in Fig. 7b was presented in Fig. 8, which represented the top and bottom regions in the residue obtained at $G = 1000$, $T = 510$ °C and $t = 180$ s, respectively. The top region (Fig. 8a) mainly consisted of sinuous

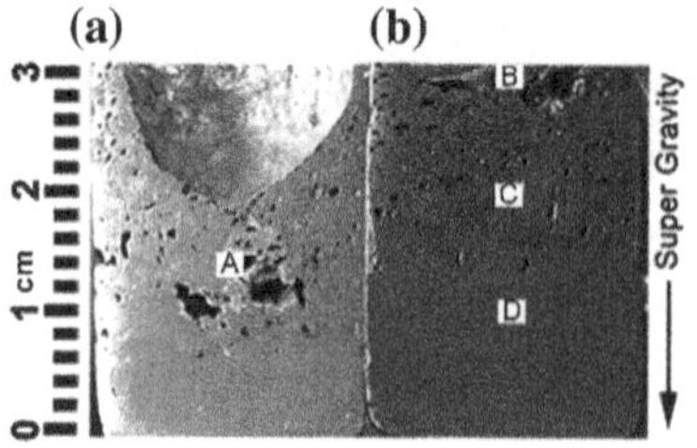

Fig. 4 Macrographs of the samples obtained by super-gravity enrichment at various gravity coefficients, $t = 180$ s and $T = 510$ °C. **a** $G = 1$; **b** $G = 600$

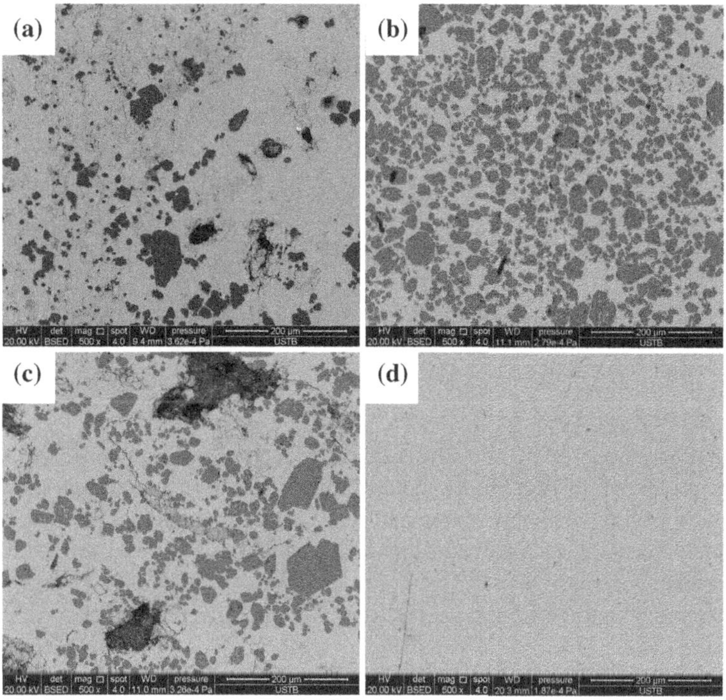

Fig. 5 Micrographs of four areas in the samples obtained by super-gravity enrichment (**a**, **b**, **c**, **d** refer to areas A, B, C, D marked in Fig. 4, respectively)

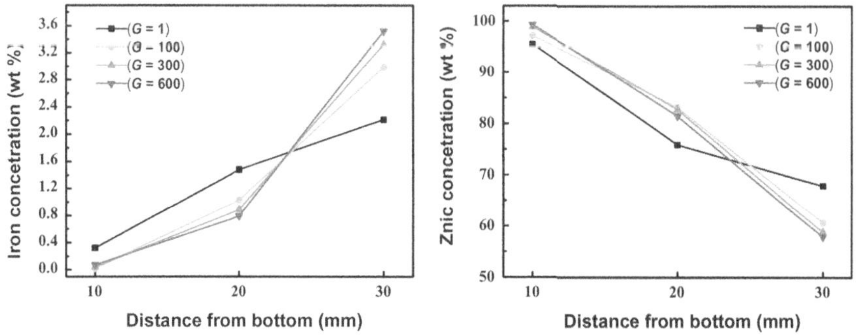

Fig. 6 The distribution of Fe and Zinc inside samples solidified under different super-gravity enrichment processed at t = 180 s and T = 510 °C

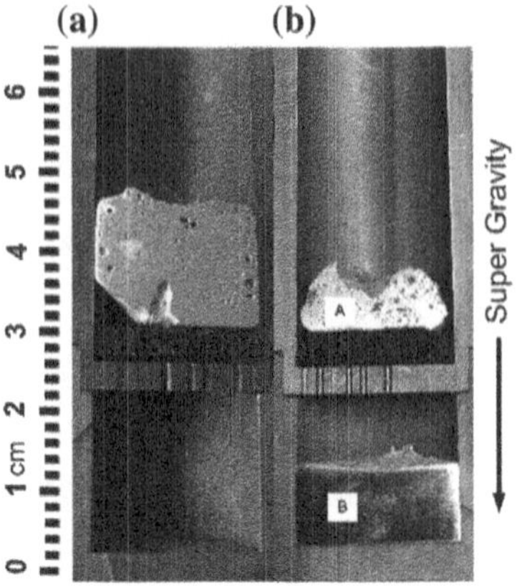

Fig. 7 Macrographs of the samples obtained by super-gravity separation at various gravity coefficients, t = 180 s and T = 510 °C. **a** G = 1; **b** G = 600

ZnO and an amount of Al–Fe–Zn particles, which was also remained zinc in this region. As shown in Fig. 8b, the bottom region had a relatively high-purity zinc (99.9 wt%). With much observation and analysis by SEM at a higher magnification, only a few fine Al–Fe–Zn particles (generally smaller than 5 μm) were found in the filtered zinc.

The removal ratio of Fe and the recovery ratio of Zn were calculated via Eqs. (2) and (3).

$$R_{\text{Znic}} = \frac{m_{\text{fz}} \times \omega_{\text{Zn\_fz}}}{m_{\text{fz}} \times \omega_{\text{Zn\_fz}} + m_{\text{r}} \times \omega_{\text{Zn\_r}}} \tag{2}$$

$$\eta_{\text{iron}} = 1 - \frac{m_{\text{fz}} \times \omega_{\text{iron\_fz}}}{m_{\text{fz}} \times \omega_{\text{iron\_fz}} + m_{\text{r}} \times \omega_{\text{iron\_r}}} \tag{3}$$

where R_{Znic} and η_{iron} are the recovery ratio of Zn and the removal ratio of Fe; $\omega_{\text{Zn\_fz}}$ and $\omega_{\text{Zn \_r}}$ denote the Zn content (wt%) in the filtered zinc and residue, respectively;

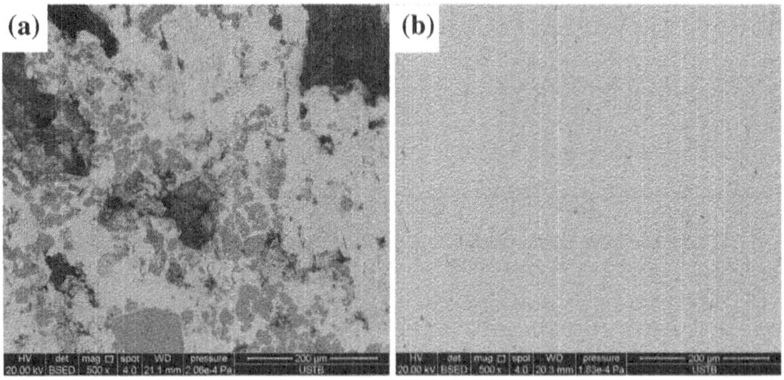

Fig. 8 BSE images of three different regions (areas ‘A and B’ in Fig. 7b) in the residue obtained at G = 600, T = 510 °C and t = 180 s. **a** Top; **b** Bottom

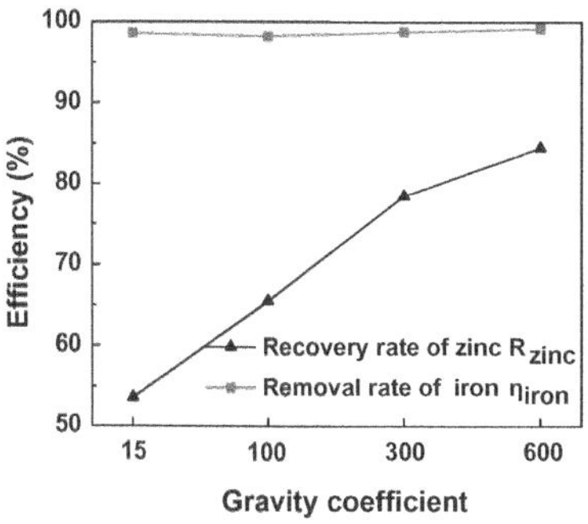

Fig. 9 Effect of gravity coefficient on the separation efficiencies processed at t = 180 s and T = 510 °C

$\omega_{\text{iron\_fz}}$ and $\omega_{\text{iron\_r}}$ denote the iron content (wt%) in the filtered zinc and residue, respectively.

As shown in Fig. 9, in general, R_{Znic} increased with increasing the gravity coefficient, mainly since more molten zinc was filtrated into the lower crucible when the gravity coefficient was improved. R_{Znic} increased quickly from 53.6 to 84.5 wt% with increasing the gravity coefficient from 15 to 600. When the gravity coefficient exceeded 600, R_{Znic} still increased but at a relatively low rate. In comparison, increasing the gravity coefficient, η_{iron} was not with obvious fluctuation. The η_{iron} fluctuated in the range of 98.2–99.2 wt% with changing the gravity coefficient from 15 to 600.

Discussion

As for super-gravity enrichment, when the temperature reached 510 °C, the Al–Fe–Zn precipitated phase appeared a relatively uniform distribution in zinc liquid at G = 1. Due to the high viscosity, the Al–Fe–Zn phase cannot be separated from residual melt under the normal gravity [11]. The moving equation of a spherical solid particle in the viscous liquid can be expressed as follows according to Stokes' law:

$$\frac{\pi}{6}d^3(\rho_1 - \rho_p)Gg - 3\pi\eta d\frac{\mathrm{d}r}{\mathrm{d}t} = \frac{\pi}{6}d^3\rho_p\frac{\mathrm{d}^2r}{\mathrm{d}t^2} \quad (4)$$

where d is the diameter of the particle, m; ρ_1 and ρ_p are the densities of the liquid and solid particles, respectively, kg m^{-3}; η is the dynamic viscosity of the liquid, Pa s; and r is the displacement of the particle, m.

When the particle's movement reaches equilibrium, the right side of Eq. (4) becomes equal to zero, and the particle's moving velocity can be expressed as in the following equation:

$$Vr = \frac{d^2(\rho_1 - \rho_P)}{18\eta} Gg \tag{5}$$

where Vr is the particle's moving velocity in the viscous liquid.

The particle's moving velocity in equilibrium is in direct proportion to the square of its diameter and gravity coefficient according to Eq. (5). Thus, by applying super gravity, the moving velocity is increased, leading to the pile up of Al–Fe–Zn phase in the upper area. It can be also speculated that the larger Al–Fe–Zn particles should move faster than the finer ones, resulting in the appearance of larger particles in the top area and the finer ones in the lower positions theoretically after the super-gravity enrichment treatment.

As for super-gravity separation, the filtration processes can be discussed as Fig. 10. The capture of the solids is achieved by the adhesion of solids on the surface of the filter medium due to interfacial or surface forces.

To achieve the filtration of molten zinc, the centrifugal pressure (P_c) caused by centrifugal force acting on the surface of filter medium is required to be larger than the filtration resistance offered by the filter medium (i.e. the pressure drop across the filter-molten metal interface, $\Delta\rho$) [12]. P_c can be calculated by

$$P_c = \frac{\rho\omega^2(L_2^2 - L_1^2)}{2} \tag{6}$$

where ρ denotes the density of the molten zinc, ω is the angular velocity, and L_2 and L_1 denote the molten zinc levels measured from the centre of rotation. More detailed description can be found in Ref. [13].

$\Delta\rho$ can be described by the Young–Laplace equation:

$$\Delta\rho = -\frac{4\sigma\cos\theta}{d} \tag{7}$$

where σ denotes the surface tension of molten zinc, θ is the wetting angle, and d is the mean pore size of the filter medium.

At the beginning of super-gravity separation, only the filtration resistance is required to be overcome. But as the super-gravity separation proceeds, reticulated

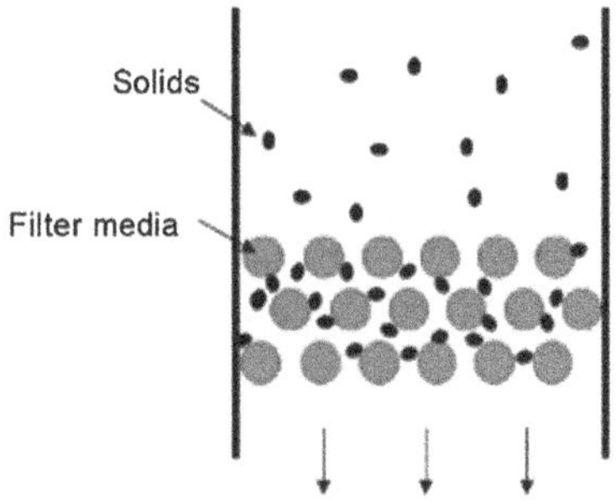

Fig. 10 Schematics of filtration mode in super-gravity separation

ZnO and Al–Fe–Zn particles accumulate above the CFF to form the cake of which the thickness increases gradually. This progressively decreases the effective opening diameter [d in Eq. (7)] and increased $\Delta\rho$. Increasing gravity coefficient can improve the P_c and overcome higher $\Delta\rho$; therefore, more molten zinc can be flowed through the filter medium, leading to the higher R_{zinc} and η_{iron} (Fig. 9).

Conclusions

In this study, detailed characterization of a hot-dip galvanizing dross was conducted and super gravity was proven to be a promising method for removing iron impurities and recovering zinc from galvanizing dross. In super-gravity enrichment, the distribution of Al–Fe–Zn particles had an obvious gradient, and with the increase of gravity coefficient, the gradient became steeper. In super-gravity separation, at $G = 600$, $t = 180$ s and $T = 510$ °C, over 84.5 wt% zinc was recovered with a high purity of about 99.9 wt%, and about 99.2 wt% iron impurity was removed.

Acknowledgements This work was supported by the National Natural Science Foundation of China (No. 51704022), the Fundamental Research Funds for the Central Universities (FRF-TP-16-036A1) and the project of State Key Laboratory of Advanced Metallurgy (41617002).

References

1. Arioka T, Hori M, Toki T, Yakawa A, Morikawa M (2001) A hydrometallurgical process for the recovery of zinc. Sep Purif Technol 161:172–186
2. Dong AP, Shu D, Wang J, Cai X, Sun B, Cui J, Shen J, Ren Y, Yin X (2008) Continuous separation of Fe–Al–Zn dross phase from hot dip galvanised melt using alternating magnetic field. Mater Sci Technol 24:40–44
3. Liu YZ, Zhou X, Chen Y, Zhang Z, Zhou LL (2015) Clostridium butyricum can be used as a potential alternative for the antibiotic in Cherry Valley ducks. J Anim Plant Sci 25:99–106
4. Li DG (2005) Selective precipitation and separation of valuable constituent in blast furnace slags. *Northeastern University* 25–32
5. Gopala A, Kipphardt H, Matschat R, Panne U (2010) Process methodology for the small scale production of purity zinc using a resistance heated vacuum distillation system. Mater Chem Phys 122:151–155
6. Zheng H, Gu Z, Zheng Y (2008) Electrorefining zinc dross in ammoniacal ammonium chloride system. Hydrometallurgy 90:8–12
7. Ren X, Wei Q, Hu S, Wei S (2010) The recovery of zinc from hot galvanizing slag in an anion-exchange membrane electrolysis reactor. J Hazard Mater 181:908–915
8. Wang Z, Gao JT, Shi AJ, Meng L, Guo ZC (2018) Recovery of zinc from galvanizing dross by a method of super-gravity separation. J Alloys Compd 735:1997–2006
9. Zhao LX, Guo ZC, Wang Z, Wang MY (2010) Removal of low-content impurities from Al by super-gravity. Metall Mater Trans B 48:93–97
10. Zhao LX, Wang Z, Guo ZC (2013) Purification of metallurgical-grade silicon by Sn–Si refining system with calcium addition. Sep Purif Technol 21:1185–1192

11. Wang Z, Gao JT, Shi AJ, Meng L, Guo ZC (2017) Recovery of zinc from Zn–Al–Fe melt by supergravity. ISIJ Int 10:735–738
12. Reuter M (2015) Metal recycling: opportunities, limits. Infrastructure Miner Eng 79:2–5
13. Damoah L, Zhang L (2010) Removal of inclusions from aluminum through filtration. Metall Mater Trans B 41:886–907

Parameters of the Metallic Calcium Reduction from Magnesium Production Residues

K. C. Tasyurek, M. Bugdayci and O. Yucel

Abstract In this paper, we presented an investigation of the theoretical and the industrial definitions of the metallic calcium production by metallothermic process in the vacuum atmosphere. In the studies, Al is the only reductant that is used for the metallothermic calcium production. In experiments, effects of percentage of Al stoichiometry as a reductant, time variances of reduction and temperature of reduction were investigated. In this study, experiments were carried out at 1200, 1250, 1300 °C and 100% Al, 125% Al and 150% Al stoichiometry to produce metallic calcium from the residue of metallic magnesium production residue. All raw materials and residue phases were characterized by using atomic absorption spectrometry (AAS), X-Ray Diffraction Spectrometry (XRD) and chemical analysis techniques. From the results of these experiments, it was understood that the use of Al with the increasing stoichiometry increases the efficiency of the metallic calcium production with the increase of reduction time.

Introduction

Calcium metal is a kind of a shiny silver white metal, and physical properties are quite ductile and soft. Metallic calcium is rapidly oxidizing in the atmosphere above 300 °C [1, 2]. The calcium as a metal reacts with water to form calcium hydroxide. Calcium hydroxide formed melts at 845 °C and boils at 1420 °C. Calcium metal is highly reactive with oxygen due to this it is never found in elemental form naturally. Primary calcium sources are the three carbonates—calcite, aragonite and

---

K. C. Tasyurek (✉) · M. Bugdayci · O. Yucel
Metallurgical and Materials Engineering Department, Faculty of Chemical and Metallurgical Engineering, Istanbul Technical University, 34469 Maslak, Istanbul, Turkey
e-mail: kctasyurek@gmail.com

M. Bugdayci
Chemical and Process Engineering Department, Faculty of Engineering, Yalova University, 77100 Yalova, Turkey

© The Minerals, Metals & Materials Society 2019

T. Jiang et al. (eds.), *10th International Symposium on High-Temperature Metallurgical Processing*, The Minerals, Metals & Materials Series,
https://doi.org/10.1007/978-3-030-05955-2_45

dolomite—the sulphates—gypsum and anhydrite—and fluorites in the form of fluorite or fluorspar [2, 3].

Main calcium producer countries are Russia, China, USA and France. Russia's metallic calcium production factories capacity is 4000 tons. In 2011, China increased the quantity of calcium production, and the common reason for this increasing is related with there were only two companies producing the calcium with the electrolytic way in the 1960s but after production by aluminothermic process has started and many companies has begun the metallic calcium production. China's yearly production capacity is 30,000–35,000 tons of calcium with aluminothermic production method, and they still produce 4000–6000 tons of calcium using electrolytic process. USA produces 1000–2000 tons of calcium by aluminothermic method. Also, France has some facilities for the production of calcium metal but they are not producing for now. The sources of the calcium are almost unlimited because of the presence of living organisms [4–6].

The first metallic calcium reduction is related to Pidgeon process where aluminum was used as a reducing material. The reaction is done at 1170 °C in a steel retort under vacuum atmosphere. The stoichiometry of aluminum had taken %100–110–120, respectively. And calcium reduction recoveries have been observed %85.8, %92.0 and %96. The experiment's heating cycles are 12, 17 and 24 h, respectively [7].

Experimental

Domestic sources of Mg reduction residue were provided as raw material for using calcium source as a raw material. Chemical analysis and AAS (Perkin Elmer Analyst 800) techniques were applied to residue. Chemical analysis of Mg production residue is given in Table 1, and Al (min. %99.5 by wt.) was also employed as a reductant.

In Fig. 1, the XRD pattern of residue was obtained by using PANalytical PW 3040/60 XRD. XRD pattern is given. The residue includes mainly CaO and SiO_2 according to the XRD pattern.

In the experimental studies, cylindrical shape of stainless steel retort was used. Metallic calcium was vapourized; then, the residue was inside of the ceramic pod. The vacuum atmosphere inside the retort was provided by dual-stage rotary vane mechanical vacuum pump, Edwards 8 E2M8 Rotary Vane Dual Stage Mechanical Vacuum Pump, which can hold a final pressure of 1×10^{-3} mbar in Fig. 2. By the SiC resistance furnace, the retort was externally heated at maximum temperature of 1350 °C. All experiments were related at 1200 °C, 1250 °C and 1300 °C and

Table 1 Chemical analysis of Mg production residue

CaO	MgO	Al_2O_3	Fe_2O_3	SiO_2	Na_2O	K_2O
57.65	3.68	0.35	5.55	28.35	0.57	0.18

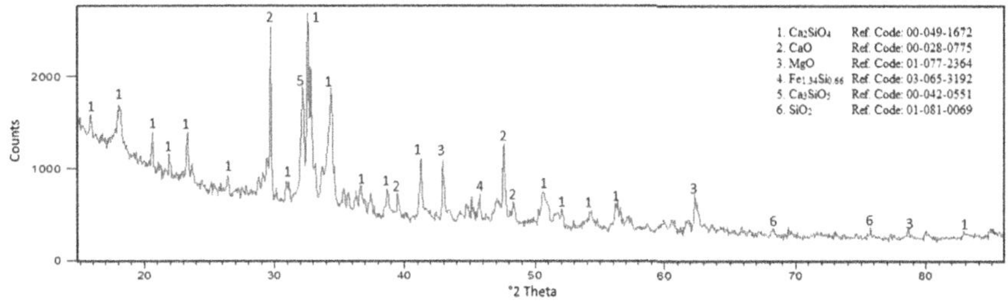

Fig. 1 X-ray diffraction (XRD) pattern of raw materials of residue

conducted with 100% Al, 125% Al and 150% Al stoichiometry of reductant and the times depends on 60–120–180–240 min.

The steel retort was left for cooling up at the furnace with the same vacuum values with the reduction and it was continued until the room temperature. Then cover was opened it up and the residue left in the ceramic base were weighted and analyzed. The rate of Ca metal recovery was calculated from residue by using the equation given below.

$$\text{Ca recovery}\ \%\ (\text{from residue}) = \{[(\text{Ca}_{0\%} \times W_0) - (\text{Ca}_{1\%} \times W_1)]/(\text{Ca}_{0\%} \times W_0)\} \times 100 \quad (1)$$

where W_0 is the weight of raw materials, $\text{Ca}_{0\%}$ is the weight percentage of calcium in raw materials, W_1 is the weight of residue, and $\text{Ca}_{1\%}$ is the weight percentage of calcium in residue.

The research studies of the thermodynamic parameters were calculated by using the FactSage 6.4 database. Under 1 mbar and 1 bar vacuum pressures, Fig. 3 is plotted to understand the changing of the reduction conditions of MgO and CaO

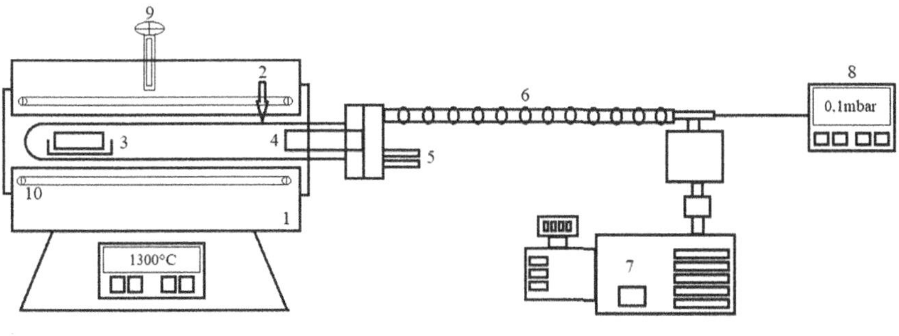

Fig. 2 Schematic sketch of experimental setup. (1) Furnace, (2) stainless steel retort, (3) charge, (4) condensation section, (5) cooling water in and out, (6) vacuum connection, (7) vacuum pump, (8) digital pressure gauge, (9) PtRh30/PtRh6 thermocouple and (10) SiC resistance

using the Al reductant. For 1 bar pressure, it can be seen that the reaction temperature needs to be 2332 °C for CaO reduction with an Al reductant.

Production of metallic calcium vapour pressure is important for condensing the calcium as a metal in the process. Figure 3 and Table 2 give reduction temperatures of MgO, CaO with Al reductant, and it is clearly seen that CaO reduction starts at 1298 °C. Figure 4 shows the probable phases of the CaO reduction process. $Ca_{(g)}$ phase increased dramatically after 1150 °C. At 1300 °C and 2.25 mol, the $Ca_{(g)}$ occurred. Due to the thermodynamic investigations, $Ca_{(g)}$ occurred between 1150 and 1200°C. The figures are plotted for the 100% stoichiometry of the reductant Al at 1 mbar.

Pidgeon process reactions occur at lower temperatures at 1 mbar process duration than that of 1 bar [8]. Figure 5 shows the effect of CaF_2 addition and $Ca_{(g)}$ products were decreased above 1200 °C. Figure 6 shows the effect of FeSi as a reductant and $Ca_{(g)}$ starting to occur above 1400 °C. Neither of these additions have positive effect on CaO reduction.

In Fig. 7, Ca vapour pressure started and increased dramatically at 650 °C. In this study of this project, an average process pressure of 0.5 mbar was used and the process temperatures were 1200, 1250 and 1300 °C. Therefore, the formation of the gas phase of calcium is possible with Al used as a reductant and 1 mbar under vacuum conditions.

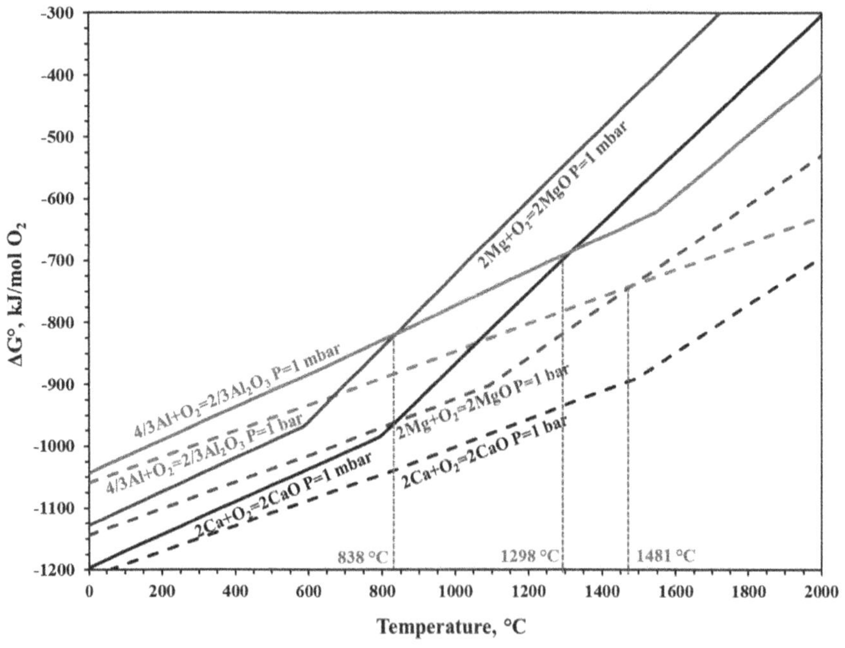

Fig. 3 Change of Gibbs free energy for the reduction of CaO and MgO using Al reductant at 1 mbar and 1 bar

Table 2 Minimum reduction temperatures of CaO and MgO for Al reductant at 1 bar and 1 mbar

Minimum reduction temperature (°C)		
	1 bar	1 mbar
CaO	2332	1298
MgO	1481	838

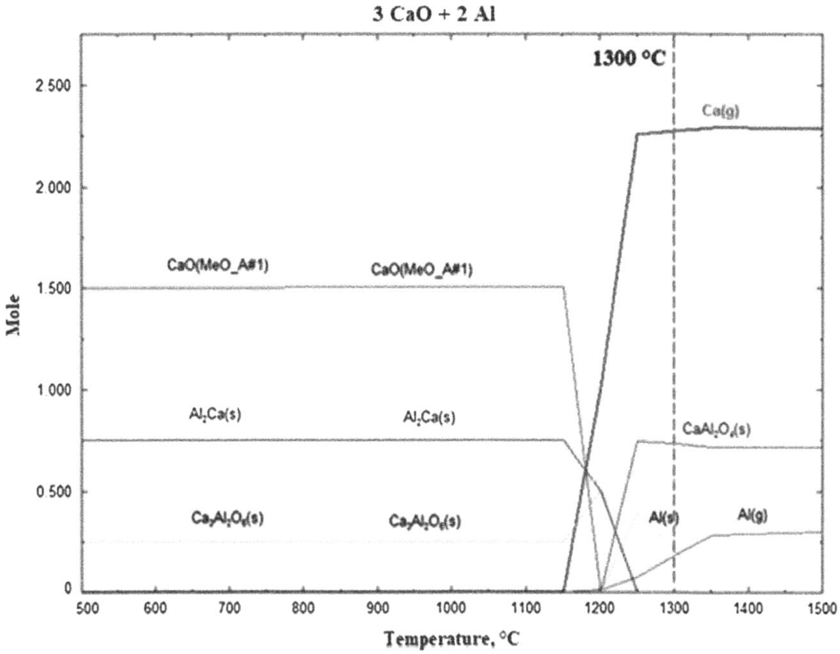

Fig. 4 Change of reaction products with increasing process temperature for Al reductant (1 mbar)

In this study, experiments were set up to investigate the effects of reductants on the Pidgeon process with the changing of the stoichiometry of Al and the time. The change of Ca recovery was investigated for 100, 125 and 150% Al reductant stoichiometry while increasing the reduction time. All experiments were done with 1 L retort, and the experiment's process duration is started with 60 min and increased to 240 min step by step. Average process pressure was about 0.5 mbar.

Results and Discussion

Experimental studies began with the theoretical requirement 100% Al stoichiometric reductant, and time effect was investigated for calcium reduction at 1200 °C. Raw materials for metallic calcium production residue and 100% Al stoichiometric

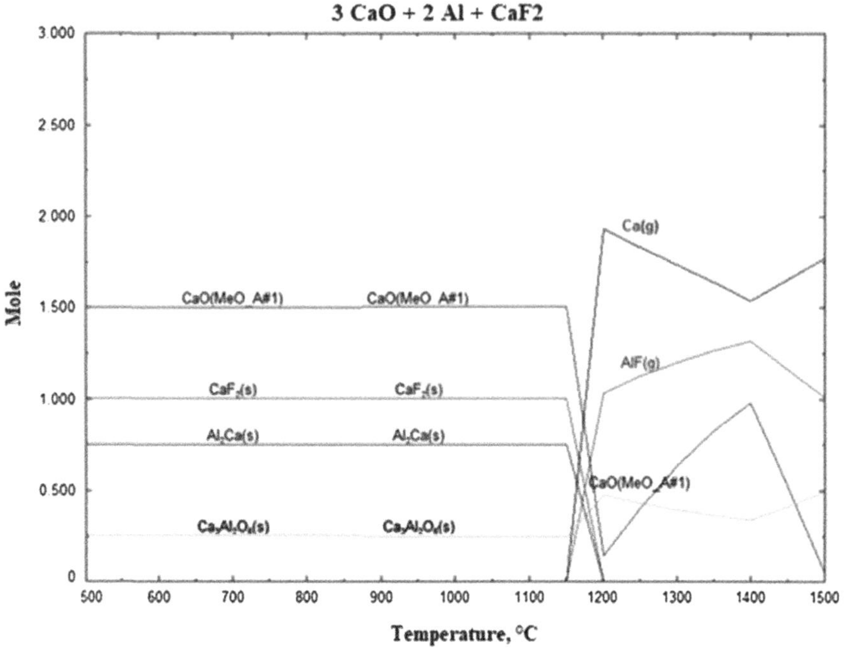

Fig. 5 Change of reaction products with increasing process temperature for Al and CaF_2 reductant (1 mbar)

powder mixtures were prepared, and experiments were conducted for the same amount of mixtures value of 9 g. Recovery rate of calcium was calculated from the residue. It was observed that the 100% Al reductant and the extended time did not affect the recovery rate of calcium.

The recovery rate results were 3–19% in the first experiments. Then, Al stoichiometry was increased to 125%, and the recovery rate results were 5–21% in this experiment. After 150% Al stoichiometry was tried, the recovery rate results were 5–20%. This showed us the amount of the reductant was not enough to get Ca reduction with high recovery rates at Fig. 8.

In the second set of experiments, the reaction temperature was fixed at 1250 °C. This time, the highest recovery detected for the total charge amount of 9 g, 150% Al stoichiometry, 1250 °C and 240 min was 22%. Figure 8 shows the Ca recovery rates for Al reductant with increasing stoichiometry and process duration at 1250 °C. The differences evaluated between the Ca recovery rates increased with the time and Al stoichiometry.

The recovery results were increased from 7 to 20% at 100% Al stoichiometry. Then, Al stoichiometry was increased to 125%, and the recovery results were 6–20%. After the experimental set was repeated at 150% Al stoichiometry, the recovery rate results were 11–22%. This showed us the amount of the reductant was not enough for getting Ca reduction with high recovery rates.

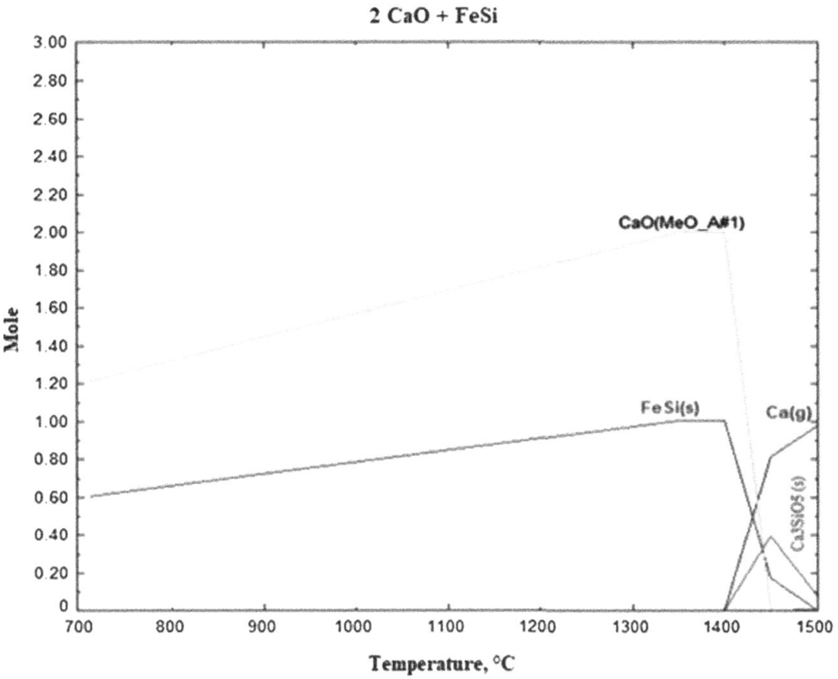

Fig. 6 Change of reaction products with increasing process temperature for FeSi reductant (1 mbar)

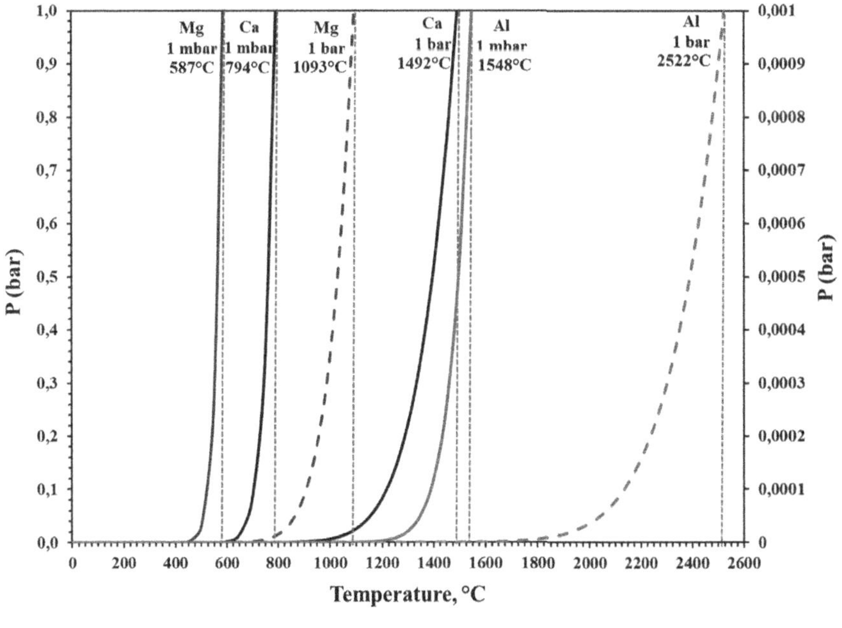

Fig. 7 Change of vapour pressure of the Ca, Mg and Al at 1 mbar and 1 bar

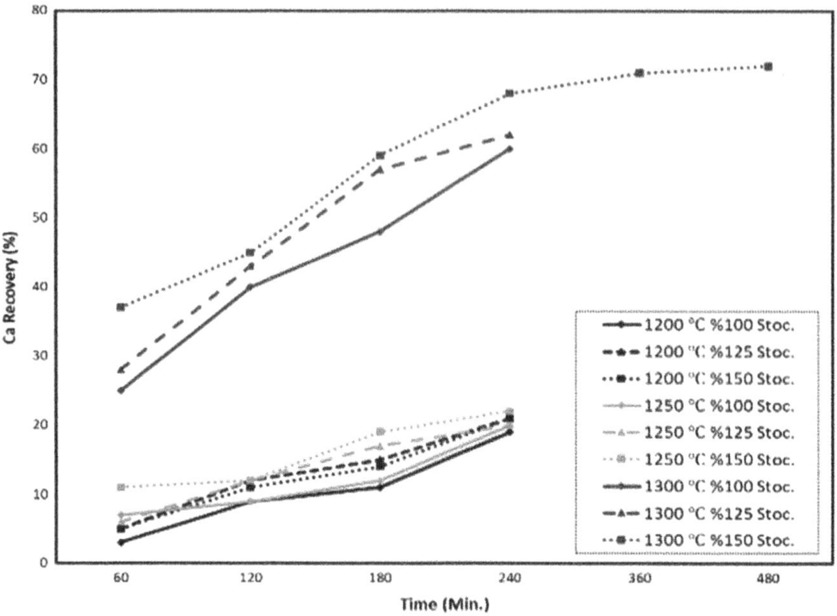

Fig. 8 Ca recovery rates for Al reductant with increasing stoichiometry and process duration at 1200, 1250 and 1300 °C for 60–480 min under 1 mbar

In the third experimental set, the reduction temperature was set up at 1300 °C, and the highest CaO amount in residue was obtained in the experiment conducted with 150% Al and 240 min with 68% recovery rate. The recovery results were 25–60% at 100% Al stoichiometry. When Al stoichiometry was changed to 125%, the recovery results were 36–67%.

Finally, Al stoichiometry was fixed at 150% and the recoveries were 37–72%. This showed us the amount of the reductant was enough for getting Ca reduction with high recovery rates. In brief, increasing time and the reductant stoichiometry likewise increases the recovery of Ca.

Conclusion

The reduction of residue and raw materials were investigated by using different reductant stoichiometries of 100% Al, 125% Al and 150% Al at 1200, 1250 and 1300 °C while varying the process durations between 60 and 240 min. With these results of experimental studies, the highest Ca recovery from residue is obtained at the experiment with 9 g charge weight, 150% stoichiometric Al at 1300 °C with 480 min reduction time; therefore, highest calcium reduction recovery rate was investigated 72%.

Despite the high cost of Al as a reductant for the metallothermic reduction of calcium, the optimization of the process is too important for the mass production expenses. The experimental results showed us that the time increased from 60 to 480 min, and recovery rate of Ca increased. For instance, 150% Al stoichiometry 60 min has recovery ratio about 37%; meanwhile, 150% Al 480 min Ca recovery ratio increased to 72% at 1300 °C.

References

1. Cardarelli F (2008) Materials handbook: a concise desktop reference, 2nd edn. Springer, London, pp 620–622
2. Mantell CL (1973) The alkaline earth metals: calcium, barium, and strontium. In: Hampel CA (ed) Rare metals handbook, 2nd edn. Reinhold, New York, pp 15–25
3. Jia X, Cheng-Fu G, Yong-Gang J, Yong-He W, Cent J (2010) South Univ Technol 17:918–923
4. Emsley J (2011) Nature's building blocks, an A-Z guide to the elements. Oxford University Press, Oxford
5. Gray Theodore (2009) The elements, a visual exploration of every known atom in the universe. Black Dog & Leventhal Publishers Inc, New York
6. Stwertka A (2012) A guide to the elements, 3rd edn. Oxford University Press, Oxford
7. Pidgeon LM, MacCatty SA (1949) Production of calcium, USA Patent Office, 15 Mar 1949
8. Beruto D, Searcy AW, Kim MG (2004) Thermochim Acta 424:99–109

Smelting Studies for Recovery of Iron from Red Mud

Ender Keskinkilic, Saeid Pournaderi, Ahmet Geveci and Yavuz A. Topkaya

Abstract Red mud can be regarded as a by-product of aluminium extraction process since it contains a significant amount of iron and some valuable elements. Therefore, the treatment of red mud has been a hot topic for some decades. The authors have recently started a laboratory-scale project dealing with stepwise recovery of valuable elements from red mud of Seydisehir Aluminum Plant, Turkey. The first step is related to recovery of iron and pyrometallurgical methods (smelting and solid-state reduction) will be performed. Nonferrous metals will then be selectively leached in the second step. In the extent of the present work, a literature review relevant to the smelting studies for recovery of iron from red mud was presented.

Keywords Red mud · Pyrometallurgy · Smelting · Iron

Introduction

Red mud is also called as bauxite residue, which is the waste material and more correctly by-product obtained from Bayer process for alumina production. The amount of red mud is at least equal to the amount of alumina produced, and most of the time, it is significantly more [1]. World bauxite residue production is more than 70 Mt annually [2]. Therefore, a huge amount of material is being accumulated continuously near the aluminium production plants, causing allocation of

E. Keskinkilic (✉)
Department of Metallurgical and Materials Engineering, Atilim University, 06830 Ankara, Turkey
e-mail: ender.keskinkilic@atilim.edu.tr

S. Pournaderi
Agri Ibrahim Cecen University, Ataturk Mah, 04500 Patnos, Agri, Turkey

A. Geveci · Y. A. Topkaya
Department of Metallurgical and Materials Eng, Middle East Technical University, 06531 Ankara, Turkey

© The Minerals, Metals & Materials Society 2019

T. Jiang et al. (eds.), *10th International Symposium on High-Temperature Metallurgical Processing*, The Minerals, Metals & Materials Series,
https://doi.org/10.1007/978-3-030-05955-2_46

substantially large area for disposal and more importantly environmental issues related to toxicity of bauxite residue [3]. As well as their amounts in red mud are concerned, iron oxide and aluminum oxide are the two primary components, so efforts have been made to recover iron and aluminum from red mud for more than fifty years [4]. Red mud has also been regarded as a source for recovery of certain elements, such as rare-earth elements (REEs). A typical bauxite residue may contain the following phases like hematite, goethite, gibbsite, diaspore, calcite and cancrinite. [5].

Recovery of iron from red mud is important for two respects: first, iron oxide is the major oxide component, source of iron, although it is not as high as the one appearing in economical iron ores. Second, removal of iron from bauxite residue is helpful to increase the recoveries of the elements in leaching step. Leaching without prior iron removal results in significant iron dissolution, which badly affects the leaching performance of REEs to be recovered [5]. There are many studies related to either the first and/or the second target. Both studies can be classified mainly in three categories: (a) solid-state reduction, (b) magnetic separation and (c) smelting.

The use of red mud in iron and steelmaking was one of the older smelting examples for iron recovery [6]. Mozharenko and Noskov proposed the use of red mud as a charge material to blast furnaces [7]. On the other hand, the direct use of bauxite residue in blast furnaces causes a significant decrease in refractory life due to its high alkaline content [8]. High alumina content of red mud increases viscosity of blast furnace slag, so this cannot be tolerated for successful furnace operation. Even after smelting of red mud for pig iron production, the slag's alkali composition and higher viscosity of slag remain big problems for the vessel where smelting is conducted. To overcome the problem associated with high viscosity, great amount of CaO-containing fluxes (lime or limestone) are to be used to increase CaO/SiO_2 ratio from the values less than unity to nearly two [2]. Therefore, high slag volume and alkali problem are the two bottlenecks of smelting operations. High variability of red mud compositions from plant to plant also prevents the establishment of a well-defined red mud smelting process [9]. Due to those factors, smelting part of red mud treatment still has important problems to be solved for successful industrial operation.

The authors have recently started a laboratory-scale project dealing with stepwise recovery of valuable elements from red mud of Seydisehir Aluminum Plant, Turkey. The first step in the project is related to recovery of iron and pyrometallurgical methods (smelting and solid-state reduction) will be performed. Nonferrous metals will then be selectively leached in the second step. In the extent of this work, smelting studies for recovery of iron from red mud were presented. The investigations dealing with solid-state reduction and magnetic separation are outside the scope of this paper.

Smelting Studies for Recovery of Iron

Smelting studies for recovery of iron from red mud was dating back to the end of sixties and much was performed at the beginning of seventies. In one of these studies, electric arc furnace (EAF) smelting of Jamaican and Surinam red mud were performed for production of iron and alumina [10]. The major components of Jamaican and Surinam red mud are shown in Table 1.

Pellets made up of red mud and ground limestone were charged to EAF together with the coke. After complete melting, the bath temperature was increased to ~1500 °C, and then the furnace was shut down after about 15 min. After cooling of around one day, pig iron and slag were removed from the furnace. The slag was then subjected to leaching with Na_2CO_3 to obtain soluble sodium aluminate. After a number of tests conducted with both Jamaican and Surinam red mud, it was found that iron recoveries could reach 98% and alumina recoveries of 84% could be obtained. Slightly better iron recoveries were reported for Jamaican red mud [10].

It was outlined in a paper dealing with metal recovery from scrap and waste that in 1971, Mc Dowell-Willman Engineering Co. developed a process for the smelting of prereduced red mud pellets in EAF to obtain pig iron followed by LD converter process to produce steel. It was reported that it could be an alternative way for steel production for the countries where there are significant number of aluminum plants but no integrated iron and steelmaking industry [1]. Other than soda–lime–carbon sinter process investigated by Fursman et al. [10], complex separation–melting process (Hungary) and smelting–slag disintegration process (Yugoslavia) were the pilot plant studies conducted in the past [11]. It was indicated in the literature that ROMELT process [12] established for iron making by Moscow Institute of Steel and Alloys (MISA) was considered by Indian companies NALCO and RSIL to smelt bauxite residue [13]. In relation to the smelting of red mud, patents can easily be found in the literature, so some examples are presented in the references for the reader's attention [14–16].

Kauβen and Friedrich [8] outlined the pig iron compositions obtained by Logomerac [17] and Ziegenbalg [18] in the red mud smelting experiments performed in EAF using red mud as a charge material (Table 2). As it can be seen from these literature data, it seems that the variabilities of elements in pig iron are very high.

Kumar et al. reviewed the innovative methodologies for the utilisation of blast furnace slag, red mud and fly ash. In the scope of that work, it was outlined that

Table 1 Chemical composition of Jamaican and Surinam red mud [10]

Material	Al_2O_3	SiO_2	TiO_2	Fe_2O_3	Na_2O	CaO
Jamaican red mud	14.6	3.40	6.87	50.9	3.18	4.96
Surinam red mud	19.0	11.9	12.15	24.81	9.29	3.6

Table 2 Pig iron compositions in EAF smelting tests from the literature [8]

Element (wt%)	Logomerac [17]	Ziegenbalg [18]
C	3.8–4.5	4.2–5
Si	0.5–3	0.3–0.6
Ti	0.1–0.3	0.2–0.7
Cr	0.2–0.4	–

Thakur and Das [6] smelted red mud in EAF with the presence of coke at 1600–1700 °C and that iron alloy having recoveries of 90% was obtained. It was further indicated that slag was subjected to leaching with sulphuric acid to recover iron oxide and titanium dioxide [11].

Kauβen and Friedrich [8] studied reductive smelting of red mud for iron recovery using laboratory-scale EAF. The average composition of red mud is given in Table 3. In each run, approximately 200 g of lignite coke was smelted together with 3200 g of red mud in dried condition. Temperature of the EAF was reported in the range of 1600–1700 °C. It was found that commercial pig iron having composition of ~4 wt% carbon, less than 0.5 wt% silicon and 0.2 wt% titanium could be obtained. The authors also discussed the alkali problem created by the presence of Na_2O. The boiling temperature of metallic sodium is fairly low (897 °C) compared to smelting temperatures. Therefore, a significant amount of sodium vapour formed during reduction of Na_2O with solid carbon. It was indicated that sodium vapour then showed reoxidation in the off-gas chamber [19]. It was reported that total alkali content after reduction was below 0.1 wt% [18].

Balomenos et al. targeted a zero-waste process using EAF technology. Greek red mud having composition of 16.22 wt% Al_2O_3, 47.74 wt% Fe_2O_3, 10.73 wt% CaO, 6.08 wt% SiO_2, 5.93 wt% TiO_2 and 2.51 wt% Na_2O was smelted in a dust treating EAF to obtain pig iron and viscous slag. Pig iron was reported to contain 93.206 wt % Fe, 4.802 wt% C, 0.737 wt% Si and 0.566 wt% Ti. CaO, Al_2O_3, SiO_2 and TiO_2 were reported as the major components of the viscous slag. As a part of ENEXAL project, the viscous slag was subjected to high-speed air/water jet and therefore fiberized aiming to be used as mineral wool. Insulating mineral wool necessitates lower radioactive content, so the authors had to use at least 220 kg of fluxes per ton of red mud to decrease the parameter related to radioactivity called as "activity concentration index, I" below 6. Actually, one and a half time greater amount of fluxes were added to achieve the desired CaO/SiO_2 ratio, in the range of 0.8–1 [20]. More detail about fiberization tests were published in another paper. According to SEM analysis results, inorganic fibres having diameter 20 μm were obtained and some thicker fibres were also reported. The authors indicated that commercial

Table 3 Average composition of red mud used in EAF smelting experiments [8]

Component (wt%)	Fe_2O_3	Al_2O_3	SiO_2	TiO_2	Na_2O	CaO
Red mud	30–50	10–20	5–20	3–15	3–7	1–8

vitreous wool products have diameters in the range of 3–10 μm. The authors concluded that uniform fibres could be produced from slag melts by melt blowing or melt spinning [21].

Ercag and Apak investigated smelting of Turkish bauxite residue from Seydisehir to obtain pig iron and slag, which was then leached to recover TiO_2 and Al_2O_3. Average composition of red mud from Seydisehir Aluminum Plant is shown in Table 4 [22]. With the addition of bentonite, 100 g red of red mud, 100 g of dolomitic limestone and 16.16 g of coke were pelletized and sintered in an ash furnace at 1100 °C for 1 h. Afterwards, the pellets were placed in a graphite crucible and charged to a high-temperature furnace heated from room temperature to 1550 °C and kept for 1 h at this temperature. The furnace was then cooled down nearly one day. With magnetic separation, 11.89 g of pig iron was collected. Chemical composition of pig iron was reported as 95.68 wt% Fe, 1.10 wt% Ti and 0.0066 wt% Al. The amount of slag was indicated as 100 g. Slag TiO_2 and Fe_2O_3 compositions were measured as 1.25 wt% and 0.2776 wt%, respectively [23]. The slag was then subjected to leaching and solvent extraction processes to obtain Al_2O_3 and pigment-grade TiO_2 [22].

Borra et al. studied smelting reduction of red mud for iron and subsequent REE's recovery. The chemical analysis of the bauxite residue used in that study is illustrated in Table 5. The red mud was mixed with calcium silicate (CaO: 51.2 wt% and SiO_2: 46.4 wt%), and graphite powder and handmade pellets were prepared. After drying at 105 °C for one day, pellets were placed in a graphite crucible and were heated to 1500–1600 °C in a high-temperature vertical alumina tube furnace under argon atmosphere. The crucible waited at the set temperature for 1 h. It was found that around 15 wt%, calcium silicates in the furnace charge were sufficient to form a liquid slag at 1500 °C. Smelting at 1600 °C was not found as suitable due to the increased reaction of sample and graphite crucible. The authors reported that for a clear-cut metal–slag separation, the amount of calcium silicate in the charge should be at least 20 wt%. Excess carbon practice was not successful and the optimum graphite amount was determined as 5 wt%, which corresponded to stoichiometric carbon requirement for reduction of Fe_2O_3. Impurities present in the pig iron were reported as C: 5.1 wt%, Si: 0.19 wt%, Ti: 0.33 wt%, P: 0.12 wt% and S: 0.004 wt%. Recovery of iron was found to be larger than 85%. The remaining of the work was dedicated to leaching experiments using the slag obtained from smelting, and the recoveries of REEs were discussed deeply [5].

Ning et al. targeted to smelt the red mud to obtain pig iron and slag, which was aimed to be used in steelmaking as a desulfurization flux. The chemical composition of the red mud used in smelting experiments is shown in Table 6. Pellets were formed with the addition of coke powder, and the reduction roasting

Table 4 Average composition of red mud obtained from Seydisehir Aluminum Plant [22]

Component (wt%)	Fe_2O_3	Al_2O_3	SiO_2	TiO_2	Na_2O	CaO	CO_2	P_2O_5	S	V_2O_5	LOI
Red mud	37.6 ±0.3	19.1 ±0.5	15.7 ±0.2	4.9 ±0.1	9.5 ±0.5	2.4 ±0.3	1.8 ±0.4	0.18 ±0.1	0.07 ±0.005	0.045 ±0.001	7.8 ±0.2

Table 5 Chemical composition of the bauxite residue used in the smelting experiments [5]

Component (wt%)	Fe_2O_3	Al_2O_3	SiO_2	TiO_2	Na_2O	CaO	Major rare-earth elements		
							Ce (g/ton)	La (g/ton)	Sc (g/ton)
Red mud	44.6	23.6	10.2	5.7	2.5	11.2	368	114	121

experiments were conducted in a vertical tube furnace at 1100–1200 °C for 12–20 min. For smelting reduction experiments, $CaO–Al_2O_3–SiO_2$ slag system (Fig. 1 [24]) was selected as a basis and charges were prepared with the addition of CaO and Al_2O_3 to obtain slag compositions in the region of the lowest melting temperature compound, $12CaO.7Al_2O_3$. All materials were thoroughly mixed and placed into a graphite crucible, which was charged into a horizontal tube furnace at 1550 °C and the crucible was remained in the furnace for 40 min under argon atmosphere. Afterwards, the furnace was cooled down slowly, and pig iron and slag phases were removed from the furnace for chemical analysis. The typical composition of the metal was reported as (i.e. one of the experiments) 5.34 wt% C, 0.016 wt% S, 0.16 wt% P and 0.013 wt% Si. The composition of the major components of the slag was given as 53.89 wt% CaO, 35.74 wt% Al_2O_3, 4.57 wt% SiO_2, 3.62 wt% TiO_2 and 1.41 wt% Na_2O [25]. As well as CaO, Al_2O_3 and SiO_2 components of the slag are concerned; the composition is similar to the ladle furnace slags in secondary steelmaking (Table 7 [26]). As it can be seen from Table 7, very small amount of Na_2O is present in ladle furnace slags. On the other hand, Na_2O present in the red mud was also transferred to the slag phase during red mud smelting. There are a number of laboratory studies showing good desulfurization ability of Na_2O slags, and some examples are given in the references [27–30]. Moreover, the use of Na_2CO_3 for desulfurization of hot metal was studied, in detail [31, 32]. On the other hand, the use of alkali-containing agents in hot metal desulfurization or in ladle metallurgy reduces the life of the refractories due to its corrosive nature, so it is not industrially preferred. In a study related to hot metal desulfurization, Turkdogan pointed out that due to the corrosive nature of Na_2O containing slags formed during desulfurization by Na_2CO_3, refractory wear in the process is tremendous. Gaseous reaction products lead to environmental problems. Moreover, as the amount of Na_2O decreases by time, the possibility of sulphur going back to metal phase increases. It was indicated that it was crucial to take desulfurization slag away from metal as quickly as possible [33].

Rath et al. investigated the use of plasma arc technology to recover iron from red mud. Fine red mud of the following composition was used in the smelting

Table 6 Chemical composition of the red mud used in the smelting experiments [25]

Component (wt%)	Fe_2O_3	Al_2O_3	SiO_2	TiO_2	Na_2O	CaO	MgO	K_2O	LOI
Red mud	67.01	7.90	5.01	3.92	1.725	0.35	0.09	0.078	13.24

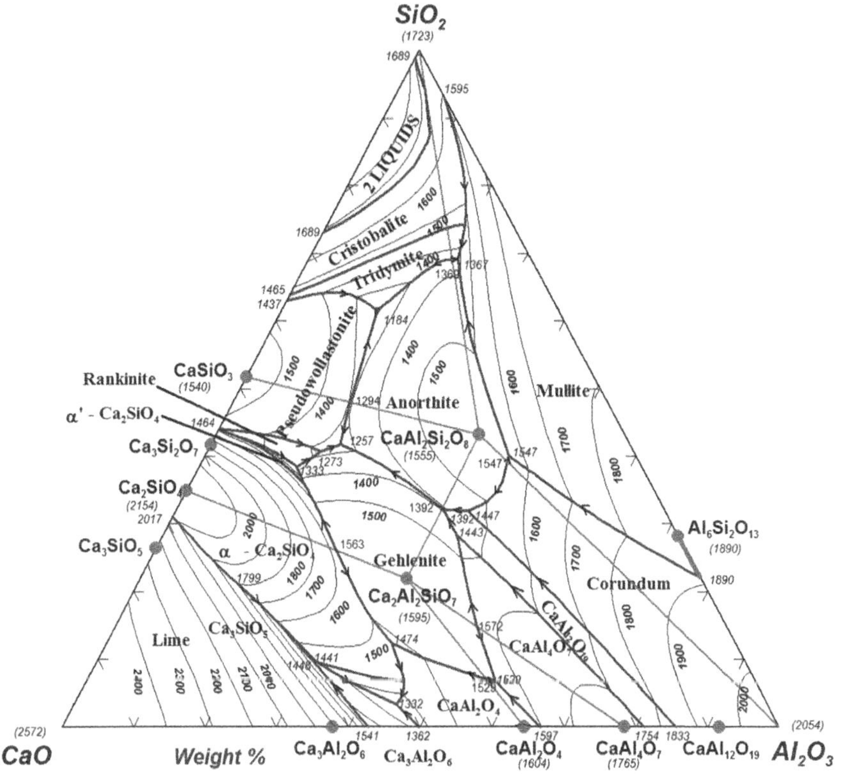

Fig. 1 $CaO–Al_2O_3–SiO_2$ ternary phase diagram [24]

experiments: 55.2 wt% Fe_2O_3, 16.5 wt% Al_2O_3, 6.3 wt% SiO_2, 3.1 wt% Na_2O, 1.4 wt% CaO, 4.5 wt% TiO_2, 0.1 wt% S, 0.08 wt% P and 12.5 wt% LOI. After drying at 383 K, red mud was mixed with graphite and certain fluxes including limestone, dolomite and fluorspar. The batches were adjusted in such a way that 350 g of dried red mud was charged in each run. The charge materials fed into a graphite crucible were smelted in a plasma arc furnace for periods of 5–18 min under argon atmosphere. The authors studied the variation of metal recovery with time, reductant amount and basicity. It was reported that metal recovery increased up to 15 min, whereas it showed a decrease beyond this point. It was indicated that this decrease might be attributed to the formation of complex compounds at elevated temperatures. The amount of reductant in the batches was in the range of 9–14 wt%, and it was found that metal recovery decreased beyond 11 wt%. It was further reported that metal recovery showed a decreasing trend beyond the basicity value of 0.25. Moreover, increased amount of reductant and basicity resulted in an increase in power consumption. Furthermore, the authors dealt with the kinetics of the process upon investigation of metal recovery and found that the reduction reaction followed the first-order kinetics, surface diffusion being the rate controlling

Table 7 Chemical composition of ladle furnace slags of a low-carbon steel grade (all the samples were taken just at the end of ladle furnace operations) [26]

Heat#	Fe_tO	SiO_2	MnO	Al_2O_3	CaO	MgO	P_2O_5	S	Na_2O	K_2O	TiO_2	Cr_2O_3
1	1.95	6.33	2.77	30.99	49.50	8.11	0.0010	0.074	0.128	0.0145	0.138	0.002
2	2.19	6.58	3.02	31.93	48.71	7.34	0.0010	0.054	0.045	0.0010	0.106	0.015
3	1.86	7.40	2.43	30.13	49.34	8.54	0.0009	0.064	0.093	0.0063	0.137	0.001
4	1.84	4.50	1.54	26.41	56.29	9.14	0.0135	0.108	0.052	0.0103	0.099	0.001
5	1.05	2.29	0.31	33.99	53.36	8.54	0.0009	0.424	0.017	0.0009	0.031	0.001

step. Multi-linear regression analysis (MLRA) was also conducted considering the four factors, namely effect of red mud, energy efficiency, effect of basicity and effect of flux. It was concluded that those four components accounted for 75.4% of the variation. Faster reaction time, handling of fine red mud particles without the necessity of agglomeration and CO formation at high temperatures were indicated as the main advantages of plasma arc process [34]. Statistical modelling of iron recovery from red mud was studied in another paper [35].

Conclusions

Since the end of the sixties, smelting of red mud has been deeply investigated for recovery of iron and/or enhancement of subsequent leaching recoveries of aluminium and other elements including the rare earths. In relation to smelting studies, the following main conclusions can be drawn:

- High variation of red mud compositions from plant to plant prevents establishment of well-defined smelting process. Those variations also cause high variability in pig iron compositions.
- Smelting experiments conducted using industrial furnaces and pilot or laboratory-scale works all necessitate the use of significant amount of CaO-containing fluxes to form a suitable slag, causing an increase in slag volume. The presence of Na_2O in red mud not only leads to evaporation/reoxidation of sodium, but also decreases the refractory life of the smelting vessels.
- Plasma arc furnace is given as an example of new technologies. Some smelting investigations suggested zero-waste processes in such a way that smelting slags were used for mineral wool production and evaluated as a flux material, e.g. desulphurization flux. Although some drawbacks are seemingly present in them, those can be regarded as promising attempts for the treatment of red mud.

Acknowledgements The authors would like to thank The Scientific and Technological Research Council of Turkey (TUBITAK) for the financial support given under Project No: 117M185.

References

1. Shamsuddin M (1986) Metal recovery from scrap and waste. J Met 38(2):24–31
2. Bruckard WJ, Calle CM, Davidson RH, Glenn AM, Jahanshahi S, Somerville MA, Sparrow GJ, Zhang L (2010) Smelting of bauxite residue to form a soluble sodium aluminum silicate phase to recover alumina and soda. Mineral Process Extr Met (Trans Inst Min Metall C) 119(1):18–26
3. Hammond K, Mishra B, Apelian D, Blanpain B (2013) CR^3Communication: Red mud—a resource or a waste? JOM 65(3):340–341

4. Klauber C, Gräfe M, Power G (2009) Review of bauxite residue “Re-use” options. CSIRO Document DMR-3609 Project ATF-06-3: Management of bauxite residues
5. Borra CR, Blanpain B, Pontikes Y, Binnemans K, Gerven TV (2015) Smelting of bauxite residue (red mud) in view of iron and selective rare earths recovery. J Sustain Metall. https://doi.org/10.1007/s40831-015-0026-4
6. Thakur RS, Das SN (1994) Red mud analysis and utilisation. PID and Wiley Eastern, New Delhi
7. Mozharenko NM, Noskov VA (2001) Possible directions in the use of red mud in metallurgical production. Metallurgicheskaya I Gornorudnaya Prom 2:127–128
8. Kauβen F, Friedrich B (2015) Reductive smelting of red mud for iron recovery. Chem Ing Tech 87(11):1535–1542
9. Mishra B, Gostu S (2017) Materials sustainability for environment: red-mud treatment. Front Chem Sci Eng 11(3):483–496
10. Fursman OC, Mauser JE, Butler MO, Stickney WA (1970) Utilization of red mud residues from alumina production. U.S. Bureau of Mines Report of Investigations, RI-7454
11. Kumar S, Kumar R, Bandopadhyay A (2006) Innovative methodologies for the utilization of wastes from metallurgical and allied industries. Resour Conserv Recycl 48:301–314
12. Valavin VS, Romenets VA, Saluja JS (2002) Romelt process-the tested single stage ironmaking technology. In: Jouhari AK, Galgali RK, Misra VN (eds) Smelting reduction for iron making. Allied Publishers Pvt. Ltd., New Delhi, pp 100–113
13. Paramguru RK, Rath PC, Misra VN (2004) Trends in red mud utilization—a review, mineral processing & extractive. Metall Rev 26(1):1–29
14. Dobos G, Felfoldi Z, Horvath G, Kaptay G, Osvald Z, Solymar K (1976) Method for the treatment of red mud. U.S. Patent 3989513A. 2 Nov 1976
15. АнатолийАнатольевич ГОЛУБЕВ, ЮрийАлександрович ГУДИМ (2013) Pyrometallurgical red mud processing method. WO2013070121A1. 16 May 2013
16. Red sludge pyrometallurgical processing method, http://russianpatents.com/patent/247/2479648.html. Accessed 17 Aug 2018
17. Logomerac VG (1975) Complex treatment of red mud. Neue Hutte 20(3):145–148
18. Ziegenbalg S, Rudort M, Löve D, Horvath G, Siklosi P, Solymär K, Felföldi Z (1985) Erzmetall 38(4):200–204
19. Horvath G (1974) Acta techn Acad Sci Hung 79:413–449
20. Balomenos E, Gianopoulou I, Panias D, Paspaliaris I (2011) A novel red mud treatment process: process design and preliminary results. Trav 36(40):255–266
21. Balomenos E, Gianopoulou I, Gerogiorgis D, Panias D, Paspaliaris I (2013) Resource-efficient and economically viable pyrometallurgical processing of industrial ferrous by-products. Waste Biomass Valor. https://doi.org/10.1007/s12649-013-9280-5
22. Ercag E, Apak R (1997) Furnace smelting and extractive metallurgy of red mud: recovery of TiO_2, Al_2O_3 and pig iron. J Chem Technol Biotechnol 70:241–246
23. Ercag E (1995) The recovery of iron, titanium dioxide and rare earth concentrate from red mud. PhD thesis, Istanbul University
24. Fact Sage 5.3.1. http://www.crct.polymtl.ca/fact/download.php Accessed 17 Aug 2018
25. Ning G, Zhang B, Liu C, Li S, Ye Y, Jiang M (2018) Large-scale consumption and zero-waste recycling method of red mud in steel making process. Minerals https://doi.org/10.3390/min8030102
26. Keskinkilic E (2007) Examination of desulfurization behaviour of ladle furnace slags of a low-sulfur steel. PhD thesis, Middle East Technical University
27. Iwai H, Kunisada K (1989) Desulfurization and simultaneous desulfurization and dephosphorization of molten iron by Na_2O–SiO_2 and Na_2O–CaO–SiO_2 fluxes. ISIJ Int 29(2):135–139
28. Hernandez A, Romero A, Chavez F, Angeles M, Morales RD (1998) Dephosphorization and desulfurization pretreatment of molten iron with CaO–SiO_2–CaF_2–FeO–Na_2O Slags. ISIJ Int 38(2):126–131

29. Choi J, Kim D, Lee H (2001) Reaction kinetics of desulfurization of molten pig iron using $CaO–SiO_2–Al_2O_3–Na_2O$ slag systems. ISIJ Int 41(3):216–224
30. Van Niekerk WH, Dippenaar RJ (1993) Thermodynamic aspects of Na_2O and CaF_2 containing lime-based slags used for the desulphurization of hot metal. ISIJ Int 33(1):59–65
31. Suito H, Ishizaka A, Inoue R, Takahashi Y (1981) Simultaneous dephosphorization and desulfurization of carbon-saturated iron by sodium carbonate-sodium sulfate flux. Trans ISIJ 21:156–164
32. Moriya T, Fujii M (1981) Dephosphorization and desulfurization of molten pig iron by Na_2CO. Trans ISIJ 21:732–741
33. Turkdogan ET (1985) Slags and fluxes for ferrous ladle metallurgy. Ironmaking Steelmaking 12(2):64–78
34. Rath SS, Jayasankar K, Satapathy BK, Mishra BK, Mukherjee PS (2011) Kinetics and statistical behaviour of iron recovery from red mud using plasma arc furnace. High Temp Mater Proc. https://doi.org/10.1515/htmp.2011.031
35. Rath SS, Pany A, Jayasankar K, Mitra AK, Satish Kumar C, Mukherjee PS, Mishra BK (2013) Statistical modeling studies of iron recovery from red mud using thermal plasma. Plasma Sci Technol 15(5):459–464

Optimization on Drying of Acid Leaching Slag by Microwave Heating Using Response Surface Methodology

Xuemei Zheng, Aiyuan Ma, Hairong Gao, Xiaoling Li, Xianzhu He, Ming Sun and Fengjiao Gu

Abstract In this paper, the technology that acid leaching slag was dried by microwave heating was studied. The influence of microwave power, microwave drying time, material mass and their interaction on the relative dehydration rate of acid leaching slag were investigated using response surface methodology (RSM). Based on the central composite design (CCD), a quadratic model was developed to correlate the preparation variables to the dehydration rate. The optimum conditions for drying acid leaching slag has been identified to be a drying temperature of 100 °C, drying duration of 12 min, microwave power of 750 W and material mass of 50 g. The relative dehydration rate of microwave drying acid leaching slag could reach 97.87% which is very close to the predicted dehydration rate of 98.81%, indicating that the response surface methodology optimization of process parameters for dehydration rate of acid leaching slag by microwave roasting are reliable.

Keywords Acid leaching slag · Microwave drying · Dehydration rate Response surface methodology

Introduction

Acid leaching slag is presently obtained from zinc leach residue in a zinc hydrometallurgy process, which contains significant amounts of Zn. The acid leaching slag consists of lead (Pb), cadmium (Cd), chromium (Cr), nickel (Ni), other non-ferrous metals, and a small amount of rare indium (In) and germanium (Ge) metals [1–3]. The valuable metals, such as zinc and indium can be concentrated in rotary kiln by roasting pretreated. Additionally, acid leaching slag contains

X. Zheng · A. Ma (✉) · H. Gao · X. Li · X. He · M. Sun · F. Gu
School of Chemistry and Materials Engineering, Liupanshui Normol University, Liupanshui 650093, China
e-mail: may_kmust11@163.com

X. Zheng
e-mail: zxm_lpssy19@163.com

© The Minerals, Metals & Materials Society 2019

T. Jiang et al. (eds.), *10th International Symposium on High-Temperature Metallurgical Processing*, The Minerals, Metals & Materials Series,
https://doi.org/10.1007/978-3-030-05955-2_47

higher content of water and chlorine, which dissolved in water to form corrosive acid, even lead to kiln corrosion perforation. Therefore, drying-based pretreatment of acid leaching slag has to be carried out at early stage before the industrial use.

Compared with the traditional heating style, the microwave heating is a highly efficient, clean, and green metallurgy technology [4]. The unique microwave heating characteristics include the low processing time, the selective and volumetric heating, and the controllable heating process. Water is a bipolar molecule, and water has a relatively high dielectric constant (about 76.79 F/m) [5]. Microwave drying is a new technique and is widely used in mineral [6–8], food [9, 10] and tobacco [11], etc.

In this study, the removal of water from acid leaching slag by microwave drying was investigated. The paper intends to investigate the influences of microwave power, microwave drying time, material mass on the dehydration rate, and find the optimum conditions using response surface methodology (RSM) based on the central composite design (CCD).

Materials and Methods

Materials

The acid leaching slag used in the experiments was received from zinc hydrometallurgical plant in Yunnan Province in China. The moisture in acid leaching slag is almost all free water, with the initial moisture content 18.6%. The sample composition was characterized by X-ray diffraction and fluorescence (XRF) measurements. The main chemical composition of the acid leaching slag is listed in Table 1, which reveals that acid leaching slag has a high contents of iron and zinc, as well as Pb, Si, Ca and S.

Experimental Equipment and Methods

The drying experimental equipment was a power of 3 kW box-type microwave reactor designed by Kunming University of Science and Technology. The device was made up of microwave reactor and off-gas absorption system, as shown in Fig. 1.

Table 1 The chemical compositions of acid leaching slag (mass fraction, %)

Composition	Zn	Pb	Ca	Fe	Si	S	Mn	Mg	Cu	H_2O
Content (%)	14.59	5.45	4.65	20.70	7.59	12.13	2.27	1.70	0.88	18.6

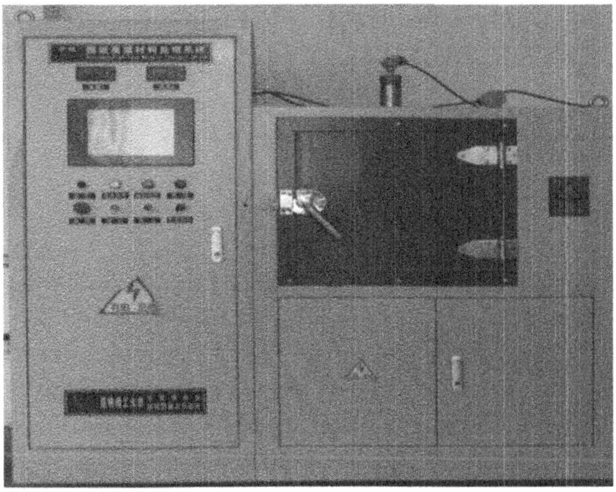

Fig. 1 Microwave drying equipment

In drying experiments, firstly, a certain amount of samples was placed into a mullite crucible, which was surrounded with thermal insulation material. Then, all of these were then transferred to the microwave reactor. A thermocouple was used to measure the temperature, in the range of 0 to 1300 °C. Secondly, the starting of the experiment was marked by activating the microwave and off-gas absorption systems. The off-gas absorption system was developed to collect the flue dust generated in the experimental process.

The water content of the sample after microwave drying was calculated. The dehydration rate (η) of acid leaching slag was expressed as,

$$\eta = \frac{M - M'}{M} \times 100\% \tag{1}$$

where M and M' represent the contents in the initial materials and samples after drying for water, respectively.

Results and Discussion

Experimental Design

In this study, the microwave power (X_1, W), material mass (X_2, g), and drying time (X_3, min) are considered independent variables. The dependent variable is the dehydration rate of acid leaching slag (Y). Table 2 shows the independent variables and levels of the design model.

A total of 20 runs of CCD experiments consist of 8 factor points, 6 pivot points and 6 center points. The responses to each of the experimental conditions concerning the dehydration rate are listed in Table 3. Considering the operability of

Table 2 Experimental range and level of variables

Variables	Level code		
	−1	0	1
Microwave power, X_1 (W)	500	750	1000
Material mass, X_2 (g)	40	50	60
Drying time, X_3 (min)	6	12	18

experiments, drying experiments, the microwave power of run 10, 14 and 19 was set as 1170, 750 and 330 W, the material mass of run 11 and 16 was set as 33 and 67 g, and the drying time of run 13 and 14 was set as 2 and 22 min, respectively.

The ANOVA result of the Quadratic model for the dehydration rate are listed in Table 4. The model *F*-value of 109.36 implies that the model is significant. Values of "Prob> *F*" of less than 0.0001 indicate that the model terms are significant. In this case, X_1, X_3, X_1 X_3, X_2 X_3, X_1^2, X_2^2 and X_3^2 are the significant model terms. For a good fitness of a model, the correlation coefficient should be at least 0.80. The "Pred R-squared" of 0.9234(R_1^2) is in reasonable agreement with the "Adj R-squared" of 0.9809(R_{1adj}^2) plotted in the Table 4. Hence, the model can be used to navigate the design space.

Table 3 Central composite design arrangement and results

Run	Microwave power X_1 (W)	Material mass X_2 (g)	Drying time X_3 (min)	Dehydration rate *Y* (%)
1	500.00	40.00	6.00	30.6344
2	1000.00	40.00	6.00	61.2366
3	500.00	60.00	6.00	40.1147
4	1000.00	60.00	6.00	78.8530
5	500	40.00	18.00	85.3333
6	750.00	50.00	12.00	97.8578
7	500.00	60.00	18.00	70.3978
8	1000.00	60.00	18.00	88.1900
9	750.00	50.00	12.00	97.8578
10	1170.45	50.00	12.00	97.1720
11	750.00	33.18	12.00	82.6382
12	750.00	50.00	12.00	97.8578
13	750.00	50.00	1.91	34.6774
14	750.18	50.00	22.09	98.0968
15	750.00	50.00	12.00	97.8578
16	750.00	66.82	12.00	70.0003
17	750.00	50.00	12.00	97.8578
18	1000.00	40.00	18.00	98.5060
19	329.55	50.00	12.00	44.0108
20	750.00	50.00	12.00	97.8578

Table 4 Dehydration rate analysis of variance for response surface quadratic model

Source	Sum of square	df	Mean square	F value	Prob> F
Model	10,739.93	9	1193.33	109.36	<0.0001
X_1	2635.34	1	2635.34	241.52	<0.0001
X_2	27.58	1	27.58	2.53	0.1429
X_3	4156.26	1	4156.26	380.90	<0.0001
X_1X_2	20.34	1	20.34	1.86	0.2021
X_1X_3	184.09	1	184.09	16.87	0.0021
X_2X_3	342.54	1	342.54	31.39	0.0002
X_1^2	1357.97	1	1357.97	124.45	<0.0001
X_2^2	850.48	1	850.48	77.94	<0.0001
X_3^2	1805.70	1	1805.70	165.49	<0.0001
Residual	109.12	10	10.91	–	–

$R_1^2 = 0.9234$, $R_{1adj}^2 = 0.9809$

The constants and coefficients were obtained by fitting the data listed in Table 4 in to Eq. (2). The equation in terms of coded factors is obtained:

$$\begin{aligned} Y = {} & -366.08238 + 0.29502X_1 + 8.37035X_2 + 18.22138X_3 + 6.37780E^{-4}X_1X_2 \\ & - 3.19797E^{-3}X_1X_3 - 0.10906X_2X_3 \quad 1.55315E^{-4}X_1^2 - 0.076821X_2^2 - 0.31093X_3^2 \end{aligned} \tag{2}$$

It is important to confirm that the selected model provides an adequate approximation to the real system. By using the diagnostic plots, the predicted vs. actual valuc, thc model adequacy can be judged.

Figure 2 shows the comparison of the predicted values versus the experimental values of dehydration rate. As can be seen, the experimental data are distributed on the both sides of the model prediction, indicating the model is valid and successfully fits the experimental data.

Response Analysis

To achieve better understanding of the interactions of the variables and to determine the optimum level of each variable for the maximum dehydration rate, three-dimensional response surfaces plots of the relationship between X_1 and X_2, X_1 and X_3, X_2 and X_3 were calculated and illustrated in Fig. 3.

Figure 3a shows the effect of the microwave power and material mass on the dehydration rate for a constant drying temperature and drying time. The dehydration rate increased with the microwave power. Compared with the material mass, the microwave power has a more significant effect on the dehydration rate.

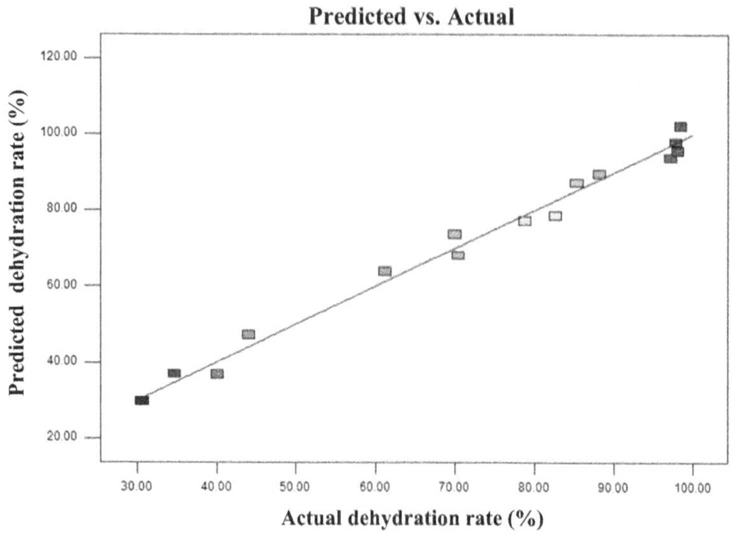

Fig. 2 Linear correlation between actual and predicted dehydration rate

Figure 3b shows the dehydration rate of the acid leaching slag as a function of the microwave power and drying time. The efficiency increased significantly as the microwave power and drying time increased. This can be easily explained by the fact that the heating rate increased rapidly with the microwave power. However, the higher the dehydration rate would be obtained with the time extending.

As shown in Fig. 3c, compared with material mass, the drying time has a relatively bigger effect on the dehydration rate. Within a certain range of time, extended the drying time improves the dehydration rate significantly.

Process Optimization

The experimental value was compared with the predicted in order to determine the validity of the model. From the model, optimized conditions were obtained and given in Table 5. The predicted value of 98.81% and actual experimental value of 97.87% for the dehydration rate of acid leaching slag are obtained. It is obviously to observe that experimental result is very accord with the predicted result. The relative error is only 0.94% with the predicted value.

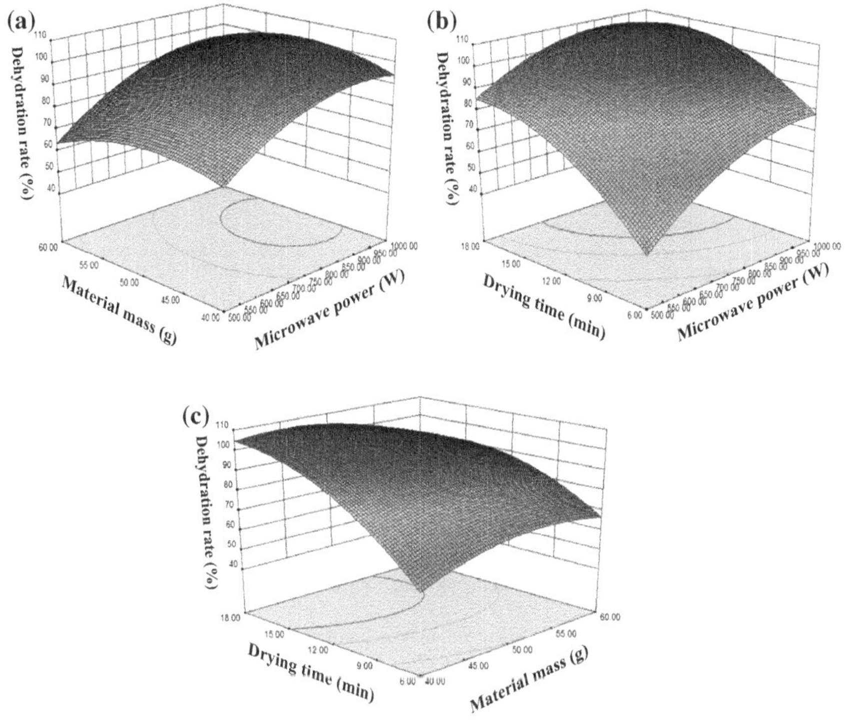

Fig. 3 Response surface plot of dehydration rate for material mass vs microwave power (**a**), drying time versus microwave power (**b**), drying time versus material mass (**c**)

Table 5 Optimization process parameters of regression model

Microwave power X_1 (W)	Material mass X_2 (g)	Drying time X_3 (min)	Drying temperature (°C)	Dehydration rate *Y* (%)	
				Predicted	Experimental
750	50	12	100	98.81	97.87

Conclusions

Dehydration rate of acid leaching slag using microwave drying by RSM, The results indicated that the dehydration rate is significant affected by microwave power and drying time compared with material mass. The optimized calcination conditions are as follows: microwave power 750 W, drying time 12 min and material mass 50 g. The dehydration rate of acid leaching slag are 98.81%, which coincides well with experiments value of 97.87% under the optimized conditions, suggesting that regressive equation fits the dehydration rate perfectly. The unique microwave heating characteristics showed the efficiency and economy of removing water from acid leaching slag by microwave drying.

Acknowledgements The authors are grateful for the financial support from the key cultivation disciplines of Liupanshui Normal University (LPSSYZDPYXK201708), high-level talent research start-up fund of Liupanshui Normal University (LPSSYKYJJ201707).

References

1. Ma AY, Zheng XM, Li SW et al (2018) Zinc recovery from metallurgical slag and dust by coordination leaching in NH_3–CH_3COONH_4–H_2O system. R Soc open sci 5:180660
2. Rabah MA, El-Sayed AS (1995) Recovery of zinc and some of its valuable salts from secondary resources and wastes. Hydrometallurgy 37(1):23–32
3. Li XH, Zhang YJ, Qin QL et al (2010) Indium recovery from zinc oxide flue dust by oxidative pressure leaching. Trans Nonferrous Metals Soc of China 20:s141–s145
4. Ma AY, Zheng XM, Peng JH et al (2017) Dechlorination of zinc oxide dust derived from zinc leaching residue by microwave roasting in a rotary kiln. Braz J Chem Eng 34(1):193–202
5. Ma AY, Zhang LB, Peng JH et al (2014) Dielectric properties and temperature increase of zinc oxide dust derived from volatilization in rotary kilns. J Microw Power Electromagn Energy 48:25–34
6. Li C, Liao JJ, Yin Y et al (2018) Kinetic analysis on the microwave drying of different forms of water in lignite. Fuel Process Technol 176:174–181
7. Le T, Ju SH, Lu LM et al (2018) Microwave drying of CuCl residue from hydrometallurgical zinc recovery process. Drying Technol 2:1–12
8. Feng KL, Chen G, Chen J, Ruan R et al (2018) Experimental investigation on drying characteristic of titanium slag using microwave heating. Drying Technol 36(13):1525–1530
9. Cuccurullo G, Giordano L, Metallo A et al (2018) Drying rate control in microwave assisted processing of sliced apples. Biosyst Eng 170:24–30
10. Wray D, Ramaswamy HS (2015) Novel concepts in microwave drying of foods. Drying Technol 33(7):769–783
11. Zi WH, Zhang XL, Peng JH et al (2013) Optimization of microwave drying biomass material of stem granules from waste tobacco using response surface methodology. Drying Technol 31 (11):1234–1244

Part X
Utilization of Complex Ores

Effect of Core Diameter on the Compressive Strength and Porosity of Itakpe Iron Ore Pellets

JU Odo and VU Nwoke

Abstract The effect of core diameter on the compressive strength and porosity of Itakpe iron ore pellets has been studied. The holes at the centre of the pellets were introduced using balls of combustible materials (palm kernel husk, rice husk and sugar cane husk) of various sizes (2, 3, 4, 5 and 6 mm). The produced green pellets were dried and fired, chemical composition, physical and mechanical properties were determined using standard methods. The results showed that the chemical composition of the pellets produced is acceptable and can be used commercially and the palm kernel husk has the best compressive strength with 7.16 KN/mm^2 while the porosities were in the range of 25–31%. The result also showed that good porosity and maximum compressive strength were achieved with hollow pellets formed with combustible materials of diameter 4 mm for all the samples.

Keywords Iron ore · Core diameter · Compressive strength

Introduction

Iron, along with its generic products, is currently the most widely utilized metal in various sectors of the world's economy. Many factors contribute to this, and ranging from the good mechanical properties it possesses to the low cost associated with its production. Iron is mainly produced through two methods: the blast furnace, BF, route (pig iron) and the direct reduction, DR, route (sponge iron). According to the World Steel Association, 2011 [1], crude steel production was standing at 1.4 billion tones by the end of 2010. Of these, 70% was produced via

J. Odo (✉) · V. Nwoke
Department of Metallurgical and Materials Engineering,
Nnamdi Azikiwe University, Awka, Nigeria
e-mail: odojus@yahoo.com

V. Nwoke
e-mail: uv.nwoke@unizik.edu.ng

© The Minerals, Metals & Materials Society 2019
T. Jiang et al. (eds.), *10th International Symposium on High-Temperature Metallurgical Processing*, The Minerals, Metals & Materials Series,
https://doi.org/10.1007/978-3-030-05955-2_48

basic oxygen furnace (BOF), which uses pig iron from the blast furnace and 28% through the electric arc furnace (EAF), which uses sponge iron and scrap.

Iron ore can be used directly in its natural form as a raw material for processing iron, or it can be upgraded through beneficiation before it is charged into the BF or DR furnaces. The feedstock is evaluated for physical and metallurgical properties [2]. Physical properties give an indication of the material behaviour during handling and descent in the furnace. Metallurgical properties on the other hand indicate the material behaviour during the reduction process. In selecting iron ore for iron and steel industries, some of the properties which need to be considered include porosity, tumbler, abrasion and shatter indices, chemical composition, reduction behaviour and loss on ignition [3].

The performance of a blast furnace depends greatly on the physical and chemical characteristics of the materials. The burden materials, which are charged through the throat, are coke, lump ores and agglomerated ores in the form of sinter or pellets. Lump ores are significant than pellets and sinters. However, they are inferior, particularly with respect to softening and melting, they affect the smooth running of the blast furnace, and this increases coke consumption [4]. Swelling and disintegration of iron ore have been two major drawbacks in their acceptance as feed for blast furnace and direct reduction furnace [5].

Increasing demand for iron and steel has led to the design and construction of blast furnace of increasing size. The huge amount of ores required to feed those large capacity furnaces led to the development of mechanical mining and handling of ores. Transportation and handling of ores also produced further amount of finer fractions, and screening of ores to remove these fines becomes difficult [6]. Pelletization of iron ores concentrates for use as charge material in the blast furnace has been gaining importance in the steel industry, and this is the result of an attempt to meet the increased demand for iron and steel with lower-quality ores and ores extracted from beneficiation plants, all of which are usually in the form of finely divided particles, too finely for direct processing in the blast furnace. Pelletization is the newest of the agglomeration processes, differing from sintering, in that a green unbaked pellet or ball is first formed and then hardened by heating and was first done at the University of Minnesota by Davis and his associates [7, 8]. Their work on concentrates and agglomeration of low-grade iron ores showed that it was possible to ball or pelletize fine magnetite concentrate in a balling drum and that if the ball were fired at sufficiently high temperature, hard, indurated pellets well adapted for use in the blast furnace could be made. Since then, intense interest in pelletization has been developed in recent years because of the outstanding performance achieved by steel producers in extended operations with pellets as the principal iron-bearing material in the blast furnace burden. The products of pelletization process (pellets) go into blast furnaces for direct iron making by reducing the pellets with gas. This is because pellets offer good reducibility than sinter during reduction smelting in blast furnace due to more favourable distribution of porosity in the pellets. Having strong pellets with adequate compressive strength, porosity and other desired properties will increase productivity, improve the working condition and ensure a long lifespan of the equipment used in our steel industries. This not only will save cost but will enhance technological and industrial development.

Some of the major problems encountered using pellets in the blast furnace and direct reduction iron making processes include: low mechanical strength and inadequate porosity at high temperatures. These negatively affect the efficient working condition of the furnace and greatly reduce productivity. Sometimes, they create dangers due to inadequate permeability of gases at the reaction zone of the blast furnace.

The iron ore pellets which are charged into the blast furnace and direct reduction plants are classified into two major types: acid pellets and self-fluxing pellets containing limestone, magnesium oxide, dolomite, etc. These pellets are produced particularly:

1. To avoid blowing off of iron ore particles, thereby loosing heat and increasing total heat demand
2. To achieve fuel economy
3. To increase tumble index, compressive strength and porosity which will lead to improved reducibility.

However, the conventional pellets do not necessarily possess satisfactory reduction characteristics at high temperature range in the lower part of the blast furnace because of hindered gas flow into the inner portion of the pellets. This is attributed to the sizes of the pores in the pellets which are less than 0.1 mm in diameter and the metallic iron in the outer peripheral portion of the pellets which form shell layers. As a result, the reduction of the pellets occur retarded and low melting point slag are formed in the pellets leading to the clogging of the pores, softening and sticking to each other causing various operational problems which include: increased consumption of fuel, non-uniform flow of gases, hanging and sliding of pellets, broken tuyeres and damage to other essential parts of the furnace. These subsequently may lead to loss in productivity, increased maintenance cost and eventual total breakdown of the plant.

Production and usage of pellets with enough porosity and adequate compressive strength will possibly eliminate the problem of poor permeability and inadequate compressive strength reportedly experienced in the usage of conventional pellets (partially solid pellets). Addition of combustible materials to the iron ore concentrate and creation of internally hollow pellets will enhance the total porosity range, between 25 and 30% with uniform quality and sufficient compressive strength.

The production of pellets with adequate porosity, uniform quality and sufficient compressive strength will reduce the problems of cracking and breaking of pellets, poor permeability and non-uniform flow of the reducing gases which are the major causes of softening and sticking of the pellets. These problems cause damage to the tuyeres and other major parts of the blast furnace, and finally result in total breakdown of the plant.

Experimental Procedure

The raw materials used in this work include but not limited to iron ore concentrate, bentonite, coke breeze, sodium silicate, palm kernel husks, sawdust, rice husks, sugar cane husks, water. There are various stages of production of the hollow pellets, and they include raw material sourcing, chemical composition analysis, raw material preparation, mixing operations, pellet production, green ball induration, chemical analysis and physical/mechanical test (porosity and compressive strength). The chemical composition of all the raw materials used in this research was done using the ultraviolet spectrometer at Geology Department, University of Nigeria, Nsukka, Enugu State, Nigeria. Each raw material was soaked in water inside a beaker for 48 h, and the product formed a solution with water which was fed into the equipment and analysed. The results were read off from the computer screen and used to calculate the compositions/percentages of oxides, elements/component in each material. The iron ore stone was gotten from National Iron Ore Company Itakpe (NIOMCO) in Kogi State, Nigeria, and processed to iron ore concentrate at the National Development Center (NDC), Jos, Plateau State. The iron ore was washed thoroughly with water and dried under the sun for about 72 h. The dried ore concentrate was ground and screened to the desired size ranges to upgrade it to about 60–65% of the raw pellet mix. The pulverized iron concentrate, additives and binders were further screened to adjust particle sizes using sieves of different mesh sizes (<85 and <52 mm) placed on an analysette for a period of six minutes. The combustible materials (sawdust, rice husk, sugar cane husk and palm kernel husk) were screened, sorted and washed with water to remove dirt and fine sand particles; thereafter, they were also dried for 48 h under the sun.

Mixing operation was carried out in two stages. The combustible materials were separately mixed with sodium silicate, and small balls of diameters 2, 3, 4, 5 and 6 mm were prepared using hand moulding. Secondly, 1000 g of iron ore concentrate from the <52 and <85 mm (mesh size) was measured out using an electronic weighing balance and charged into a clean mixer. Three percentage bentonite (binder), 5% limestone and 2% coke breeze were also measured out in proportion to the iron ore concentrate and fed to the mixer. Calculated volume of water about 5–6% of the concentrate was measured out and added to the mixture, and the disc was gradually rotated to complete a homogenous mix ready for pelletization. Pellet production is normally done using the balling drum or pelletizing disc, but in this work hand pelletizing was adopted considering the quantity of raw materials and number of pellets required which will not be economical for balling drum or pelletizing disc production. The formed balls of combustible materials were then embedded in the concentrate mix and rolled into balls of the desired diameter and shape: 13-mm diameter. This process was carried out for the production of pellets with core sizes of 2-, 3-, 4-, 5- and 6-mm diameters from the combustible materials. The samples were produced and labelled as follows:

Cx, PKC_2, PKC_3, PKC_4, PKC_5 and PKC_6; RHC_2, RHC_3, RHC_4, RHC_5 and RHC_6; SDC_2, SDC_3, SDC_4, SDC_5 and SDC_6; and SCC_2, SCC_3, SCC_4, SCC_5 and SCC_6

where Cx is the control sample (ore concentrate), PKC is palm kernel husk core, RHC is rice husk core, SDC is sawdust core, SCC is sugar cane husk core, and the subscripts represent the core diameter. There are six samples in each composition above. This is followed by green ball induration process carried out in three stages (drying, preheating and firing). The green balls were dried in air at room temperature for 72 h followed by preheating which was carried out at temperatures of 0–350 °C in 30 min and held for 20 min. The pellet temperature was quickly raised to 900 °C in the next 30–45 min and held for about 5–15 min. In the firing stage, the temperature was increased to 1200 °C in 120 min and allowed to cool with the furnace temperature. Both preheating and firing were carried out simultaneously in a calcination furnace. Chemical analysis of the produced pellets was done using atomic absorption spectrometer (AAS) followed by the physical and mechanical tests.

The compressive strength of the fired pellets was carried out using California bearing ratio (CBR) testing machine. The fired pellets were inserted between the compressive heads of the compression test one after the other. Load was applied to the pellets by an electrically operated screw thread until the specimen collapsed. The value of the compressive strength was read off from an attached circular counter, six pellets of the same composition and firing temperatures were used for the test, and the mean compressive strength of the six specimens was recorded as the compressive strength. In the case of porosity, the pellets were heated to 200 °C in an oven and oven cooled. Six pellets from each composition were weighed to obtain the dry weight and then soaked in water for 24 h and their weight recorded after removal from water. This gave the wet or soaked weight, the pellets were finally re-soaked in water and suspended with string and the suspended weight recorded. All the weighing was carried out using a digital weighing balance. Percentage apparent porosity and other physical properties like apparent density, bulk density and percentage water absorption were calculated using the following formula.

$$\%\text{ apparent porosity} = \frac{\text{Wet Weight} - \text{Dry Weight}}{\text{Wet Weight} - \text{Suspended Weight}} \times \frac{100}{1} \times \text{density of the immersed liquid} \tag{1}$$

$$\text{Apparent density} = \frac{\text{Dry Weight}}{\text{Dry Weight} - \text{Suspended Weight}} \tag{2}$$

$$\text{Bulk density} = \frac{\text{Dry Weight}}{\text{Wet Weight} - \text{Suspended Weight}} \tag{3}$$

$$\text{Percentage water absorption} = \frac{\text{Wet Weight - Dry Weight}}{\text{Dry Weight - Suspended Weight}} \times \frac{100}{1} \tag{4}$$

Results and Discussion

The chemical composition of the ore concentrate from Itakpe iron ore after beneficiation and ore processing to remove impurities as shown in Table 1 falls within the Japanese Industrial Standard (JIS) and American Institute of Testing Materials (ASTM) accepted composition of iron ore concentrate for the production of iron using both blast furnace process and direct reduction method.

The chemical composition of the combustible materials used in the production of hollow pellets is presented in Table 2. The palm kernel husks have the highest percentage of calcium oxide (CaO) and magnesium oxide (MgO): 4.28 and 1.68%, respectively. The alkaline oxide contents are seen to be higher in sawdust and lowest in sugar cane husks. Also, the two major oxides of the combustible materials silica (SiO_2) and alumina (Al_2O_3) were seen to have highest concentrations in rice husks: 27.94 and 12.87%, respectively.

A careful study of Table 3 shows that there is an increase in the percentage of oxides (SiO_2, Al_2O_3 and CaO) in the produced hollow pellets when compared with the control sample, thereby influencing the final properties of the pellet. This was as a result of diffusion of oxides during burning at high temperature from the combustible materials. The chemical compositions of the pellets produced after the final firing temperature of 1350 °C are of high grade and suitable for blast furnace and direct reduction operations when compared to JIS and ASTM standards.

Compressive Strength

The compressive strength of pellets determines the pellet resistance to breakage during handling and transportation to the furnace plant. This also determines how the pellets will resist distortion or breakage after charging, during gradual descending from the stack to the hearth of the furnace. Pellets having compressive strength between 5.0 and 8.0 KN/mm^2 are internationally considered to be ideal for use in the blast furnace pig iron production process (Table 4).

Figure 1 shows the effect of core diameter on the compressive strength of hollow pellet using all the combustible materials. The compressive strength of hollow pellets formed by palm kernel husk concentrate increased with core diameter up to 4 mm and thereafter decreased with further increase in core diameter. The highest compressive strength of 7.16 KN/mm^2 was obtained in hollow pellets formed with core formation of materials of 4-mm diameter. The compressive strength of hollow pellets formed by rice husk concentrate was low when compared with that of PKC. However, the compressive strength of pellets produced with SDC was second to PKC and this is attributed to the high percentage/concentration of alkali oxides (Na_2O and K_2O) which readily melt on heating to high temperature and reacting with ore particles of the pellets, thereby increasing their bond strength. The lowest compressive strength was found in the pellets produced with SCC. The difference in

Table 1 Chemical composition of the standard bulk, Itakpe bulk and Itakpe concentrate

Composition (%)	Fe_T	FeO	Fe_2O_3	TiO_2	SiO_2	Al_2O_3	CaO	MgO	MnO	P	S	Others
Standard bulk	31.6	3.7	–	0.16	40.0	0.40	5.50	2.1	0.12	0.1	0.18	15.70
Itakpe bulk	37.6	–	–	–	24.0	20.51	5.50	2.10	–	0.12	0.18	10.50
Itakpe concentrate	68.31	0.18	3.51	–	9.37	15.35	1.20	0.48	–	0.05	0.05	1.50

Table 2 Chemical composition of the combustible materials

Oxides/compounds (%)	CaO	MgO	SiO_2	Al_2O_3	FeO	PO_4	Na_2O	K_2O	L.O.1
Palm kernel husk	4.28	1.68	24.40	12.74	0.26	0.04	1.82	2.75	1.12
Sawdust	0.79	0.28	20.17	10.74	0.06	0.05	2.56	3.33	0.81
Rice husk	3.03	0.73	27.94	12.87	0.19	0.03	1.67	1.56	0.94
Sugar cane husk	2.05	0.81	15.27	14.18	0.09	0.015	1.07	1.15	1.25

Table 3 Chemical composition of the produced pellets and pellets after successful pelletization tests

Sample oxides/ elements	Fe_T	Fe_2O_3	SiO_2	Al_2O_3	CaO	MgO	P	S	Others
Control	70.14	2.60	3.17	0.59	0.32	0.22	0.04	0.02	22.9
PKC	69.31	2.60	3.32	0.95	0.77	0.22	0.04	0.02	22.8
RHC	70.43	2.57	3.42	0.93	0.68	0.22	0.03	0.018	21.7
SDC	68.81	2.55	3.27	0.59	0.63	0.18	0.04	0.02	23.9
SCC	69.70	2.58	4.0	0.64	0.67	0.25	0.02	0.01	22.1

Note *PKC* Palm kernel husks + concentrate; *RHC* rice husks + concentrate; *SDC* sawdust + concentrate; *SCC* sugar cane husks + concentrate

Table 4 Average dry compressive strength of produced hollow pellets

Core diameter (mm)	Compressive strength of the final pellet with vacant cores (KN/mm^2)			
	PKC (avg)	RHC (avg)	SDC (avg)	SCC (avg)
2	6.94	6.88	6.92	6.70
3	7.05	7.00	7.01	7.00
4	7.16	7.10	7.13	7.06
5	6.68	6.63	6.68	6.62
6	5.68	5.41	5.54	5.41

Note The compressive strength of the control sample is 6.69 KN/mm^2

percentages of oxide constituents of the combustible materials was found to have remarkable influence on the strength as well as other properties. The oxides at high temperatures of about 1200–1350 °C tend to burn of and diffuse into the pellets, thus assisting in sintering and increase in the bond strength which in turn increased the compressive strength. Also, it could be seen that maximum compressive strength was achieved with pellets formed with combustible materials of diameter 4 mm for all the samples. After those pellets produced from combustible materials with 4-mm diameter, strength decreased and becomes too poor at about 6-mm diameter. The pellets at this stage become too weak and may disintegrate both during handling processes and in the blast furnace producing fine particles that are not appropriate for the smooth operation of the blast furnace. On the other hand,

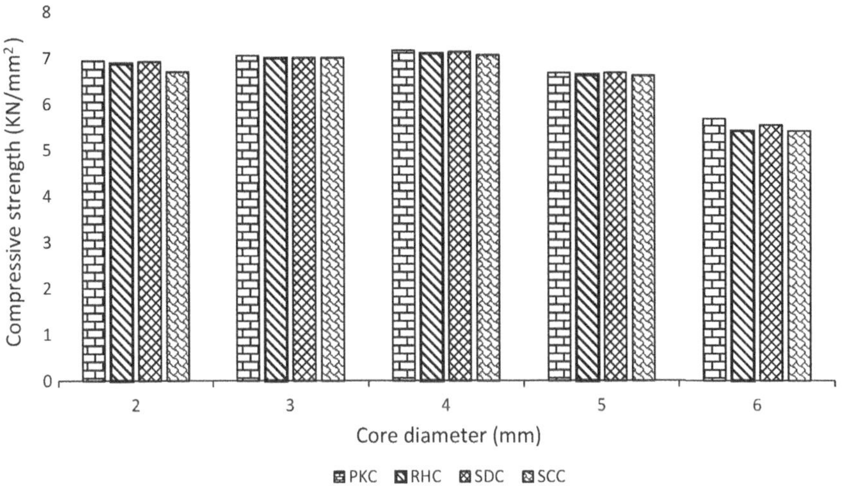

Fig. 1 Effect of core diameter on the compressive strength of hollow pellets produced using different combustible materials

those pellets produced with less than 3-mm diameter have low strength, poor porosity and permeability that they may easily crack and disintegrate during handling and transportation; also in the furnace melt and clog the pores and obstruct uniform flow of the reducing gases and retard reducibility. It can also cause sticking of the pellets within the belly, heart and other high temperature zones of the furnace which affect productivity.

Porosity

Porosity is a very important factor in determining the quality of pellets. Increase in porosity of the pellets can contribute immensely to the improvement in reducibility at high temperature regions, for example, furnace belly leading to decrease in the amount of FeO and reduction in the hanging and sticking phenomena caused by poor permeability and non-uniform flow of reducing gases. It also helps in the reduction of clogging of the pores of the pellets by low melting slag. The porosity of the control pellet sample was 20.62 which is within the acceptable range (15–25%) for conventional pellet porosity which has been proven not to be very good for efficient blast furnace operation. The results of the porosity of the hollow pellets are shown in Table 5 and represented graphically in Fig. 2. It shows that the porosity of the pellets increased with increase in core diameter and those hollow pellets formed with combustible materials of diameter 3 mm and 4 mm have mean porosities in the range of 25.56–30.52% which is within the acceptable range of 25–30% total porosity by JIS and International Standard Organization (ISO).

Table 5 Average porosity of the produced hollow pellets

Core diameter (mm)	Porosity (%)			
	PKC (avg)	RHC (avg)	SDC (avg)	SCC (avg)
2	23.75	23.04	23.52	23.04
3	26.68	25.56	25.60	26.06
4	30.44	30.52	29.98	30.52
5	32.18	32.26	31.62	32.22
6	34.38	34.24	34.12	34.86

Note The porosity of the control sample is 20.62

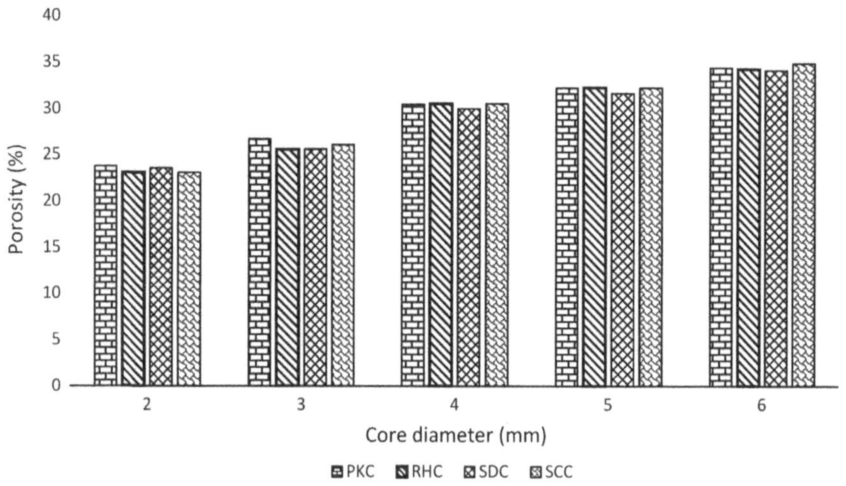

Fig. 2 Effect of core diameter on the porosity of pellets produced using different combustible materials

Therefore, pellets having total porosity above 30% are not considered good for use in the blast furnace pig iron production because they have insufficient compressive strength and easily break into particles during handling operations or in the furnace, resulting in various operational problems.

Conclusion

This research work has been able to produce iron ore pellets with improved porosity, compressive strength and other physical properties when compared to conventional solid pellets. The results of the chemical composition of the produced pellets showed they have acceptable chemical composition and could make commercially quality pellets capable of withstanding handling and transportation. The pellets produced will promote high gas efficiency and excellent reducibility at elevated temperature zones of the blast furnace leading to high productivity as well as reduction in coke ratio. Due to high porosity and compressive strength, the pellets will resist softening and sticking phenomena at high temperature zones of the blast furnace and eliminate operational problems like non-uniform flow of gases, hanging slipping and cracking.

Acknowledgements The lead author gratefully acknowledges the financial support from Nigeria Tertiary Education Trust Fund (TETFund). The authors also wish to thank all academic and technical staff that made input in this work.

References

1. Morris G (2011) World steel in figures. World Steel Association, Published online 01 July 2011 www.worldsteel.org
2. Mashao MS (2009) Evaluation of iron ore pellets and sinters for BF and DR use. The Southern African Institute of Mining and Metallurgy, http://www.saimm.co.za/events/0809pelsint/downloads/Mashao.pdf
3. Kumar M, Jena S, Pate SK (2008) Characterization of properties and reduction behavior of iron ores for application in sponge ironmaking. Mineral Process Extr Metall Rev 29(2):118–129
4. Wu SL, Xu HF, Tian YQ (2009) Evaluation of lump ores for use in modern blast furnaces as part of mixed burden practice. Ironmaking Steelmaking 36(1):19–23
5. Sharma T, Gupta RC, Prakash B (1990) Effect of gangue content on the swelling behavior of iron ore pallets. Miner Eng 3(5):509–516
6. Jones GK (1979) Industrial minerals. In: Austria institutes of metals and minerals. New Castle pellets and granules symposium 1979
7. Eklund N, Dahlstedt A (2002) The choice of pellets in a mixed blast furnace burden and how it affects the process control. Paper presented at 14th conference on hungarian pig iron and steel making Balatonzéplak, Lulea, Sweden, 12–13 Sept 2002
8. Kawatra SK, Elisele TC, Gartler JA, Hardison CA, Lewandoski K (2004) Report on novel binders and methods for agglomeration of ore. Semiannual technical progress report 2004

Correlation Between Reduction Degree and Softening and Melting Properties of Pellets

Yu-Zhu Pan, Ai-Jun Zhang, Lin Lin, Jing-Song Wang, Huai-Xuan Feng and Qing-Shan Lin

Abstract With increasing application of oxygen-enriched or even oxygen blast furnace ironmaking technologies, more intensive indirect reduction, followed by softening and melting, of iron ore pellets may take place in the blast furnace. In this study, the correlation between reduction degree and softening and melting properties of pellets was investigated. The results showed that with the increased of the reduction degree, the softening start temperature (T_s) of pellets did not change significantly and melting start temperature (T_m) obviously increased. The dropping temperature of the pellets slightly increased with the reduction degree of the pellets increase. Since the increased temperature range of the softening zone interval (ΔT_S/°C) of the pellet was greater than the reduced temperature range of the melting zone interval (ΔT_M/°C), the range of the cohesive of the pellet was increased. The permeability of the pellets was obviously improved as the reduction degree increased. This study provided important guidance for blast furnace production under high reduction degree conditions.

Y.-Z. Pan · J.-S. Wang (✉) · H.-X. Feng · Q.-S. Lin
State Key Laboratory of Advanced Metallurgy, University of Science and Technology Beijing, Beijing 100083, China
e-mail: wangjingsong@ustb.edu.cn

Y.-Z. Pan
e-mail: 591130214@qq.com

H.-X. Feng
e-mail: fenghuaixuan@foxmail.com

A.-J. Zhang
Yong Feng Group, Qihe, Shandong 251100, China
e-mail: lgzhangai@sohu.com

L. Lin
School of Energy and Environmental Engineering, University of Science and Technology Beijing, Beijing 100083, China
e-mail: linlin@ustb.edu.cn

© The Minerals, Metals & Materials Society 2019
T. Jiang et al. (eds.), *10th International Symposium on High-Temperature Metallurgical Processing*, The Minerals, Metals & Materials Series,
https://doi.org/10.1007/978-3-030-05955-2_49

Keywords Blast furnace · Reduction degree · Pellets · Cohesive zone · Permeability

Introduction

Pellet is the main iron-containing burden for blast furnace ironmaking. It has the characteristics of high iron content, reducibility and low harmful elements content. However, the softening and melting properties of pellets are worse than sinter, which limits the proportion of pellets used in blast furnace ironmaking. With the advancement of blast furnace ironmaking technology, a large number of new technologies had been applied in blast furnace production, such as high oxygen enriched with PIC [1–6], oxygen blast furnace [7–10], hot reducing gas injection [11, 12] and utilization of metallization burdens [13–16]. The application of these techniques will lead to an increase of the reduction degree and change the softening and melting properties of the pellets. However, the change of the permeability of the pellets layer and the cohesive zone were caused by the reduction degree had not been systematically investigated in previous studies.

The main purpose of this study is to investigate a correlation between reduction degree and softening and melting properties of pellets. The study was carried out by using softening, melting and dropping experiments of pre-reduced pellets; the reduction degree of the pellets is 50, 60, 70 and 80%, respectively. Specific experimental methods are introduced in part 2.

Experimental

Preparation of Pre-reduced Pellets

Pellets with a diameter of 10–12.5 mm were selected and the chemical composition *as* shown in Table 1. The reduction furnace was used to reduce the pellet is shown in Fig. 1a. The pellets were placed in a stainless steel reaction tube; reduction gas was introduced into the reaction tube from the gas inlet. The pellets were heated to 900 °C under a nitrogen atmosphere, the nitrogen gas flow was 5 L/min and the heating rate of the furnace was 5 °C/min. When the temperature of the pellets reached 900 °C, reduction gas ($CO:N_2$ = 30:70) was introduced from the gas inlet. The target reduction degree of mixed burden is 50, 60, 70, and 80%. Pellets will lose weight in the reduction progress. The weight loss of pellets during the reduction process was recorded by the balance. When the weight loss of the pellets reached the target reduction degree, the heating was stopped and the pellets were cooled to room temperature under a nitrogen flow rate of 5 L/min. The relationship between the reduction degree and the weight loss of the pellets is given by Eq. (1):

$$R = \left[\frac{0.1W_1}{0.43W_2} + \frac{M_1 - M_t}{M_0 \times 0.43W_2} \times 100\right] \times 100 \quad (1)$$

In Eq. (1), R is the reduction degree of the mixed burden at time t; M_0 is the initial weight of the mixed burden, g; M_1 is the weight of mixed burden at the beginning of reduction, g; M_t is the weight of the mixed burden at time t, g; W_1 is the initial FeO content of pellet, %; W_2 is the initial total iron content of the mixed pellet, %; 0.11 is the conversion factor of oxygen required for the oxidation of FeO to Fe_2O_3; 0.43 is the conversion factor of oxygen required when all of the iron is oxidized to Fe_2O_3.

Softening, Melting and Dropping Experiments of Pre-reduced Pellets

Softening, melting and dripping experiments of pre-reduced pellets were carried out by softening and melting furnace (Fig. 1b). The pre-reduction pellets were placed in the graphite reaction tube. Coke (20 g) with the particle size of 5–10 mm was placed on the upper part and the lower part of the pellets. Throughout the experiment, 12 L/min of N_2 was introduced from the inlet at the bottom of the furnace. Heating was stopped when the first drop of burden was seen in the peephole and cooled to room temperature under a nitrogen atmosphere of 5 L/ min. The droplets of burden were collected by the lower graphite crucible. Throughout the experiment, the burden layer was subjected to a pressure of 1 kg/cm^2, the distance of burden shrinks and the pressure difference are recorded by a displacement metre and a record U-differential pressure gauge, respectively. The index and physical significance of softening, melting and dripping of the pre-reduced pellets were included in Table 2.

Results and Discussion

The Effect of Reduction Degree on the Cohesive Zone and Permeability

The results of softening, melting and dripping experiments of the pellets are shown in Table 3. In order to more intuitively show changes in the cohesive zone, the data in Table 3 were plotted as a histogram, as shown in Fig. 2. As can be seen from

Table 1 Composition (mass %) of pellets

Burden	TFe	FeO	Al_2O_3	CaO	MgO	SiO_2	Basicity
Pellet	67.26	0.64	0.27	1.08	<0.05	1.49	0.72

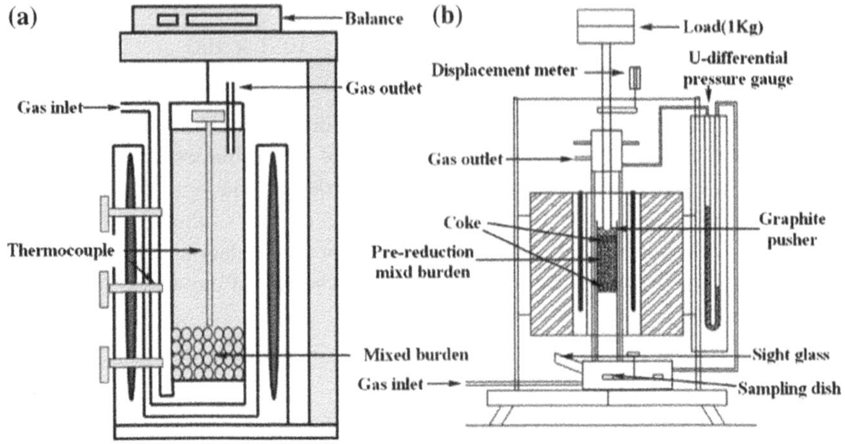

Fig. 1 **a** Reduction furnace. **b** The softening and melting furnace

Table 2 Index and physical significance of softening, melting and dripping of the pre-reduction mixed burden

Parameters	Physical significance
T_S/°C	Softening start temperature (°C), the burden lays contraction of 10%
T_m/°C	Melting start temperature (°C), pressure drop steeply of burden lays
ΔT_S/°C	Softening zone interval ($T_m - T_S$) (°C)
T_d/°C	Dripping temperature (°C), the temperature at which dripping starts.
ΔT_M/°C	Melting zone interval ($T_d - T_m$) (°C)
ΔT_C/°C	cohesive zone Interval ($T_d - T_S$) (°C)
ΔP_{max}/Pa	Maximum differential pressure (Pa)
S/KPa °C	Permeability index (kPa °C) namely $S = \int_{T_s}^{T_d} (\Delta P_{max} - \Delta P_T) dT$ ΔP_T means the differential pressure at T

Fig. 2, as the reduction degree increased, the softening start temperature of the pellets was maintained at about 1100 °C. The melting start temperature increased as the reduction degree increases. When the reduction degree was 50%, the melting start temperature was 1280 °C, and melting start temperature was 1367 °C when the reduction degree increased to 80%. As the melting start temperature increased, the softening interval of the pellet increased, when the reduction degree increases from 50 to 80%, the softening interval increases from 115 to 248 °C. The dropping temperature was increased from 1416 to 1435 °C when the reduction degree increases from 50 to 80%. Since the increase of the melting start temperature was larger than the increase of the drip temperature, the temperature interval of ΔTM/°C was decreased from 136 to 68 °C.

It can be seen from Table 3 that the maximum pressure difference occurred when the reduction degree was 60%. When the reduction degree increased from 50% to

Table 3 Experimental results of softening, melting and dropping of pre-reduced pellets

R (%)	$T_{10\%}$/°C	$T_{40\%}$/°C	ΔT_B/°C	T_s/°C	T_d/°C	ΔT_M/°C	ΔP_{max}/Pa	S/KPa °C
50	1125	1212	87	1280	1416	136	4292.4	517.13
60	1107	1200	93	1317	1426	109	5194	512.74
70	1114	1215	101	1339	1439	100	4802	431.20
80	1119	1248	129	1367	1435	68	1783.6	87.96

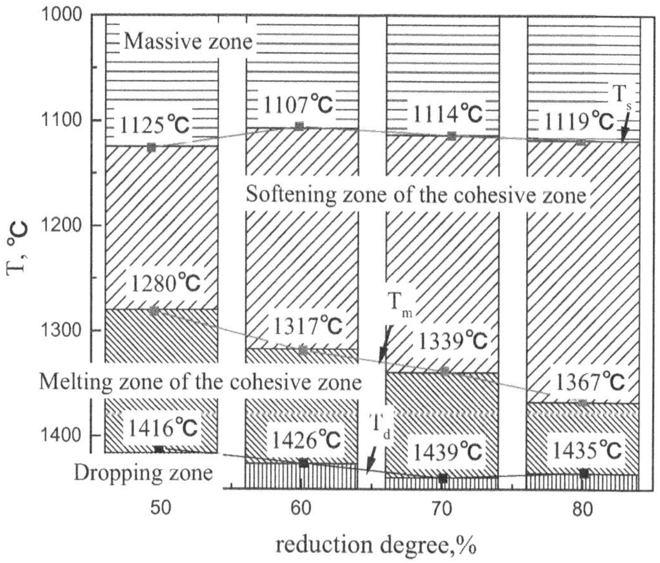

Fig. 2 Interval of the blast furnace cohesive zone under different reduction degrees

60%, the maximum pressure difference increased from 4292.4 to 5194 Pa. When the reduction degree exceeded 60%, the maximum pressure difference decreased as the reduction degree increased. When the reduction degree increased from 60 to 80%, the maximum pressure difference was reduced from 5194 to 1783.6 Pa. In general, the maximum pressure difference of the pellets decreased as the reduction degree of the pellets increased. Figure 3 shows the variation of the permeability index of the pellets layer with the reduction degree. It can be seen from Fig. 3 and Table 3 that as the reduction degree increased, the permeability index of the layer decreased. When the reduction degree was 50%, the permeability index of the pellets layer was 517.13 S/KPa °C. When the reduction degree increased to 80%, the permeability index of the pellets layer was only 87.96 S/KPa °C.

In summary, as the reduction degree of the pellets increased, the range of the cohesive zone of the pellets expands. The softening zone of the cohesive zone was increased and the melting zone is reduced, the gas permeability of the pellets layer is improved.

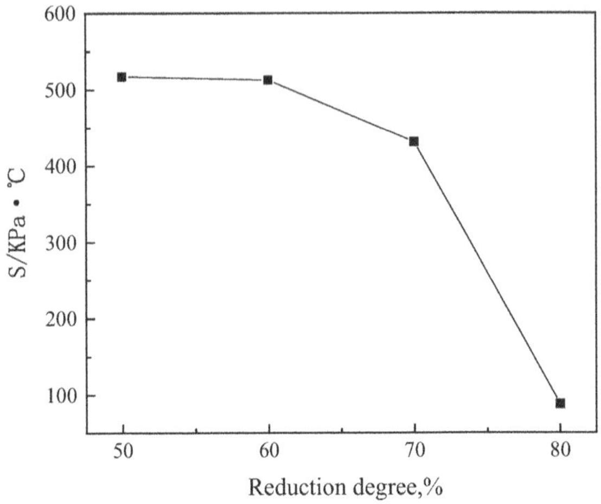

Fig. 3 Variation of the permeability of *the pellet* layer with the reduction degree

Influence Mechanism of Reduction Degree on Changes of Cohesive Zone

According to the research results of Wu et al. [17], in the acidic pellets, the gangue phase is mainly Fe_2SiO_4, and its melting point is only 1205 °C. The increase of the reduction degree means that more iron is reduced to metallic iron, which increases the resistance of the pellet; on the other hand, the iron in the pellet is more reduced, and FeO content in the gangue phase is lowered, which reduces the content of Fe_2SiO_4. The production of the cohesive zone includes liquid iron and slag, and the melting point of iron is affected by the carbon content. A large number of studies have shown that there are three ways of carburizing metal iron, namely direct carburizing [18, 19], gaseous carburizing [20–22] and molten carburizing [23, 24]. Direct carburizing refers to the carbon of the coke or coal direct dissolution into iron in form solid–liquid or solid–solid. According to the research results of Deng, the amount of carburization from solid–solid form to metallic iron before melting is very small, and it has little effect on the softening and melting of metallic iron, this carburizing only occurs on the contact surface of coke and metallic iron. Comparing the reduction conditions of the experiment and the conditions described in the literature, it is known that the carburization reaction occurring during the reduction has little effect on the carbon content of the metallic iron in the pellet, because gaseous carburizing requires high CO concentration and temperature. During the contact of iron-containing slag with carbon, the iron element in the slag will form Fe–C particles. Due to the change of slag composition, the Macranganin effect drives the molten Fe–C particles to carburize into the metallic iron. In this experiment, molten carburizing is considered to be the main carburizing method. As the reduction degree increases, FeO in the slag decreases, and the manner of molten carburization is suppressed. Due to the increase in iron phase, the carbon content of iron is reduced. Therefore, under conditions of high reduction degree, the melting and dripping of the iron phase require a higher temperature.

Through the experiments in this study, it is found that increasing the reduction degree can improve permeability and the softening and melting properties of the pellets, which has important guiding significance for the blast furnace. First, it is possible to achieve a higher reduction degree of reduction in the massive zone by improving the reducibility of the iron-containing burden in the furnace. In addition, the reduction atmosphere of the furnace is increased to improve the degree of reduction before softening and melting. Finally, adding the appropriate proportion of metallization burden to the blast furnace can be applied.

Conclusions

By using softening, melting and dropping experiments of different reduction degrees of pellets, the correlation between reduction degree and softening and melting properties of pellets were systematically studied. The results of the study showed that the softening interval of the pellet became wider as the reduction degree increased, the melting interval became narrower, the temperature range of the softening interval increased, and the permeability was significantly improved. Based on the results of previous studies, the reasons for obtaining the results of this study were given. The increase of the reduction degree increased the amount of metallic iron in the pellets, resulting in increasing deformation resistance of the pellets, the formation of low-melting slag phase and molten carburization was suppressed. According to the experimental results, some guidance was provided for the production of the blast furnace in order to improve the permeability of the blast furnace cohesive zone, which improving the reducibility of the iron-containing burden, improving the reduction potential of the furnace and using the metallization burden.

References

1. Du SD, Chen WH, Lucas J (2007) Performances of pulverized coal injection in blowpipe and tuyere at various operational conditions. Energy Convers Manag 48:2069–2076
2. Wijayanta AT, Alam MS, Nakaso K, Fukai J, Kunitomo K, Shimizu M (2014) Combustibility of biochar injected into the raceway of a blast furnace. Fuel Process Technol 117:53–59
3. Guo BY, Zulli P, Rogers H, Mathieson JG, Yu AB (2005) Three-dimensional simulation of flow and combustion for pulverized coal injection. ISIJ Int 45:1272–1281
4. Shen YS, Guo BY, Yu AB, Maldonado D, Austin P, Zulli P (2008) Three-dimensional modelling of coal combustion in blast furnace. ISIJ Int 48:777–786
5. Shen YS, Guo BY, Yu AB, Zulli P (2009) A three-dimensional numerical study of the combustion of coal blends in blast furnace. Fuel 88:255–263

6. Babich A, Gudenau HW, Formoso A, Mavrommatis K, Froehling C, Garcia L (2002) Choice of technological regimes of a blast furnace operation with injection of hot reducing gases. Rev Metal 38:288–305
7. Meijer K, Denys M, Lasar J, Birat JP, Still G, Overmaat B (2009) ULCOS: ultra-low CO_2. Ironmaking Steelmaking 36:249–251
8. Zuo GQ, Hirsch A (2009) The trial of the top gas recycling blast furnace at LKAB's EBF and scale-up. Metall Res Technol 106:387–392
9. Danloy G, Berthelemot A, Grant M, Borlée J, Sert D, van der Stel J, Jak H, Dimastromatteo V, Hallin M, Eklund N, Edberg N, Sundqvist L, Sköld B-E, Lin R, Feiterna A, Korthas B, Müller F, Feilmayr C, Habermann A (2009) ULCOS-pilot testing of the low-CO_2 blast furnace process at the experimental BF in Luleå. Metall Res Technol 106:1–8
10. An XW, Wang JS, Lan RZ, Han YH, Xue QG (2013) Softening and melting behavior of mixed burden for oxygen blast furnace. J Iron Steel Res Int 20:11–16
11. Desai B, Ramna RV, Sathaye JM (2008) Effect of hot reducing gas (HRG) injection on blast furnace operational parameters: theoretical investigation. Ironmaking Steelmaking 35:43–50
12. Qi YH, Yan DL, Gao JJ, Zhang JC, Li MK (2011) Study on industrial test of the oxygen blast furnace. Iron Steel 46:6–8
13. Kaushik P, Fruehan RJ (2006) Mixed burden softening and melting phenomena in blast furnace operation part 2—mechanism of softening and melting and impact on cohesive zone. Ironmaking Steelmaking 33:520–528
14. Kaushik P, Fruehan RJ (2007) Mixed burden softening and melting phenomena in blast furnace operation part3—mechanism of burden interaction and melt exudation phenomenon. Ironmaking Steelmaking 34:10–22
15. Chung SH, Kim KH, Sohn I (2015) DRI from recycled iron bearing wastes for lower carbon in the blast furnace. ISIJ Int 55:1157–1164
16. She XF, Wang JS, Liu JZ, Zhang XX, Xue QG (2014) Increasing the mixing rate of metalized pellets in blast furnace based on the high-temperature interactivity of iron bearing materials. ISIJ Int 54:2728–2736
17. Wu SL, Han HL, Xu HF, Yan LJ (2010) Research on mechanism of interaction between sinter and lump ores in blast furnace. Chin J Process Eng 10(Suppl):37–42
18. Deng SQ, Che CR, Yan LJ (1993) Study on carburization of iron in upper part of blast furnace. Iron Steel 28:12–16
19. Deng SQ, Shi YX, Wang Q, Li CY (1992) Carburization of hot metal in low part of BF. Iron Steel 27:5–14
20. Kaspersma JH, Shay RH (1981) Carburization of iron by CO-based mixtures in nitrogen at 925 C. Metall Mater Trans B 12:77–83
21. Murakami T, Fukuyama H, Nagata K (2001) Mechanisms of carburization and melting of iron by CO gas. ISIJ Int 41:209–215
22. Asano R, Sasaki Y, Ishii K (2002) Carburization of Iron by Ar–CO–H_2 at 1523 K. ISIJ Int 42:121–126
23. Murakami T, Ohno M, Suzuki K, Owaki K, Kasai E (2017) Acceleration of carburization and melting of reduced iron in iron ore-carbon composite using different types of carbonaceous materials. ISIJ Int 57:1928–1936
24. Ohno K, Miki T, Hino M (2004) Kinetic analysis of iron carburization during smelting reduction. ISIJ Int 44:2033–2039

Effect of TiO_2 on the Viscous Behaviour of the $CaO–SiO_2$–14 Mass% Al_2O_3–8 Mass% $MgO–TiO_2$ Slag

Zhengde Pang, Yuyang Jiang, Xuewei Lv, Zhiming Yan and Wenchao He

Abstract In the present work, the effect of TiO_2 on the viscosity and the free running temperature of the $CaO–SiO_2–Al_2O_3–MgO–TiO_2$ slag system were studied. At fixed CaO/SiO_2 ratio of 1.1, 14 mass pct Al_2O_3 and 8 mass pct MgO, the viscosity decreases with an increase in TiO_2 content between 20 and 30 mass pct due to depolymerization of slag structure, which means that TiO_2 behaves as a network modifier. The free running temperature decreases with the increase of TiO_2 content. The change of the apparent activation energy is in accordance with the change of viscosity. Five different viscosity models and FactSage software were employed to predict the viscosity. Yan's model agrees well with experimental data.

Keywords TiO_2 · Viscosity · Free running temperature · Viscosity models

Introduction

After rapid development of iron and steel industry in recent decades,high-grade iron resources have been consumed gradually. The TiO_2-rich iron ores are quite abundant and basically have not been explored. Such resources have attracted increasing attention and are gradually being used in the steel industry [1–3]. As a result of the consumption of TiO_2-bearing iron ores, the content of TiO_2 in the blast furnace slag has significantly increased. The TiO_2 significantly changes slag physicochemical properties such as the free running temperature, viscosity and stability. The free running temperature is related to the fluidity of slag indicating temperature at which slag can flow freely.

Z. Pang · Y. Jiang · X. Lv (✉) · Z. Yan · W. He
College of Materials Science and Engineering,
Chongqing University, Chongqing 400044, China
e-mail: lvxuewei@163.com

© The Minerals, Metals & Materials Society 2019

T. Jiang et al. (eds.), *10th International Symposium on High-Temperature Metallurgical Processing*, The Minerals, Metals & Materials Series,
https://doi.org/10.1007/978-3-030-05955-2_50

Many studies [4–8] have been done to research the effect of TiO_2 on the viscosity of TiO_2-bearing blast furnace slag (see Fig. 1). In the $CaO–SiO_2–Al_2O_3–TiO_2$ slag system, Ohon and Ross [4] found out that the slag viscosity decreases with the addition of TiO_2 content under inert condition at various Al_2O_3 contents between 10 mass pct and 20 mass pct. Work done by Park et al. [5] showed the influence of TiO_2 in the $CaO–SiO_2–Al_2O_3–MgO$ system, where TiO_2 content addition up to 10 mass pct decreased the slag viscosity under inert condition. This result is consistent with the studies carried out by Liao et al. [6] and Sohn et al. [7]. Yan et al. [8] also indicated that under inert atmosphere the increased TiO_2 addition decreases the viscosity of liquid phase. However, the effect of TiO_2 on the free running temperature in combination with the viscosity is not well understood or controlled for blast furnace slag. In this work, the influence of TiO_2 addition on the viscous behaviour of the $CaO–SiO_2–Al_2O_3–MgO–TiO_2$ slag has been studied by using a rotating viscometer.

Experimental

The chemical composition of slag used in the present study is shown in Table 1. The samples were prepared by analytical pure reagent CaO, SiO_2, Al_2O_3, MgO and TiO_2 powder. To remove any carbonate, hydroxide or H_2O, these chemical reagents

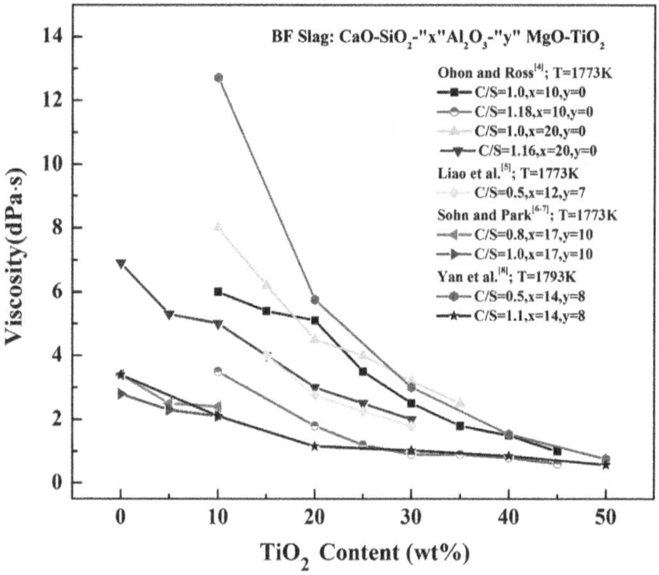

Fig. 1 Effect of TiO_2 on the viscosity of $CaO–SiO_2–Al_2O_3–MgO$-based slag at various CaO/SiO_2 ratios at 1773 K

were calcined at 1273 K in a muffle furnace. The cylindrical pellet samples were placed in a molybdenum crucible (inner diameter of 46 mm, outer diameter of 48 mm and height of 120 mm). Viscosity measurements were carried out by the rotating cylinder method [9] using a Brookfield digital viscometer. The experimental arrangement including the dimensions of the crucibles and spindle is schematically shown in Fig. 2. Weighted analytic reagent powders (total weight: 200 g) were thoroughly pressed into cylindrical pellets. The sample and the crucible were placed in the hot zone of the furnace, which was heated with $MoSi_2$. The crucible and probe, made of molybdenum, were used for the viscosity measurement. To avoid the oxidation of the Mo crucible and probe, purified argon gas was blown into the furnace chamber at a constant flow rate of 1.0 L/min. The sample was heated to 1793 K and held for 3 h to form a homogeneous slag for viscosity measurement. After the slag sample melted, the Mo probe was immersed in the melt up to a depth of 2.5 mm via controlled position of the loading platform. The viscosity of slag was measured every 20 K from 1773 to 1693 K.

Results and Discussion

Viscosity

Figure 3 shows change of the viscosity of the $CaO–SiO_2$–14 mass% Al_2O_3–8 mass % $MgO–TiO_2$ slag as a function of TiO_2 and temperature (1693–1773 K) at constant CaO/SiO_2 ratio of 1.1. The increase of TiO_2 decreases the viscosity, which is consistent with previous studies [10–12]. According to work done by Shankar et al. [11], the decrease in the viscosity suggests that TiO_2 behaves as a basic oxide, which depolymerizes the slag network structure. Park et al. [5] have studied the effect of TiO_2 content on slag structure using Fourier transform infrared spectroscopy and found out that the corresponding vibration band of Si–O and Al–O complex structure decreased with increased TiO_2 content. This indicates that the complex structure of Si–O and Al–O is reduced, and the slag viscosity decreased.

Table 1 Chemical compositions of studied slag

	CaO	SiO_2	MgO	Al_2O_3	TiO_2	CaO/SiO_2
A1	30.38	27.62	8	14	20	1.1
A2	29.33	26.67	8	14	22	1.1
A3	28.29	25.71	8	14	24	1.1
A4	27.24	24.76	8	14	26	1.1
A5	26.19	23.81	8	14	28	1.1
A6	25.14	22.86	8	14	30	1.1

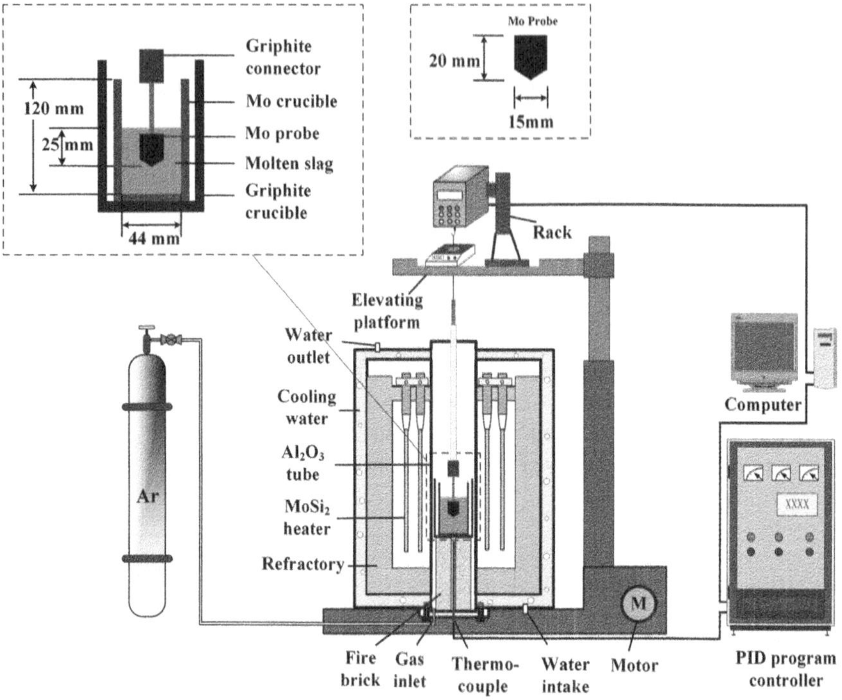

Fig. 2 Experimental set-up for the viscosity measurement and dimensions of the crucible and probe (dotted line box)

Fig. 3 Effect of TiO_2 on the viscosity of the $CaO–SiO_2$–14 mass% Al_2O_3–8 mass% $MgO–TiO_2$ slag at CaO/SiO_2 ratio of 1.1

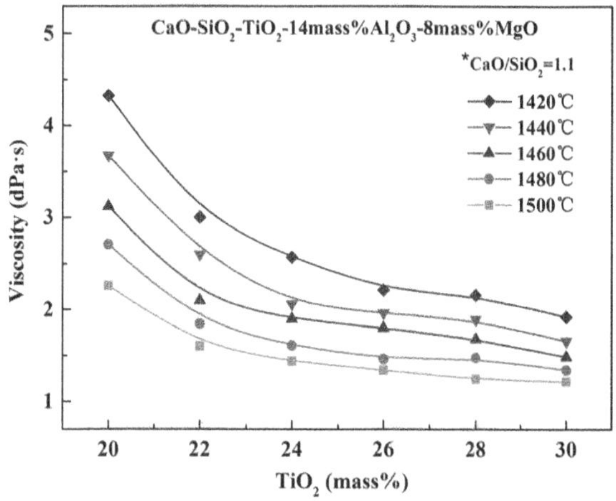

Free Running Temperature

The free running temperature of slag is another important indicator of viscous behaviour [13, 14]. It depends on the free flow of slag after melting. It could be determined by using the viscosity–temperature curve. Figure 4 shows effect of TiO_2 on the viscosity–temperature curves, free running temperature of the $CaO–SiO_2–14$ mass% $Al_2O_3–8$ mass% $MgO–TiO_2$ slag at CaO/SiO_2 ratio of 1.1. As expected, the viscosity increases with decreased temperature. It is evident from that the free running temperature decreases slightly when TiO_2 content increases (see Fig. 4).

The viscosity data over the entire temperature region can be described by an Andrade equation [15] (see Eq. 1):

$$\eta = \eta_0 \exp\left(\frac{E_\eta}{RT}\right) \tag{1}$$

where η_0, E_η and R are constant, the apparent activation energy of viscous flow and the universal gas constant, and T is the thermodynamic temperature. Variations of $E\eta$ can reveal the change of frictional resistance for viscous flow indicating a change in the structure of the molten slag or, more directly, a change in the cohesive flow units forming the slag structure. Figure 5a shows viscosities of slag as a function of reciprocal temperature, and Fig. 5b shows the relationship between the content of TiO_2 and the apparent activation energy of viscous flow calculated according to Eq. (1). The apparent activation energy (Eg) decreases with increasing TiO_2 content, which is very similar to the change of the free running temperature. This result also indicates that the structural unit for viscous flow of slag becomes smaller with increasing TiO_2 content.

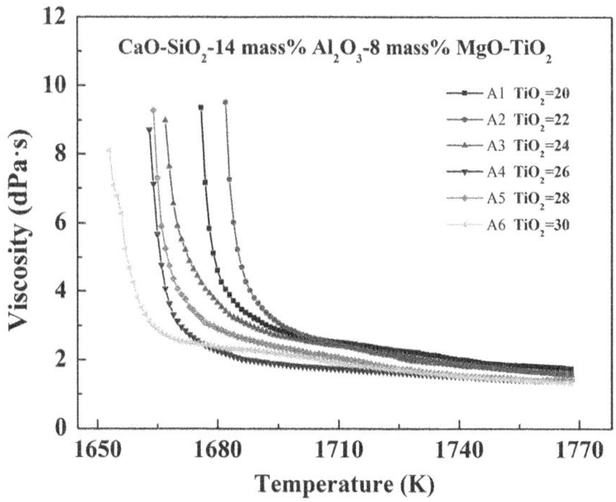

Fig. 4 Effect of TiO_2 on the viscosity–temperature curves of the $CaO–SiO_2–14$ mass% $Al_2O_3–8$ mass% $MgO–TiO_2$ slag at CaO/SiO_2 ratio of 1.1

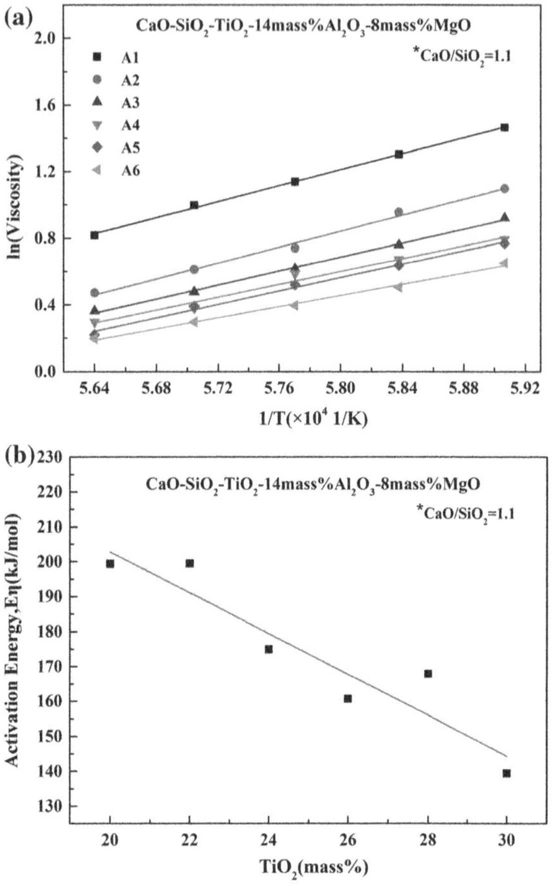

Fig. 5 Temperature dependence of viscosity (**a**) and apparent activation energy (**b**) of the $CaO–SiO_2$–14 mass% Al_2O_3–8 mass% $MgO–TiO_2$ slag at CaO/SiO_2 ratio of 1.1

Models and Prediction of Viscosity

The measured viscosities were compared with the estimated viscosities from FactSage software and five different models, which are Riboud's model, Urbain's model, Zhang's model, optical model and the modified Reddy model by Yan et al. [16–20]. Figure 6 shows the comparison between the measured and the estimated viscosity for the $CaO–SiO_2–MgO–Al_2O_3–TiO_2$ slag, from which it can be seen that Yan's model agrees well with experimental data (the average deviation is 11.3%). This model is structural-based, and the temperature dependence of viscosity is calculated by the Arrhenius equation. The temperature compensation effect was considered in the calculation. For the $CaO–SiO_2–MgO–Al_2O_3–TiO_2$ slag system, the viscosity values can be accurately predicted by the modified Reddy model.

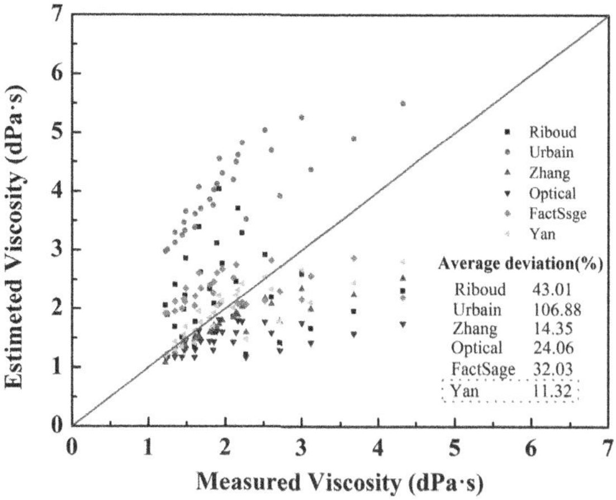

Fig. 6 Comparison between measured viscosity and estimated viscosity

Conclusions

In present work, the viscous behaviour of the CaO–SiO_2–14 mass% Al_2O_3–8 mass % MgO–TiO_2 slag system was experimentally studied. The viscosity and the free running temperature were measured, and the apparent activation energy was calculated using Arrhenius equation. The increase of TiO_2 addition decreases the slag viscosity and the free running temperature. The apparent activation energy decreases with increasing TiO_2 content, which is similar to the change of the free running temperature. The estimated viscosity values by modified Reddy model agree well with experimental values.

Acknowledgements This study was supported by the Fundamental Research Funds for the Central Universities (Grant No. 2018CDQYCL0026) and the Fundamental Research Funds for the Central Universities (2018CDPTCG0001/32).

References

1. Zhang G, Chou K, Zhang J (2014) Influence of TiO_2 on viscosity of aluminosilicate melts. Ironmaking Steelmaking 41(1):47–50
2. Zhang S, Zhang X, Bai C, Wen L, Lv X (2013) Effect of TiO_2 content on the structure of CaO–SiO_2–TiO_2 system by molecular dynamics simulation. Isij Int 53(7):1131–1137
3. Sohn I, Wang W, Matsuura H, Tsukihashi F, Min DJ (2012) Influence of TiO_2 on the viscous behavior of calcium silicate melts containing 17 mass% Al_2O_3 and 10 mass% MgO. ISIJ Int 52(1):2012
4. Ohno A, Ross H (1963) Can Metall Q 2:259
5. Park H, Park JY, Kim GH, Sohn I (2012) Steel Res Int 83:150
6. Liao JL, Li J, Wang XD, Zhang ZT (2012) Ironmaking Steelmaking 39:133
7. Sohn I, Wang W, Matsuura H, Tsukihashi F, Min DJ (2012) ISIJ Int 52:158
8. Yan Z, Lv X, He W, Xu J (2017) ISIJ Int 57(1):31–36

9. Yan Z, Lv X, Liang D, Zhang J, Bai C (2017) Transition of blast furnace slag from silicates-based to aluminates-based: viscosity. Metall Mater Trans B 48(2):1092–1099
10. Colf JVD, Howat DD (1979) J S Afr Inst Min Metall 79:255–263
11. Shankar A, Gornerup M, Lahiri AK, Seetharaman S (2007) Ironmaking Steelmaking 34:477–481
12. Nakamura T, Morinaga K, Yanagase T (1977) Nippon Kinzoku Gakkaishi 4:1300–1304
13. Qiao R, Du H (1988) Influence of TiO_2 on the viscosity and free running temperature of the slag containing fluorine. J Iron Steel Res 10:1–4
14. Liang L, Hao Z, Wen D, Jia X, Deng Y (2017) Study on viscosity and melting temperature of no. 7 blast furnace slag from Baotou steel. Hunan Nonferrous Metals
15. Joo Hyun Park (2013) Structure-property relationship of CaO–MgO–SiO^2, slag: quantitative analysis of raman spectra. Metall Mater Trans B 44(4):938–947
16. Riboud PV, Roux Y, Lucas LD, Gaye H (1981) Fachber Hut-tenpraxis Metall 19:859–869
17. Urbain G, Cambier F, Deletter M, Anseau MR (1981) Trans J Br Ceram Soc 80:139–141
18. Zhang G, Kuochih C (2012) Modelling viscosities of CaO–MgO–Al_2O_3–SiO_2 molten slags. ISIJ Int 52(3):355–362
19. Mills K (2011) The estimation of slag properties. Southern African Pyrometallurgy
20. Yan Z, Reddy RG, Lv X, Pang Z, He W (2018) Structure-based viscosity model development for titania aluminosilicate slags. Ironmaking Steelmaking: 1–7

Granulation of Semisteel by Rotary Disc Atomizer

Wenchao He, Xuewei Lv, Feifei Pan, Xueqin Li, Zhiming Yan and Zhengde Pang

Abstract Semisteel, the main by-product of titania slag smelting process of ilmenite concentrate with an electric furnace, is of high sulphur and low carbon. As a result, the semisteel cannot be widely used in the steeling process and the market value is relatively low. In order to enhance the added value of semisteel, a novel process which was based on a rotary disc atomizer was suggested on semisteel to produce iron powders in the present study. The oxidation degree of the obtained iron granules was studied by thermogravimetric analysis (TG) in the mixture gas of nitrogen and hydrogen.

Keywords Granulation · Semisteel · Rotary disc atomizer

Introduction

Natural rutile, one of the important resources for producing titanium dioxide, is mainly deposited in Australia, South Africa and India with a global reserve of about 45 million tons [1]. Ilmenite has been used in titanium industry as the raw material due to the gradual depletion of natural rutile resource. Generally, the ilmenite concentrate is smelted by the electric furnace to obtain Ti-rich material with high grade [2, 3]. High-titania slag and semisteel are the main product and by-product of titania slag smelting process of ilmenite with electric furnace, respectively. The obtained high-titanium slag becomes the main raw material for titanium dioxide production. However, the semisteel cannot be widely used in the steelmaking

W. He · X. Lv (✉) · F. Pan · X. Li · Z. Yan · Z. Pang
Chongqing Key Laboratory of Vanadium-Titanium Metallurgy and New Materials, Chongqing University, No. 174 Shazheng Street, Shapingba District, Chongqing 400044, China
e-mail: lvxuewei@163.com

W. He · X. Lv · F. Pan · X. Li · Z. Yan · Z. Pang
College of Materials Science and Engineering, Chongqing University, No. 174 Shazheng Street, Shapingba District, Chongqing 400044, China

© The Minerals, Metals & Materials Society 2019

T. Jiang et al. (eds.), *10th International Symposium on High-Temperature Metallurgical Processing*, The Minerals, Metals & Materials Series,
https://doi.org/10.1007/978-3-030-05955-2_51

process because of its high sulphur–low carbon content. As a result, the market value of the semisteel is relatively low. A series of post-treatment processes such as adding carburizing agent, desulphurizing agent and ferroalloy are generally conducted on the semisteel to produce cast iron in the steelmaking plant [4]. While the cast iron production process has a number of drawbacks, like the long flow and high energy consumption, an alternative process is needed to enhance the added value of semisteel.

Titanium dioxide is mainly produced by the sulphate process in China. The ferric iron attached to the titanium dioxide's surface will affect the whiteness of the titanium dioxide. Therefore, iron powders are needed to reduce the ferric iron (Fe^{3+}) in the titanyl sulphate solution during the reducing step of the sulphate process [5]. A novel iron powder production process using semisteel was proposed in the present work according to the component characteristics of the semisteel. The iron powders produced by semisteel were used in the sulphate process, so there is no need to remove the sulphur.

In China, the iron powder is mainly produced by Hoganas process and water atomization process [6, 7]. The Hoganas process with forging scale or iron concentrate powders as raw materials includes the following steps: carbon reduction in the tunnel, crushing, hydrogen reduction in the belt furnace, screening and blending. In particular, the carbon reduction in the tunnel and the hydrogen reduction in the belt furnace last for nearly 74 and 3 h, respectively, resulting in high energy consumption and low production efficiency. The water atomization with hot metal or molten steel as raw materials can be divided into a number of steps: water atomization by high-pressure water, dying, hydrogen reduction in the belt furnace, crushing, screening and blending. Although the water atomization process is of high production efficiency, it has obvious drawbacks like water consumption and energy consumption. Therefore, an alternative process is required to obtain iron powders.

Previously, a number of atomization processes have been proposed and carried out in the investigation of molten slag, including the rotary cup atomizer process, rotary disc or drum atomizer process, rotary cylinder atomizer process and rotary multi-nozzle cup atomizer process [8–12]. In comparison with the water atomization process, the rotary atomization consumes less water and electrical energy. Therefore, the rotary disc atomizer process was adopted to produce iron powders in the present study.

Experimental

Materials

The semisteel was obtained from one steel company in China. The main chemical composition of the semisteel is shown in Table 1.

Table 1 Chemical composition of the semisteel

Composition	Fe	S	C
Content (wt%)	97.4	0.35	0.15

Apparatus and Procedures

Figure 1 shows the schematic diagram of the rotary disc atomizer apparatus used in the present study. The experimental apparatus mainly includes four parts: (1) a graphite crucible with a hole at the bottom with an induction coil in the outer layer; (2) a rotary disc atomizer; (3) a rotation-driving equipment—The rotating speed can be adjusted through the frequency converter, and the maximum value is 3000 revolutions per minute (rpm); (4) the iron granules collector.

The semisteel was melted in a graphite crucible and kept at approximately 1723 K (1450 °C). The rotary disc was preheated for 15 min using a burner with nature gas to avoid the breakage due to the thermal shock. Rotation speed was fixed at a desired value, and the molten alloy was poured onto the centre of the rotating disc. The molten semisteel was spread out from the edge of the disc and disintegrated into droplets. The droplets were cooled and satisfied to granules.

The flow rate of molten semisteel in the experiment was nearly 1.3×10^{-5} m^3/s, and the rotating speed was 800 rpm.

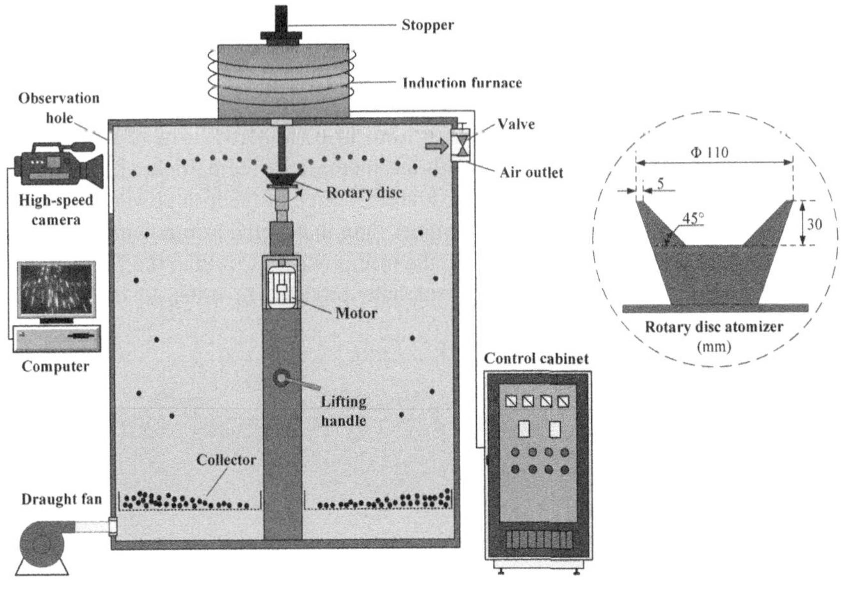

Fig. 1 Schematic of experimental apparatus and multi-nozzles rotary cups

According to the water atomization process, a series of post-treatments were conducted on the obtained iron granules including hydrogen reduction in the belt furnace. In the present study, the iron granules were reduced in a hydrogen atmosphere with the Netzsch STA 449C thermal analyzer.

Results and Discussion

Morphology of Iron Granules

The obtained iron granules with different diameters at 800 rpm is shown in Fig. 2. By contrast, the iron granule with smaller size has a better spherical morphology. Besides, the surface colour of the majority of iron granules is grey. It is speculated that a layer of iron oxides was formed on the surface of the iron granules during the atomization process. Therefore, the TG measurements are required to remove the iron oxide layer and study the oxidation degree of the obtained iron granules.

TG Analysis

The TG measurements on the iron granules with different size were conducted in the Netzsch STA 449C thermal analyzer at 1173 K (900 °C). The reducing atmosphere used in the present study was H_2-75% and N_2-25% gas mixture. During the TG experiments, the weight loss of iron granules can be approximately taken as a constant when the reaction time was 1 h. Figure 3 shows the weight loss of iron granules, and the corresponding oxidation degree of iron powders of different size. It can be found that when the particle size is in the range from 0.3–0.35 mm to 0.6–0.55 mm, the oxidation degree varies from 11.67 to 8.13%. The results indicate that the oxidation degree decreases with the increase of the particle size.

According to the TG analysis, the oxidation degree of the iron powders obtained by the rotary disc atomizer process is higher than that of the iron powders obtained by the water atomization. Additionally, the iron powders less than 0.45 mm should be more than 95% in the industrial sulphate process. In order to decrease the

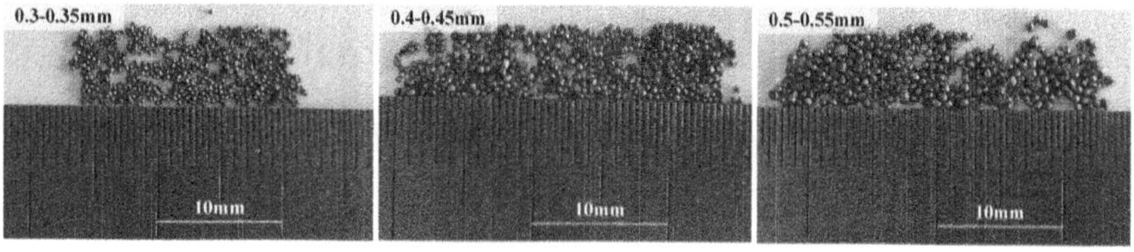

Fig. 2 The obtained iron granules with different sizes

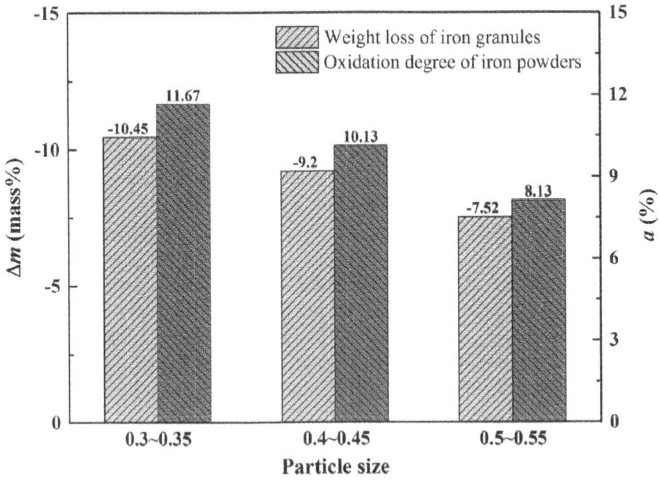

Fig. 3 The weight loss of iron granules and oxidation degree of iron powders with different size

oxidation degree of the iron powders and investigate the distribution of iron granules, the effect of rotating speed and flow rate of the semisteel should be further studied in the future.

Conclusions

In order to enhance the added value of the semisteel, a novel process for producing iron powders is described in this paper. The iron granule obtained by rotary disc atomizer process with smaller size has a better spherical morphology. The oxidation degree of iron powder decreases with the increase of the particle size and varies from 11.67 to 8.13% when the particle size is in the range from 0.3–0.35 mm to 0.6–0.55 mm.

Acknowledgements The authors are especially thankful to the Fundamental Research Funds for the Central Universities (Grant No. 2018CDQYCL0026), the National Natural Science Foundation of China (Grant No. 51674053) and Graduate Research and Innovation Foundation of Chongqing (Grant No. CYB17002).

References

1. Chen K, Liu X, Sun X (2013) The supply-demand situation of rutile resources and the prospect analysis of development and exploitation in Xinyi area. Mod Min 29(4):126–127
2. Yan S (1999) Utilization and construction model of titanium resources in Panzhihua. Nonferrous Met Des 4:66–69

3. Lu R, Jia H (2014) The production technology status and development analysis of high titanium slag in China. China Metal Bull 10:43–45
4. Yu W (2005) Experimental study of carburisation and desulfuration for semisteel. Sichuan Metall 27(2):29–32
5. Tang W, Zhang J (2011) Energy consumption analysis and comments of manufacturing titanium dioxide by sulfuric acid process and chloride process. Inorg Chem Ind 43(6):7
6. Liu S, Zhao J (2007) Development and prospect of iron and steel powder China. In: 2007 CSM annual meeting proceedings 1:361–364
7. Yu X (1997) Sponge iron and reduced iron powder produced by Hoganas process. Powder Metall Ind 3:31–39
8. Pickering SJ, Hay N, Roylance TF et al (1985) New process for dry granulation and heat recovery from molten blast-furnace slag. Ironmaking Steelmaking 12(1):14–21
9. Kashiwaya Y, Yutaro IN, Akiyama T (2010) Development of a rotary cylinder atomizing method of slag for the production of amorphous slag particles. ISIJ Int 50(9):1245–1251
10. Kashiwaya Y, Yutaro IN, Akiyama T (2010) Mechanism of the formation of slag particles by the rotary cylinder atomization. ISIJ Int 50(9):1252–1258
11. Wang D, Peng H, Ling X (2014) Ligament mode disintegration of liquid film at the rotary disk rim in waste heat recovery process of molten slag. Energy Proc 61:1824–1829
12. Qin Y, Lv X, Bai C et al (2013) Dry granulation of molten slag using a rotating multi-nozzle cup atomizer and characterization of slag particles. Steel Res Int 84(9):852–862

Dissolution Kinetics of Titanium in Carbon-Saturated Iron

Leizhang Gao, Tongxiang Ma, Zhiming Yan and Meilong Hu

Abstract Adding titanium ore in the blast furnace is an efficient method to prolong its hearth life. The content of titanium is important for prolonging blast furnace life. In this study, the dissolution kinetics of titanium in carbon-saturated iron was studied. The carbon-saturated pig iron was heated to 1200, 1250 and 1300 °C, respectively, and then an appropriate amount of titanium-iron alloy powder was added. After a certain period of time, a quartz tube was used to take out the samples from the molten iron and the solubility of titanium was measured by an inductive coupled plasma emission spectrometer (ICP). By fitting the experimental results with different solid–liquid reaction kinetics models, the limiting step of the whole process was determined. The results show that in the process of dissolution the mass diffusion is the limiting step. According to the Arrhenius empirical equation, the activation energy of the dissolution is about 450.96 kJ/mol.

Keywords Dissolution kinetics · Carbon-saturated pig iron · Activation energy

Introduction

Since the development of the blast furnace, the smelting intensity became more and more important and increasing emphasis has been placed on extending the campaign life of the blast furnace. The two main restrictions on campaign life are shell cracking and hearth refractory erosion. Of these, the restriction of shell cracking was overcome using a cooling system and repair engineering [1]. To date, premature hearth failure is one of the main factors limiting blast furnace campaign life. Before 1890, metallurgists [2] agreed that titanium was detrimental to ironmaking process. After 1914, Bachman [3] detailed a series of extended blast furnace trials using titaniferous ores sourced in an attempt to determine their commercial value.

L. Gao · T. Ma · Z. Yan · M. Hu (✉)
College of Materials Science and Engineering, Chongqing University, Chongqing 400044, China
e-mail: hml@cqu.edu.cn

© The Minerals, Metals & Materials Society 2019

T. Jiang et al. (eds.), *10th International Symposium on High-Temperature Metallurgical Processing*, The Minerals, Metals & Materials Series,
https://doi.org/10.1007/978-3-030-05955-2_52

Until 1977, Narita et al. [4] concluded that titanium compounds were effective in protecting against erosion. After nearly 180 years of development, using titanium-bearing iron ores as raw materials in the blast furnace has been determined to be effective and is the main strategy for extending campaign life [5–9]. Many studies have been published about protecting the hearth using titanium dioxide-bearing ores as raw materials [10–12], most of which focused on the thermodynamics of the relationship between the titanium solution and temperature. The content of titanium is very low in carbon-saturated iron, so it is necessary to obtain dissolution rate of titanium.

Insert the metal titanium rod directly into the molten steel, two separate dissolution periods were identified: a steel shell period and a free dissolution period [13]. When the rod was in contact with the melt, a solidified shell was formed around the rod, which was not conducive to the determination of the dissolution time. Afterwards, the method of studying the dissolution kinetics was improved in the experiment, using rotation or stirring. When exploring the factors affecting the dissolution rate, the dissolution temperature, the rotation speed and the particle size are generally considered [14, 15]. In this experiment, the free dissolution method was used. In order to prevent the surface of the sample from forming an outer shell, ferrotitanium alloy powder was added at the experimental temperature to obtain a free dissolution rate of titanium.

Experimental

Raw Materials

Pig iron was obtained from Steel Corp in China. The composition of the ferrotitanium alloy and iron block was given in Table 1.

Preparation of Samples with Different Titanium Contents

The desired titanium content in pig iron was obtained by adding ferrotitanium alloy and pig iron in an alumina crucible (73 mm diameter × 63 mm height) in this study. To prevent the alumina crucible from breaking, the crucible containing a mixture of pig iron was placed in another big alumina crucible. Finally, the crucibles were placed in a furnace, as shown in Fig. 1.

Table 1 The iron block and ferrotitanium compositions (wt%)

Element	Fe	C	Ti	Si	Mn
Iron block	94	4.57	0.04	0.53	0.47
ferrotitanium	48	0.10	35	4.5	2.5

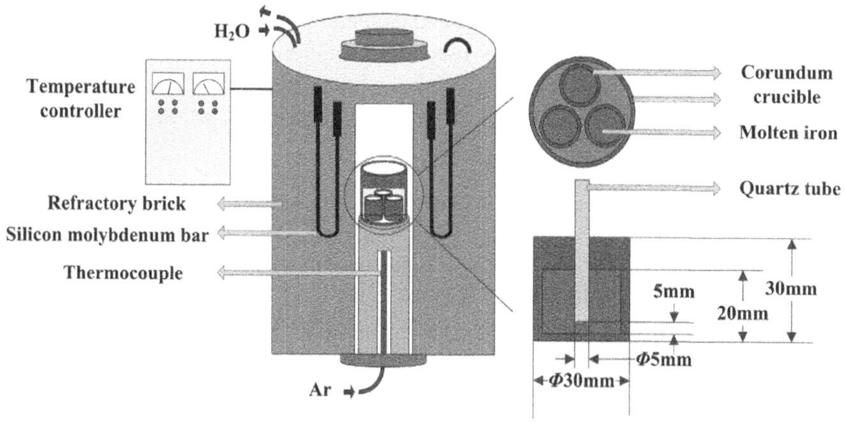

Fig. 1 Schematic of the equipment used to prepare the sample of pig iron with saturated carbon and different titanium contents

The whole process was performed in an argon atmosphere. The temperature was raised at a normal heating rate and finally held at the target temperature for 4 h to ensure that the entire molten iron reaches the target temperature. At the end of the holding time, the furnace cover was opened and molten iron was absorbed using a quartz tube (inner diameter of 5 mm) to obtain the original titanium content in the pig iron. Then using the same specification quartz tube to add ferrotitanium alloy (after the alloy is added, the titanium dissolves and just reach saturation), simply stir it to disperse the alloy on the surface of the molten iron, and no longer stir in the subsequent sampling. After the alloy was added, samples were taken at 0.5, 1.5, 3.5, 7.5, 15.5, 23.5 and 31.5 h. The extracted sample was obtained by ICP to determine the solubility of titanium.

Results and Discussion

Titanium Dissolution Rate

The ferrotitanium alloy has a small particle size (80 mesh), and it is considered that the temperature immediately reaches the target temperature after the powder was in contact with the molten iron (1 kg), and the molten iron cooling layer is not formed on the surface. Define the dissolution rate $X_{(Ti)}$ of titanium as shown in Eq. (1) [16]:

$$X_{(Ti)} = \frac{w_t - w_0}{w_\infty - w_0} \quad (1)$$

where w_t is titanium solubility at time t, w_0 is titanium solubility at time 0, w_∞ is titanium solubility at equilibrium time.

The relationship between solubility and temperature of titanium has been studied by many scholars, but the samples used were different from those of on-site pig iron. The relationship between the solubility of titanium in the pig iron on site and the temperature has been obtained before this experiment, as shown in Eq. (2) below:

$$[W_{Ti}] = \frac{5099.88}{T} + 2.48 \tag{2}$$

where W_{Ti} is titanium solubility, T is temperature.

At the target temperature of 1200, 1250 and 1300 °C, the sample obtained was tested by ICP, and the content of titanium and dissolution rate was shown in Fig. 2. Dissolution can be divided into two stages, rapid dissolution and slow dissolution. After the alloy was added into the molten pig iron, the alloy was in the rapid dissolution stage for the first 3 h. The concentration of the titanium metal dissolved and detached around the alloy was higher than that of the molten iron. The diffusion resistance was relatively small, so it could be quickly diffused out. After 3 h, the differential concentration of titanium around the alloy and titanium in the molten iron resulted in slow dissolution.

The Dynamics Model of Solid–Liquid Reaction

The Dynamics Model

At present, the solid–liquid reaction kinetics model of mineral dissolution mainly includes first-order reaction model, solid–liquid interface dissolution model and internal diffusion-type model.

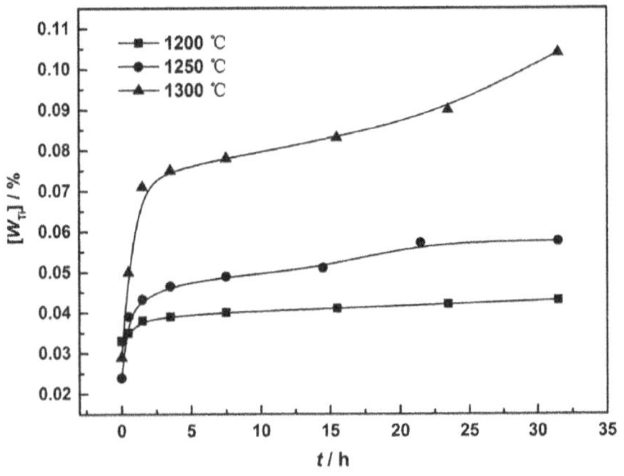

Fig. 2 The titanium content in the molten iron with time increasing at 1200, 1250, 1300 °C

(1) First-order reaction dynamics
The reaction rate is related to the concentration formability of the reactants. The first-order reaction model reaction rate Eq. (3) [17] is:

$$\frac{dM}{dt} = -KM^m \tag{3}$$

where M is the content of titanium in the experimental sample, K is the reaction rate constant, m is the number of reaction stages. Also, where m = 1, integrate both sides of the Eq. (3) and get:

$$\ln(1 - X) = Kt \tag{4}$$

(2) Solid–liquid interface dissolution model
When the solid–liquid interface dissolution reaction is controlled, the interface between Fe in the melt and alloy, Fe acts on the surface of the particles, so that titanium falls off and dissolves. The reaction rate Eq. (5) [18] is:

$$\frac{dX}{dt} = K\left(\frac{3V}{r_0^3}\right)(1 - X)^{2/3} \tag{5}$$

where V is the molar volume of the reactant and r_0 is the mineral particle size. Integrate both sides of the Eq. (5) and get:

$$1 - (1 - X)^{1/3} = Kt \tag{6}$$

(3) Internal diffusion-type model
When the internal diffusion reaction of solid–liquid reaction is controlled, as the dissolution reaction, the product layer of dissolved titanium thickens, and the resistance to the diffusion of Fe atoms in the molten iron to the interface becomes larger and larger. The kinetic Eq. (7) [19] is:

$$1 - 3(1 - X)^{2/3} + 2(1 - X) = Kt \tag{7}$$

At the temperatures corresponding to those in Fig. 1, the dissolution rate X of titanium and the time t were brought into Eqs. (4), (6) and (7) to obtain the linear correlation R^2 and degree of fitting Q. This data was processed using MATLAB programming. The results were shown in Fig. 3.

It can be seen from Fig. 3 that the values of the degree of fitting Q increase with the increasing of the temperature of the three control models, indicating that the temperature rises and the numerical fitting error becomes larger. The more high the temperature, the more intense the diffusion of the titanium, resulting in greater fluctuations in the experimental data obtained. Compared with the linear correlation R^2 and the degree of fitting Q of each model. At every temperature, the fitting linearity R^2 of the internal diffusion-type model $1–3(1 - X)^{2/3} + 2(1 - X)$ is the

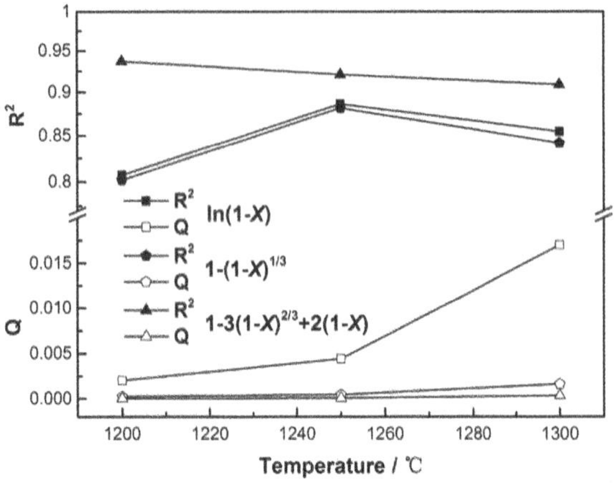

Fig. 3 The fitting result of solid–liquid reaction kinetics equations (R^2: linear correlation, Q: degree of fitting)

best, and the degree of fitting error Q is the smallest. Therefore, the internal diffusion-type model most conforms to ferrotitanium alloy dissolved in carbon-saturated molten pig iron.

Activation Energy *E*

In the process of seeking the dissolution activation energy *E*, using the Arrhenius equation, as shown in Eq. (8) [20]:

$$K = \mathrm{A} \cdot \mathrm{e}^{-\frac{E}{\mathrm{R}T}} \tag{8}$$

The dissolution activation energy *E* can be calculated by deforming the Eq. (8), as shown in Eq. (9):

$$\ln K = \ln \mathrm{A} - \frac{E}{\mathrm{R}T} \tag{9}$$

where *E* is the dissolution activation energy, *K* is the slope of the fitting function, *T* is the temperature, R is the gas constant and A is fixed parameters.

The *K* value and the temperature *T* corresponding to the internal diffusion-type model at 1200, 1250 and 1300 °C are brought into the Eq. (9), and the fitting result is shown in Fig. 4. Where the linear correlation value of R^2 is 0.9894, the relative error is less than 2% (generally between 2 and 5%). From the slope of the line, the dissolution activation energy *E* is 450.96 kJ/mol. This indicates that the titanium dissolves slowly.

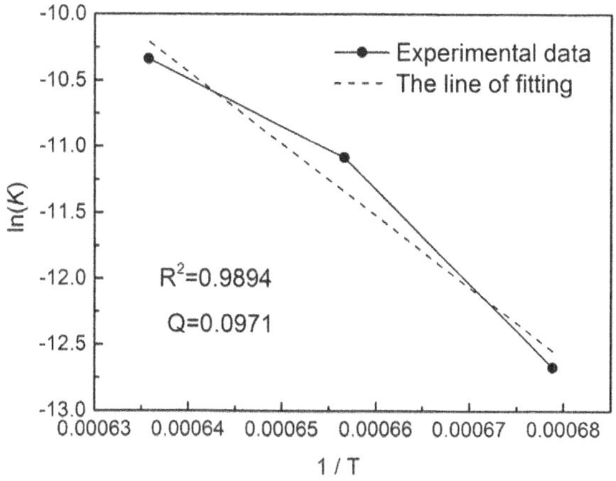

Fig. 4 The fitting result of the experimental data based on the Arrhenius equation

Conclusions

Dissolution of titanium in molten pig iron can be divided into two stages, rapid dissolution and slow dissolution. The rapid dissolution makes up about 10% of the whole dissolution time. The dissolution and diffusion of titanium dissolved in carbon-saturated molten pig iron is the main limiting step in the whole process.

The dissolution activation energy of the titanium element dissolved in carbon-saturated molten pig iron was about 450.96 kJ/mol showed that the dissolution speed of titanium is very slow.

Acknowledgements The authors are especially grateful to the project 51674054 supported by National Natural Science Foundation of China and supported by the National Key R&D Program of China (2017YFB0603801).

References

1. Nouchi T, Sato M, Takeda K (2006) Extension of blast furnace campaign at JFE steel. In: 34 th McMaster University symposium on iron and steelmaking, Hamilton, Ontario, Canada, 24–38 May 2006
2. Rossi AJ (1890) Titanium in blast furnaces. J Am Chem Soc 12(4):91–117
3. Bachman FE (1914) The use of titaniferous ore in the blast furnace. American Iron and Steel Institute Year Book, pp 371–419
4. Narita K, Maekawa M, Onoye T (1977) Formation of titanium compounds, so-called titanium-bear, in the blast furnace hearth. ISIJ 62(8):525–534
5. Hisada K, Jyomoto Y, Etou B (1964) 25 On the properties of Ti-bears sampled from the blast furnace hearth. In: 68th grand lecture meeting of the iron and steel institute of Japan, Tetsu, Hagane, Japan, pp 1616–1619, September 1964
6. Ollno A, Ross HU (1963) Liquidus-temperature measurements in the lime-titania-alumina-silica system. Can Metall Q 2(3):243–258

7. Ohno A, Ross HU (1963) Optimum slag composition for the blast-furnace smelting of titaniferous ores. Can Metall Q 2(3):259–279
8. Ohno A, Ross HU (1964) The reduction of titanium dioxide in titaniferous slags. Can Metall Q 3(3):257–267
9. Street S, Copeland C, Worral E (2014) Curse to cure: the utilization of titanium-bearing products in blast furnace ironmaking. Iron Steel Technol 11(3):71–87
10. Li Y, Fruehan RJ (2001) Thermodynamics of TiCN and TiC in Fe-Csat melts. Metall Mater Trans B 32(6):1203–1205
11. Morizane Y, Ozturk B, Fruehan RJ (1999) Thermodynamics of TiO_x in blast furnace-type slags. Metall Mater Trans B 30(1):29–43
12. Zhao YF (2014) The impact of titanium on skull formation in the blast furnace hearth. Iron Steel Technol 11(3):95–103
13. Argyropoulos SA, Guthrie RIL (1984) The dissolution of titanium in liquid steel. Metall Trans B 15(1):47–58
14. Morales GV, Capretto ME, Fuentes LM (2000) Dissolution kinetics of hydroboracite in water saturated with carbon dioxide. Hydrometallurgy 58(2):127–133
15. Lu WH, Yin ZL, Ding ZY (2017) Dissolution kinetics of copper from multi-metal copper alloy roasted in oxygen. J Central South Univ 24(2):335–340
16. Keng WU, Zhao Y, Qian W (2002) Foam behavior parameter in foaming process originated from reducing TiO_2 in synthetic blast furnace slag. Chin J Nonferrous Metals 12(4):817–821
17. Xu ZM, Huang RQ, Tang ZG (2005) Sili-cate mineral dissolution kinetics and the significance of research on landslide. Chin J Rock Mechan Eng 24(9):1480–1491
18. Hua QX, Han YW, Tang JW (2016) Study on dissolution kinetics of single superphosphate granules in water. Phosphate Compd Fertil 31(5):10–12
19. Aracena A, Sanino A, Jerez O (2018) Dissolution kinetics of molybdite in KOH media at different temperatures. Trans Nonferrous Metals Soc China 28(1):177–185
20. Juan M, Rodrtguez DM, Teresa SM (2009) Study of the best designs for modifications of the Arrhenius equation. Chemometr Intell Lab Syst 95(2):199–208

Research on Mineral Structure and Compositions of Peru Raw Ore

Wen Pan, Lei Liu, Ya-peng Zhang, Xia Zhao and Zhi-xing Zhao

Abstract In this study, mineral structures and compositions of Peru raw ore (PRO) were investigated by optical microscope, scanning electron microscope and X-ray diffraction. The results show that PRO is primarily composed of magnetite, haematite, gangue and sulphide. A small amount of chalcopyrite and calcium phosphate also exist in PRO. Gangues are mainly quartz minerals. Sulphides exist as FeS, Fe_2S_3 and FeS_2, respectively. Sulphides, most of which accrete with gangues, are embedded among ferrous bodies in the form of banding, schistose, or porphyritic structures. The particle sizes of sulphides and gangues are between 200 and 500 μm. Some of them are over 1000 μm. It can be concluded that high liberation degree of sulphides and gangues will be obtained after crushing. After flotation and magnetic separation in Shougang Jingtang dressing plant, the Fe content of PRO was improved from 56.68 to 65.75%. The sulphur content dropped to 0.48% from 3.44%.

Keywords Peru raw ore · Mineral compositions · Microstructure · Beneficiation

Peru is rich of iron ore resources with a reserve of 882 million ton and an average grade of more than 50%, which are mainly distributed in Ica province, Arequipa province and Andes mountains [1]. Shougang Hierro Peru S.A.A, a subsidiary of Shougang group, is the largest Chinese mining investment company in South America. The PRO studied in this paper, with low price and high original grade, were from the iron ore company. Study on the microstructures and mineral compositions of the iron ore was carried out by the mineralogical research methods

W. Pan (✉) · Y. Zhang · Z. Zhao
Beijing Key Lab of Green Recyclable Process for Iron & Steel Production Technology, Beijing 100043, China
e-mail: panwen@shougang.com.cn

L. Liu
Beijing Shougang Co., Ltd.,, Beijing, Hebei 064400, China

X. Zhao
Shougang Institute of Technology, Beijing 100144, China

© The Minerals, Metals & Materials Society 2019

T. Jiang et al. (eds.), *10th International Symposium on High-Temperature Metallurgical Processing*, The Minerals, Metals & Materials Series,
https://doi.org/10.1007/978-3-030-05955-2_53

[2–4], which provides technical references for the subsequent usage of the raw iron ore in Shougang Group.

Samples and Experiments

From the perspective of macroscopic morphology, the PRO was compact. Bright striped, flaky metallic luster minerals were observed scattering in the ore matrix after cutting, rough grinding and polishing, as shown in Fig. 1. Table 1 presents the chemical compositions of PRO. The PRO has relatively high original grade, which is nearly 57%, but it also contains ultra-high sulphur content, which is up to 3.44%.

In order to further understand the internal structure of PRO, the microstructure and mineral compositions of PRO were investigated with optical microscope, scanning electron microscope and X-ray diffraction phase analysis (XRD). The optical microscopy used in the test was the German Leica optical microscopy, the scanning electron microscopy was the Japanese electron jsm-7001f scanning electron microscopy and the XRD analysis was conducted using the German company brock X-ray diffraction instrument.

Fig. 1 Macro morphology of PRO

Table 1 Chemical composition of PRO (wt%)

TFe	SiO_2	CaO	MgO	Al_2O_3	K_2O	Na_2O	S	ZnO	LOI
56.68	9.42	2.17	2.98	1.42	0.43	0.19	3.44	0.17	−0.50

Microstructures of PRO

Figures 2a–d are the microscopic appearance of the PRO under reflected light (50 times) of optical microscopy. It can be seen that PRO is mainly composed of iron-bearing minerals, metallic luster minerals and gangue; among which the metallic luster minerals need to be analysed and identified by scanning electron microscopy and XRD.

Most of the above-mentioned metallic luster minerals are exhibited as porphyric structures, bands or sheets. And most of the grains were between 200 and 500 mm, and a few grains were larger than 1000 mm in diameter. A part of the gangue was clustered between the magnetite grains (Figs. 2a and b), and the other part was patchy and scattered among the metallic luster minerals (Fig. 2d). Both the metallic luster minerals and the gangue minerals are clustered and have a high degree of dissociation. Therefore, the separation of the gangue minerals, the metallic luster minerals and the iron bearing minerals can be realized by crushing the PRO below −90 meshes (180 μm).

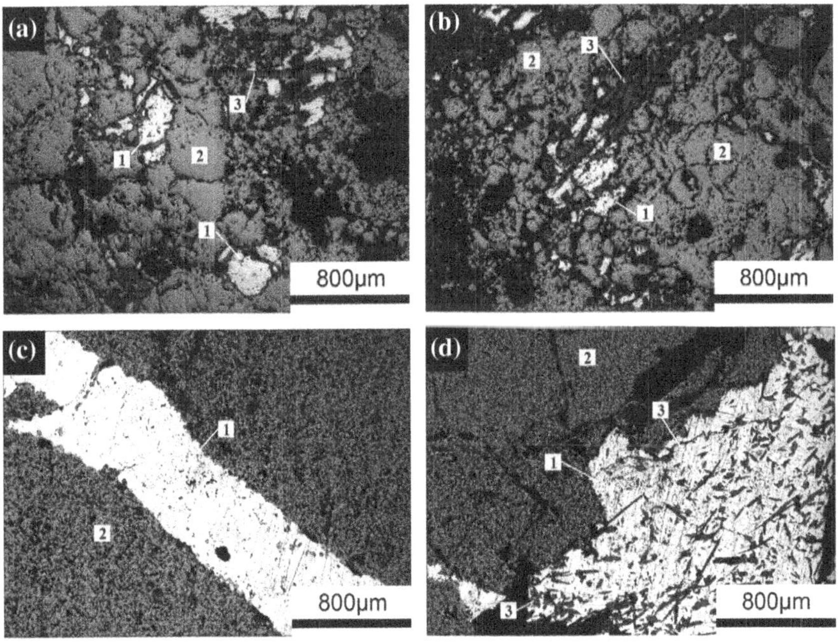

Fig. 2 Optical microscope images of PRO. 1—Metallic luster minerals; 2—Ferric oxide; 3—Gangue

Mineralogy Analysis of PRO

Scanning electron microscope and elemental energy spectrum analysis were applied to investigate the mineral types in PRO. The results are shown in Fig. 3.

Energy Spectrum Analysis of Mineral Elements

Figures 4a–d are some scanning electron microscopy photographs of PRO. Combined with the observation results of the optical microscope and the analysis data of elemental energy spectrum, the element compositions of minerals under the scanning electron microscopy were classified and sorted. The results are shown in Table 2, 3, 4, 5 and 6.

(A) metallic Luster Minerals

According to the results of energy spectrum analysis in Table 2, the main compositions of metallic luster minerals were iron sulphide which included FeS_2, FeS and Fe_2S_3. It can be seen that the Fe/S molar ratio of iron sulphide in PRO was between 0.99 and 1.97. Thus, it can be inferred that the metallic luster minerals in PRO contained all three kinds of iron sulphides. In addition, solid solution of element boron can be occasionally seen in some iron sulphide minerals, as shown in Table 2, position 3.

(B) Ferric Oxides

It can be seen from Table 3 that Fe content of iron bearing minerals in PRO was between 60.34 and 71.80%. All the measured points of ferric oxides contain sulphur, and the mass percentage was in the range of 1.0–3.0%.

(C) Gangues

According to the results of the energy spectrum analysis of gangue minerals in Table 4, Si, Mg and Fe are the main elements of gangue minerals, which were similar to ferric oxides. All the monitoring points of gangue minerals contain sulphur, and the content range was between 1 and 10%. In addition, Al is present in many of the observation points of gangue minerals with concentration between 0.5 and 4 wt%.

(D) D) Unknown mineral #1

A large area of dark grey unknown mineral, called unknown mineral #1, was observed under the scanning electron microscope in PRO. According to the energy spectrum analysis of the elements in Table 5, this mineral was mainly composed of P and Ca elements. The molar fraction ratios of P and Ca elements of the three measurement points are, respectively, 1.50, 1.49 and 1.51, which are very close to

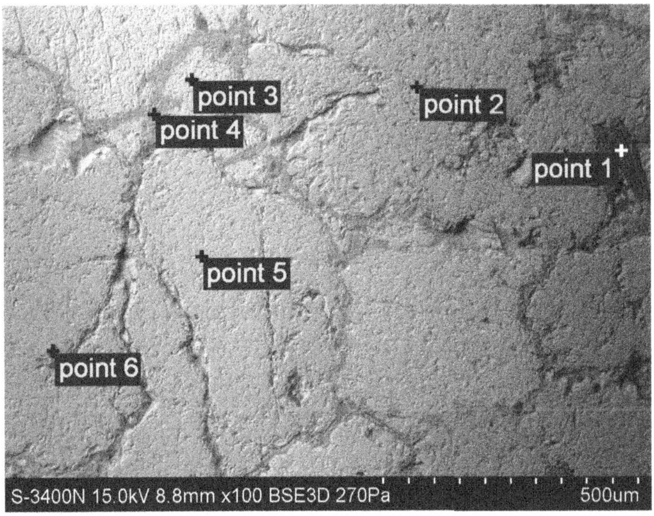

Point	O	Na	Mg	Al	Si	S	Cl	Ca	Fe
Point 1	49.14		10.86	0.58	20.63	2.41		0.32	16.07
Point 2	33.32		0.46		1.26	1.02			63.93
Point 3	9.94	0.48			0.84	31.99			56.74
Point 4	24.71	2.07			1.03	31.09			41.10
Point 5	24.44					1.19			74.37
Point 6	28.90	0.92	1.43		19.48	2.28	1.37		45.61

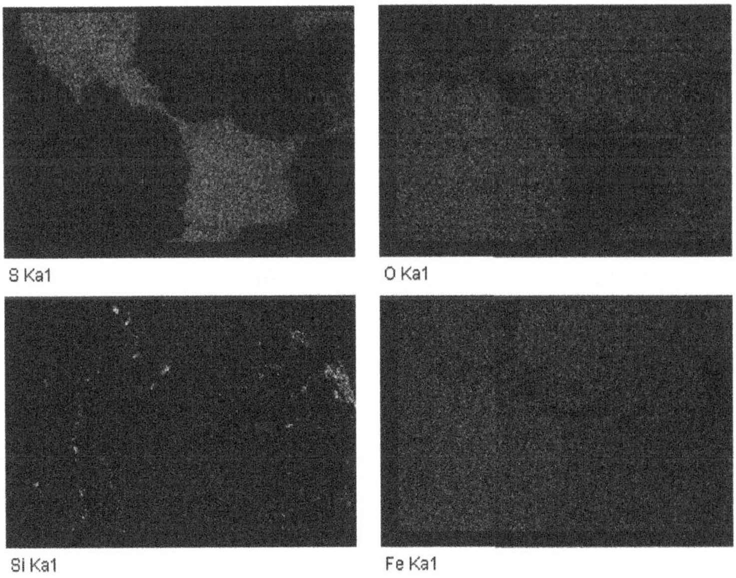

Fig. 3 SEM images and EDS analysis results of PRO

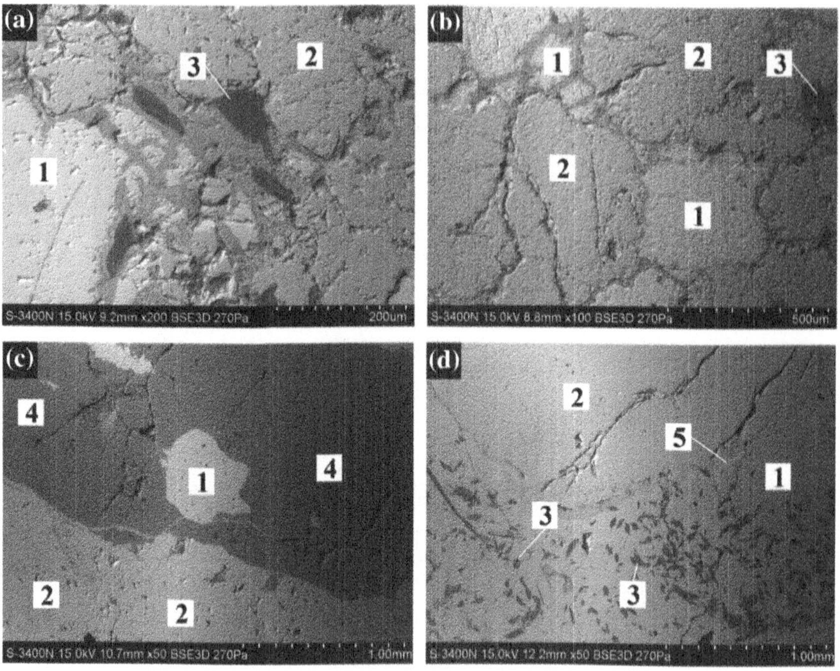

Fig. 4 SEM images of PRO. 1—metallic luster minerals, 2—ferric oxides, 3—gangue, 4—unknown mineral #1, 5—unknown mineral #2

the molar fraction ratios of P and Ca in $Ca_3(PO_4)_2$ (1.50). Therefore, it was determined that this mineral was calcium phosphate.

(E) Unknown mineral #2

The bright sheet unknown minerals, marked as unknown mineral #2, were found in all of the scanning electron microscopy (SEM) photographs of several PROs. According to energy spectrum analysis in Table 6, the mineral was mainly composed of Cu, S and Fe elements. The Fe, Cu, S mole fraction ratio are: 1. 93:1.02, 1:1. 93:1.20, 1:2. 75:3.21 in three points. In addition to point 3, the mole fraction ratios of the other two points were closely to chalcopyrite ($CuFeS_2$). So, the mineral is presumed to be chalcopyrite

Distributions of Main Chemical Elements

According to the energy spectrum analysis data of the above elements, the distributions of S, Fe, Si, Mg, Ca, Al, Na and K in sulphide, iron oxide, gangue, calcium phosphate and chalcopyrite were calculated.

Table 2 EDS analysis results of minerals with metallic luster (%)

Position	S	O	Fe	Mg	Si	Ca	Cu	B	Na	K	P	Fe: S (molar fraction)
1	31.99	9.94	56.74		0.84				0.48			1:0.99 (FeS)
2	31.09	24.71	41.10		1.03				2.07			1:1.32 (FeS. Fe_2S_3)
3	27.08	23.21	32.77	0.58	1.59	0.46	1.28	12.11	0.58	0.34		1:1.45 (Fe_2S_3)
4	34.42	20.84	41.42	0.32	1.80	0.40			0.80			1:1.45 (Fe_2S_3)
5	37.78	4.57	57.27		0.39							1:1.15 (FeS)
6	49.75	4.49	45.75									1:1.90 (FeS_2)
7	47.75	6.15	44.88	0.49	0.74							1:1.86 (FeS_2)
8	47.07	5.28	41.74			2.32			0.62		1.20	1:1.97 (FeS_2)
9	46.72	5.55	42.02			2.90					1.08	1:1.95 (FeS_2)
10	46.41	6.58	47.01									1:1.73 (Fe_2S_3. $2FeS_2$)

Table 3 EDS analysis of ferric oxides (%)

Position	Fe	O	S	Na	Mg	Si	Cl	Ca	K	P
1	63.93	33.32	1.02		0.46	1.26				
2	74.37	24.44	1.19							
3	69.37	28.17	1.82			0.63				
4	63.31	24.97	2.75	4.12	0.57	0.98	1.09	0.63	1.57	
5	69.24	28.99	1.16			0.61				
6	69.65	27.30	1.67	0.60		0.79				
7	70.58	26.02	2.81			0.59				
8	60.34	19.89	1.01	9.68		1.33	6.08	1.00		0.68
9	67.72	31.15	1.13							
10	71.80	27.09	1.11							

(1) S element

Figure 5 showed the distribution of S elements in various minerals. S elements were widely distributed in all minerals of PRO, and the sequence of content from high to low was as follows: sulphide, chalcopyrite, gangue, calcium phosphate and ferric oxides.

Table 4 EDS analysis of gangues (%)

Position	Si	Mg	S	O	Fe	Na	Al	Cl	Ca	Cu
1	20.63	10.86	2.41	49.14	16.07		0.58		0.32	
2	19.48	1.43	2.28	28.90	45.61	0.92		1.37		
3	21.74	10.25	3.49	43.08	10.15		1.47		7.51	2.32
4	21.61	11.45	2.38	44.31	13.41		0.55		6.29	
5	19.31	8.85	4.74	43.47	16.42		1.17		6.03	
6	11.03	8.39	1.31	37.96	29.64	3.70	4.00	3.97		
7	19.58	9.69	9.59	39.04	14.84		0.83		6.42	
8	11.48	7.73	5.23	38.33	33.15		4.07			

Table 5 EDS analysis of unknown mineral #1 (%)

Position	P	Ca	O	S	Na	Si	Cl	Fe	F
1	**19.02**	36.88	34.79	1.64	0.58	0.49	2.00	4.61	
2	**15.33**	29.56	35.91	5.04		0.52		9.32	4.32
3	**16.58**	32.40	39.13	0.70		0.83	0.42	5.54	4.40

Table 6 EDS analysis of unknown mineral #2 (%)

Position	Cu	S	Fe	O	Mg	Si	Ca	K
1	**31.29**	30.47	28.11	5.84	1.10	2.07	0.64	0.48
2	**30.54**	29.70	32.45	6.84		0.46		
3	**16.68**	23.15	47.26	10.86		2.05		

(2) Fe element

Figure 6 showed the distribution of Fe in various minerals. Similar to sulphur, Fe was distributed in all minerals, and the content of Fe is successively from low to high: iron oxides, sulphide, chalcopyrite, gangue and calcium phosphate. The type of ferric oxides was determined by XRD analysis in the following part.

(3) 3) Si, Mg, Ca, Al

Figures 7a–d, respectively, showed the distribution of Si, Mg, Ca and Al in each mineral. Generally, the above four elements were mainly distributed in gangue minerals, especially Al element. Ca was the highest in calcium phosphate, followed by gangue.

(4) Na, K

The alkali metal elements Na and K have a small amount of distribution in each mineral, as shown in Figs. 8a and b. In comparison, Na element was more widely

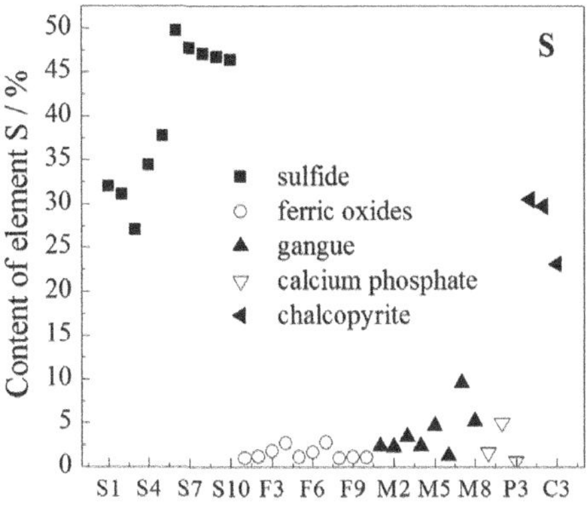

Fig. 5 Distribution map of element sulphur

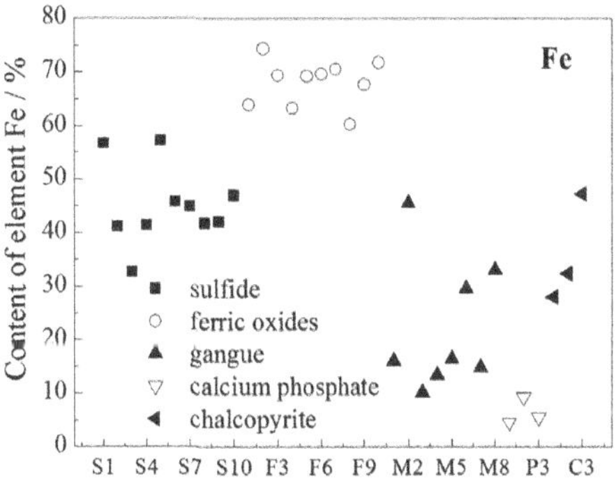

Fig. 6 Distribution map of element ferrum

distributed, existing in all minerals except calcium phosphate, and the contents fluctuated greatly. For example, in ferric oxides, Na content was close to 10% at the highest level and less than 1% at the lowest level. K is only occasionally found in sulphides, ferric oxides and calcium phosphate (less than 2%).

XRD Analysis of PRO

XRD analysis was carried out to further confirm the mineral species and quantitatively analyse the percentage content of each mineral. The test samples were grinded to −200 meshes. The results were shown in Table 7.

According to the X-ray diffraction results, the ferric oxides in PRO were mainly magnetite and a few haematites. The main mineral in gangue was quartz. However, no sulphide, chalcopyrite and calcium phosphate minerals were found in the X-ray diffraction results of the test samples, mainly because the contents of these three minerals in the sample were lower than the measured range (Fig. 9).

As the sulphide mineral types in PRO cannot be confirmed with the power sample, the sample of PRO used in the optical microscope (See in Fig. 1) was put in

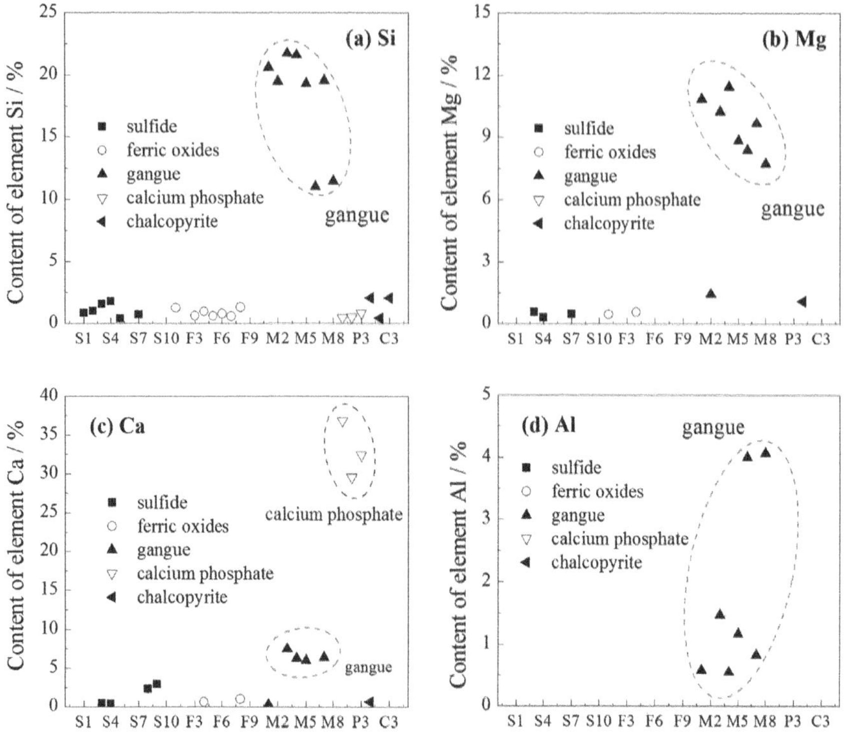

Fig. 7 Distribution map of element Si, Mg, Ca and Al

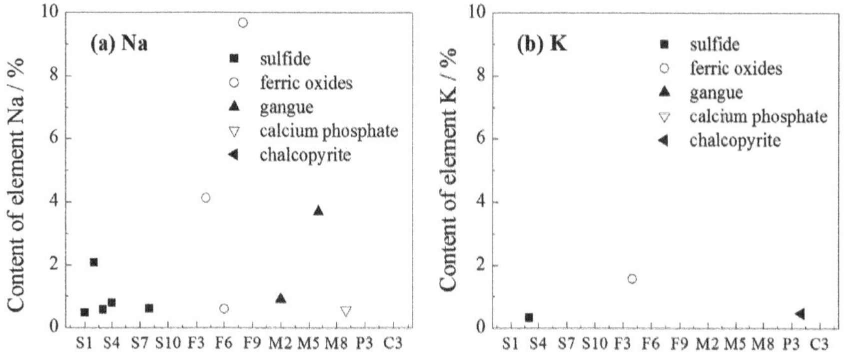

Fig. 8 Distribution map of element Na, K

X-ray scanner. And the X-ray beam was directly located to the sulphide mineral (metallic luster minerals). Two points were selected (See in Fig. 10a and b).

All the mineral types in the scanning area of point 1 were magnetite (Fe_3O_4), and the sulphide mineral was still not detected due to the low content of sulphide in the

Table 7 XRD analysis result of powder PRO (%)

Fe_3O_4	Fe_2O_3	SiO_2
61.7	5.6	32.7

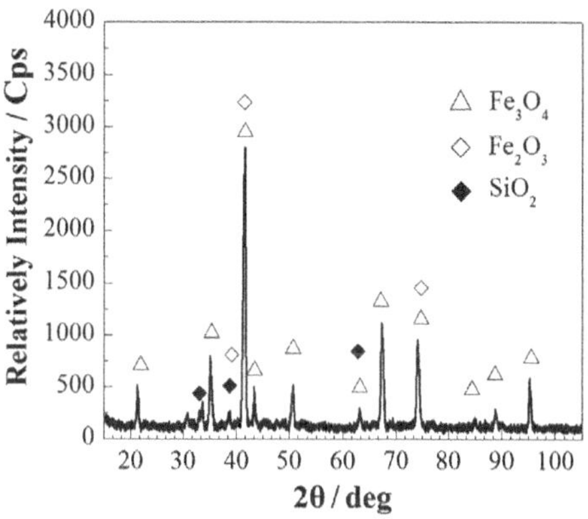

Fig. 9 X-ray diffraction pattern of PRO

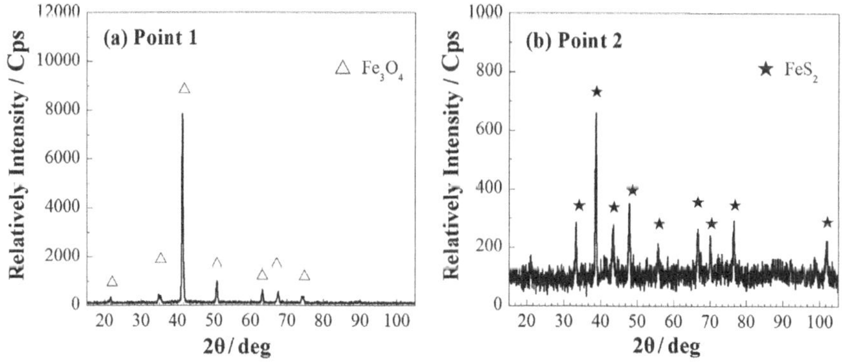

Fig. 10 X-ray diffraction pattern of lump PRO

region. According to Fig. 10b, the mineral type in the scanning area of measuring point 2 was pyrite (FeS_2), which was consistent with the conclusions of scanning electron microscopy and energy spectrum analysis above. Therefore, it can be confirmed that pyrite existed in PRO.

Table 8 Chemical compositions of PRO before and after beneficiation (wt%)

	TFe	SiO_2	CaO	MgO	Al_2O_3	K_2O	Na_2O	S	ZnO	LOI
Before	56.68	9.42	2.17	2.98	1.42	0.43	0.19	3.44	0.17	−0.50
After	65.75	4.13	0.47	1.59	0.84	0.16	0.11	0.48	0.01	−1.57

Industrial Application of PRO

In the case that the load of sintering S is not excessive, a part of PRO was directly broken and then used for sintering by shougang jingtang to take the price advantage of PRO [5].

As the sulphide minerals, gangue minerals and ferric oxides in PRO had a good degree of dissociation, a part of PRO was processed with combined separating technology (flotation and magnetic separation) after breaking and grinding, which would effectively reduce the content of K, Na and some other harmful elements such as zinc, S. Chemical compositions before and after processing were shown in Table 8.

Conclusions

(1) The main minerals in PRO were magnetite, haematite, gangue and sulphide, as well as a small amount of chalcopyrite and calcium phosphate. In the gangue, quartz was the main mineral. And the sulphide includes FeS_2, FeS and Fe_2S_3.
(2) Element S was widely distributed in all minerals of PRO, and the content was in the order from high to low: sulphide, chalcopyrite, gangue, calcium phosphate and ferric oxides.
(3) Si, Mg, Ca and Al are mainly distributed in gangue minerals, especially Al elements. No Al element was found in other minerals.
(4) Elements Na and K had a small amount of distribution in each mineral. In comparison, Na elements are more widely distributed and existed in all minerals except calcium phosphate. K was only occasionally found in sulphide, ferric oxides and calcium phosphate, and its content was not high (less than 2%).
(5) The grains of gangue mineral and sulphide in PRO were between 200 and 500 mm in diameter. There was high degree of dissociation between ferric oxides and gangue, which was easy to separate after crushing.
(6) After the combined separating processing at Jingtang iron & steel co., ltd. in Shougang Group, the grade of Fe increased from 56.68 to 65.75%, S content decreased from 3.44 to 0.48%, K_2O content decreased from 0.43 to 0.16%, Na_2O content decreased from 0.19 to 0.11% and ZnO content decreased from 0.17 to 0.01%.

Acknowledgements Supported by National Key R&D Program of China (2017YFB0304300 & 2017YFB0304302)

References

1. Guo Juan, Liu Shuchen (2010) Prospect of Peru mineral resources and China-Peru mining cooperation. Land Resour Inf 7:33–38
2. Huang Z, Leng S, Sun J (1996) Study on the mineralogy of phosphorus ore process in huangmaling mine. Ind Min Process 25(3):27–29
3. Zhang W (1996) Study on mineralogy of tin deposit ore in west Yunnan iron plant. J Kunming Univ Sci Technol (Natural Science Edition) 21(3):6–12
4. Chen Z (1991) The technical mineralogical characteristics of the original ore of the phosphate rock deposit in Hubei province. Ind Min Process 20(2):20–23
5. Yang S, Cheng Z, Pei Y et al (2014) Research about application of Peru iron ore in sintering production in Jingtang. Hebei Metall 7:10–15

Part XI
Poster Session

A Literature Review of Heat Capacity Measurement Methods

Guishang Pei, Junyi Xiang, Gang Li, Shanshan Wu, Feifei Pan and Xuewei Lv

Abstract Heat capacity is the fundamental thermodynamic parameter, which always be used to obtain enthalpy, entropy and Gibbs energy. Those thermodynamic parameters are critical for chemical reaction, phase equilibrium, material synthesis and design. Various calorimetric methods for the measurement of the heat capacity have been reviewed and compared. The measurement methods of the heat capacity can be divided into two types: low-temperature and high-temperature heat capacity measurement, according to the temperature. The principle and equipment have been introduced systematically. It is recognized that the Drop method is becoming increasingly important, as it can provide more precise calorimetric data at high temperature that was attributed to its special 3D heat sensor.

Keywords Calorimetric methods · Heat capacity · DSC · Drop method

Introduction

Thermodynamics is the branch of physical chemistry concerned with heat and temperature, which deals with energy and the transformation of energy in various ways [1]. Thermodynamics developed out of a desire to increase the efficiency of early steam engines, particularly through the work of French physicist [2]. Subsequently, Scottish physical Lord Kelvin [3] was the first to formulate a concise definition of thermodynamics in 1854. The principle of thermodynamic lay the foundation for phase transformation, phase equilibrium, chemical reaction, material synthesis and design. In thermodynamics, various functions, which are deduced from three laws of thermodynamics, were proposed to describe the present state according to different conditions, such as internal energy, entropy, enthalpy, Helmholtz function and Gibbs free energy et al [4]. Moreover, as for the several

G. Pei · J. Xiang · G. Li · S. Wu · F. Pan · X. Lv (✉)
School of Materials Science and Engineering, Chongqing Univesity,
Chongqing 400044, China
e-mail: lvxuewei@163.com

© The Minerals, Metals & Materials Society 2019

T. Jiang et al. (eds.), *10th International Symposium on High-Temperature Metallurgical Processing*, The Minerals, Metals & Materials Series,
https://doi.org/10.1007/978-3-030-05955-2_54

state functions, Gibbs free energy [5] (ΔG) is extensively used to judge the direction of the chemical reaction under constant pressure conditions, the value of ΔG is obtained by integrating heat capacity at specific temperature range. Therefore, calorimeter technology is critical tools to obtain the thermodynamic data.

In principle, all of the thermodynamics quantities associated with pure substances can be computed by means of calorimetric technique which measures heat content evolution and absorption during heating and cooling process of a pure substance, phase transformation or a chemical reaction. The practicability and reliability of the specific calorimeter are affected by number of crucial factors, such as heat conduction, its own heat capacity and the efficiency by which heat is transferred between sample and calorimeter. And it cannot be sufficiently emphasized that the reasonable design and correct use of a calorimeter are also significant when considering the reliability of the final results.

According to the different temperature interval, the low temperature and high temperature is the main calorimeter type. Generally, in order to calculate the thermodynamics functions at 298 K, the low-temperature heat capacity measurement is one of absolutely necessary ways, while for high-temperature heat capacity is determined the high-temperature stability of compound and solution. In this review, we mainly focus the solution calorimeter for low temperature, while DSC and Drop calorimeter for high-temperature.

Low-Temperature Calorimetric Technology

Generally, the heat capacity is contributed by lattice vibrations and the thermal excitations of electrons, while at low temperature, almost all lattice vibrations cease to contribute, leaving the thermal excitations of electrons dominant [6]. Especially below 5 K, low-temperature heat capacity is linearly related to temperature. The low-temperature heat capacity of the whole temperature interval can be determined adiabatic calorimetry.

Adiabatic calorimetry is the main quantitative method to obtain the values of low-temperature heat capacity, also the primary principle and equipment, including the experimental procedure and data processing will be systematically presented as follow.

As to the most frequently used low-temperature heat capacity measurement method, adiabatic calorimetry always be employed at vacuum conditions. The pivotal apparatus for vacuum adiabatic calorimetry generally consists of three main parts: vacuum adiabatic calorimeter, data acquisition and control system and personal computer [7–10]. The low-temperature conditions should be created for the first step for this kind of method. Generally, for the temperature range from 100 K to room temperature, the liquid nitrogen was used in order to reach this temperature, while, for blow 50 K, the only available material is liquid helium at present; high vacuum atmosphere is obtained by means of a vacuum pump, which is the device that removes gas molecules from a sealed tube to the outer conditions. Data acquisition and control system are the significantly essential part to record and process the data between the experimental process, such as heat content, temperature, time and so on; besides, personal computer is the terminal device in order to efficiently reach human-computer interaction.

The heat capacity is measured in the temperature range by a step heating method that involves a determination of the enthalpy increment and corresponding temperature change of the container with the sample [8]. The molar heat capacity of the sample is calculated as shown in Eq. (1):

$$Cp = \left(\frac{\Delta H}{T_1 - T_2} - Cp, m\right) \times \frac{M}{m} \tag{1}$$

where ΔH is enthalpy increment in the heating period t, T_1 and T_2 denote start and final temperatures of this period, $C_{p,m}$ is the heat capacity of the empty container, m and M are the mass and molar mass of the sample in a vacuum conditions, which are determined by data acquisition and control system and recorded by personal computer system.

Heat capacity of materials at low temperature has been measured using vacuum adiabatic calorimeter, indicating that the low-temperature heat capacity amounts are available and accurate. Paukov et al. [11] has measured the low-temperature heat capacity of natural zinnwaldite, heat capacity and other thermodynamic functions at 298 K for zinnwaldite have be calculated and good agree will with the theoretical values, which were estimated using additive method of calculation. Berezovskii [12] also determined low-temperature heat capacity of polynuclear and trinuclear spin crossover coordination compounds in the temperature range 80–300 K by means of vacuum adiabatic calorimeter. Based on the obtained heat capacity data, the values of enthalpy, entropy and Gibbs energy at 298 K were also calculated using the relationship of those thermodynamic functions and heat capacity values, its greatly enriched the thermodynamic database.

According to a lot of experimental results based on vacuum adiabatic calorimetry, which demonstrates that vacuum adiabatic calorimeter is the most widely used and accurate method to obtain the low temperature, especially the temperature range below 30 K, the measurement accuracy and the ability can be significantly affected by operating process.

High-Temperature Calorimetric Technology

At present, the most frequently used high-temperature calorimeter technology can be classified into two kinds: Differential Scanning Calorimetry (DSC) method and Drop method. There is a big difference between those two kinds of means about the measurement principle and process equipment. As for DSC method, a sustained heat flow can be input as the addition energy to offset the heat content with the heating or cooling process, therefore the heat capacity can be obtained by integrating the heat flow changes. While as regards Drop method, before the calorimeter test, the powdery sample should be pressed into cylinder in order to keep the shape consistent with the standard sapphire sample, the studied sample and standard sapphire sample are placed in the sample holder at room temperature and then dropping it directly into a calorimeter held at a higher temperature. The calorimeter measures the heat evolved while the sample heats to the temperature of the calorimeter. The measurement and evaluation of this two kinds of calorimeter method were presented.

DSC Method

There are two types of DSC, which were always be used to measure heat capacity and heat of phase transition, can be divided into heat flux type and power compensation type, the basic principle of two kinds type is a little bit difference, as for heat flux type the measure process keeps heat flux constant, while, as for power compensation type it keeps power supply constant, the measure process can be described as follow in details.

According to American Society for Testing Material (ASTM) [13] standard, measurements of heat capacity were operated in a dynamic mode with heating rate of 10 K min^{-1} in the argon flow (99.998%, 50 ml min^{-1}) and in isothermal period during 15 and 10 min before and after the dynamic mode, correspondently. Sapphire (α-Al_2O_3) was used as the standard reference sample [14]. The heat capacity of sample was determined by the comparison method. Before the calorimeter process, the weight of standard sapphire sample and studied sample should be taken for the first step. Measurement method of heat capacity included the measurement of DSC thermal flows of the empty crucibles (baseline), the standard sapphire sample (reference) and the studied sample. The measurements of heat capacity were made for reproducible data of DSC thermal flows of the empty sapphire crucibles (baseline) and the standard sample (reference) and the measured sample. Calculation of the specific heat capacity (*Cp*) of the studied material was performed by the formula (2):

$$Cp = \frac{m_r}{m_s} \times \frac{M_s}{M_s} \times \frac{DSC_s - DSC_b}{DSC_r - DSC_b} \times Cp_r \qquad (2)$$

where m_r and m_s was the weight of sapphire (mg) and sample (mg), respectively. DSC_r was the DSC signal of sapphire (mV), DSC_s was the DSC-signal of the sample (mV), DSC_b was the DSC-signal of a baseline (mV), and Cp_r was the heat capacity of a sapphire (J mol^{-1} K^{-1}). The numerical values of heat capacity then were implemented using standard functions and settings of special thermal analysis software package.

The measured sensitivity of the DSC sensor was estimated by DSC measurement of standard sapphire sample. Accuracy of "weight" measurements in the temperature range was equal to ±0.01 mg, temperatures of phase transformations and fusion 10 K and enthalpies of phase transformations and fusion ± 5% approximately.

In previous studies, DSC is widely used to measure heat capacity of variety of compounds such as ceramics, organic materials, polymers, and ionic liquids. [15–21], Venkata et al. [19] measured the heat capacity of La_6UO_{12}, Sm_6UO_{12} and Eu_6UO_{12} with a heat flux type differential scanning calorimetry in the temperature range 298–800 K, $NaTeO_4$(s) and $TiTe_3O_8$(s) [22] in the temperature range 300–700 K, based on the obtained values, the other thermodynamic data were calculated, it reveals that the high-temperature heat capacity are in adequate agreement with the values estimated by using Neumann-Kopp's law from the literature data.

Just as the phase enthalpy computed by DTA, the measured heat capacity values are not enough accurate due to the baseline changes at the heating process. The sample was inputted the bottom of crucible, therefore, the heat content only can be determined by 2D delivery, the quantities were absolutely under the actual values; this drawback limits the usage of DSC for taking heat capacity especially at high temperature to some extent.

Drop Method

Heat capacity can be measured accurately by a number of advanced techniques based on Drop method. The basic principle of drop calorimetry is the measurement of the enthalpy increments of the sample that is being dropped from the ambient temperature to the detector which is kept at higher temperature. When the studied sample at the ambient temperature reaches the additional heat must be delivered to the system in order to maintain the programmed setting target temperature. The calorimeter then measures the heat evolved while the sample cools to the temperature of the calorimeter. The measurement at which the sample is initially heated is varied and a plot of H_T-H_{298} to temperature is drawn, as shown in Fig. 1, which is heat content of

TiC obtained by Naylor [23] and Levinson [24]. It indicated that the heat content values of TiC are fitted well with each other. A popular calorimeter for this is the diphenyl either calculated by Hultgren et al. [25] in 1958, Davies and Pritchard [26] in 1972, but its temperature range is limited below about 1050 K. Through processing the experiment for several temperatures, enthalpy as a function of temperature is obtained and the heat capacity is calculated from its derivative under atmospheric conditions as shown the formula (3):

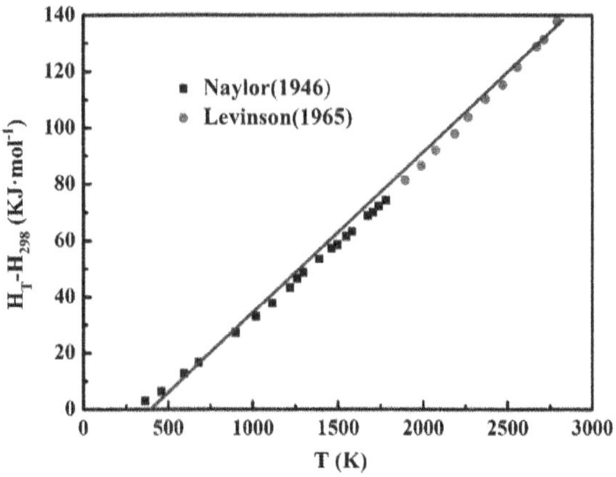

Fig. 1 Heat content of TiC with experimental data of Naylor [23] and Levinson [24]

$$Cp = \left(\frac{dH}{dT}\right)_P \tag{3}$$

where H and T were the enthalpy (J) and temperature (K), respectively. The subscript p said the heat capacity values based on the formula (3) were determined under atmospheric conditions.

Drop sensor, which is the necessary device for Drop method, is installed in Multi-detector High Temperature Calorimeter. After the drop calorimetry instruments, it is required to have compacted pieces of the material using a hydraulic press. The mass of samples ranged from 80 mg to 100 mg in order to keep the weight of a standard Sapphire consistent.

Typical heat flow curve of one measurement sequence is shown in Fig. 2 consisting of four drops of studied samples and four drops of Sapphire reference. Each peak of the heat flow signal was integrated individually using a data analysis software due to obtain the Q_s and Q_r terms from equations, respectively. By the heat flow curve enthalpy changes as a function of temperature is obtained as shown formula (4):

$$\Delta_{T_a}^{T_m} H_m = \frac{Q_s}{Q_r} \times \frac{m_r}{m_s} \times \frac{M_s}{M_r} \times \int_{T_a}^{T_m} Cp(T)dT_r \tag{4}$$

where T_a and T_m are the ambient and detector temperatures, m and M are the mass and the molar mass of the standard material or sample, respectively. And $Cp(T)$ is the temperature dependent function of the sapphire reference molar heat capacity in literature, Q_s and Q_r are heat content by integrated individually each peak of the heat flow signal using Origin Software.

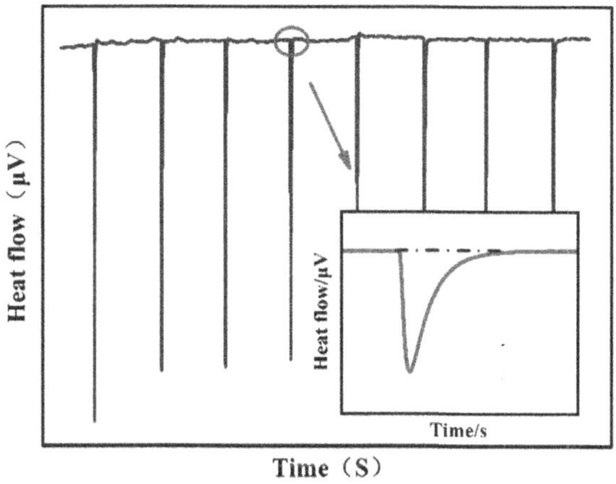

Fig. 2 A typical output of the enthalpy increment measurement of the sample using the drop calorimeter

At present, it is recognised that the Drop method is becoming increasing important, as it can provide more precise calorimetric data at high temperature that attributed to its special 3D heat sensor, especially about 1700–1800 K. AC. Macleod et al [27] have measured the heat capacity of $UO_{2.25}$ using an adiabatic vacuum drop calorimeter and covering a temperature range from 800 to 1600 K, indicating that the high-temperature heat capacity is in excellent agreement with the previous study.

However, time-consuming and expensive equipment also limit the development of this technology to some extent. In addition, the Drop techniques suffer from the disadvantage that heat capacity value is derived from the heat content curves and are not directly measured. There are not sensitive enough to accurately measure small changes in energy associated with, for example, order/disorder transformations or magnetic transitions.

Conclusion

Heat capacity and other corresponding thermodynamic parameters are critical to chemical reaction, phase equilibrium, material synthesis and design. In this study, the two kinds of measurement methods of the heat capacity: low-temperature and high-temperature heat capacity measurement have been reviewed and compared. Adiabatic calorimetry is the main quantitative method to obtain the values of low-temperature heat capacity, while high-temperature heat capacity always be measured by DSC method and Drop method. The principle and equipment have been introduced systematically for each method. It is recognised that the Drop method is becoming increasingly important, as it can provide more precise calorimetric data at high temperature that was attributed to its special 3D heat sensor.

Acknowledgements This work was supported by the project funded by China Postdoctoral Science Foundation and the Fundamental Research Funds for the Central Universities (2018CDYJSY0055).

References

1. Lukas H, Fries SG, Bo S (2007) Computational thermodynamics. Cambridge University Press
2. Dunning-Davies J (2011) 11–Thermodynamic equilibrium and stability. In: Concise thermodynamics. Elsevier, New York, pp 71–74
3. Thomson W, LLD (2009) On his new navigational sounding machine and depth-gauge. Roy Unit Servic Insti J 25(110):374–386
4. Dugdale JS (1967) Entropy and its physical meaning, 2nd edn. Taylor, Francis
5. Gibbs JW (1879) On the equilibrium of heterogeneous substances. Trans. Connecticut. Acad
6. Morishita M (2004) Standard gibbs energy of formation of $Mg_{48}Zn_{52}$ determined by solution calorimetry and measurement of heat capacity from near absolute zero kelvin. Metall Mater Trans B 35(5):891–895
7. Malyshev VM (1986) Automatic low-temperature calorimeter. Instrum Exp Tech 2(6):1456–1459
8. Varushchenko RM, Druzhinina AI, Sorkin EL (1997) Low-temperature heat capacity of 1-bromoperfluorooctane. J Chem Thermodyn 29(6):623–637
9. O'Neal HE, Gregory NW (1959) Vacuum adiabatic heat capacity calorimeter. Rev Sci Instrum 30(6):434–438
10. Tan ZC (2008) A fully automated adiabatic calorimeter for heat capacity measurement between 80 and 400 K. J Therm Anal Calorim 92(2):367–374
11. Paukov IE (2010) A low-temperature heat capacity study of natural lithium micas. J Therm Anal Calorim 99(2):709–712
12. Berezovskii GA (2008) Heat capacity of polynuclear $Fe(HTrz)_3(B_{10}H_{10})\cdot H_2O$ and trinuclear $[Fe_3(PrTrz)_6(ReO_4)_4(H_2O)_2](ReO_4)_2$ complexes (HTrz=1,2,4-triazole, PrTrz=4-propyl-1,2,4-triazole) manifesting 1A1⇔5T2 spin transition. J Therm Anal Calorim 93(3):999–1002
13. Board CA (1999) American society for testing and materials (ASTM). Trf5 Jus Br 55(5):32–37
14. Pilař R (2014) Modified stepwise method for determining heat capacity by DSC. J Therm Anal Calorim 118(1):485–491
15. Carvalho PJ (2009) High pressure phase behavior of carbon dioxide in 1-alkyl-3-methylimidazolium bis(trifluoromethylsulfonyl)imide ionic liquids. J Supercrit Fluid 50(2):105–111
16. Schick C (2009) Differential scanning calorimetry (DSC) of semicrystalline polymers. Anal Bioanal Chem 395(6):1589
17. Yoshida T (2009) Heat capacity at constant pressure and thermodynamic properties of phase transitions in $PbMO_3$ (M =Ti, Zr and Hf). J Therm Anal Calorim 95(2):675–683
18. Leitner J (2009) Heat capacity, enthalpy and entropy of calcium niobates. J Therm Anal Calorim 95(2):397–402
19. Krishnan RV (2008) Heat capacity of La_6UO_{12}, Sm_6UO_{12} and Eu_6UO_{12} by DSC. Thermochim Acta 472(1):95–98
20. Krishnan RV, Nagarajan K (2006) Heat capacity measurements on uranium–cerium mixed oxides by differential scanning calorimetry. Thermochim Acta 440(2):141–145
21. Atanasova L, Baikusheva-Dimitrova G (2012) Heat capacity and thermodynamic properties of tellurites $Yb_2(TeO_3)_3$, $Dy_2(TeO_3)_3$ and $Er_2(TeO_3)_3$. J Therm Anal Calorim 107(2):809–812
22. Cobble JW (1963) The Thermochemical properties of uranium compounds. Oliver & Boyd (24):4056

23. Naylor BF (2002) High-temperature heat contents of titanium carbide and titanium nitride[1]. J Am Chem Soc 68(3):370–371
24. Levinson LS (1965) High-temperature heat contents of TiC and ZrC. J Chem Phys 42 (8):2891–2892
25. Margolin H (1958) Constitution of binary alloys. J Electrochem Soc 105(12):260
26. Davies JV, Pritchard HO (1972) The properties of diphenyl-ether calorimeters. J Chem Thermodyn 4(1):9–22
27. Macleod AC (1972) Enthalpy of $UO_{2.25}$ to 1600 K by drop calorimetry. J Chem Thermodyn 4 (5):699–708

Application of Offgas Analysis on Predicting Carbon Content at End Point During Steelmaking Process

Rong Cheng, Jiongming Zhang, Liangjin Zhang and Haitao Ma

Abstract Using gas analysis to predict the carbon content at the end point is a new technology within steelmaking. The offgas data of a 100 ton BOF during steelmaking were analysed. The changing rules of the CO, CO_2 and O_2 volume fractions in the offgas played an important role in judging the status of decarburization in the converter during steelmaking. The carbon deposition reaction error, the CO_2 dissolution in the west dedusting and the added CO_2 from air into offgas had little influence on the predicted carbon content. The carbon integral model had serious errors regarding the predicted results. A new model based on the correlation between the volume fraction of CO and the carbon content in converter was built to predict the carbon content near the end point. The hit rate of the carbon prediction model was 95.79% for the end point carbon content $\leq 0.088\%$ with a precision of $\pm 0.03\%$.

Keywords Carbon content · Prediction · CO · Decarburization

Introduction

Steelmaking is an important link during the metallurgical processes. The product quality is heavily influenced by the end point of carbon content in steelmaking. The carbon content at the end point during steelmaking is predicted by the static model, the assistant lance or the offgas analysis.

R. Cheng · J. Zhang (✉) · H. Ma
State Key Laboratory of Advanced Metallurgy, University of Science and Technology Beijing, Beijing 100083, China
e-mail: jmz2203@sina.com

L. Zhang
China National Construction Steel Quality Supervision and Test Centre, Central Research Institute of Building and Construction Co., Ltd, Mcc Group, Beijing 100083, China

© The Minerals, Metals & Materials Society 2019

T. Jiang et al. (eds.), *10th International Symposium on High-Temperature Metallurgical Processing*, The Minerals, Metals & Materials Series,
https://doi.org/10.1007/978-3-030-05955-2_55

The static model was based on the material balance between the flow of oxygen blowing, the quantity of auxiliary raw materials and the quantity of liquid iron [1, 2]. The material balance calculation was only based on the initial and end states of steelmaking. The lack of blowing process analysis resulted in the hit ratio of the carbon content at end point prediction was often less than 60% during steelmaking.

Predicting the end point of carbon content by the assistant lance is typically used during steelmaking [3–8]. The molten steel temperature and the carbon content that are determined by the assistant lance before the end point are applied during steelmaking to adjust the blowing parameters. The hit ratio of predicting the carbon content at the end point based on the assistant lance is higher than the static model. The assistant lance is generally suitable for large tonnage converters, the one-time probes of assistant lance should not be an ignored cost during steelmaking, and the assistant lance could not monitor the blowing during steelmaking in real time.

Predicting the carbon content during steelmaking via the offgas analysis is a new technology. The decarburization rate and the carbon content in the bath were calculated by analysing the composition and flow of the offgas during steelmaking [9–12]. The offgas analysis method is suitable for converters with various tonnages, the probe of the offgas analysis could be reused, and the carbon content could be monitored in real time during the blowing process. Some companies developed the systems of offgas analysis to predict the end point during steelmaking [13–16].

In this paper, the influences that the carbon deposition reaction, the CO_2 dissolution in west dedusting and the added CO_2 from air into the offgas had on predicting the carbon content at the end point were analysed. The carbon integral model for the prediction was discussed. A novel method, based on the correlation between the volume fraction of the CO and the carbon content in the bath, was built to predict the carbon content at the later stage of decarburization.

Experiment

Figure 1 shows the offgas analysis system. The sampling probes of the laser gas were between the dust-removal system and the converter. The flowmeter was near the exhaust fan. About 100 tons of liquid iron with a carbon content of 4.3 mass% and 24 ton scrap steel with a carbon content of 0.05 mass% were added in the converter before blowing. The initial and the final chemical compositions of the molten steel are shown in Table 1. Parameters of the gas analysers are shown in Table 2.

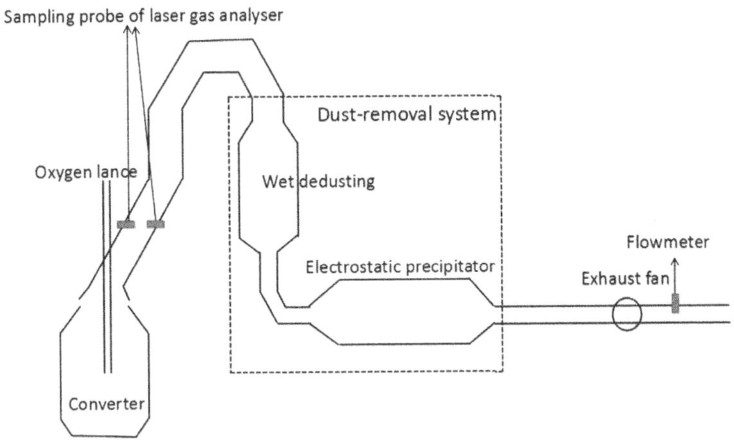

Fig. 1 Sketch of the offgas analysis system

Table 1 Initial and final chemical compositions of the molten steel

Composition(%)	C	Si	Mn	P	S	Ti	V
Initia	4.3	0.4	0.32	0.11	0.002	0.078	0.019
End point	0.028	–	0.075	0.009	0.006	–	–

Table 2 Parameters of gas analysers

Equipment of gas analysis	Laser gas analyser
Sampling position	3 m above gas hood
Types of analysed gases	CO, CO_2, O_2
Measuring range (volume fraction)	0–100%
Temperature of the offgas	1000–1300 °C
Interval time of analysis	2 s

Results and Discussion

Composition and Flow of Offgas

Figure 2 shows the changes in the offgas composition and the flow during the steelmaking processes. The oxygen blowing time ranged from 75th second to 1200th second. The flow of offgas was nearly stable during the oxygen blowing. During the earlier stage of the oxygen blowing (75th second–300th second), the volume fractions of CO_2 and CO increased and then decreased during the later stage of oxygen blowing (1020th second–1200th second). The volume fraction of O_2 was around 0% during the middle stage of oxygen blowing (300th second–1020th second). The volume fraction of O_2 was about 21%, which was the same as the volume fraction of O_2 in pure air.

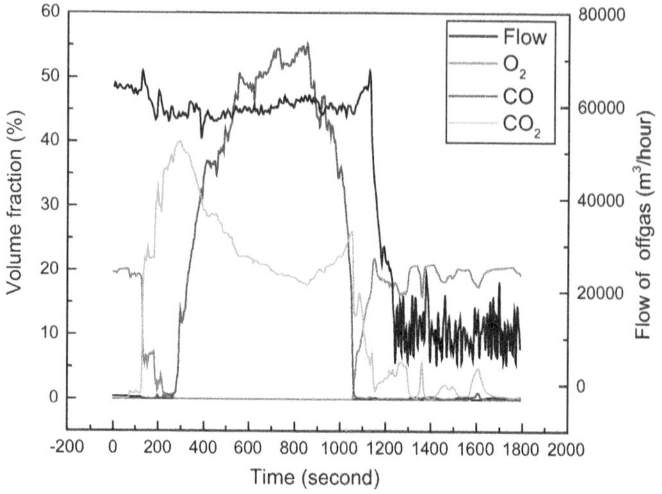

Fig. 2 Change of the offgas compositions and the flow

Influence of CO_2 Dissolution

The wet dedusting system was between the sampling probe of laser gas analyser and the offgas flowmeter. A small quantity of CO_2 in the offgas could be dissolved into water when the offgas passed through the wet dedusting system. The solubility of CO_2 in the water is shown in Table 3. The water temperature in the wet dedusting system was between 60 and 80 °C. The pressure in the wet dedusting system was lower than the standard atmospheric pressure. The low temperature and the high pressure could increase the solubility. The most dissolved quantity of CO_2 at 60 °C in standard atmospheric pressure was calculated via Eq. (1).

$$\frac{\alpha \times Q_{water} \times (t_{end} - t_0) \times 12 \times 1000}{60 \times 22.4 \times 100,000} \times 100\% = 0.01592\% \quad (1)$$

α the solubility of CO_2 in water at 60 °C in the standard atmospheric pressure, 0.361(ml/g),
Q_{water} the flow of dusting water (380), ton/h,
t_0 the start time of oxygen blowing, seconds and
t_{end} the end time oxygen blowing (13 × 60), seconds

Table 3 Solubility of CO_2 in water within the standard atmospheric pressure

Temperature (°C)	0	25	50	60	75	100
Solubility (ml/g)	1.79	0.752	0.423	0.361	0.307	0.231

The flow velocity of the offgas was 10–30 m/s. The time required to pass through the wet dedusting system was about three seconds. The water was not saturated with dissolved CO_2 within that short time. The dissolved CO_2 was less than 0.01592%, which was negligible for the carbon content at the blowing end point (about 0.03%).

Influence of Carbon Deposition Reaction

Carbon deposition reaction of the offgas is shown in Eq. (2). The carbon could be changed from the gas phase carbon (CO) into the solid phase (C) when the carbon deposition reaction degree was serious. The accuracy of the offgas composition and the flow determined by analysis was influenced by the carbon deposition reaction.

$$2CO = CO_2 + C \tag{2}$$

$$\Delta G^{\theta} = -166,550 + 171T \tag{3}$$

T temperature of the offgas

The carbon deposition reaction was exothermic. Equation (3) shows that the carbon deposition reaction could not occur when the temperature was higher than 700 °C. Liu et al. [17] found that the carbon deposition reaction was slow and was negligible when the temperature was lower than 500 °C. A volume fraction of CO_2 greater than 8% could effectively suppress the carbon deposition reaction. The CO_2 volume fraction of the offgas was often greater than 20%. The effect that the carbon deposition reaction had on the accuracy of the offgas composition and flow was negligible.

Influence of CO_2 from Air

There was a large gap between the converter mouth and the offgas hood. There could be air inhaled into the offgas. The influence of the CO_2 from the air on the accuracy of the carbon content at the blowing end point was analysed. The volume fraction of the CO_2 in air was about 0.03%. The amount of CO_2 from the air is calculated in Eq. (4). The calculated result showed that the CO_2 from air was about 0.001816%, which was negligible for the carbon content at the blowing end point (0.03%).

$$\frac{\int_{t_0}^{t_{end}} (1 - \varphi_{CO} - \varphi_{CO_2} - \varphi_{O_2}) \times 0.03 \times Q_{offgas} \times 12 \times 1000}{79 \times 22.4 \times 3600 \times 100,000} \times 100\% = 0.001816\% \tag{4}$$

φ_{CO}	volume fraction of CO in offgas, %,
φ_{CO_2}	volume fraction of CO_2 in offgas, %,
φ_{O_2}	volume fraction of O_2 in offgas, %,
$(1 - \varphi_{CO} - \varphi_{CO_2} - \varphi_{O_2})$	volume fraction of N_2 in offgas, %, and
Q_{offgas}	flow of the offgas, m^3/h

Carbon Integral Model

Equation (5) shows the carbon integral model that was based on the principle of mass conservation. The carbon contents in the bath and the offgas were conserved.

$$W_t = W_0 - W_{gas} = W_0 - \int_{t_0}^{t} v_C dt \tag{5}$$

W_t	the carbon content in the bath at time t second, kg,
W_O	the carbon content in bath before oxygen blowing, kg,
W_{gas}	the carbon content in offgas from t_0–t second, kg,
V_c	decarburization rate, kg/s, and
t	oxygen blowing time, second

Equation (6) shows the v_c:

$$v_C = \frac{12 \times Q_{offgas} \times (\varphi_{co} + \varphi_{CO_2})}{1000 \times 3.6 \times 22.4} \tag{6}$$

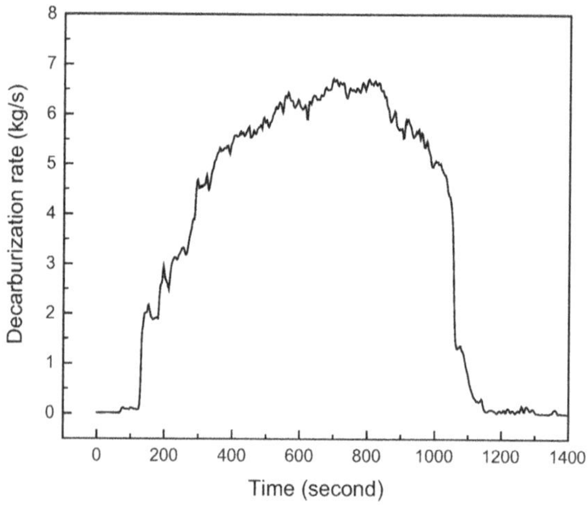

Fig. 3 Decarburization rate during steelmaking

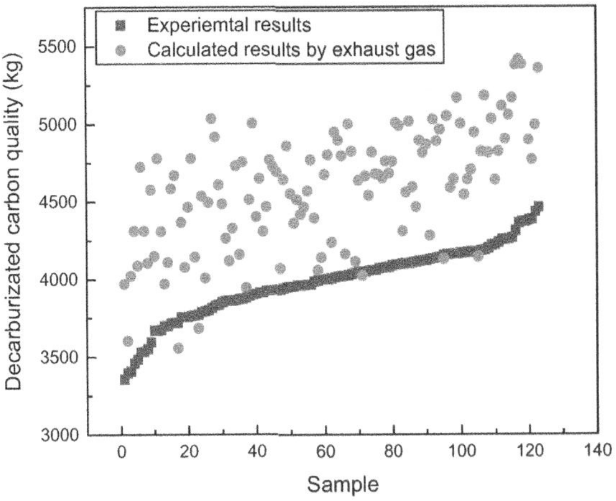

Fig. 4 Decarburized carbon qualities

Figure 3 shows the decarburization rate during steelmaking. In Fig. 3, the decarburization rate contained the earlier stage, the middle stage and the later stage.

Figure 4 shows the 95 group decarburized carbon qualities that were calculated by the offgas analysis and the difference between the carbon content from the initial and end points. The experimental results that were calculated between the difference of carbon content from the initial and end points in the bath were different from the calculated results of the offgas. The large difference could be caused by the errors of composition, the flow of offgas and the error of carbon content in the steel scrap.

$$W_{\Delta offgas} = \int_{0}^{t_{end}} v_C dt \tag{7}$$

$$W_{\Delta C} = \frac{W_{Fe} \times w_0 - (W_{Fe} + W_{steelscrap}) \times w_{end}}{100} \tag{8}$$

$W_{\Delta offgas}$ decarburized carbon quality calculated by the offgas, kg,
$W_{\Delta C}$ decarburized carbon quality calculated by the difference between the carbon in the initial and end points, kg,
W_{Fe} weight of molten steel, kg,
W_{scrap} weight of steel scrap, kg,
w_0 carbon content of molten steel, %, and
w_{end} carbon content in bath end point, %

The experimental results were different than the calculated results of the offgas, so the calculated results from the offgas required correction. The correction, based

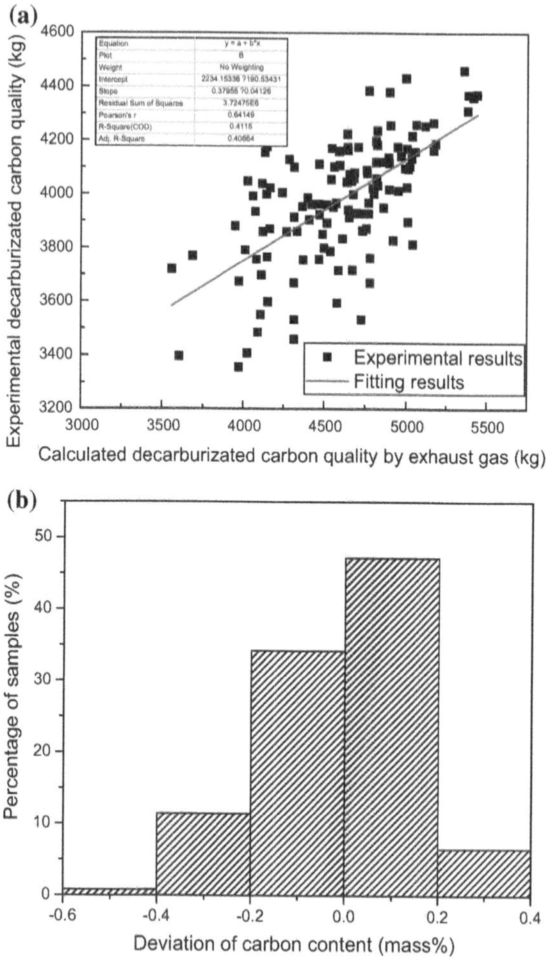

Fig. 5 Correction for $W_{\Delta offgas}$ and deviation after correction, a) correction for $W_{\Delta offgas}$ based on the least square method and b) deviation between $W_{\Delta offgas2}$ and $W_{\Delta C}$

on the least square method, is shown in Fig. 5a and Eq. (9). Figure 5b shows the deviation of the carbon content between $W_{\Delta offgas2}$ and $W_{\Delta C}$. The deviation was calculated by Eq. (10). Figure 5b shows that the hit rate of the deviation was about 82% with a precision of ±0.2% carbon content.

$$W_{\Delta offgas2} = 2234.15 + 0.37955 W_{\Delta offgas} \tag{9}$$

$$w_{deviation} = \frac{W_{\Delta C} - W_{\Delta offgas2}}{W_{Fe} + W_{steelscrap}} \times 100\% \tag{10}$$

$W_{\Delta offgas\ 2}$ corrected $W_{\Delta offga}$,

$w_{deviation}$ deviation between $W_{\Delta offgas2}$ and $W_{\Delta C}$

Carbon Content Prediction at the Later Stage

The carbon integral model failed to precisely predict the carbon content at the end point. Figure 6a shows the volume fractions of CO and CO_2 at the later stage of oxygen blowing. Figure 6a shows that the curve of CO volume fraction was smooth and regular. A new method was proposed to predict the carbon content during the later stage of decarburization, shown in Fig. 6b. The method was based on the relationship between the carbon content and the volume fraction of CO during the later stage of decarburization. As shown in Fig. 6b, the end point was the initial

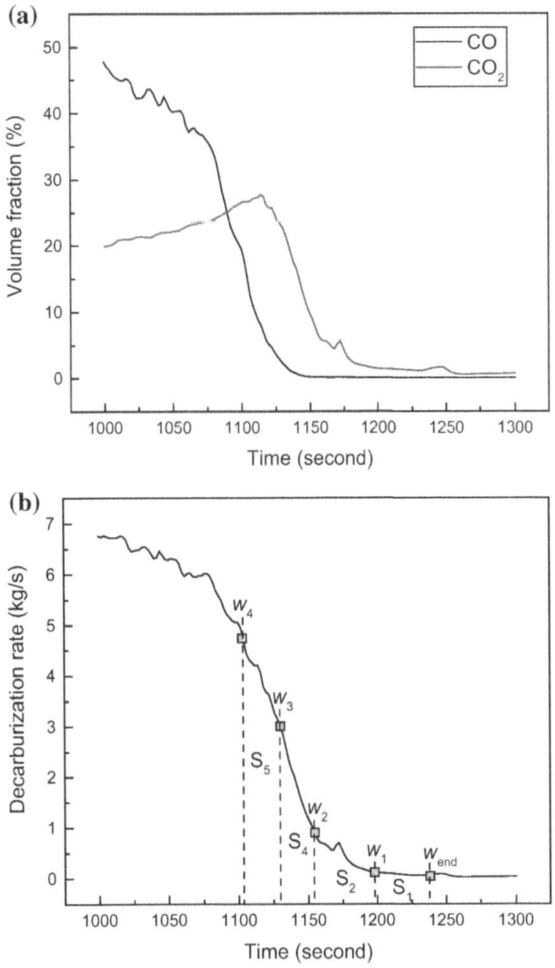

Fig. 6 Method for predicting the carbon content during the later stage of decarburization, a) CO volume fraction and b) the method

Table 4 Parameters of the nodes

No.	Initial node (%)	Second node (%)	Third node (%)	Fourth node (%)
Volume fraction of CO	0.06	2	10	3
Carbon content	0.02	0.088	0.1479	0.24835
Precision	±0.02	±0.03	±0.06	±1
Hit rate	98.94	95.79	95.79	89.47

node with the coordinate (volume fraction of CO, carbon content), the second node was before the initial node on the timeline, and the carbon content of the second node could be obtained by integration with Eq. (11).

$$\begin{aligned} w_1 &= w_{end} + 100\left(\int_{t_1}^{t_{end}} v_C dt\right) / \left(W_{Fe} + W_{steelscrap}\right) \\ &= w_{end} + 100 S_1 / \left(W_{Fe} + W_{steelscrap}\right) \end{aligned} \quad (11)$$

$$\begin{aligned} w_2 &= w_{end} + 100\left(\int_{t_2}^{t_{end}} v_C dt\right) / \left(W_{Fe} + W_{steelscrap}\right) \\ &= w_{end} + 100(S_1 + S_2) / \left(W_{Fe} + W_{steelscrap}\right) \end{aligned} \quad (12)$$

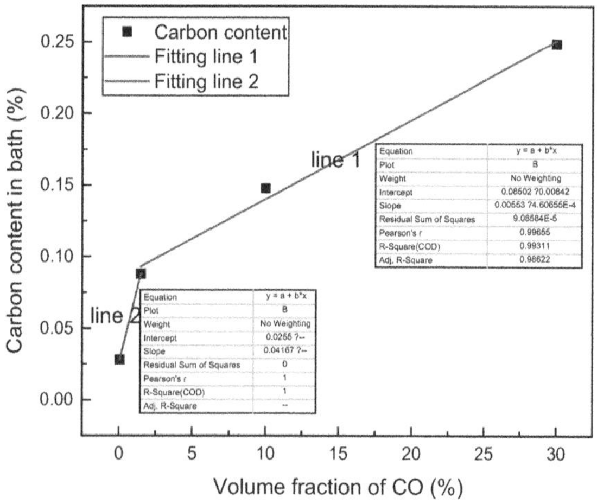

Fig. 7 Relationship between the carbon content and the volume fraction of CO during the later stage of decarburization

$$w_3 = w_{end} + 100\left(\int_{t_3}^{t_{end}} v_C dt\right) / \left(W_{Fe} + W_{steelscrap}\right) \\ = w_{end} + 100(S_1 + S_2 + S_3) / \left(W_{Fe} + W_{steelscrap}\right) \quad (13)$$

$$w_n = w_{end} + 100\left(\int_{t_n}^{t_{end}} v_C dt\right) / \left(W_{Fe} + W_{steelscrap}\right) \\ = w_{end} + 100(S_1 + S_2 + \cdots S_n) / \left(W_{Fe} + W_{steelscrap}\right) \quad (14)$$

t_1, t_2, t_3…t_n — the time of second node, third node, fourth node… n-th node, s,
w_1, w_2, w_3…w_n — the carbon content of second node, third node, fourth node…… n-th node, s,
t_{end} — the time of the initial node (the end point of oxygen blowing), s,
w_{end} — carbon content at the initial node, %, and
S_1, S_2, S_3… S_n — decarburized carbon content between t_{end} and t_1, between t_1 and t_2, between t_2 and t_3… between t_{n-1} and t_n, kg

Table 4 shows the parameters of the four nodes. The carbon contents of the nodes were the average value of 95 group samples. The hit rate was for the average carbon content with the precision shown in Table 4. Figure 7 shows the relationship between the carbon content and the volume fraction of CO at the later stage of decarburization. Their relationship was a positive correction.

Conclusions

The feasibility of predicting the carbon content at the end point through analysing the offgas was discussed. The main conclusions are as follows:

The carbon deposition, the CO_2 dissolution in west dedusting and the added CO_2 from air into the offgas had little influence on the predicted carbon content at the end point. The carbon integral model failed to precisely predict the carbon content at end point. The method based on the positive correction relationship between the carbon content and the volume fraction of CO during the later stage the relationship could precisely predict the carbon content at the later stage.

Acknowledgements This work was supported by the National Natural Science Foundation of China under Grant 51474023.

References

1. Pacciarelli D, Pranzo M (2004) Production scheduling in a steelmaking-continuous casting plant. Comput Cheml Eng 28(12):2823–2835
2. Litvinenko EF, Vakula VS (2003) Development of systems for controlling steel-making in oxygen-converter shops. Metall Gornorud Promst 8:4390–4395
3. Boom R (1994) Sublance technology for oxygen steelmaking. Steel Technol Int 3:91–94
4. Naito K, Kitamura S, Ogawa Y (2002) Effects of BOF top blowing and bottom stirring conditions on suppressing excessive oxidation. Ironmak Steelmak 29(3):208–213
5. Friedl E, Kaiser HP, Heinz W (1990) Automatic blowing process in BOF and direct tapping using the sublance system. Metall Plant Technol Int 13(1):28–32
6. Robson W J, Rimes P (1993) Successful sublance implementation at BHP, New Castle Works. In: 76th steelmaking conference, pp 163–168
7. Zonneveld P (1995, April) How to boost BOF output and reduce cost per ton of steel. In: Steelmaking conference proceedings, vol 78, pp 147–152
8. Boom R, Zonneveld P, Parker J (1989) Benefits of a sublance system in oxygen steelmaking. Steel Times Int 13(4):36–37
9. Kanemoto M, Okajima M (1990) Phosphprus and manganese control technologies in basic oxygen furnace and their future view. Tetsu-to-Hagané 76(11):1964–1971
10. Iso H, Ueda Y, Yoshida T et al (1988) Refining control of top and bottom blowing converter by manipulating bottom blown flow rate. Trans ISIJ 28(5):372–381
11. Hara T, Nakamura H, Tawara M, Yamagami T (1989, April) Development of direct tapping at no. 2 steelmaking shop in Kure works. In: Steelmaking conference proceedings, vol 72, pp 155–158
12. Kato Y, Takahashi Y, Sakuraya T, Fujii T, Omiya S (1990) Optimization of the blowing sequence in a top and bottom blowing converter containing pretreated hot metal. Tetsu-to-Hagané 76(4):560–567
13. Takawa T, Tatayama K, Katohgi K (1988) Analysis of converter process variables from exhaust gas. Trans ISIJ 28(1):59–67
14. Merriman D (1997) Mass spectrometry for oxygen steelmaking control. Steel Times 225 (11):439
15. Sun S, Liao D, Pyke N et al (2008) Development of an offgas/model technology to replace sublance operation for KOBM endpoint carbon control at ArcelorMittal dofasco. Iron Steel Technol 5(11):36–42
16. Scipolo V, Maiolo J (2008) Application of EFSOP holistic optimization TM technology to oxygen steelmaking. In: AISTech proceedings, AIST, Pittsburgh
17. Liu JZ, Xue QG, Zhang SY (2004) Carbon deposition reaction based on heating recycling gas of oxygon blast furnaces. J Univ Sci Technol Beijing 36(5):592–596

Calcination of Strontium Carbonate in Rotary Kiln Furnace

Raşit Sezer, Emre Yilmaz, Selim Ertürk and Cüneyt Arslan

Abstract The calcination of the strontium carbonate in the rotary kiln is usually carried out at temperatures as high as 1300 °C in practical applications. During the calcination process, partial melting of powders occurs at about 1150 °C, due to the partial pressure of CO_2. Because of adhesion of the melts to the furnace walls, a ring formation can be seen in the progressive stages. Nevertheless, carbon additions to the $SrCO_3$ before calcination can prevent the agglomerations by reducing the partial pressure of CO_2. In this study, calcination of $SrCO_3$ is determined by adding metallurgical coke (0–30%) in a laboratory-scale rotary kiln furnace. Experiments were run at the temperatures between 1100 and 1300 °C. The calcined powders were analysed by XRD. The maximum calcination efficiency was obtained with 15% coke addition at 1200 °C.

Keywords $SrCO_3$ · Calcination · Rotary kiln furnace

Introduction

The main strontium source is strontium carbonate in the production of colour TV tubes, UV barrier glasses, hexaferrite magnets, strontium aluminate cement and phosphorescence materials. Strontium carbonate ($SrCO_3$) is produced by pyrometallurgical processes in many applications [1].

$SrCO_3$ is decomposed to strontium oxide (SrO) and carbon dioxide (CO_2) either before or during the application of these production processes. $SrCO_3$ is calcined at temperatures above 1000 °C in atmospheric conditions and is decomposed into SrO and CO_2 (Reaction 1) [2].

R. Sezer (✉)
Karadeniz Technical University, Trabzon 61000, Turkey
e-mail: rasitsezer@gmail.com

R. Sezer · E. Yilmaz · S. Ertürk · C. Arslan
Istanbul Technical University, Istanbul 34469, Turkey

© The Minerals, Metals & Materials Society 2019

T. Jiang et al. (eds.), *10th International Symposium on High-Temperature Metallurgical Processing*, The Minerals, Metals & Materials Series,
https://doi.org/10.1007/978-3-030-05955-2_56

$$SrCO_{3(s)} = SrO_{(s)} + CO_{2(g)} \quad (1)$$

Calcination of $SrCO_3$ begins with the separation of CO_2 from the surface layer and formation of the SrO shell, in accordance with the shrinking core model [3]. At this stage, the reaction rate is controlled by the chemical reaction and the transfer of CO_2 in the environment by diffusion. As the reaction proceeds to the interior, the diffusion of the reaction product CO_2 into the crusted porous SrO layer. This is the slowest step that determines the reaction rate.

In industrial applications, the calcination of $SrCO_3$ is usually carried out in rotary kilns. As the calcination progresses, $SrCO_3$ is transformed from α phase (in rhombohedral structure) to β phase (in hexagonal structure) at above 950 °C [4].

Ptacek et al [5]. have investigated the mechanism and kinetics of calcination of α-$SrCO_3$ and β-$SrCO_3$ powders by TG-DTA. The α-$SrCO_3$ transforms first to the β phase during calcination, which results in a lower activation energy for calcination with the instant nuclei resulting from this phase transformation. As a result of the calculations made, the activation energy for calcination of α-$SrCO_3$ was found to be 227 kJ/mol, while β-$SrCO_3$ was found to be 255 kJ/mol [5].

Although the calcination of $SrCO_3$ in the rotary kiln is theoretically carried out at 1150 °C, it is practically up to 1300 °C. During calcination at temperatures between 1100 and 1150 °C, some parts of $SrCO_3$ melts, due to the partial pressure of CO_2. Adhesions and ring formation are observed in the kiln walls due to these melts. When the CO_2 partial pressure exceeds the critical value (0.4 atm), material starts to melt, while the calcination yield is more than 85%. Systems such as vacuum or hydrogen gas are used to reduce the CO_2 pressure in the environment. Instead of such expensive systems, a method is developed by Coatney and his team (1975) to overcome this problem, that is, adding carbon to $SrCO_3$. They found that the $SrCO_3$ powder size (d_{85}) should be less than 44 microns and that the amount of carbon to be used should be greater than 8% in order to prevent the formation of a liquid phase and sticking to the furnace walls. Very fine $SrCO_3$ powders were pelleted shortly after entering the furnace and then the combustion of carbon-prevented dust and pellets from sticking to the furnace [6].

In this study, effect of metallurgical coke addition on the calcination of $SrCO_3$ was investigated. The weight ratio of metallurgical coke to $SrCO_3$ is between 0 and 30%. Calcination processes were done in laboratory-scale rotary kiln furnace at temperatures between 1100 and 1300 °C. Calcination efficiency is determined as a function of kiln temperature and the coke ratio.

Experimental Study

The starting material, $SrCO_3$, is produced from celestite ore (Barit Maden Türk A.Ş.) by black ash method in previous studies [7, 8]. Calcination experiments were carried out using a laboratory-scale rotary kiln. The alumina–mullite reactor tube of the kiln is 1200 mm long and has an inner diameter of 90 mm. The furnace can run at a slope of 0°–8°, and the reactor tube can rotate continuously at 0–30 rpm. The 400-mm-long middle section of the tube is the heating zone surrounded by the molybdenum-disilicide ($MoSi_2$) resistances. The heating regime of the furnace is shown in Fig. 1.

The calcination experiments in the rotary kiln were carried out at 1100, 1200 and 1300 °C with constant rotational speeds (2 rpm) and slope (3°). To increase the calcination yield, finely ground (<250 μm) metallurgical coke was added to $SrCO_3$ with a weight ratio of 15% or 30%. Analysis of the powders obtained as a result of the calcination experiments in the rotary kiln was carried out by XRD.

Results and Discussion

In the DSC–TG curves, shown in Fig. 2, the phase transition is observed with a sharp endothermic peak at 950 °C. Calcination starting at approximately 900 °C (Reaction 1) accelerates after phase transformation to reach maximum speed at 1067 °C. The weight reduction (TG curve), which is consistent with the broad endothermic peak in the DSC curve indicates that the CO_2 released as a result of the calcination is removed from the structure.

In the calcination experiments with the laboratory rotary kiln, neither ring formation nor any adhesion was observed due to the fact that the CO_2 pressure in the furnace environment did not reach the critical value. However, because of the short

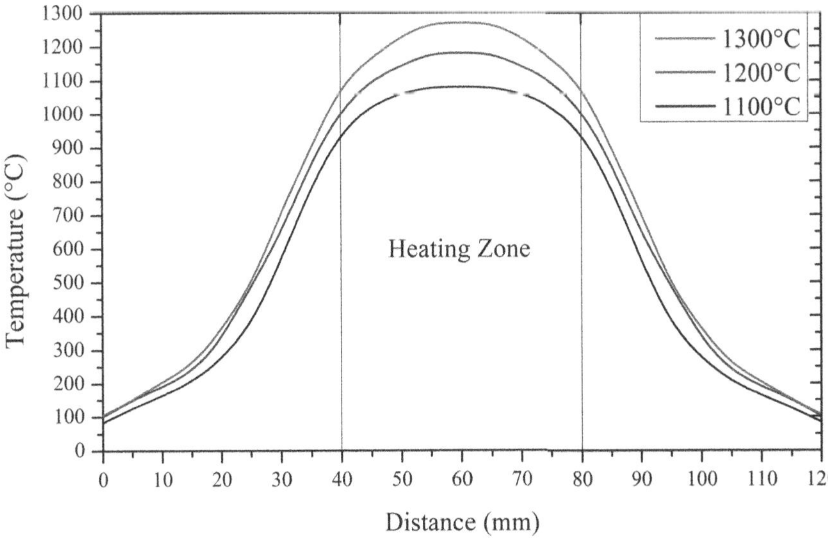

Fig. 1 Heating regime rotary kiln furnace

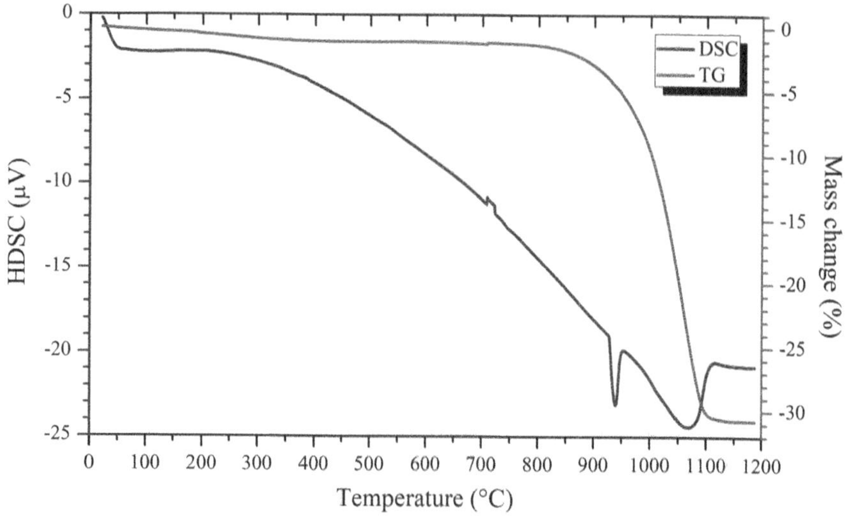

Fig. 2 DSC–TG curves of $SrCO_3$

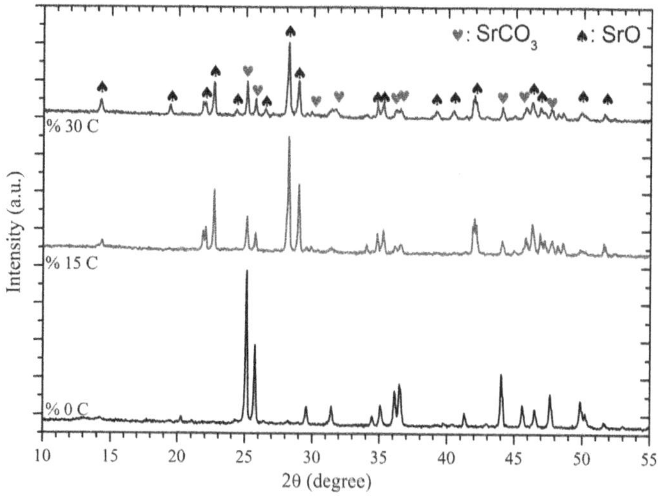

Fig. 3 XRD analysis of powders calcined at 1100 °C

length of the kiln, dwelling time of the powders (calcination period) is not sufficient. In experimental studies, dwelling time was determined as 14 min, with 3° oven slope and 2 rpm rotational speed. Uncalcined $SrCO_3$ was detected in the powders obtained at different temperatures (1100–1300°C) (Fig. 3).

$$CO_{2(g)} + C_{(s)} = 2CO_{(g)} \tag{2}$$

$$CO_{2(g)} + C_{(s)} = 2CO_{(g)} \tag{3}$$

$$C_{(s)} + {}^{1}\!/_{2}O_{2(g)} = CO_{(g)} \tag{4}$$

To increase the calcination efficiency, $SrCO_3$ powders were mixed with finely ground (<250 μm) metallurgical coke (15–30%) before feeding to the furnace. The calcination product, CO_2, is reduced with the metallurgical coke (Reaction 2) to decrease the partial pressure of CO_2 in the kiln. Thanks to this, the calcination reaction (1) contributes to the progress of the products. Moreover, thanks to the combustion of metallurgical coke with O_2 in the kiln (comes from the atmosphere) (Reaction 3 and 4), spaces are formed in the dust heaps. In these spaces, reaction gases can easily move away from the dust pile.

XRD analyses of the powders obtained from the calcination experiments in the rotary kiln are given in Figs. 3, 4 and 5. In the rotary kiln at 1100 °C, $SrCO_3$ cannot be fully calcined. However, the addition of metallurgical coke at both 15% and 30 increases the calcination efficiency of $SrCO_3$ (Fig. 3). Without the addition of coke, the heat energy in the rotary kiln at 1100 °C is not sufficient for the calcination of $SrCO_3$. The energy released by the burning of the carbon in the coke added at this temperature ensures the calcination of the carbonate.

By increasing the temperature of the rotary kiln to 1200 and 1300 °C, carbonate powders can be calcined, whether by adding coke or not. The $SrCO_3$ peaks are found to be the least intensive (the highest calcination efficiency) with mixtures to which 15% metallurgical coke is added (Figs. 4 and 5).

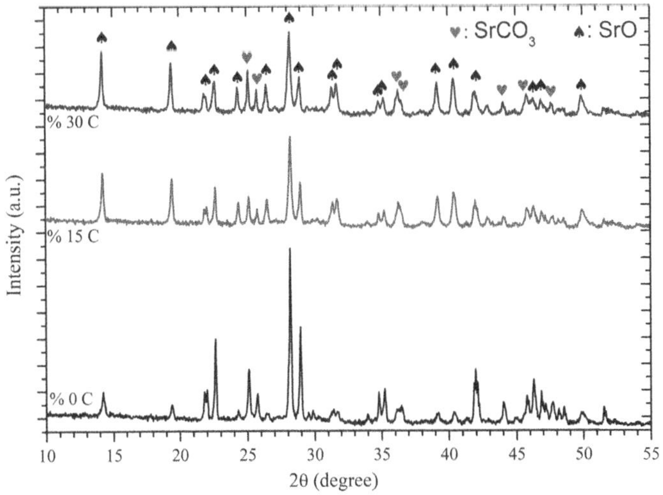

Fig. 4 XRD analysis of powders calcined at 1200 °C

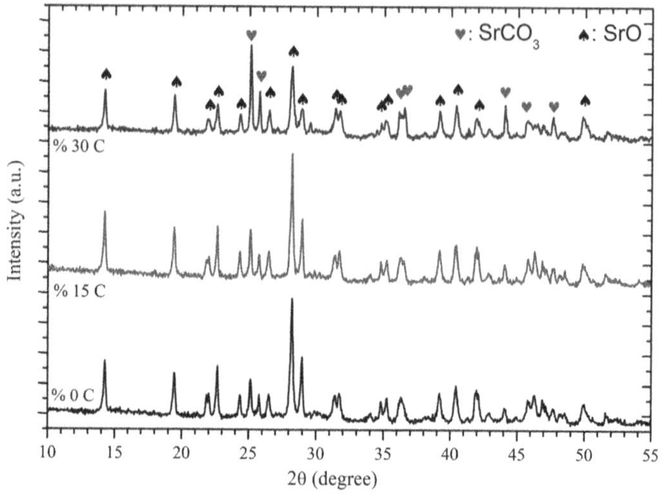

Fig. 5 XRD analysis of powders calcined at 1300 °C

If metallurgical coke is added in high proportion (30%), the CO_2 gas, which is released as a result of the combustion of carbon, causes the recarbonation of SrO in the cooling zone of the furnace. At temperatures below the calcination temperature, the SrO can be converted back to $SrCO_3$ due to the calcination reaction (Reaction 1) in the presence of the CO_2 gas in the opposite direction (left). For this reason, as a result of experiments with 30% coke depletion (especially at 1300 °C), the intensity of $SrCO_3$ peaks increases.

Conclusion

Calcination of strontium carbonate occurs relatively slowly at temperatures up to 1067 °C, although it begins at temperatures above 900 °C. Calcination of the rotary kiln did not yield full-scale calcination of $SrCO_3$ at 1100 °C. However, 15 and 30% metallurgical coke added to $SrCO_3$ prior to calcination ensures that the calcination efficiency to increase and a complete calcination.

Acknowledgements The authors greatly acknowledge the financial support provided by Scientific and Research Council of Turkey (TUBITAK) under the contract no 115M631.

References

1. Hibbins SG (2000) Strontium and strontium compounds In: Ley C (ed) Kirk-othmer encyclopedia of chemical technology. Wiley, New York
2. Torres T, Almanza R, Flores V, de Castro R, Herrera T (2007) Thermal behavior of $SrSO_4$–$SrCO_3$ and $SrSO_4$–$SrCO_3$–Al_2O_3 mixtures. Mater Character 58(10):859–863
3. Arvanitidis I, Xiao X, Seetharaman S (1999) Effect of heat and mass transfer on the thermal decomposition of $SrCO_3$ compacts. Metall Mater Trans B 30(5):901–908
4. Arvanitidis I, Du SC, Sohn HY, Seetharaman S (1997) The intrinsic thermal decomposition kinetics of $SrCO_3$ by a nonisothermal technique. Metall Mater Trans B 28(6):1063–1068
5. Ptacek P, Bartonickova E, Svec J, Opravil T, Soukal F, Frajkorova F (2015) The kinetics and mechanism of thermal decomposition of $SrCO_3$ polymorphs. Ceram Int 41(1):115–126
6. Coatney RL, Housh LM, Van Dreser ML (1975) Calcination of strontium carbonate. US. Patent 3,875,298
7. Erturk S, Sezer R, Hizli IG, Bilen A, Arslan C (2017) Roasting of celestite in laboratory scale rotary furnace. Paper presented in 8th international symposium on high-temperature metallurgical processing, San Diego, California, 26 Feb–2 Mar 2017
8. Bilen, A, Hizli IG, Sezer R, Erturk S, Arslan C (2017) Strontium carbonate precipitation from strontium sulfide solution. Paper presented in XVIIth Balkan Mineral Processing Congress—BMPC 2017, Antalya, Turkey, 1–3 Nov 2017

Determination of Effect of Li_2O on the Structure of CaO-Al_2O_3-Based Slag by Molecular Dynamics Simulation and Raman Spectrum

Sai Wang, Bo Ran Jia, Sheng Ping He and Qian Wang

Abstract In this study, the structural properties of the CaO-Al_2O_3-Li_2O-based slag with varying Li_2O content have been investigated via molecular dynamics simulation and Raman. The results showed the average bond lengths of Al-O remained at 1.75 Å. The addition of Li_2O contributed to the decrease in the bridging oxygen proportion and the network connectivity (Q^n) degree, Meanwhile, the five-fold coordinated Al was decreased when mass fraction of Li_2O was increased. The results of the Raman spectrum analysis show that the proportion of the complex structural unit Q^4 decreases with the increase of the Li_2O content, the decrease of the degree of polymerization of the slag network (DOP) indicates that the melt structure becomes simple and consistent with the results obtained by the molecular dynamics simulation.

Keywords Molecular dynamics · Microstructure · Raman spectroscopy · DOP

Introduction

The properties of slag are determined by its chemical composition and microstructure. Numerous empirical models have been suggested to predict the physical and chemical properties of actual slag, such as the viscosity, crystallinity and so on. Wu et al. [1] have studied the effects of Na_2O, K_2O, BaO and CaO on R_xO-Al_2O_3 (R = Na, K, Ba, Ca) binary slag system by molecular dynamics simulation. The results demonstrate that basic oxides not only have the function of depolymerization network structure but also can stabilize the $[AlO_4]^{5-}$ tetrahedral structure. Mirhadi and Mehdikhani [2] studied the crystallization behaviour of Li_2O-CaO-SiO_2-Al_2O_3 by DTA and XRD experiments and observed the rod-like

S. Wang · B. R. Jia · S. P. He (✉) · Q. Wang
College of Materials Science and Engineering, Chongqing University,
Chongqing 400044, China
e-mail: heshp@cqu.edu.cn

© The Minerals, Metals & Materials Society 2019

T. Jiang et al. (eds.), *10th International Symposium on High-Temperature Metallurgical Processing*, The Minerals, Metals & Materials Series,
https://doi.org/10.1007/978-3-030-05955-2_57

growth morphology of $LiAlO_2$ glass by electron microscope. Since the addition of Li_2O can optimize the performance of mold flux, domestic researchers [3, 4] mainly discussed the impact of different components such as Al_2O_3/SiO_2, Na_2O, B_2O_3 and Li_2O on the performance of low-reactive mold flux based on the idea of substituting SiO_2 with Al_2O_3. On the basis of previous studies, Li_2O plays an important role in regulating slag viscosity and physical and chemical properties, but the mechanism of Li_2O on the microstructure of slag is not clear. In this paper, molecular dynamics simulation and Raman spectroscopy were used to study the structural transformation of slag and the mechanism of Li_2O action with different Li_2O contents.

Materials and Methods

Simulating Method

All MD simulations were performed using the pairwise potential model. The interatomic Born-Mayer-Huggins (BMH) potential form was applied and is given as follows:

$$U_{ij}(r) = \frac{q_i q_j}{r_{ij}} + A_{ij}\exp(-\mathrm{B}_{ij}\mathrm{r}) - \frac{C}{r_{ij}} \tag{1}$$

where $U_{ij}(r)$ is the interatomic potential, q_i and q_j are the charges of ions i and j, respectively, r_{ij} is the distance between i and j ions, and A_{ij}, B_{ij}, and C_{ij} are parameters for the BMH potentials. The interatomic potential parameters of all the systems were taken from Hirao and Kawamura [5]. The first term represents Coulomb interactions, the second term represents the inter-core short-range repulsion, and the last term represents the van der Waals attractions. We calculated the Coulomb potential using the Ewald sum method. The charge of Ca, Al, Li and O is 2, 3, 1 and -2, respectively. The detailed mass fraction, atomic number, density [6] and simulated cell length of the CAL system are shown in Table 1.

Table 1 Composition and atomic number of CaO-Al_2O_3-Li_2O samples

Sample number	Mass fraction			Atomic number					Density (g/cm^3)	Length (Å)
	CaO	Al_2O_3	Li_2O	Ca	Al	Li	O	Total		
CAL1	45	45	10	668	734	554	2046	4002	2.679	37.31
CAL2	44	44	12	638	700	650	2013	4001	2.654	37.05
CAL3	43	43	14	609	670	738	1983	4000	2.630	36.86
CAL4	42	42	16	582	638	828	1953	4001	2.670	36.72
CAL5	41	41	18	555	610	910	1925	4000	2.585	36.54

In this study, we investigated the microstructure of CAL melts with a total number of atoms around 4000 for each sample. The time interval step was 1 fs, and a cubic unit cell with periodic boundary conditions was chosen. The initial temperature was set at 5000 K in 30,000 steps to agitate the atoms and eliminate the effects of the initial atomic distribution. Next, the temperature was decreased down to 1723 K in 50,000 steps. Last, the simulated melt was equilibrated at 1673 K in 20,000 steps. All simulations were performed in a NVT ensemble, at which the number of ions, the volume and the temperature remained constant.

Roman

Reagent-grade CaO, Al_2O_3 and highly purified Li_2O were used as raw materials, placing 50 g of pre-melted slag in a Mo crucible until melted in a resistance furnace at 1723 K temperatures in Ar atmosphere. The temperature was kept constant at 1723 K for 1 h to homogenize the liquid slag, followed by quenching the slag with water to form solidified glassy samples. After quenching, the microstructure of the samples was investigated using Roman spectrometry. The measurement was conducted at infrared bands of corresponding to 2000–100 cm^{-1}.

Results and Discussion

Partial RDFs and CNs

The abscissa of the RDF curve's first peak represents the average bond length between first-neighbour atom pairs. The coordination numbers (CNs) of an atom *i* surrounding an atom *j* are determined by the numerical integration of the RDFs in the cut-off radius that is obtained from the first minima of the corresponding $g_{ij}(r)$. The cut-off radius is uniquely determined at each state. The ordinate value corresponding to the CN platform is the average CN. Figure 1 shows the corresponding RDFs and CNs of different atomic pairs for the CAL4 system.

Some studies [7, 8] show that the higher the initial peak of RDFs, the better the stability. The first peak corresponds to the average distance between atoms. According to Fig. 1a, the radial distribution function of Al-O RDFs is sharp and narrow, indicating that its stability is very good. The average distance between Al-O is maintained at 1.75, while the first peak of Li-O RDF is not as sharp as that of Al-O RDFs, indicating that Li-O bonds are not stable and easy to break, Al-O tetrahedral structure plays a supporting role in the network.

Table 2 shows the average bond lengths of different atomic pairs under different Li_2O. With the increase of Li_2O, the bond lengths of Al-O and Li-O almost do not change, which indicates that the bond lengths of Al-O and Li-O are not affected by

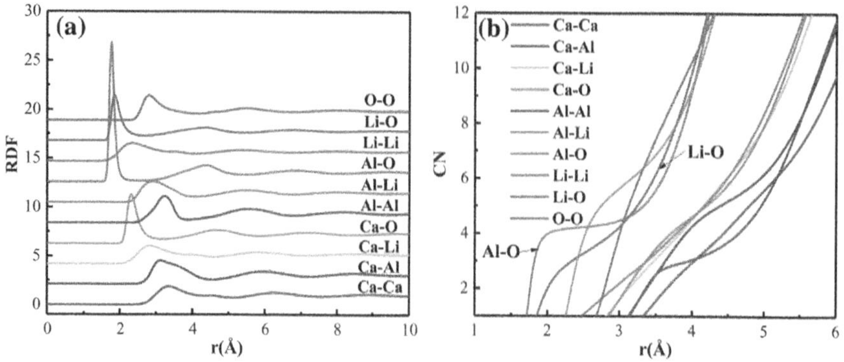

Fig. 1 **a** Partial radial distribution function (RDFs) and **b** average coordination number (CNs) of different atom pairs

Li_2O. As for the average bond length for O-O, it is somewhat longer than 2.6 Å, which is consistent with other published studies [9]. With the increase of Li_2O content, the average length of O-O bond increases, indicating that the bond becomes weaker and Li_2O has a depolymerization effect on Al-O network structure.

As can be seen from Fig. 1b, the Al-O CNs has a wider platform with a slight slope. The average Al CNs is thought to reflect the presence of Al^{3+} in several different aluminate complexes whose CNs range between 4 and 5. Li-O coordination number curve does not have a platform, indicating that the structure of Li-O is unstable. At the same time, the coordination number curve of Li-O does not have an obvious platform, mainly acting as network modifiers.

Table 3 shows the average coordination between different atoms, With the increase of Li_2O content, the average CN of Al-O decreases from 4.26 to 4.14, indicating that the increase of Li_2O content enhances the stability of $[AlO_4]^{5-}$ tetrahedron structure; the average CN of Ca-O fluctuates between 5.96 and 6.07; the average CN of Al–Al decreases from 3.29 to 2.84, indicating that the connection ratio between $[AlO_4]^{5-}$ tetrahedron decreases and the structure becomes simplified. In order to further analyze the effect of Li_2O on the coordination number of Al-O, as shown in Fig. 2.

Figure 2 reveals that for all systems, the presence of primary four-fold coordinated Al^{3+}, indicating that $[AlO_4]^{5-}$ in the slag structure acts as the network skeleton. With the increase of Li_2O content, the proportion of four-fold coordinated

Table 2 Average bond lengths of different atomic pairs of CAL system

Pair	Ca-O	Al–Al	Al-O	Li-O	O-O
Rij/Å(CAL1)	2.31	3.25	1.75	1.84	2.78
Rij/Å(CAL2)	2.30	3.26	1.75	1.85	2.78
Rij/Å(CAL3)	2.28	3.25	1.75	1.84	2.79
Rij/Å(CAL4)	2.27	3.25	1.75	1.84	2.77
Rij/Å(CAL5)	2.25	3.26	1.75	1.84	2.79

Table 3 Coordination numbers of different atomic pairs of CAL system

Pair	Ca-O	Al–Al	Al-O	Li-O
CAL1	6.01	3.29	4.26	3.63
CAL2	5.96	3.14	4.22	3.63
CAL3	6.07	3.02	4.18	3.63
CAL4	6.05	2.95	4.17	3.62
CAL5	6.02	2.84	4.14	3.62

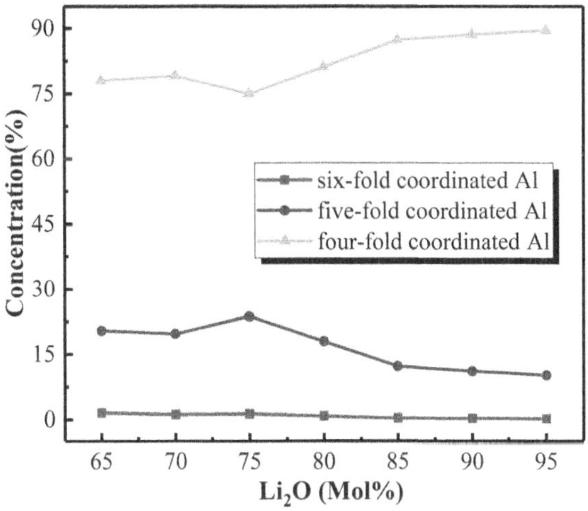

Fig. 2 Al-O coordination numbers under different Li_2O contents

Al increases gradually, and the five-fold coordinated Al was decreased, indicating that Li_2O can promote the stabilization of $[AlO_4]^{5-}$ tetrahedral network structure.

Distribution of Oxygen Types and Q^n Species

The calculated fractions of the oxygen types in the different systems show the same variation trend in that the O_f and O_{nb} concentrations increase and the proportion of O_b decreases with increasing Li_2O content. With the increase of Li_2O, Fig. 3a shows O_f with a continuous increasing trend, and O_{nb} gradually increases from 41.8 to 51.3%. Oxygen Tricluster(OT) content is very small and basically unchanged, so that the increase of Li_2O simplifies the network structure of the system.

Another key aspect of this short-range network structure is the distribution of Q_n species, meaning Al with various numbers n of O_b. The distribution of these Q_n species is related to the connectivity of the melt network. According to Fig. 3b, Q^3 and Q^4 of Al depolymerize into Q^0, Q^1 and Q^2 with increasing Li_2O contents. The proportion of Q^3 and Q^4 decreases with the increase of Li_2O content. The existence of Q^5 of Al is due to the OT in the aluminate melts. The results show that the

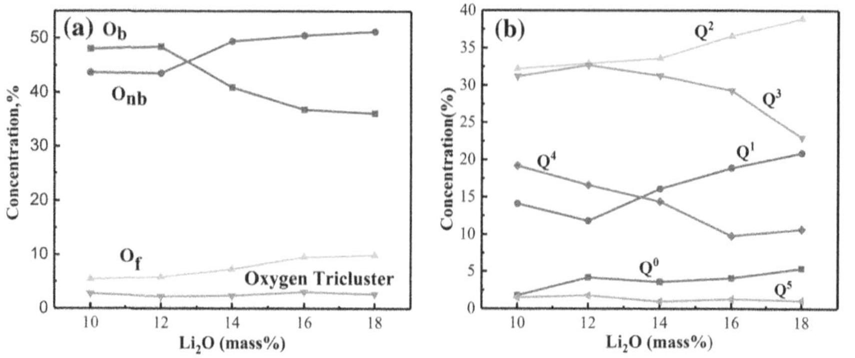

Fig. 3 **a** Concentrations of different oxygen types; **b** Concentrations of different Q^n species for Al of the CaO-Al_2O_3-Li_2O system

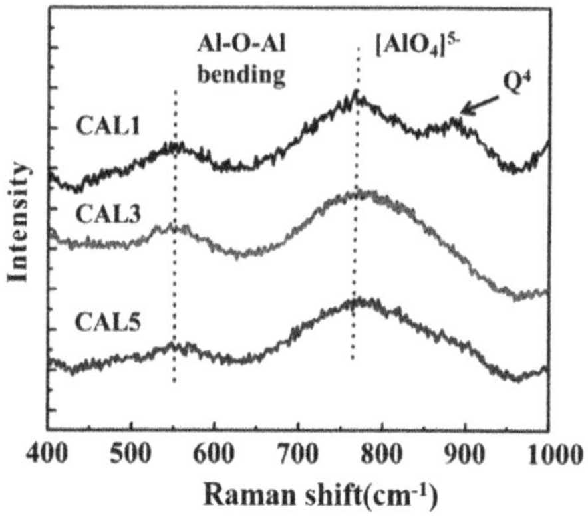

Fig. 4 Raman spectra of melt with different Li_2O contents

addition of network modifier Li_2O can increase the content of Q^0, Q^1 and Q^2 more easily and simplify the structure of $[AlO_4]^{5-}$ tetrahedron (Fig. 4).

Raman Spectroscopy

According to the results of previous studies [10, 11], the characteristic bands of different structural units in the Raman band of aluminate system can be determined. It can be seen from Fig. 4 that with the increase of Li_2O content, the centre of the high-wave section 600–950 cm^{-1} is shifted towards the direction of the low-wave number, indicating that the slag structure gradually overturns from complex structure to simple structure. In CAL1 system, there is an obvious peak around 880 cm^{-1}, which corresponds to the completely polymerized structure unit (Q^4) in the aluminum oxygen network structure. With the increase of Li_2O, Raman strength

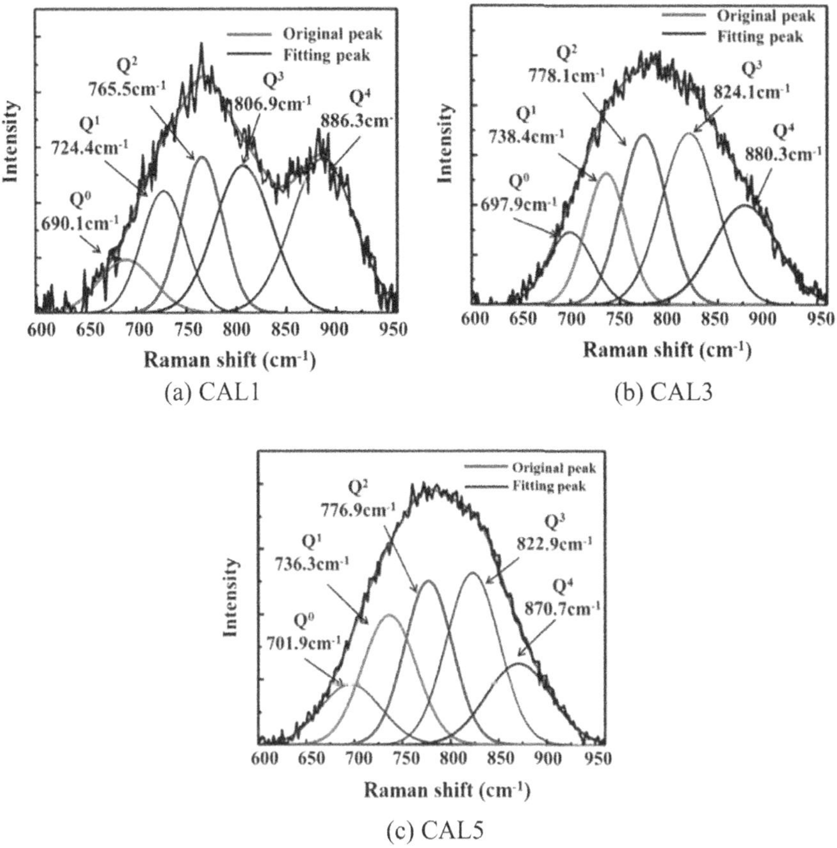

Fig. 5 Raman spectrum peak fitting: **a** CAL1, **b** CAL3 and **c** CAL5

of Q^4 structural unit with high degree of polymerization in the aluminum oxygen network structure gradually weakens and disappears, indicating that the slag structure gradually becomes simplified. At 550 cm^{-1}, the Raman strength of Al-O-Al gradually decreases with the increase of Li_2O, indicating that the proportion of bridge oxygen in the system keeps decreasing.

In order to further analyze the aluminate network structure at 600–950 cm^{-1} of the high-wave number section, the spectral peak in the range 600–950 cm^{-1} of the high-wave number section was subjected to Gaussian fitting to obtain a more intuitive semi-quantitative relationship. The results obtained by fitting are shown in Fig. 5 and Table 4.

Comparing the results of Raman spectroscopy experiment and MD simulation, we find that the trend of Q^n is basically the same. In order to further analyze the variation of slag polymerization degree, the complexity of slag was characterized by the degree of slag polymerization proposed by Zhang [12]:

Table 4 The centre of gravity and area fraction corresponding to different structural units obtained by Raman spectra fitting under different Li_2O

Structure units	CAL1		CAL3		CAL5	
	Centre (cm^{-1})	Area (%)	Centre (cm^{-1})	Area (%)	Centre (cm^{-1})	Area (%)
Q^0	690.1	8.76	697.9	9.99	701.9	11.14
Q^1	721.4	16.23	738.4	16.77	736.3	21.26
Q^2	765.6	19.73	778.1	24.08	776.9	23.92
Q^3	806.9	25.78	824.1	29.40	822.9	27.63
Q^4	886.3	29.50	880.3	19.75	870.7	16.05

$$\mathrm{DOP} = \frac{x(Q^3) + x(Q^4)}{x(Q^0) + x(Q^1)} \tag{2}$$

where DOP represents the ratio of the complex structural units (Q^3 and Q^4) to the simple structural units (Q^0 and Q^1) in the slag structure. As shown in Fig. 6, with the increase of Li_2O, the proportion of Q^4 in the system gradually decreased. The proportion of Q^3 to Q^2 increased first and then decreased. The proportion of Q^0 to Q^1 increased continuously, indicating that the melt structure was simplified with the addition of Li_2O. In addition, a sharp decrease in the DOP value can be seen from the diagram, which also shows that the structure of the slag becomes simple.

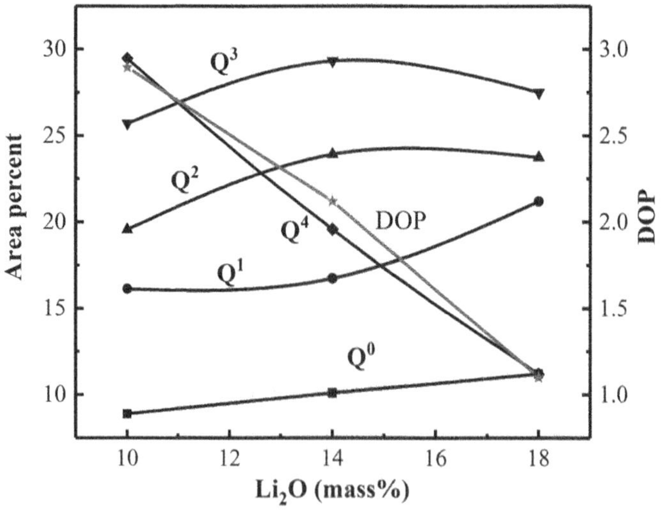

Fig. 6 The area fraction of Qn and the change of DOP under different Li_2O content

Conclusion

In this study, we investigated the microstructure of CaO-Al_2O_3-Li_2O (CAL) molten slag at different Li_2O contents using MD simulation and Raman measurements. The conclusions are the following:

It can be obtained that the average bond lengths of, Ca-O, Al–Al, Al-O, Li-O, and O-O are 2.31, 3.25, 1.75, 1.84 and 2.78Å,the average CNs for Ca-O, Al–Al, Al-O and Li-O pairs are 6.01, 3.29, 4.26, and 3.63.

With the increase of Li_2O, Li^+ polarizes the bridging oxygen, while the number of network forms decreases and the number of non-bridging oxygen increases. Li_2O as a network modifier enhances the stability of $[AlO_4]^{5-}$ tetrahedron structure. According to the statistical analysis of $[AlO_4]^{5-}$ tetrahedron, it is found that the proportion of Q^0, Q^1 and Q^2 shows an increasing trend, while the proportion of Q^3 and Q^4 shows a decreasing trend, indicating that the structure of Al-O is depolymerized.

The influence of Li_2O on CaO-Al_2O_3-Li_2O slag structure was studied by Raman spectrum, the change of area ratio of different Q^n structural units and the decrease of DOP value indicate that the degree of polymerization of slag network decreases with the increase of Li_2O, and the complex network structural units tend to simplify.

Acknowledgements The authors deeply appreciate the funding support from the National Natural Science Foundation of China (project no. 51874057) and the Key projects of national natural science foundation of China (project no. U1660204).

References

1. Wu T, Wang Q, Yao TH, He SP (2016) Molecular dynamics simulations of the structural properties of Al_2O_3-based binary systems. J Non-Cryst Solids 435:17–26
2. Mirhadi B, Mehdikhani B (2011) The effect of compositional changes on the crystallisation behaviour and mechnical properties of Li_2O-CaO-SiO_2-Al_2O_3. Adv Mater Sci 11:11–21
3. Kubicki JD, Toplis MJ (2002) Molecular orbital calculations on aluminosilicate tricluster molecules: Implications for the structure of aluminosilicate glasses. Am Mineral 87:668–678
4. Allwardt JR (2005) Aluminum coordination and the densification of high-pressure aluminosilicate glasses. Am Mineral 90:1218–1222
5. Hirao K, Kawamura K (1994) Materials design using personal computer. Shokabo, Tokyo. 12:52–54
6. Huang XH (2013) Metallurgy principle of iron and steel. Metallurgical Industry Press, China, pp 315–316
7. Wu T, He SP, Liang Y, Wang Q (2015) Molecular dynamics simulation of the structure and properties for the CaO-SiO_2 and CaO-Al_2O_3 systems. J Non-Cryst Solids 411:145–151
8. Zhang S, Zhang X, Liu W, Lv X, Bai C, Wang L (2014) Relationship between structure and viscosity of CaO-SiO_2-Al_2O_3-MgO-TiO_2 slag. J Non-Cryst Solids 402:214–222
9. Hannon AC, Vessal B, Parker JM (1992) The structure of alkali silicate glasses. J Non-Cryst Solids 150:97–102

10. Licheron M, Montouillout V, Millot F, Neuville DR (2011) Raman and 27Al NMR structure investigations of aluminate glasses: (1-x)Al_2O_3-xMO, with M = Ca, Sr. Ba J Non-Cryst Solids 357:2796–2801
11. Neuville DR, Cormier L, Massiot D (2006) Al coordination and speciation in calcium aluminosilicate glasses: effects of composition determined by ^{27}Al MQ-MAS NMR and Raman spectroscopy. Chem Geol 229:173–185
12. Zhang Z, Xie B, Zhou W, Diao J, Li HY (2016) Structural characterization of FeO–SiO_2–V_2O_3 slags using molecular dynamics simulations and FT-IR spectroscopy. ISIJ Int 56: 828–834

Dissolution Reaction of Earthy Graphite in Liquid Steel

Hongyan Yan, Xiaojun Hu, Chao Luo, Jinglong Liang and KuoChih Chou

Abstract Earthy graphite is an abundant natural resource with a high content of fixed carbon. It can be considered as a carbon source for steel carburizers. Also, it can reduce the production cost of enterprises. The study is aimed to clarify the kinetics of dissolution reaction and the carbon increase effect. The experiments were carried out by quartz tube sampling at 1873 K, with 0.1 and 0.4% theoretical carbon increase. Effect of added amount on S, N and O content in steel was studied. The results indicated that carbon increase effect was equivalent to some company carbon additives. The added amount of earthy graphite has little influence on S, N and O content. Considering the carbon oxidation reaction consumption during carbon dissolution process, a dynamic model was obtained and the reaction rate constant of the carbon dissolution reaction was calculated.

Keywords Earthy graphite · Dissolution · Kinetics · Liquid steel

Background

At present, most enterprises mainly use the carbonization method to control the carbon content in liquid steel. The carbon additives used in the carbon-increasing method require high carbon content, low sulphur and phosphorus, and low ash. Usually, the carbon additives are electrode blocks, coke powder and calcined anthracite. The electrode block recarburizer has high carbon content and strong

H. Yan · J. Liang
College of Metallurgy and Energy, North China University of Science and Technology, Tangshan 063210, China

X. Hu (✉) · K. Chou
State Key Laboratory of Advanced Metallurgy, University of Science and Technology Beijing, No. 30, College Road, Haidian District, Beijing 100083, China
e-mail: huxiaojun@ustb.edu.cn

C. Luo
Hesteel Group Tangsteel Company, Tangshan 063000, China

© The Minerals, Metals & Materials Society 2019

T. Jiang et al. (eds.), *10th International Symposium on High-Temperature Metallurgical Processing*, The Minerals, Metals & Materials Series,
https://doi.org/10.1007/978-3-030-05955-2_58

oxidation resistance, the production process is complicated and the cost is high, the price of coke powder is low, the ash content and sulphur content are high, the carbon content is low, and the use effect is not good [1]. At present, the most commonly used anthracite recarburizer needs to be purified by beneficiation and then graphitized. Although it basically meets the needs of steelmaking production, the cost is still high [2]. China has abundant natural graphite resources. Among them, earthy graphite (cryptocrystalline graphite) has abundant reserves and high grade, and the fixed carbon can reach more than 80%. The sulphur content and phosphorus content are low, and the relative density is relatively large. It has been mentioned that the composition and physical and chemical properties of earthy graphite can basically meet the requirements of steelmaking recarburizers [3, 4]. There are few reports on the dissolution kinetics of earthy graphite in liquid steel. In this paper, earthy graphite was used to increase the carbonization in liquid steel. The dissolution kinetics of earthy graphite was studied.

Experiments

The carbon-increasing experiment proceeded by increasing the carbon in the steel mill in the laboratory simulation steel, melt the steel sample in the silicon-molybdenum rod furnace, and then added the prepared recarburizer to carry out the carbon-increasing experiment. During the experiment, the molten steel carbonation experiments of different types of recarburizers were carried out with different theoretical carbon additions.

(1) The samples were placed in a silicon molybdenum rod furnace and heated to 1600 °C. High-purity Ar with a flow rate of 1 L/min was introduced as a shielding gas during the heating process (Table 1).
(2) After reaching the reaction temperature, the first sample is taken with a quartz tube, and the extracted steel sample is cooled and weighed. The mass of the remaining steel sample plus the mass of the iron piece was calculated by the theoretical carbon addition amount of 0.1 and 0.4%, respectively. The carbon-reinforcing agent was wrapped in a steel tube and inserted into the molten steel. The mixture was continuously stirred and timed, and sampled and weighed at 0, 5, 10, 20 and 40 min, respectively. The specific conditions are shown in Table 2.

Table 1 Chemical composition of steels (mass fraction)

C	Mn	Al	Ti	S	O	N	P
0.58–0.63	0.15–0.17	0.039–0.045	0.031–0.068	0.0098–0.0109	0.0037–0.0060	0.0089–0.0149	0.009–0.013

Table 2 The experimental and conditions

Carbon additives	m_{steel}/g	m_{foil}/g	$m_{additives}$/g	Theoretical carbon increasement/%
Earthy graphite	350.87	3.35	0.37	0.10
	337.05	5.39	1.38	0.40
Company 1	351.57	2.68	0.36	0.10
	560.02	6.28	2.29	0.40
Company 2	380.37	3.64	0.39	0.10
	612.91	5.89	2.55	0.40

(3) The water-cooled experimental samples were subjected to surface treatment, and the C, S content and N and O contents of the samples were, respectively, tested. The instruments tested were EMIA-820V infrared carbon-sulphur metre and TCH-600 O-N-H analyser imported from Japan.

Results

Effect on Carbon Concentration

Figure 1 showed the change of mass fraction $w[C]$ in steel samples over time. The dissolution reaction took two stages. In the first 10 min, $w[C]$ increased significantly. The maximum value was reached in about 10 min and then decreased. The theoretical variation of carbon concentration with 0.1 and 0.4% is basically same.

According to the change of carbon concentration, combined with Eq. (1), the yield η of recarburizer can be obtained.

$$\eta = \frac{w[C]_{\%,d} - w[C]_{\%,0}}{w[C]_{\%,t} - w[C]_{\%,0}} \times 100\% \quad (1)$$

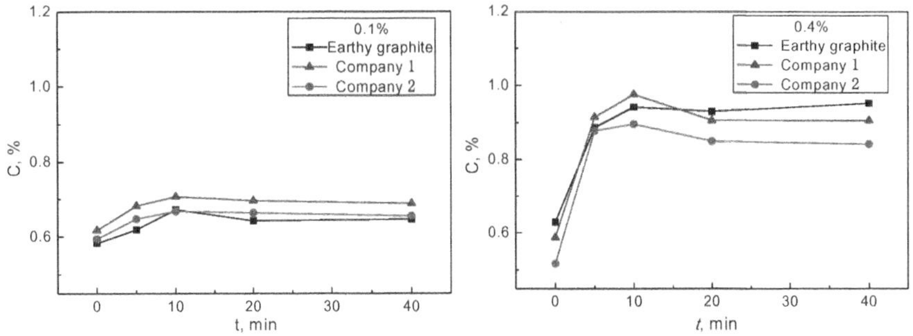

Fig. 1 Variation of carbon mass fraction in steels with time for different carburants amounts

η represented the yield of recarburizer, $w[\mathrm{C}]_{\%,d}$ represented the detected carbon content. $w[\mathrm{C}]_{\%,0}$ represented carbon content of raw steel, $w[\mathrm{C}]_{\%,t}$ was calculated target with carbon increasement. $w[\mathrm{C}]_{\%,t}$ could be calculated by Eq. (2).

$$w[\mathrm{C}]_{\%,t} = \left(\frac{w[\mathrm{C}]_{\%,d} \times m_C}{m_{\text{steel}} + m_{\text{foil}} + m_C} \right) \tag{2}$$

$w[\mathrm{C}]_{\%,d}$ could be obtained by thermogravimetric combustion. m_{steel}, m_{foil}, m_C represented the quality of steel, iron and recarburizer. The yield of the carbonation reaction can be obtained by Eqs. (1) and (2).

Figure 2 showed the variation of carbonation yield with time. In the first 10 min, yield increased significantly. The maximum value was reached in about 10 min and then decreased. When the theoretical carbon addition is 0.1 and 0.4%, the curve of the earthy graphite was basically at the same level as that of other enterprises. This indicated that the carbonation effect of earthy graphite was comparable to that of other enterprise recarburizers. The fixed carbon content of earthy graphite was lower than that of other enterprises, and its yield was comparable to enterprise carbonizers. It indicated that the carbon addition effect of earthy graphite was better than that of other enterprises.

Effect on Sulphur

Figure 3 showed the effect of carbonation on S mass fraction in liquid steel. The S mass fraction hardly changes, and the S mass fraction used earthy graphite was lower than that in other enterprises.

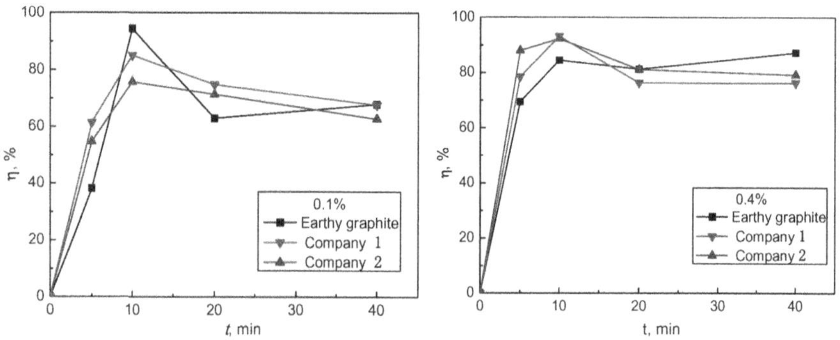

Fig. 2 Variation of carbon yield with time for different carburants

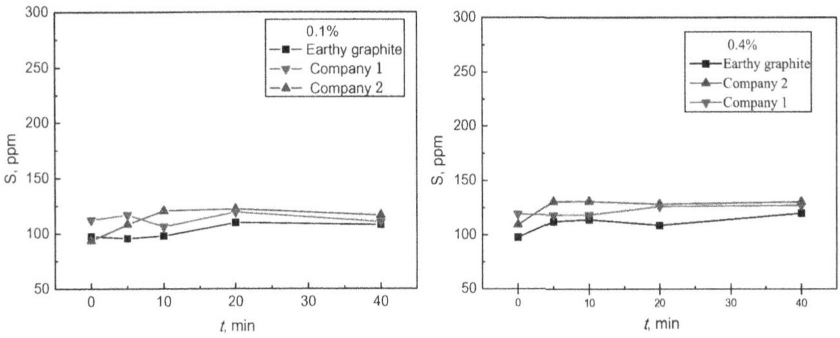

Fig. 3 Effect on S mass fraction in the liquid steel

Effect on Nitrogen

Figure 4 showed the effect of carbonation on N mass fraction in liquid steel. When the theoretical carbon addition is 0.1 and 0.4%, the N mass fraction increased with time. The N mass fraction both varied from 0.01 to 0.02%. It indicated that the earthy graphite carbonation experiment had little effect on the N mass fraction. It was equivalent to the effect of other enterprises' recarburizers.

Effect on Oxygen

Figure 5 showed the effect of the carbonation on the total oxygen mass fraction in the liquid steel. The oxygen mass fraction varied from 0.005 to 0.010%. It indicated that carbonation used earthy graphite had little effect on the total oxygen content. There were several points with high oxygen content in the figure, because the

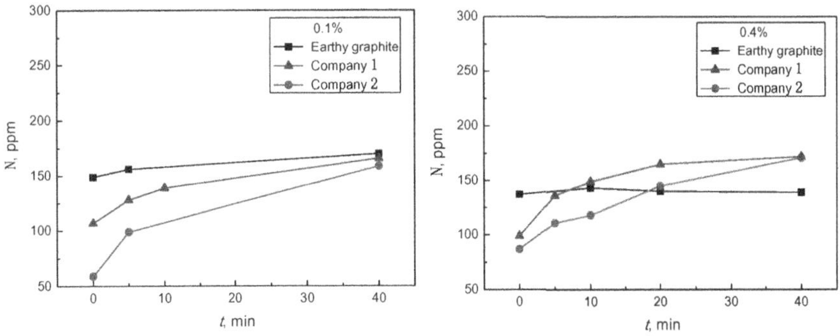

Fig. 4 Effect on N mass fraction in the liquid steel

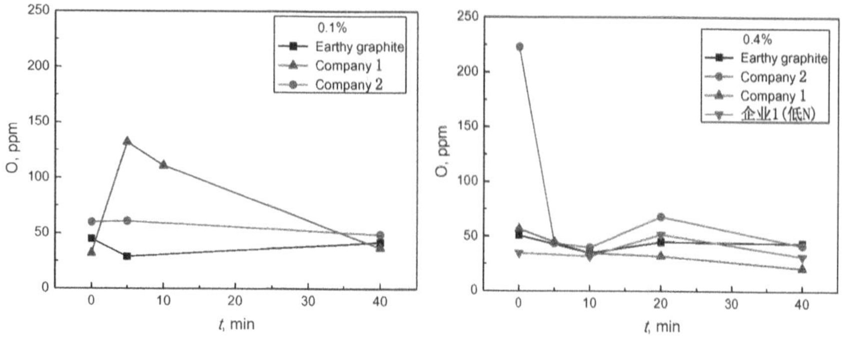

Fig. 5 Effect on total oxygen mass fraction in the liquid steel

samples drawn with quartz tube were oxidized by air during cooling. Compared with other manufacturers' recarburizers, the oxygen mass fraction was relatively lower. This indicated that the addition of earthy graphite had no obvious influence on the total oxygen content in liquid steel. It demonstrated that the earthy graphite could be used for carbonation in steelmaking.

Discussion

Some researchers [6–9] believed that the dissolution of carbon in steelmaking was mass transfer rate controlled. It can be expressed as first-order reaction kinetics equation.

$$\frac{dC_t}{dt} = k(C_s - C_t), \quad k = \frac{Ak_m}{V_m} \tag{3}$$

C_t was the carbon concentration in the liquid steel; C_s was the saturated solubility for carbon. k, k_m, A and V_m represented dissolution reaction rate constant, mass transfer coefficient, the surface area of molten steel and the volume of molten steel, respectively. The theoretical carbon increase in this experiment is at most 0.4%; the saturated solubility was not obtained. Then, C_s can be expressed by ideal carbon dissolved concentration C_i.

After integral (3) and took initial conditions as $t = 0$, $C_t = C_0$, Eq. (4) was obtained.

$$C_t = C_i + (C_0 - C_i)e^{-kt} \tag{4}$$

If the carbon dissolution process satisfied first-order reaction kinetics equation, the concentration of carbon with time should be increased until the desired carbon concentration was reached. It was found that the concentration and yield of the carbon 10 min before the dissolution of carbon increased continuously, and the reaction showed a slow decline after 10 min. It indicated that after 10 min, the carbon in liquid steel was depleted, resulting in a decrease in carbon concentration and yield. The above phenomenon was due to the oxidation loss of carbon caused by the reaction of carbon and oxygen in liquid steel. It can be expressed as Eq. (5).

$$[\mathrm{C}] + \frac{1}{2}\mathrm{O}_{2(g)} \rightarrow \mathrm{CO}_{(g)} \tag{5}$$

Research [10–13] showed that the carbon-oxygen reaction according to Eq. (5) was controlled by mass transfer. In the low-carbon region (w[C] less than 1%), the kinetics of carbon and oxygen reaction can be expressed as follows [10, 11, 13, 14],

$$j'_{\mathrm{C}} = k'C_t \tag{6}$$

k' represented the reaction rate constant of carbon-oxygen reaction in liquid steel. When the theoretical carbon addition is 0.1 and 0.4%, the mass fraction of carbon in the liquid steel is in the range of 0.59 to 1%. It belongs to the low-carbon region. Therefore, the kinetics of carbon loss reaction in the carbonization process can be expressed as Eq. (6). The mass transfer process of carbon in molten steel is expressed as follows,

$$j''_{\mathrm{C}} = k(C_i - C_t) \tag{7}$$

$$j_C = k(C_i - C_t) - k'C_t \tag{8}$$

After integral (8) and took initial conditions as $t = 0$, $C_t = C_0$, Eq. (9) was obtained.

$$C_t = \frac{k}{k+k'}C_i + (C_0 - \frac{k}{k+k'}C_i)e^{-(k+k')t} \tag{9}$$

Thus, a kinetic equation to describe the process of carbon addition in the liquid steel was obtained.

Figure 6 showed the results of data fitting by using Eqs. (4) and (9), respectively. It found that the fitting result with Eq. (9) was significantly better than Eq. (4). If there was no carbon oxidation reaction, the carbon dissolution curve in Fig. 6 should be along the direction of the dotted line. The actual dissolution curve followed the trajectory of the solid line in Fig. 6. Above phenomenon indicated that the first 10 min of dissolution, $j''_{\mathrm{C}} > j'_{\mathrm{C}}$. The carbon dissolution rate increased with time; the carbon concentration in the liquid steel increased continuously. Around 10 min, $j''_{\mathrm{C}} = j'_{\mathrm{C}}$,carbon concentration reached the maximum. After 10 min, $j''_{\mathrm{C}} < j'_{\mathrm{C}}$, carbon concentration dropped slightly. In summary, Eq. (9) can be used to describe the carbonation of recarburizer in liquid steel.

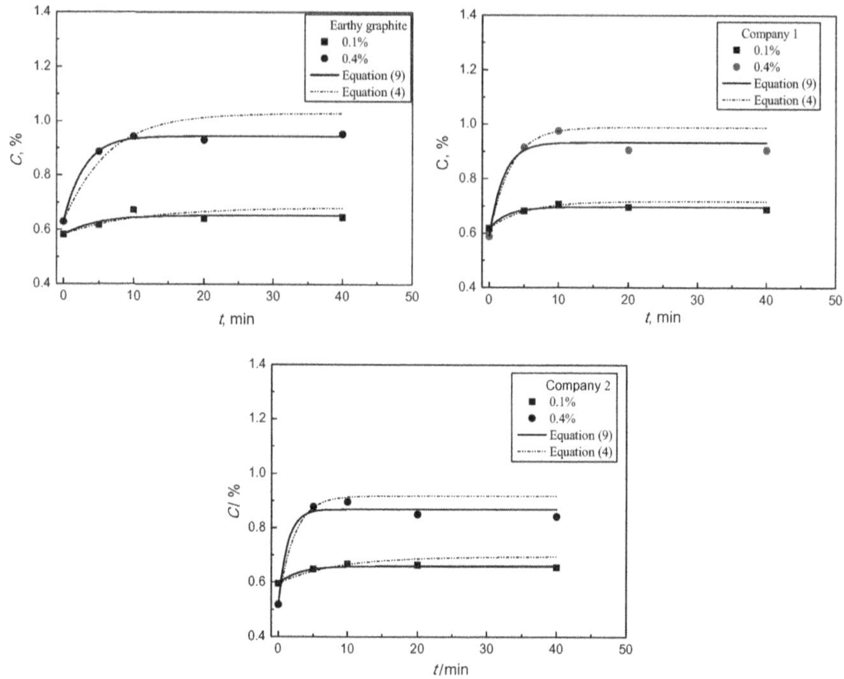

Fig. 6 Fitted results with different kinetic models

Table 3 Calculated results of reaction rate constants of k and k'

Carbon additives	Theoretical carbon increasement/%	a	$b \times 10^2$	$k \times 10^3$/s^{-1}	$k' \times 10^3$/s^{-1}
Earthy graphite	0.10	0.96	0. 34	3.26	0.14
	0.40	0.92	0. 58	5.32	0.48
Company 2	0.10	0.95	0. 57	5.39	0.29
	0.40	0.95	1.27	12.02	0.68
Company 1	0.10	0.97	0. 65	6.32	0.19
	0.40	0.94	0. 83	7.84	0.46

Table 3 showed the values of reaction rate constants k. k' under different experimental conditions. a and b represented $\frac{k}{k+k'}$ and $k + k'$, respectively. a, b can be obtained by fitting Eq. (9) to the carbon concentration curve. Then, the values of k and k' can be calculated by the values of a and b.

Conclusions

(1) Earthy graphite and other enterprises' recarburizers dissolved faster in the first 10 min, increased rapidly and reached a maximum at about 10 min, and had a downward trend after 10 min. The carbonation effect of earthy graphite was equivalent to that of other enterprises. The effect of earthy graphite on S, N and O in liquid steel cannot affect the quality of molten steel.

(2) A carbonation kinetic model was obtained,

$$C_t = \frac{k}{k+k'}C_i + (C_0 - \frac{k}{k+k'}C_i)e^{-(k+k')t}$$

Compared with the model that does not consider the carbon oxidation reaction, this model can better describe the carbonation process of the recarburizer in the molten steel.

Acknowledgements This work was supported by the National Natural Science Foundation of China (No. 51334001, No. 51804126, No. 51474019), Natural Science Foundation of Hebei Province (No. E2018209266) and Natural Science Foundation of Tangshan city (No. 17110208a).

References

1. Wang YM, Han CH, Wang F (2005) Study on molding techniques of graphite-like carbon additive for steel-making. Ganshu Technol 27(1):9–10
2. Sun XC, Qu YF (2000) Development of a new carburizing agent for steel-making-a kind of semi-graphite. Rock Miner Anal 19(2):112–115
3. Zhou WP (1989) EAF steelmaking used earthy graphite carbon. Southern Steel 4:9–21
4. Chen SH, Wei B, Long LX (2008) Application of carburization material on production of cast iron. Foundry Technol 29(6):823–826
5. Li CE, Brown TC (2001) Carbon oxidation kinetics from evolved carbon oxide analysis during temperature-programmed oxidation. Carbon 5(39):725–732
6. Khanna R, McCarthy F, Sun H et al (2005) Dissolution of carbon from coal-chars into liquid iron at 1550 °C. Metall Mater Trans B 36(6):719–729
7. Dahlke VO, Knacke O (1955) The dissolution of carbon in liquid iron. Arch Eisenhuttenwes 26:373–378
8. Wu C, Sahajwalla V (2000) Dissolution rates of coals and graphite in Fe-CS melts in direct ironmaking: Influence of melt carbon and sulfur on carbon dissolution. Metall Mater Trans B 31(2):243–251
9. Kosaka M, Minowa S (1968) On the rate of dissolution of carbon into molten Fe-C alloy. Trans Iron Steel Inst Jap 8(6):392–400
10. Swisher JH, Turkdogan ET (1967) Decarburization of iron-carbon melts in CO_2-CO atmospheres-kinetics of gas-metal surface reactions. Aime Met Soc Tran 239(5)
11. Baker LA, Warner NA, Jenkins AE (1967) Decarburization of a levitated iron droplet in oxygen. Aime Met Soc Trans 239(6):857–864

12. Robertson DGC, Jenkins AE (1970) Reaction of liquid iron and its alloys in pure oxygen. Plenum Press, New York, pp 393–408
13. Zong JH, Yoon JK (1990) Theoretical interpretation of the decarburization mechanism in convective oxygen steelmaking. Metall Trans B 21(1):49–57
14. Meyer HW, Porter WF, Smith GC et al (1968) Slag-metal emulsions and their importance in BOF steelmaking. J Met 20(7):35–42

Distribution Behaviour of Metals in Copper Alloy Under Super-Gravity Field

Long Meng, Zhe Wang, Yiwei Zhong, Kuiyuan Chen and Zhancheng Guo

Abstract In the present study, the influences of super-gravity field with different cooling rates on the distribution behaviour and the morphology of metallic elements of Cu alloy were investigated. The Cu alloy was obtained from waste printed circuit boards by super-gravity separation, and it contained precious and common metals. The different solubilities of metallic elements in liquid and solid phases lead to component segregation during solidification of alloys. The results showed that super-gravity technology can enrich and distribute metallic elements in the Cu alloy according to different densities, and low cooling rate was favourable for the precipitation and enrichment of metals in the Cu alloy. When the cooling rate was 2 °C/min, Fe was enriched in the top of the Cu alloy, while Pb and Sn were enriched in the bottom of the alloy. The precious metals were combined with Pb and Sn, and they were enriched with segregation of Pb and Sn.

Keywords Copper alloy · Super-gravity field · Metal · Distribution · Enrichment

Introduction

Nowadays, the manufacturing of electronic and electrical equipment (EEE) and the associated generation of electronic waste (e-waste) are experiencing a rapid growth due to the high consumption of electronic devices and their early obsolescence throughout the world [1]. Printed circuit boards (PCBs) are the key central components of electronic equipment and composed of polymers, ceramics and metals, which make the process more difficult to recycle. The disposal of PCBs has several health issues, such as the release of toxic and hazardous compounds, contamination of water, soil and air, and impacts on regions surrounding recycling facilities.

L. Meng · Z. Wang · Y. Zhong · K. Chen · Z. Guo (✉)
State Key Laboratory of Advanced Metallurgy, University of Science and Technology Beijing, 30 Xueyuan Road, Beijing 100083, People's Republic of China
e-mail: zcguo@ustb.edu.cn

© The Minerals, Metals & Materials Society 2019

T. Jiang et al. (eds.), *10th International Symposium on High-Temperature Metallurgical Processing*, The Minerals, Metals & Materials Series,
https://doi.org/10.1007/978-3-030-05955-2_59

However, several precious metals (Ag, Au, Pd and Pt) are present in PCBs in much higher concentrations than typically found in corresponding ores. These precious metals are frequently used in the electronics industry, particularly as contact materials, for electroplating, or as connectors because of their high chemical stability, corrosion resistance and electrical conductivity [2]. For example, Ag is used in switches, contacts and solders, as well as bonding wire in integrated circuits; Pd is used in multilayer capacitors and connectors; and Pt is used in hard disks, in proton-exchange membrane fuel cells and in thermocouples [3, 4]. The environmentally sustainable recovery of precious metals from e-waste is expected to have a significant economic impact. With appropriate recycling, e-waste could be an important secondary resource of precious metals for modern resource conservation and environmental protection. Many technologies have been investigated for the recovery and concentration of precious metals. The conventional approaches of recycling precious metals were hydrometallurgy and pyrometallurgy. Park et al. used aqua regia as a leachant to recover 98 wt% of the input Ag, liquid–liquid extraction to obtain 93 wt% of the input Pd, and dodecanethiol and sodium borohydride to recycle 97 wt% of the input Au [5]. Alzate et al. developed a novel methodology to recover Au from waste EEE (WEEE) using ammonium persulfate ($(NH_4)_2S_2O_8$), and the results indicated that more than 98% of the Au were recovered [6]. Sahin et al. used an iodine–hydrogen peroxide (I_2-H_2O_2) solution system to recover precious metal, and the leaching efficiency of Au was 99.98% in shake flask tests and 99.85% in semi-pilot tests using 3% iodine concentration, 1% H_2O_2 concentration and 15% solids-% [7]. Li et al. suggested a thiourea leaching process to extract about 90% Au and 50% Ag from 100 mesh PCBs at room temperature using solutions containing 24 g/L thiourea and an Fe^{3+} concentration of 0.6% [8]. Rashchi et al. utilized nitric acid to recover Ag (87 wt%) and Cu (98 wt%) from PCB scraps under the optimized condition [9]. Fogarasi et al. developed an eco-friendly chemical–electrochemical process for the simultaneous recovery of Cu and separation of Au-rich residue from waste PCBs (WPCBs). The results showed that a high-purity Cu deposit (99.04 wt%) and 25 times Au concentration higher than that in the initial WPCB samples were achieved [10]. Cayumil et al. used high-temperature pyrolysis to concentrate and recover precious metals in PCBs. This process was successful in concentrating several precious metals in a small volume fraction and reduced volumes for further processing/refinement by up to 75% [1]. In this study, a novel approach was reported to significantly concentrate precious metals present in WPCBs. This approach is cost-effective and produce minimal secondary waste. Super-gravity is a new technique for strengthening multiphase flow transfer and reaction process. It uses a rotating centrifuge to produce a stable, adjustable centrifugal field instead of a conventional gravitational field. The technology has been successfully applied to the enrichment of valuable elements from different slags [11, 12], the fabrication of functionally graded materials [13, 14], the treatment of waste [15, 16] and the removal of impurities from alloys or metals [17, 18].

In a super-gravity field, different melting points between the solid-particle and liquid melt would result in particles distributing and separating gradually along the centrifugal direction. Based on the characteristics, different metals (Pb, Sn, Zn and Cu) were selectively separated and recovered from WPCBs by super-gravity in the previous studies [19, 20]. In this research, precious metals were concentrated in the Cu alloy and PCB residues by super-gravity technology. The solidification process of Cu alloy was studied under the super-gravity field, and the influences of super-gravity field with different cooling rates on the distribution behaviour and the morphology of metallic elements of Cu alloys were investigated.

Experimental Section

Materials and Methods

The WPCBs were purchased from an enterprise in Hunan province of China. Prior to their receipt, the PCBs were crushed, sorted and sieved to obtain a particle size of about 1.0 mm, and the majority of nonmetallic materials were removed from the PCBs. The obtained PCBs contained a range of metals and alloys with different melting points, and the molten metals and alloys were separated rapidly from the solid particles using a super-gravity field. Most of the low-melting-point metals in the PCBs were preferentially recycled by super-gravity separation, and the high-melting-point metals (Cu, Zn, Au, Ag and Pd) were remained in the residues. Therefore, the residues need to be reheated, rotated and centrifuged to allow the recovery of Cu by using super-gravity technology. The common metals (Pb, Sn, Cu and Zn) were selectively separated and recovered from WPCBs by super-gravity separation. The precious metals were concentrated in the Cu alloys and the final residues, respectively.

Equipment

The super-gravity field was generated by a centrifugal apparatus, the schematic and physical diagrams of which are illustrated in Fig. 1, with the heating furnace and the counterweight fixed symmetrically onto the horizontal rotor. The temperature of the separation tank was controlled by a program controller with an R-type thermocouple, with a precision range of ±3 °C. The gravity coefficient (G) was calculated as the ratio of super gravitational acceleration to normal gravitational acceleration via Eq. (1).

$$G = \frac{\sqrt{g^2 + (\omega^2 R)^2}}{g} = \frac{\sqrt{g^2 + (\frac{N^2 \pi^2 R}{900})^2}}{g} \quad (1)$$

where ω is the angular velocity (rad/s), N is the rotating speed of the centrifuge (rev/min), R is the distance from the centrifugal axis to the centre of the sample (R = 0.25 m) and g is normal gravitational acceleration.

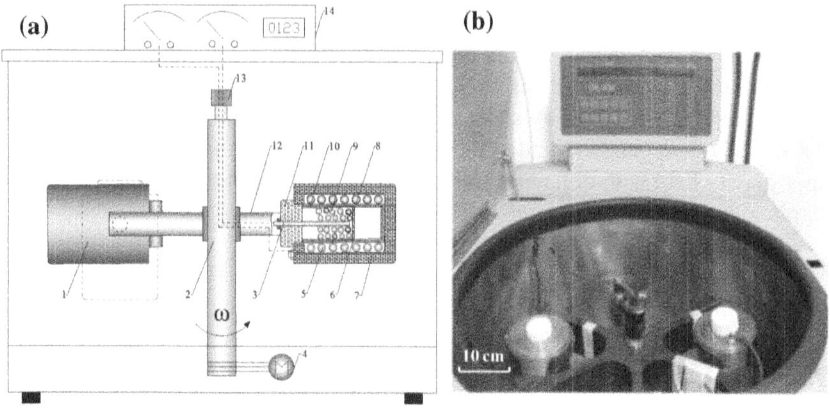

Fig. 1 Schematic and physical diagrams of centrifugal separation apparatus (1—counterweight; 2—centrifugal axis; 3—thermocouple; 4—motor; 5—resistance coil; 6—furnace chamber; 7—refractory materials; 8—filter hole; 9—PCB particles; 10—graphite crucible; 11—refractory brick; 12—horizontal rotor; 13—conductive slipping; 14—dashboard)

Experimental Procedures

The PCB particles were placed into a set of graphite crucibles of inner diameter 19 mm. The bottom of the upper graphite crucible included small holes (1 mm), through which metals in the liquid state could pass. Graphite felt (thickness 3 mm) was laid at the bottom of the upper crucible to retain small solid particles. Similarly, graphite felt (thickness 5 mm) was laid on the PCB particles to avoid oxidization during the separation process. Figure 2 shows the flow chart of super-gravity separation of metals from PCB particles [16]. In the third step process, the 2# PCB residues were heated to 1200 °C, and then the Cu-Zn alloy was separated and solidified under the super-gravity field. The different solubility of metallic elements in liquid and solid phases leads to component segregation during the solidification of alloys. Due to the different densities of metals, the enrichment behaviour of metals in the Cu alloy was present in the super-gravity field.

Characterization

The alloys obtained from PCB residues were sectioned longitudinally along the centre axis. They were characterized by scanning electron microscope and energy-disperse X-ray spectrum (SEM/EDS, MLA 250, FEI Quanta, USA), and electron probe X-ray micro analyzer (EPMA-1720H, Shimadzu, Japan).

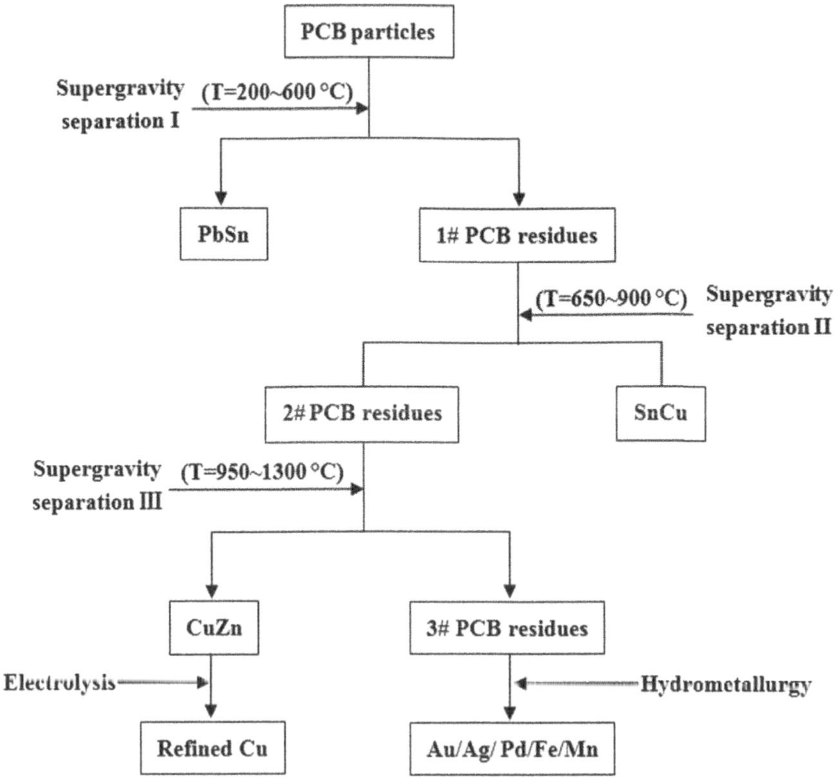

Fig. 2 Flow chart of super-gravity separation of metals from PCB particles

Results and Discussion

After the separation of Pb-Sn and Sn-Cu, the 2# PCB residues were reheated to 1200 °C. The centrifugal device was rotated and centrifuged under a gravity coefficient of 800. The effects of cooling rates on the metal distribution behaviours in the Cu-Zn alloys were investigated.

First, a cooling rate of 10 °C/min was conducted in the solidification process under the super-gravity field. Figure 3 shows the SEM mapping of common metals in the top and bottom of Cu-Zn alloy, respectively. As indicated, Pb and Sn were evenly distributed, and no macro-segregation was observed in the sample. In the solidification process, large cooling rate cannot effectively enrich common metals in the Cu-Zn alloy under a super-gravity field. Figure 4 shows the EPMA mapping of precious metals in the top and bottom of Cu-Zn alloy, respectively. The results showed that the precious metals with large density were combined with Pb and Sn in the Cu-Zn alloy, and they were uniformly distributed in the Cu-Zn alloy, without any enrichment.

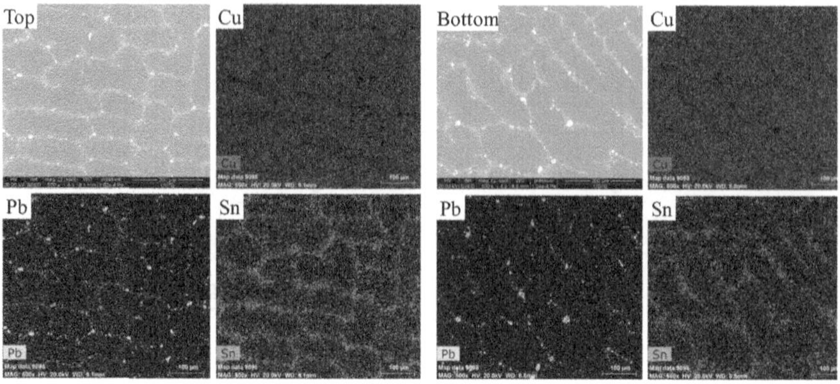

Fig. 3 SEM mapping of common metals in the top and bottom of Cu-Zn alloy

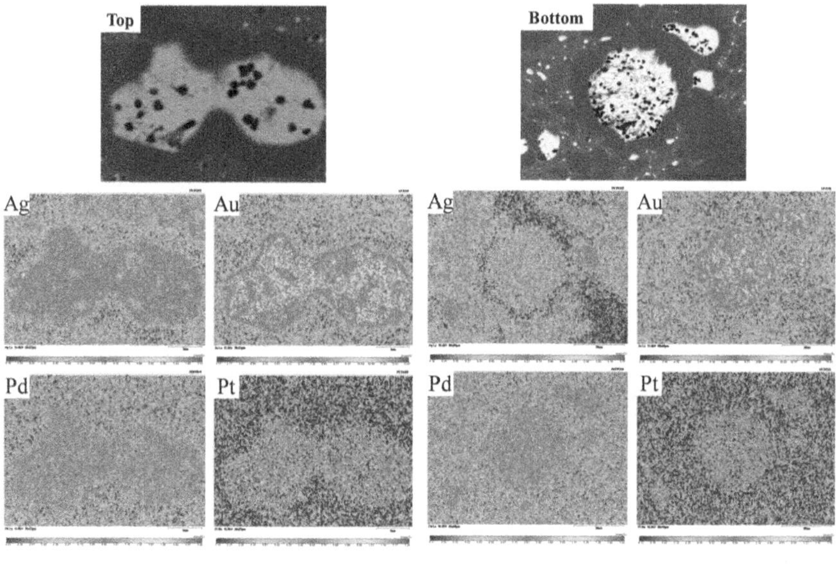

Fig. 4 EPMA mapping of precious metals in the top and bottom of Cu-Zn alloy

Decreasing the cooling rate may cause the precipitation of metallic elements in the alloy. Therefore, the centrifugal solidification was carried out with a cooling rate of 5°C/min. Figure 5 shows the microstructure of common metals in the top and bottom of Cu-Zn alloy. As shown, the contents of Pb and Sn in the top of Cu-Zn alloy decreased, while the distribution of Pb and Sn in the bottom of Cu-Zn alloy. Sn is easy to combine with Pb, and its alloy density is larger than that of Cu. Therefore, they were concentrated in the bottom of Cu-Zn alloy along the centrifugal direction with the temperature decreased. At 5 °C/min, metallic elements

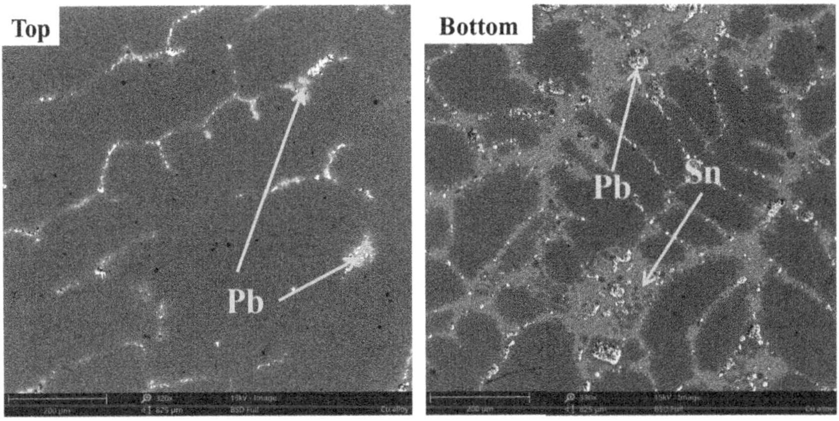

Fig. 5 SEM mapping of common metals in the top and bottom of Cu-Zn alloy

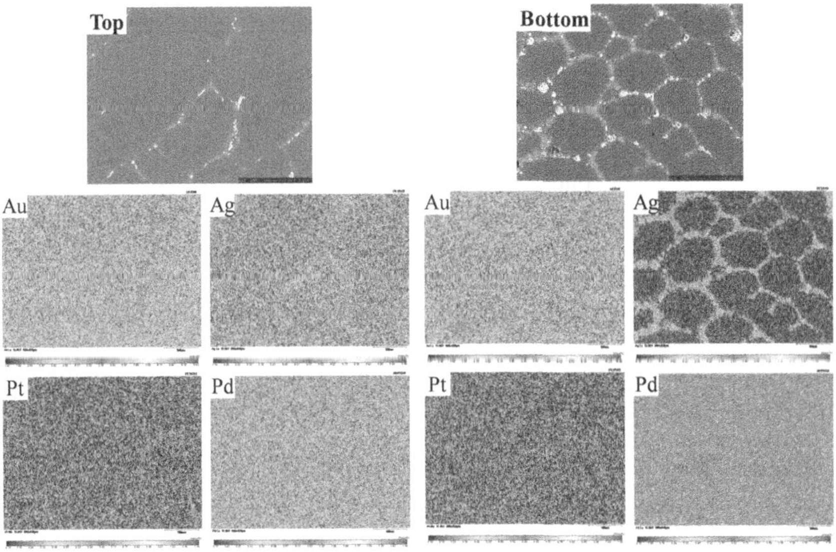

Fig. 6 EPMA mapping of precious metals in the top and bottom of Cu-Zn alloy

present an enrichment distribution in the alloy. Figure 6 shows the EPMA mapping of precious metals in the top and bottom of Cu-Zn alloy, respectively. The results showed that the precious metals are combined with Pb and Sn and distributed in the bottom of Cu-Zn alloy. It is indicated that low cooling rate was beneficial to the enrichment of metallic elements in the alloy under the super-gravity field.

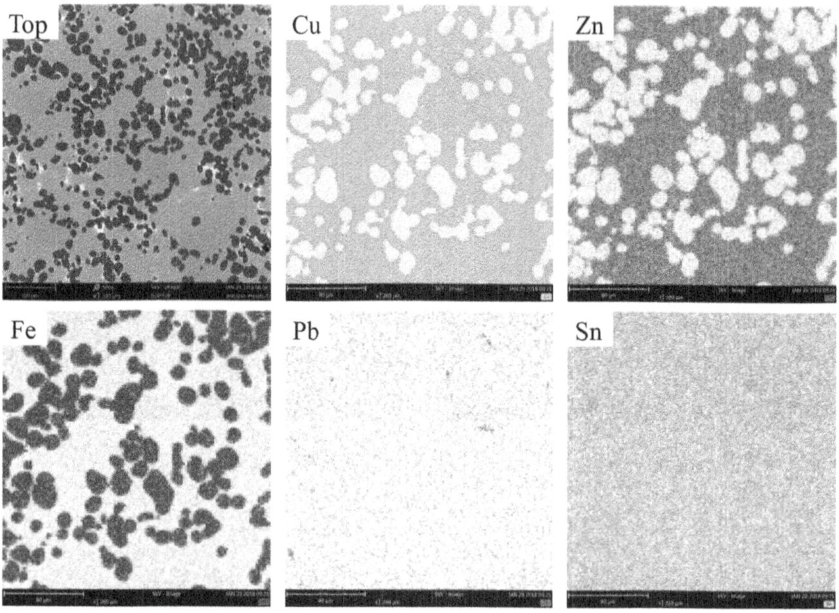

Fig. 7 SEM mapping of common metals in the top of copper alloy

The cooling rate was further reduced, and the centrifugation solidification was carried out at 2 °C/min. Figures 7 and 8 show the SEM mapping of common metals in the top and bottom of Cu-Zn alloy, respectively. As shown, the contents of Pb and Sn in the top of Cu-Zn alloy decreased obviously, and they were mostly distributed in the bottom of the Cu-Zn alloy. Fe, as a low-density metal, was concentrated in the top of Cu-based alloy. Figure 9 shows the EPMA mapping of precious metals in the top and bottom of Cu-Zn alloy, respectively. The results showed that the precious metals were distributed in the bottom of Cu-Zn alloy, while they were not distributed in the top of Cu-Zn alloy. At 2 °C/min, metallic elements present obvious segregation and enrichment in the Cu alloy.

The concentrated precious metals will be recycled by electrolytic refining of the Cu alloy and subsequent treatment by hydrometallurgy. Our findings will improve the complete super-gravity separation and concentration process of metals from PCBs and increase the utilization of secondary resources. Future studies will focus on scale-up experiments and the assessment of the economic costs and environmental impact of the proposed process.

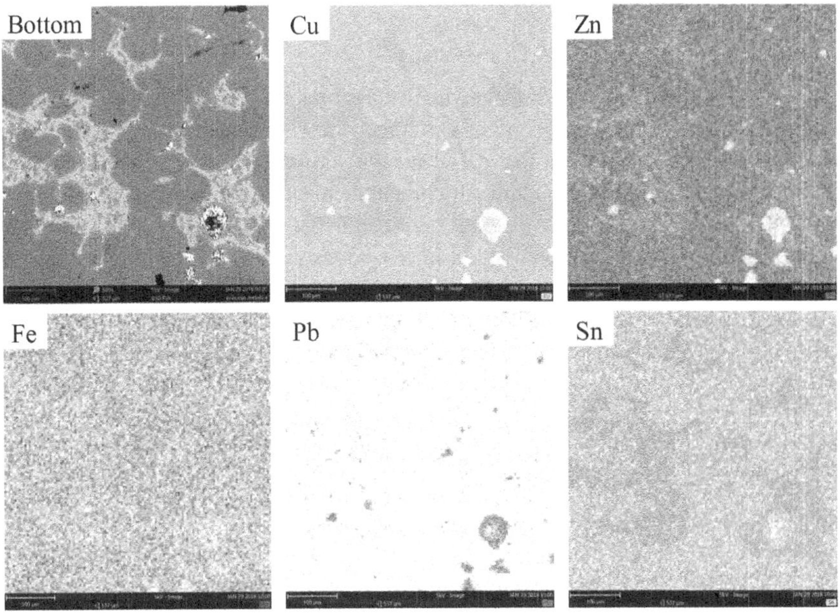

Fig. 8 SEM mapping of common metals in the bottom of copper alloy

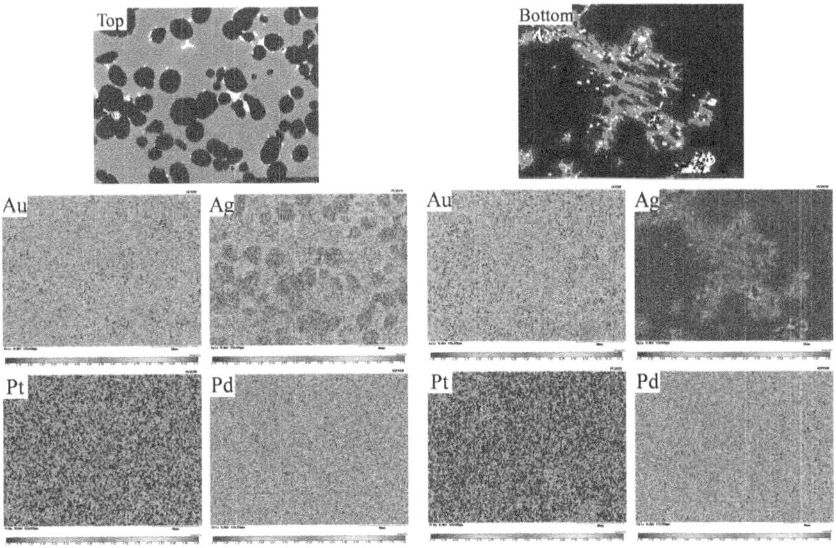

Fig. 9 EPMA mapping of precious metals in the top and bottom of Cu-Zn alloy

Conclusions

The Cu-Zn alloy was obtained from waste PCBs by a third super-gravity separation process. Super-gravity solidification can affect the enrichment and distribution of metallic elements in the Cu alloy. The results showed that low cooling rate was favourable for the distribution and enrichment of metallic elements in the Cu alloy. At 2 °C/min, a large amount of Fe was enriched in the top of the Cu alloy, while the majority of Pb and Sn were enriched in the bottom of the alloy. Precious metals were combined with Pb and Sn and enriched in the bottom of the Cu-Zn alloy. This paper provides a reference to control the enrichment and purification of metallic elements whose physical properties are different in alloys under a super-gravity field.

Acknowledgements This work was financially supported by the project of National Natural Science Foundation of China (Grant No. 51704022).

References

1. Cayumil R, Khanna R, Rajarao R, Mukherjee PS, Sahajwalla V (2015) Concentration of precious metals during their recovery from electronic waste. Waste Manage 57:121–130
2. Cui J, Roven HJ (2011) Electronic waste. Waste 281–296
3. Kogan V (2006) Recovery of precious metals from electronic scrap by hydrometallurgical processing. WO
4. Schluep M (2009) Recycling: from E-waste to resources
5. Park YJ, Fray DJ (2009) Recovery of high purity precious metals from printed circuit boards. J Hazard Mater 164:1152–1158
6. Alzate A, López ME, Serna C (2016) Recovery of gold from waste electrical and electronic equipment (WEEE) using ammonium persulfate. Waste Manage 57:113–120
7. Sahin M, Akcil A, Erust C, Altynbek S, Gahan CS, Tuncuk A (2015) A potential alternative for precious metal recovery from E-waste: iodine leaching. Sep Sci Technol 50:2587–2595
8. Li JY, Xu XL, Liu WQ (2012) Thiourea leaching gold and silver from the printed circuit boards of waste mobile phones. Waste Manage 32:1209–1212
9. Joda NN, Rashchi F (2012) Recovery of ultra fine grained silver and copper from PC board scraps. Sep Purif Technol 92:36–42
10. Fogarasi S, Imre-Lucaci F, Imre-Lucaci Á, Ilea P (2014) Copper recovery and gold enrichment from waste printed circuit boards by mediated electrochemical oxidation. J Hazard Mater 273:215–221
11. Li JC, Guo ZC, Gao JT (2014) Isothermal enriching perovskite phase from CaO-TiO_2-SiO_2-Al_2O_3-MgO melt by super gravity. ISIJ Int 54:743–749
12. Li JC, Guo ZC (2014) Innovative methodology to enrich britholite (Ca3Ce2[(Si, P)O4]3F) phase from rare-earth-rich slag by super gravity. Metall Mater Trans B 45:1272–1280
13. Rahimipour MR, Sobhani M (2013) Evaluation of centrifugal casting process parameters for in situ fabricated functionally gradient Fe-TiC composite. Metall Mater Trans B 44:1120–1123
14. Xie Y, Liu C, Zhai Y, Wang K, Ling X (2009) Centrifugal casting processes of manufacturing in situ functionally gradient composite materials of Al-19Si-5 Mg alloy. Rare Met 28:405–411

15. Cambiella A, Benito JM, Pazos C, Coca J (2006) Centrifugal separation efficiency in the treatment of waste emulsified oils. Chem Eng Res Des 84:69–76
16. Meng L, Wang Z, Zhong YW, Guo L, Gao JT, Chen KY, Cheng HJ, Guo ZC (2017) Supergravity separation for recovering metals from waste printed circuit boards. Chem Eng J 326:540–550
17. Li JW, Guo ZC, Tang HQ, Wang Z, Sun ST (2012) Si purification by solidification of Al–Si melt with super gravity. Trans Nonferrous Met Soc China 22:958–963
18. Zhao LX, Guo ZC, Wang Z, Wang MY (2010) Removal of low-content impurities from Al by super-gravity. Metall Mater Trans B 41:505–508
19. Meng L, Gao JT, Zhong YW, Wang Z, Chen KY, Guo ZC (2018) Supergravity separation for recovering Pb and Sn from electronic waste. Sep Purif Technol 191:375–383
20. Meng L, Zhong YW, Wang Z, Chen KY, Guo ZC (2018) High-temperature centrifugal separation of Cu from waste printed circuit boards. J Clean Prod 199:831–839

Effect of H_2/CO Ratio on Gas Consumption and Energy Utilization Rate of Gas-Based Direct Reduction Process

C. Y. Xu, A. Y. Zheng, J. L. Zhang, R. R. Wang, Y. Li, Y. Z. Wang and Z. J. Liu

Abstract Direct reduced iron (DRI) is an indispensable raw material for producing high-quality steel and an alternative to scrap. The Midrex process and the HYL process are the main gas-based direct reduction processes for producing DRI. In this paper, the theoretical analysis of the thermal mass balance in the shaft furnace under different H_2/CO ratios is carried out. The reduction of gas consumption of different H_2/CO ratios was calculated by the formulas. These calculated values are fitted with curves to obtain the relationship between the reducing gas consumption and the H_2/CO ratio. Effects of reducing temperatures on the consumption of reducing gas and the influence of H_2/CO ratio on the energy utilization rate were also investigated. The calculation results show that the consumption of reducing gas decreases with increasing H_2/CO ratio and temperature, and the energy utilization rate of the shaft furnace presents the same trend.

Keywords Gas-based shaft furnace · Direct reduction · Gas consumption · Heat balance · Mass balance

Introduction

In order to get rid of the fetters of coking coal resource shortage to the development, adapt to the increasing requirements of environmental protection and reduce the energy consumption of steel production, non-blast furnace ironmaking technology, which uses non-coking coal as energy source, has become a research hotspot in the

---

C. Y. Xu · A. Y. Zheng · J. L. Zhang · R. R. Wang · Y. Li
Y. Z. Wang · Z. J. Liu (✉)
School of Metallurgical and Ecological Engineering,
University of Science and Technology Beijing, Beijing 100083, China
e-mail: liuzhengjian@ustb.edu.cn

J. L. Zhang
School of Chemical Engineering, The University of Queensland,
St Lucia, QLD 4072, Australia

© The Minerals, Metals & Materials Society 2019
T. Jiang et al. (eds.), *10th International Symposium on High-Temperature Metallurgical Processing*, The Minerals, Metals & Materials Series,
https://doi.org/10.1007/978-3-030-05955-2_60

steel industry. Direct reduction, which is the major part of non-blast furnace ironmaking, has also become a focus on iron and steel metallurgy.

According to the data published by Midrex Technologies, Inc, worldwide production of direct reduced iron was 72.76 million tons in 2016, among which the direct reduction iron produced by Midrex and HYL-III/Energiron process accounts for 64.8 and 17.4%, respectively [1]. Compared with Midrex process, HYL process has the following differences: (1) Midrex process adopts nickel as catalyst, while HYL process adopts H_2O; (2) Midrex process has strict requirements on sulphur content of raw materials, while high sulphur ore can be used in HYL process; (3) H_2/CO in the reducing gas of HYL process is higher than that of Midrex process, and operating temperature of HYL process is also high because the reduction of H_2 is endothermic process; (4) HYL process adopts high-pressure operation, and Midrex process is normal pressure.

The calculation of energy balance is an important part of different ironmaking processes, which has been studied by many investigators [2, 3]. On the basis of energy balance calculation, Qin et al. discussed the influence of various factors on the temperature of the top gas by the calculation of heat balance [4]. Results show that the temperature of the top gas increases with the consumption, CO ratio and the inlet temperature of reducing gas. Wang et al. discussed the influence of a single factor on the temperature of top gas on the basis of energy balance [5]. Results show that the temperature of the top gas decreases with the increase of the iron temperature and the heat loss of the furnace wall. Top temperature of the furnace increases with the amount of carburizing in DRI.

In this paper, the production conditions of DRI under two typical shaft furnace processes (Midrex and HYL-III) are simulated, and the energy and mass balance of two processes are also calculated on the basis of the conservation law of energy. Moreover, energy utilization ratio of the two gas-based shaft furnace reduction processes is analysed under different H_2/CO ratios, which aims to understand the main difference between the two processes from the perspective of energy balance.

Actual Production Parameters

Both Midrex and HYL-III processes use pellets and natural lump ore as raw materials. In this paper, the pellets produced by a steel plant are used as main charge material for the shaft furnace, and the composition of pellets is shown in Table 1.

Midrex and HYL-III use different natural gas reforming technologies, and the corresponding composition and temperature of reducing gas of each process are

Table 1 Composition of charging pellets, wt%

TFe	Fe_2O_3	FeO	SiO_2	CaO	Al_2O_3	MgO	P	S	TiO_2	Other	Total
68.750	97.660	0.570	1.270	0.068	0.500	0.130	0.001	0.004	0.110	0.061	100

Table 2 Composition and temperature of reducing gas, wt%

No.	CO	CO_2	H_2	CH_4	N_2 + Ar	H_2O	Total
1	34.45	2.60	53.47	3.00	0.65	5.83	100.00
2	13.00	3.30	74.10	7.00	1.20	1.40	100.00

also different. The H_2/CO ratio and temperature of the reducing gas is low in MIDREX process, which is due to the reduction of iron oxide by CO would release a lot of heat, thus the temperature of reducing gas is decreased in order to prevent the sticking behavior of burden. The H_2/CO ratio and temperature of the reducing gas is relative higher in HYL-III process, which is due to the reduction of iron oxide by H_2 would absorb heat, thus the temperature of reducing gas should be increased in order to ensure adequate heat in the furnace and high production efficiency. In this paper, two kinds of reducing gas with different H_2/CO ratios were selected and the specific parameters are shown in Table 2, in which No. 1 simulated the reduction conditions of Midrex shaft furnace [6] and No. 2 simulated HYL-III shaft furnace [7]. The temperature of the reducing gas of No. 1 and No. 2 is 1174 and 1203 K, respectively.

In actual production, the top gas of shaft furnace contains dust. Due to the small amount of dust brought in by reducing gas which can be neglected, this paper proposed that the furnace dust is 30 kg/tHM, and all of them come from pellets, so the composition of the dust is the same with pellets and is shown in Table 3.

DRI is produced by reducing iron ore at a temperature below the melting temperature, and there is no slagging reaction in the smelting process; thus, phase composition of gangue components in the iron ore did not change much before and after the reaction. Also, as the raw material for high-quality steel, the sulphur content in the direct reduction iron should be less than 0.02%. Moreover, the reduction process is accompanied by carburizing reaction, so the carbon element is also contained in the DRI. Combining theory and actual production, the carbon content of DRI is proposed as 2%. On the basis of the above conclusions, the main composition of the DRI is shown in Table 4.

Mass and Energy Balance

Mass Balance

Calculation of Amount of Charge Material

This calculation is based on the production of 1t DRI.

According to the iron balance and the consideration of dust discharging amount, the mass of charging pellets could be calculated by Eqs. (1) and (2):

Table 3 Composition of the dust in the furnace

Components	TFe	Fe_2O_3	FeO	SiO_2	CaO	Al_2O_3	MgO	P	S	TiO_2	Others	Total
Proportion (%)	68.75	97.66	0.57	1.27	0.07	0.50	0.13	0.001	0.004	0.11	0.06	100.37
Mass (kg)	950.63	1350.37	7.88	17.56	0.94	6.91	1.80	0.01	0.06	1.52	0.84	1382.73

Table 4 Main composition of DRI, wt%

Components	TFe	MFe	P	S	C	Metallization
Proportion	93.00	84.00	0.02	0.02	2.00	90.32

$$G_{\rm pe} \times (TFe\%)_{\rm pe} = G_{\rm DRI} \times (TFe\%)_{\rm DRI} \quad (1)$$

$$G_{\rm dust} \times (TFe\%)_{\rm dust} = (G_{\sum \rm pe} - G_{\rm pe}) \times (TFe\%)_{\rm pe} \quad (2)$$

where $G_{\sum \rm pe}$, $G_{\rm pe}$, $G_{\rm DRI}$ and $G_{\rm dust}$ are the mass of total pellet, pellet, DRI and dust, respectively; $(TFe\%)_{\rm pe}$, $(TFe\%)_{\rm DRI}$ and $(TFe\%)_{\rm dust}$ are the percentage of TFe in pellets, DRI and dust, respectively. By inserting the data in Tables 1, 3 and 4 into Eqs. (1) and (2), the required amounts of total pellet could be calculated out and the result is 1382.73 kg.

Calculation of the Reducing Gas Volume

The reaction in the shaft furnace is more complicated under high temperature, which includes the reduction reaction of iron oxide, carburizing reaction, carbon emission reaction, water–gas conversion reaction and methane conversion reaction, as shown in Eqs. (3)–(8).

$$3Fe + 2CO = Fe_3C + CO_2 \quad (3)$$

$$3Fe + CH_4 = Fe_3C + 2H_2 \quad (4)$$

$$2CO = C + CO_2 \quad (5)$$

$$CO + H_2O = CO_2 + H_2 \quad (6)$$

$$CH_4 + CO_2 = 2H_2 + 2CO \quad (7)$$

$$CH_4 + H_2O = 3H_2 + CO \quad (8)$$

The temperature in shaft furnace is continuously increasing from top to bottom, and the ore suffered different reduction processes with the increase of temperature. Studies have shown that no matter what reducing agent was used, the iron oxides were changed from high-price oxides to low-price oxides, and the order is as follows: T > 570°C, $Fe_2O_3 \rightarrow Fe_3O_4 \rightarrow FeO \rightarrow Fe$; T < 570 °C, $Fe_2O_3 \rightarrow Fe_3O_4 \rightarrow Fe$. The main role of the reducing gas is to reduce the iron oxide as a reducing agent and secondly is to supply heat. In addition, in the shaft furnace reduction process, the reduction of each stage is carried out in series, and the reduction from FeO to Fe has the highest demand for the reducing atmosphere.

Therefore, the amount of reducing gas used in this stage is the theoretical amount of reducing gas required for reducing iron ore, as given in Eq. (9) [5, 8].

$$\begin{aligned} V_{\text{theory}} &= \frac{V_{H_2/CO(FeO\to Fe)} + V_{CO(\text{carburizing})} + V_{H_2+CO(\text{methane conversion})}}{\eta} \\ &\quad \times \frac{1}{(H_2\%)_{\text{reducing gas}} + (CO\%)_{\text{reducing gas}}} \\ &= \frac{400(MFe\%)_{DRI} + 3730(C\%)_{DRI} + 4V_{\text{reducing gas}}\left[(CH_4\%)_{\text{top gas}} - (CH_4\%)_{\text{reducing gas}}\right]}{\frac{(H_2\%)_{\text{reducing gas}}\times KH_2}{1+KH_2} + \frac{(CO\%)_{\text{reducing gas}}\times K_{CO}}{1+K_{CO}} - \left[\frac{(H_2O\%)_{\text{reducing gas}}}{1+KH_2} + \frac{(CO2\%)_{\text{reducing gas}}}{1+K_{CO}}\right]} \\ &\quad \times \frac{1}{(H_2\%)_{\text{reducing gas}} + (CO\%)_{\text{reducing gas}}} \end{aligned} \tag{9}$$

where K_{H2} and K_{CO} represent the reaction equilibrium constants of the third stage under the atmosphere of H_2 and CO, respectively; η represents the comprehensive utilization rate of the mixed reducing gas in the third stage of reduction. Because the CH_4 content in the top gas and the reducing gas are small and the change is not large, the influence of methane conversion on the gas utilization rate could be ignored in Eq (9).

According to the references, value of the equilibrium constant K of the two reactions is as follows [9]:

$$K_{CO} = \exp(-2.35 + \frac{1942.51}{T}) \tag{10}$$

$$K_{H_2} = \exp(-1.943 + \frac{2818.14}{T}) \tag{11}$$

As the equilibrium constant is related to the temperature, the demand for the theoretical amount of reducing gas is affected by the reduction temperature. By inserting the data in Table 2 into Eqs. (9), (10) and (11), the actual demand of reducing gas for the Midrex and HYL-III processes is calculated to be 1700 and 1605 m^3, respectively (it is usually 1.05–1.2 times of the theoretical value [5, 10]).

Calculation of the Composition and Content of Top Gas

Based on the mass conservation of C and H before and after the reduction, the volumes of CO, CO_2, H_2 and H_2O in the top gas are as follows:

$$V_{H_2(\text{top gas})} = V_{H_2(\text{reducing gas})} - V_{H_2(Fe_2O_3\to Fe)} - V_{H_2(Fe_2O_3\to FeO)} \tag{12}$$

$$V_{H_2O(\text{top gas})} = V_{H_2O(\text{reducing gas})} - V_{H_2O(Fe_2O_3 \to Fe)} - V_{H_2O(Fe_2O_3 \to FeO)} - V_{H_2O(\sum pe)} \quad (13)$$

$$V_{CO(\text{topgas})} = V_{CO(\text{reducing gas})} - V_{CO(Fe_2O_3 \to Fe)} - V_{CO(Fe_2O_3 \to FeO)} - V_{CO(\text{methane conversion})} \quad (14)$$

$$V_{CO_2(\text{topgas})} = V_{CO_2(\text{reducing gas})} - V_{CO_2(Fe_2O_3 \to Fe)} - V_{CO_2(Fe_2O_3 \to FeO)} - V_{CO_2(\text{methane conversion})} \quad (15)$$

According to the mass conservation of sulphur element, after the desulfurization reaction, volume of H_2S in top gas is as follows:

$$V_{H_2S(\text{top gas})} = \frac{22.4 \times G_{S(\text{topgas})}}{32} = \frac{22.4 \times \left[G_{S(\sum pe)} + G_{S(\text{reducing gas})} - G_{S(DRI)} - G_{S(\text{dust})}\right]}{32} \quad (16)$$

N_2 and other gases do not take part in the chemical reaction, and thus the volume remains unchanged before and after the reaction:

$$V_{N_2+Ar(\text{reducing gas})} = V_{N_2+Ar(\text{topgas})} \quad (17)$$

Take the data of Table 2 into Eqs. (12)–(17), the volume and content of each component of top gas of the Midrex and HYL-III shaft furnace could be calculated, and the results are shown in Tables 5 and 6, respectively.

Mass Calculation of Each Component in DRI

Since the smelting of DRI is carried out at a lower temperature and there is no slagging reaction, the gangue components in the ore almost unchanged during the

Table 5 Composition of top gas of Midrex shaft furnace

Components	CO	CO_2	H_2	CH_4	N_2 + Ar	H_2O	H_2S	Total
Volume (m^3/t)	327.28	273.58	580.24	51.00	11.05	427.86	1.47	1672.48
Proportion (%)	19.57	16.36	34.69	3.05	0.66	25.58	0.09	100.00

Table 6 Composition of top gas of HYL-III shaft furnace

Components	CO	CO_2	H_2	CH_4	N_2 + Ar	H_2O	Total
Volume (m^3/t)	67.82	180.97	759.17	112.35	19.26	452.60	1592.17
Proportion (%)	4.26	11.37	47.68	7.06	1.21	28.43	100.00

smelting process. The content of each component in DRI could be calculated according to the material conservation, as shown in Eqs. (18)–(24).

$$G_{\mathrm{TFe}} = (TFe\%) \times G_{\mathrm{DRI}} \tag{18}$$

$$G_{M\mathrm{Fe}} = (MFe\%) \times G_{\mathrm{DRI}} \tag{19}$$

$$(FeO\%) = \frac{[(TFe\%) - (MFe\%)] \times 72}{56} \tag{20}$$

$$G_{\mathrm{FeO}} = (FeO\%) \times G_{\mathrm{DRI}} \tag{21}$$

$$G_{\mathrm{C}} = (C\%) \times G_{\mathrm{DRI}} \tag{22}$$

$$G_{\mathrm{SiO_2}} = G_{\mathrm{SiO_2(pe)}} - G_{\mathrm{SiO_2}(dust)} \tag{23}$$

$$G_{\mathrm{S}} = G_{\mathrm{S(pe)}} + G_{\mathrm{S(reducing\ gas)}} - G_{\mathrm{S(dust)}} - G_{\mathrm{S(topgas)}} \tag{24}$$

Based on the above calculation process and results, the mass balance of the Midrex and HYL-III process is shown in Tables 7 and 8, respectively.

Heat Balance

Calculation of shaft furnace heat balance is to study the energy budget during smelting process, which is based on the conservation law of energy. Heat balance

Table 7 Mass balance for the Midrex furnace, kg

Input item		Output item	
Total pellet	1382.73	DRI	1000.00
Reducing gas (dry)	950.28	Top gas (dry)	1014.34
Moisture in reducing gas	79.64	Moisture in top gas	343.82
		Dust	30.00
Total	2412.65	Total	2388.16

Table 8 Mass balance for the HYL-III furnace, kg

Input item		Output item	
Total pellet	1382.73	DRI	1000.00
Reducing gas (dry)	575.36	Top gas (dry)	532.10
Moisture in reducing gas	18.06	Moisture in top gas	363.70
		Dust	30.00
Total	1976.15	Total	1925.80

calculation methods include the whole furnace heat balance and regional heat balance. The first method calculates the amount of heat generated and consumed according to Hess law of thermochemistry and is based on the initial shape as well as the final shape of the charging material in furnace, but it neglects the actual chemical reaction in the shaft furnace. Therefore, the second thermal equilibrium calculation method, which could consider water–gas shift reaction and carburization reaction and better reflect the essence of heat exchange in shaft furnace, is used in this paper. Based on the results of material balance calculation, this paper takes 298 K as the reference temperature and calculates the energy budget of shaft furnace.

Calculation of Heat Input

The charging temperature of pellets is assumed as 298 K in this paper. The relationship between the specific heat and the temperature of each component in pellets could be obtained by references. The amount of each component in the pellets can be obtained from Table 1. The physical heat brought by the pellets can be derived from Eq. (25).

$$\begin{aligned} Q_{\sum \mathrm{pe}} &= Q_{\mathrm{Fe_2O_3\,brought}} + Q_{\mathrm{FeO\,brought}} + Q_{\mathrm{SiO_2brought}} + Q_{\mathrm{CaObrought}} + Q_{\mathrm{Al_2O_3\,brought}} \\ &\quad + Q_{\mathrm{MgO\,brought}} + Q_{\mathrm{MnO\,brought}} + Q_{\mathrm{H_2O\,brought}} + Q_{\mathrm{TiO_2\,brought}} \\ &= \int_{273}^{298} \left[\begin{array}{l} \mathrm{n}_{\mathrm{Fe_2O_3}} \cdot C_{p(\mathrm{Fe_2O_3})} + \mathrm{n}_{\mathrm{FeO}} \cdot C_{p(\mathrm{FeO})} + \mathrm{n}_{\mathrm{SiO_2}} \cdot C_{p(\mathrm{SiO_2})} + \mathrm{n}_{\mathrm{CaO}} \cdot C_{p(\mathrm{CaO})} + \mathrm{n}_{\mathrm{Al_2O_3}} \\ \cdot C_{p(\mathrm{Al_2O_3})} + \mathrm{n}_{\mathrm{MgO}} \cdot C_{p(\mathrm{MgO})} + \mathrm{n}_{\mathrm{MnO}} \cdot C_{p(\mathrm{MnO})} + \mathrm{n}_{\mathrm{H_2O}} \cdot Cp(\mathrm{H_2O}) + \mathrm{n}_{\mathrm{TiO_2}} \cdot C_{p(\mathrm{TiO_2})} \end{array} \right] \cdot dt \end{aligned} \tag{25}$$

where $Q_{\sum \mathrm{pe}}$ represents the total physical heat brought by pellets, kJ; Q_{brought} represents the physical heat brought by each component in the pellets, kJ; C_p represents the specific heat of each component in the pellet, kJ mol^{-1} K^{-1}; n represents the molar mass of each component in the pellet, mol. The physical heat of the Midrex and HYL-III shaft furnace brought by pellets can be calculated, and the result is 0 kJ. (This paper takes 298 K as the reference temperature, and the temperature of pellets is 298 K too.)

The relationship between the molar heat capacity at constant pressure and temperature of the components in the reducing gas can be obtained by references. The amount of each component in the reducing gas can be determined from Table 2. The physical heat introduced by the reducing gas is given by Eq. (26).

$$\begin{aligned} Q_{\mathrm{reducing\,gas}} &= Q_{\mathrm{H_2brought}} + Q_{\mathrm{N_2brought}} + Q_{\mathrm{H_2Obrought}} + Q_{\mathrm{CObrought}} + Q_{\mathrm{CO_2brought}} + Q_{\mathrm{CH_4\,brought}} + Q_{\mathrm{H_2Sbrought}} \\ &= \int_{273}^{T_{\mathrm{reducing\,gas}}} \left[\begin{array}{l} \mathrm{n}_{\mathrm{H_2}} \cdot C_{p(\mathrm{H_2})} + \mathrm{n}_{\mathrm{N_2}} \cdot C_{p(\mathrm{N_2})} + \mathrm{n}_{\mathrm{H_2O}} \cdot C_{p(\mathrm{H_2O})} + \mathrm{n}_{\mathrm{CO}} \cdot C_{p(\mathrm{CO})} \\ + \mathrm{n}_{\mathrm{CO_2}} \cdot C_{p(\mathrm{CO_2})} + \mathrm{n}_{\mathrm{CH_4}} \cdot C_{p(\mathrm{CH_4})} + \mathrm{n}_{\mathrm{H_2S}} \cdot C_{p(\mathrm{H_2S})} \end{array} \right] \cdot dt \end{aligned} \tag{26}$$

where $Q_{\text{reducing gas}}$ represents the physical heat brought by the reducing gas, kJ; Q_{brought} represents the physical heat introduced by each substance in the reducing gas, kJ; C_p represents the specific heat of each component in the reducing gas, J mol^{-1} K^{-1}; $T_{\text{reducing gas}}$ represents the temperature of reducing gas, K; n represents the molar mass of each component in the reducing gas, mol. By inserting the data from Table 2, the physical heat introduced by reducing gas could be calculated. The results of calculation show that the physical heat introduced into the Midrex and HYL-III shaft furnaces by the reducing gas is 2,334,454.9 and 1,930,108.5 kJ, respectively.

The carbon content of DRI assumed in calculation is 2%, and the needed amount of CO and the carburizing reaction energy consumption are calculated by Eq. (27) and (28) [11]:

$$V_{\text{CO carburizing}} = \frac{(C\%)_{DRI} \times G_{DRI} \times 2 \times 22.4}{12} \tag{27}$$

$$Q_{\text{carburizing}} = \mathrm{n_{CO}} \times \Delta \mathrm{H}_{\mathrm{T}}^{\Theta} \tag{28}$$

where $V_{\text{CO carburizing}}$ represents the volume of CO consumed by the carburizing reaction, m^3; $(C\%)_{\text{DRI}}$ represents the carburizing rate, %; G_{DRI} represents the quality of DRI, kg; $Q_{\text{carburizing}}$ represents the heat release by the carburizing reaction, kg; n_{CO} represents the molar mass of CO, mol; ΔH represents the thermal effect of the standard reaction, kJ/mol.

Calculation of Heat Expenditure

The heat transfer and reduction reaction are completed during the countercurrent contact process of reducing gas and pellets in the reduction section. Although the temperature of reducing gas is different under two conditions, while there is a large amount of H_2 in the reducing gas, a lot of heat is absorbed by pellets during the reduction process of the wustite. The physical heat brought out by DRI can be obtained by Eq. (29).

$$\begin{aligned} Q_{\text{DRI}} &= Q_{\text{Fe}_2\text{O}_3\text{ex}} + Q_{\text{SiO}_2\text{ex}} + Q_{\text{CaOex}} + Q_{\text{Al}_2\text{O}_3\text{ex}} \\ &\quad + Q_{\text{MgO ex}} + Q_{\text{MnOex}} + Q_{\text{H}_2\text{O ex}} + Q_{\text{TiO}_2\text{ex}} \\ &= \int_{273}^{\mathrm{T_{topgas}}} \left[\begin{aligned} &\mathrm{n_{Fe_2O_3}} \cdot C_{p(\mathrm{Fe_2O_3})} + \mathrm{n_{SiO_2}} \cdot C_{p(\mathrm{SiO_2})} + \mathrm{n_{CaO}} \cdot C_{p(\mathrm{CaO})} + \mathrm{n_{Al_2O_3}} \cdot C_{p(\mathrm{Al_2O_3})} \\ &+ \mathrm{n_{MgO}} \cdot C_{p(\mathrm{MgO})} + \mathrm{n_{MnO}} \cdot C_{p(\mathrm{MnO})} + \mathrm{n_{H_2O}} \cdot C_{p(\mathrm{H2O})} + \mathrm{n_{TiO_2}} \cdot C_{p(\mathrm{TiO2})} \end{aligned} \right] \cdot dt \end{aligned} \tag{29}$$

where Q_{DRI} represents the physical heat taken away by DRI, kJ; Q_{ex} represents the amount of heat taken away by each component of DRI, kJ; C_p represents the specific heat of each component in DRI, J mol^{-1} K^{-1}; $T_{\text{top gas}}$ represents the discharging temperature of DRI from reduction section, K; n represents the molar mass of each component in DRI, mol.

The physical heat taken away by the top gas can be calculated by Eq. (30).

$$Q_{\text{topgas}} = Q_{H_2\text{ex}} + Q_{N_2\text{ex}} + Q_{H_2O\text{ex}} + Q_{CO\text{ex}} + Q_{CO_2\text{ex}} + Q_{CH_4\text{ex}} + Q_{H_2S\text{ex}}$$
$$= \int_{273}^{T_{\text{topgas}}} \begin{bmatrix} n_{H_2} \cdot C_{p(H_2)} + n_{N_2} \cdot C_{p(N_2)} + n_{H_2O} \cdot C_{p(H_2O)} + n_{CO} \cdot C_{p(CO)} \\ + \ n_{CO_2} \cdot C_{p(CO_2)} + n_{CH_4} \cdot C_{p(CH_4)} + n_{H_2S} \cdot C_{p(H_2S)} \end{bmatrix} \cdot dt \tag{30}$$

where $Q_{\text{top gas}}$ represents the physical heat taken away by the top gas, kJ; Q_{ex} represents the physical heat taken away by each component of the top gas, kJ; C_p represents the specific heat of each component of the top gas, J mol^{-1} K^{-1}; n represents the molar mass of each component in the top gas, mol.

The heat consumption of iron oxide reduction reaction includes: the reduction of iron oxides by H_2 and CO produces metallic iron and residual iron oxide, which consumes heat. The heat consumption of each process can be calculated by Eq. (31).

$$\begin{aligned} Q_{Fe_2O_3 \rightarrow Fe/FeO} = {} & n_{H_2(Fe_2O_3 \rightarrow Fe)} \times \Delta H_T^{\Theta} + n_{CO(Fe_2O_3 \rightarrow Fe)} \times \Delta H_T^{\Theta} \\ & + n_{H_2(Fe_2O_3 \rightarrow FeO)} \times \Delta H_T^{\Theta} + n_{CO(Fe_2O_3 \rightarrow FeO)} \times \Delta H_T^{\Theta} \end{aligned} \tag{31}$$

where $Q_{Fe_2O_3-Fe/FeO}$ and $Q_{Fe2O3\text{-}Fe/FeO}$ represents the heat taken away by the reduction of iron oxide, kJ; n represents the molar mass by each reduction reaction, mol; ΔH represents the thermal effect of the standard reaction, kJ/mol.

The physical heat brought out by dust can be calculated by Eq. (32).

$$\begin{aligned} Q_{\text{dust}} = {} & Q_{Fe_2O_3\text{ex}} + Q_{SiO_2\text{ex}} + Q_{CaO\text{ex}} + Q_{Al_2O_3\text{ex}} \\ & + Q_{MgO\text{ex}} + Q_{MnO\text{ex}} + Q_{H_2O\text{ex}} + Q_{TiO_2\text{ex}} \\ = {} & \int_{273}^{T_{\text{topgas}}} \begin{bmatrix} n_{Fe_2O_3} \cdot C_{p(Fe_2O_3)} + n_{SiO_2} \cdot C_{p(SiO_2)} + n_{CaO} \cdot C_{p(CaO)} + n_{Al_2O_3} \cdot C_{p(Al_2O_3)} \\ + n_{MgO} \cdot C_{p(MgO)} + n_{MnO} \cdot C_{p(MnO)} + n_{H_2O} \cdot C_{p(H2O)} + n_{TiO_2} \cdot C_{p(TiO_2)} \end{bmatrix} \cdot dt \end{aligned} \tag{32}$$

where Q_{dust} represents the physical heat taken away by dust, kJ; Q_{ex} represents the physical heat taken away by each component in the dust, kJ; $T_{\text{top gas}}$ represents the temperature of top gas, K; C_p represents the specific heat of each component in the dust, J mol^{-1} K^{-1}; n represents the molar mass of each component in the dust, mol.

Finally, the heat quantity used for the heat loss should also be taken into account. The final heat balance results are shown in Tables 9 and 10.

Table 9 Heat balance for the Midrex shaft furnace

Source of heat			Expenditure of heat		
Item	Heat (kJ)	Percentage (%)	Item	Heat (kJ)	Percentage (%)
Physical heat of reducing gas	2,334,454.9	89.7	Enthalpy of top gas	1,344,475.0	51.7
Physical heat of total pellet	0.0	0.0	Physical heat of DRI	793,221.2	30.5
Heat released by the carburizing reaction	267,015.9	10.3	Reduction heat of iron oxide	207,673.1	8.0
			Physical heat of dust	14,309.3	0.6
			Heat loss	241,792.2	9.3
Total	2,601,470.8	100.0	Total	2,601,470.8	100.0

Table 10 Heat balance for the HYL-III shaft furnace

Source of heat	Expenditure of heat				
Item	Heat (kJ)	Percentage (%)	Item	Heat (kJ)	Percentage (%)
Physical heat of reducing gas	1,930,108.5	87.8	Enthalpy of top gas	1,342,689.3	51.2
Physical heat of total pellet	0.0	0.0	Physical heat of DRI	670,356.7	25.6
Heat released by the carburizing reaction	267,015.9	12.2	Reduction heat of iron oxide	364,765.8	13.9
			Physical heat of dust	14,983.4	0.6
			Heat loss	227,552.1	8.7
Total	2,197,124.4	100.0	Total	2,620,347.2	100.0

Results and Discussion

Reducing gas consumption and reducing gas utilization rate are two basic indicators to evaluate the energy consumption of gas-based reduction shaft furnace. Thus, decreasing the reducing gas consumption and improving the utilization rate are the keys to improve the energy utilization rate.

Effect of H_2/CO Ratio on the Gas Consumption

Gas consumption of two processes (Midrex and HYL-III) was calculated under the same indexes of DRI, and the influence of methane conversion reaction on reducing gas consumption was neglected. The effect of H_2/CO ratio on the gas consumption was studied by changing the H_2/CO ratio and maintaining other components in reducing gas unchanged. Theoretical reduction gas volume needed to produce 1t of DRI under different H_2/CO conditions is calculated, and curve fitting on the calculated values is performed by MATLAB, as shown in Fig. 1.

As can be seen from Fig. 1, the amount of theoretical reducing gas needed to produce 1t DRI decreases with the increase of H_2/CO ratio. That means when producing 1 t sponge, the theoretical reducing gas consumption of HYL-III process is less than that of the Midrex process. This is mainly due to the fact that under 1173 K, the utilization rate of H_2 is higher than that of CO, and thus the increase of H_2 ratio in the reducing gas would increase the total gas utilization rate and decrease the demand amount of reducing gas.

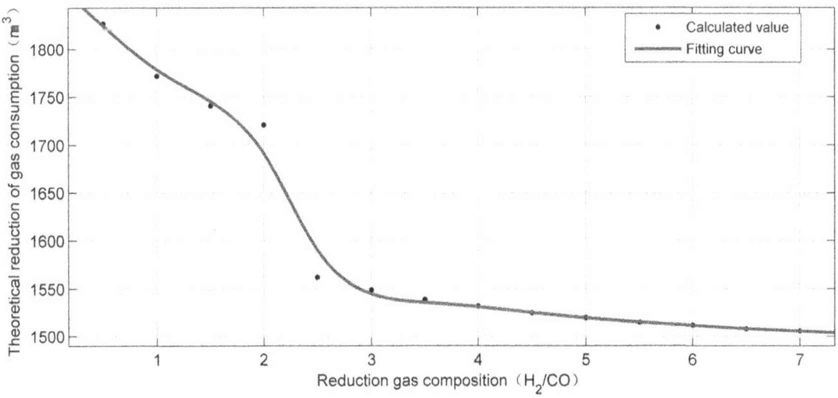

Fig. 1 Effect of reducing gas composition on the theoretical gas consumption

Effect of Reducing Temperature on the Gas Consumption

Theoretical gas consumption to produce 1 t DRI under different temperatures was calculated by changing H_2/CO ratio and reducing temperature, and results are shown in Fig. 2.

It can be seen from Fig. 2 that under ratios of $H_2/CO = 1.6$ (Midrex) and $H_2/CO = 5.7$(HYL-III), the theoretical reducing gas consumption decreases with the increase of temperature, and the greater the H_2/CO ratio is, the faster the reducing gas consumption decreases. When the temperature rises from 900 to 1000 °C, the theoretical gas consumption of Midrex ($H_2/CO = 1.6$) process decreases from 1723 m^3/DRI(t) to 1625 m^3/DRI(t), which would decrease by 98 m^3/DRI(t), while the value is 142 m^3/DRI(t) for HYL-III. That means the temperature has a bigger influence on the gas consumption of HYL-III than MIDREX, the main reason is that the reducing gas of HYL-III process contains more H_2, the reduction of iron oxides by which is an endothermic reaction, thus the gas utilization rate increases rapidly with the increase of temperature and obviously would decrease the reducing gas consumption.

Effect of H_2/CO Ratio on the Energy Utilization Rate

The energy of shaft furnace mainly comes from reducing gas, and the energy of reducing gas includes two parts: physical heat and chemical energy in which the physical heat is proportional to the temperature of the reducing gas and the

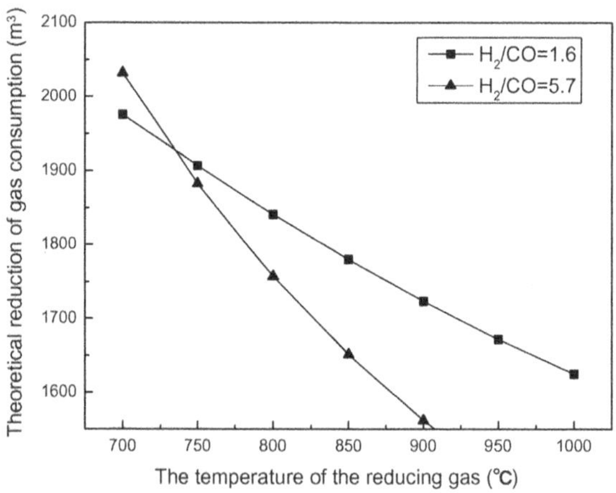

Fig. 2 Effect of reduction temperature on the theoretical gas consumption

chemical energy is mainly the heat released by the oxidation reaction of H_2 and CO in reducing gas. The utilization rate is mainly determined by H_2/CO ratio of the reducing gas. According to the physical heat utilization ratio and chemical energy utilization ratio, the total energy utilization rate of the shaft furnace could be calculated as follows [12]:

$$\eta_t = \frac{E_{ph}\eta_{ph} + E_{ch}\eta_{ch}}{E_{ph} + E_{ch}} \tag{33}$$

where η_t represents total energy utilization rate of shaft furnace,%; η_{ph} represents vertical furnace physical heat utilization,%; η_{ch} represents shaft furnace chemical heat utilization,%; E_{ph} represents reducing gas physical heat total, GJ/t; E_{ch} represents shaft furnace total energy utilization, GJ/t.

The total energy consumption distribution and its utilization in the Midrex and HYL-III processes can be calculated according to the mass balance and Eq. (33).

Figure 3 shows the energy consumption distribution of shaft furnace under $H_2/CO = 1.6$ (Midrex) and $H_2/CO = 5.7$ (HYL-III), which indicated that with the increase of H_2/CO ratio, the proportion of heat absorbed by the oxidation of H_2 increases. Apply the data to Eq. (27), and energy utilization rate of two processes can be calculated, which are 48.97 and 43.60%, respectively. Thus, the energy utilization rate of the gas-based reduction shaft furnace process decreases with the increase of H_2/CO ratio in the reducing gas.

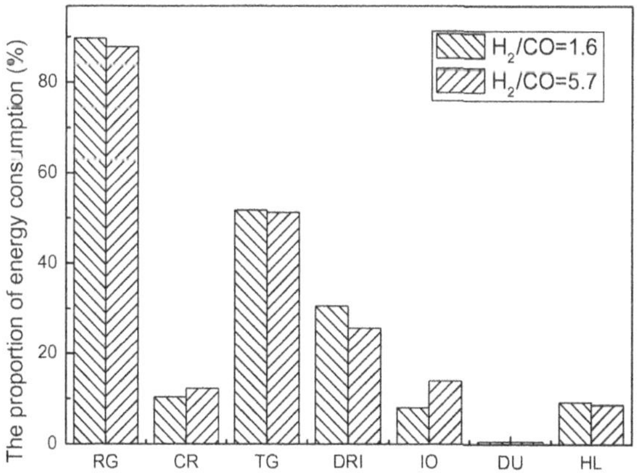

Fig. 3 Energy consumption distribution ratio of gas-based reduction shaft furnace. RG—physical heat of reducing gas; CR—heat released by the carburizing reaction; TG—enthalpy of top gas; DRI—physical heat of DRI; IO—reduction heat of iron oxide; DU—physical heat of dust; HL—heat loss

Conclusion

The theoretical analysis of the thermal mass balance in the shaft furnace reduction process under different H_2/CO ratios is carried out in this paper based on the method of regional heat balance calculation, the reducing gas consumption and energy utilization rate of each process are also evaluated, and the results are shown as follows:

(i) With the increase of the H_2/CO ratio, the demand of reducing gas for the production of 1t DRI in the gas-based reduction shaft furnace decreases. The higher the H_2/CO ratio, the higher the decrease in reducing gas consumption of the process is more significant. It can be seen that the reducing gas composition and the reduction temperature determine the maximum possible reducing gas utilization under the thermodynamic conditions.

(ii) The consumption of reducing gas decreases with the increase of temperature. With the increase of H_2/CO ratio, the effect of temperature on the consumption of reducing gas is more significant, which is due to the increasing utilization rate of reducing gas.

(iii) The energy utilization rate of shaft furnace decreases with the increase of H_2/CO ratio, which decreases by 5.37% when H_2/CO increases from 1.6 to 5.7. This is inversely proportional to the theoretical demand of reducing gas. The thermodynamic conditions can only solve the problem of limit; thus, it is necessary to rely on sufficient dynamic conditions to further make the actual reducing gas utilization close to this limit.

Acknowledgements The authors would like to express their thanks for the support by National Key R&D Program of China (2017YFB0304302-01).

References

1. Midrex Technologies, Inc. (2017) Available from: https://www.midrex.com/assets/user/news/MidrexStatsBook2016.pdf
2. You XM, Du T (1986) Energy Analysis and material balance of melt reduction process. In: China Metal Society Fourth Melt Reduction Conference
3. Yi LY (2013) Fundamental research on gas-based direct reduction of CO-H_2 mixtures of iron ore pellets. Central South University
4. Qin J (2008) Study on reduction characteristics and carbon behavior of DRI production in shaft furnace. Chongqing University
5. Wang ZC (2013) Basic research on direct reduction process of coal gas-air shaft furnace. Dongbei University
6. Zare Ghadi A, Valipour MS, Biglari M (2017) Int J Hydrogen Energy 42:103–118
7. Heinrich P (1991) Sintering and pelletizing, pp 30–34
8. Wu CB, Qin J, Gao Y et al (2008) The calculation and influencing factors of the minimum reduction gas amount of DRI in shaft furnace. In: National conference on metallurgical physics and chemistry

9. Huang XG (1981) Principles of steel and metallurgy. Metallurgical Industry Press
10. Lin ZQ (1977) The calculation method and application of reducing gas minimum requirement in the direct reduction process of shaft furnace. Ironmaking Steelmaking 3:91–97
11. Ye DL, Hu JH (2002) Thermodynamic data manual for inorganic substances. Metallurgical Industry Press
12. Fang J (2010) Non-blast furnace ironmaking process and theory. Metallurgical Industry Press

Effects of Particle Size of Coke on Iron Ore Sintering Process

Huaiying Ma, Wen Pan, Lei Liu, Zhidong Zhang and Chunlai Wang

Abstract In order to increase the combustion efficiency of coke breeze in sintering process, thermogravimetry-differential scanning calorimetry tests of coke breeze and sintering pot tests were carried out. The effects of particle size of coke on combustion and sintering process were examined. The surface air velocity and the exhaust gas composition of a sintering machine were also tested in a sintering strand. The results showed that the fine coke breeze burned quickly, but high ratio of fine particle in coke breeze was not beneficial for the improvement of the sinter quality. The low permeability of the front section of the sintering machine caused the low combustion efficiency of coke breeze which could be improved by increasing the coarse powder ratio of coke breeze.

Keywords Coke breeze · Particle size · Combustion efficiency Sintering

Introduction

Coke breeze is the main heat source in sintering process. Its combustion yields enough amounts of heat which melts the sinter mix. One of the factors that significantly affect the course of coke breeze combustion is its grain size [1–3]. It is known that fine-grained coke reduces permeability of the sinter mix and the use of coke breeze with particle size below 1 mm is avoided on the sintering strand [4, 5]. Most of the sintering plant in China controls the ratio of −3 mm coke breeze over 80% to get the best sintering performance. In this study, experiments were carried out to evaluate effects of particle size of coke on iron ore sintering and to obtain the proper proportion of 1–3 mm coke breeze.

H. Ma (✉) · W. Pan
Shougang Group Corporation, Research Institute of Technology, Beijing, China
e-mail: captainma@163.com

L. Liu · Z. Zhang · C. Wang
Shougang Qian'an Steel Company, Beijing, China

© The Minerals, Metals & Materials Society 2019
T. Jiang et al. (eds.), *10th International Symposium on High-Temperature Metallurgical Processing*, The Minerals, Metals & Materials Series,
https://doi.org/10.1007/978-3-030-05955-2_61

Sample and Experiment

Table 1 summarizes the chemical analyses of the component ores used in the tests. Coke breeze of −0.1, 0.1–1, 1–3 mm were tested by STA449C TG-DSC apparatus made by Netzsch, the air flow was 20 L/min with a heating rate of 20 °C/min. First, all the raw materials were weighed and mixed manually, the second manual mixing was carried after a certain amount of water was added. Then, the granulated mix was transferred into a drum for granulation for 5 min. After that, they were charged into a sinter pot (diameter: 250 mm, height of 500 mm) with 2 kg hearth materials which were 10–16 mm of sinter product. Underpressure of ignition and sintering were controlled at constant level of 6 and 12 kPa separately. The sintering was finished when the waste gas was cooled down to 200 °C from the maximum temperature. The sinter was crushed by a single toothed roller crusher and then subjected to a standard testing.

Size distribution of coke breeze in each sintering test was listed in Table 2. The first 4 tests were carried out using coke breeze of one range.

Table 1 Chemical compositions and blending ratio of sintering materials (mass%)

Materials	TFe	SiO_2	CaO	MgO	Al_2O_3	LOI	Ratio
Ore A	61.31	3.94	0.05	0.10	1.99	5.37	17.88
Ore B	65.46	6.75	0.31	0.65	0.26	−2.83	15.55
Ore D	57.24	4.95	0.07	0.05	1.34	10.11	3.89
Ore E	61.01	4.16	0.062	0.07	2.09	6.81	13.22
Ore F	64.52	4.38	0.00	0.00	0.83	1.62	3.89
Return fine	54.45	6.06	11.94	2.09	1.75	−0.65	30.32
Limestone		2.51	47.55	5.20	0.00	42	6.17
Dolomite		3.57	31.04	19.93	0.00	45	1.15
Lime		4.01	78.54	9.67	0.00	10	3.89
Coke breeze							4.05

Table 2 Size distribution of coke breeze in each test (mass%)

Tests	<1 mm	1–3 mm	3–5 mm	5–8 mm
Test 1	100			
Test 2		100		
Test 3			100	
Test 4				100
Test 5	45.6	38.0	16.4	
Test 6	45.6	28.0	26.4	
Test 7	45.6	22.8	21.5	10.1
Test 8	45.6	20.2	19.1	15.1

Results

TG-DSC Tests

Figure 1 shows the thermogravimetry-differential scanning calorimetry (TG-DSC) curves of coke breeze with different particle size. From the TG curves in the upper part of the figure, coke breeze size −0.1 mm showed the lowest initial weight loss temperature and maximum mass loss rate. The DSC curves showed that the −0.1 mm coke breeze had the highest exothermic rate below 700 °C, followed by 0.1–1 and 1–3 mm coke breeze. Above 700 °C the exothermic rate of −0.1 mm coke breeze became lower and showed a small endothermic peak.

Sintering Pot Tests

Effects of particle size of coke breeze on flame front speed (FFS) and productivity were shown in Fig. 2. Coke breeze of 1–3 and 3–5 mm showed a positive effect on increasing FFS and productivity. Figure 3 shows that yield was increased and coke consumption was decreased using coke breeze of the same size.

The maximum tumbler index (TI) and mean diameter have been observed in Fig. 4 when 1–3 mm coke breeze was used.

Figures 5 and 6 show the results for FFS, productivity, yield and fuel consumption. With the increasing ratio of 1–3 mm coke breeze, productivity, FFS and fuel consumption increased. Figure 7 shows the TI and mean diameter, respectively. In case of 22.8% coke breeze size 1–3 mm bypass, the maximum TI and mean diameter were reached.

Fig. 1 TG-DSC curves of coke in different size

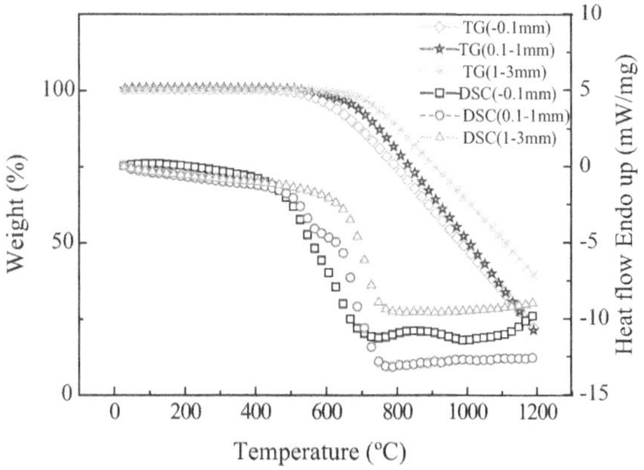

Fig. 2 Effect of coke size on FFS and productivity

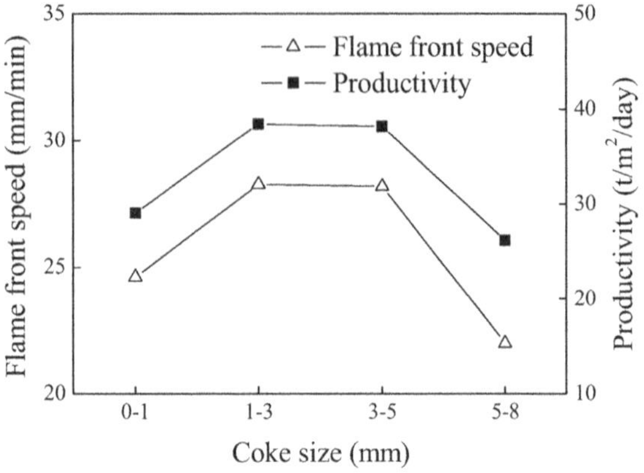

Fig. 3 Effect of coke size on yield and fuel consumption

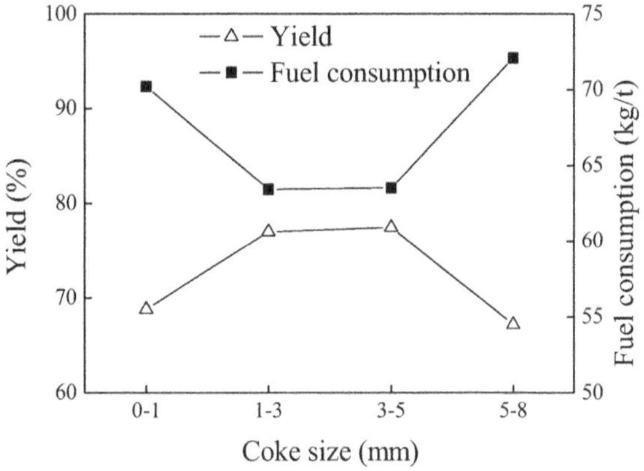

Fig. 4 Effect of coke size on TI and mean diameter of sinter

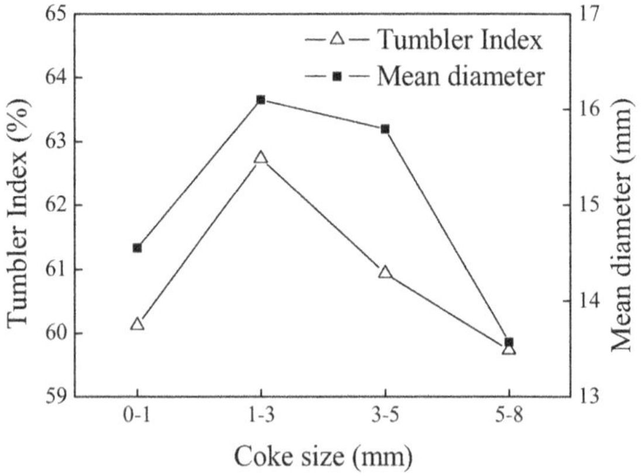

Fig. 5 Effect of ratio of 1–3 mm coke breeze on FFS and productivity

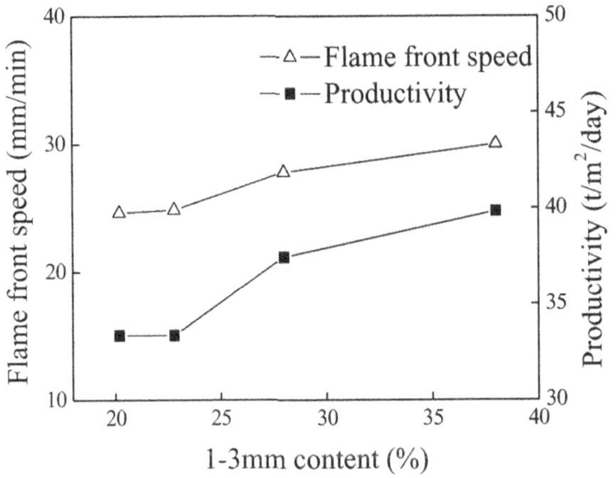

Fig. 6 Effect of ratio of 1–3 mm coke breeze on yield and fuel consumption

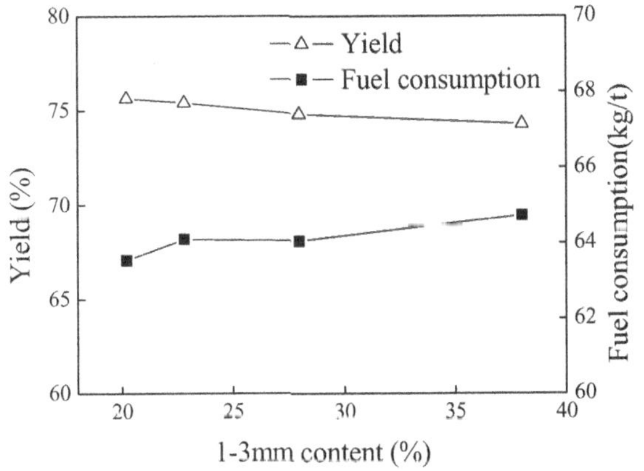

Fig. 7 Effect of ratio of 1–3 mm coke breeze on TI and mean diameter of sinter

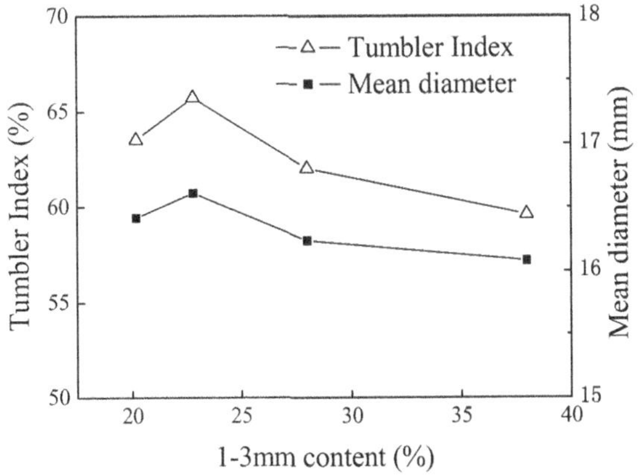

Application at a Commercial Plant

The wind speed above the sintering bed is shown in Fig. 8. Along the length direction of sintering strand, the wind speed firstly descended then ascended. It indicates that less oxygen flowed across the bed until the permeability was increased.

In order to improve the bed permeability, interval between the coke crusher rollers was increased. The increasing ratio of coke breeze size 1–3 mm with the decreasing ratio of coke breeze size −1 mm has been shown in Fig. 9. For convenience, the primary coke was named fine coke and coke with more 1–3 mm particles was called coarse coke.

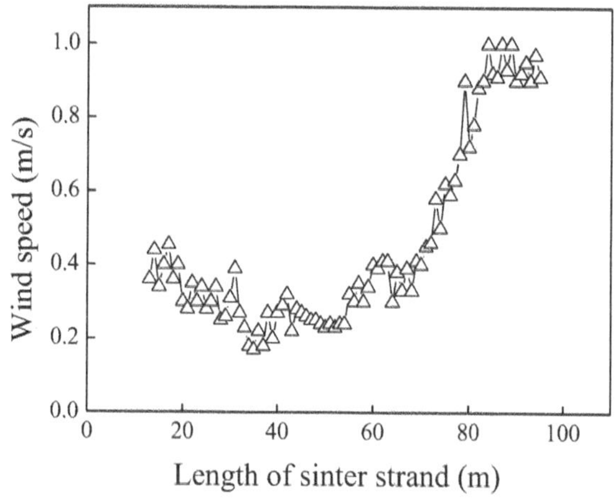

Fig. 8 Wind speed of the sintering bed surface

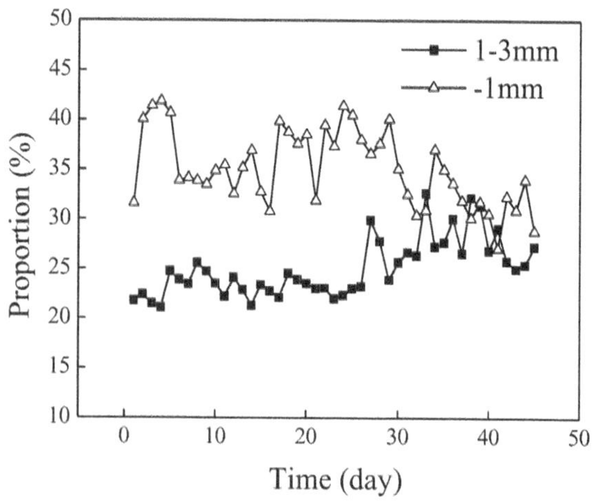

Fig. 9 Proportion of 1–3 mm particles and −1 mm particles in coke breeze

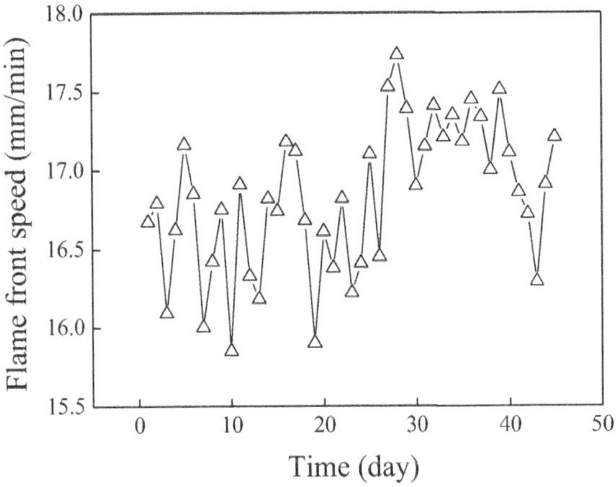

Fig. 10 FFS changes of the sintering strand

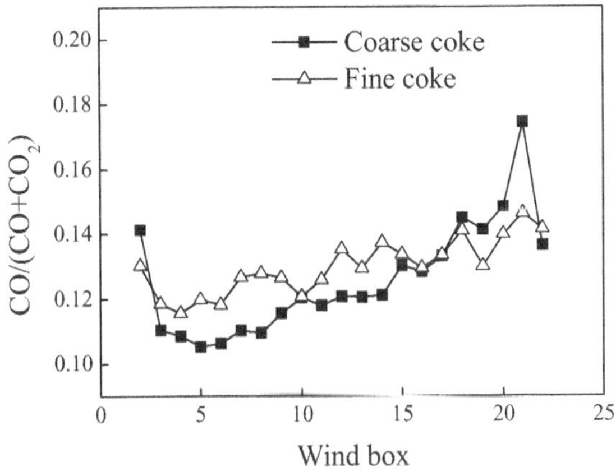

Fig. 11 $CO/(CO + CO_2)$ in each wind box using coarse and fine coke breeze

Figure 10 shows that the FFS increased after the coarse coke breeze had been used. Ratios of $CO/(CO + CO_2)$ in each wind box before and after the coke size changed were shown in Fig. 11. Ratios of $CO/(CO + CO_2)$ decreased after coarse coke breeze had been used, which indicates a higher combustion efficiency of coke.

Discussion

Coke breeze of smaller size burns fast because of its large specific surface area. FFS is influenced by the combustion speed of coke, the distribution of coke in quasi-particles and the permeability of the sintering bed. In the sintering process, by using coke breeze size −1 mm the mean diameter of raw materials before granulation decreased and the quasi-particle diameter became smaller [6]. Thus, a slower

FFS of adding coke breeze size −1 mm has been observed. Because of the lower combustion speed and the decline in numbers under the same mass, adding coke breeze of larger size (+5 mm) also decreased the FFS. Results of sintering pot tests showed that FFS became faster when the ratio of coke breeze size 1–3 mm increased. It was considered that the permeability of the packed bed increased with the increasing ratio of coke particles of 1–3 mm because the latter fraction played a role as nuclei particles in the granulation process [7].

Conclusions

Laboratory sintering tests were conducted to study the effects of particle size of coke breeze on the sintering process. The results showed that both the coke breeze of −1 mm and +5 mm could cause a slow sintering speed and inferior sintering performance. With the increasing ratio of coke breeze size 1–3 mm, productivity and FFS increased. Effects of coke breeze size on sintering were also investigated in a sintering plant. The FFS and combustion efficiency increased with the increasing ratio of 1–3 mm coke breeze.

References

1. Zhou MS, Zhao DM, Li JJ (2018) Effects of size distribution of coke breeze on sintering process. China Metall 28(02):11–16
2. Loo CE (1991) Role of coke size in sintering of hematite ore blend. Ironmak Steelmak 18 (1):33–40
3. Tobu Y, Nakano M, Nakagawa T (2013) Effect of granule structure on the combustion behavior of coke breeze for iron ore sintering. ISIJ Int 53(9):1594–1598
4. Damien O, Sun Q, Shen HB (2008) Effect of coke size on iron ore sintering. Iron Steel 43 (10):8–12
5. Nielser M, Stecko J, Blacha L (2014) Application of fine-grained coke breeze fractions in the process of iron ore sintering. Metalurgija 53(1):37–39
6. Takayuki M, Ryota K, Ko-ichiro O (2013) Effect of particle size of iron ore and coke on granulation property of quasi-particle. ISIJ Int 53:1503–1509
7. Yan L (1994) Development status of iron ore sintering method in the world. Sinter Pelletizing 2:49–50

Experimental Study on Dechlorination of Cold-Rolling Sludge at High-Temperature Roasting

Yi Li, Hongwei Cheng, Guangshi Li, Xiaoyong Mei, Xionggang Lu and Qian Xu

Abstract Hitherto, the iron content after cold-rolling sludge dehydration in iron and steel enterprises has reached up to 60%. However, it contains a lot of chlorine, which is the source of dioxins. And the presence of chlorine makes the cold-rolling sludge possess highly corrosivity, causing great trouble in the reproduction process. Therefore, dechlorinating experiments on cold-rolling sludge are of great significance. In this paper, we proposed an efficient way to dechlorinate the cold-rolling sludge by using high-temperature roasting method. XRD, SEM and other analytical methods were carried out to characterize the sludge. For comparison, the samples were heated from 200 to 500 °C for 2 h under air atmosphere. The dechlorinating efficiency reaches a very high value of 99%, and the main phase of final product is Fe_2O_3. In addition, we analysed the reaction mechanism and clarified the removal form of chlorine in the roasting process. These findings open up a promising avenue for steel enterprises to deal with cold-rolling sludge.

Keywords Cold-rolling sludge · Dechlorination · Roasting

Introduction

After dewatering of cold-rolled sludge, the total iron content is as high as 60%. It also contains various metal elements and has good resource recycling value. Due to the production process, the obtained cold-rolling sludge contains 10–20% chlorine, wherein the chlorine-containing compound is a source of dioxins [1, 2]. And the

---

Y. Li · H. Cheng (✉) · G. Li · X. Mei · X. Lu (✉) · Q. Xu
State Key Laboratory of Advanced Special Steel & School of Materials Science and Engineering, Shanghai University, Shanghai 200444, China
e-mail: hwcheng@shu.edu.cn

X. Lu
e-mail: luxg@shu.edu.cn

© The Minerals, Metals & Materials Society 2019

T. Jiang et al. (eds.), *10th International Symposium on High-Temperature Metallurgical Processing*, The Minerals, Metals & Materials Series,
https://doi.org/10.1007/978-3-030-05955-2_62

sludge is highly corrosive, causing difficulties in its return to production. Every year, the total amount of cold-rolling sludge produced by various steel mills in China is more than 100,000 tons [3]. Its complex composition, large output and serious damage have become difficult points for environmental protection of stainless steel enterprises. However, there is no safety, economical and practical recovery process at home and abroad. Baosteel once explored the making of sludge into pellets and returned to the blast furnace for metallurgical production [4]. However, the chlorine element was enriched in the blast furnace, and the adverse effects on the stable production of the blast furnace could not be ignored. In addition, the pollution of the toxic waste gas also needed to be resolved. Thus, it is very significative to explore a dechlorination method for cold-rolling sludge, which can cost-effectively reduce the disposal cost, recycle valuable resources and control the generation of pollution.

In this paper, cold-rolling sludge was transformed into metal oxides by oxidation roasting method. Meanwhile, the chloride compounds with a higher evaporation gradually ran into the vapour phase and then separated from the cold-rolling sludge. Based on previous studies, the reaction mechanism of dechlorination process was further established.

Experimental Materials and Methods

Experimental Materials

The cold-rolling sludge was collected from a stainless steel enterprise (China), and the main chemical composition of the sludge was given in Table 1. It showed that the cold-rolling sludge was composed of many complex components. Besides, Fe element content was much higher than any other elements, followed by Cl element. If the chlorine was not removed, it would come into the metallurgical system and result in a high Cl content in blast furnace, which is harmful to the metallurgical process [5–7].

According to the X-Ray diffraction pattern of cold-rolling sludge shown in Fig. 1, the main mineral phases were ferrous chloride ($FeCl_2{\cdot}4H_2O$) and goethite (FeOOH). And all reagents used in this study were of analytical grade.

Table 1 Main chemical composition of cold-rolling sludge (mass, %)

Element	Fe	Cl	O	Al	Cr	Si
Content/%	60.08	19.67	17.05	1.2	0.68	0.36

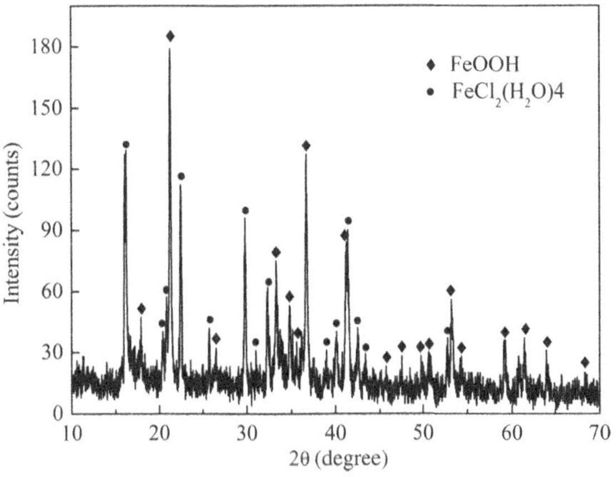

Fig. 1 X-ray diffraction of cold-rolling sludge

Methods

10 g of samples was placed into the corundum crucible. The corundum crucible was transferred into the tube after the tube furnace was heated to the set roasting temperature.

At the same time, air was passed into the reactor. The rate of air flow into the reactor was controlled by gas flowmeter. The tail gas was promptly discharged from the tube. In addition, we also investigated the effect of different roasting temperatures on Cl removal rate. After high-temperature roasting, the cold-rolling sludge samples were cooled to room temperature. Potentiometric titration [8] was used to determine chlorine content. Cl removal rate in cold-rolling sludge was mathematically expressed as:

$$\eta_{(Cl)} = \frac{M_{(Cl)} - M'_{(Cl)}}{M_{(Cl)}} \times 100\% \quad (1)$$

where $M_{(cl)}$ represents initial chloride content in cold-rolling sludge; $M'_{(Cl)}$ denotes Cl content in cold-rolling sludge after roasting; and $\eta_{(Cl)}$ represents % Cl removal.

The composition of the cold-rolling sludge was determined by XRF (XRF-1800, SHIMADZU LIMITED) with full spectrum scanning at 8°/min. Mineral compositions of the cold-rolling sludge and the dechlorination products were characterized by X-ray diffraction on a Rigaku model 3 kW D/MAX2200V PC. The XRD data were collected at 50 kV/50 mA and with a 2 theta range from 10 to 80° (0.02°/step, 0.33 s/step).

Experiment

Oxidation Roasting

The purpose of oxidation roasting was to transform the chloride into oxide. And great efforts were devoted to investigating the oxidation process of $FeCl_2{\cdot}4H_2O$ [9, 10]. According to their studies, the oxidation of ferrous chloride could take place above 200 °C, including the hydrolysis of Ferrous chloride forming HCl, the decomposition of goethite and the oxidation of Ferric chloride forming Cl_2. Therefore, the cold-rolling sludge was roasted at 200, 300, 400 and 500 °C respectively in a tube furnace with air flow rate of 50 ml/min for two hours.

Reaction of Pure $FeCl_2{\cdot}4H_2O$ with Air

The reaction process of pure ferrous chloride was performed in air to study the dechlorination roasting mechanism. The differences of dechlorination effects between pure substance and sludge were also investigated. The calcination was carried out at 500 °C for 2 h. The sample mass was 10 g, and the air flow rate was 50 ml/min. Then the roasting product was subjected to chlorine content determination and phase analysis.

Results and Discussion

Effect of Roasting Temperature on Cl Removal Rate

The chlorine removal rate at different temperatures (200–500 °C) was plotted in Fig. 2. It showed that the low reaction rate at the low temperature, resulting a low Cl removal rate. The Cl removal rate improved gradually along with the roasting temperature increment, indicating the roasting temperature affects the reaction rate of compounds. In addition, the higher temperature, the bigger reaction rate and higher Cl removal rate would be obtained. Specifically, the chlorine removal rate reached 3.98% at the roasting temperature of 200 °C and 99% at the roasting temperature of 500 °C. Besides, the chloride compounds with a higher evaporation gradually ran into the vapour phase and then separating from the cold-rolling sludge along with a rising of roasting temperature (Fig. 3).

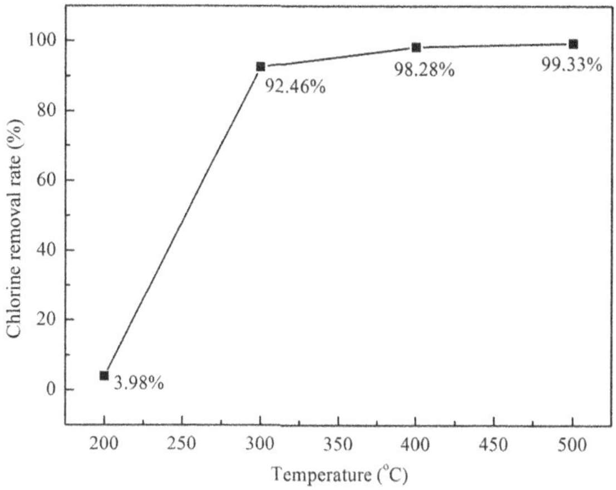

Fig. 2 Effect of roasting temperature on the chlorine removal rate of cold-rolling sludge

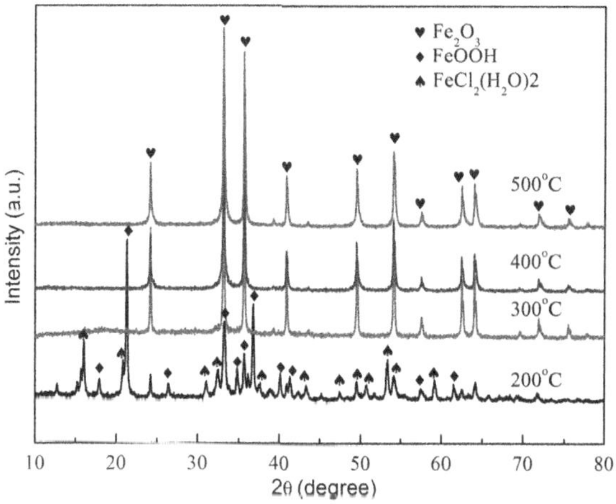

Fig. 3 X-ray diffractions of the cold-rolling sludge after oxidation roasting at 200, 300, 400 and 500 °C for 2 h

The Reaction Mechanism of Dechlorination Process

SEM image of the cold-rolling sludge after oxidation roasting at 500 °C for 2 h was shown in Fig. 4. The spherical particles were Fe_2O_3 with a radius between 20 and 30 nm. To have a better understanding of dechlorination roasting reaction mechanism, the chlorination of pure $FeCl_2 \cdot 4H_2O$ was conducted in air. Oxygen in the air reacts with ferrous chloride to form iron oxide, and there were various side reactions that produce ferric chloride [11, 12], which can be volatilized or oxidized to form chlorine. In the air, the main phase of the condensation on the tube after roasting at 500 °C is ferric chloride. In addition, sufficient moisture is essential to the dechlorination reaction of ferrous chloride. Compared with the residual chlorine content of the products, it can be found that the residual chlorine content of the dechlorination product of sludge is much lower than pure ferrous chloride roasting

Fig. 4 Micromorphology of cold-rolling sludge after oxidation roasting at 500 °C for 2 h

product. This is because the presence of goethite in the cold-rolling sludge, which decomposed at high temperature to produce H_2O, directly or indirectly involved in the dechlorination process.

Based on the preceding analogous experimental results, the main chemical reactions in the roasting can be summarized as follows:

$$2FeOOH(s) = Fe_2O_3(s) + H_2O(g) \tag{2}$$

$$4FeCl_2(s) + O_2(g) + 4H_2O(l) = 2Fe_2O_3(s) + 8HCl(g) \tag{3}$$

$$FeCl_2(s) + 1/4O_2(g) = 2/3FeCl_3(s) + 1/6Fe_2O_3(s) \tag{4}$$

$$FeCl_3(s) = FeCl_3(g) \tag{5}$$

$$FeCl_3(s) + 3/4O_2(g) = 1/2Fe_2O_3(s) + 3/2Cl_2(g) \tag{6}$$

Conclusions

The main dechlorination of sludge was derived from the high-temperature hydrolysis of $FeCl_2$ to form HCl, which will be discharged along with the tail gas. At the same time, the roasting dechlorination experiments were adopted to make the halides volatilize from the cold-rolling sludge to the gas phase under high-temperature conditions and achieved the purpose of dechlorination. Optimization experiments showed that cold-rolling sludge dechlorination rate could reach more than 99%. The dechlorination process of cold-rolling sludge at high-temperature roasting was high efficiency and presented a series of advantages such as cleaner production, easy operation, short reaction time and so on.

References

1. Anderson DR, Fisher R (2002) Sources of dioxins in the United Kingdom: the steel industry and other sources. Chemosphere 46(3):371–381
2. Addink RF, Espourteille Altwicker ER (1998) Role of inorganic chlorine in the formation of polychlorinated dibenzo-p-dioxins/dibenzofurans from residual carbon on incinerator fly ash. Environ Sci Technol 32(21):3356–3359
3. Wang L (2007) Recycling use of dust and chromium containing sludge from stainless steel-maker. Masters thesis, Xi'an University of Architecture and Technology
4. Hu S, Chen RH, Wang RY (2012) Study on metallurgical properties and pellet proportioning technology of stainless steel cold-rolling sludge. Baosteel Technol 6(2):21–26
5. Fu YK, Tang XS (2012) Chlorine balance in blast furnace and chlorine distribution in gas pipeline. J. of Anhui Univ Technol (Nat Sci) 29(3):198–201
6. Liu XJ, Lu Q, Zhang SH (2012) Migration principle of chlorine in BF production. Adv Mater Res 402:107–110
7. Zhang W et al (2015) Harm of chlorine in blast furnace and analysis on its reaction behavior. Angang Technol 6:6–10
8. Li XY, Hu JK (2014) Determination of high content chlorine in copper concentrates by automatic potentiometric titration. Chinese J Anal Chem 4:11–13
9. Tang XM, Dai WZ (1993) $FeCl_2 \cdot 4H_2O$ kinetics of roasting process. J Chem Ind Eng 44:19–25
10. Wang L (2017) Research on fluidized calcination process of $FeCl_2 \cdot 4H_2O$ and its Kinetics. Masters thesis, Zhejiang University of Technology
11. Elliott AC (1970) Effluent Water Treat J 10:385–390
12. Mellor JW (1935) A comprehensive treatise on inorganic and theoretical chemistry, vol 14. Longmans Green and Co., London, pp 20–21

Extraction Process of Antimony from Stibnite by Electrothermal Volatilization

Dongbo Li and Xiaohua Yang

Abstract A new pyrometallurgical process to extract antimony from stibnite concentrate was proposed. Thermodynamic analysis of extraction process was conducted. Based on the thermodynamic analysis, the influence of volatilization temperature, volatilization time, blowing oxygen, low melting point slagging on the volatilization rate of antimony and the residual rate of antimony in the slag was investigated. The results show that high temperature benefits for Sb_2S_3 and Sb_2O_3 volatilization reaction and the appropriate temperatures is 1623–1723 K, while Sb_2S_3 volatilization trend is stronger than that of Sb_2O_3 at the same temperature. While the stibnite concentrated content is 75% (mass fraction, similarly hereinafter), FeO content is 16%, CaO content is 16%, the volatilization time is 60 min, the volatilization temperature is 1673 K, low melting point slagging and non-blowing operation are necessary, the primary volatilization rate of antimony from stibnite concentrate is 94.78%, the residual rate of antimony from melting residue is 0.2%, and the effect of volatilization antimony from stibnite concentrate by electrothermal volatilization is better.

Keywords Electrothermal volatilization · Stibnite · Thermodynamic
Saturated vapour pressure

Introduction

There are two main types of modern antimony production methods—pyrometallurgy and hydrometallurgy [1–3]. At present, the former is the main. The vast majority of antimony smelting enterprises have been using the traditional blast furnace volatile–reverberatory furnace reduction process, which is of low envi-

D. Li · X. Yang (✉)
China ENFI Engineering Corporation, Beijing 100038, China
e-mail: yangxh@enfi.com.cn

D. Li
e-mail: lidb@enfi.com.cn

© The Minerals, Metals & Materials Society 2019

T. Jiang et al. (eds.), *10th International Symposium on High-Temperature Metallurgical Processing*, The Minerals, Metals & Materials Series,
https://doi.org/10.1007/978-3-030-05955-2_63

ronmental protection, high energy consumption, low concentration of SO_2 smoke being difficult to control and high level of antimony in slag. With the continuous strengthening of China's environmental pollution supervision, the continuous improvement of the three waste discharge standards, transformation of smelting technology and industrial upgrading become a key issue for the survival of antimony enterprises [4].

For the problem of antimony industrial, many researchers have devoted themselves to the development of new metallurgical process. The fixed sulphur reduction smelting process [5–12] was solved effectively the problem of low concentration SO_2 flue gas by fixing the sulphur to the smelting product, which the sulphur-fixing agents include soda ash, zinc oxide, pyrite cinder, hydrometallurgical ferric cake, oxidized iron ore or copper oxide. However, there are some problems such as high soda ash consumption, high cost and complex treatment process of sulphur fixation products, its industrial applications in the field of antimony smelting are limited. Leiting et al. [13] used air-volatilized molten pool of low-grade antimony smelting learn from the technique of fume furnace. Duan [14] carried out a semi-industrial test for the smelting of oxygen–sulphide concentrate, and obtained high-quality crude antimony and high concentration of SO_2 flue gas, but the slag contained antimony was high and unstable. It can be seen that the existing new technology of antimony refining is still difficult to meet the needs of the process, so the development of new antimony refining technology is still an urgent need for the development of the industry.

In this paper, the author aimed at the problems existing in the molten pool volatile melting process, on the basis of theory analysis combined with the melting process and slag process mineralogy, which explores to reduce the amount of antimony in slag and the effective ways to improve crude antimony oxide volatilization rate the author proposed a new technology of antimony smelting—electrothermal volatilization separating antimony process and achieved good results. It combines the volatilization smelting and the oxidation smelting in the electrothermal evaporation stage to form the crude antimony oxide by one step, and then electrothermally reducing the coarse antimony to the metal antimony, thereby greatly increasing the concentration of SO_2 fume, meeting the acid requirement and solving the environmental protection problem. The process was significantly reduced with the amount of flux, which greatly reduced the content of antimony in slag, and obtained high-quality crude antimony and high direct rate of antimony. On the basis of thermodynamic analysis, the effects of volatilization temperature, volatilization time, oxygen blowing operation and low melting point slag system on antimony volatilization rate and antimony residues in slag were analysed.

Experimental

Experimental Material

(1) The analysis result of major chemical elements of stibnite concentrate is shown in Table 1.
(2) The XRD analysis of antimony concentrate is shown in Fig. 1, and the antimony phase analysis is shown in Table 2.
(3) The additives required for this test are haematite, calcium oxide and coke powder, and the chemical analysis results are listed in Tables 3, 4 and 5.

Testing Device and Analysing Equipment

Intermediate frequency induction heating material used in the experiment, rated power of 100 kw, realize the control of furnace heating rate by adjusting the current;

Table 1 Main chemical element composition of stibnite concentrate (%)

Name	Sb	S	Fe	O	Si
Content (%)	48.55	15.33	5.14	16.02	9.92

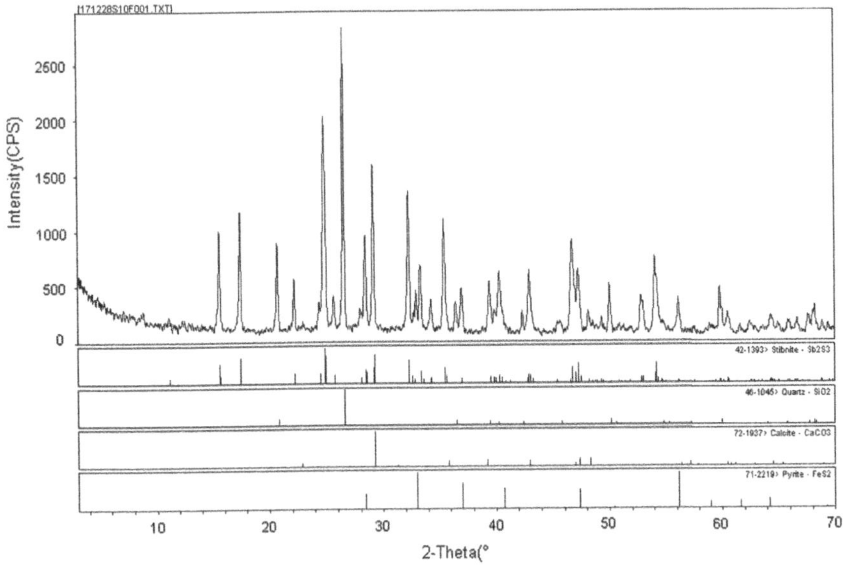

Fig. 1 XRD pattern of stibnite concentrate

Table 2 Analysis of antimony phase, silica phase and iron phase in antimony concentrate (%)

Element	SiO_2	FeS	Sb_2O_3	Sb_2S_3	Sb_2O_4
Content (%)	21.26	6.61	3.25	54.31	8.65

Table 3 Chemical composition of haematite (%)

Element	SiO_2	Fe_2O_3	CaO	Al_2O_3	MgO
Content (%)	11.31	61.46	12.36	0.50	10.75

Table 4 Chemical composition of calcium oxide (%)

Element	SiO_2	FeO	CaO	Al_2O_3	MgO
Content (%)	2.35	0.93	88.00	1.05	3.02

Table 5 Chemical composition of coke powder (%)

Element	C	SiO_2	FeO	CaO	Al_2O_3	MgO
Content (%)	79.91	7.05	1.02	0.33	4.01	0.10

With silicon carbon crucible (outer) and corundum crucible (inner) combination form as smelting furnace, the effective size of Ø 200 × 370 mm, for bag filter dust collecting equipment, bag dust collector front with a settling chamber, after the cloth bag dust collector is equipped with a suction fan; The tail gas is treated with limewater.

The analytical instruments include Japanese Neo-TTRIII X-ray diffractometer for structural analysis and atomic absorption spectrophotometer AA320N for elemental analysis.

Methods of Testing

The effects of volatilization temperature, volatilization time, low melting point slag system and oxygen blowing operation on antimony volatilization rate and residual antimony rate of slag were investigated in this paper. The purpose of high efficiency volatilization of antimony in antimony concentrates was achieved by optimizing technological parameters. The experimental steps are as follows: (1) according to certain ratio according to the quality of antimony concentrate, haematite, calcium oxide and coke powder, will add a small amount of binder after four blending for granulating, and drying the standby; (2) started a medium frequency induction furnace, and the program was warmed up to a set of values, and then you add a mixture of the ingredients, and start the timing reaction; (3) the rising flue was equipped with the air outlet, and the volatile matter was oxidized by oxygen to the air to enter the dust collecting device; (4) after the reaction, the crucible was cooled naturally, and then the reaction products were sampled one by one.

The formula for the calculation formula of antimony volatility and residual antimony rate of slag is shown in Formulas (1) and (2):

$$\mathbf{V_r = (M_t - E_t) \times \eta / M_t} \tag{1}$$

$$\mathbf{E_r = E_t / M_s} \tag{2}$$

Type: V_r is the element antimony volatilization rate; M_t is the quality of antimony elements in ores. E_t is the quality of antimony elements in slag. η is the production rate of crude antimony trioxide; E_r is the residual antimony rate of slag. M_s is the quality of slag.

Principle of Test

In the process of vapour oxidation, the main reaction is as shown in Eqs. (3)–(11):

$$2Sb_2S_3 + 9O_2(g) = 2Sb_2O_3 + 6SO_2(g) \tag{3}$$

$$2FeS_2 = 2FeS + S_2(g) \tag{4}$$

$$S_2(g) + 2O_2(g) = 2SO_2(g) \tag{5}$$

$$2Sb_2O_3 + O_2 = 2Sb_2O_4 \tag{6}$$

$$Sb_2O_4 + C(g) = Sb_2O_3 + CO(g) \tag{7}$$

$$4Sb_2O_4 + S_2(g) = 4Sb_2O_3 + 2SO_2(g) \tag{8}$$

$$Sb_2S_3 + 2Sb_2O_3 = 6Sb + 3SO_2 \tag{9}$$

$$2Sb_2S_3 + 3Sb_2O_4 = 10Sb + 6SO_2 \tag{10}$$

$$2Sb_2O_4 = 2Sb_2O_3 + O_2 \tag{11}$$

$$Sb_2S_{3(s)} = Sb_2S_{3(l)} \tag{12}$$

$$Sb_2S_{3(l)} = Sb_2S_{3(g)} \tag{13}$$

The saturated vapour pressure of antimony and its sulphide and oxide at different temperatures are shown in Fig. 2. As can be seen from Fig. 2, as the temperature increases, the saturated vapour pressure of Sb_2S_3, Sb_2O_3 and Sb increases gradually, and the growth trend of Sb_2S_3 is the most rapid. When the temperature is below 1201 K, the saturated vapour pressure of Sb_2O_3 is slightly greater than Sb_2S_3

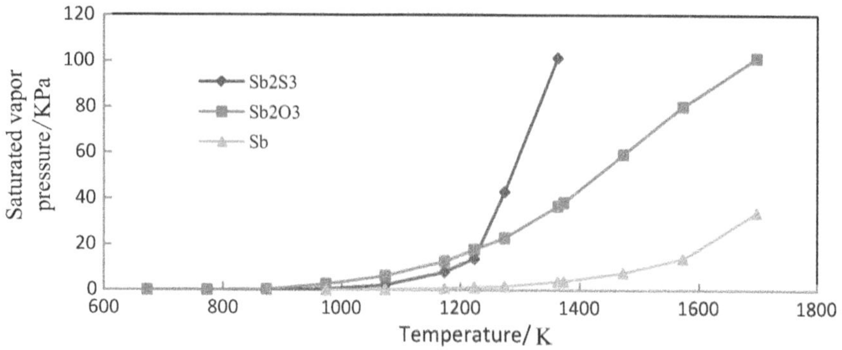

Fig. 2 Relationships between saturated vapour pressure and temperature

saturated vapour pressure. In the 1201–1363 K (Sb_2S_3 boiling point is 1363 K) and at the same temperature, Sb_2S_3 saturated vapour pressure of saturated vapour pressure than Sb_2O_3 is much larger. In 1273–1698 K (Sb_2O_3 boiling point is 1698 K) and at the same temperature, Sb_2O_3 saturation vapour pressure is far bigger than Sb.

When the antimonite mixture is added to the crucible to form a liquid melting pool (temperature greater than 1473 K), on the one hand, the mixture of Sb_2S_3 and Sb_2O_3 rapid evaporation first, followed by Sb_2O_4 first decomposed into Sb_2O_3 again volatile (beginning decomposition temperature 1203 K), and other volatile material is very small. On the other hand, some antimony oxides (Sb_2O_3 and Sb_2O_4) in the mixture were reduced to metallic antimony, accompanied by a small amount of single sulphur into antimony mattes, which were separated by volatilization and antimony mattes, and the separation and extraction of antimony in antimony concentrate were completed.

The structure of molten slag and antimony matte formed by the evaporation process is shown in Fig. 3.

Fig. 3 Structure of antimony matte and slag after the antimony volatilization

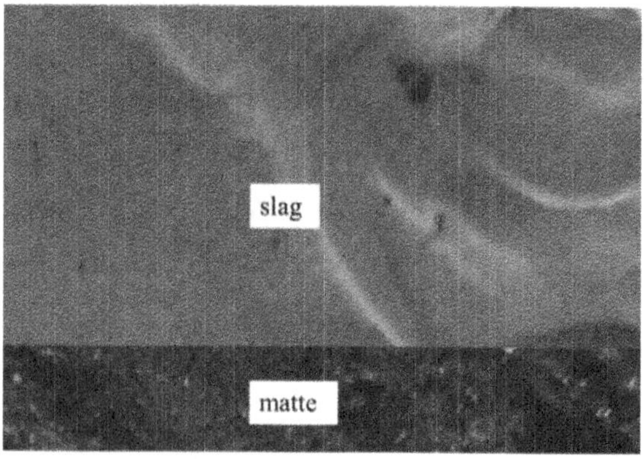

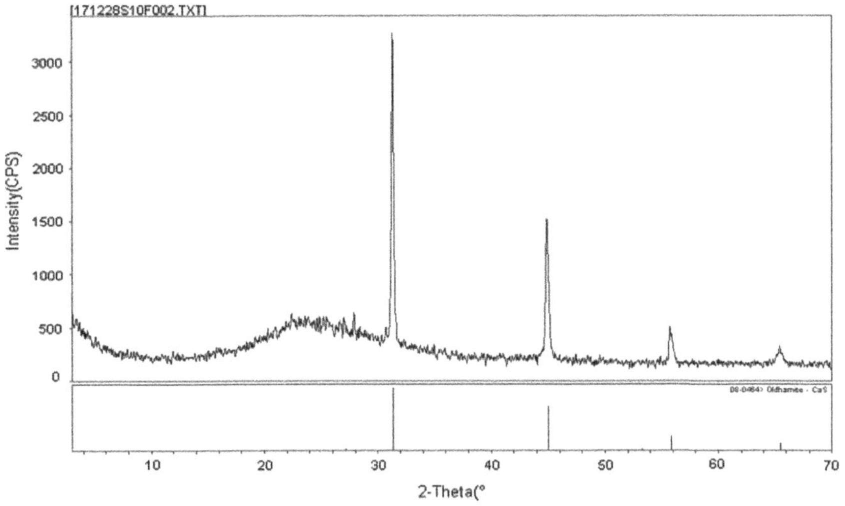

Fig. 4 XRD pattern of slag

Table 6 Composition of slag (%)

Element	SiO_2	CaS	Al_2O_3	TFe	MgO
Content (%)	57.1	27.0	13.0	0.7	2.1

Table 7 Composition of antimony matte (%)

Name	TSb	TS	Fe	FeO	FeS
Content (%)	15.7	12.4	32.9	21.4	17.2

Fe—Simple iron; TS—Total Sulfur; TSb Total antimony

The XRD of the slag is shown in Fig. 4.

The crystal phase in the XRD test slag is mainly CaS; at the same time, there are also a large number of amorphous phases in the slag. The chemical analysis results are shown in Table 6:

Because of the melting point of the slag, the melting point is higher than the antimony matte, and it is less dense than the antimony matte, which is why the slag is cooled down, and the slag first solidifies and floats at the top of the antimony.

Typical antimony matte phase composition is as follows (Table 7):

In the antimony matte, the phase consisting of Fe–FeS has low melting point (1262 K) eutectic, the system to dissolve the low melting point metal antimony and elemental sulfur, and the whole system with low melting point and specific gravity is opposite bigger, so the final solidification, and is located in the lower part of slag. Dissolved metal antimony in antimony matte after converting crude antimony oxide powder can be obtained directly, and converting process is easy to control higher yield (for comparison, antimony in antimony matte can think all transformed into crude antimony oxide powder, loss ignored).

From the above composition, it can be seen that Sb_2S_3 and Sb_2O_3 (Sb_2O_4) in stibnite concentrate are reduced to antimony mattes in a small amount, and the rest are volatilized. The slag and antimony sulphide have almost no antimony sulphide and antimony oxide. The experimental results show that the electrothermal volatilization process makes antimony in stibnite concentrate to the maximum degree of volatilization.

Results and Discussion

Analysis of Volatile—Oxidative Thermodynamics in Stibnite Concentrate

Consult the thermodynamic data manual [15] and refer to the related data [16] for the thermodynamic analysis of the volatilization process of stibnite concentrate. Figure 5 shows the relationship between the standard Gibbs free energy and temperature of the chemical equations (3)–(13).

Figure 5 shows that from the thermodynamics point of view, the calculation of the temperature range, type (4), (9), (10), (11) the standard free energy in 1023, 1543, 1173, 1413 K, changed from positive to negative, thermodynamic trend

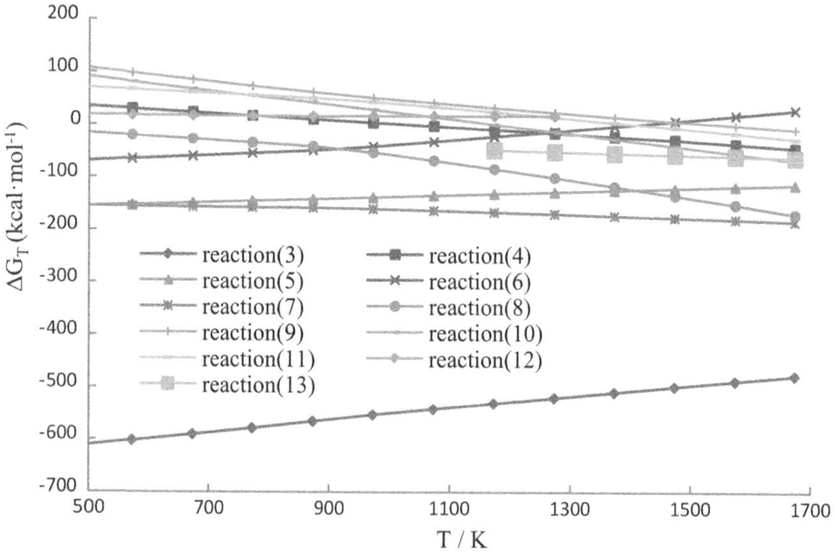

Fig. 5 Relationships between ΔG_T and temperature

gradually strengthen, and shows that only when the corresponding reaction temperature to their temperature, type (4), (9), (10), (11) to spontaneous. The standard freedom of formulas (3), (5), (7), (8) and (13) can be negative, indicating that these five reactions can be spontaneous. With the increase of temperature, (3), (5), (7), (8), (13) trend of thermodynamic order from large to small is: type (3), (7), (8), (5), (13), (5) to (6) and the type of thermodynamics trend abate showed that type (5) to (6) as the exothermic reaction. In formula (12), the standard freedom can be positive, and the Formula (12) cannot be carried out spontaneously. In conclusion, the increase in temperature is helpful for the volatilization of Sb_2S_3 and Sb_2O_3, and the volatilization of Sb_2S_3 is stronger than that of Sb_2O_3. The volatilization temperature of 1413 K is more conducive to inhibiting the generation of Sb_2O_4.

Influence of Volatilization Temperature

According to the results of thermodynamic analysis, it is more conducive to the volatilization of antimony oxides in high temperature conditions. Influence of volatilization temperature (1473–1723 K) on antimony volatilization rate (V_r) and antimony residual rate in slag (E_r) was investigated in the case of the ratio of stibnite concentrate to 75%, ferrous oxide 16%, calcium oxide 9% and volatilization time 60 min; the result of which is shown in Fig. 6.

It can be seen from Fig. 6 that, with the increase of temperature, the volatility of antimony oxide and antimony sulphide in the molten pool gradually increases, and the trend decreases after 1623 K. The residual rate of antimony in slag decreases with the increase of temperature. When the temperature is greater than 1623 K, the trend decreases. The increase in temperature is beneficial for both antimony oxide and antimony sulphide volatilization, as well as the reduction of the residual rate of antimony in slag, thus increasing the melting temperature within a range and

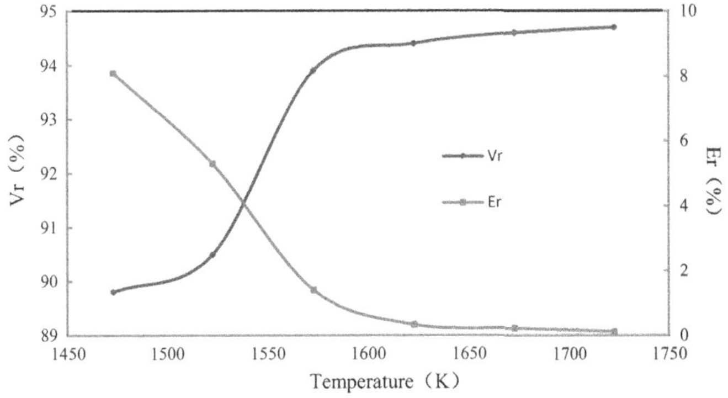

Fig. 6 Effect of volatilization temperature on antimony volatilization rate and residual rate of antimony in slag

facilitating volatilization. However, considering the comprehensive cost of smelting, the temperature is suitable for 1623–1723 K.

In stibnite concentrate, the volatilization of antimony is gas–solid reaction, and the reaction steps are: (1) the solid state containing antimony is diffused outward in the material; (2) on the gas–solid reaction interface containing antimony phase stripping; (3) Sb_2S_3 and Sb_2O_3 are diffused through the porous product layer to the surface of the solid reactants. (4) the gas products Sb_2S_3 and Sb_2O_3 diffuse through the gas phase diffusion boundary layer to the gas phase body. It can be seen from Fig. 6 that, within that range of 1573–1723 K, the volatilization rate of antimony increases with increasing temperature, is approximately linear, which shows that the stibnite concentrate volatilization process is diffusion control. Therefore, in order to reduce the rate of residual antimony in slag, and increase the volatilization rate of the concentrate, the volatile temperature should be increased as far as possible.

The Influence of Volatile Time

Test in stibnite ratio of 75%, 16% ferrous oxide, calcium oxide is 9%, the evaporating temperature is 1673 K condition, volatile test time (30–80 min) on the volatilization rate of antimony (V_r) and the influence of antimony residual rate in slag (Er), the results are shown in Fig. 7.

It can be seen from Fig. 7 that, with the increase of volatilization time, antimony volatilization rate tends to rise and then becomes stable after 60 min. However, the residual rate of antimony in slag was higher before, and it also tended to be stable after 60 min. In general, the residual rate of antimony was not very large.

At the beginning of the reaction, the amount of reactants is greater, the reaction rate is faster, and as the reaction progresses, the stibnite concentrate and the ferrous oxide are gradually consumed; therefore, at the later stage of the reaction, the reaction rate is decreased, and the volatility of the antimony and the residual rate of

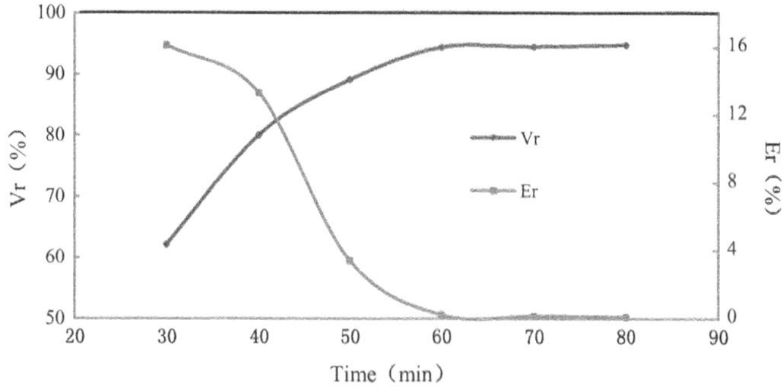

Fig. 7 Effect of volatilization time on evaporation rate of antimony and residual rate of antimony in slag

antimony in the slag tend to be stable. The volatility of smelting 80 min and the volatility of 60 min, the volatility of the antimony, and the residual rate of antimony, is a little change, so choosing the right time of evaporation, the antimony volatilization rate and the residual rate of the antimony in slag, is very important, and the results show that the time of evaporation is appropriate for 60 min.

Influence of Low Melting Point Slag on the Volatilization of Antimony

The kinetics conditions for the diffusion of reactants can be enhanced under certain liquid medium conditions. The experiment selected 75% of antimony concentrate, 16% of ferrous oxide, and 9% of calcium oxide (the volatilization pool is the low melting point slag system, hereinafter called the molten pool) and the two volatile conditions of the pure antimony concentrate (not suitable for other additives, hereinafter referred to as the non-molten pool), and the effects of volatilization time (30 ~ 80 min) on antimony volatilization rate (V_r) and antimony residual rate in slag (Er) were investigated, and the results were shown in Fig. 8.

It can be seen from Fig. 8 that with that increase of volatilization time, both the molten bath and the non-molten bath have a tendency to increase the volatilization rate of the smelting process, and the volatilization of the volatilize 60 min is stable, but the volatilization of the molten pool is much higher than that of the non-potable bath. The residual rate of antimony in slag was higher before and after 60 min; it was stable, but the residual rate of antimony was not significant, but the residual rate of antimony was much lower than that of non-molten pool.

Early reaction, it compared to non-molten pool, the volatilization rate of antimony is higher, antimony residual rate in slag is low, under the condition of molten pool, this is because at 1673 k, volatile in the molten pool formed into liquid slag

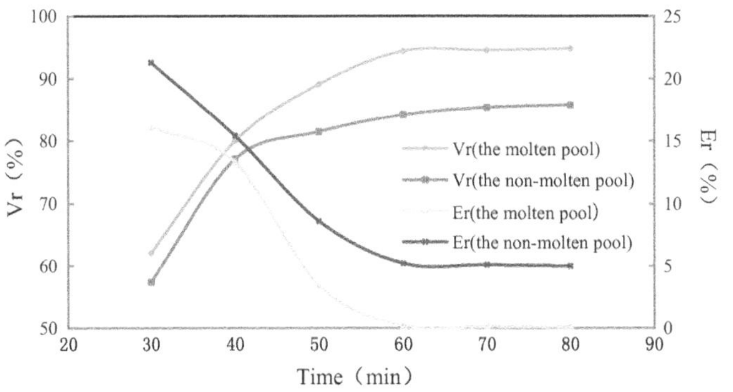

Fig. 8 Effect of low melting point slagging on evaporation rate of antimony and residual rate of antimony in slag

(slag melting point about 1503 k), and have sufficient heat and good liquidity, there is a better accelerate the volatilization of antimony, thermodynamics and kinetics conditions remain the liquid slag, antimony residual rate in slag lower nature; If don't make molten pool, antimony oxide and antimony sulfide after evaporation, evaporation system formed, which is mainly composed of SiO_2 slag, containing a small amount of FeS, sulfur monomer, due to their high SiO_2 melting point (1996 K), evaporation process under solid state, volatile matter diffusion or stripping resistance is big, the final steam down because of the volatiles in the diffusion and stripping resistance and stopped the volatilization, it is inevitable that antimony volatilization rate of stibnite concentrate is low and the residual rate of antimony is higher in solid slag system. Therefore, the ideal volatile effect should be based on the formation of a low melting point slag system (i.e. molten pool).

The Effect of Oxygen Blast on Antimony Volatilization

The dynamic conditions of the chemical reaction can also be enhanced under oxygen blowing conditions. Test selection in stibnite ratio of 75%, 16% ferrous oxide, calcium oxide is 9%, the evaporating temperature is 1673 k and under the condition of two kinds of the oxygen blast and the non-oxygenation, volatile time (30–80 min) for antimony volatilization rate (Vr) and the influence of antimony residual rate in slag (Er); the result is shown in Fig. 9.

It is shown that in the two conditions of oxygen and non-oxygenation, as the volatile time increases, the volatility of antimony is increasing, and the volatility of the volatile 60 min is stable. However, the residual rate of antimony in slag was higher before, and it also tended to be stable after 60 min. In general, the residual rate of antimony was not very large.

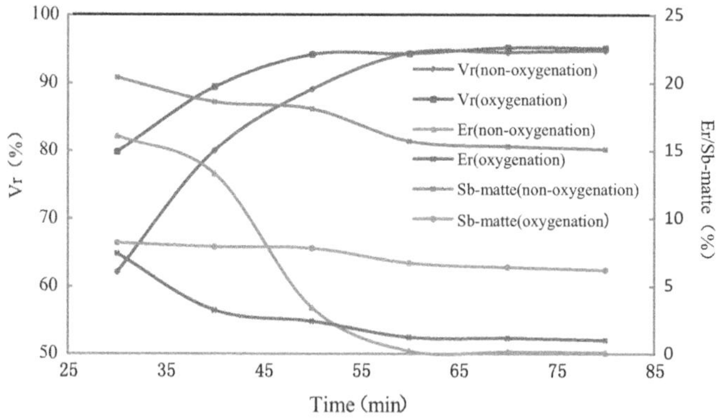

Fig. 9 Effect of blowing oxygen on evaporation rate of antimony and residual rate of antimony in slag

In the initial stage of reaction, the volatilization reaction under oxygen condition is higher than that of non-oxygenation condition, so the volatility of antimony is high, and the residual rate of antimony in slag is relatively low. Following the reaction, the volatilization trend and the tendency of antimony remained to be basically consistent after 60 min, but the antimony residual rate in slag under the oxygen blowing condition was higher, 1.2% at 60 min and 1.0% at 80 min. At the same time, the ratio of non-oxygenated antimony is to 0.2 and 0.1%, and the ratio of antimony to oxygen is six to ten times higher. It is shown that the oxygen blowing operation is detrimental to the reduction of the antimony residue in the molten slag, because the molten pool will produce the Sb_2O_4 and FeO as the oxygen blowing operation, which will form the $FeSbO_4$ at a certain temperature, which is high melting point (~ 1346 K), stable and less volatile, thus remain in the slag, resulting in an increased slag and antimony residue.

On the other hand, under the condition of oxygen blowing, the ratio of antimony matte to antimony was 6.7%, which was 15.7% when the oxygen was not blown. This is because the majority of antimony in antimony matte was oxidized to Sb_2O_3, and the residual antimony ratio in antimony matte was significantly reduced. However, whether or not an oxygen blowing operation is performed, the antimony content of the antimony matte is high and recovery must be carried out.

Trend Analysis of Antimony in Smelting Process

As shown in Fig. 10, the trend analysis of the antimony element in the smelting process showed that, in the condition of blowing oxygen and not blowing oxygen,

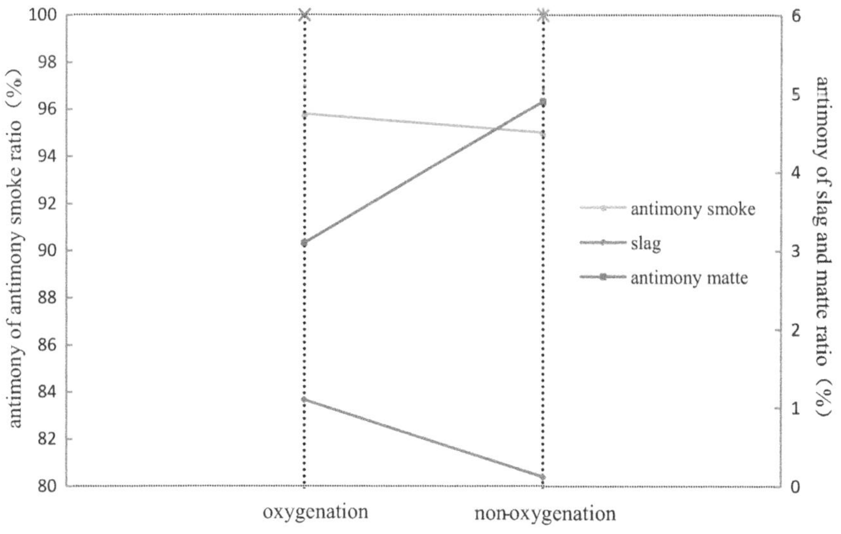

Fig. 10 Distribution on antimony in the process of smelting evaporation

the amount of antimony in the slag was 0.12% of the total amount of antimony in the ore, and 5.10% of antimony in the proportion of antimony matte. The proportion of antimony in the antimony smoke was 94.78% (i.e. the volatile ratio of antimony element in the concentrate was 94.78%, the same as the lower one). In the oxygen blowing operation, the slag contains antimony ratio of 1.10%, antimony matte containing 3.10% and antimony smoke with antimony 95.80%.

Under the condition of no oxygen blowing, antimony matte were made by electrothermal volatilization, and the antimony matte were further recovered by blowing oxygen. At this time, the amount of antimony in the tail slag was 0.12%, and the amount of antimony volatilization was 99.88% (for the comparison, the total conversion of antimony in antimony matte was calculated by the calculation of crude antimony powder, and the loss was neglected, and the same was the same). However, the antimony matte produced by volatilization of antimony concentrate in the oxygen blowing condition still need to be recovered by oxygen volatilization, and the amount of antimony in the tail slag is 1.1%, and the proportion of antimony volatiles is only 98.9%.

Conclusion

From the thermodynamic point of view, the temperature rise was helpful to the volatilization of Sb_2S_3 and Sb_2O_3, and at the same temperature, Sb_2S_3 was more volatile than Sb_2O_3. The evaporation temperature was suitable for 1623–1723 K. Under the conditions of low melting point slag system, non-blowing operation, volatilization temperature of 1673 K and volatilization time of 60 min, the residual amount of antimony in the slag was 0.2%, the one-time volatility of antimony is 94.78%, and the effect of antimony volatilization was better in the ore.

Electrothermal volatilization process had opened up a new way to extract antimony from stibnite, which had a good environmental protection compared with that of traditional blast furnace volatilization and separation antimony technology, and the gas SO_2 concentration was high (can make acid), high yield of antimony volatilization and wide application prospect.

Acknowledgements Foundation item: Project (2013FJ1009) is supported by the Science and Technology Major Special Fund of Hunan Province, China, and Project (2014ZX00X03) is supported by the Science and Technology Projects Fund of China Minmetals Corporation, China.

References

1. Zhao T (1987) Antimony. Metallurgical Industry Press, Beijing, pp 54–104
2. Wang C, Qiu D, Jiang P (2002) Status and development of antimony metallurgy in China. Nonferrous Metals: Extr Metall 5:6–12

3. Wang D, Zhang J, Wang Z et al (1995) The designation handbook of heavy nonferrous metal smelting. Metallurgical Industry Press, Beijing, pp 252–366
4. Liu F, Le X-C, Mcknight-Whitford A (2010) Antimony speciation and contamination of waters in the Xikuangshan antimony mining and smelting area, China. Environ Geochem Health 32(5):401–413
5. Chen Y, Huang C, Tang M et al (2005) Production of antimony by directly reducing-matting smelting of stibnite concentrate. The Chin J Nonferrous Metals 15(8):1311–1316
6. Liu X, Yang J, Wu Y et al (2012) Separation of needle antimony from stibnite through low-temperature sulfur-fixing smelting process. The Chin J Nonferrous Metals 22(10):2896–2901
7. Ye L, Tang C, Tang M et al (2012) Separation antimony from stibnite concentrate through a low temperature smelting. J Central South Univ (Sci Technol), 43(9):
8. Tang C (2003) New process studies on reducing-making matte smelting of lead and antimony. Central South University, Chang sha, pp 58–68
9. Tang M, Tang C, Zhang D (2000) Nonferrous metals smelting process without SO_2 formation-reduction-matte-making smelting of nonferrous metal sulfide ores and sulfur bearing materials. Nonferrous Metals 52(4):58–60
10. Tang C, Tang M, Yao W et al (2003) Reducing-matting smelting of jamesonite concentrate. J Central South Univ (Sci Technol) 34(5):502–505
11. Tang C, Tang M, Yao W et al (2003) The laboratory research on reducing—matting smelting of jamesonite concentrate. In: Proceedings of 132th annual meeting and exhibition. Boston: TMS, 2003, pp 689–697
12. Tang M, Chen Y, Tang C, et al (2010) A low temperature molten salt of antimony clean metallurgy method: China, 101914693 A[P]. 2010-09-01
13. Lei T, Wang J (2001) Industrial experiment study on the treatment of low-grade antimony ore by bath melting—continuous fuming method. Nonferrous Metals (Extr Metall) (2):15–17, 20
14. Duan F (2010) Approach to new technology of oxygen-enriched top blowing bath smelting of antimony sulfide concentrate. Nonferrous Metals Eng Res 31(6):12–16
15. Ye D, Hu J (2002) Handbook of thermodynamic data for applied inorganic material. Mechanical Industry Press, Beijing, pp 175–948
16. Li H (2005) Metallurgical principle. Science Press, Beijing, pp 128–218

Influence Factors Analysis on Scavenging of Chlorine Impurity from Crude Titanium Sponge

Liang Li, Fuxing Zhu, Dachun Liu, Kaihua Li, Zhuo Sheng and Baoqiang Xu

Abstract Titanium is exceptional value to industry, military and consumer applications due to the attractive combination of properties exhibited by metallic titanium and its alloys. The Kroll reduction reaction is a complex heterogeneous exothermic reaction with the possibility of the formation of $TiCl_2$ and $TiCl_3$ as reaction intermediates. Chlorine impurity has been a significant technical challenge in titanium production ever since the inception of the Kroll processes themselves. In this investigation, an attempt was made to check the influence factors of titanium sponge structure, raw materials, times of distillation and vacuum distillation process parameters on chorine impurity scavenging. The results indicate that the decrease of Ti sponge porosity increased the difficulty of chloride impurities removing. Chlorine content increased rapidly from 0.012 to 0.121% when the system pressure grew from 0.1 to 12 Pa in the vacuum distillation process. The chlorine content in titanium sponge was also increased with the increase of NaCl, $MgCl_2$ and $CaCl_2$ impurities in Mg reductant.

Keywords Chlorine impurity · Titanium sponge · Vacuum distillation
Kroll process

L. Li · D. Liu (✉) · B. Xu
National Engineering Laboratory for Vacuum Metallurgy, Kunming University of Science and Technology, Kunming 650093, Yunnan, China
e-mail: lcd_2002@sina.com

L. Li
e-mail: 254324378@qq.com

L. Li · F. Zhu · K. Li · Z. Sheng
State Key Laboratory of Vanadium and Titanium Comprehensive Utilization, Pangang Group Research Institute Co. Ltd., Panzhihua 617000, Sichuan, China

F. Zhu
College of Materials and Chemistry & Chemical Engineering, Chengdu University of Technology, Chengdu 610059, Sichuan, China

© The Minerals, Metals & Materials Society 2019

T. Jiang et al. (eds.), *10th International Symposium on High-Temperature Metallurgical Processing*, The Minerals, Metals & Materials Series,
https://doi.org/10.1007/978-3-030-05955-2_64

Introduction

Titanium and its alloys are widely used in aerospace, military and other industries due to the attractive combination of properties, which include excellent good fatigue resistance, corrosion resistance, a high strength-to-weight ratio and bio-compatibility [1, 2]. Titanium is also one of the most difficult to extract metals mainly due to its high chemical reactivity, high melting point (1668 °C) and high thermodynamic stability of its oxide, TiO_2 [3]. Many attempts, such as metallothermic (Mg, Al, Si, etc.), carbothermic and hydrogen reduction of its oxide, have not been successful as the product had always been contaminated with impurities [4]. Until 1937, Kroll developed magnesium reduction of titanium tetrachloride ($TiCl_4$) for producing titanium sponge with high purity. This process has been widely used over the last 70 years for the commercial production of titanium extraction [5]. $TiCl_4$ is reduced by magnesium at high temperature in a closed reactor under an inert gas atmosphere. The reduction reactions are a complex heterogeneous exothermic reaction with the possibility of the formation of lower-valence titanium chlorides as reaction intermediates [6]. Some chloride salts' impurity remains within the sponge mass and must be removed either by evaporation or by leaching. The residual chlorides in titanium sponge cause a variety of detrimental effects during subsequent processing, such as microinclusions and poor weldability [7]. So far, vacuum distillation is generally considered to obtain higher quality titanium sponge. But it is a slower, more costly procedure and energy-intensive batch process, where $MgCl_2$ and Mg entrapped in the titanium sponge are vaporized and removed at about 950–1000 °C over several days [8]. An attempt was made in this study to improve the scavenging efficiency of chlorine impurity from titanium sponge. Some factors were investigated in an attempt to check the influence of titanium sponge structure, magnesium raw materials and vacuum distillation system pressure on scavenging efficiency of chorine impurities.

Experimental Principle and Method

Vacuum distillation always has been used to separate excess Mg, $MgCl_2$, $TiCl_2$ and $TiCl_3$ in the titanium sponge industrial production process. The fundamental condition to determine whether the separation of impurities in titanium sponge is the differences of the vapour pressure in the pure state [9] can be calculated with the Van Laar Equation (1). Meanwhile, the boiling point of Mg and $MgCl_2$ also can be reduced under vacuum which will enlarge their relative volatility.

$$\lg p^0 = AT^{-1} + B\lg T + CT + D \tag{1}$$

Table 1 Evaporation constant of main substances in crude titanium sponge

	A	B	C	D	Temperature/°C
Ti	−23200	−0.66	——	13.86	1670–3289
Mg	−7550	−1.41	——	14.915	650–1090
$MgCl_2$	−10,840	−5.03	——	27.65	714–1418
$TiCl_2$	−9470	——	——	8.8	680–1300
$TiCl_3$	−9420	−2.52	——	19.68	25–730

where p^0 is the saturated pressure of pure substances, Pa; *A*, *B*, *C* and *D* are evaporation constants, and *T* is the absolute temperature in K. Table 1 lists the evaporation constants of main substances in crude titanium sponge [10].

According to the constants in Table 1, the saturated vapour pressures of Mg, $MgCl_2$, $TiCl_2$ and $TiCl_3$ can be calculated with Eq. (1), and the results are presented in Fig. 1, respectively. Table 2 lists the relative volatility of Mg and $MgCl_2$ with Ti.

It can be observed in Fig. 1 that the value of $\lg p^0$ for $MgCl_2$, $TiCl_2$, $TiCl_3$ and Mg increases with the rise of temperature. They are positive at the temperature range of 650–1300 °C. The $\lg p^0$ of $TiCl_2$ and $TiCl_3$ is less positive than that of Mg at the same temperature. In addition, according to Table 2, the saturated vapour pressure of Mg and $MgCl_2$ is much higher than Ti between 900 and 1000 °C. The relative volatility of Mg and $MgCl_2$ with Ti is 4.37×10^{12} and 3.12×10^{11} when temperature is 900 °C. That means the volatilization ability of Mg, $MgCl_2$, $TiCl_2$ and $TiCl_3$ is much stronger than titanium. Similarly, the boiling temperatures of Mg and $MgCl_2$ can be reduced by the temperature less than 750 °C when system pressure was less than 100 Pa. And we can expect a better separation between titanium and Mg, $MgCl_2$, $TiCl_2$ and $TiCl_3$ when vacuum distillation temperature is controlled between 900 and 1000 °C.

Chlorine impurity in titanium sponge is mainly because of the residue of $MgCl_2$ and the adsorption of chloride ions by titanium sponge surface. Figure 2 shows the

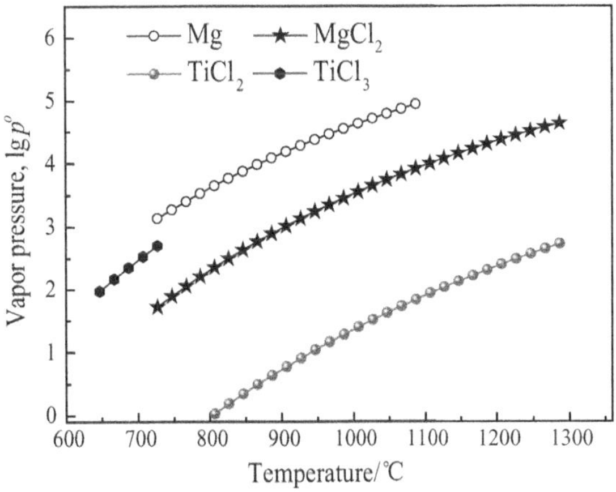

Fig. 1 $\lg p^0$-T diagram of $MgCl_2$, $TiCl_2$, $TiCl_3$ and Mg

Table 2 Relative volatility of Mg and $MgCl_2$ with Ti

Temperature/°C	Saturated vapour pressure of Mg/Pa	Saturated vapour pressure of $MgCl_2$/Pa	Relative volatility of Mg/Ti	Relative volatility of $MgCl_2$/Ti
900	13,119	937	4.37×10^{12}	3.12×10^{11}
1000	38,628	3310	2.58×10^{11}	2.21×10^{11}

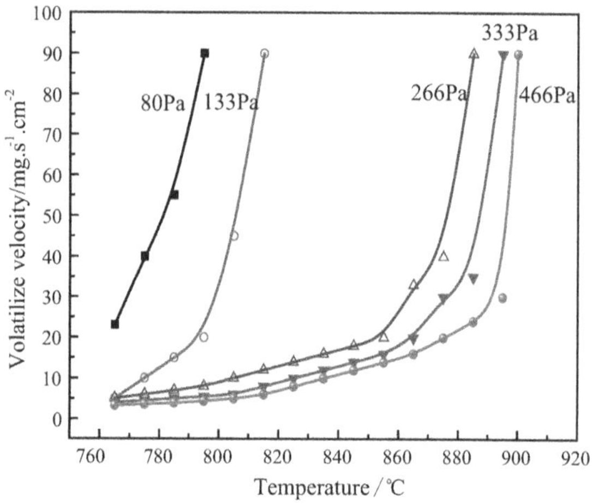

Fig. 2 Volatilization velocity of $MgCl_2$

volatilization velocity of $MgCl_2$ at different pressure. As shown in Fig. 2, it is a significant rise of $MgCl_2$ volatilization velocity along with system pressure reduction from 466 to 80 Pa. That means the lower the pressure is, the less $MgCl_2$ residues and less chlorine impurity in titanium sponge.

Experiment materials of crude titanium sponge are from products of Kroll reduction process. The Ti fraction was around 55–60%, Mg fraction was around 25–30%, and $MgCl_2$ fraction was around 10–15%. In addition, a small amount of $TiCl_2$ and $TiCl_3$ was detected in the materials.

The distillation device for titanium sponge purification is presented in Fig. 3. The bottom distillation tank can be used to collect pure titanium sponge after distillation. And top condensation recovery tank, with outside installed water jacket, is used to collect Mg and $MgCl_2$ during distillation process. Distillation tank and condensation recovery tank are connected by the reduction of the tank cap and centre channel with a magnesium plug. The furnace was heated with heating rate of 5 °C/min to 900–1050 °C during distillation process. After that, titanium sponge lump was cooled and crushed.

The concentration of chlorine element analysis was performed by ICP-OES (ICAP6300 The-rom, USA). Morphologies and element content of titanium sponge

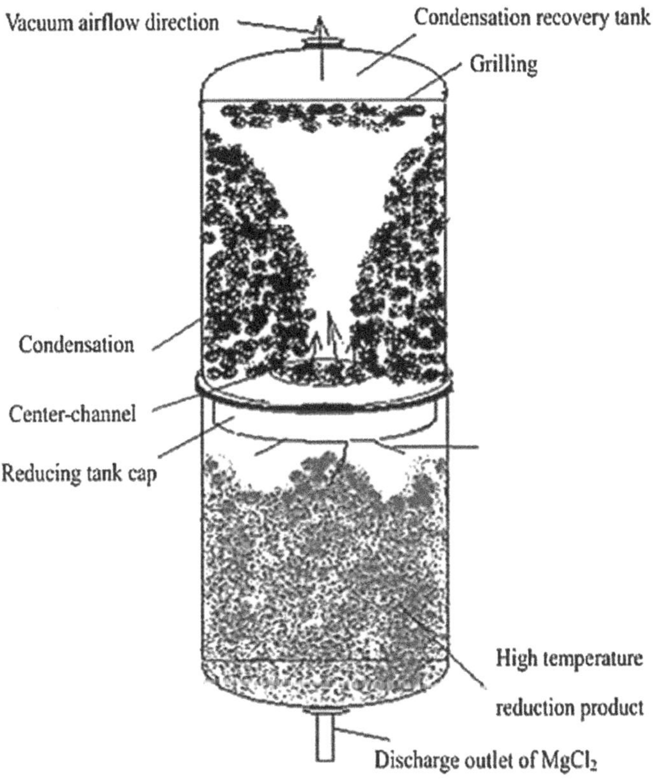

Fig. 3 Reactor for Ti sponge purification

and magnesium were characterized by scanning electron microscope (SEM) and energy-dispersive spectrometer (EDS) (TM3030Plus, Japan), respectively.

Results and Discussion

Influence of Titanium Sponge Structure

As is well known, the distribution of pore in titanium sponge is non-uniform in different bulks. It is very difficult to measure the concrete size in the different titanium sponge bulk. However, from lots of macroscopic and microscopic measurement results, most of the size of pores in regular titanium sponges, which was list in Fig. 4, was distributed from 30 μm to 1 cm. But for the sintered and hard titanium sponge that shown in Fig. 5, there was almost no pore or the size of pore was less than 30 μm, which will prevent the scavenging of chloride impurities.

Figure 6 shows the SEM and EDS images of sintered titanium sponge.

Fig. 4 SEM image of porous regular Ti sponge

Fig. 5 Macrophotograph of sintered Ti sponge

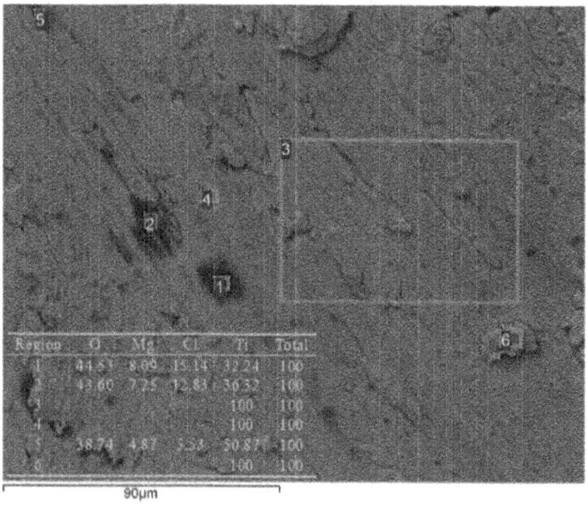

Fig. 6 SEM and EDS of sintered titanium sponge

As seen from Fig. 6, there was almost no pore can be detected. Part of by-products $MgCl_2$ was wrapped in sintered or hard titanium sponge, which will also hydrolyse in moist air and increase the oxygen content in titanium sponge. Thus, higher porosity of the titanium sponge is beneficial to the removal of chlorine impurities, especially for the scavenging of $MgCl_2$.

Influence of Raw Materials

Actual production process indicates that chlorine content of titanium sponge increases significantly with the increasing chlorine content in raw material of magnesium reductant. The chlorine content of titanium sponge was increased from 0.045 to 0.065 wt% with the chlorine content of magnesium increased from 0.005 to 0.0075 wt% when the distillation process was conducted at 1000 °C for 157 h under 2–3 Pa. Figure 7 shows the SEM and EDS results of Mg with high chlorine impurities which was used in the Kroll process to produce titanium sponge with 0.065wt% chlorine impurity.

As seen in Fig. 7, chloride electrolyte impurities, such as NaCl and $CaCl_2$, were observed obviously in the case of high chlorine impurities in the magnesium raw materials. As seen in Fig. 8, the vapour pressures of NaCl and $CaCl_2$ are much lower than that of $MgCl_2$ [10]. That means NaCl and $CaCl_2$ are more difficult to vaporize than $MgCl_2$ under the distillation temperature range of 900–1050 °C. Parts of NaCl and $CaCl_2$ will remain in the titanium. Thus, high pure magnesium is an important condition to ensure the production of titanium sponge with low chlorine impurity.

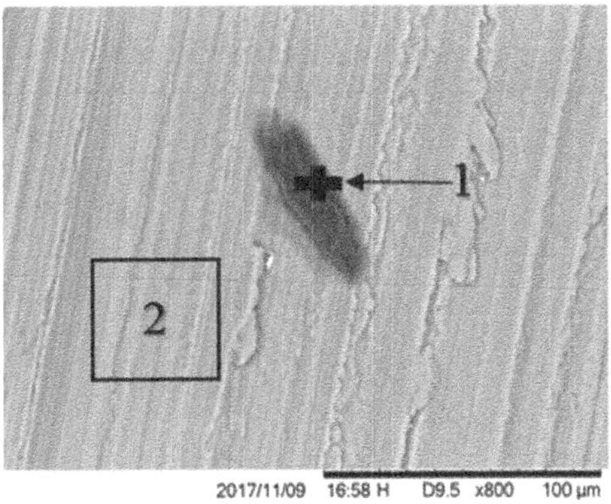

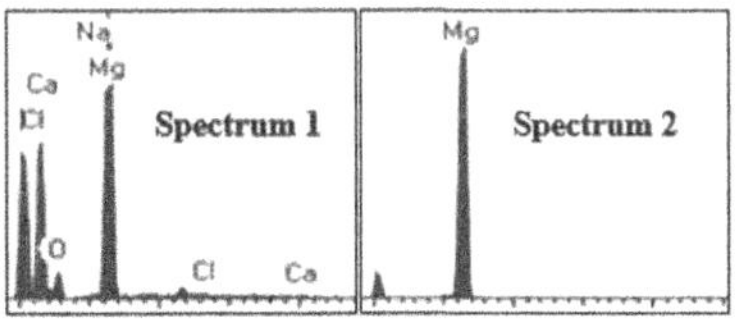

Element	Spectrum 1 -Weight/%	Spectrum 2 -Weight/%
Oxygen	23.59	
Sodium	0.867	
Magnesium	73.90	100
Chlorine	0.989	
Calcium	0.652	

Fig. 7 SEM and EDS of Mg with high Cl impurities

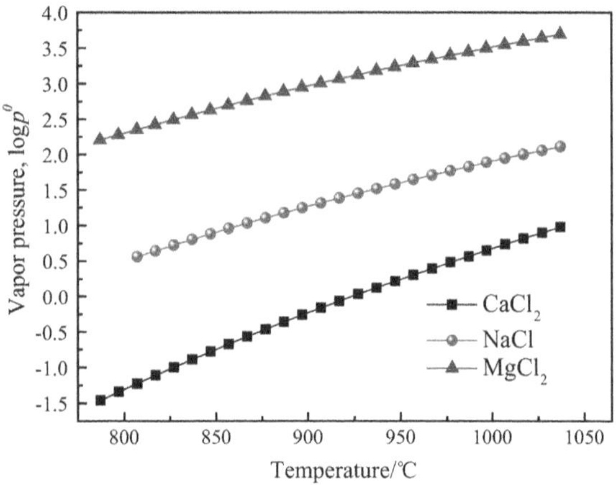

Fig. 8 $\lg p^0$-T of NaCl and $CaCl_2$

Influence of Vacuum Distillation Process System Pressure

Evaporation rate of the chlorine impurity components is determined directly by the vacuum distillation process system pressure when the distillation temperature is the constant. The degree of $MgCl_2$ evaporation will impact more on the product quality of titanium sponge. In order to investigate the influence of the system pressure on $MgCl_2$ evaporation, vacuum distillation separation experiments of the reduction products were conducted at different pressure conditions between 0.1 and 12 Pa. Results of experiments are shown in Fig. 9.

According to Fig. 9, chlorine content increased rapidly from 0.012 to 0.121% when the system pressure grew from 0.1 to 12 Pa in the vacuum distillation process. Titanium sponge with chlorine impurity under 0.06wt% can be obtained under lower distillation system pressure, especially when the system pressure was less than 3–4 Pa. The chlorine content in titanium sponge increased quickly when the system pressure is more than 4 Pa. Therefore, a better system pressure parameters

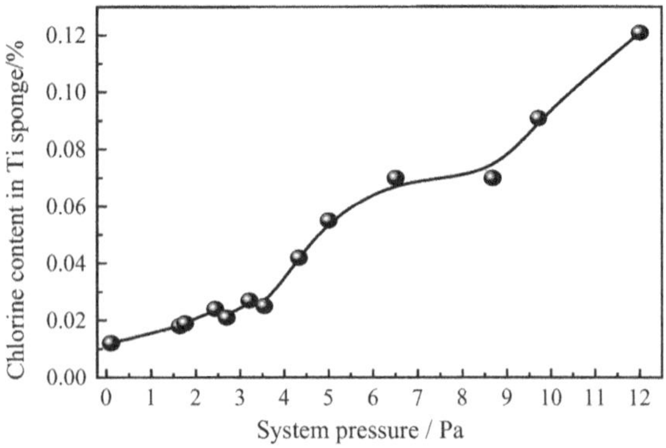

Fig. 9 Relationship between contents of chlorine with system pressure

should be controlled below 4 Pa at the vacuum distillation process for obtaining lower chlorine impurity titanium sponge products. The system pressure influence on the content of chlorine in titanium sponge is outstanding. We can have a rough estimate of titanium sponge products by knowing the distillation system pressure.

Influence of Twice Vacuum Distillation

The vacuum distillation process of crude titanium sponge will be half-stopping because of the blockage of the distillation centre channel and the breakage of resistance wire of electric furnace. In this case of half-stopping furnace, it needs to conduct the twice vacuum distillation operation. Figure 10 shows the macrophotograph of titanium sponge after twice vacuum distillation at 950–1000 °C for 160 h under 4 Pa. As shown in Fig. 10, some undistilled chloride impurities were observed directly at the bottom of titanium lump. The shrinkage of titanium sponge was also exacerbated, which delayed the smooth volatilization of chloride impurities.

The titanium sponge product details of normal and twice vacuum distillation experiments are summarized in Table 3. It is seen that the chlorine impurity content in normal once distillation titanium sponge product was much higher than that of twice distillation titanium sponge. The value of chlorine content in twice distillation titanium sponge may reach 0.121–0.150%, which would seriously reduce the quality of titanium sponge. The experiments also verified that the substandard of titanium sponge increased from 7.69 to 12.17% when the twice distillation process was conducted. Thus, effective measures for avoiding blockage of the distillation centre channel and the breakage of resistance wire are needed during the crude titanium sponge vacuum distillation process.

Fig. 10 Macrophotograph of titanium sponge after twice vacuum distillation. **a** Bottom of titanium lump and **b** top of titanium lump

Table 3 Product details of normal and twice vacuum distillation

Items	Part	Grade	Batch weight/t	Cl impurity content/%	Substandard of Ti/%
A Normal once distillation	1	0_A	7.29	0.035–0.050	7.69
	2	0_A		0.013–0.042	
B Twice distillation	1	0	7.23	0.055–0.060	12.17
	2	1		0.071–0.081	
C Twice distillation	1	3	7.59	0.095–0.153	10.14
	2	3		0.121–0.150	

Conclusions

The cases studied in this paper show that scavenging of chlorine impurity from crude titanium sponge was greatly influenced by titanium sponge structure, magnesium raw material, vacuum distillation process parameters and the times of vacuum distillation of titanium sponge. To this end, higher porosity of the titanium sponge is beneficial to the removal of chlorine impurities. High pure magnesium raw material is an important condition to ensure the production of titanium sponge with low chlorine impurity. Chlorine content increased rapidly from 0.012 to 0.121% when the system pressure grew from 0.1 to 12 Pa in the vacuum distillation process. Meanwhile, the chlorine impurity content in normal once distillation titanium sponge product was much higher than that of twice distillation titanium sponge, and the value of chlorine content in twice distillation titanium sponge may reach 0.121–0.150%.

Acknowledgements This work was supported in part by National Science Foundation of China (No. 51874156), Transfer Payment Program of Sichuan Province of China (No. 2017GZYZF0039) and the Scientific Research Item of Panzhihua City (No. 2017CY-G-5). The encouragement of academician Dai Yongnian is greatly appreciated.

References

1. Nagesh CR, Kumar GB, Saha B, Gokhale AA (2017) Titanium sponge production and processing for aerospace applications. Aerospace materials and material technologies. Springer, Singapore, pp 73–89
2. Dehghan-Manshadi A, Bermingham MJ, Dargusch MS, StJohn DH, Qian M (2017) Metal injection moulding of titanium and titanium alloys: challenges and recent development. Powder Technol 319:289–301
3. Jiao H, Tian D, Wang S, Zhu J, Jiao S (2017) Direct preparation of titanium alloys from ti-bearing blast furnace slag. J Electrochem Soc 164(7):D511–D516
4. Ono K, Suzuki RO (2002) A new concept for producing Ti sponge: calciothermic reduction. JOM 54(2):59–61
5. Zhang W, Zhu Z, Cheng CY (2011) A literature review of titanium metallurgical processes. Hydrometallurgy 108(3–4):177–188

6. Low RJ, Qian M, Schaffer GB (2012) Sintering of titanium with yttrium oxide additions for the scavenging of chlorine impurities. Metall Mater Trans A 43(13):5271–5278
7. Fan Z, Niu HJ, Cantor B, Miodownik AP, Saito T (1997) Effect of Cl on microstructure and mechanical properties of in situ Ti/TiB MMCs produced by a blended elemental powder metallurgy method. J Microsc 185(2):157–167
8. Gao F, Nie Z, Yang D, Sun B, Liu Y, Gong X, Wang Z (2018) Environmental impacts analysis of titanium sponge production using Kroll process in China. J Clean Prod 174:771–779
9. Zhang Y, Deng J, Jiang W, Mei Q, Liu D (2018) Application of vacuum distillation in refining crude lead. Vacuum 148:140–148
10. Liang L, Dachun L, Heli W, Kaihua L, Juhai D, Wenlong J (2018) Removal of chloride impurities from titanium sponge by vacuum distillation. Vacuum 152:166–172

Kinetic Study on Decomposition and Dissolution of Limestone in Converter Slag

Haohua Deng, Nan Wang, Min Chen and Guangzong Zhang

Abstract The kinetic behaviour of limestone decomposition and dissolution in converter slag at temperatures in the range of 1300–1400 °C was studied. The influence of decomposition reaction on the lime dissolution rate was also investigated. The rate-limiting steps of decomposition and dissolution of limestone in converter slag were determined on the basis of shrinking core hypothesis. The experimental results indicate that a mixed controlling mechanism of chemical reaction and heat conduction through the produced lime layer should be considered for the limestone decomposition in converter slag. The rate-controlling mechanisms of the dissolution of produced lime differ during the various sub-processes. In addition, examination of the lime microstructure at higher slag temperature revealed that cracks are formed in decomposed lime layer; lime dissolution is promoted by molten slag and penetrates through these cracks.

Keywords Kinetic behaviour · Limestone decomposition · Dissolution rate
Converter slag

Introduction

Although concerns and debates exist, direct use of limestone to replace lime as converter slagging flux has been the subject of much attention in recent years [1, 2]. The concerns relate to decomposition and dissolution behaviours of limestone during the steelmaking process, dissolution rate in converter slag, reasonable addition amount and proper adding moment. In China, many industrial trials [3–5] on limestone slagging mode have been conducted in some steel plants with the key motivation of lower steelmaking cost. Compared with the lime slagging mode, adding limestone to the converter is thought [6, 7] to be favourable to CO_2 emission reduction by eliminating the calcination process of limestone in shaft or rotary kiln,

H. Deng · N. Wang · M. Chen (✉) · G. Zhang
School of Metallurgy, Northeastern University, Shenyang 110819, China
e-mail: chenm@smm.neu.edu.cn

© The Minerals, Metals & Materials Society 2019

T. Jiang et al. (eds.), *10th International Symposium on High-Temperature Metallurgical Processing*, The Minerals, Metals & Materials Series,
https://doi.org/10.1007/978-3-030-05955-2_65

and at the same time, the sensible heat loss of calcined lime could be avoided. In addition, numerous CO_2 bubbles generated from decomposition reaction will stir the melt bath and promote refining reactions.

Reports, mainly based on the industrial observations, state that the decomposition reaction in converter would proceed intensively at higher steelmaking temperature and the limestone lump would break down into small particles due to the higher CO_2 partial pressure. Therefore, it is easier for smaller limestone particles to be heated and undergo a continuous decomposition reaction, and thus, the freshly decomposed lime particulate would dissolve into molten slag quickly. On the other hand, the replacement of lime with limestone in converter was not advisable according to the experimental results at 1600 °C and the decomposition reaction controlled by heat transfer was thought to proceed slowly, resulting in a much longer dissolution time [1]. Some researchers express concern about the lower bath temperature and slagging rate caused by endothermic decomposed reaction [8]. In the limestone slagging mode, the two reactions for decomposition reaction and slagging of the decomposed lime would occur simultaneously; thus, decomposition reaction would have significant influences on the dissolution rate, for example, local temperature drop in slag bath. Furthermore, microstructure of the decomposed lime is supposed to be varied under different temperature conditions, which is considered to affect the dissolution rate greatly.

In the present work, the decomposition and dissolution behaviours of limestone in converter slag were studied by using the laboratory-scale experiment at 1300–1400 °C and the effect of slag temperature was investigated. Moreover, the rate-controlling mechanisms of limestone decomposition and dissolution were discussed based on the unreacted core model. The aim of the present work is to secure an improved understanding of the limestone slagging mode during converter steelmaking process.

Experimental

Limestone lumps were cut into 10 × 10 × 10 mm cubic. Synthetic slag containing 30.0 wt% CaO, 30.0 wt% SiO_2 and 40.0 wt% FeO_x was prepared in a platinum crucible by melting the oxide mixture under an argon atmosphere. The synthetic slag composition was based on the slag formation route with high FeO_x content.

A tubular electric furnace with $MoSi_2$ heating elements was employed. The prepared slag sample was placed in a platinum crucible and charged to the high-temperature zone of the tube furnace. A PtRh30–PtRh6 thermocouple was placed at the crucible bottom to measure the slag temperature. A corundum rod with the rotating speed of 150 rpm was adapted to stir the molten slag continuously. High-purity argon gas flowed through the reaction chamber at a constant rate throughout each experiment. After reacting for the required time, the sample was taken out quickly and quenched in liquid nitrogen. Time and temperature were the variables—the temperature range of molten slag ranged from 1300 to 1400 °C, and the reaction time ranged from 30 to 900 s. The quenched crucible was cut along the longitudinal axis, embedded in polyester resin, and the cross section of the sample was grinded and polished. Dimensions of the residual limestone were measured by light microscope, and the cross section was observed by scanning electron microscope (SEM) after gold coating. The elemental distribution within each sample was analysed by energy-dispersive spectrometer (EDS). In addition, the apparent porosity of the produced lime was determined based on the image analysis using Image-Pro Plus software.

Results and Discussion

Limestone Decomposition in Converter Slag

The microstructural evolution of the limestone samples decomposing at various slag temperatures for 180 s is shown in Fig. 1. The apparent porosities of the lime layer produced at 1300–1400 °C for 180 s were determined by the image analysis method, and the results indicate that the apparent porosity is about 27.8–30.2% at 1300 °C, while those are approximately 30.9–31.7% at 1350 °C and 31.7–32.6% at 1400 °C.

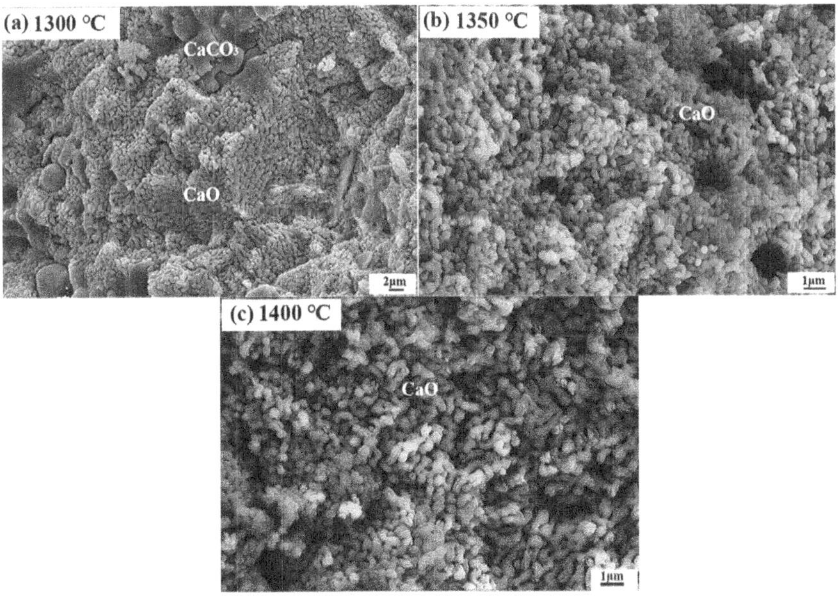

Fig. 1 SEM images of the cross section of limestone at three temperatures for 180 s

The decomposition rate of limestone in the converter slag can be described by the decomposition degree X (%), expressed as in Eq. (1).

$$X = 100\left[1 - \left(\frac{l}{l_0}\right)^3\right] \quad (1)$$

where l is the side length of undecomposed limestone (m); l_0 is the initial side length of limestone (m).

Figure 2 shows the relationship between the decomposition degree X and reaction time t at 1300–1400 °C. Slag temperature has an obvious effect on the decomposition degree of limestone, and the decomposition degree increases with the slag temperature.

The classical shrinking core model was employed to investigate the kinetics of limestone decomposition in the slag [9], and the experimental results were so analysed. In the shrinking core model, it is assumed that the decomposition reaction initially takes place at the outer surface of the limestone sample, and as the decomposition reaction proceeds, exterior surface of the sample is covered by a lime layer and a limestone core remains in the interior region. The overall rate is presumably determined by the interrelationships among three major rate processes: heat transfer, mass transfer and interfacial chemical reaction [10]. Under the assumption of the heat conduction through the lime layer controls the process, a relationship between the decomposition degree X and reaction time t is expressed by Eq. (2).

By analysing the experimental data with model fitting method according to Eq. (2), the plot of $1 - 2/3X - (1 - X)^{2/3}$ and t presents a good linear relationship with the correlation coefficient R^2 over 97.00%, as shown in Fig. 3a. Then, the effective thermal conductivity of lime layer λ_e is obtained by the slopes of these lines.

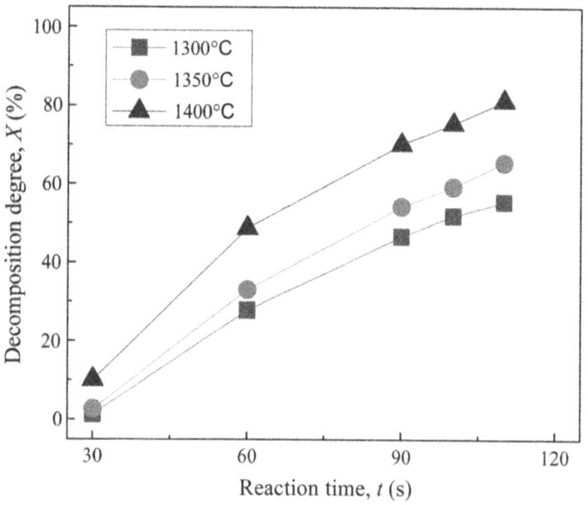

Fig. 2 Profiles of decomposition degree at three temperatures

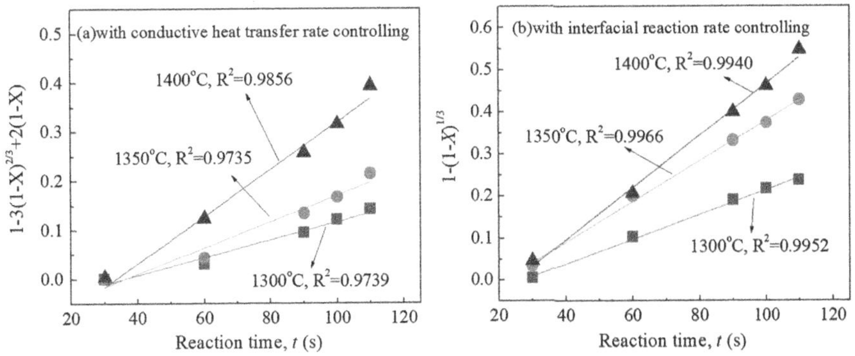

Fig. 3 Kinetic plots at three temperatures

$$1 - 3(1 - X)^{2/3} + 2(1 - X) = \frac{24\lambda_e (T_b - T_c)}{\Delta H \rho_{CaCO_3} l_0^2} t \quad (2)$$

where t is the reaction time (s), ρ_{CaCO_3} is the limestone density (mol m^{-3}), λ_e is the effective thermal conductivity of lime layer (W m^{-1} K^{-1}), T_b is the slag temperature (K), T_c is the reaction surface temperature (K), and ΔH is the reaction enthalpy of limestone decomposition (J mol^{-1}).

Furthermore, as for the mass transfer during decomposition process, the following three elementary steps take place in succession: interfacial decomposition reaction, internal diffusion of CO_2 through the lime layer and external mass transfer of CO_2. Since the mass transfer resistance within the external boundary layer can be negligible, the rate-controlling step could be either the interfacial decomposition reaction or the internal diffusion.

Under the assumption of the interfacial decomposition reaction controls the entire process, a relationship between the decomposition degree X and the reaction time t is expressed by Eq. (3).

$$1 - (1 - X)^{1/3} = \frac{2k_r (C_e - C_b)}{\rho_{CaCO_3} l_0} t \quad (3)$$

where k_r is the chemical reaction rate constant (m s^{-1}), C_e is the equilibrium concentration of CO_2 (mol m^{-3}), and C_b is the bulk concentration of CO_2 (mol m^{-3}); high-purity argon gas was used in the decomposition experiment, so the partial pressure of CO_2 in bulk gas is thought to be zero.

Under the assumption of the internal diffusion controlling process, a relation between the decomposition degree X and the reaction time t is expressed by Eq. (4).

$$1 - 3(1 - X)^{2/3} + 2(1 - X) = \frac{24D_e(C_e - C_b)}{\rho_{CaCO_3} l_0^2} t \quad (4)$$

where D_e is the effective diffusion coefficient ($m^2\ s^{-1}$).

The rate-controlling step in the limestone decomposition process within the slag could be determined by Eqs. (3) and (4). Compared with the internal diffusion mechanism, the plot of $1 - (1 - X)^{1/3}$ versus t, as shown in Fig. 3b, presents a better linear relationship with the correlation coefficient R^2 over 99.0%, indicating that the interfacial reaction could be considered as the rate-controlling step. In addition, the chemical reaction rate constant k_r can be obtained from the slopes of these lines in Fig. 3b.

Based on the above analysis, the decomposition rate of limestone is controlled by both the heat conduction through the lime layer and interfacial decomposition reaction. The effective thermal conductivity of lime layer λ_e and the chemical reaction rate constant k_r are listed in Table 1.

Limestone Dissolution in Converter Slag

According to the kinetic model of limestone decomposition, the complete decomposition time of the limestone samples at 1300–1400 °C is about 320 s, 240 s and 180 s, respectively. So, the dissolution process of limestone is considered to consist of two sub-processes. The first sub-process is the coexisting process composed of limestone decomposition and dissolution of the produced lime, and the second one is the sole dissolution of lime after the complete decomposition of limestone.

Figure 4 presents the interface region between the slag and the lime produced at 1350 °C for different time. Figure 4a shows the SEM microphotograph of the interface region at 1300 °C for 30 s; some cracks formed inside the decomposed lime layer due to a greater pressure gradient are caused probably by CO_2 generation. At higher slag temperature (1300–1400 °C), the greater decomposition rate of limestone would increase the generation rate of CO_2, resulting in a high partial pressure of CO_2 [7, 11]. The formation of these cracks makes it easier for the molten slag to penetrate into the decomposed lime and promote the dissolution process, as shown in Fig. 4b.

Dissolution rate of the produced lime in the slag can also be depicted by the dissolution degree X' (%), as expressed in Eq. (5).

Table 1 Effective thermal conductivity of lime layer λ_e and chemical reaction rate constant k_r

Slag temperature (°C)	λ_e ($W\ m^{-1}\ K^{-1}$)	k_r ($m\ s^{-1}$)
1300	9.20×10^{-2}	7.39×10^{-4}
1350	1.19×10^{-1}	8.04×10^{-4}
1400	1.87×10^{-1}	8.37×10^{-4}

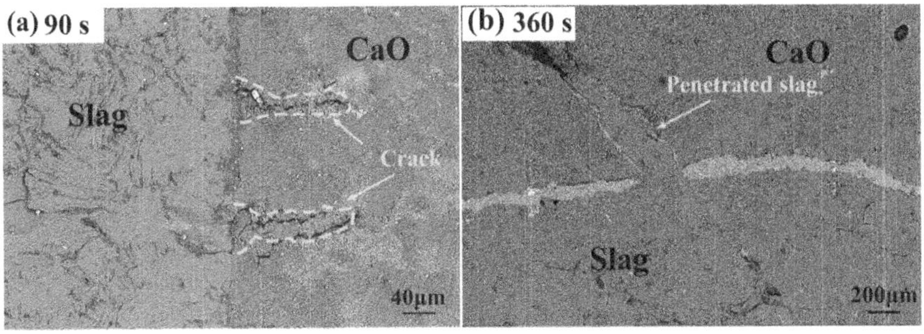

Fig. 4 SEM images of the lime–slag interface at 1350 °C for different time

$$X' = 1 - \left(\frac{a}{a_0}\right)^3 \quad (5)$$

where a is the side length of undissolved lime (m); a_0 is the initial side length of lime (m).

Figure 5 shows the relationship between the dissolution degree X' and reaction time t at 1300–1400 °C. It indicates that the slag temperature has an obvious effect on the dissolution degree of limestone and the dissolution degree increases with increasing the slag temperature.

According to the decomposition degree of limestone, the maximum temperature drop of slag bath can be calculated and the results are shown in Fig. 6. Obvious temperature drop of slag bath caused by the intensive endothermic decomposition reaction is noted especially at higher slag temperature, which is considered to lower the dissolution degree. Therefore, sufficient heat supply to the slag bath is required for the practical limestone slagging mode.

Fig. 5 Dissolution degree curves at three temperatures

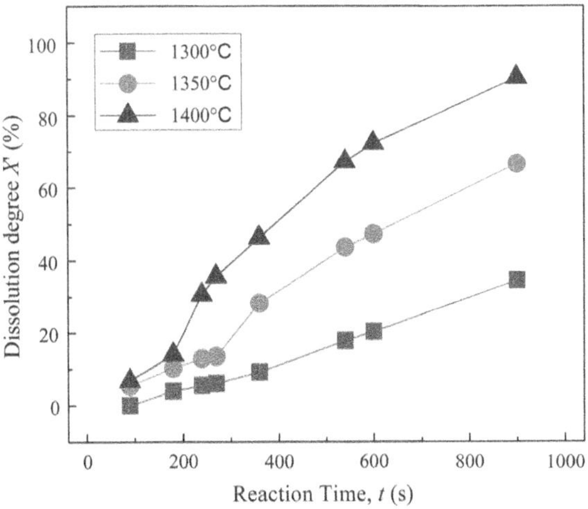

Fig. 6 Maximum temperature drop of slag bath at three temperatures

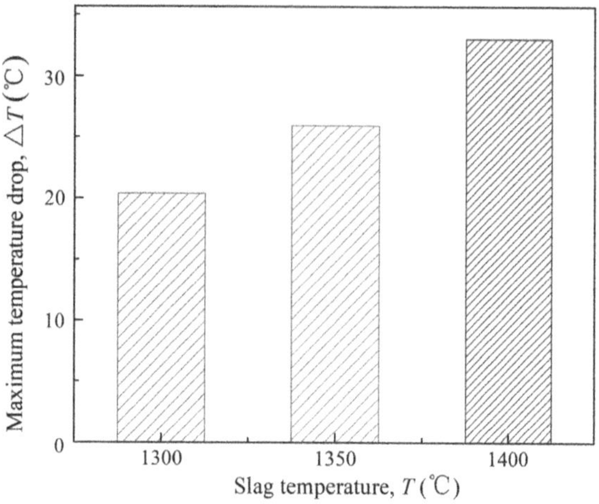

According to the shrinking core model, the dissolution of limestone in the slag is thought to include two elementary steps, i.e. interfacial chemical reaction of the decomposed lime ($CaO_{(s)}$ = (CaO)) and mass transport through the boundary layer.

Under the assumption of the interfacial reaction controlling process, a relationship between the dissolution degree X' and the reaction time t is expressed as in Eq. (6).

$$1-(1-X')^{1/3}=\frac{2k_c(C^{*}_{CaO}-C^{b}_{CaO})}{\rho_{CaO}a_0}t \tag{6}$$

where k_c is the chemical reaction rate constant (m s^{-1}), C^{*}_{CaO} is the saturation concentration of CaO in the slag (mol m^{-3}), and C^{b}_{CaO} is the concentration of CaO in the slag (mol m^{-3}).

Under the assumption that the process is controlled by mass transfer, a relationship between the dissolution degree X' and the reaction time t is expressed as in Eq. (7) [12].

$$1-(1-X')^{2/3}=\frac{2k_D(C^{*}_{CaO}-C^{b}_{CaO})}{\rho_{CaO}a_0}t \tag{7}$$

where k_D is the mass transfer rate constant (m s^{-1}).

The kinetic plot during the coexisting process of dissolution at 1300–1400 °C is shown in Fig. 7. Compared with the kinetic analysis results, the plot of $1-(1-X)^{1/3}$ versus t shown in Fig. 7a represents an improved linear relationship with the correlation coefficient R^2 greater than 99.0%, indicating that the interfacial reaction could be considered as the rate-controlling step in the coexisting process of dissolution. Furthermore, the chemical reaction rate constant k_c can be obtained from the slopes of these lines in Fig. 7a.

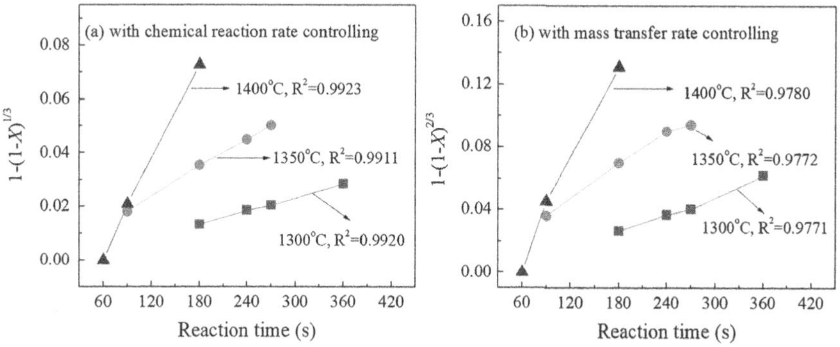

Fig. 7 Kinetic plot during the coexisting process of dissolution at three temperatures

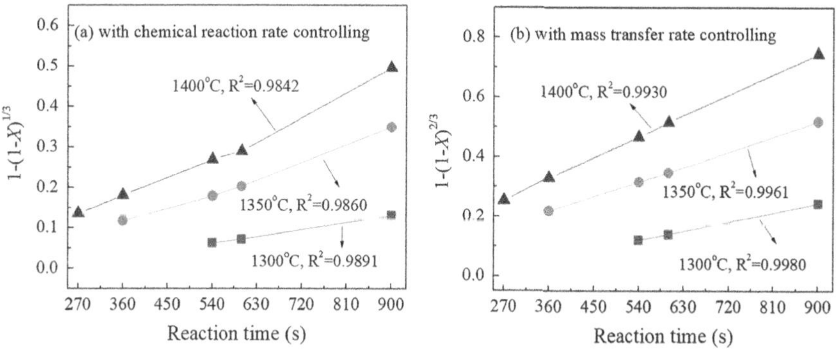

Fig. 8 Kinetic plot during the sole dissolution process of lime produced at three temperatures

Figure 8 presents the kinetic analysis results during the sole dissolution process of the produced lime. It can be noted that the plot of $1 - (1 - X)^{2/3}$ versus t, as shown in Fig. 8b, represents an improved linear relationship with the correlation coefficient R^2 greater than 99.0%, indicating that the mass transfer through boundary layer could be considered as the rate-controlling step.

Based on the above analysis, during the coexisting process of limestone decomposition and dissolution of produced lime, the dissolution rate is governed by the chemical dissolution reaction, and during the sole dissolution process, the mass transfer through the boundary layer is the rate-controlling step. The chemical reaction rate constant k_c and mass transfer rate constant k_D are listed in Table 2.

Table 2 Chemical reaction rate constant k_c and mass transfer rate constant k_D

Slag temperature (°C)	k_c (m s^{-1})	k_D (m s^{-1})
1300	6.43×10^{-6}	2.31×10^{-5}
1350	1.10×10^{-5}	3.43×10^{-5}
1400	3.25×10^{-5}	4.69×10^{-5}

Conclusions

Decomposition and dissolution of limestone in converter slag were studied, and the effect of temperature was examined. The influence of decomposition reaction on the dissolution rate of the decomposed lime was also investigated. The following conclusions were drawn.

(1) Kinetic analysis based on shrinking core model shows that the limestone decomposition in converter slag at 1300–1400 °C involves a combination of the dissociation reaction rate and heat conduction through the produced lime layer. Slag temperature has a significant influence on the degree of decomposition
(2) Apparent porosity of the lime decomposed from limestone in converter slag increases with increasing the slag temperature in the temperature range of 1300–1400 °C
(3) The temperature drop of slag bath caused by decomposition reaction is considered to lower the dissolution degree, and sufficient heat supply is required for a practical limestone slagging mode.
(4) The rate-controlling mechanism of the limestone dissolution process in converter slag at 1300–1400 °C is confirmed, and the corresponding kinetic parameters are obtained.

Acknowledgments The authors gratefully acknowledge the National Natural Science Foundation of China (51574065, 51574066, 51774072 and 51774073) and the National Key R&D Program of China (No. 2017YFB0304201, 2017YFB0304203, 2016YFB0300602) which have made this research possible.

References

1. Deng TF, Patrice N, Mattias E, Du SC (2013) Limestone dissolution in converter slag at 1873 K (1600 °C). Metall Mater Trans B 44B(1):98–105
2. Zhu YX, Zhong LC (2017) Melting slagging mechanism and application of lime and limestone in BOF steelmaking. Steelmaking 33(1):12–17
3. Qin D, Yang J, Wei S, Zhang J, Sun J, Liu J (2014) Analysis and process practice on steelmaking technology of 100t top-blown oxygen converter by limestone for slag making. Spec Steel 35(5):34–36
4. Chen Z, Lu ZJ, Tang J (2013) Practice on using limestone to replace partial lime for slagging in high hot metal ratio production. Sci. Techno. Liuzhou Steel 6:4–6
5. Dong D, Feng J, Nian W, Li H, Liang M, Liu R (2013) Experimental study on using limestone instead of lime for slagging in 60t converter. China Metall 23(11):58–61
6. Li H, Qu Y (2010) Discussion on limestone addition instead of lime for energy–saving and emission reduction in BOF steelmaking. China Metall 20(9):45–48
7. Feng J, Li H, Li CX, Song WC, Nian W, Guo LF (2013) Research on using limestone for slagging during BOF steelmaking process. Adv Mater Res 690–693:237–245
8. Wu W, Dai SF (2016) Influence of limestone on metallurgical effect of converter. China Metall 26(5):52–56

9. Amiri A, Ingram GD, Maynard NE, Livk I, Bekker AV (2015) An unreacted shrinking core model for calcination and similar solid-to-gas reactions. Chem Eng Commun 202(9): 1161–1175
10. Wang N, Zhang GZ, Chen M, Deng HH (2017) Decomposition behavior of limestone in early converter slag. Powder Techno 320:73–79
11. Johnston J (1910) The thermal dissociation of calcium carbonate. J Am Chem Soc 32(8): 938–946
12. Guo X, Sun Z, Dyck VJ, Guo M, Blanpain B (2014) In situ observation on lime dissolution in molten metallurgical slags—kinetic aspects. Ind Eng Chem Res 53:6325–6333

Low-Grade Phosphorus-Containing Iron Ore for the Removal of Cu(II) Ion from Wastewater

Xiaoli Yuan, Dongshan Zhou, Wentang Xia and Qingyun Huang

Abstract The possible use of low-grade phosphorus-containing iron ore (LGPIO), a raw and wasted industrial solid material as an adsorbent to remove Cu(II) ion from wastewater, was investigated. Studies were carried out as a function of initial pH value, adsorption time, initial Cu(II) ion concentration, adsorption temperature and LGPIO dosage. The results show that the Cu(II) ion removal efficiency exceeds 99.65% and Cu(II) ion concentration is less than 0.30 mg/L under the conditions of pH value 6.1, temperature 25 °C, adsorption time 30 min, initial Cu(II) ion concentration 100 mg/L and adsorbent dosage 10 g/L. After Cu(II) ion removal reaction, the Cu(II) ion concentrations completely accord with the requirement of national discharge standard of water pollutants for iron and steel industry (GB 13456-2012) (TCu(II) = 0.5 mg/L) in China. Therefore, it can be concluded that LGPIO is a promising low-cost adsorbent which is suitable for the removal of Cu (II) ion from wastewater.

Keywords Low grade phosphorus-containing iron ore (LGPIO) · Cu(II) ion removal · Wastewater · Adsorption

Introduction

Copper is an useful engineering materials which is widely used in several important alloys, ceramics and pesticides [1]. However, copper becomes one of the most toxic pollutants released into the environment when its ingestion beyond the permissible level, which can cause harmful biochemical and many health hazards effect on human being, such as causing anaemia, stomach and intestinal disorder, kidney and liver damage[2, 3]. Especially the discharge of Cu(II) ion-containing wastewater

X. Yuan (✉) · D. Zhou · W. Xia · Q. Huang
School of Metallurgy and Material Engineering, Chongqing University of Science and Technology, Chongqing 401331, China
e-mail: 68267523@qq.com

© The Minerals, Metals & Materials Society 2019

T. Jiang et al. (eds.), *10th International Symposium on High-Temperature Metallurgical Processing*, The Minerals, Metals & Materials Series,
https://doi.org/10.1007/978-3-030-05955-2_66

has caused serious environmental pollution. Therefore, the removal of Cu(II) ion from wastewater is a very urgent task in the world.

Some techniques including solvent extraction, precipitation, membrane filtration, evaporation, coagulation, ion exchange and adsorption have been developed for Cu(II) ion removal from wastewaters at different concentrations [4]. Of all these methods, adsorption is a quite useful technique to wastewater treatment, especially for wastewater with relative low Cu(II) concentrations. In recent years, more and more studies have been focused on seeking out economical and high-affinity adsorbents including activated carbon, activated alumina, clay minerals, fly ash, coal, waste wool, peat moss, saw-dust, serpentine mineral and blast-furnace slag [5–10]. Among these, mineral wastes have been recognized the favourable characteristics of environmental friendliness, chemical stability (low solubility), easy accessibility and low cost.

Low-grade phosphorus-containing iron ore (LGPIO) is an industrial solid waste, which comes from mining high phosphorus iron ores. LGPIO is becoming a serious problem in the world due to the fact that it not only causes landslide and dust pollution, but occupies the land resources. However, LGPIO contains metal oxides (haematite), silicate, dolomite and apatite. All of which are the effective ingredient for Pb(II) ion and phosphate ion removal from wastewater [11, 12].

The aim of this work was to investigate the feasibility of using LGPIO as an adsorbent for Cu(II) ion removal from wastewater. The effects of initial pH value, initial Cu(II) concentration, adsorption temperature, particle size, adsorbent dosage and adsorption time on the Cu(II) ion removal efficiency of LGPIO were studied.

Materials and Methods

Materials

The sample of LGPIO (mass percent, TFe 25.49%, P 1.26%, $SiO_2$16.13%, CaO 20.76%, MgO 7.99%, Al_2O_3 5.55%, MnO 0.36%) used in this research was obtained from Wushan, Chongqing, China. LGPIO was crushed and screened to <0.25 mm in particle size before being used.

Experimental Methods

A defined volume of the Cu(II) ion stock solution with a concentration of 1 g Cu(II)/L was diluted to the initial concentration by adding deionized water. About 100 mL of Cu(II) ion solution was poured into the glass beaker of 200 mL, and its pH value was adjusted to desired value with HCl or NaOH in moderate doses. Then the LGPIO was added into the wastewater. The sample was taken out by filtration after equilibrium, and the pH of the wastewater was measured by pH meter (Mettler Toledo). The Cu(II) cation concentrations of all samples including wastewater were analysed by flame atomic absorption spectrophotometry method (TAS-986, Beijing Puxi Science and Technology Instrument Co Ltd China).

Results and Discussion

Effect of Initial pH

Figure 1 showed the Cu(II) ion removal efficiency at different initial pH value as adsorption temperature, the stirring speed, the range of LGPIO particle size, LGPIO dosage, initial Cu(II) ion concentration, adsorption time were fixed in 25 °C, 250 r/min, <0.25 mm, 10 g/L, 100 mg/L and 30 min, respectively. The data in Fig. 1 show that the initial pH value is a very important parameter for the removal of Cu(II) ion from wastewater. With the initial pH value increasing from 2 to 6.1, the Cu(II) ion removal efficiency increased rapidly from 94.37 to 99.89%, and the equilibrium decreased from 5.63 to 0.11 mg/L. However, when the initial pH value ranged from 6.1 to 10, pH value almost had no effect on the Cu(II) ion removal efficiency, and the Cu(II) removal efficiency kept in above 99%. The maximum (99.89%) removal of copper was found at pH 6.1. The lower Cu(II) removal efficiency at low pH (<6.1) can be attributed to the combination of adsorbent with hydronium ion, which made it protonated and rejected Cu^{2+} due to the presence of high content of H^+, namely the electron repulsion. At higher pH(>6.1), the influence of H^+ is negligible. The Cu(II) ion may be removed from solution by precipitation. Similar results have been reported by the adsorption Cu(II) ion on the fly ash [2].

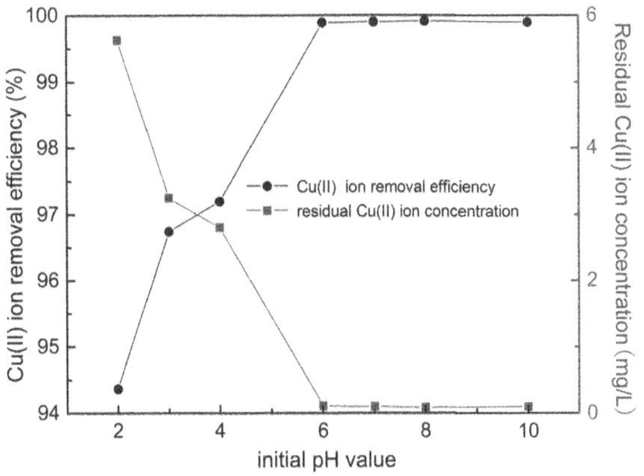

Fig. 1 Effect of initial pH on Cu(II) ion efficiency

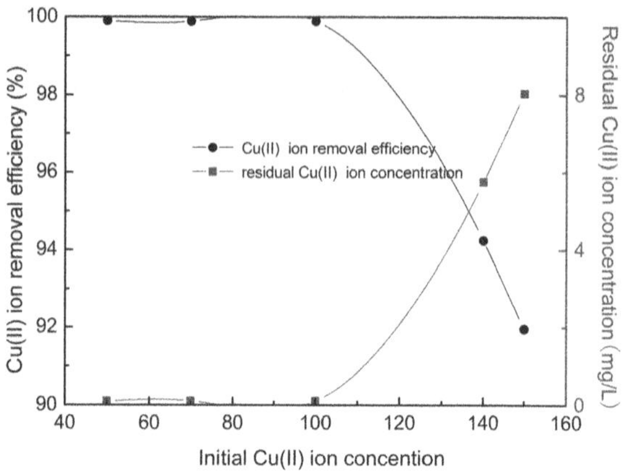

Fig. 2 Effect of initial Cu(II) ion concentration on Cu(II) ion efficiency

Effect of Initial Cu(II) Ion Concentration

Figure 2 is the plot of relationship between residual Cu(II) ion concentration and the Cu(II) ion removal efficiency on the conditions of reaction temperature 25 °C, reaction time 30 min, initial pH value 6.1, LGPIO dosage 10 g/L, the range of LGPIO particle size <0.147 mm. From Fig. 2, when the initial Cu(II) ion concentration increased from 50 mg/L to 100 mg/L, the Cu(II) ion removal efficiency kept in above 99.89%. But, with further increases in the initial Cu(II) ion concentration, the Cu(II) ion removal efficiency decreased and the residual Cu(II) ion concentration increased rapidly. So the appropriate initial Cu(II) ion concentration is 100 mg/L under the above condition.

Effect of Adsorption Temperature

Figure 3 showed the Cu(II) ion removal efficiencies and residual Cu(II) ion concentrations at different adsorption temperatures as initial pH, the stirring speed, reaction time, LGPIO dosage, the range of LGPIO particle size were fixed to 6.1, 250 r/min, 30 min, 10 g/L, <0.147 mm, respectively. Figure 3 shows that when reaction temperature was increased from 25 to 60 °C, the Cu(II) removal efficiency and the residual Cu(II) ion concentration have almost no change. It indicated that the adsorption temperature almost had no effect on the Cu(II) ion removal efficiency. Considering the factors including Cu(II) removal efficiency and processing cost, 25 °C would be appropriate for the industrial application.

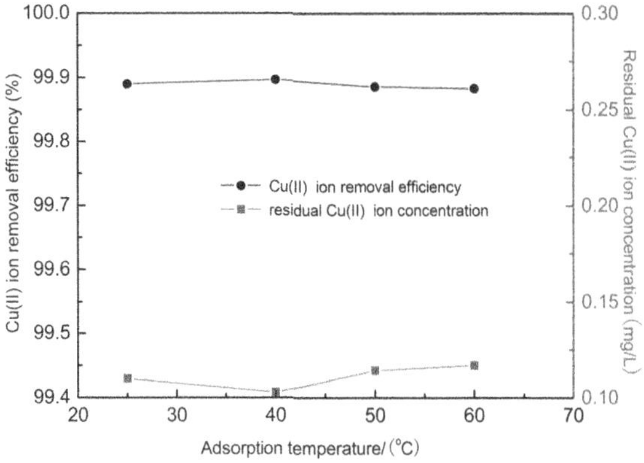

Fig. 3 Effect of adsorption temperature on Cu(II) ion removal efficiency

Effect of Particle Size

The effect of particle size on the Cu(II) ion removal efficiency was examined on the conditions of reaction temperature 25 °C, initial pH value 6, initial Cu(II) concentration 100 mg/L, stirring speed 250 r/min, LGPIO dosage 10 g/L and adsorption time 30 min, and the results are presented in Fig. 4. When the particle size decreased, the Cu(II) ion removal efficiency increased. It is obvious that the Cu (II) ion removal efficiency increased slightly when the particle size range from 0.147–0.075 to <0.075 mm. Synthetically considering the factors including Cu(II) removal efficiency, processing cost and the cycle utilization of adsorbent, the range of LGPIO particle size <0.147 mm would be appropriate for the industrial application.

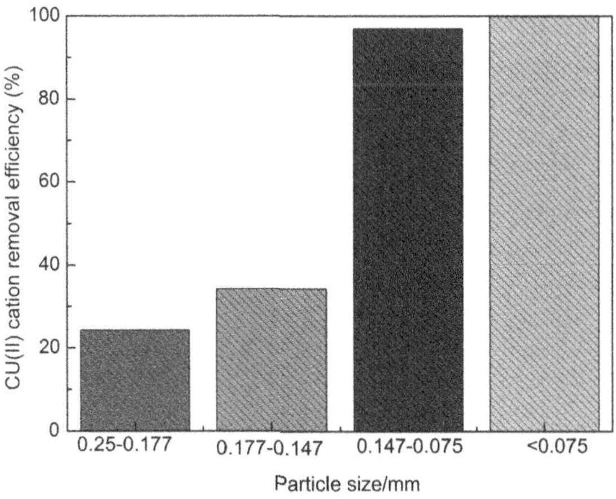

Fig. 4 Effect of particle size on Cu(II) ion removal efficiency

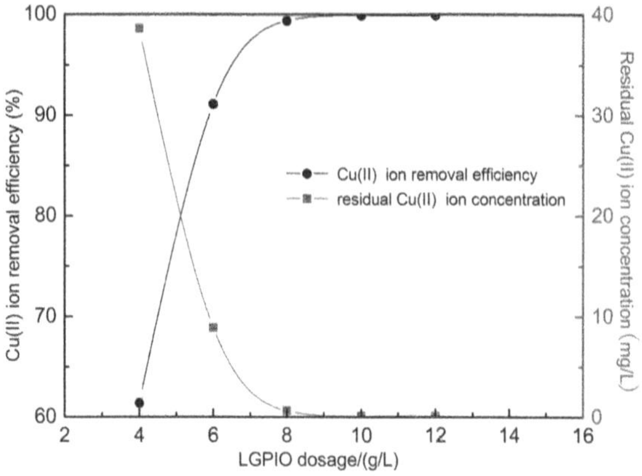

Fig. 5 Effect of LGPIO dosage on Cu(II) ion removal efficiency

Effect of LGPIO Dosage

The plots of Cu(II) ion removal efficiency and Cu(II) ion equilibrium concentration versus LGPIO dosage were presented in Fig. 5. The experiments were conducted on the conditions of reaction temperature 25 °C, stirring speed 250 r/min, the range of LGPIO particle size <0.147 mm, reaction time 30 min and pH value 6.1. From Fig. 5, the LGPIO dosage notably influenced the process of Cu(II) ion removal from wastewater. When LGPIO dosage was ranged from 4 to 8 g/L, the Cu(II) ion removal efficiency increased remarkably with increasing LGPIO dosage. As LGPIO dosage was 10 g/L, the Cu(II) removal efficiency was above 99.89%. When LGPIO dosage was further increased, the Cu(II) ion removal efficiency and residual Cu(II) concentration nearly kept constant.

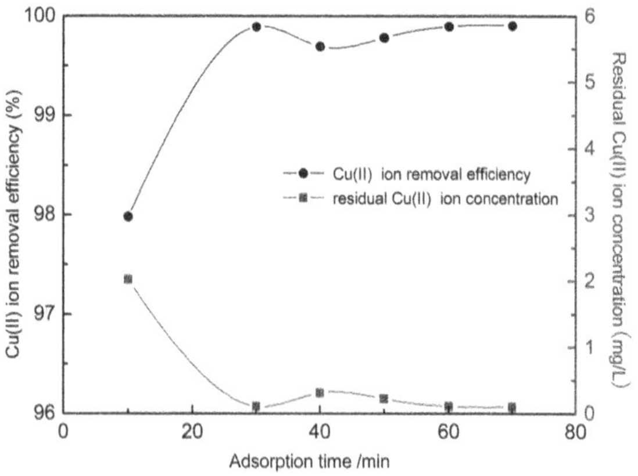

Fig. 6 Effect of reaction time on Cu(II) ion removal efficiency

Table 1 Effect of wastewater volume on Cu(II) ion removal

Solution volume/mL	Residual Cu(II) ion concentration after adsorption(mg/L)	Cu(II) ion removal efficiency (%)
100	0.072	99.93
300	0.084	99.92
500	0.0956	99.94
700	0.1032	99.90

Effect of Adsorption Time

The effect of adsorption time on the Cu(II) ion removal efficiency was examined on the conditions of adsorption temperature 25 °C , the range of LGPIO particle size <0.147 mm, stirring speed 250 r/min, LGPIO dosage 10 g/L and initial pH value 6.1, and the result was shown in Fig. 6. From Fig. 6, when the reaction proceeded for 30 min, the Cu(II) ion removal efficiency reached above 99.5%. With the reaction time further increasing, the Cu(II) ion removal efficiency increased slowly. The results indicated that Cu(II) ion in wastewater was removed by LGPIO quickly.

Verifying Experiment

In order to further verify the most favourable conditions, the favourable reaction conditions were determined as follows: pH value 6.1, temperature 25 °C, adsorption time 30 min, initial Cu(II) ion concentration 100 mg/L, particle size lower than 0.147 mm, adsorbent dosage 10 g/L and stirring speed 250 r/min. Four verifying experiments were executed under favourable conditions. The experimental results are showed in Table 1.

From Table 1, the results indicated that these optimal reaction conditions were credible. The Cu(II) removal efficiencies all exceeded 99.9% in the four verification experiments, and the Cu(II) concentrations completely accorded with the requirement of national discharge standard of water pollutants for iron and steel industry (GB 13456-2012) (TCu(II) = 0.5 mg/L) in China.

Conclusions

The adsorption experiments results of present work indicate that the low-grade phosphorus-containing iron ore (LGPIO) has a good performance on Cu(II) ion removal from wastewater. The optimal experimental parameters were determined as follows: initial pH value 6.1, temperature 25 °C, reaction time 30 min, the range of

LGPIO particle size <0.147 mm, initial Cu(II) concentration 100 mg/L, adsorbent dosage 10 g/L and stirring speed 250 r/min. Under the optimal experimental parameters, the Cu(II) removal efficiency exceeded 99%, and the Cu(II) concentrations of wastewater were lower than 0.5 mg/L, which completely meet the requirement of national discharge standard of water pollutants for iron and steel industry (GB 13456-2012) (TCu(II) = 0.5 mg/L) in China. From environmental and economical concerns, the results show that LGPIO, an industrial solid waste, could be as a promising adsorbent for Cu(II) ion removal from wastewater.

Acknowledgements This research was financially supported by the Scientific and Technological Research Program of Chongqing Municipal Education Commission (Grant No. KJQN201801507). This work was also financially supported by the National Natural Science Foundation of China (Grant Nos. 51504051, 51174246 and 51574053) and Chongqing Research Program of Basic Research and Frontier Technology (Grant No. cstc2015jcyjA90001).

References

1. Shrivastava AK (2009) A review on copper pollution and its removal from water bodies by pollution control technonogies. Indial J Environ Prot 29(6):552–560
2. Alinnor IJ (2007) Adsorption of heavy metal ions from aqueous solution by fly ash. Fuel (86):853–857
3. Feng X, Gao JS, Wu ZC (2008) Removal of copper ions from electroplating rinse water using electrodeionization. J Zhejiang Univ Sci A 9(9):1283–1287
4. Solisio C, Lodi A, Torre P, Converti A (2006) Copper removal by dry and re-hydrated biomass of Spirulina platensis. Biores Technol 97:1756–1760
5. Ho YS, Porter JF, McKay G (2002) Equilibrium isotherm studies for the sorption of divalent metal ions onto peat: copper, nickel and lead single component systems. Water Air Soil Pollut 141:1–33
6. Foo KY, Hameed BH (2010) Review: insights into the modeling of adsorption isotherm systems. Chem Eng J 156:2–10
7. Radke CJ, Prausnitz JM (1972) Sorption of organic solutes from dilute aqueous solutions on activated carbon. Indial Eng Chem Found 11:445–451
8. Korngold E, Aronov L, Kedem O (1998) Novel ion-exchange spacer for improving electrodialysis I. Reacted spacer. J Membr Sci 138(2):165–170
9. Virarghavan T, Dronamraju MM (1992) Utilization of coal ash in water pollution control. Int J Environ Stud 40:79–85
10. Singh DK, Jyotsana L (1992) Removal of toxic metal ions from wastewater by coal based adsorbent. Pollut Res 11(1):37–42
11. Yuan XL, Bai CG, Xia WT (2014) Acid-base properties and surface complexation modeling of phosphate anion adsorption by wasted low grade iron ore with high phosphorus. J Colloid Interface Sci 428:208–214
12. Yuan XL, Xia WT, Yin JG (2015) Adsorptive removal of Pb(II) from aqueous solution using raw and wasted low grade iron ore with phosphorus as adsorbent. In: 6th International Symposium on High-Temperature Metallurgical Processing, pp 635–642

Mechanism of the Chlorination Roasting of Nickel Sulphide Concentrate with Ammonium Chloride

Xiaoyong Mei, Hongwei Cheng, Cong Xu, Guangshi Li, Xionggang Lu and Qian Xu

Abstract Nickel resources are mainly in the form of sulphides in China, and chlorination roasting is one of the most important means of treating nickel sulphide ore. This paper reports the results of an investigation of the reaction mechanism of ammonium chloride roasting of flotation concentrate. The effect of roasting temperature on the calcine composition is discussed on the basis of roasting products analysis by SEM and vibrational spectra. The role of ammonium chloride in the reaction process was investigated in depth by X-ray micro-area analysis. The results show that the chlorination of ammonium chloride is achieved through the solid–solid chlorination and the chlorine-forming reaction, instead of chlorination by HCl formed by the decomposition of NH_4Cl. Metal chlorides and double sulphate salts are formed during the chlorination roasting process, and the formation of double salts benefits the thermal stability of the product.

Keywords Sulphide · Roasting · Ammonium chloride · Reaction mechanism

Introduction

Nickel is an important non-ferrous metal element widely used in stainless steel, alloys, electroplating and other industries. The world's nickel resources exist mainly in the form of nickel sulphide or laterite nickel ore. Traditional nickel sulphide ore processing technology involves complicated processes, low metal yield due to loss of cobalt and nickel by hydrometallurgy, SO_2 flue gas emissions and other shortcomings [1]. In view of the sharp reduction of the reserves of rich mineral resources combined with strict concentrate grade and magnesium content

X. Mei · H. Cheng (✉) · C. Xu · G. Li · X. Lu (✉) · Q. Xu
State Key Laboratory of Advanced Special Steel & School of Materials Science and Engineering, Shanghai University, Shanghai 200444, China
e-mail: hwcheng@shu.edu.cn

X. Lu
e-mail: luxg@shu.edu.cn

© The Minerals, Metals & Materials Society 2019

T. Jiang et al. (eds.), *10th International Symposium on High-Temperature Metallurgical Processing*, The Minerals, Metals & Materials Series,
https://doi.org/10.1007/978-3-030-05955-2_67

restrictions in pyrometallurgical processes, a means for efficiently processing of low-grade complex mixed mines are an urgent necessity.

Chlorination roasting is a promising pretreatment method for ore raw materials [2]. Solid chlorinating agents such as NaCl and $CaCl_2$ are commonly used in chlorination roasting. Ammonium chloride, a by-product of the soda ash industry, is also a good source of chlorine. However, the question of how to use ammonium chloride efficiently is an important issue that restricts the production of the combined alkali process [3].

At present, chlorination roasting research reported in the literature mainly focuses on the roasting of oxidized ore. Terakado et al. [4] of Nagoya University in Japan successfully used ammonium chloride to extract indium from metal sludge. Researchers at the Tomsk Polytechnic University in Russia studied chlorination roasting of mixed ores, oxidized ores and sulphides [5, 6]. Bayer et al. [7] compared the chlorination effects of several solid chlorinating agents on chalcopyrite and indicated that the reaction temperature of NH_4Cl is lower than that of NaCl and KCl. In fact, the ammonium salt has poor thermal stability and is susceptible to thermal decomposition. The mechanism of action of ammonium chloride and sulphide is more complicated than that of NaCl and $CaCl_2$.

This study employed nickel sulphide flotation concentrate as the research object and used ammonium chloride as the chlorinating agent to explore the effects of calcination temperature and chlorinating agent addition on the composition of the calcined product. Detailed vibrational spectroscopy and micro-area analysis were carried out to elucidate the calcination reaction mechanism.

Materials and Methods

The concentrate used in the experiment was from Jinchuan Mining Company. The main components and X-ray diffraction spectra of flotation concentrate are shown in Fig. 1. The flotation concentrate is mainly composed of pentlandite, chalcopyrite, pyrrhotite and pyrite, as well as a small amount of magnetite. The concentrate size is mainly between 10 and 100 μm, and the particle size of less than 74 μm accounts for about 81 vol.%. The NH_4Cl used in the experiment is of analytical grade. For uniform mixing, the ammonium chloride was ground to 100% passing 0.28 mm.

The chlorination roasting experiment was carried out in a muffle furnace. Concentrate and ammonium chloride were blended according to a predetermined mass ratio and mixed evenly in the crucible. When the furnace was raised to the preset temperature, the crucible was placed in the muffle furnace and calcined for 120 min. The crucible was retrieved after the roasting was complete, and the furnace cooled to room temperature.

The calcine samples were analysed by X-ray diffraction spectroscopy, and the product microstructure was inspected by scanning electron microscopy. The spectral structure was analysed by Fourier transform infrared spectroscopy and laser micro-Raman spectroscopy. The FTIR spectral scanning range was 4000–400 cm^{-1},

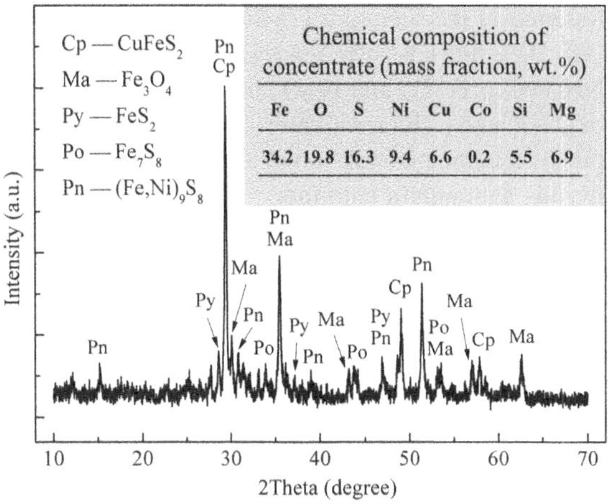

Fig. 1 X-ray diffraction spectra and the chemical analysis of flotation concentrate

and the sample was prepared by KBr pelleting. The room temperature Raman spectral scanning range is 100–800 cm^{-1} and 100–1800 cm^{-1}, respectively, and the laser wavelength used is 633 nm.

Results and Discussion

The roasting temperature range was selected in the range of 200–325 °C to reduce the thermal decomposition loss of ammonium chloride. The X-ray diffraction pattern of the calcined product at different temperatures is shown in Fig. 2.

The XRD patterns in Fig. 2 that at 200 and 225 °C reveal that the NH_4Cl peak is present, accompanied by a large amount of unreacted sulphide ore. It has been

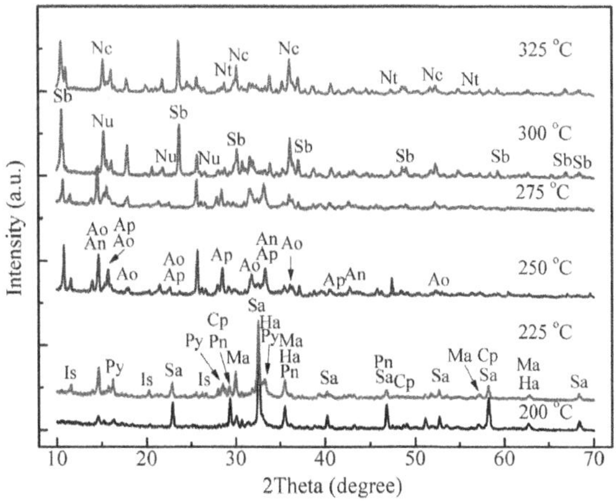

Fig. 2 XRD patterns of products roasted at various temperatures: Sa–NH_4Cl, Ha–Fe_2O_3, Is-$(NH_4)_3Fe(SO_4)_3$, An–NH_4NiCl_3, Ap–$(NH_4)_2CuCl_4$, Ao–$(NH_4)_2FeCl_5$, Sb–$NH_4Fe(SO_4)_2$, Nu–$(NH_4)_2Ni_2(SO_4)_3$, Nc–$NiCl_2$, Nt–CuCl

reported in the literature [7] that NH_4Cl and chalcopyrite react at 200 °C, but, in fact, they did not fully react at 225 °C. The presence of the diffraction peaks of NH_4FeCl_3, Fe_2O_3 and Fe_3O_4 indicates that the iron in the mineral is preferentially oxidized and a chlorination reaction occurs. Since $FeCl_2$ is not stable above 200 °C, the formation of ammonium chloride double salt may improve the stability of ferrous ammonium chloride.

The main phases in the sample roasted at 250 °C are NH_4NiCl_3, $(NH_4)_2CuCl_4$ and $(NH4)_2FeCl_5$, as well as $(NH_4)_3Fe(SO_4)_3$, $(NH_4)_2Fe(SO_4)_2$, $NH_4Fe(SO_4)_2$ and $(NH_4)_2Ni_2(SO_4)_3$. In the samples roasted at temperatures greater than 300 °C, simple chloride or sulphate salts such as $NiCl_2$, $CuCl_2$, CuCl and $FeSO_4$ replaced the NH_4NiCl_3, $(NH_4)_2CuCl_4$ and $(NH_4)_2FeCl_5$. The ferrous sulphate double salt is mainly in the form of $(NH_4)_2Fe(SO_4)_2$ between 250 and 275 °C and decomposes to $(NH_4)_2Fe_2(SO_4)_3$ above 300 °C. The trivalent iron sulphate double salt is mainly in the form of $(NH_4)_3Fe(SO_4)_3$ between 250 and 275°C. The results also show that the presence of the double salt phase is beneficial to maintain the stability of chlorides and sulphates. In addition, whether it is ammonium chloride double salt or sulphate double salt, its decomposition is carried out step by step [8], mainly to gradually remove NH_4Cl or $(NH_4)_2SO_4$.

Figure 3 presents an image of the microscopic surface topography of the calcine. Starting at 225 °C, ore particles attach to the surface of ammonium chloride and react to form corresponding chlorides and sulphates. The local structure of the product is so dense that it cannot be distinctly distinguished from the topography. At 275 and 300 °C, the particles formed tend to be regular and have a certain geometry. At a roasting temperature of 325 °C, as shown in Fig. 3f, the geometrical morphology of the surface of the product is destroyed and is porous. Considering the previous XRD analysis, the decomposition of some double salts and the escape of gas is suspected.

Next, the influence of the ammonium chloride addition was studied, and the results are shown in Fig. 4. The concentrate phase composition is almost unchanged when the ammonium chloride is not added. As the amount of ammonium chloride increases, the chalcopyrite and pentlandite diffraction peak intensities gradually diminish and completely disappear at 80% ammonium chloride addition, indicating that the sulphide was substantially reacted. Ammonium chloride addition was increased until the diffraction peaks of various metal double salts appeared in the product, including both sulphate and chloride. When the ammonium chloride is excessive, the diffraction peak of ammonium sulphate appears at the same time. It is noteworthy that iron oxide and iron sulphate are detected at 50 wt% ammonium chloride addition, ostensibly due to the preferential oxidation of iron.

Spectral analysis was carried out to further investigate the changes in the flotation concentrate and ammonium chloride during the roasting process. Figure 5a is a FTIR chart of calcine produced at different temperatures. The sample shows an absorption peak of water at 3100–3500 and 1600–1620 cm^{-1}, indicating that the sample is hygroscopic. The infrared spectra at 200 and 225 °C exhibited NH_4^+ stretching vibration peaks at 2812 and 3042 cm^{-1}, which were not observed after 250 °C due to the low intensity. The N–H bond stretching vibration near

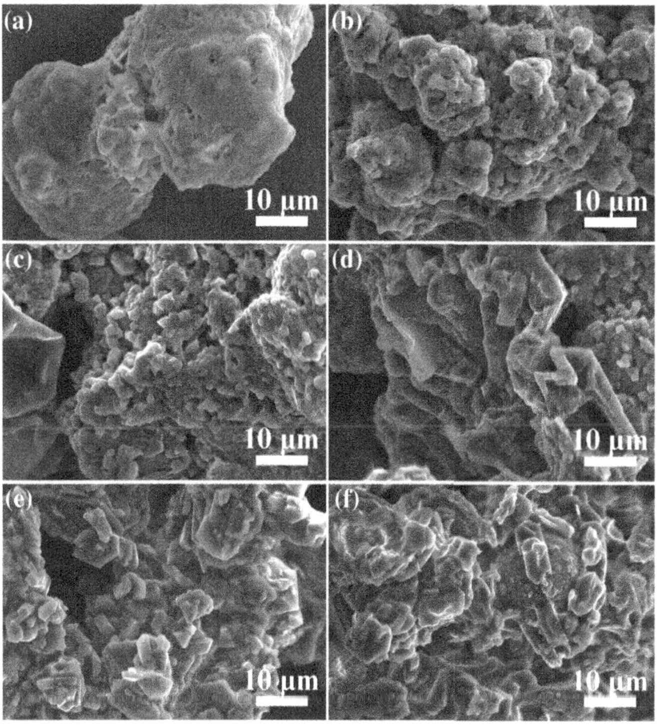

Fig. 3 SEM images of calcine samples produced at various temperatures: **a** 200 °C, **b** 225 °C, **c** 250 °C, **d** 275 °C, **e** 300 °C, **f** 325 °C

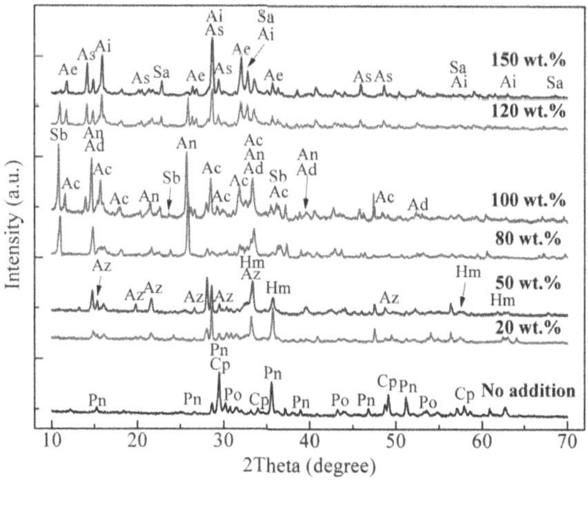

Fig. 4 XRD patterns of calcine samples produced by roasting at 250 °C for 120 min with various additions of NH_4Cl Az–$(NH_4)_2Fe_2(SO_4)_3$, Ae–$(NH_4)_3Fe(SO_4)_3$, Ac–$(NH_4)_2CuCl_4$, Ad–NH_4FeCl_3, As–$(NH_4)_2SO_4$, Ai–$(NH_4)_2NiCl_4$

1406 cm^{-1} has always existed. As the roasting temperature increases, the peak position shifts towards the high wave number direction and appears broadened, indicating that the electrostatic interaction of the N–H bond decreases as the temperature increases. This result is consistent with the fact that the ammonium chloride salt of the metal becomes unstable after the temperature rises. The 1300–400 cm^{-1} region in the IR spectrum is the characteristic peak region of sulphate and the fingerprint region of metal–oxygen bonds. The characteristic peak of sulphate is clearly observed at 250 °C and gradually disappears after 300 °C. Between 225 and 275 °C, 1261 cm^{-1} and 1111 cm^{-1} are two antisymmetric stretching vibration peaks of $\nu_3(SO_4^{2-})$, 970 cm^{-1} is the ν_1 vibrational peak of sulphate, 655 and 599 cm^{-1} are the ν_4 bending vibration peak of sulphate [6], and its vibrational peak of ν_2 at 450 cm^{-1}. The distribution of peak positions indicates that the two triple degeneracy modes of sulphate radicals are split due to the decrease in symmetry, and $\nu_3 > \nu_1$, $\nu_4 > \nu_2$, which indicates that the sulphate is mainly dominated by the C_{2v} configuration in the calcined product [9].

Figure 5b is a Raman diagram of the calcine samples produced at different temperatures. All bands less than 400 cm^{-1} are assigned to the internal vibration mode of the Me–Cl bond and the Fe–O bond [9]. Among them, the strong band at 294 cm^{-1} is related to the ν_1 vibration of the Fe–Cl bond, and the two medium-strength bands at 226 cm^{-1} and 413 cm^{-1} are also classified as the stretching vibration of the Fe–Cl bond. In the Raman spectrum at 200 and 225 °C, the broadband appearing at 470–498 cm^{-1} is assigned to the symmetric stretching mode of the S–S bond of metal sulphide [10]. This band disappeared after 250 °C, indicating that the sulphide completely reacted, consistent with the results of the previous XRD analysis. In the Raman spectrum at 250 °C, the vibration mode of Ni–Cl appears at 134 and 199 cm^{-1}. At 325 °C, the structure of the metal chloride bond will be destroyed due to poor thermal stability, resulting in a simple chloride, where 112, 217 and 409 cm^{-1} are vibrations of the Cu–Cl bond. The band, while at 172 and 266 cm^{-1}, is the vibration mode of $NiCl_2$. The bending vibration of SO_4^{2-}

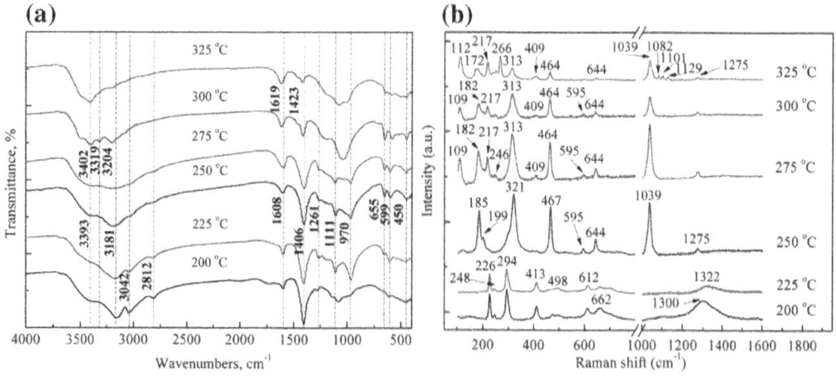

Fig. 5 Vibrational spectra of products at roasted various temperatures: **a** FTIR and **b** Raman

is 464 cm^{-1}, while the strongest band at 1039 cm^{-1} belongs to the v_1 vibration mode of sulphate, and the v_3 vibration band appears at 1275 cm^{-1}, while 595 and 644 cm^{-1} are two v_4 bands, which also indicate that the sulphate is not a T_d point group in the calcined product, but a less symmetric C_{2v} point group [11, 12]. As for the v_3 band, only one appears, probably because the other band strength is too low to be observed.

In order to investigate the role of ammonium chloride in the roasting process, X-ray micro-area analysis was performed on different calcined product layers, and the results are shown in Fig. 6.

Figure 6a, b is different product layers close to the solid ammonium chloride particles and away from the solid ammonium chloride particles, respectively. Among them, the product layer (a) close to the solid ammonium chloride particles is green and relatively loose, and the main phases are $(NH_4)_2SO_4$, $(NH_4)_2Ni_2(SO_4)_3$, $(NH_4)_3Fe(SO_4)_3$ and $(NH_4)_2Fe_2(SO_4)_3$, mainly composed of sulphate double salt. The appearance of the ammonium sulphate diffraction peak demonstrated that the chlorine formation reaction was carried out during the calcination. The product layer (b) far from the ammonium chloride particles is black red, and the particles are fine, and the main phase is the formed iron oxide and the unreacted sulphide ore.

Comparison of Fig. 6a, b reveals that direct contact between the concentrate and the ammonium chloride particles has a great influence on the calcine morphology. The comparison shows that ammonium sulphate is mainly a by-product of the chlorine formation reaction, and the formation of metal sulphate double salt may be related to ammonium sulphate [13]. It is generally believed that sulphate is formed by direct contact with iron oxide and SO_2, O_2, etc. [14], but the temperature of conventional sulfation can reach about 500 °C [15], a temperature much higher than those evaluated in this research. Moreover, in combination with the blank calcination experiment, in the case of not adding ammonium chloride in Fig. 4, it can be concluded that almost all of the sulphide in the concentrate cannot be directly oxidized by air at the experimental calcination temperature. That is to say, the formation of iron oxide is not the direct oxidation of iron sulphide in the

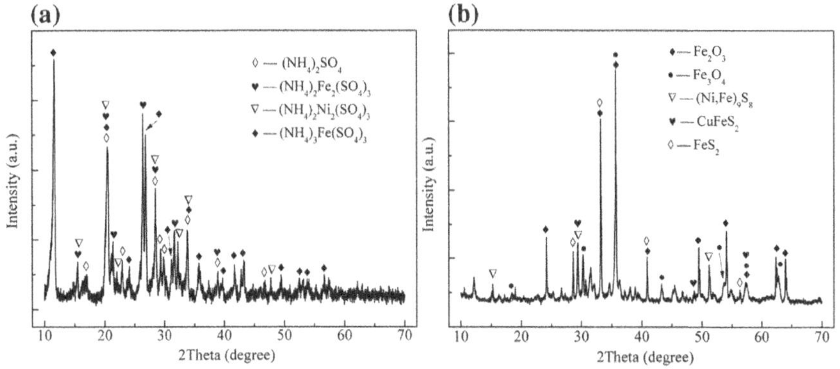

Fig. 6 X-ray micro-area analysis of different product layers at 250 °C

concentrate, but is achieved by oxidation of the chlorination product. According to the results of the micro-area analysis, the roasting process includes both the chlorination reaction of ammonium chloride in direct solid–solid phase contact with the sulphide, and the gas–solid phase reaction of ammonium chloride to form chlorine by the chlorine-forming reaction.

Conclusions

The following conclusions are drawn from the results of this chlorination roast study:

1. Chlorination of the sulphide ore using ammonium chloride as the chloridizing agent is not achieved through HCl formed by thermal decomposition of ammonium chloride; it is the result of the direct contact of solid particles and the indirect chlorination of chlorine.
2. The reaction products of the roasting of ammonium chloride and sulphide include chloride and sulphate, not only in the presence of a large number of iron sulphate double salts of different valence states, but also a small amount of nickel sulphate double salt. Copper is present in the form of $(NH_4)_2CuCl_4$.
3. Ammonium chloride or ammonium sulphate can be effectively combined with metal chlorides to form various forms of soluble double salt compounds, thereby avoiding the production of metal oxides.

Acknowledgements The project was funded by the National Basic Research Program of China (2014CB643403), National Natural Science Foundation of China/Joint Fund for Iron and Steel Research (U1860203) and CAS Interdisciplinary Innovation Team.

References

1. Liu YQ, Chang QZ (2004) Review and prospect of Jinchuan company's comprehensive treatment of smelter off-gas. Sulphuric Acid Ind 5:10–14
2. Mukherjee TK, Gupta CK (1983) Base metal resource processing by chlorination. Miner Process Extr Metall Rev 1(1–2):111–153
3. Sun MS, Wang FM, Cai WF et al (2014) Research on the application of ammonium chloride. Chem Ind Eng Prog 33(4):999–1005
4. Terakado O, Saeki T, Irizato R et al (2010) Pyrometallurgical recovery of indium from dental metal recycling sludge by chlorination treatment with ammonium chloride. Mater Trans 51 (6):1136–1140
5. D'Yachenko AN, Kraidenko RI (2010) Processing oxide-sulfide copper ores using ammonium chloride. Russ J Non-Ferrous Met 51(5):377–381
6. Borisov VA, D'Yachenko AN, Kraidenko RI (2011) Reaction of ammonium chloride with the copper(II) sulfide and oxide, and identification of the reaction products. Russ J Gen Chem 81(7):1430–1433

7. Bayer G, Wiedemann HG (1992) Thermal analysis of chalcopyrite roasting reactions. Thermochim Acta 198(2):303–312
8. Borisov VA, D'Yachenko AN, Kraidenko RI (2012) Mechanism of reaction between cobalt (II) oxide and ammonium chloride. Russ J Inorg Chem 57(7):923–926
9. Nakamoto K (1997) Infrared and Raman spectra of inorganic and coordination compounds: part B: applications in coordination, organometallic, and bioinorganic chemistry. Wiley, New York, pp 184–188
10. Reddy GU, Seshamaheswaramma K, Nakamura Y et al (2012) Electron paramagnetic resonance, optical absorption and Raman spectral studies on a pyrite/chalcopyrite mineral. Spectrochim Acta A Mol Biomol Spectrosc 96:310–315
11. Marinova D, Georgiev M, Stoilova D (2009) Vibrational behavior of matrix-isolated ions in Tutton compounds. II. Infrared spectroscopic study of and ions included in copper sulfates and selenates, J Mol Struct 938(1–3): 179–184
12. Chio CH, Sharma SK, Muenow DW (2007) The hydrates and deuterates of ferrous sulfate ($FeSO_4$): a Raman spectroscopic study. J Raman Spectrosc 38(1):87–99
13. Nagaishi T, Ishiyama S, Matsumoto M et al (1984) Reactions between ammonium sulphate and metal oxides (metal = Cr, Mn and Fe) and thermal decomposition of the products. J Therm Anal 29(1):121–129
14. Shi JF, Wang ZX, Hu QY et al (2013) Recovery of nickel and cobalt from nickel laterite ore by sulfation roasting method using ammonium bisulfate. Chin J Nonferrous Met 2:510–515
15. Dahlstedt A, Seetharaman S, Jacob KT (1992) Thermodynamics of salt roasting of sulphide ores. Scand J Metall 21(6):242–245

Numerical Simulation Investigation on the Flow and Temperature Fields in Tundish with Gas Injection into Ladle Shroud

Wang Zhou, Tao Zhang and San-Xing Chen

Abstract Injecting gas into a ladle shroud to produce fine bubbles is an effective way to promote inclusion removal from molten steel in tundish. The flow and temperature fields in tundish with gas injection into ladle shroud have been investigated based on a *k*-ε model, Euler two-phase flow model and population balance model (PBM), and the relationships between the distribution of bubble diameter and the conditions of gas injection were investigated. The results indicate that gas injection into a ladle shroud significantly improves the flow and temperature fields in tundish. However, with the increase of gas flow rate, surface disturbance of the molten steel obviously increases. In addition, bubbles gather near the exit of shroud with the increase of gas flow rate, which is not conducive to the collision of bubbles and inclusions. Therefore, the gas flow rates of less than 15.13 m^3/h are recommended.

Keywords Numerical simulation · Tundish · Fine bubble · Population balance model

Introduction

As the last vessel for molten steel before it enters the mould, the tundish plays an important role in improving the steel purity, especially for the removal of inclusions [1–3]. It is understood that larger inclusions can be transferred from the steel to the

W. Zhou (✉) · S.-X. Chen
Chongqing CEPREI Industrial Technology Research Institute, Chongqing 401332, China
e-mail: wangzhou@cqu.edu.cn

W. Zhou
College of Materials Science and Engineering, Chongqing University, Chongqing 400044, China

T. Zhang
Department of Science and Technology, Chongqing University of Education, Chongqing 400065, China

© The Minerals, Metals & Materials Society 2019
T. Jiang et al. (eds.), *10th International Symposium on High-Temperature Metallurgical Processing*, The Minerals, Metals & Materials Series,
https://doi.org/10.1007/978-3-030-05955-2_68

overlaying slag film by employing appropriate flow control techniques. However, smaller inclusions particles are difficult to remove because their transfer velocities are so slow [4]. Consequently, reducing the number of small inclusions in molten steel is one of the main objectives in steelmaking.

Gas injection [5–7] is an effective method to promote the removal of inclusions, especially for those smaller than 50 μm. Inclusion removal by bubble flotation occurs by two phenomena: adherence of inclusion on the bubble surface [8] and inclusion capture by the wake of the bubbles [9]. Moreover, fine bubbles performed better in the removal of inclusions. Evans et al. [10] indicated that bubble diameter decreases as the fluid turbulence intensity increases. In steel continuous casting, the strong turbulence intensity can be obtained at both the submerged entry nozzle (SEN) and the ladle shroud. By comparison, gas injection into the ladle shroud is a better way to form fine bubbles because the float-out time is sufficient in the tundish. However, the flow and temperature fields are bound to change with the gas injection, and these factors directly relate to the quality of molten steel. Hence, an investigation of the effects of the flow and temperature fields in the tundish with gas injection into ladle shroud is highly significant.

This work was financially supported by Scientific and Technological Research Program of Chongqing Municipal Education Commission (Grant No. KJ1601403) and Chongqing Research Program of Basic Research and Frontier Technology (No. cstc2017jcyjAX0258).

Mathematical Model

The mathematical model of tundish flow is based on the following assumptions:

(1) molten steel is an incompressible Newtonian fluid.
(2) no slag layer is considered; instead a free surface is assumed.
(3) atmospheric pressure of the liquid surface in the tundish.
(4) bubbles are assumed to have a spherical shape.
(5) initial bubble size is the same as the gas nozzle size.
(6) argon temperature is the same as molten steel prior to injection into the ladle shroud.
(7) argon behaves as an ideal gas (the ideal gas state equation is satisfied).

Based on this series of assumptions, four governing equations were solved in the mathematical model: the continuity equation, the momentum equations, the turbulence equation and the energy equation. The collapse and agglomeration of argon bubbles are simulated via PBM. The interaction between bubbles and molten steel is simulated via Euler two-phase flow model. Typical non-slipping conditions were applied to all solid surfaces. A pressure boundary condition is applied at the outlet, and the pressure at this level is set to be atmospheric.

Water Model

In order to verify the accuracy, rationality and applicability of the mathematical model, a water model with a scale of 1:3 was established, and the structure schematic is shown in Fig. 1. The flow field in tundish is displayed by adding a blue ink tracer in an instant.

Results and Discussion

Fine Bubbles Generation

Effect of Nozzle Size

The effect of nozzle size on fine bubble generation while pouring 85.74 m^3/h steel with a gas flow rate of 21.43 m^3/h was investigated. Gas injection into shroud was examined at five nozzle sizes: 2, 2.5, 3, 3.5 and 4 mm. The computational results indicate that the formation bubbles are mainly concentrated in the range of 0.25–0.75 mm in diameter, and the ratio of 0.25–0.75 mm fine bubbles is in the range of 70–80%. According to Wang et al. [11], the physical modelling results indicate that the formation bubble sizes are less than 0.5 mm with the water flow rate of 43.2 m^3/h and gas flow rate of 4.32 m^3/h. The relationship between the ratio of fine bubbles with diameters from 0.25 to 0.75 mm and nozzle size is shown in

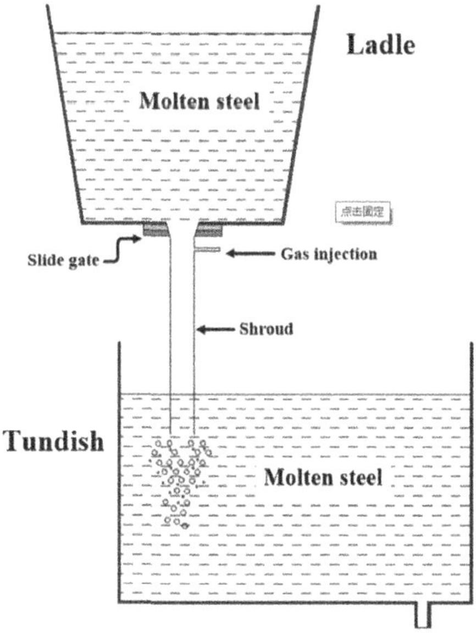

Fig. 1 Structure schematic of water model

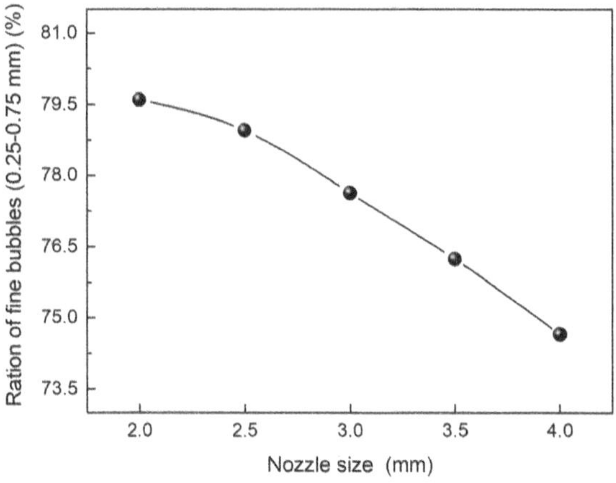

Fig. 2 Relationship between ratio of fine bubbles (0.25–0.75 mm) and nozzle size

Fig. 2. The ratio of fine bubbles (0.25–0.75 mm) decreases with increasing nozzle size. Smaller size nozzle obviously favours fine bubble generation, which is consistent with the study by Zhang et al. [6].

Effect of Gas Flow Rate

The effect of varying gas flow rate on the fine bubble generation using a 3 mm nozzle while pouring 85.74 m^3/h steel was investigated, and the computational model results are shown in Fig. 3. The ratio of fine bubbles (0.25–0.75 mm) decreases from 84.51 to 77.46% as the gas flow rate increases from 4.51 to 28.58 m^3/h. Low gas flow rate favours fine bubble generation because the number of bubbles formed per unit time increases with the increase of gas flow rate. Thus, coalescence and growth of bubbles are more obvious than that in low gas flow rate.

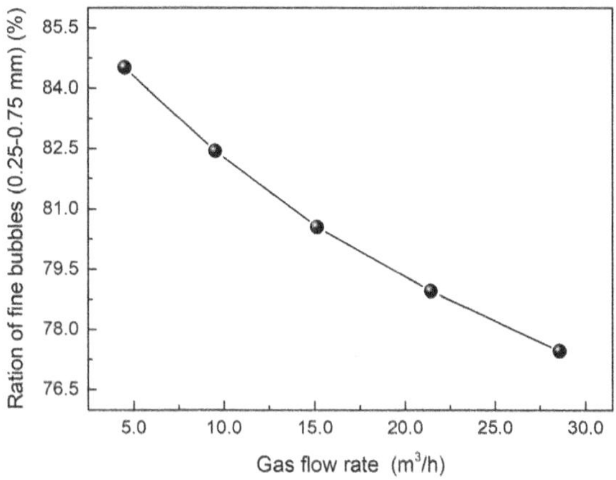

Fig. 3 Relationship between ratio of fine bubbles (0.25–0.75 mm) and gas flow rate

Figure 4a–d shows the distribution of volume fraction of argon bubbles at different gas flow rates. The aggregation extent of bubbles increases sharply with the increase of gas flow rate, especially when the gas flow rate exceeds 9.53 m^3/h. With increasing gas flow rate, the increase in total volume of formation bubbles per unit time increases the buoyancy of the gas phase. The impulse generated by buoyancy greatly reduces the downward momentum of the bubbles, thus the formation bubbles aggregate around the outlet of shroud, which is not conducive to the uniform distribution of bubbles.

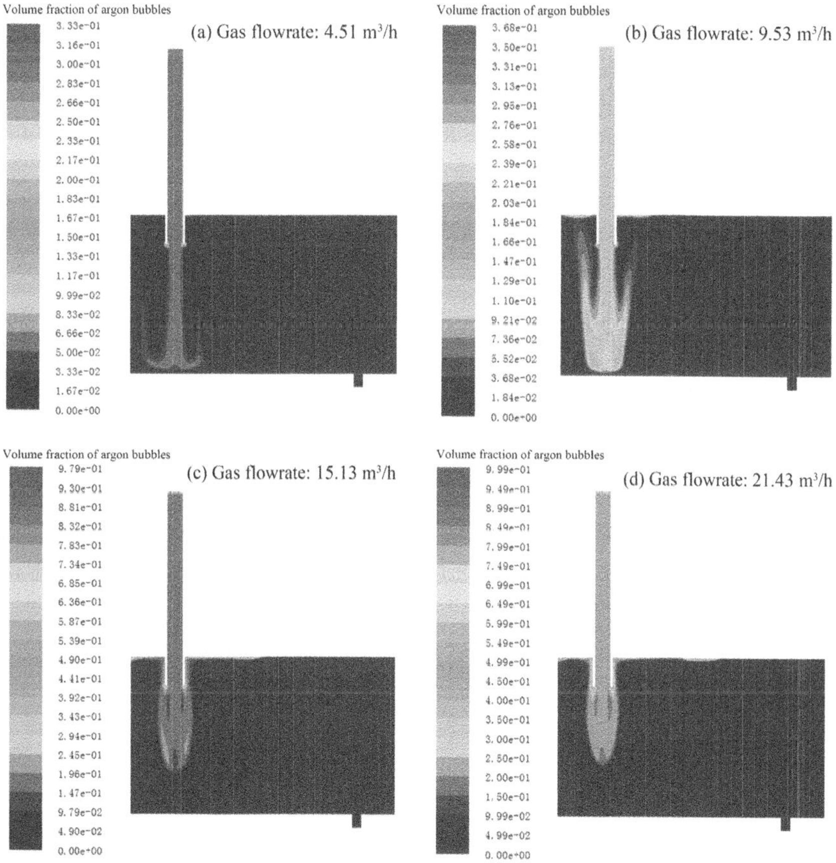

Fig. 4 Distribution of volume fraction (argon bubbles) with different gas flow rates

Influence of Gas Injection on Flow Fields in Tundish

The influence of gas injection on the flow field in tundish with a 3 mm nozzle while pouring 85.74 m^3/h steel was investigated.

As shown in Fig. 5a–b, the flow tracks of fluid in tundish were investigated via mathematical and physical models. There is a short flow in the tundish without gas injection, and the flow tracks of fluid could be improved by gas injection because the buoyancy of bubbles can reduce the downward velocity of molten steel and promote the molten steel to flow upward. Gas injection reduces the short flow, thereby improving the flow field in the tundish.

With increased gas flow rate, the resulting increase in turbulent kinetic energy dissipation of molten steel is transformed into the interfacial energy of bubbles. Comparison of Fig. 6a with Fig. 6b–d shows that the turbulent kinetic energy of molten steel is lowered by the gas injection, and the maximum of turbulent kinetic energy without gas injection is larger than five times that with gas injection. Therefore, gas injection could reduce the erosion of the bottom refractory by molten steel.

Although gas injection is conductive to improving the flow field in tundish, Fig. 6b–d show that the turbulent kinetic energy of molten steel surface will be increased with the increase of gas flow rate. This finding means that the gas flow

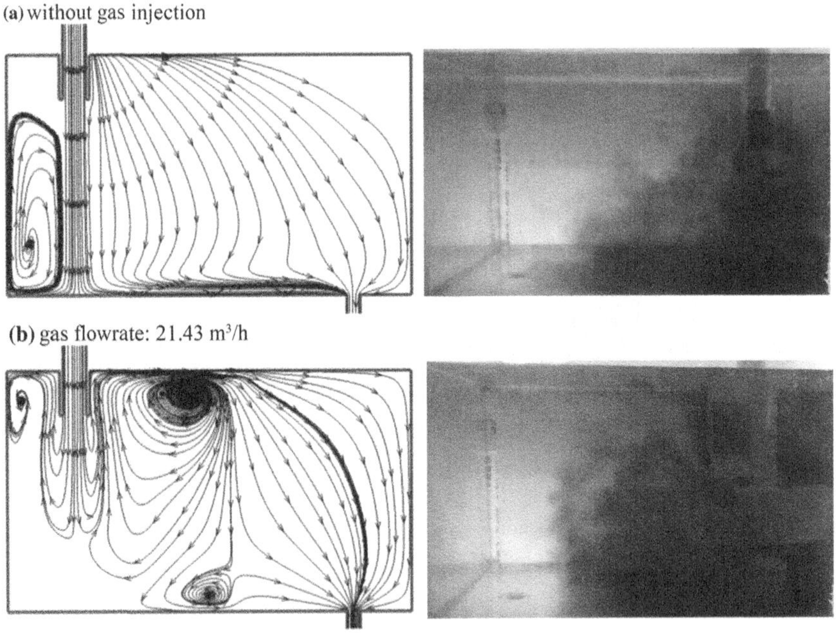

Fig. 5 Flow track of fluid in tundish **a** without gas injection, **b** gas flow rate: 21.43 m^3/h

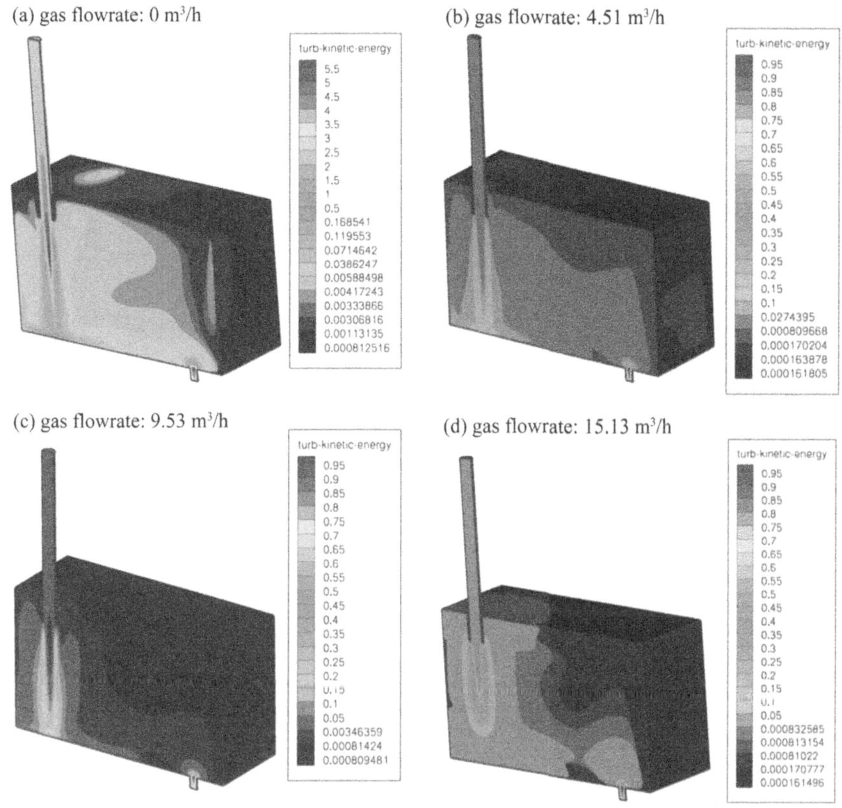

Fig. 6 Distribution of turbulent kinetic energy in tundish. Gas flow rates: **a** 0 m^3/h, **b** 4.51 m^3/h, **c** 9.53 m^3/h, **d** 15.13 m^3/h

rate should be controlled. Combined with the distribution of bubbles in the tundish, the gas flow rate lower than 15.13 m^3/h should yield better results.

Influence of Gas Injection on Temperature Fields in Tundish

The influence of gas injection with a 3 mm nozzle on the temperature field in tundish while pouring 85.74 m^3/h of steel was investigated, and the results are shown in Fig. 7a–d. The data show that gas injection benefits the uniform of temperature in the tundish. The minimum temperature of the symmetry face of the tundish increases from 1818.5 to 1820.51 K with the gas flow rate increase from 4.51 to 15.13 m^3/h. This increase results from the fact that the stir of bubbles

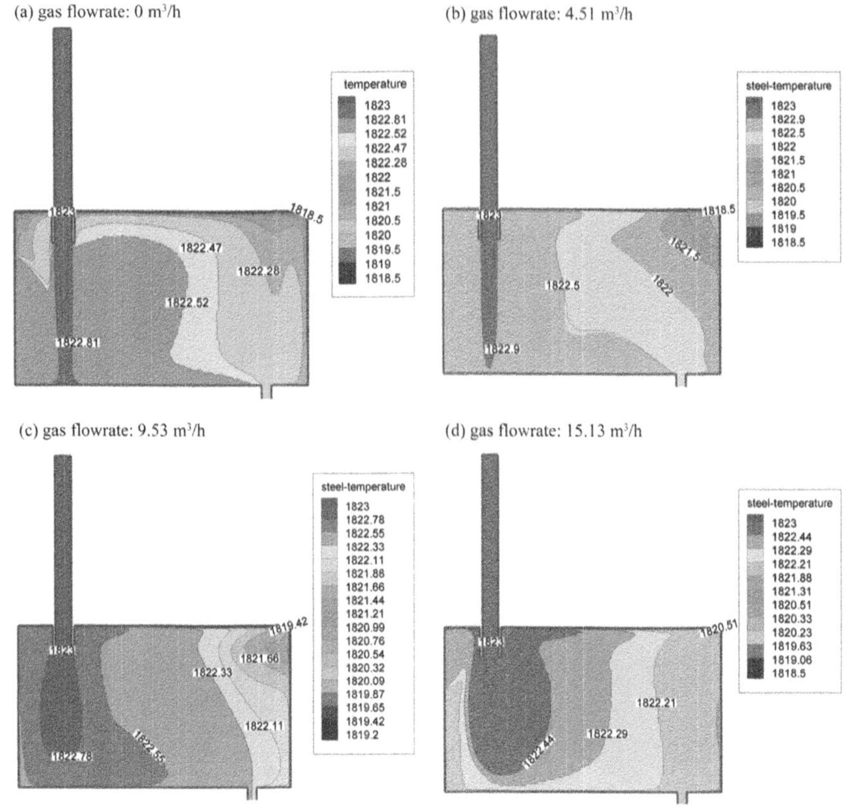

Fig. 7 Temperature fields in tundish. Gas flow rates: **a** 0 m^3/h, **b** 4.51 m^3/h, **c** 9.53 m^3/h, **d** 15.13 m^3/h

promotes the mass transfer of the molten steel in tundish; moreover, the short flow can be reduced by gas injection. Hence, another benefit of gas injection is improved temperature field in the tundish.

Conclusions

In this work, mathematical models have been developed to describe the bubble formation behaviour, flow field and temperature field in the tundish. The collapse and agglomeration of argon bubbles were simulated via PBM, and the interaction between bubbles and molten steel was simulated via Euler two-phase flow model. The results indicate that small nozzle size and low gas flow rate benefit the formation and distribution of fine bubbles. Gas injection can improve the flow and temperature fields in the tundish, especially for reduction of short flow and uniform of temperature. In order to avoid the increase of surface disturbance of the molten steel and obtain the uniform distribution of bubbles, the gas flow rate should be maintained below 15.13 m^3/h while pouring 85.74 m^3/h of steel.

References

1. Zhang LF, Taniguchi S, Cai K (2000) Fluid flow and inclusion removal in continuous casting tundish. Metall Mater Trans B 31(2):253–266
2. Miki Y, Thomas BG (1999) Modeling of inclusion removal in a tundish. Metall Mater Trans B 30(4):639–654
3. Seshadri V, Silva CAD, Silva IAD, Júnior ADSE (2012) A physical modelling study of inclusion removal in tundish using inert gas curtain. Tecnol Metal Mater Miner 9:22–29
4. Guthrie RI (2009) A review of fluid flows in liquid metal processing and casting operations. ISIJ Int 49(10):1453–1467
5. Chang S, Cao X, Zou Z (2018) Regimes of micro-bubble formation using gas injection into ladle shroud. Metall Mater Trans B 49(3):953–957
6. Zhang QY, Wang LT, Xu ZR (2006) A new method of removing inclusions in molten steel by injecting gas from the shroud. ISIJ Int 46(8):1177–1182
7. Rogler JP, Heaslip LJ, Mehrvar M (2004) Inclusion removal in a tundish by gas bubbling. Can Metall Quart 43(3):407–415
8. Zhang LF, Aoki J, Thomas BG (2006) Inclusion removal by bubble flotation in a continuous casting mold. Metall Mater Trans B 37(3):361–379
9. Yang HL, He P, Zhai YC (2014) Removal behavior of inclusions in molten steel by bubble wake flow based on water model experiment. ISIJ Int 54(3):578–581
10. Evans GM, Jameson GJ, Atkinson BW (1992) Prediction of the bubble size generated by a plunging liquid jet bubble column. Chem Eng Sci 47(13/14):3265–3272
11. Wang L, Lee HG, Hayes P (1996) A new approach to molten steel refining using fine gas bubbles. ISIJ Int 36(1):17–24

Recovery Indium from Indium-Enriched Cementation and Co-precipitation Simultaneously Fe-As

Hui Huang, Zhigan Deng, Chang Wei, Xingbin Li, Gang Fan, Cunxiong Li and Minting Li

Abstract A process for extracting of indium from indium-enriched cementation produced in zinc hydrometallurgy process was investigated by oxygen pressure leaching and normal pressure and oxygen-rich leaching. The indium was leached in sulphuric acid medium, at the same time of iron and arsenic were co-precipitated in the form of $FeAsO_4$. The conditions of selective leaching were explored in the presence of oxidant (H_2O_2). The experimental results indicated that high efficiency of extracting indium and co-precipitation of iron and arsenic was obtained with the appropriate operation conditions. Which were determined as the initial sulphuric acid concentration of 0.26 mol/L, liquid–solid ratio of 10 mL/g, at 80 °C for 6 h. And in the process controlled the pH between 1.3 and 1.8, the sulphuric acid total amount of 0.75 mol/L and add H_2O_2 of 0.6 mL/g. Under the optimum conditions, the leaching per cent of indium was more than 92%, precipitation per cent of iron and arsenic was more than 58 and 60%. Compared with traditional methods of leaching, this process has the characteristics of low acid consumption high leaching rate of indium and removed simultaneously iron and arsenic.

Keywords Indium · Acid leaching · Co-precipitation · Ferric arsenate

Introduction

Indium is an important rare metal with several applications due to its excellent physical and chemical properties. It has been widely used in coating, electrical components and semiconductors, solders and alloys and other research [1–3]. It is

H. Huang · Z. Deng (✉) · C. Wei · X. Li · G. Fan · C. Li · M. Li
Faculty of Metallurgical and Energy Engineering, Kunming University of Science and Technology, 253 Xuefu Road, Kunming, Yunnan 650093, China
e-mail: dengzhigan83@163.com

H. Huang
Metallurgy & Materials Engineering Faculty, Kunming Metallurgy College, Kunming 650033, China

© The Minerals, Metals & Materials Society 2019

T. Jiang et al. (eds.), *10th International Symposium on High-Temperature Metallurgical Processing*, The Minerals, Metals & Materials Series,
https://doi.org/10.1007/978-3-030-05955-2_69

worth nothing that indium does not occur in the native state and it is a trace element in some zinc, lead, copper and tin minerals, particularly in sphalerite and marmatite [4–8]. Indium is recovered upon processing base metal concentrates and it is recovered mainly as by-product of zinc [9, 10]. The typically traditional process for extracting indium from indium-enriched cementation includes the following steps: acid leaching, extraction and stripping, and replacement and electro refining [11].

China is rich of zinc resource, most of which exist in forms of sphalerite of high iron and indium. The indium-enriched cementation was got from the zinc hydrometallurgy process [12–14]. However, the traditional leaching procedure presents the character of high content of impurities (Fe and As) and low leaching rate of indium. To address the problem, the effects of reaction time, pressure, final pH in the presence of oxidant on indium leaching were studied in this work.

Experimental

Materials

The indium-enriched cementation was got from a zinc production plant in Wenshan, Yunnan and China, which was obtained by indium precipitation. The indium-enriched cementation was dried to a constant weight at 60 °C for 24 h. The samples were ground and sieved to an average particle size of 75 μm was used for further studies. The chemical components of cementation are given in Table 1. X-ray diffraction (XRD, Rigaku model D/max-2500) result is shown in Fig. 1. The composition of the cementation is very complex, including Cu_2O, FeO, $ZnSO_4{\cdot}7H_2O$ and Cu_2In. And including maybe $In(OH)_3$, Cu_3As, which are cementation and have no diffraction peak in Fig. 1. Using the hydrogen peroxide (30% H_2O_2, analytical grade) as the oxidant.

Experiment Principle

The indium, zinc, copper and others were dissolved from the indium-enriched cementation with the sulphuric acid media under oxidizing condition. Indium leaching was performed by controlling the acid concentration, at the same time, iron and arsenic co-precipitation in the form of $FeAsO_4$. The reactions as follow shown:

Table 1 Main chemical components of indium-enriched cementation (mass fraction, %)

In	Zn	Cu	Fe	As	SiO_2	F	Cl	Al
2.26	10.25	6.39	11.38	4.23	17.39	0.11	0.32	8.32

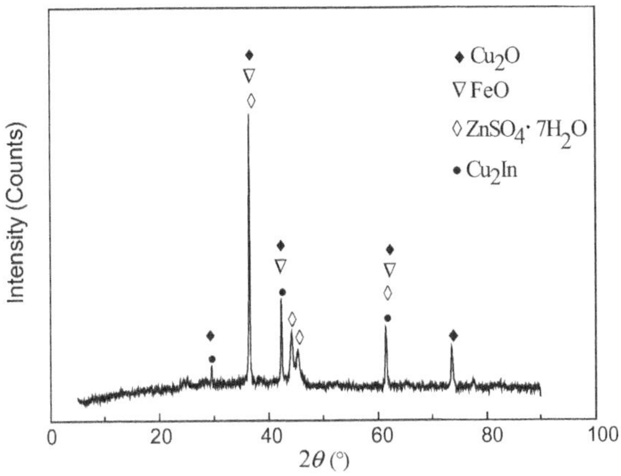

Fig. 1 X-ray diffraction pattern of indium-enriched cementation

$$FeO + H_2SO_4 \rightarrow FeSO_4 + H_2O \quad (1)$$

$$2In(OH)_3 + 3H_2SO_4 \rightarrow In_2(SO_4)_3 + 6H_2O \quad (2)$$

$$Cu_2O + 2H_2SO_4 + H_2O_2 \rightarrow 2CuSO_4 + 3H_2O \quad (3)$$

$$FeSO_4 + Cu_3As + 2H_2SO_4 + 6H_2O_2 \rightarrow FeAsO_4 \cdot 2H_2O \downarrow + 3CuSO_4 + 6H_2O \quad (4)$$

Method

Acidic leaching was performed using sulphuric acid solution, according to the liquid–solid ratio of 10 mL/g at the stirring velocity of 500 r/min with oxidant, the effect of reaction time, pressure, acid concentration and reaction temperature on the leaching were examined. The oxidative pressure leaching tests were performed in a 0.5 L autoclave (GS-0.5) with TA2 lining, equipped with PID temperature controller, a heating mantle and a variable speed stirrer. The atmospheric pressure leaching experiments were carried out in a 500 mL three-neck flask immersed in thermostatically controlled water bath, equipped with a condenser pipe, mechanical stirrer and thermometer.

After reaching the reaction time, the slurry was filtered and the residue was washed adequately with acid water (pH at 1–1.5, Shanghai INESA Scientific Instrument Co., Ltd. PHSJ-5 PH Meter). Iron, zinc, indium and arsenic in the experimental samples were analyzed by means of the oxidation–reduction titration and atomic absorption spectrum.

Results and Discussion

Effect of Pressure

The oxidative pressure leaching was carried out at 4 h, 130 °C, liquid–solid ratio of 10 mL/g, the stirring velocity of 500 r/min, the partial pressure of oxygen at 0.7 MPa with the initial sulphuric acid concentration of 0.26, 0.36, 0.46 and 0.51 mol/L. The results are shown in Figs. 2 and 3.

It is shown in Fig. 2 that the leaching rate of Zn and Cu is increased with increasing initial H_2SO_4 concentration. For the indium leaching, the leaching rate is increased at first, and then decreased with increasing initial H_2SO_4 concentration. In Fig. 3, it is shown that the precipitation rate of Fe, As and SiO_2 is decreased with

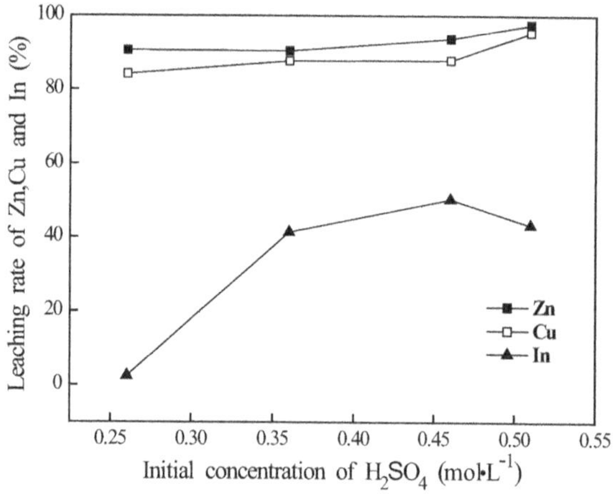

Fig. 2 Effect of initial H_2SO_4 concentration on Zn, Cu and In leaching rate in oxidative pressure leaching

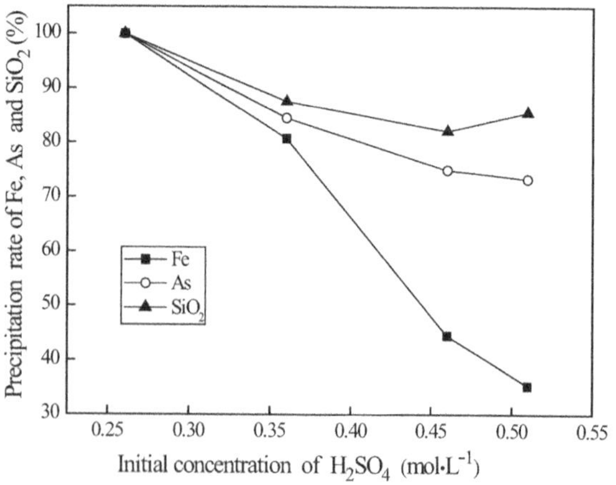

Fig. 3 Effect of initial H_2SO_4 concentration on Fe, As and SiO_2 precipitation rate in oxidative pressure leaching

Table 2 Atmospheric leaching results

Leaching rate (%)			Precipitation rate (%)		
Zn	Cu	In	Fe	As	SiO_2
97.6	96.6	62.57	65.79	84.44	82.19

increasing initial H_2SO_4 concentration. However, the leaching rate of indium cannot achieve the desired effect in the oxidative pressure leaching.

So, the further experiments are conducted in the atmospheric pressure under the conditions as follows. The leaching temperature is 80 °C, liquid–solid ratio of 10 mL/g, the stirring velocity of 500 r/min, reaction time is 4 h, H_2O_2 dosage of 0.6 mL/g with the initial sulphuric acid concentration of 0.51 mol/L. The results are shown in Table 2.

By contrast Table 2, Figs. 2 and 3 show that the leaching of Zn, Cu and In is more efficiency in the atmospheric leaching than the oxidative pressure leaching. There is little difference between the atmospheric leaching and oxidative pressure leaching for the precipitation rate of Fe, As and SiO_2. Therefore, atmospheric leaching is chosen for the recovery of indium and Fe–As co-precipitation.

Effect of Reaction Time

In the atmospheric conditions at 80 °C, liquid–solid ratio of 10 mL/g, the stirring velocity of 500 r/min, H_2O_2 dosage of 0.6 mL/g with the initial sulphuric acid concentration of 0.51 mol/L, the effect of reaction time(4 and 6 h) on the leaching is conducted. After reaction 6 h, the leaching rate of indium reached 70.38%; it is more than 4 h (62.57%).

Effect of Final pH

In the atmospheric conditions at 80 °C, reaction time 6 h, liquid–solid ratio of 10 mL/g, the stirring velocity of 500 r/min, H_2O_2 dosage of 0.6 mL/g with the initial sulphuric acid concentration of 0.26 mol/L, the effect of final pH on the leaching is conducted. During the experiment, the pH of slurry was controlled to 0.8–2.4 using sulphuric acid. The results are shown in Figs. 4 and 5.

It is shown in Figs. 4 and 5 that the leaching rate of indium is decreased, but the precipitation rate of Fe, As and SiO_2 is increased with the rising of final pH in slurry. In order to obtain a high leaching rate of indium and appropriate precipitation rate of Fe, As and SiO_2, the better final pH of slurry is 1.4. Under these conditions, the total amount of acid is 0.75 mol/L, X-ray diffraction (XRD, Rigaku model D/max-2500) for residue and the result shown in Fig. 6. Compared with another acid leaching process [15–18], this process has the characteristics of low

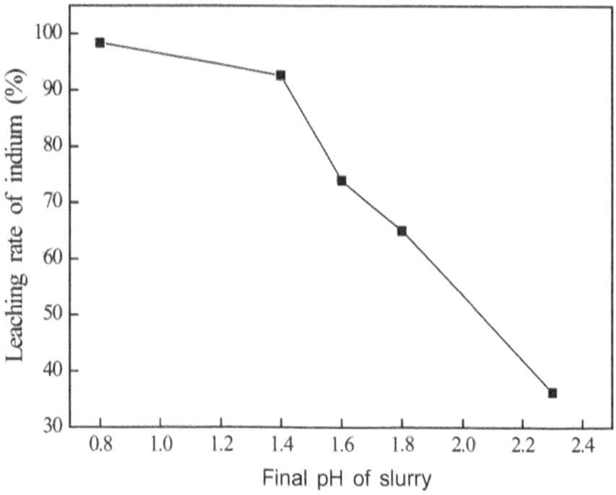

Fig. 4 Effect of final pH of slurry on indium leaching rate in atmospheric leaching

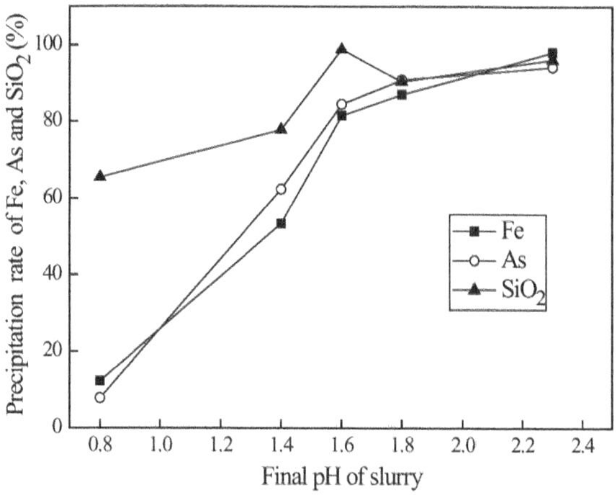

Fig. 5 Effect of final pH of slurry on Fe, As and SiO_2 precipitation rate in atmospheric leaching

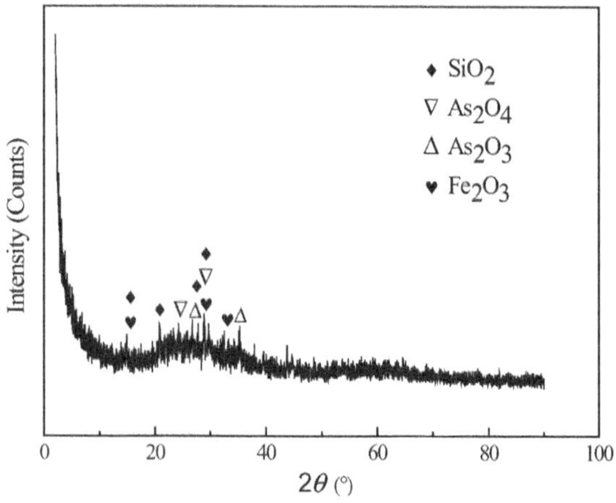

Fig. 6 X-ray diffraction pattern of leaching residue

acid consumption and high leaching rate of indium, at the same time, iron and arsenic co-precipitation in the form of $FeAsO_4$.

Conclusions

(1) Compared with traditional hydrometallurgical process of indium, by controlling the final pH of slurry, the new process can achieve low acid consumption and high leaching rate of indium, and the co-precipitation of Fe-As.
(2) The optimum conditions of atmospheric leaching are initial concentration of sulphuric acid 0.26 mol/L, controlled the final pH of slurry about 1.4, at 80 °C for 6 h, liquid–solid ratio 10 mL/g, the total amount of acid 0.75 mol/L and H_2O_2 dosage of 0.6 mL/g. Under these conditions, the leaching rate of indium is more than 92%, precipitation rate of iron and arsenic was more than 58 and 60%.

Acknowledgements We acknowledge the National Natural Science Foundation of China (No. 51804146, 51564030, 51664030, 51664029, 51664038), National Basic Research Program of China No.2014CB643404), Yunnan Province Applied Foundation Research Programs (No.2016FB097) and Analysis and Testing Center of Kunming University of Science and Technology, China.

References

1. Zhang Y, Li Y, Li C, Wang W, Zhang J, Wang R (2012) Effects of dopant content on optical and electrical properties of In_2O_3: W transparent conductive films. Rare Met 32(2):168–171
2. Li Y, W Wenwen, Zhang J, Wang R (2012) Preparation and properties of tungsten-doped indium oxide thin films. Rare Met 32(2):158–163
3. Wang S (2008) Progress of extraction and application of indium. Eng Sci 10(5):85–94
4. Gao J (2005) Recent progress in the determination of gallium, indium, and thllium. Rare Met 24(1):1–7
5. Hsieh SJ, Chen CC, Say WC (2009) Process for recovery of indium from ITO scraps and metallurgic microstructures. Mater Sci Eng B158(1):82–87
6. Terakado O, Saeki T, Irizato R, Hirasawa M (2010). Pyrometallurgical recovery of indium from dental metal recycling sludge by chlorination treatment with ammonium chloride. Mater Trans 51(6):1136–1140
7. Chen N, Fan Y, Gui W, Yang C, Jiang ZH (2014) Hybrid modeling and control of iron precipitation by goethite process. Chin J Nonferrous Met 24(1):254–261
8. Wu XL, Qin WQ., Wu SK, Ma XH., Niu YJ & Yang CR (2013) Recovery of gallium from zinc concentrate by pressure oxygen leaching. Rare Met 32(6):622–626
9. Yan GU, Zhang TA, Yan LIU, Mu WZ, Zhang WG, Dou ZH, Jiang XL (2010) Pressure acid leaching of zinc sulfide concentrate. Trans Nonferrous Met Soc China 20(S1):136–140
10. Li XH, Zhang YJ, Qin QL, Yang J, Wei YS (2010) Indium recovery from zinc oxide flue dust by oxidative pressure leaching. Trans Nonferrous Met Soc China 20:s141–s145
11. Zeng-ling WU (2011) Indium resources application and the studying progress of separation and recovery technology. Copper Eng 1:25–30

12. Qiang Feng, Chang Wei, Zhigan Deng, Gang Fan (2014) Enriching indium from zinc sulfate solution bearing high iron and indium containing. Chin J Rare Met 38(4):687–692
13. Yang F, Deng Z, Wei C, Li C, Li X (2014) Iron-removal by hematite from leaching liquor of high iron sphalerite. Chin J Nonferrous Met 24(9):2387–2392
14. Li X, Deng Z, Li C, Wei C, Li M, Fan G, Rong H (2015) Direct solvent extraction of indium from a zinc residue reductive leach solution by D2EHPA. Hydrometallurgy 156:1–5
15. Yao JH, Li XH, Li YW (2011) Study on indium leaching from mechanically activated hard zinc residue. J Min Metall, Sect B Metall 47(1): 63–72
16. Guocai ZHU, Z Mingdan, Fan G, Zhao Y (2007) Recovering indium with sulfating roasting from copper-smelting ash. Rare Met 26(5):488–491
17. Li S, Tang M, Jing HE, Yang S, Tang C, Chen Y (2006) Extraction of indium from indium-zinc concentrates. Trans Nonferrous Met Soc China 16(6):1448–1454
18. Liu D, Yang B, Dai Y, Du G, T W, Long HAN (2005) Extracting indium metal from indium-enriched slag. Chin J Rare Met 29(4):574–577

Research on Comprehensive Recovery and Harmless Treatment Process of Copper Smelting Slag

Dongbo Li, Yaguang Guo, Shuaibiao Liang and Deng Ma

Abstract Taking copper smelting slag as raw material, and anthracite and quartzite as additives, the research on comprehensive recovery and harmless treatment of copper smelting slag is carried out by using a 120 kW submerged arc furnace. Theoretical analysis shows that when the temperature is increased from 1200 to 1450 °C, the slag viscosity decreases from 0.283 Pa s to 0.033 Pa s. In the range of 1300–1450 °C, selective reduction can be used to comprehensively recover Cu, Pb, Zn, Sb and other valuable metals in the slag and deeply remove arsenic(As) from the slag. The reduction ability of C is better than that of CO. The results show that addition of quartz can effectively reduce the copper content in the final slag. Under the conditions of holding temperature, quartz content, anthracite content and holding time, 1450 °C, 5%, 10% and 120 min, respectively, Zn volatilizes and gets oxidized, in the same time Cu settles and gets enriched in matte, and the contents of Cu, Pb, Zn, Sb and As in the final slag are reduced to 0.31, 0.043, 0.12, 0.0045 and 0.0014%, respectively. The copper in the final slag mainly exists as sulphide droplets, and arsenic in copper matte is mainly distributed in the region where metallic iron exists.

Keywords Copper smelting slag · Comprehensive recovery · Harmless treatment
Pilot experiment

At present, two processes mainly used for treating copper smelting slag are slag flotation process and electric furnace cleaning process [1–3]. The traditional electric furnace cleaning process, with low recovery rate of copper (about 0.7–1.0% copper content in waste slag) [4], has been less used and the flotation process [5, 6] is widely used in production, because it can deeply recover copper from slag (about 0.3–0.4% copper content in waste slag). However, the flotation process has defects of long flowsheet, large capital construction and big investment [7, 8]. With the increasing complexity of copper smelting raw materials and the increasing smelting

D. Li · Y. Guo · S. Liang · D. Ma (✉)
China ENFI Engineering Corporation, Beijing 100038, China
e-mail: billpuma@163.com

© The Minerals, Metals & Materials Society 2019

T. Jiang et al. (eds.), *10th International Symposium on High-Temperature Metallurgical Processing*, The Minerals, Metals & Materials Series,
https://doi.org/10.1007/978-3-030-05955-2_70

intensity, the contents of Zn, Pb, Sb and other valuable metals in the raw materials are increased [9–12]. In fact, most of the above elements are oxidized and get into slag phase and harmful substances contained in the raw material such as As are also enriched in copper slag under the high oxygen potential conditions [13]. It is possible neither to recover valuable metals such as Zn in the slag, resulting in waste of resources, nor to remove harmful substances such as arsenic in the slag, resulting in secondary pollution hazards from storage of tailings in the slag beneficiation process [14–16]. In order to solve the technical issues arising from recovery of valuable metal elements and remove harmful elements in copper smelting slag, the authors put forward a new technique for comprehensive recovery of copper smelting slag—CR (comprehensive recovery) technology—carry out a series of experiments. The new technology consists of a special furnace—CR furnace—and a relatively high-temperature process. This paper mainly introduces result of the pilot-scale experiment.

Experimental Scheme and Theoretical Analysis

Experimental Materials

The experimental materials are copper smelting slag and additives, among which the copper smelting slag is produced by oxygen bottom-blown smelting furnace in a Chinese company, and its composition is shown in Table 1.

The microstructure of copper smelting slag was analysed by scanning electron microscope, as shown in Figs. 1 and 2. The results of electron probe microanalysis of the marked points in Fig. 1 are shown in Tables 2 and 3. The results of electron probe microanalysis are contents of a single element and a high-valence oxide of a single element, which correspond to slag phase and matte/metal phase, respectively. It is shown from element's surface scanning analysis that Cu and S are basically distributed in the same area. Pb and Zn contents are higher at the position where S content is higher, so this position is presumed to be Pb–Zn sulphide. Zn in slag phase is dispersed, and Zn content is higher in the region where Si content is higher. Part of Sb exists in the Cu–S phase, and Sb in the slag phase is enriched in the region where Si content is high. It is shown from electron probe microanalysis that point 1 and point 2 in Fig. 1 are an iron oxide-rich region and a fayalite phase-rich region, respectively. Point 3 represents a copper matte region with Cu and S contents being 73.25 and 23.28%, respectively, and the main component at point 4 is the metallic elemental antimony.

Table 1 Chemical composition of copper smelting slag, %

Cu	Fe	Pb	Zn	Sb	S	As	Al_2O_3	CaO	MgO	SiO_2
4.55	42.11	0.89	3.44	0.95	2.55	0.44	3.18	1.96	2.09	22.66

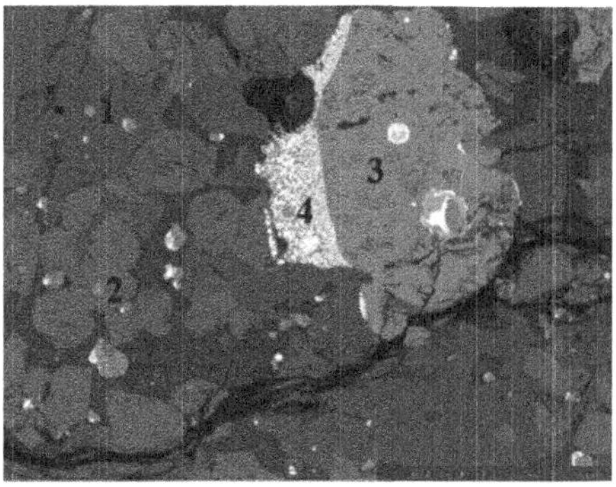

Fig. 1 Microstructure of copper smelting slag (×200 times)

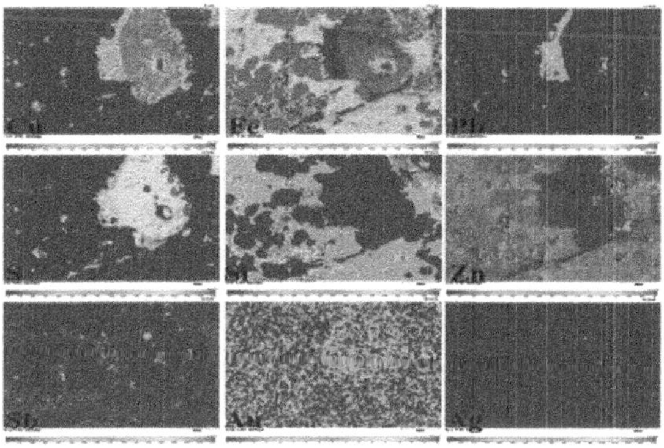

Fig. 2 Analysis of element surface scanning shown in Fig. 1

The Analysis and Testing Center of Changsha Research Institute of Mining and Metallurgy was commissioned to analyse the phase partition of the main metal elements in the copper smelting slag, and the results are shown in Tables 4, 5, 6, 7, 8 and 9. It can be seen from Table 4 that the copper in the slag mainly exists in sulphide phase and metal phase, which will be recycled through settling separation. It can be seen from Table 5 that the zinc in the slag mainly exists in zinc ferrite

Table 2 Composition analysis of points 1–2 by electron microprobe in Fig. 1, %

No.	MgO	Al_2O_3	SiO_2	TiO_2	MnO	Fe_2O_3	ZnO	CaO
1	0.26	1.42	0.35	0.85	0.39	93.94	2.79	
2	2.09		30.14		1.22	62.1	4.11	0.33

Table 3 Composition analysis of points 3–4 by electron microprobe in Fig. 1, %

No.	S	Fe	Cu	As	Sb	Pt	Pb	Bi
3	23.28	3.46	73.25					
4		0.44	2.16	3.46	88.3	3.14	1.31	1.19

Table 4 Copper phase partition, %

Metallic copper	Copper in copper oxide	Copper in copper sulphide	Others	Total
0.8	0.1	3.53	0.12	4.55

Table 5 Zinc phase partition, %

Zinc in zinc oxide	Zinc in zinc sulphate	Zinc in zinc sulphide	Zinc in zinc silicate	Zinc in zinc ferrite	Total
0.028	0.023	0.51	1.25	1.61	3.44

Table 6 Lead phase partition, wt%

Lead in lead oxide	Lead in lead sulphate	Lead in lead sulphide	Lead in lead jarosite	Total
0.16	0.023	0.64	0.066	0.89

Table 7 Iron phase partition, %

Metallic iron	Iron in iron sulphide	Iron in ferrous oxide	Iron in ferric oxide	Iron in magnetic iron	Other insoluble iron	Total
2.14	1.52	14.73	1.55	19.53	2.64	42.11

Table 8 Antimony phase partition, %

Metallic antimony	Antimony in antimony sulphide	Antimony in antimony trioxide	Antimony in other metal solid solutions	Total
0.06	0.04	0.18	0.64	0.95

Table 9 Arsenic phase partition, %

Arsenic in arsenic oxide	Arsenic in arsenic sulphide	Arsenic in arsenate	Arsenopyrite and metal-coated arsenic	Total
0.003	0.024	0.133	0.280	0.440

phase and zinc silicate phase, which will be recycled by deep reduction and volatilization.

Anthracite and quartzite are used as additives in the experiments, of which the compositions are shown in Tables 10 and 11, respectively.

Table 10 Anthracite composition, %

Fixed carbon	Ash	Volatile
75.86	16.16	7.98

Table 11 Quartzitecomposition, %

SiO_2	CaO	Al_2O_3	MgO	Fe
96.83	0.08	0.95	0.016	0.079

Experimental Apparatus

The experiment was carried out by using a 120 kW submerged arc furnace, and the experimental apparatus is shown in Fig. 3.

The experimental steps are as follows:

1. Start the submerged arc furnace, feed copper smelting slag to the furnace, and melt the slag.
2. When the copper smelting slag was completely melted, takes out samples, measures the temperature, adds additives, starts the time and keeps experimental temperature constant. In the holding process, take samples and measure the temperature every 20 min.

Fig. 3 Experimental apparatus. a: Feeding system; b: electric furnace system; c: dust collection and desulphurization system; and d: ladle

3. After reaching the preset holding time, cut off the power to shut down the furnace. Put the products into the ladle through the discharge port at the bottom of the furnace body to cool down. Finally, sample the final slag and copper matte for chemical analysis by the Analysis and Testing Center of Changsha Research Institute of Mining and Metallurgy when the products are cooled to room temperature.

Theoretical Analysis

The objective of this study was to recover valuable metals from the copper smelting slag and realize harmless treatment of the slag, of which the key points are the recovery of copper by matte settling and the recovery of zinc by volatilization.

Two main factors affecting the settling and separation of matte are:

(a) viscosity of molten slag and (b) interfacial tension between copper matte and slag.

The two main factors affecting the recovery of zinc in slag are:

(a) Types of reducing agents and (b) temperature.

Based on the above theory, the reaction thermodynamics and viscosity of copper smelting slag were analysed.

The mixed viscosity of copper smelting slag was calculated using modules Equilib and Viscosity in the FactSage thermodynamic calculation software. Figure 4 shows that the viscosity of the raw slag is 0.283 Pa s at 1200 °C and 0.033 Pa s at 1450 °C. According to Stokes formula (Formula 1), lowering viscosity can reduce the resistance force of matte settling and increase the settling speed of matte in slag, thus facilitating the settling of matte droplets in the slag.

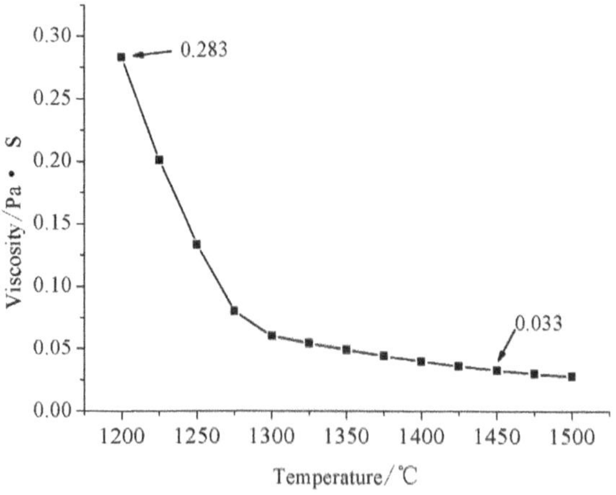

Fig. 4 Viscosity of copper smelting slag

$$v = \frac{2r^2 g(\rho_0 - \rho_1)}{9\mu_s} \tag{1}$$

where

- r Metal droplet radius (m)
- ρ_1 Molten slag density (kg/m^3)
- ρ_0 Copper matte (metal) droplet density (kg/m^3)
- g Gravity acceleration (9.8 N/kg)
- μ_s Molten slag viscosity (Pa s)
- v Metal droplet speed (m/s).

Increasing either the temperature or the SiO_2 content in slag can increase the interfacial tension between copper smelting slag and matte, promoting the separation of slag and matte and facilitating the settling separation of matte droplets [17]. Moreover, the increase of SiO_2 content in molten slag results in the conversion of siloxane polymerization anions to make up the molten slag to a more complex structure, and thus increases the radius or chain length of the polymerization anions, thereby leading to an increase in molten slag volume and a decrease in density, which can promote the settling of copper matte.

The Gibbs free energy of reduction reaction of the main metal oxides in the slag by C was calculated using the module reaction of FactSage software, as shown in Fig. 5. As can be seen from Fig. 5, the reduction temperature and reduction potential required for Cu_2O, PbO and $PbSiO_3$ to be reduced by carbon are relatively low, but the initial reduction temperature of $ZnFe_2O_4$ is high. When the temperature rises above 1300 °C, the sequence of main oxides with Gibbs free energy of C reduction reaction from low to high is as follows: Cu_2O, $CuFe_2O_4$, PbO, $PbSiO_3$, As_2O_3, Sb_2O_3, $ZnFe_2O_4$, Fe_3O_4, ZnO, Zn_2SiO_4, FeO and Fe_2SiO_4.

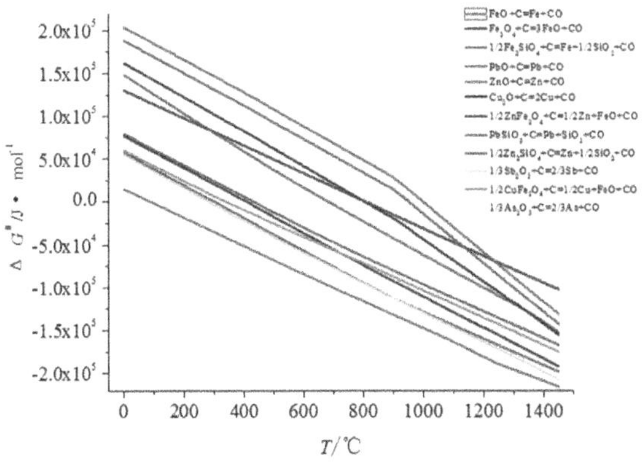

Fig. 5 Gibbs free energy of C reduction reaction of valuable metal oxides in copper smelting slag

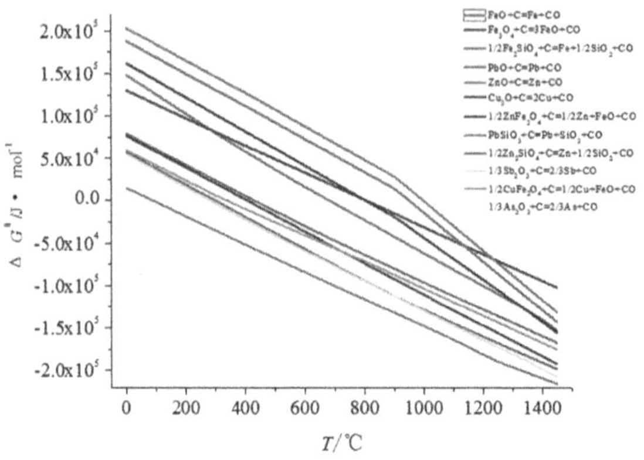

Fig. 6 Gibbs free energy of CO reduction reaction of valuable metal oxides in copper smelting slag

At the same time, the Gibbs free energy of CO reduction reaction of the main oxides in the slag is calculated and listed in Fig. 6. As can be seen from Fig. 6, when the temperature exceeds 1300 °C, the sequence of main oxides in the slag with the Gibbs free energy of CO reduction reaction from low to high is as follows: Cu_2O, $CuFe_2O_4$, PbO, $PbSiO_3$, As_2O_3, Sb_2O_3, $ZnFe_2O_4$, Fe_3O_4, ZnO and Zn_2SiO_4. However, the FeO and Fe_2SiO_4 cannot be reduced.

It can be deduced by comparing the Gibbs free energy calculation results of the reduction of main metal oxides in the copper slag by C and CO that the reduction potential of C is better than that of CO at 1300–1450 °C.

Experimental Scheme

Based on the results of theoretical analysis, the influence of temperature and the contents of anthracite and quartzite on comprehensive recovery of copper smelting

Table 12 Experimental materials and conditions of each heat

No.	Copper smelting slag (kg)	Anthracite (kg)	SiO_2 (kg)	Holding temperature (°C)	Holding time (min)
1	300	30	0	1250	120
2	300	30	0	1350	120
3	300	30	0	1450	120
4	300	30	15	1450	120
5	300	15	15	1450	120

Note During the temperature-holding process, the operating parameters are adjusted by temperature measurements, and the temperature is controlled at ±25 °C slag was investigated. The experimental materials and conditions of each heat are shown in Table 12.

Experimental Results and Analysis

Experimental Results

Using the experimental apparatus shown in Fig. 3 and the above experimental methods, the experiment was carried out according to the experimental scheme shown in Table 12. The compositions of the final slag and copper matte obtained from the experiment are shown in Tables 13 and 14. From Table 13, it can be seen that the contents of Cu, Pb, Zn and Sb in the final slag of the fourth heat are all reduced to the lowest, i.e. 0.31, 0.043, 0.12 and 0.0045%, respectively.

Discussion on Chemical Analysis of Experimental Products

In order to make a detailed analysis on the experimental results, the samples taken during the experiment were analysed chemically and the variation trend of each element's content during the experiments was summarized.

Table 13 Composition of final slag, %

Heat	Cu	Pb	Zn	Sb	As
1	0.69	0.17	0.85	0.011	
2	0.51	0.11	0.42	0.0080	
3	0.45	0.051	0.082	0.0054	0.0025
4	0.31	0.043	0.12	0.0045	0.0014
5	0.35	0.046	0.16	0.0050	0.0032

Table 14 Composition of copper matte, %

Heat	Cu	Pb	Zn	Sb	As
1	52.38	3.01	0.13	7.32	
2	38.41	2.43	0.098	5.89	
3	27.32	1.65	0.064	4.81	1.43
4	28.13	1.54	0.055	4.01	1.39
5	30.97	1.72	0.061	4.43	1.52

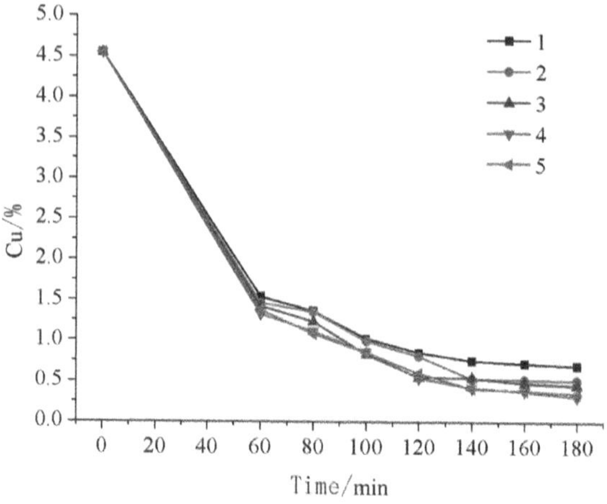

Fig. 7 Relationship between Cu content in slag and smelting time in different heats

Analysis on Change of Cu Content in Slag

Figure 7 shows the variation trend of copper content in slag with smelting time under different experimental schemes. It can be seen from the figure that the copper content in the slag decreases from 4.55% to less than 2.0% after the slag is completely melted and the copper content in slag gradually decreases with the settlement of copper matte inclusions and the reduction and settlement of copper oxides during the temperature-holding process. Upon completion of the experiment, the copper content in the final slag of each heat can be reduced to less than 0.7%.

Holding temperature, contents of anthracite and quartz in raw material can all have an effect on the copper content in the final slag. It is deduced by comparing the results of three heats numbered from 1 to 3 that copper content in the final slag decreases from 0.69 to 0.45% when the holding temperature rises from 1200 to 1450 °C, which is attributed to the fact that the slag viscosity decreases from 0.283 Pa s to 0.033 Pa s.

It is concluded by comparing the results of third heats with fourth heats that copper content in the final slag decreases from 0.45 to 0.31% after addition of 5 wt % quartzite, which is attributed to the fact that the addition of quartzite not only increases the interfacial tension between copper smelting slag and matte but also promotes the conversion of silicon and oxygen polymerization anions in slag to a more complex structure.

It is deduced by comparing the results of fourth heats with fifth heats that copper content in the final slag increases from 0.31 to 0.35% when the anthracite content decreases from 10 to 5%, which is mainly attributed to the fact that increase of carbon amount leads to reduction of a large number of iron oxides and larger density difference between matte and slag.

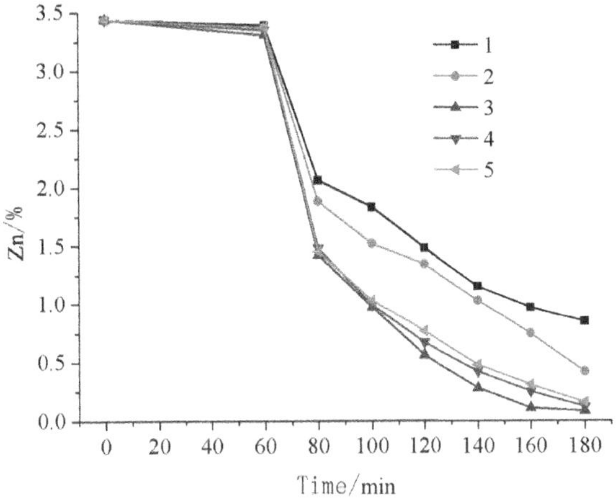

Fig. 8 Relationship between Zn content in slag and smelting time in different heats

Analysis on Change of Zn Content in Slag

Figure 8 shows the change trend of Zn content in slag with smelting time under different experimental schemes. As can be seen from the figure, the zinc content in the slag is reduced from 3.44% to about 3.3%, during the temperature-holding process after the slag is completely melted and gradually decreased during the holding heat-preserved process. After the experiment, the zinc content in the final slag per heat can be reduced to less than 1%.

The holding temperature and amount of anthracite and quartz can all have an effect on the zinc content in the final slag. It is deduced by comparing the results of three heats numbered from 1 to 3 that Zn content in the final slag decreases from 0.85 to 0.082% when the holding temperature rises from 1200 to 1450 °C, which is mainly attributed to the fact that higher temperature brings to better reducibility of zinc oxide. It is seen by comparing the experimental results of the 3rd and 4th heats that Zn content in final slag increases from 0.082 to 0.12% after the addition of quartzite into the copper smelting slag, which is mainly due to the fact that addition of quartzite leads to the generation of $ZnSiO_3$ which is difficult to be reduced. By comparing the experimental results of the 4th and 5th heats, it can be seen that when anthracite content is reduced from 10 to 5%, the zinc content in final slag is increased from 0.12 to 0.16%. This is mainly due to the fact that reduction and volatilization of Zn oxides is more difficult because of the decrease of anthracite addition at the late stage of the reaction when the zinc oxide content and activity are lower.

Analysis on Change of Pb Content in Slag

Figure 9 shows change of Pb content in slag with smelting time under different experimental schemes. It can be seen from the figure that, Pb content in the slag

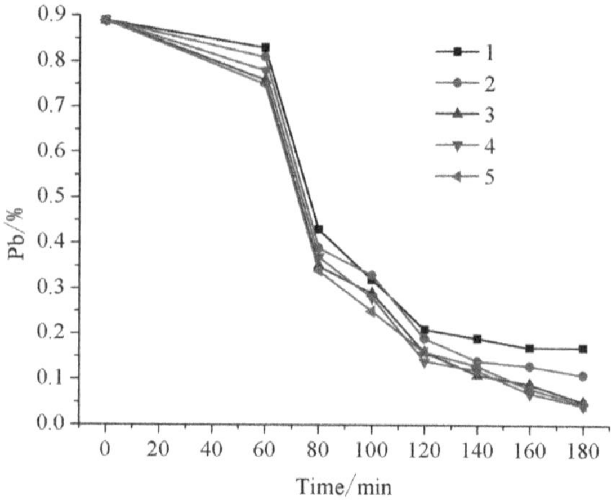

Fig. 9 Relationship between Pb content in slag and smelting time in different heats

decreases from 0.89% to about 0.8% after the slag is completely melted and gradually decreases with the settlement of lead sulphide and the reduction and volatilization of lead oxide during the temperature-holding process. At the end of experiment, Pb content in the final slag per heat can be reduced to less than 0.2%.

It is deduced by comparing the results of three heats numbered from 1 to 3 that Pb content in the final slag decreases from 0.17 to 0.051%. when the holding temperature rises from 1200 to 1450 °C. This is mainly because the increase in temperature results in an increase in reduction of lead oxide and a decrease in slag viscosity, which is beneficial to the settlement of lead sulphide and the mass transfer of lead oxide. By comparing the experimental results of the 3rd and 4th heats, it can be seen that adding 5% quartzite reduces the lead content in the final slag from 0.051 to 0.043%, which is consistent with the variation of matte settlement. Adding quartzite can increase the interfacial tension between slag and lead sulphide, and it is beneficial to the settlement of lead sulphide. By comparing the experimental results of the 4th and 5th heats, it can be seen that when anthracite content is reduced from 10 to 5%, the lead content in final slag is increased from 0.043 to 0.046%. This is mainly due to the fact that reduction and volatilization of Pb oxides is more difficult because of the decrease of anthracite addition at the late stage of the reaction when the lead oxide content and activity are lowered.

Analysis on Change of Sb Content in Slag

Figure 10 shows the change trend of antimony content in slag with smelting time under different experimental schemes. As can be seen from the figure, Sb content in the slag decreases from 0.95% to about 0.7% after the slag is completely melted and gradually decreases during the temperature-holding process. At the end of the experiment, Sb content in the slag per heat can be reduced to less than 0.01%.

The holding temperature and amount of anthracite and quartz can all have an impact on Sb content in the final slag. By comparing the experimental results of the

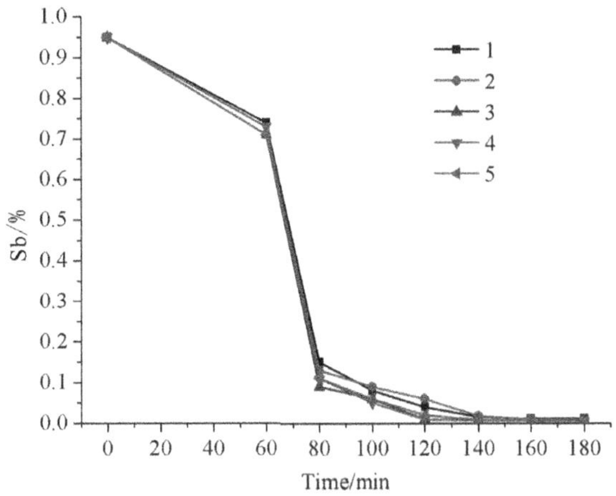

Fig. 10 Relationship between Sb content in slag and smelting time in different heats

first to third heats, it can be seen that Sb content in the final slag decreases from 0.011 to 0.0054% when the holding temperature is increased from 1250 to 1450 °C. It is mainly because that the increase in temperature results in an increase in reduction of antimony oxide and a decrease in slag viscosity, which is beneficial to the settlement of metal solid antimony and the mass transfer of antimony oxide. By comparing the experimental results of the third and fourth heats, it can be seen that addition of 5% quartzite reduces Sb content in the final slag from 0.0054 to 0.0045%, which is consistent with variation of matte settlement. The addition of quartzite can increase the interfacial tension between slag and antimony-containing metal solid solution. By comparing the experimental results of the fourth and fifth heats, it can be seen that the antimony content in the final slag increases from 0.0045 to 0.0050% when the anthracite content decreases from 10 to 5%. This is mainly because that reduction and volatilization of Sb oxides are more difficult because of the decrease of anthracite addition at the late stage of the reaction when the antimony oxide content and activity are lowered.

Analysis of Arsenic Removal Effect

Arsenic is a toxic element, and the existence of arsenic causes serious pollution hazards in slag storage. This problem cannot be effectively solved by the existing slag floatation process and traditional slag cleaning process in an electric furnace. Sb content in the final slag can be reduced to 0.0014% and kept below 0.005% by adding appropriate amount of anthracite and quartzite at 1450 °C to realize harmless treatment of the final slag in this study.

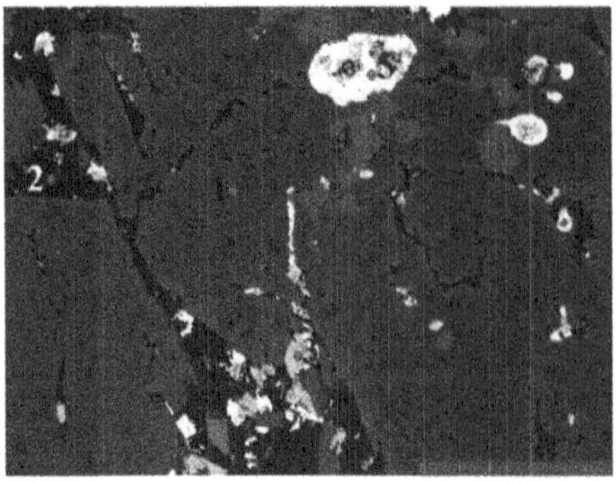

Fig. 11 Microstructure of final slag (× 150 times)

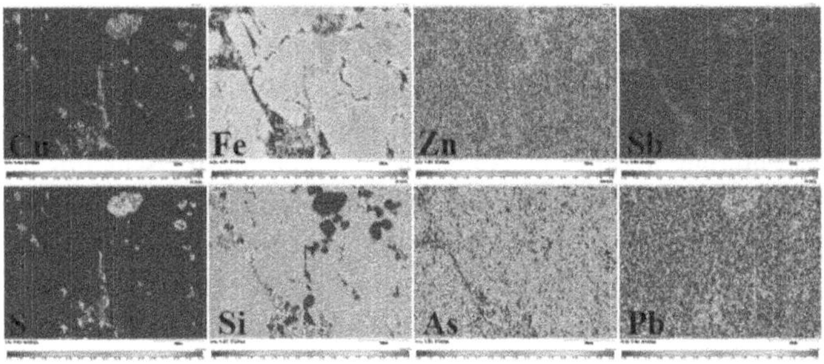

Fig. 12 Field-of-view element surface scanning shown in Fig. 11

Table 15 Analysis of points 1, 2 and 4 by electron microprobe in Fig. 11, %

No.	Fe_2O_3	SiO_2	CaO	MgO	Al_2O_3	Cr_2O_3	TiO_2	Pb	SO_3
1	67.12	30.13	0.38	2.38					
2	23.68	52.11	14.41		8.72				1.09
4	45.69			0.74	4.92	47.48	1.17		

Discussion on Physical Analysis of the Products

Physical Analysis of Typical Final Slag

The final slag obtained from the fifth heat experiment was analysed by scanning electron microscope, with the microstructure and element surface scanning as shown in Figs. 11 and 12, and composition analyses by electron microprobe, as shown in Tables 15 and 16, were carried out for the main phases as shown in Fig. 11.

Table 16 Analysis of points 3 and 5 by electron microprobe in Fig. 11, %

No.	Fe	Cu	S
3	62.58	1.67	35.75
5	3.25	73.62	23.13

Table 17 Copper phase in final slag

Phase	Metallic copper	Copper in copper oxide	Copper in copper sulphide	Others	Total
Content/%	0.05	0.005	0.28	0.015	0.35

The analysis of the results shows that the existing arsenic and zinc in the final slag are dispersed and Pb and Sb are mainly enriched in sulphide inclusion regions. The phases at point 1, point 2, point 3, point 4 and point 5 are mainly fayalite phase, fayalite and silicon oxide-rich phase, the iron sulphide inclusion, the iron chromate formed by slag erosion of lining and the copper sulphide inclusion.

The Analysis and Test Center of Changsha Research Institute of Mining and Metallurgy was commissioned to carry out phase analysis of the residual copper in the final slag. The results, as shown in Table 17, show that 80% of the copper in the final slag is in sulphide inclusion and 14.3% of the copper is in metal state.

Physical Analysis of Typical Copper Sulphide

The copper matte obtained from the experiment of the fifth heat was analysed by scanning electron microscopy, with its microstructure and element surface scanning as shown in Figs. 13 and 14, and composition analyses by electron microprobe, as shown in Table 18, were carried out for the main phase as shown in Fig. 13.

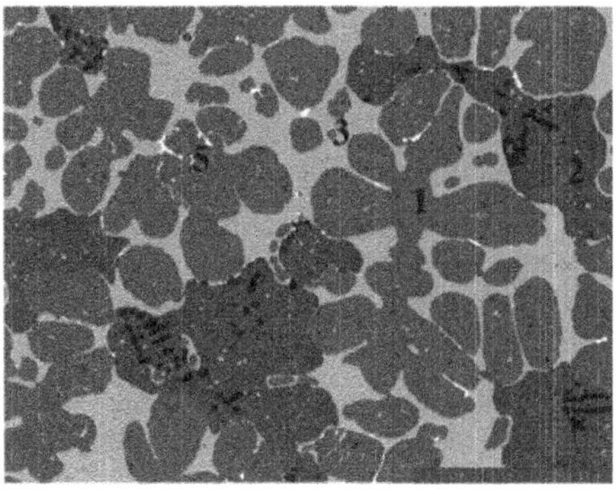

Fig. 13 Microstructure of copper matte (× 200 times)

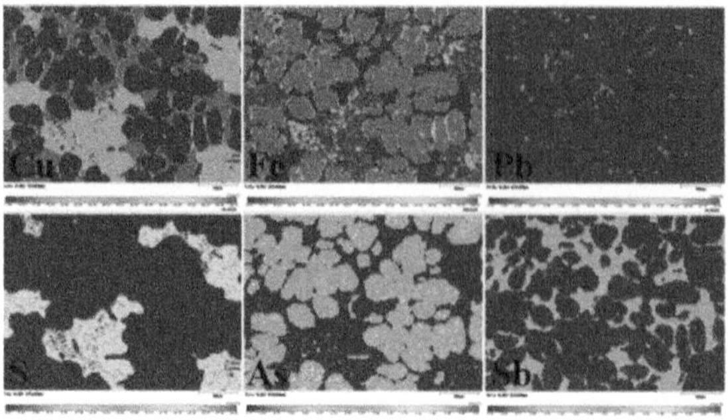

Fig. 14 Field-of-view element surface scanning shown in Fig. 13

Table 18 Analysis of points 1–5 by electron microprobe in Fig. 13, %

No.	Fe	Co	Cu	Sb	Zn	Sn	As	Pb	S
1	93.68	2.71	2.49				1.12		
2	6.35		72.23						21.42
3	2.34		95.2	2.12		0.35			
4	25.57		2.98		35.33				36.12
5	2.31		0.38					97.3	

As shown in the result analysis, the main phase is metallic iron at point 1 where arsenic is mainly distributed in the metallic iron-rich region. Copper matte is the main phase at point 2 where sulphur is distributed in the same region with most copper. Copper mainly exists at point 3 where antimony is mainly distributed in the same region with metallic copper. Small-sized inclusions exist at point 4 with iron and zinc sulphide as the main components, and the main phase at point 5 is metallic lead.

Conclusion

(1) A new treatment technique of comprehensive utilization for copper smelting slag—CR (comprehensive recovery)—is proposed through theoretical analysis and experimental research. This technique not only overcomes many disadvantages existing in the traditional treatment process but also realizes comprehensive recovery of valuable metal elements such as copper, lead, zinc and antimony in slag, deeply removes arsenic in slag and eliminates the pollution hazards in final slag storage.

(2) Under the conditions that anthracite content is 10%, quartz content is 5%, holding temperature is 1450 °C, and holding time is 120 min, the copper, lead, zinc, antimony and arsenic contents in the final slag are reduced to 0.31, 0.043,

0.12, 0.0045 and 0.0014%, while the contents of copper, lead, zinc, antimony, and arsenic in the matte phase are 28.13, 1.54, 0.055, 4.01 and 1.39%, respectively.

(3) It is shown by experimental result that the addition of quartz can effectively reduce the copper content in the final slag, but it has a negative impact on the deep recovery of zinc.

(4) It is seen from microanalysis that the residual copper in the final slag mainly exists as sulphide droplet inclusions, and arsenic in the matte is mainly distributed in the region where iron phase aggregates.

References

1. Zhang J, Qi Y, Yan D, Xu H (2015) A new technology for copper slag reduction to get molten Iron and copper matte. J Iron Steel Res Int 22(5):396–401
2. Zhang H, Shi X, Zhang B, Hong X (2013) Behaviors of the molten copper slags in the vertical electric field. ISIJ Int 53(10):1704–1708
3. Sarrafi A, Rahmati B, Hassani HR et al (2004) Recovery of copper from reverberatory furnace slag by flotation. Miner Eng 17(3):457–459
4. Jiang M, Li C (2009) Depletion and resourceful utilization of the copper smelting slag. Chin Nonferrous Metall 3:57–60
5. Bian R, Yuan J, Chne J (2012) Depletion of copper smelting slag and technological economy analysis. Nonferrous Met Ext Metall 3:14–17
6. Liao Y, Ye C, Wang Y, Cao L (2017) Resource utilization of copper smelter slag-a state-of-the-arts review. Chem Ind Eng Prog 36(8):3066–3073
7. Liu Y (2012) Techniques contrast on slag dressing with slag depletion of bottom blowing furnace copper smelting slag. Nonferrous Min Metall 28(6):33–35
8. Jiang F, Huang H, Sun W, Liu R, Xiong Y (2013) Comparative experimental research on flotation of copper from different copper slag. Nonferrous Met (Mineral Process Sect) 6:60–63
9. Chen G, Yang H (2016) Distribution of associated metals in double top blown copper smelting process and its behavior. Chin Nonferrous Metall 2:68–71
10. Roghani G, Takeda Y, Itagaki K (2000) Phase equilibrium and minor element distribution between FeO_X-SiO_2-MgO–based slag and Cu_2S-FeS matte at 1573 K under high partial pressures of SO_2. Metall Mater Trans B 31B:705–712
11. Chaubal PC, Sohn HY, George DB, Bailey LK (1989) Mathematical modeling of minor-element behavior in flash smelting of copper concentrates and flash converting of copper mattes. Metall Trans B 20B:39–51
12. Tan P (2011) Modeling and control of copper loss in smelting slag. J Miner Met Mater Soc 63 (12):51–57
13. He X (2013) Research on distribution of impurity elements in copper smelting. Nonferrous Met (Ext Metall) 2:55–57
14. Cao Z, Sun T, Xue X, Liu Z (2017) Recovery of iron and zinc from copper slag by rotary hearth furnace with anthracite as reductant. Min Metall Eng 37(2):74–78
15. Zhao K, Cheng X, Qi Y, Gao J, Shi X (2012) Analysis of copper slag treatment process and new technology of comprehensive utilization. Chin Nonferrous Metall 1:56–60
16. Jiang P, Wu P, Hu X, Zhou G (2016) Copper slag comprehensive utilization development and new technology is put forward. Chin Min Mag 25(2):76–79
17. Fu C (1984) Nonferrous metallurgy principle, 1st edn. Metallurgical Industry Press, Beijing

Structure–Property Correlations of $Al_2O_3 \leftrightarrow SiO_2$ Substitution in Blast Furnace Slag

Zhiming Yan, Xue-wei Lv, Ramana G. Reddy, Zhengde Pang and Wenchao He

Abstract Knowledge of the structure and properties of aluminosilicate slags is of great importance in metallurgical processes. The quantitative structural information could be obtained from Raman spectra of the quenched slag samples. The relationship between the properties and structure in the CaO-MgO-SiO_2-TiO_2-Al_2O_3 system with various Al_2O_3/SiO_2 ratio were studied in present work. Various properties such as viscosity, density, and electrical conductivity can be expected as a simple linear function of structural complexity. It is can be concluded that these physical properties are strongly dependent on the structural complexity of aluminosilicate melts. The methodology outlined in the current study can be used to predict the physicochemical properties of aluminosilicate slags when using the high alumina iron ore in ironmaking process.

Keywords Aluminosilicate slags · Structure · Properties

Introduction

As a result of the consumption and huge global demand for high-quality steel products, and the increasing cost of raw materials for ironmaking, steel producers have increased their use of low-cost, low-grade raw materials. The usage of high alumina ore will result in the blast furnace slag system translation to aluminosilicate based [1–7]. The aluminosilicate slags are very important not only in the metallurgical processes but also in the glass manufacture and magmatic processes in petrology. Structure of alumonosilicate slags is the dominant factor affecting the

Z. Yan · X. Lv (✉) · Z. Pang · W. He
College of Materials Science and Engineering, Chongqing University, Chongqing 400044, China
e-mail: lvxuewei@163.com

Z. Yan · R. G. Reddy
Department of Metallurgical and Materials Engineering, The University of Alabama, Tuscaloosa 35487, AL, USA

© The Minerals, Metals & Materials Society 2019

T. Jiang et al. (eds.), *10th International Symposium on High-Temperature Metallurgical Processing*, The Minerals, Metals & Materials Series,
https://doi.org/10.1007/978-3-030-05955-2_71

high temperature properties such as viscosity, density, surface tension, and electrical conductivity. Therefore, many researchers have tried to measure and predict these macroscopic thermo-physical properties as a function of composition of silicates based on the microscopic view of structures. Raman spectroscopy was originally used to probe the local anionic structure of alumonosilicates and provide a quantitative measurement of the various chemical species involved in the melts network structure [8–10].

We previously reported the properties such as viscosity, density, surface tension, sulfide capacity, and electrical conductivity of the CaO-SiO_2-MgO-Al_2O_3-TiO_2 slag system when substituted silica by alumina [1–3]. In the present work, the structural information of the slags was investigated by Raman spectroscopic analysis for melt-quenched samples. Furthermore, the relationships among the structural information and thermos-physical properties are developed.

Experimental Procedure

Preparation of Samples

The chemical compositions of the slags investigated in the present study are shown in Table 1. The slags for the thermos-physical measurements were made from synthetic slags, which were synthetized with the known chemical composition slag from the analysis pure grade regent. The analysis pure grade regent was first claimed at 1273 K (1000 °C) in a muffle furnace to remove the impurities such as carbonate, hydroxide, and moisture. Weighted chemicals were thoroughly pressed into cylindrical pellets before being used in the experiments. About 200 g of the sample was used in the Mo crucible, which gave a slag bath of at least 40 mm deep contained in crucibles. After completing the thermos-physical measurements, the melts were reheated to 1773 K (1500 °C) and quickly quenched on a water-cooled plate, but the cooling rate of water quenching is not precisely known. Slag samples were analyzed using X-ray fluorescence spectroscopy to confirm the composition, and the results were in good agreement with the anticipated compositions.

Analysis of Raman Spectra

Raman spectroscopy (*Lab. RAM HR Evolution; HORIBA Scientific, France*) with a 532-nm laser was used for further analysis of the slag structure. The analysis of the Raman spectra was constrained from 1200 to 400 cm^{-1}, and the area fraction and quantitative amounts of the individual silicate and aluminate bonds were ascertained by deconvoluting the Raman spectra with the Peak Fit V4 software. X-ray diffraction (XRD) (Model D/max2500/PC (Cu Kα)) analysis was carried out to

Table 1 Composition of slag samples (wt%) and the peak area ratio obtained from a Gaussian deconvolution of Raman spectra

Slag	SiO_2	CaO	Al_2O_3	MgO	TiO_2	A/S	$Al\text{-}O^c$	$Al\text{-}O^-$	Si-O-Al	Q^0	Q^1	Q^2	Q^3
1	32	40	18	9	1	0.56	0.189	0.062	0.151	0.178	0.206	0.112	0.103
2	30	40	20	9	1	0.67	0.169	0.050	0.162	0.176	0.205	0.138	0.100
3	28	40	22	9	1	0.79	0.156	0.072	0.189	0.271	0.178	0.091	0.043
4	26	40	24	9	1	0.92	0.141	0.077	0.204	0.264	0.213	0.070	0.031
5	24	40	26	9	1	1.08	0.179	0.097	0.192	0.246	0.197	0.061	0.028
6	22	40	28	9	1	1.27	0.185	0.127	0.162	0.191	0.219	0.064	0.051

ensure the phase of samples. Scanning was carried out at an angular range of 10°–90° and scan rate of 4°/min. The absence of any characteristic peaks in the XRD pattern indicates that water quenching imparts sufficient cooling to obtain an amorphous structure.

Figure 1 shows a typical deconvolution result carried out in the present study, e.g. for the 40 mass% CaO-SiO_2-Al_2O_3-9 mass% MgO-1 mass% TiO_2 slags system, $Al_2O_3/SiO_2 = 0.92$. Bands marked at 876, 905, 978, and 1059 cm^{-1} correspond to Q^0 (850–880 cm^{-1}), Q^1 (900–920 cm^{-1}), Q^2 (950–1000 cm^{-1}), and Q^3 (1050–1100 cm^{-1}) units, respectively [10–12]. The stretching Al-O-Si vibration is observed near 820–850 cm^{-1}, which the peak at 834 cm^{-1} [13]. The $Al\text{-}O^-$ and $Al\text{-}O^0$ stretching vibrations are separately observed near 700–800 cm^{-1} and 530–610 cm^{-1}, which the peak is 721, 567 cm^{-1}, respectively [8, 14, 15]. Here, the Raman scattering for Q^4 unit (1200 and 1060 cm^{-1}) could not generally be resolved due to its very low intensity in these aluminosilicate systems.

Hence, from the quantitative analysis of Raman spectra, the obtained configuration of each structural units is shown in Table 1, wherein the relative area fractions of the fitted Gaussian bands are listed.

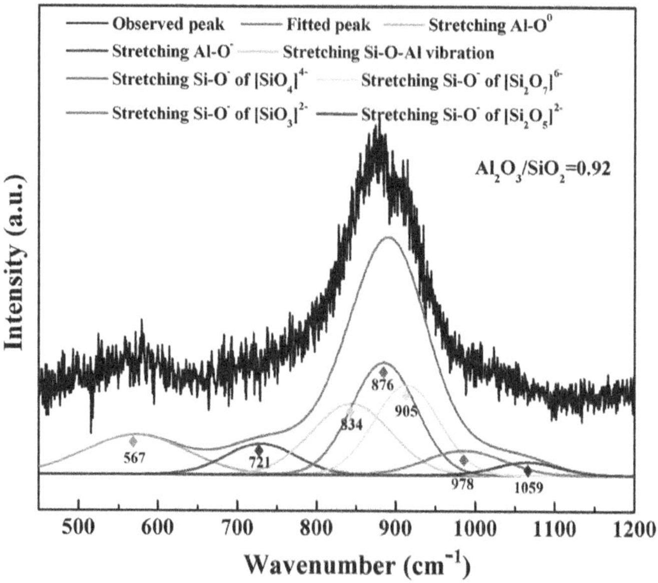

Fig. 1 Raman spectroscopy of as-quenched samples and deconvoluted spectra for the CaO-SiO_2-Al_2O_3-MgO-TiO_2 system; $Al_2O_3/SiO_2 = 0.92$

Results and Discussion

Effect of $Al_2O_3 \leftrightarrow SiO_2$ on the Structure

Figure 2 shows the fraction of structural units as a function of the Al_2O_3/SiO_2 ratio obtained from the deconvoluted Raman spectra. The $(Q^0 + Q^1)$ and Si-O-Al increase when the Al_2O_3/SiO_2 ratio is less than 0.92 and decreases when the alumina content further increases. By contrast, the $(Q^2 + Q^3)$ and Al-O^0 structural units demonstrate contradictory results. Notably, the separation of the variation for Al_2O_3/SiO_2 is nearly 0.92, such that the atoms of Al are close to those of Si. This phenomenon may explain why the Si-O-Al concentration reaches its maximum value. The structural unit of Al-O^0 increases with increasing Al_2O_3/SiO_2 content. According to the Raman analysis, the ratio of complex structures and simple structures was defined as structural complexity (Sc) to characterize the structure of the slags, which shown as Eq. (1):

$$Sc = \frac{f_{Q^2} + f_{Q^3} + f_{(Si-O-Al)} + f_{(Al-O^0)} + f_{(Al-O^-)}}{f_{Q^0+Q^1}} \quad (1)$$

where f_i is the fraction of structure unit i.

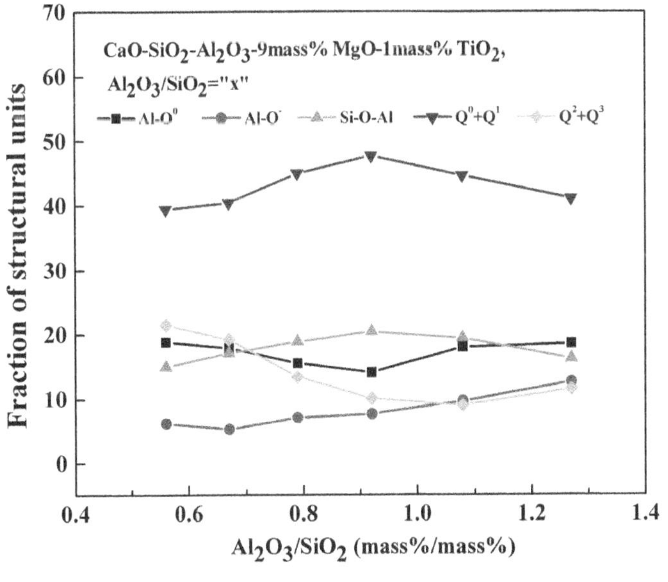

Fig. 2 Fraction of the characteristic structural units with varying Al_2O_3/SiO_2

Structure–Property Relationship

Viscosity

The viscosity of molten slags has played an essential role in determining the performance and productivity of the ironmaking process. The viscosity of the studied slag system has been reported in the previous work [3]. Figure 3a shows the relationship between the viscosity and the logarithm of structural complexity (Sc). The viscosity of the aluminosilicate slags increases linearly as ln (Sc) increases at 1773 K. The activation energy of aluminosilicate melts was obtained from an Arrhenius relationship (Eq. 2) between viscosity and temperature.

$$\eta = \eta_0 \exp\left(\frac{E_\eta}{RT}\right) \tag{2}$$

In Fig. 3b, the activation energy for the viscous flow of $CaO\text{-}SiO_2\text{-}MgO\text{-}Al_2O_3\text{-}TiO_2$ slag is a linear function of structure complexity, which is defined as ln (Sc.) in the present study as follows.

$$E_\eta = (\text{kJ}) = 84.20 \times \text{In}(Sc.) + 186.56\left(\text{r}^2 = 0.92\right) \tag{3}$$

Consequently, the ln (Sc.) can be experimentally obtained and act as a good polymerization index, which could be used to quantify the effect of structure on the viscosity of aluminosilicate melts.

Density

Density (q) is an useful variable for structural interpretation of molten slags. The relationship between density of the $CaO\text{-}SiO_2\text{-}MgO\text{-}Al_2O_3\text{-}TiO_2$ slag and ln (Sc.)

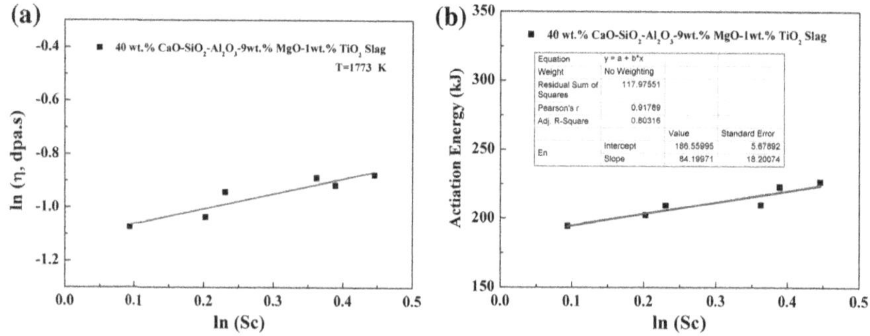

Fig. 3 Relationship between the viscosity (**a**) and activation energy (**b**) with the structural complexity at 1773 K

value at 1773 K is shown in Fig. 4. The density of slags linearly decreases by increasing the ln (Sc.) at a given temperature with the following relationship:

$$\ln \rho\left(\mathrm{g/cm^3}\right) = -0.048 \times \ln\left(Sc.\right) + 2.69\left(r^2 = 0.92\right) \quad (4)$$

In the alumonosilicate systems, the charge compensation is required of $[AlO_4]^{5-}$ tetrahedra due to the difference in the valence comparing with $[SiO_4]^{4-}$ which are usually achieved by alkali metals and alkaline earths. In the case of $Al_2O_3 \leftrightarrow SiO_2$ substitution for the slag with fixed CaO content (40 mass pct), the bridging oxygens will increase. Richet et al. [9]. concluded that bridging oxygens have a larger partial molar volume than nonbridging oxygens. The results are shown in Eq. (4) and Fig. 4 are consistent with their conclusion.

Electrical Conductivity

Electrical conductivity of molten slag is a significant physical property in the metallurgical process. Such property is crucial in the comprehensive utilization of blast furnace slag with an electric arc furnace, such as the preparation of rock wool from the slag through electric arc furnace. The electrical conductivity (k) follows an Arrhenius-type relationship, (Eq. 5), similar to viscosity:

$$\kappa = A\ \exp\left(\frac{E_K}{RT}\right) \quad (5)$$

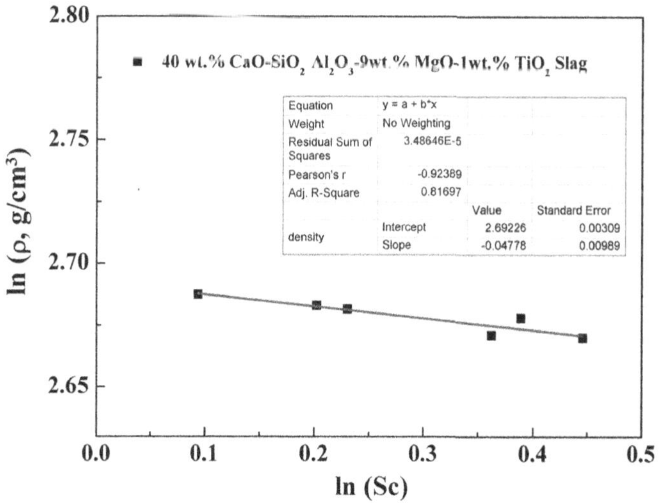

Fig. 4 Density of studied slag as a function of structural complexity

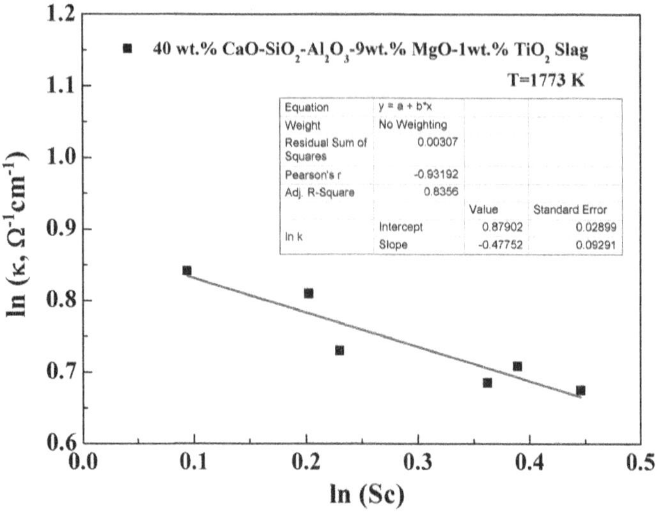

Fig. 5 Electrical conductivity of studied slag as a function of structure complexity

The relationship between the electrical conductivity (ln k) and the structure complexity, viz. ln (Sc.), of the $CaO\text{-}SiO_2\text{-}MgO\text{-}Al_2O_3\text{-}TiO_2$ slags at 1773 K is shown in Fig. 5. The electrical conductivity obtained by Zhang's model [16]. ln (k) decreases linearly as ln (Sc.) increases with the following correlation:

$$\ln \kappa \left(\Omega^{-1}\mathrm{cm}^{-1}\right) = -0.48 \times \ln (Sc.) + 0.88 \left(\mathrm{r}^2 = 0.93\right) \qquad (6)$$

Conclusions

The structure information of the $CaO\text{-}MgO\text{-}SiO_2\text{-}TiO_2\text{-}Al_2O_3$ system with various Al_2O_3/SiO_2 ratio was studied in present work by Raman spectroscopic analysis. Furthermore, the relationship between the properties and structure in this system was evaluated. The high temperature properties (viscosity, density, and electrical conductivity) are expected as a simple linear function of structural complexity. It is can be concluded that these physical properties are strongly dependent on the structural complexity of aluminosilicate melts. It is should be noted that the measurements of high temperature physicochemical properties of aluminosilicate melts are high cost and time-consuming tasks, and the methodology outlined in the current study can be used to predict the physicochemical properties of aluminosilicate slags when using the high alumina iron ore in ironmaking process.

Acknowledgements This study was supported by Graduate Research and Innovation Foundation of Chongqing (Grant No. CYB17002) and China Scholarship Council. We also thank Department of Metallurgical and Materials Engineering, The University of Alabama for supporting this research work.

References

1. Yan Z, Lv X, Pang Z, Lv X, Bai C (2018) Transition of blast furnace slag from silicate based to aluminate based: density and surface tension. Metall Materi Trans B 49:1322–1330
2. Yan Z, Lv X, Pang Z, He W, Liang D, Bai C (2017) Transition of blast furnace slag from silicate based to aluminate based: sulfide capacity. Metall Materi Trans B 48:2607–2614
3. Yan Z, Lv X, Liang D, Zhang J, Bai C, Jie Z (2016). Transition of blast furnace slag from silicates-based to aluminates-based: viscosity. Metall Materi Trans B 48:1092–9
4. Yan Z, Lv X, Pang Z, Bai C (2018) Physicochemical properties of high alumina blast furnace slag. In: 9th international symposium on high-temperature metallurgical processing. Phoenix, Arizona
5. Liang D, Yan Z, Lv X, Zhang J, Bai C (2017) Transition of blast furnace slag from silicate-based to aluminate-based: structure evolution by molecular dynamics simulation and raman spectroscopy. Metall Materi Trans B 48:573–581
6. Yan Z, Lv X, Zhang J, Qin Y, Bai C (2016) Influence of MgO, Al_2O_3 and CaO/SiO_2 on the viscosity of blast furnace type slag with high Al_2O_3 and 5 wt-% TiO_2. Can Metall Q 55:186–194
7. Zhang J, Lv X, Yan Z, Qin Y, Bai C (2016) Desulphurisation ability of blast furnace slag containing high Al_2O_3 and 5 mass% TiO_2 at 1773 K. Ironmaking Steelmaking 43:378–384
8. Mysen BO, Michael J. Toplis (2007) Structural behavior of Al^{3+} in peralkaline, metaluminous, and peraluminous silicate melts and glasses at ambient pressure. Am Miner 92:933–46
9. Mysen BO, Richet P (2005) Silicate glasses and melts: properties and structure: Elsevier
10. Mysen BO, Finger LW, Virgo D, Seifert FA (1982) Curve-fitting of Raman spectra of silicate glasses. Am Miner 67:686–695
11. Mysen BO, Virgo D, Seifert FA (1982) The structure of silicate melts: implications for chemical and physical properties of natural magma. Rev Geophys 20:353–383
12. Mysen BO, Virgo D, Scarfe CM (1980) Relations between the anionic structure and viscosity of silicate melts—a Raman spectroscopic study. Am Miner 65:690–710
13. Huang C, Behrman E (1991) Structure and properties of calcium aluminosilicate glasses. J Non-Cryst Solids 128:310–321
14. Mysen BO, Virgo D, Kushiro I (1981) The structural role of aluminum in silicate melts; a Raman spectroscopic study at 1 atmosphere. Am Miner 66:678–701
15. Mcmillan P, Piriou B (1983) Raman spectroscopy of calcium aluminate glasses and crystals. J Non-Cryst Solids 55:221–242
16. Zhang GH, Chou KC (2012) Correlation between viscosity and electrical conductivity of aluminosilicate melts. Metall Mater Trans B 43:849–855

The Study of Hot Metal Dephosphorization by Replacing Part of Lime with Limestone

Haohua Deng, Min Chen, Nan Wang and Guangzong Zhang

Abstract The study of hot metal dephosphorization by replacing part of lime with limestone at temperature of 1350–1400 °C has been carried out in this work, and the effects of slag temperature and substitution proportion of limestone were examined. Moreover, the influence of limestone decomposition and dissolution on the dephosphorization rate of hot metal was also investigated. It was found that adding limestone instead 50% in mass of lime and the intensive endothermic reaction of limestone decomposition would reduce the slag temperature and increase the slag viscosity. The dynamic condition of dephosphorization was obviously deteriorated, and the dephosphorization rate of hot metal decreased at 1350 °C, while the thermodynamic condition of dephosphorization could be improved and the dephosphorization rate of hot metal increased at 1400 °C.

Keywords Dephosphorization · Limestone · Hot metal · Decomposition

Introduction

The dephosphorization plays an important role in the converter steelmaking process. In view of this importance, many studies on the hot metal dephosphorization have been carried out [1–3], including reaction mechanism and the factors affecting the dephosphorization rate such as lime property, slag composition and temperature and so on [4, 5]. Ono et al. investigated the dephosphorization kinetics during lime injection with oxygen and found that increasing the slagging rate of solid lime is necessary to increase the dephosphorization rate of hot metal [6]. Lime is usually used as the slagging material for slagging and dephosphorization in the converter. It is well known that active lime is produced in a rotary kiln by calcination of limestone at 1000–1300 °C for a few hours, and then, the cooled lime is transported to the converter for slagging [7]. During this process, the loss of sensible heat of the

H. Deng · M. Chen · N. Wang (✉) · G. Zhang
School of Metallurgy, Northeastern University, Shenyang 110819, China
e-mail: wangnan@mail.neu.edu.cn

© The Minerals, Metals & Materials Society 2019

T. Jiang et al. (eds.), *10th International Symposium on High-Temperature Metallurgical Processing*, The Minerals, Metals & Materials Series,
https://doi.org/10.1007/978-3-030-05955-2_72

calcined lime is inevitable. In recent years, with the demand for lower production cost and emission reduction in steelmaking industries, directly using limestone instead of lime as slagging material in converter steelmaking process has been paid great attention, and some industrial trials on limestone slagging mode have been conducted in China [8, 9].

It is known to all that the dephosphorization rate of hot metal is mainly dependent on the three factors, basicity of slag, slag temperature and oxidization of slag [10]. In the basic oxygen steelmaking process, fast dissolution of lime in slag would promote the quick formation of a suitable slag, which could promote the hot metal dephosphorization. By comparing with the dissolution of lime in converter slag, the dissolution of limestone is accompanied by the decomposition of limestone during the limestone steelmaking process. The decomposition of limestone is a strong endothermic reaction, which is supposed to have great influence on the slag bath temperature. Moreover, in the anthors' previous study, the limestone decomposition in converter was studied and the results indicated that the decomposition reaction of limestone could affect the dissolution of the lime [11]. The local temperature drop caused by the intensive endothermic reaction of limestone decomposition may result in a slower slagging rate, which could have an effect on the dephosphorization rate of hot metal.

In this work, the hot metal dephosphorization by replacing part of lime with limestone at temperature of 1350–1400 °C was studied, and the slag temperature and substitution proportion of limestone were examined. The motivation of this paper is to provide more scientific evidence for better understanding the hot metal dephosphorization by replacing part of lime with limestone under the converter steelmaking condition.

Experimental

Raw limestone and active lime lumps supplied by a steelmaking company in China were cut into the small particle sample with the size of 2 mm, respectively. CaO powder used in the synthesized slag was obtained by the calcination of the reagent-grade $CaCO_3$ powder at 900 °C for 24 h. Thereafter, the initial converter slag (synthesized slag) was prepared by mixing CaO, reagent-grade SiO_2 MgO and Al_2O_3 powders thoroughly. The composition of the synthesized slag was 40 wt% CaO, 36 wt% SiO_2, 9 wt% MgO and 15 wt% Al_2O_3. To ensure the melting of the synthesized slag at experimental temperature, the reagent-grade CaF_2 was added to the above composite, with external addition of 20 wt%. The hot metal was prepared by melting the pig iron. The composition of pig iron is shown in Table 1.

Table 1 Composition of pig iron (wt%.)

C	Si	Mn	P	S
4.00	0.27	0.30	0.091	0.046

The reagent-grade Fe_2O_3 was used as an oxidant during the hot metal dephosphorization process.

The electric furnace with $MoSi_2$ heater and alumina reaction tube was employed. The MgO crucible was put in the constant temperature zone of the reaction tube. The hot metal temperature was measured by a PtRh30–PtRh6 thermocouple located at the bottom of the crucible. A corundum rod with the rotating speed range of 100 rpm was used to stir the molten slag during the reaction process. About 300 g of pig iron sample and 10 g of synthesized slag were charged into the crucible and heated to the required experimental temperature, namely 1350 and 1400 °C, respectively. High-purity argon gas was introduced from the bottom of the reaction tube and led out from the top. After the pig iron and synthesized slag were melted completely, Fe_2O_3, active lime and limestone were charged into the crucible. The addition amounts of Fe_2O_3, active lime and limestone are shown in Table 2. The hot metal and slag were frequently sampled with a 4-mm inner diameter quartz tube and a steel rod, respectively, over a period of 5, 10, 20 and 30 min. Thereafter, the hot metal and slag sample were quenched. The hot metal and slag samples were analysed with chemical titration.

Results and Discussion

Figure 1 shows the variation of the content of [P] in hot metal with the reaction time. It can be observed that the lower the temperature of the molten metal, the greater became the dephosphorization effects. In addition, during the process of slagging with limestone instead 50% of lime, the content of [P] in hot metal was higher than that during the process of slagging with lime at 1350 °C, while it took on an opposite trend at 1400 °C, indicating that slagging with limestone instead 50% of lime was favourable to the hot metal dephosphorization at 1400 °C, but that was disadvantage to the hot metal dephosphorization at 1350 °C.

Figure 2 shows the variation of the ratio of phosphorus content of the slag and hot metal (%P)/[%P] with the reaction time. It can be observed that the ratio of (%P)/[%P] increased with increasing the reaction time. In addition, during the process of slagging with limestone instead 50% of lime, the ratio of (%P)/[%P] was obviously lower than that during the process of slagging with lime at 1350 °C, while the ratio of (%P)/[%P] for slagging with limestone instead 50% of lime was slightly higher than that for slagging with lime at 1400 °C, indicating that slagging with limestone instead 50% of lime was favourable to the improvement of

Table 2 Addition amounts of slagging material

Substitution proportion of limestone (%)	Limestone (g)	Lime (g)	Fe_2O_3 (g)
0	0	20.20	20.06
50	17.51	10.10	20.06

Fig. 1 Variation of content of [P] in hot metal with the reaction time

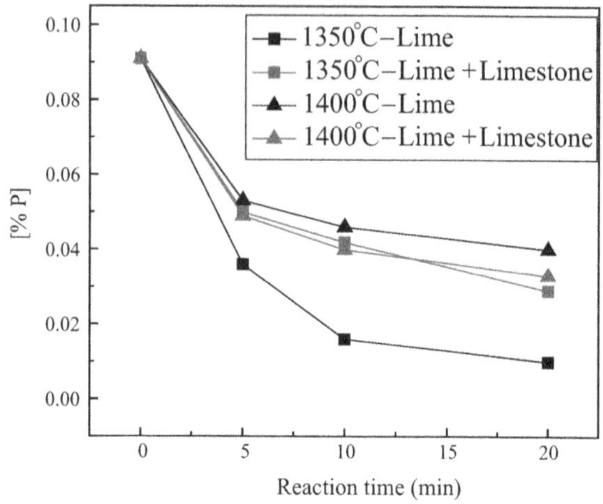

Fig. 2 Variation of the ratio of (%P)/[%P] with the reaction time

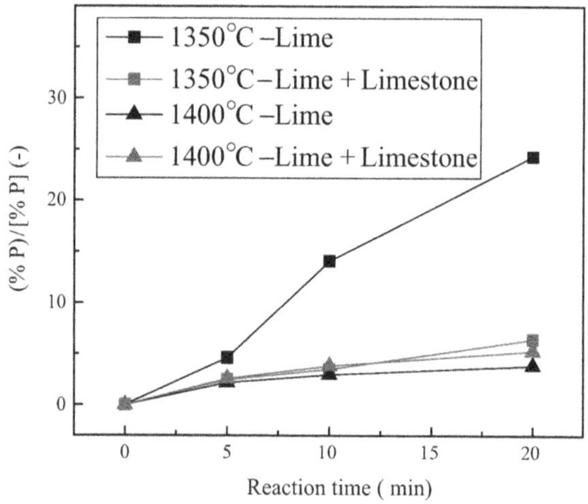

dephosphorization capacity of slag at 1400 °C. Therefore, it is considered that to ensure quick dephosphorization rate of hot metal in the steelmaking process, the hot metal should have a higher temperature while using limestone slagging.

Table 3 presents the chemical composition of slag at the end of reaction. Figure 3 shows the variation of slag basicity (CaO/SiO_2) with the reaction time. It can be observed that the slag basicity increased with increasing reaction time. During the process of slagging with limestone instead 50% of lime, the slag basicity was obviously lower than that during the process of slagging with lime at 1350 °C, and the slag basicity for slagging with limestone instead 50% of lime was less than 2.8 at the end of reaction. This is due to that the endothermic decomposed reaction of limestone decreased the slag temperature and the slagging rate. However, the slag basicity for slagging with limestone instead 50% of lime was slightly lower

Table 3 Chemical composition of slag

T (°C)	Substitution proportion of limestone (%)	SiO_2 (%)	CaO (%)	P_2O_5 (%)	FeO (%)
1350	0	14.82	44.17	1.04	39.97
	50	15.17	42.48	0.81	41.54
1400	0	14.69	44.08	0.65	40.58
	50	14.82	43.72	0.74	40.71

than that for slagging with lime at 1400 °C. Thus, the higher the temperature, the faster the dissolution rate of lime is, and the slag basicity is higher.

It is known to all that limestone decomposition is a process with strong endothermic reaction ($\Delta H_{CaCO_3} = 169120\,\mathrm{J/mol}$) [12]. The maximum heat absorption of limestone decomposition reaction could be obtained by Eq. (1).

$$\Delta H = n \cdot \Delta H_{CaCO_3} \tag{1}$$

where ΔH_{CaCO_3} is the enthalpy of decomposition reaction (J mol^{-1}) and n is the mole mass of $CaCO_3$ of limestone sample (mol).

The temperature drop of hot metal could be calculated by Eq. (2)

$$\Delta T = \frac{\Delta H}{C_{P,s} \cdot m} \tag{2}$$

where Δt is the maximum temperature drop of hot metal (°C), m is the mass of hot metal (kg), and $C_{P,s}$ is the constant pressure capacity of hot metal (J (kg °C)$^{-1}$).

During the process of slagging, it is assumed that the heat absorbed of limestone decomposition was all from the hot metal. Based on Eqs. (1–2), the maximum

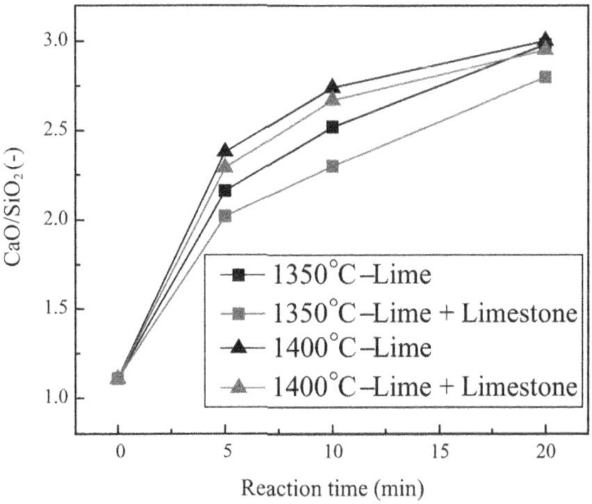

Fig. 3 Variation of the ratio of CaO/SiO_2 with the reaction time

temperature drop of hot metal was calculated as about 54 °C while using limestone instead 50% of lime slagging. Thus, the temperature of hot metal was about 1296–1350 °C for slagging with limestone instead 50% of lime at 1350 °C and that was about 1346–1400 °C for slagging with limestone instead 50% of lime at 1400 °C.

The limestone decomposition is a process with strong endothermic reaction, and thus, the temperature would inevitably be affected by the decomposition reaction. Furthermore, the slag temperature has a significant effect on the viscosity of slag. The viscosity of slag could be obtained by FactSage software. The result shows that using limestone instead 50% of lime for slagging at 1350 °C, the viscosity increased from 0.461 to 0.597 Pa s, while it increased from 0.362 to 0.462 Pa s at 1400 °C. It can be found that the increment of viscosity was less at 1400 °C. That is to say, the dynamics condition of the hot metal dephosphorization at 1400 °C was obviously better than that at 1350 °C.

The reaction of hot metal desulfurization could be expressed by Eq. (3) [10].

$$\begin{aligned}2[\mathrm{P}] + 5(\mathrm{FeO}) + 4(\mathrm{CaO}) &= 4\mathrm{CaO}\cdot\mathrm{P_2O_5} + 5[\mathrm{Fe}]\Delta G^0 \\ &= -767162 + 288.35T\ \mathrm{J/mol}\end{aligned} \tag{3}$$

The standard Gibbs free energy of desulfurization reaction of hot metal could be obtained by Eq. (3), and the results are shown in Fig. 4. It indicates that the standard Gibbs free energy of dephosphorization reaction of hot metal increased with increasing the temperature. As mentioned above, the intensive endothermic reaction of limestone decomposition would reduce the temperature of hot metal. Thus, the slagging with limestone instead 50% of lime was favourable to the improvement of the thermodynamic condition of the hot metal dephosphorization.

Slagging with limestone instead 50% of lime could decrease the temperature of hot metal and improve the thermodynamic condition of the hot metal dephosphorization, while the dissolution rate of lime was obviously slower and the basicity of

Fig. 4 Variation of ΔG with the reaction time

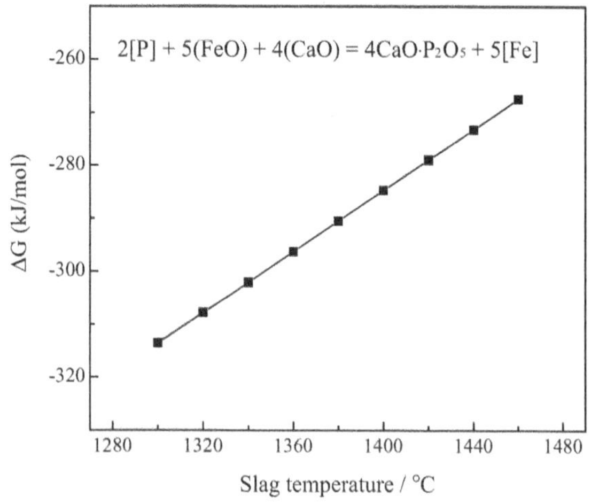

slag was lower at 1350 °C. In addition, the temperature of hot metal drop caused by the limestone decomposition could obviously increase the viscosity of hot metal and deteriorate the dynamics condition of the hot metal dephosphorization. Thus, the dephosphorization rate of hot metal decreased. Compared with the hot metal dephosphorization with limestone instead 50% of lime slagging at 1350 °C, the dissolution rate of limestone was faster and the dynamics condition of the hot metal dephosphorization was better at 1400 °C. Therefore, the dephosphorization rate of hot metal for slagging with limestone instead 50% of lime was faster than that for slagging with lime.

Based on the above analysis, the dephosphorization of hot metal is affected by the conditions of thermodynamics and dynamics. The deterioration of dynamics condition of the hot metal dephosphorization by the limestone addition hindered the dephosphorization reaction at 1350 °C, while the improvement of thermodynamics condition of the hot metal dephosphorization by the limestone addition promoted the dephosphorization reaction.

Conclusions

The hot metal dephosphorization by replacing part of lime with limestone at temperature of 1350–1400 °C was studied, and the slag temperature and substitution proportion of limestone were examined. The main results and conclusions are summarized in the following:

(1) During the process of slagging with limestone instead 50% of lime, the intensive endothermic reaction of limestone decomposition would reduce the slag temperature and increase the slag viscosity.
(2) When the use of limestone instead 50% of lime was applied for slagging, the dynamic condition of dephosphorization was obviously deteriorated and the dephosphorization rate of hot metal decreased at 1350 °C, while the thermodynamic condition of dephosphorization could be improved and the dephosphorization rate of hot metal increased at 1400 °C.

Acknowledgements The authors gratefully acknowledge the National Natural Science Foundation of China (51574065, 51574066, 51774072 and 51774073) and the National Key R&D Program of China (No. 2017YFB0304201, 2017YFB0304203, 2016YFB0300602) which have made this research possible.

References

1. Xuan XY, Shi S, Qi X, Cai JW (2015) Study on dephosphorization of high-phosphorus hot metal in $CaO–SiO_2–FeO–Na_2O–Al_2O_3$ slag system. Ming Metall 24(2):51–54

2. Nassaralla C, Fruehan RJ, Min DJ (1991) A thermodynamic study of dephosphorization using $BaO–BaF_2$, $CaO–CaF_2$ and $Bao–CaO–CaF_2$ systems. Metall Mater Trans B 22(1):33–38
3. Xu KD, Xiao LJ (2011) Theoretical analysis of hot metal dephosphorization pretreatment in converter. J Shanghai Univ 17(4):331–336
4. Bo H, Dong JJ, Ren QQ, Zhang YZ (2016) Effect of activity and microstructure of lime on dephosphorization of hot metal. Special Steel 37(5):10–13
5. Pan Y, Wen G, Zhou H, Tang P, Hou Z (2016) Study on final slag composition control for highly efficient dephosphorization of 80 t converter in CISC. Steelmaking 32(1):15–19
6. Ono H, Masui T, Mori H (2006) Dephosphorization kinetics and reaction region in hot metal during lime injection with oxygen. ISIJ Int 25(2):133–141
7. Pan C (1997) The metallurgical effects of activated lime made by rotary kiln. Steelmaking 13 (3):24–28
8. Qing DP, Yang JP, Wei SH, Zhang JS, Sun JG (2014) Analysis and process practice on steelmaking technology of 100 t top-blown oxygen converter by limestone for slag making. Special Steel 35(5):34–37
9. Dong DX, Feng J, Nian W, Li H, Liang M (2013) Experimental study on using limestone instead of lime for slagging in 60 t converter. China Metall 23(11):58–62
10. Huang XH (2010) Principles of steel metallurgy. Metallurgical Industry Press, Beijing
11. Wang N, Zhang GZ, Chen M, Deng HH (2017) Decomposition behavior of limestone in early converter slag. Powder Techno 320:73–79
12. Tang B, Wang XM, Zou ZS, Yu AB (2016) Decomposition of limestone in hot metal at 1300 °C. Steel Res Int 87(2):226–231

Study on Energy Utilization of High Phosphorus Oolitic Haematite by Gas-Based Shaft Furnace Reduction and Electric Furnace Smelting Process

Hui Sun and Miaolian Bian

Abstract Based on mass balance and heat balance calculation model, the energy utilization of pellet roasting, gas-based shaft furnace reduction and electric furnace smelting process of high phosphorus oolitic haematite was calculated, and the influence on energy consumption of the total process with different additives was investigated. Furthermore, the gas-based shaft furnace was divided into three parts, namely preheated zone, reduction zone and cooling zone. Taking chemical reactions that taken place in different parts, as well as roasting pellets and sponge iron with different temperatures into consideration, the relative mass balance and heat balance calculation model are established for energy analysis, which can provide a theoretical foundation for energy saving of gas-based shaft furnace reduction technology.

Keywords High phosphorus oolitic haematite · Mass balance · Heat balance · Gas-based shaft furnace reduction · Electric furnace smelting

Introduction

The direct reduction is a process for reducing iron ore and producing sponge iron called direct reduction iron or DRI [1]; there are several typical technologies, such as MIDREX, HYL and PERED using counter flow shaft furnace to make sponge iron [2–4]. Nowadays, gas-based shaft furnace direct reduction process is playing a dominant role in non-blast furnace iron making process. The total output of sponge iron had been more than 70 million metric tons since 2010; however, 75% of them were produced by gas-based shaft furnace. In 2013, the global output of DRI was 75.20 million metric tons, more than 78% of which was produced by gas-based shaft furnace direct reduction process [5, 6]. Especially for the recent years, the gas-based shaft furnace direct reduction process is developing faster than years

H. Sun (✉) · M. Bian
Shenwu Technology Group Corp Co., Ltd., Beijing 102200, China
e-mail: sunhui19850304@163.com

© The Minerals, Metals & Materials Society 2019

T. Jiang et al. (eds.), *10th International Symposium on High-Temperature Metallurgical Processing*, The Minerals, Metals & Materials Series,
https://doi.org/10.1007/978-3-030-05955-2_73

before, and several gas-based shaft furnace direct reduction and production facility had been built one after another in Iran, India, Egypt, the United Arab Emirates and other countries. Up to now, the design capacity of single gas-based shaft furnace direct reduction and production facility can reach 2.75 million metric tons per year, and large scale of which may become the most prominent characteristics in this century.

The high phosphorus oolitic haematite is an important resource with a huge deposit in China, which is nearly impossible to be used in blast furnace owing to its extreme difficulties of separation and concentration of iron oxide minerals by traditional magnetic, floating or gravity separation methods [7]. The phosphorus content of raw high phosphorus oolitic haematite ranges from 0.8 to 1.2% and the iron content of which is about 50%. What's more, the phosphorus and iron element are always nesting with each other and the diameter of the oolitic structure is in macrometer degree, which is hard to separate them by grinding operation [8, 9]. In recent years, a lot of efforts were made to separate phosphorus from high phosphorus oolitic haematite by gas-based shaft furnace direct reduction process and obtained low phosphorus metallic iron with phosphorus content less than 0.18% after smelting in an electric furnace. That is to say, the gas-based shaft furnace direct reduction and electric furnace process are an ideal method for extracting lowgrade phosphorus metallic iron [10].

Although the literature reports published based on the laboratory research can provide the theory and data foundation for large-scale processing the high phosphorus oolitic haematite by gas-based shaft furnace direct reduction and electric furnace process, the energy utilization of it is also deserving focusing on. The gas-based shaft furnace can be divided into three parts, namely preheated zone, reduction zone and cooling zone, and each part has its own mass and heat balance, which can help us understand the energy distribution. Moreover, based on the mass and heat balance calculating model, the energy conserve potential can also be made to serve the practical production in future.

Materials and Gases

Raw Materials

The raw materials used in this study include high phosphorus oolitic haematite and some additives. The high phosphorus oolitic haematite is from the western part of Hubei Province, named Ning-xiang type, and its chemical composition is shown in Table 1.

Table 1 Chemical composition of high phosphorus oolitic haematite, wt%

TFe	Fe_2O_3	FeO	SiO_2	CaO	Al_2O_3	MgO	K_2O	Na_2O	P
48.93	66.70	2.93	10.56	5.32	5.92	0.82	0.419	0.098	0.793

It can be seen from Table 1 that the total Fe of high phosphorus oolitic haematite is less than 50%, it is not desirable for smelting by blast furnace iron making technology because of the low degree of Fe, furthermore, more high phosphorus slag will be produced, and a hard work will be made for further utilization widely and economically.

The additives are mainly to some pure chemical reagents, like CaO, $Ca(OH)_2$, CaF_2, $CaCl_2$, $NaOH$, the proportion of which added was shown in Table 2.

There is a common sense that high phosphorus oolitic haematite is a red-brown ore with high viscosity, especially when grinding to small particles with size less than 75 μm and pelletizing with size of 10–16 mm by using disc pelletizer. Provided that no additives were added, there will be no pellets produced. Just a little water added will result in massive blocks or balls. Otherwise, the alkalinity of high phosphorus oolitic haematite is very low, so the smelting temperature will be high, which will affect the smelting process in an electric furnace. Overall, additives added to high phosphorus oolitic haematite can lower the viscosity and improve the alkalinity, not only in favour of pelletizing, but also brings convenience to smelting process.

Reducing Gas

The reducing gas is used for reducing the roasting pellets in reduction zone of gas-based shaft furnace, the temperature of it ranges from 1123 to 1223 K, and the injection quantity is about 1800 Nm^3/tDRI. The composition is shown in Table 3.

Cooling Gas

In the bottom of the gas-based shaft furnace, in order to avoid the heated sponge iron oxidized in air or carburized, it is necessary to lower the sponge iron temperature. Generally speaking, nitrogen gas and argon gas with normal temperature can be taken as cooling gas, and in view of economical efficiency and sustainable

Table 2 Proportion of additives added to 100% of high phosphorus oolitic haematite, wt%

Additives	CaO	$Ca(OH)_2$	CaF_2	$CaCl_2$	$NaOH$
Proportion	5–25	5–25	5–25	5–25	5–25

Table 3 Composition of reducing gas, vol.%

H_2	CO	CH_4	H_2O	CO_2	N_2 + Air	Σ
44.53	43.03	4.40	2.40	1.20	4.44	100.00

development, the reducing gas with normal temperature may be better suitable for cooling the heated sponge iron. In this study, the reducing gas is taken as cooling gas either, and the injection quantity is about 1000 $Nm^3/tDRI$.

Gas-Based Shaft Furnace and Calculating Model

Taking gas-based shaft furnace as a whole, the material input is cooling gas, reducing gas and roasting pellets, and the material output is cooling gas after heating, top gas and sponge iron, which can be shown in Fig. 1, and the mass balance based on chemical reactions is listed in Table 4 and shown in Table 5. It can be seen from Table 5 that the material input is mainly roasting pellets and reducing gas, and almost the same proportion. When chemical reactions are taken place in the gas-based shaft furnace under high temperature, the proportion of the products, DRI and top gas alters, and the proportion of DRI derived from roasting pellets decreases. Table 6 shows heat balance of the whole gas-based shaft blast furnace, and it also indicates the same phenomenon that the larger proportion of heat input by reducing gas and the larger proportion of heat output by top gas.

When gas-based shaft furnace is divided into preheated zone, reducing zone and cooling zone, mass balance and heat balance can be calculated, respectively.

In preheated zone, the material input is the same as that of output and is to heat the roasting pellets with no consideration of any chemical reactions at all. The reduction zone is the most important part of the whole gas-based shaft furnace; it

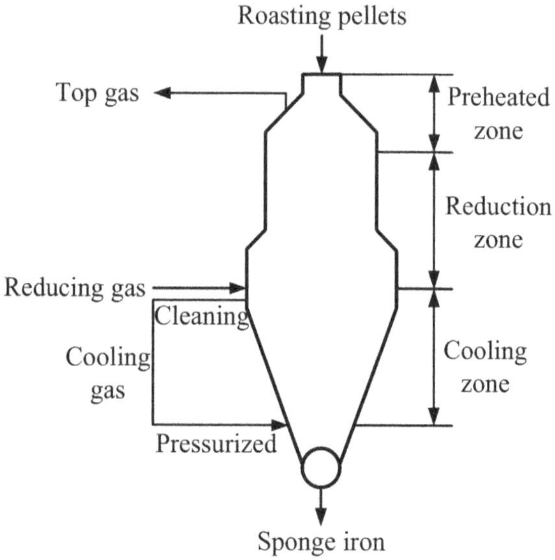

Fig. 1 Diagram of gas-based shaft furnace

Table 4 Chemical reactions involved in gas-based shaft furnace

No.	Chemical reactions	ΔH/(kJ/mol)
(1)	$3Fe_2O_3(s) + H_2(g) = 2Fe_3O_4(s) + H_2O(g)$	−2.07
(2)	$Fe_3O_4(s) + H_2(g) = 3FeO(s) + H_2O(g)$	60.45
(3)	$FeO(s) + H_2(g) = Fe(s) + H_2O(g)$	30.23
(4)	$3Fe_2O_3(s) + CO(g) = 2Fe_3O_4(s) + CO_2(g)$	−43.23
(5)	$Fe_3O_4(s) + CO(g) = 3FeO(s) + CO_2(g)$	19.29
(6)	$FeO(s) + CO(g) = Fe(s) + CO_2(g)$	−10.93
(7)	$CH_4(g) + H_2O(g) = CO(g) + 3H_2(g)$	206.12
(8)	$CH_4(g) + CO_2(g) = 2CO(g) + 2H_2(g)$	247.28

Table 5 Mass balance of whole gas-based shaft blast furnace

Material input			Material output		
Name	Mass (kg/tDRI)	Proportion (wt%)	Name	Mass (kg/tDRI)	Proportion (wt%)
Roasting pellets	1230.774	49.06	DRI	1000.000	39.86
Reducing gas	1278.031	50.94	Top gas	1508.805	60.14
Total	2508.805	100.00	Total	2508.805	100.00

[a]$Ca(OH)_2$ was taken as additive, and the proportion added to 100% of high phosphorus oolitic haematite was 10% according to a practical exploration

[b]No consideration of cooling gas because there is no mass change from beginning to end

Table 6 Heat balance of whole gas-based shaft blast furnace

Heat input			Heat output		
Name	Heat (kJ/tDRI)	Proportion (%)	Name	Heat (kJ/tDRI)	Proportion (%)
Reducing gas	2,294,341.81	100.00	Top gas	1,046,894.62	45.63
Roasting pellets	0.00	0.00	Chemical reactions	191,385.44	8.34
Cooling gas	0.00	0.00	DRI after cooling	488,239.79	21.28
			Cooling gas heated	552,509.82	24.08
			Heat loss	15,312.14	0.67
Total	2,294,341.81	100.00	Total	2,294,341.81	100.00

[a]$Ca(OH)_2$ was taken as additive, and the proportion added to 100% of high phosphorus oolitic haematite was 10% according to a practical exploration

[b]Heat loss of the gas-based furnace body is not absolutely consistent with practical production

involves all chemical reactions listed in Table 4. The cooling zone is the same with preheated zone just for heat transfer.

Results and Discussion

Total Energy Utilization Analysis

The total energy utilization of gas-based shaft furnace direct reduction and electric furnace smelting process is shown in Fig. 2.

As can be seen from Fig. 2, when the proportion of additives added to 100% of high phosphorus oolitic haematite ranging from 5 to 25%, the more the additives added, the higher the total energy consumed. When taking CaF_2 as additive, it consumes more energy than other additives in the same proportion. According to the practical study, an interesting phenomenon occurred when $Ca(OH)_2$ was taken as additive and the proportion of it was about 10%; it presented a good advantage in making raw pellets and electric furnace smelting, and that was why, the study made it as a sample in former and later for mass balance, heat balance and energy utilization analysis.

Energy Utilization Analysis of Different Systems

Figure 3 shows the energy utilization of different systems, such as pellets roasting system, reducing gas heating system and electric furnace smelting system, which occupies a significant position of gas-based shaft furnace direct reduction and

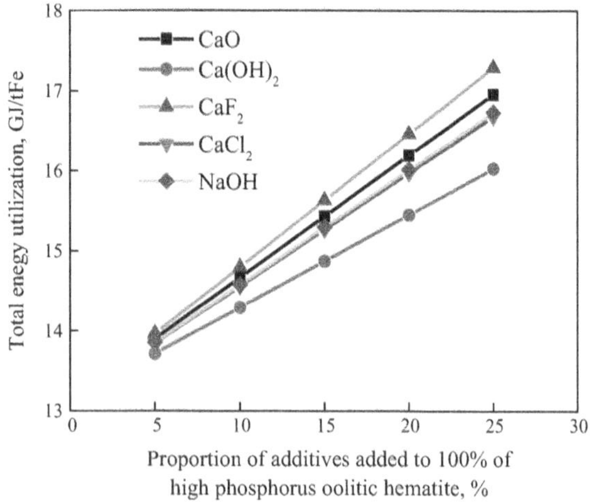

Fig. 2 Total energy utilization of gas-based shaft furnace direct reduction and electric furnace smelting process

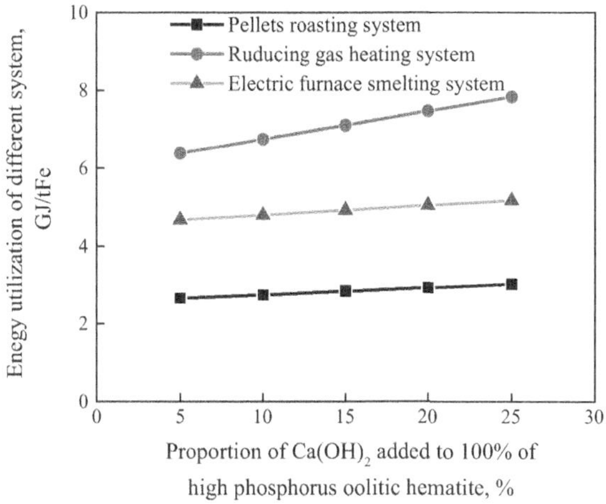

Fig. 3 Energy utilization of different systems

electric furnace smelting process. It should be pointed out that the energy utilization analysis is based on the additives of $Ca(OH)_2$ and the proportion of which added is 10 to 100% of high phosphorus oolitic haematite.

It can be seen from Fig. 3 that the reducing gas heating system consumes more energy than the other two systems, and it increases sharply along with the proportion of $Ca(OH)_2$ added. Otherwise, the pellet roasting system changes steadily with a small growth.

Potential Analysis of Energy Conservation

As well known, when transporting hot roasting pellets into gas-based shaft furnace can save energy consumption or can promote roasting pellets be reduced better with a high degree of reduction. The same as roasting pellets, hot sponge iron delivered into electric furnace can also save a lot of energy, as shown in Fig. 4.

As shown in Fig. 4a, b, when temperature of roasting pellets rises, it can save about a quarter of energy consumed by reducing gas heating system compared to Fig. 3 and that of sponge iron rises can save about a third of energy consumed by electric furnace smelting system under the same proportion of $Ca(OH)_2$ added to 100% of high phosphorus oolitic haematite, so exploring hot material transport technology and facility is imperative.

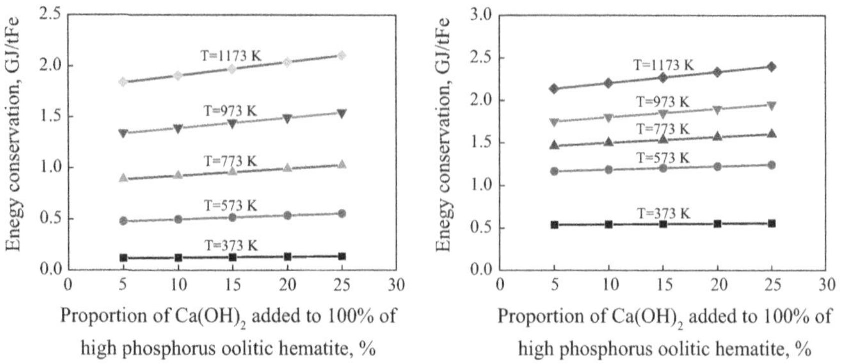

(a) Temperature of roasting pellets (b) Temperature of sponge iron

Fig. 4 Energy conservation potential analysis by using hot material. **a** Temperature of roasting pellets. **b** Temperature of sponge iron

Conclusions

(1) When the proportion of additives added to 100% of high phosphorus oolitic haematite ranging from 5 to 25%, the more the additives added, the higher the total energy consumed. According to the practical study, when $Ca(OH)_2$ was taken as additive and the proportion of it was about 10%, it presented a good advantage in making raw pellets and electric furnace smelting.

(2) The reducing gas heating system consumes more energy than pellet roasting system and electric smelting system, and it increases sharply along with the proportion of $Ca(OH)_2$ added. Otherwise, the pellet roasting system changes steadily with a small growth.

(3) When the temperature of roasting pellets rises, a quarter of energy consumed by reducing gas heating system can be saved and that of sponge iron rises can save about a third of energy consumed by electric furnace smelting system under the same proportion of $Ca(OH)_2$ added to 100% of high phosphorus oolitic haematite.

References

1. Shams A et al (2015) Modeling and simulation of the MIDREX Shaft furnace: reduction, transition and cooling zones. JOM 67(11):2681–2689
2. Wang X (2006) Iron and steel metallurgy (ironmaking part), Metallurgical Industry Press, Beijing, pp 245–267
3. Zhang Jianliang, Liu Zhengjian, Yang Tianjun (2015) Non-blast Furnace Ironmaking. Metallurgical Industry Press, Beijing, pp 54–78

4. Huitu K et al (2015) Optimization of Midrex Direct Reduced Iron Use in ore-based Steelmaking. Steel Res 86(5):456–465
5. Qi YH et al (2013) Present status and trend of direct reduction technology development in China. China Metall 23(1):9–14
6. Xu CY et al (2010) The research current situation and progress of iron ore direct reduction process and theory. Conserv Utilization Mineral Resour 4:48–54
7. Sun YS et al (2013) Recovery of iron from high phosphorus Oolitic iron ore using coal-based reduction followed by magnetic separation. Int J Mineral Metall Mater 20(5):411–419
8. Tang HQ et al (2010) Phosphorus removal of high phosphorus iron ore by gas-based reduction and melt separation. J Iron Steel Res Int 17(9):1–6
9. Li KQ et al (2011) Iron extraction from oolitic iron ore by a deep reduction process. J Iron Steel Res Int 18(8):9–13
10. Sun YS et al (2015) Exploratory tests on deep reduction of high-phosphorus Oolitic hematite ore. Min Metall Eng 35(5):68–71

Study on the Effect of Different CO_2–O_2 Mixture Gas Blowing Modes on Vanadium Oxidation

Zheng-Lei Guo, Yu Wang, Qi Lu and Shu-Chao Wang

Abstract This paper studies the effect of mixed blowing CO_2–O_2 gas on the extraction of vanadium and adopts blowing flow and stage blowing, two types of variables, to study the effect of both on the extraction of vanadium from hot iron. The initial blowing temperature of the experiment was set to 1300 °C. The experimental results showed that the whole process is sprayed with 15% CO_2, and the optimum blowing flow rate is 1.714 L/(min kg). The stage blowing has better vanadium extraction with carbon retention effect, compared with the whole blowing of 10% CO_2, and the end point temperature of the molten pool is decreased by 17 °C. The stage blowing is injecting 10% CO_2 in the initial and injecting 15% CO_2 in the middle and late period. And it has a higher vanadium oxidation rate than the whole blowing 10% CO_2, which can reach 95.14%.

Keywords Carbon dioxide · Vanadium extraction · Bath temperature

Introduction

With the steady development of society, more and more energy is consumed and more and more greenhouse gases are emitted, which makes the problem of greenhouse effect more and more serious [1, 2]. In the metallurgical industry, carbon dioxide as the main emission gas, how to effectively reduce its emissions is also one of the focuses of researchers [3, 4]. Researchers have proposed several technological methods to reduce carbon dioxide emissions. The first is to use new energy sources to reduce the use of fossil fuels. The second is carbon dioxide storage technology. The third is to recycle carbon dioxide gas—not only it can reduce carbon dioxide emissions, but it can also bring economic benefits [5, 6].

Z.-L. Guo · Y. Wang (✉) · Q. Lu · S.-C. Wang
College of Materials Science and Engineering, Chongqing University, 400044 Chongqing, China
e-mail: wangyu@cqu.edu.cn

© The Minerals, Metals & Materials Society 2019

T. Jiang et al. (eds.), *10th International Symposium on High-Temperature Metallurgical Processing*, The Minerals, Metals & Materials Series,
https://doi.org/10.1007/978-3-030-05955-2_74

In the process of vanadium extraction in the converter, the weak oxidizing property of carbon dioxide is used to replace part of oxygen and carbon dioxide participates in the molten pool reaction. The reaction of CO_2 with C and Fe is an endothermic reaction, but the reaction with Si and Mn is a weak exothermic reaction. The whole reaction process can slow down the temperature rise of the molten pool, which is beneficial to the vanadium extraction process, and does not pollute the vanadium slag and also utilizes part of the carbon dioxide [7–11].

Therefore, it is of great significance to study the participation of carbon dioxide in vanadium extraction molten pool reaction to reduce carbon dioxide emissions and environmental protection. In order to better control the temperature of molten pool, we use the stage blowing to control the temperature of molten pool. In the early stage, sufficient oxygen supply meets the requirements for vanadium extraction. The later control of the oxygen supply increases the carbon dioxide content of the mixed gas to lower the temperature of the molten pool so that it is around the carbon vanadium conversion temperature, so that the carbon remains in the hot iron.

Experimental

Experimental Materials and Equipment

The materials used in the experiment are vanadium-bearing pig iron produced by PZH Iron and Steel Group Company, and the contents of each element were chemically detected as shown in Table 1.

The gases used in the experiment are CO_2 and O_2 (industrial grade gas, purity is 99.8%).

Experimental blowpipe: corundum two-hole blowpipe (outer diameter: 8 mm, purity: 99%).

The experimental apparatus diagram is shown in Fig. 1. The heating furnace used in this experiment is a $MoSi_2$ high-temperature electric resistance furnace (5K-10-16), with a rated temperature of 1600 °C. In addition, it is equipped with a temperature controller (MTLQ-2) and a self-designed blowing system. The blowing system includes pressure reducing valve, pressure gauge, flow metre and blowpipe, which can control the gas pressure and flow rate required by the experiment. During the experiment, the control cabinet's temperature control instrument delivers a pulse signal to control the power of silicon controlled rectifier and to the heating element of electric resistance furnace. At the same time, the temperature measured by the

Table 1 Composition table of the vanadium-bearing pig iron, wt%

C	V	Si	Mn	P	S
3.73	0.325	0.053	0.14	0.075	0.145

Fig. 1 Schematic drawing of experimental apparatus

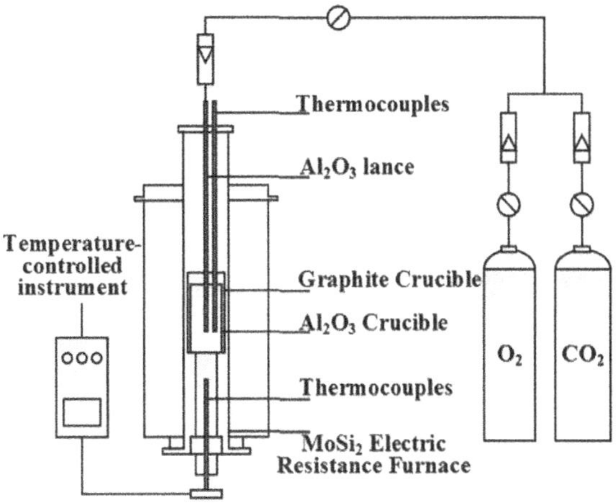

thermocouple is converted into an electrical signal which is displayed by the temperature control instrument. By Computerized Numerical Control, the $MoSi_2$ high-temperature electric resistance furnace is smoothly kept at a constant temperature, and the temperature control accuracy is ±5 °C.

Experimental Method

In this experiment, the initial blowing temperature was 1300 °C, and the total blowing time of each group experiment was 15 min. In the first experiment, the mixed gas with a CO_2 ratio of 15% is blew in the whole process of experiment, at the same time, and the blowing flow rates are changed, which are 1.428 L/(min kg), 1.714 L/(min kg), 2.000 L/(min kg). In the second experiment, we conducted experiments using a stage blowing method, and the blowing flow rate was selected to be 1.714 L/(min kg), and the mixed gas with a CO_2 ratio of 10% is blew in the first 6 min of the experiment, in the next 9 min, the mixed gas with a CO_2 ratio of 15% is blew. At the same time, set up a set of experiments, the CO_2–O_2 mixture gas with a CO_2 ratio of 10% is blew in the whole process of experiment for comparative analysis. In this experiment, the samples were taken at 60, 180, 360, 600 and 900 s, and the corresponding components were analysed.

Results and Discussion

Influence of Flow Rate

Figure 2 shows that under different blowing flow rates, the carbon content in the molten pool gradually decreases as the blowing progress. However, it has little

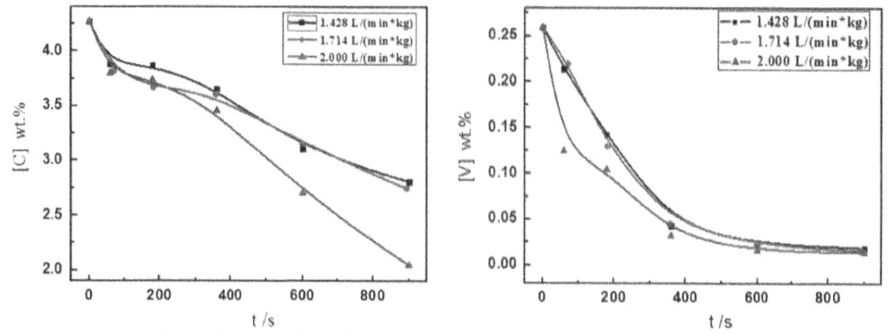

(a) Change of [C] (b) Change of [V]

Fig. 2 [C] and [V] in hot iron changes with time at different blowing flows

effect on the oxidation amount of vanadium in the end point of the hot iron and only has a great influence on the oxidation rate of vanadium in the initial stage of vanadium extraction. When the blowing flow rate was 1.714 and 1.428 L/(min kg), the changes of [C] and [V] contents were similar in the molten pool. But when the blowing flow rate is 2.000 L/(min kg), on the one hand, due to the increase of the blowing flow rate, the supply of oxygen increases correspondingly, which causes the temperature of the molten pool to rise faster, which leads to the accelerated oxidation of carbon in hot iron, and compared with the blowing flow rate of 1.428 L/(min kg), the end carbon content is reduced by 27.05%. On the other hand, because the oxygen content is increased in the initial stage of vanadium extraction, the oxidation of non-carbon elements in the hot iron is obviously accelerated, so that when the flow rate is 2.000 L/(min kg), the vanadium oxidation rate in the early stage is faster.

Through calculation, the ratio of residual carbon and residual vanadium at the end point of the reaction of vanadium extraction with carbon retention under different blowing flow rates, and we can get Table 2, and where $[C]_R$ is the end point carbon content and $[V]_R$ is the end point vanadium content.

As shown in Table 2 and Fig. 2, the blowing flow rate increased from 1.428 to 2.000 L/(min kg), and the vanadium extraction with carbon retention capacity first increased and then weakened. This is because as the gas flow rate increases, the oxygen supply increases and the temperature rise rate of the molten pool increases and promotes the oxidation of carbon in the molten pool, thereby affecting the carbon content of the end point, but it has less effect on the end point vanadium content. After comprehensive consideration, under the experimental conditions,

Table 2 Value of residual $[C]_R/[V]_R$ at different blowing flows

Blowing flow L/(min kg)	1.428	1.714	2.000
$[C]_R/[V]_R$	148	172	146

when the blowing flow rate is 1.714 L/(min kg), the task of vanadium extraction with carbon retention can be better completed. At this time, the oxidation rate of vanadium is 93.83%.

Influence of Stage Blowing

As shown in Fig. 3, due to the process of extracting vanadium from hot iron, the oxidation of carbon mainly occurs in the middle and late periods of molten pool reaction. Therefore, under the same initial blowing flow rates, after the reaction is carried out for 360 s, one of the experiments uses a CO_2–O_2 mixture gas with the CO_2 ratio of 15% for blow. In this set of experiments, the proportion of carbon dioxide in the mixed gas is increased, which suppresses the increase of the temperature of the molten pool and reduces the progress of the decarburizing reaction. So it can be seen from the end point of the vanadium extraction from hot iron, the end point carbon content of the stage blowing is higher than the end point carbon content of the constant blow CO_2–O_2 mixture gas with the CO_2 ratio of 10%. That is, the stage blowing has a better effect of protecting carbon.

It can be seen from Fig. 4 that because of the later period of stage blowing, the proportion of carbon dioxide in the mixed gas is increased, which will delay the increase of the temperature of the molten pool and can inhibit the oxidation of carbon and promote the oxidation of vanadium. Therefore, at the vanadium extraction end point, the vanadium content in the hot iron with stage blowing is lower, and after 360 s, the content of vanadium in the hot iron will always be lower than the whole process of blowing 10% CO_2, which is consistent with the experimental results.

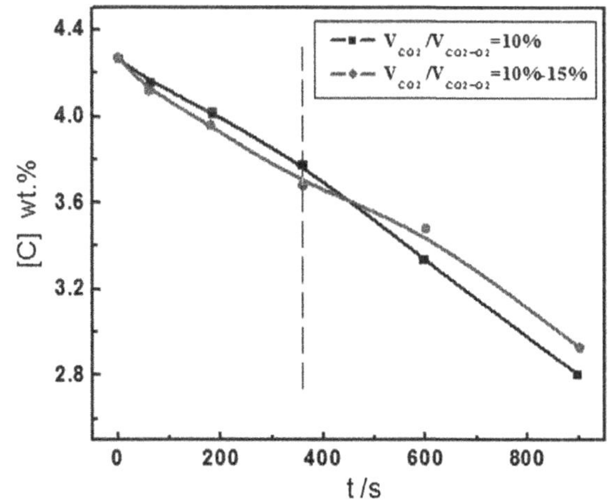

Fig. 3 Change of [C] content with time during stage blowing

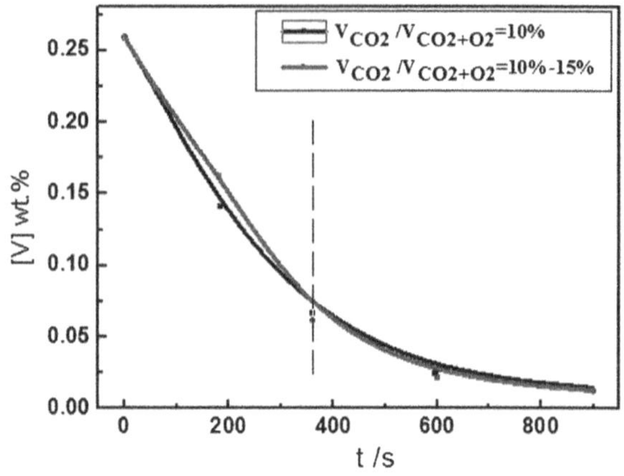

Fig. 4 Change of [V] content with time during stage blowing

As can be seen from Tables 3 and 4, comparing the contents of the end points C and V of the two tables, it can be seen that the end point C content of the stage injection is significantly higher than the whole process of blowing 10% CO_2, and the end point V content is smaller than it. And through calculation, it can be found that the value of $[C]_R/[V]_R$ at the end point of stage blowing is 233, and the oxidation rate of vanadium is 95.14%. However, with a CO_2–O_2 mixed gas having a CO_2 ratio of 10%, the value of $[C]_R/[V]_R$ at the end of the reaction is 190, and the oxidation rate of vanadium is only 94.29%. From the comparison of the two blowing methods, we can find that the sufficient oxygen is supplied in the early stage of stage blowing, and the proportion of oxygen is reduced to increase the carbon dioxide proportion in the middle and late period. In addition, it can promote the purpose of vanadium extraction and protecting carbon, and the vanadium oxidation rate is higher when using the stage blowing.

As you can see from Fig. 5, in the previous 360 s, the two blowing methods used the same blowing ratio, so the temperature of the molten pool raises basically the same. However, after 360 s, since the stage blowing uses a CO_2–O_2 mixture with a CO_2 ratio of 15%, compared with the CO_2–O_2 mixed gas with 10% CO_2, the proportion of CO_2 increases, and the proportion of O_2 decreases. In the process of extracting vanadium, the decarburization reaction mainly occurs in the middle and late periods. At this time, the proportion of carbon dioxide is increased, and the

Table 3 Change of [C] and [V] content in molten pool at stage blowing with time, wt%

CO2 ratio (%)	[C]	[V]	V oxidation rate	$[C]_R/[V]_R$
10	4.12	0.2232	95.14%	223
	3.96	0.1626		
	3.68	0.0616		
15	3.45	0.0217		
	2.93	0.0126		

Table 4 Change of [C] and [V] content in molten pool at whole blowing 10% CO_2 with time, wt%

CO2 ratio	[C]	[V]	V oxidation rate	$[C]_R/[V]_R$
10%	4.16	0.2218	94.29%	190
	4.02	0.1416		
	3.78	0.0672		
	3.34	0.0225		
	2.86	0.0148		

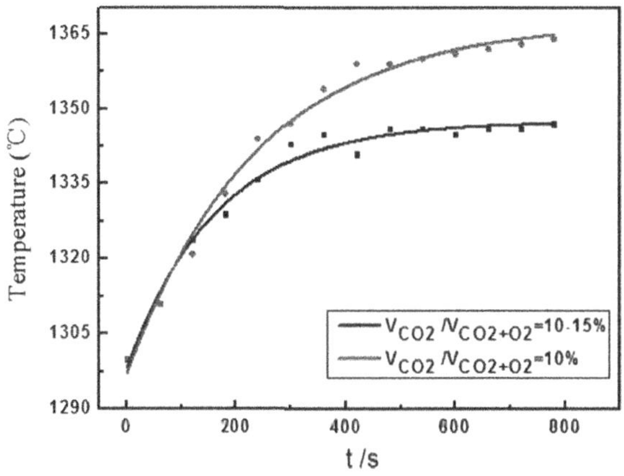

Fig. 5 Temperature changes with time during stage blowing

weak oxidation of carbon dioxide can be utilized to make more CO_2 participate in the molten pool reaction. Since CO_2 reacts with some elements in molten iron, it is a weak exothermic reaction or an endothermic reaction, so it can delay the temperature rise of the molten pool. Therefore, when using the stage blowing, the end temperature of the molten pool is reduced by 17 °C compared with 10% of CO_2.

Conclusion

Based on the above laboratory experimental research and analysis, we can obtain the following conclusions:

When the blowing temperature is 1300°C and the whole process CO_2 ratio is 15 %, by comparing three different blowing flows, which are 1.428 L/(min kg), 1.714 L/(min kg), 2.000 L/(min kg), the flow rate is 1.714 L/(min kg) has a better vanadium extraction with carbon retention effect. At this time, the oxidation rate of vanadium was 93.83%, and $[C]_R/[V]_R$ was 172, which was the highest of the three. The optimum blowing flow rate is 1.714 L/(min kg).

When the blowing temperature is 1300 °C, compared with the whole process CO_2 ratio of 10%, the stage blowing has better vanadium extraction with carbon retention effect, which can effectively reduce the end temperature of the molten pool at 17 °C, and the oxidation rate of vanadium at the end point can reach 95.14%, and the $[C]_R/[V]_R$ is also as high as 223.

Acknowledgements This work was supported by National Natural Science Foundation of China (project No.51334001) and Sharing Found of Large Scale Equipment, Chongqing University (project No.201512150031).

References

1. Grimston MC, Karakoussis V, Fouquet R et al (2001) The European and global potential of carbon dioxide sequestration in tackling climate change. Clim Policy 1(2):155–171
2. Wang K, Wang C, Lu X et al (2007) Scenario analysis on CO_2 emissions reduction potential in China's iron and steel industry. Energy Policy 35(4):2320–2335
3. Olivier JGI, Peters JAHW, Janssens-Maenhout G (2013) Trends in global CO_2 emissions. PBL Netherlands Environmental Assessment Agency
4. Hashimoto K, Habazaki H, Yamasaki M et al (2002) Advanced materials for global carbon dioxide recycling. Mater Sci Eng A 304(2):88–96
5. Brunsch R (1995) Technological ways to reduce the release of climatic-relevant emissions from the agricultural animal husbandry. Oekologische Hefte der Landwirtschaftlich-Gaertnerischen Fakultaet, Humboldt-Universitaet (Germany)
6. Kuo SJ (1997) Adapting sustainable low-carbon technologies to reduce carbon dioxide emissions from coal-fired power plants in China. Dissertation Abstr Int 58–02:0892
7. York APE, Xiao T, Green MLH et al (2007) Methane oxyforming for synthesis gas production. Cata Rev 49(4):511–560
8. Yun HH, Ruckenstein E (2002) Binary MgO—based solid solution catalysts for methane conversion to syngas. Cheminform 44(3):423–453
9. Bradford MCJ (1999) Vannice M A. CO_2 reforming of CH_4. Catal Rev 41(1):1–42
10. Nomura H, Mori K (2010) Kinetics of decarburization of liquid iron with high concentration of carbon. Tetsu-to-Hagane 58(4):29–40
11. Du WT, Wang Y, Wen G (2015) Effect of CO_2–O_2 mixed injection on C and V element oxidation in hot metal. J Chongqing Univ 38(05):66–72

Thermodynamic Analysis of Precipitation of La-O-S-As Inclusions in Steel

Congxiao Li, Hongpo Wang, Bin Bai and Lei Zhang

Abstract The formation of RE-O-S-As inclusions and corresponding arsenic content of equilibrium solid solubility were two core issues concerned when adding rare earth to restrain the harm of residual arsenic in steel. In this paper, some databases of rare earth inclusion were supplied. The influence of different La content on the formation of inclusions during solidification of As-bearing steel was calculated. The results showed that with the increase of rare earth content, the original inclusions MnS, $Al_6Si_2O_{13}$ and $Mn_2Al_4Si_5O_{18}$ in steel were replaced by $La_2Al_2O_6$, La_2O_2S, La_2S_3, La_3S_4 and LaS. The formation mechanism of As-bearing rare earth inclusions was analysed under the condition of excess rare earth, and the results showed that the As-bearing rare earth inclusions were formed during the solidification process due to enrichment of solidification frontier elements.

Keywords Residual elements · Inclusions · Rare earths · Arsenic

Introduction

Because of the lower oxidation potential than iron, the residual element As is difficult to remove in the steelmaking process, and it will continuously be enriched with the recycling of scrap steel. Although the content of As is low in steel, it usually segregates at grain boundaries and surfaces [1], which dramatically decreases the toughness and hot workability of steel products [2].

Currently, there are two main ways to control the content of residual elements in steel. On the one hand, the residual elements can be removed during the ironmaking and steelmaking processes [3]. For example, vacuum refining has the function of removing arsenic from steel, but it usually leads to a lot of iron loss. On the other hand, it is useful to add rare earth elements into steel to reduce the damage of

C. Li · H. Wang (✉) · B. Bai · L. Zhang
College of Materials Science and Engineering, Chongqing University, Chongqing 400044, China
e-mail: wanghp@cqu.edu.cn

© The Minerals, Metals & Materials Society 2019
T. Jiang et al. (eds.), *10th International Symposium on High-Temperature Metallurgical Processing*, The Minerals, Metals & Materials Series,
https://doi.org/10.1007/978-3-030-05955-2_75

residual elements. The rare earth elements La and Ce can not only react with the residual elements As, Sn and Pb to form high melting point compounds, but also reduce the contents of S and O in the steel [4].

When rare earth is added to steel, it will first react with O and S to form rare earth inclusions such as La_2O_3, La_2O_2S and LaS and then react with arsenic if the RE content is in excess. Therefore, it is important to clarify the formation mechanism of La-O-S-As inclusions.

Calculation Method

FactSage is one of the most popular computing systems in the field of thermodynamics. It has a rich database [5], which can evaluate and simulate the thermodynamic functions and thermodynamic equilibrium phase diagrams under different conditions. The whole system uses the absolute minimum Gibbs free energy as the criterion for the equilibrium point. This paper mainly used the Equilib module for calculation. Because of the incomplete rare earth element database in the FactSage, we added the thermodynamic data of La_3S_4 and La_2O_2S, shown in Table 1.

The chemical compositions for calculating in the present work are billets of grade 80 steel. The precipitation of inclusions during solidification of steel with different La contents was calculated. Table 2 shows the chemical compositions of steels.

Calculation Results and Discussion

Effect of Adding La on the Formation of Inclusions in Steel

In order to study the effect of adding rare earth on the formation of inclusions in steel, firstly, the precipitation of inclusions without La was calculated.

Table 1 Thermodynamic data of rare earth compound

No.	Reaction	$\Delta G^o = A + B \cdot T$ (J/mol)	Year	References
1	$2La_{(s)} + O_{2(g)} + S_{(s)} = La_2O_2S_{(s)}$	$-1,633,876.79 + 188.98 \cdot T$	1974	[6]
2	$3La_{(s)} + 4S_{(s)} = La_3S_{4(s)}$	$-1,826,021.82 + 220.13 \cdot T$	1986	[7]

Table 2 Chemical composition of steel (weight per cent)

Element	C	Si	Mn	P	Al	As	S	O	La
Mass/%	0.79	0.24	0.59	0.013	0.002	0.030	0.009	0.0043	0–0.10

The precipitation of inclusions during solidification of molten steel without La is shown in Fig. 1. It was calculated that 1467 and 1351 °C are the temperatures of liquidus and solidus of steel. It can be seen that the inclusions formed before the solidification are Al_2O_3 and $Al_6Si_2O_{13}$. As the temperature decreases, the main inclusion is $Al_6Si_2O_{13}$, and at the same time, SiO_2 begins to form. When the molten steel is completely solidified, the inclusions are MnS, $Al_6Si_2O_{13}$ and $Mn_2Al_4Si_5O_{18}$.

As mentioned above, elements O, Si, Mn and Al usually form composite inclusions in RE-free steel, so Al and Si play the role of deoxidation. S exists mainly in the form of MnS. In addition, there were no As-bearing inclusions.

Figures 2 and 3 show the precipitation of inclusions with the decrease of temperature in 0.03 and 0.06 mass% La steels. Figure 2 shows the precipitation of inclusions during solidification when La content is 0.03%. It can be seen that the inclusions before solidification are La_2O_2S and $AlLaO_3$, and the La_2O_2S and $AlLaO_3$ gradually reduce during the solidification process. When the molten steel is completely solidified, $AlLaO_3$ inclusions in the steel are transformed into $La_2Al_2O_6$. Finally, the main inclusions are La_2S_3, $La_2Al_2O_6$, MnS and La_2O_2S under 800 °C.

From Fig. 3, we can see that the inclusions are La_2O_2S and LaS. As the temperature decreases, the types and quantities of inclusions do not change significantly. O and S exist as sulphides and oxysulphides of La.

Figure 4 shows the formation of non-metallic inclusions with the increasing La content at 800 °C. It shows that the inclusion types were transformed from MnS, $Al_6Si_2O_{13}$ and $Mn_2Al_4Si_5O_{18}$ to $La_2Al_2O_6$, La_2O_2S, La_2S_3, La_3S_4 and LaS. When La content reaches 0.012 mass%, $Al_6Si_2O_{13}$ and $Mn_2Al_4Si_5O_{18}$ completely disappeared. Then, MnS disappeared when La content reaches 0.044 mass%. With the further increase of La content, La_2S_3 and La_3S_4 disappeared, and LaS and La_2O_2S were the final inclusions.

As mentioned above, with the La content increases, O is converted from the Si, Mn, Al composite oxides to the oxy-aluminides and oxysulphides of La. S is converted from MnS to the oxysulphides and sulphides of La.

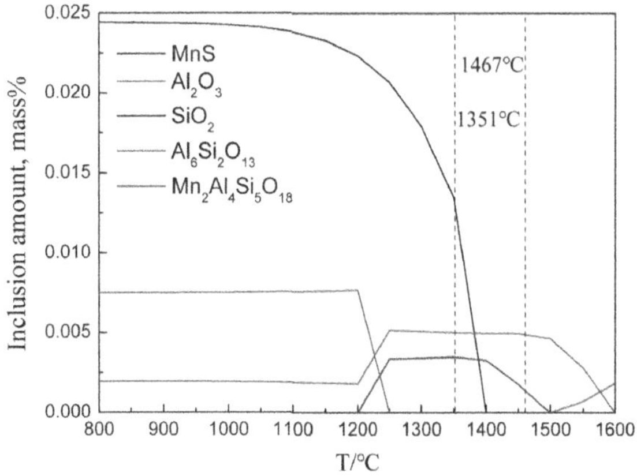

Fig. 1 Inclusion precipitation during solidification process

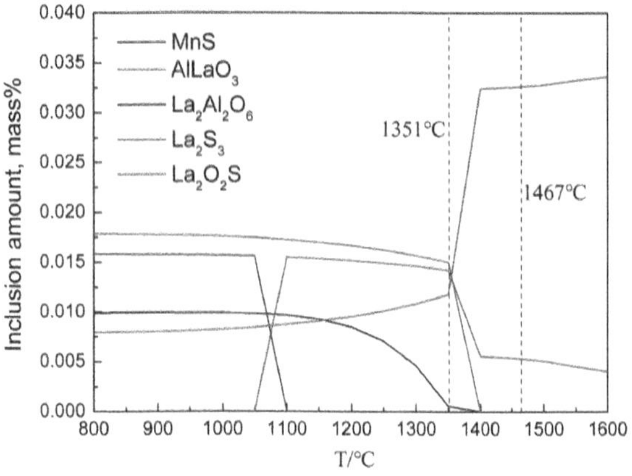

Fig. 2 Inclusion precipitation in 0.03 mass% La steel

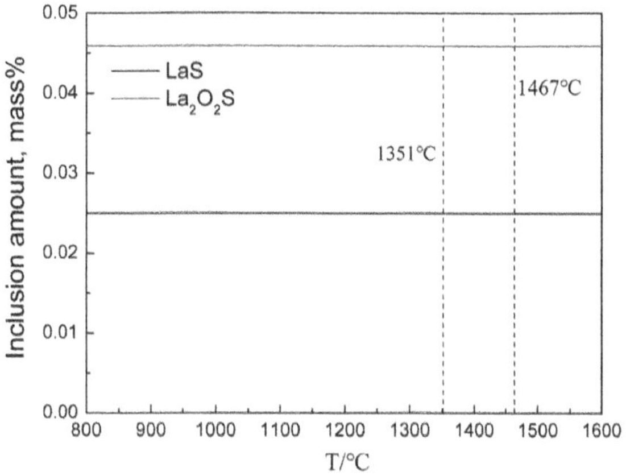

Fig. 3 Inclusion precipitation in 0.06 mass% La steel

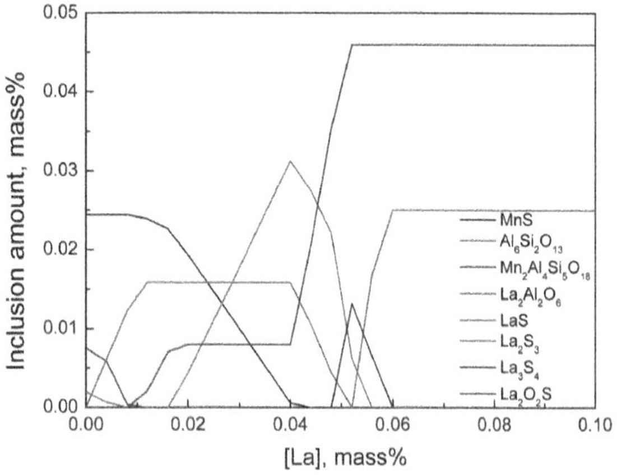

Fig. 4 Effect of La content on the precipitation of inclusions (T = 800 °C)

Formation Mechanism of As-Bearing Inclusions

Due to the strong binding ability of RE elements with oxygen and sulphur, when RE elements are added to steel to control the residual elements, they will first react with oxygen and sulphur in the molten steel. When the oxygen and sulphur contents are reduced to a certain extent, it is possible for RE elements to react with the residual elements in the molten steel. Li et al. [8] point out that RE-P-As phase will form when RE/(S + O) $\geq$ 4.2 in mass%. The experiment results of Wang et al. [9] show that in high-carbon steel containing 0.03% arsenic, when the amount of La added is from 0.003% to 0.089%, the generation sequence of La-O-S-As inclusions is La_2O_3, La_2O_2S, La_2O_2S-$LaAsO_4$ and $LaAsO_4$.

In order to clear whether As-bearing rare earth inclusions can form in steel-making process, we analysed the precipitation of inclusions at 1600 °C with different La contents. Figure 5 shows that the initial inclusion is Al_2O_3 in La-free steel. With the increase of La content, Al_2O_3 amount gradually decreases, and $AlLaO_3$ increases instead. As the consume of O, La sulphides precipitate with the increase of La content. The sequence of formation of La sulphides is La_2S_3 and LaS. However, no As-bearing inclusions form.

To further verify whether $LaAsO_4$ can form in the molten steel, the Wagner activity module was used in this section. The activity coefficient was calculated by formula (2). The activity was calculated by formula (3). The interaction coefficients are listed in Table 3. In the calculation, 0.10 mass% La was chosen (Table 4).

$$LaAsO_{4(S)} = [La] + [As] + 4[O] \quad (1)$$

$$\log f_i = \sum_{j-1}^{n} e_i^j [wt\%j] \quad (2)$$

$$a_i = f_i[\text{wt}\%i] \quad (3)$$

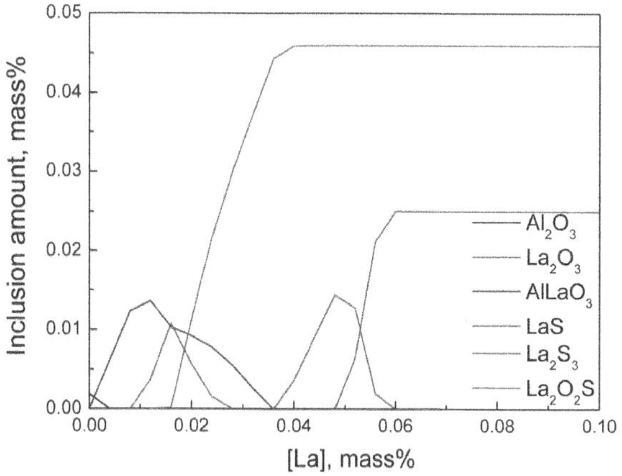

Fig. 5 Effect of La content on the precipitation of inclusions (T = 1600 °C)

Table 3 Interaction coefficients of solutes used at 1873 K (1600 °C)

					j				
i	C	Si	Mn	P	S	O	Al	As	La
O	−0.421	−0.066	−0.021	0.07	−0.133	−0.174	−1.17		-5
As	0.25	0.054	−0.031		0.0037			0.296	
La	0.03		0.28		−79	−43			−0.0078

Table 4 Activity coefficient and activity value of O, As and La

f_O	a_O	f_{As}	a_{As}	f_{La}	a_{La}
0.9803	0.0042	1.0046	0.0301	1.0027	0.01

The actual activity product of La, As and O under 1600 °C was calculated to be 9.37×10^{-14}. Because the interaction coefficient of elements is not much different between 1500 and 1600 °C, we use 9.37×10^{-14} as the activity product of the melting temperature interval.

On the other hand, Ihsan Barin found that the relationship [10] between the equilibrium activity product of La, As and O and temperature can be expressed by formula (4), and it is calculated to be 2.8×10^{-4} under 1600 °C.

$$\text{Log}\left(a_{La} \cdot a_{As} \cdot a_O^4\right)_{eq} = -56{,}046.16/\text{T} + 26.37 \quad (4)$$

We put the two activity products of La, As and O in the same picture; Fig. 6 shows that the $LaAsO_4$ cannot generate under 1600 °C.

During the solidification, La, As and O elements in the solidification front are continuously enriched due to the redistribution of solutes in semi-solid area. When they are enriched to a certain extent, the actual activity product of elements La, O

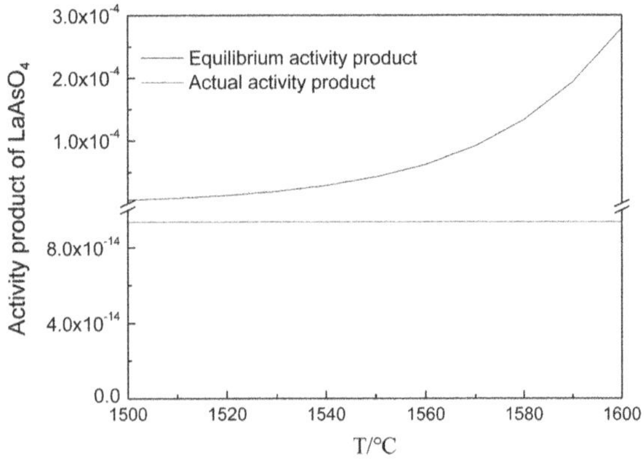

Fig. 6 The equilibrium activity product and the actual activity product of La, O and As

and As will be greater than the equilibrium activity product. Then, $LaAsO_4$ will be generated.

Therefore, the Brody-Flemings model was used to analyse the non-equilibrium solidification process. The micro-segregation equation is as follows:

$$C_L = C_O[1 - (1 - 2\alpha' k)f_S]^{(k-1)/(1-2\alpha' k)} \quad (5)$$

where C_L is the concentration of solute in the liquid phase at the solidification front, wt%; C_0 is the initial concentration of solute in the alloy, wt%; f_S is the coagulation fraction; α' is the modified Fourier number; k is the solute partition coefficient.

$$\alpha = 4D_S\tau_S/\lambda^2 \quad (6)$$

$$\alpha' = \alpha(1 - e^{\frac{-1}{\alpha}}) - \frac{1}{2}e^{\frac{-1}{2\alpha}} \quad (7)$$

where D_S is the diffusion coefficient of solute in solid phase; τ_S is the partial solidification time, s;

$$\tau_S = (T_L - T_S)/R_C \quad (8)$$

where T_L is the liquidus temperature, K; T_S is the solidus temperature, K; R_C is the cooling rate, K s^{-1}.

The relationship between the secondary dendrite arm spacing and the cooling rate is [11]:

$$\lambda = 183 \times 10^{-6}(R_C)^{-0.3616} \quad (9)$$

where λ is the secondary dendrite arm spacing, m.

The cooling rate R_c in this paper is 0.0324 K/S.

The temperature T of the solidification front liquid phase is obtained by formula (10) and then brought into formula (4) to calculate the equilibrium activity product of the solidification processes La, O and As.

$$T = T_0 - \frac{T_0 - T_L}{1 - f_S\frac{T_L - T_S}{T_0 - T_S}} \quad (10)$$

T_O is the melting point of pure iron and the temperature is 1808 K. T_L and T_S are the temperatures of liquidus and solidus of the used steel, which were calculated to be 1740 and 1624 K, respectively.

Figure 7 shows that the value of the actual activity product during the end of solidification exceeds the value of the equilibrium activity product, so $LaAsO_4$ is possibly generated.

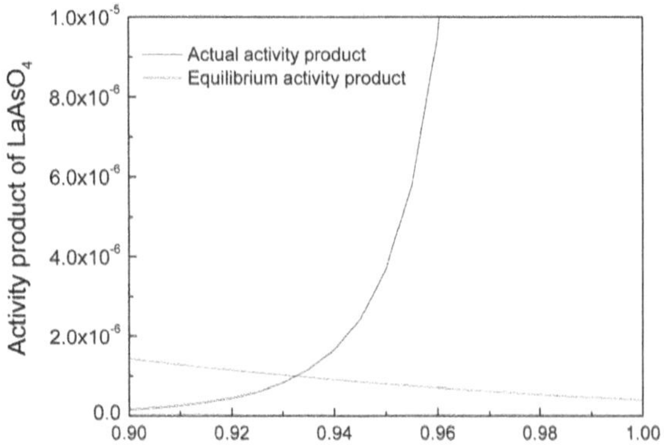

Fig. 7 Relationship between activity product and f_S in solidification process

Conclusions

(1) When no rare earth is added, elements O, Si, Mn and Al usually form composite inclusions, so Al and Si play the role of deoxidation. S is mainly in the form of MnS. After adding La, O usually forms $AlLaO_3$, $La_2Al_2O_6$ and La_2O_2S inclusions, and S usually forms La_2S_3, La_3S_4, LaS and La_2O_2S inclusions. Therefore, it can be concluded that La has a good deoxidation and desulphurization ability.

(2) Through the calculation of the non-equilibrium solidification process, it is confirmed that $LaAsO_4$ can form during the solidification process due to the enrichment of the solidification frontier elements.

Acknowledgements This research is financially supported by National Natural Science Foundation of China (Grant No. 51704051), Chongqing Postdoctoral Funds (Grant No. Xm2017192) and the Fundamental Research Funds for the Central Universities (Grant No. 106112015CDJXY130009).

References

1. Yin L, Sridhar S (2011) Residual elements arsenic, antimony, and tin on surface hot shortness. Metall Mater Trans B 42(5):1031–1043
2. Seah M, Hondros E (1973) Grain boundary segregation. Proc R Soc Lond A Math Phys Eng Sci R Soc 335(1601):191–212
3. Noboru H, Hidemi K, Teruo T (1971) Method for removing arsenic from metals or alloys: U. S. 3, 622, 302[P], 23 Nov 1971
4. Wang L, Lin Q, Ji J (2006) New study concerning development of application of rare earth metals in steels[J]. J Alloy Compd 408:384–386
5. Balecw B (2009) Factsage thermochemical software and datebases-recent development. Calphad 33(2):295–311

6. Vahed A, Kay DAR (1976) Thermodynamics of rare earths in steelmaking. Metall Trans B 7B:375–383
7. Wang L, Du T (1986) Study on the balance of [La]-[S]-[O] in molten iron. Rare Earth 2:25–32
8. Li DZ, Wang ZY (1980) Some regularities of formation and change of rare earth inclusions in steel. Iron 8:34–39
9. Wang HP, Xiong L, Zhang L et al (2017) Investigation of RE-O-S-As inclusions in high carbon steels. Metall Mater Trans B 48(6):2849–2858
10. Barin I (1995) Thermochemical data of pure substances. VCH Verlag GmbH, Weinheim
11. Won YM, Thomas BG (2001) Simple model of microsegregation during solidification of steels. Metall Mater Trans A 32(7):1755–1767

Author Index

A
Arslan, Cüneyt, 591
Augustine, Robin, 117

B
Bai, Bin, 795
Bayless, Trenin, 371
Bian, Miaolian, 777
Brooks, G.A., 297
Bugdayci, M., 479

C
Cai, Haoyu, 157
Cai, Zeyun, 381, 415
Chen, Feng, 211
Cheng, Gong, 309
Cheng, Hongwei, 221, 657, 713
Cheng, Rong, 579
Chen, Kuiyuan, 619
Chen, Min, 693, 769
Chen, San-Xing, 723
Chen, Xiaofang, 199
Chen, Xuling, 85, 257
Chen, Yuchao, 453
Chen, Zhuo, 25
Chorney, Jannette, 371
Chou, KuoChih, 609
Cui, Xiaokang, 405, 415

D
Dai, Linqing, 349
Deng, Haohua, 693, 769
Deng, Zhigan, 733
Dong, Tingliang, 289
Dong, Yijia, 145
Downey, Jerome P., 371

E
Ertürk, Selim, 591

F
Fan, Gang, 733
Fan, Kai, 65
Fan, Xiaohui, 85, 257
Feng, Fu, 443
Feng, Huai-Xuan, 523
Fu, Jianxun, 279
Fu, Pei, 25

G
Gan, Min, 85, 257
Gao, Hairong, 501
Gao, Jintao, 243
Gao, Leizhang, 545
Geveci, Ahmet, 489
Gu, Fengjiao, 501
Guo, Hanjie, 15
Guo, Lei, 467
Guo, Shenghui, 349
Guo, Yaguang, 741
Guo, Yufeng, 211
Guo, Zhancheng, 243, 467, 619
Guo, Zheng-Lei, 787

H
Hasan, M.M., 297
He, Sheng Ping, 599
He, Wenchao, 531, 539, 759

© The Minerals, Metals & Materials Society 2019

T. Jiang et al. (eds.), *10th International Symposium on High-Temperature Metallurgical Processing*, The Minerals, Metals & Materials Series,
https://doi.org/10.1007/978-3-030-05955-2

He, Xianzhu, 501
He, Yang, 427
Huang, Bing, 51, 109
Huang, Hui, 733
Huang, Qingyun, 705
Huang, Wenbin, 157
Huang, Yun, 329
Hu, Meilong, 545
Hu, Ting, 349
Hu, Xiaojun, 609
Hu, Xun, 321

J
Jia, Bo Ran, 599
Jiang, Tao, 85, 117, 145, 231
Jiang, Yuyang, 531
Jiao, Kexin, 15
Jing, Jianfa, 211
Ji, Zhiyun, 85, 257

K
Keskinkilic, Ender, 489
Kou, Ming-yin, 3
Krotov, Yury, 453
Kulkarni, Anil, 175

L
Lai, Yanqing, 321
Lan, Xi, 243
Lee, Joonho, 117
Liang, Jinglong, 609
Liang, Shuaibiao, 741
Li, Congxiao, 795
Li, Cunxiong, 733
Li, Dongbo, 665, 741
Li, Gang, 569
Li, Guanghui, 117, 145, 231
Li, Guangshi, 221, 657, 713
Li, Haokun, 15
Li, Jie, 321
Li, Kaihua, 681
Li, Liang, 681
Li, Longfei, 381, 405, 415
Li, Minting, 733
Lin, Charis I., 175
Lin, Lin, 523
Lin, Qing-Shan, 523
Lin, Tiancheng, 109
Li, Tao, 39
Liu, Bingguo, 349
Liu, Chen, 145
Liu, Dachun, 681
Liu, Hua, 349
Liu, Hui, 279
Liu, Jianhua, 427
Liu, Jun, 39
Liu, Lei, 553, 649
Liu, Mudan, 117
Liu, Shihong, 65
Liu, Wenwang, 157
Liu, Yong, 117
Liu, Z.J., 631
Liu, Zhen, 381, 405, 415
Li, Xiaoling, 501
Li, Xingbin, 733
Li, Xueqin, 539
Li, Y., 631
Li, Yi, 657
Li, Yifu, 341, 361, 393
Luo, Chao, 609
Luo, Jun, 145, 231
Lu, Qi, 787
Lu, Xionggang, 221, 657, 713
Lv, Wei, 85, 257
Lv, Xue-wei, 531, 539, 569, 759

M
Ma, Aiyuan, 501
Ma, Deng, 741
Ma, Haitao, 579
Ma, Huaiying, 649
Ma, Tengfei, 443
Ma, Tongxiang, 545
Ma, Xiaodong, 289
McLean, Alex, 427
Mei, Xiaoyong, 657, 713
Meng, Long, 619

N
Nagai, Takashi, 435
Niuman, Sebastian, 175
Nwoke, VU, 511

O
Odo, JU, 511

P
Pan, Feifei, 539, 569
Pang, Zhengde, 531, 539, 759
Pan, Wen, 553, 649
Pan, Yu-Zhu, 523
Pei, Guishang, 569
Peng, Zhiwei, 117
Pournaderi, Saeid, 489
Pu, Zhenghao, 341, 361, 393

Q
Qian, Wei, 95
Qing, Gele, 157
Qiu, Shuxing, 165

R
Rao, Mingjun, 117, 231
Reddy, Ramana G., 759
Rhamdhani, M.A., 297

S
Scheller, Piotr Roman, 309
Schumacher, Katie, 371
Sezer, Raşit, 591
Sheng, Zhuo, 681
She, Xuefeng, 443
Shi, Anjun, 467
Silaen, Armin K., 453
Singh, Jogender, 175
Song, Bo, 3, 381, 405, 415
Sun, Hu, 145, 231
Sun, Hui, 777
Sun, Ming, 501
Suo, Guangsheng, 165

T
Takamatsu, Yuichi, 435
Tang, Guangwu, 453
Tang, Huiqing, 65
Tasyurek, K.C., 479
Tian, Yunqing, 157
Tian, Zhongliang, 321
Topkaya, Yavuz A., 489

W
Wallace, Grant C., 371
Wang, Chunlai, 649
Wang, Feng, 289
Wang, Hongpo, 795
Wang, Jing-Song, 443, 523
Wang, Liancheng, 117
Wang, Nan, 693, 769
Wang, Qian, 137, 599
Wang, Qiangqiang, 309
Wang, R.R., 631
Wang, Sai, 599
Wang, Shu-Chao, 787
Wang, Wenbo, 309
Wang, Xue-Si, 137
Wang, Yijie, 15
Wang, Yu, 95, 787
Wang, Y.Z., 631
Wang, Zhe, 467, 619
Wei, Chang, 733
Wei, Lizhen, 221
Wong, Guojing, 85, 257
Wu, Chengbo, 329
Wu, Shanshan, 569
Wu, Sheng-li, 3
Wu, Tuo, 125
Wu, Xiaojiang, 157
Wu, Yue, 165

X
Xiang, Junyi, 569
Xia, Wentang, 705
Xie, Jianbo, 279
Xie, Yongfang, 199
Xiong, Xiaolu, 221
Xu, Baoqiang, 393, 681
Xu, Cong, 713
Xu, C.Y., 631
Xue, Jilai, 75, 189
Xue, Qingguo, 443
Xu, Junjie, 361, 393
Xu, Meng, 157
Xu, Qian, 221, 657, 713

Y
Yamamoto, Namiko, 175
Yang, Bin, 341, 361, 393
Yang, Kai, 321
Yang, Lingzhi, 211
Yang, Xiaohua, 665
Yang, Yindong, 427
Yan, Hongyan, 609
Yan, Jianchuan, 39
Yan, Zhiming, 531, 539, 545, 759
Yao, Shun, 3
Ye, Lei, 117
Ye, Qing, 117
Yilmaz, Emre, 591
You, Yanjun, 361
Yuan, Fang, 125
Yuan, Hua-Zhi, 137
Yuan, Xiaoli, 705
Yu, Caixiang, 221
Yucel, O., 479
Yue, Weichao, 199
Yu, Junjie, 231

Z
Zeng, Chang-Ping, 137
Zhai, Qijie, 279
Zhang, Ai-Jun, 523
Zhang, Aiping, 51, 109
Zhang, Fuming, 269
Zhang, Gaopeng, 329

Zhang, Guangzong, 693, 769
Zhang, Haitao, 329
Zhang, Huan, 341, 393
Zhang, Jianliang, 15
Zhang, Jiongming, 579
Zhang, J. L., 631
Zhang, Lei, 795
Zhang, Liangjin, 579
Zhang, Libo, 349
Zhang, Lifeng, 309
Zhang, Ning, 467
Zhang, Peng, 289
Zhang, Shengfu, 165
Zhang, Tao, 723
Zhang, Yan, 157
Zhang, Yanling, 125
Zhang, Ya-peng, 553
Zhang, Zhidong, 649
Zhao, Baojun, 289
Zhao, Lu-ming, 95
Zhao, Qiang, 75, 189
Zhao, Shiwei, 199
Zhao, Tianyang, 25
Zhao, Xia, 553
Zhao, Zheng, 125
Zhao, Zhi-xing, 553
Zheng, A.Y., 631
Zheng, Fuqiang, 211
Zheng, Xuemei, 501
Zhong, Honggang, 279
Zhong, Qiang, 145
Zhong, Yiwei, 619
Zhong, Yong, 51, 109, 329
Zhou, Chenn Q., 25, 453
Zhou, Dongshan, 705
Zhou, Heng, 3
Zhou, Ping, 25
Zhou, Wang, 723
Zhou, Xiaohu, 165
Zhou, Xunwei, 257
Zhu, Fuxing, 681
Zhu, Mingmei, 51, 109
Zhu, Rong, 15
Zhu, Rongjin, 165

Subject Index

A

Acidic and basic oxides, 299
Acid leaching, 734, 737
Acid leaching slag, 501–503, 506–508
Activation energy, 65, 70, 71, 73, 531, 535–537, 545, 550, 551, 592, 764
Adsorption, 244, 371, 372, 705–709, 711
Alkali, 15–19, 21–23, 85, 118, 165, 189, 190, 197, 299, 303, 351, 490, 492, 494, 516, 560, 714, 765
Alkali carbonate, 19, 165, 167, 173
Alkali circulation process in blast furnace, 16
Alkalinity, 355
Al-Sc master alloy, 321–327
Aluminosilicate slags, 759, 764, 766
Aluminum, 127, 128, 138, 142, 190, 199–201, 206, 207, 244, 309, 321, 325, 327, 405, 467, 480, 489–491, 493, 497, 604, 605
Ammonium chloride, 713–716, 718–720
Analysis of pore structure, 169
Analysis of Raman spectra, 760
Analysis on Change of Cu content in Slag, 749
Analysis on Change of Pb Content in Slag, 751
Analysis on Change of Sb Content in Slag, 752
Analysis on Change of Zn content in Slag, 750
Annealing, 168
Apparent viscosity, 109, 110, 112–116
Arsenic, 741, 742, 744, 753–756, 795, 796, 799
Ashes, 157–162
Assessment of Briquette Quality, 68
Average residence time and inclusion removal rate, 59

B

Basic hypothesis, 100
Basicity, 86, 115, 125, 126, 129–131, 134, 191, 196, 275, 297–303, 306, 354–356, 443, 444, 446–451, 495, 497, 525, 770, 772–774
Basicity in slag chemistry, 299
Bath temperature, 353, 491, 694, 770
Beneficiation, 244, 512, 516, 564, 742
Blast furnace, 3–6, 11, 12, 15–23, 65, 86, 157, 165, 166, 169, 189, 191, 196, 198, 244, 270, 273–277, 443, 444, 490, 491, 511–513, 516, 518–521, 523, 524, 527, 529, 531, 532, 545, 546, 631, 632, 658, 665, 678, 759, 765, 777–781
Blast furnace operation line, 15, 16, 21
Bonding strength experiment, 148
Bottom blown breathable brick, 333
Boundary conditions, 332

C

Calcination, 507, 515, 591–596, 660, 693, 714, 719, 769, 770
Calculation method, 796
Calculation of alkali balance in blast furnace, 16
Calculation of amount of charge material, 633
Calculation of carbon loss, 6
Calculation of direct reduction degree, 5
Calculation of heat expenditure, 640
Calculation of heat input, 639
Calculation of the composition and content of top gas, 636

© The Minerals, Metals & Materials Society 2019

T. Jiang et al. (eds.), *10th International Symposium on High-Temperature Metallurgical Processing*, The Minerals, Metals & Materials Series,
https://doi.org/10.1007/978-3-030-05955-2

Calculation of the reducing gas volume, 635
Calorimetric methods, 569
Carbide synthesis, 373
Carbon content, 21, 22, 65–68, 70, 72, 158, 349, 351, 352, 371, 374, 376, 377, 528, 540, 579, 580, 583–589, 633, 640, 789–791
Carbon dioxide, 11, 591, 787, 788, 791–793
Carbon integral model, 584
Carbon loss, 3–12
Carbon-saturated pig iron, 545
Carbothermic reduction, 117–119, 123, 211, 212
Carburization Modelling, 373
Carburization Precursor Production, 373
Cathodic current density, 324
Cell design and electrolysis procedure, 322
Centrifugal flow, 39, 40
CeO_2, 381–390
Ce treatment, 415, 416, 418, 421, 424
The characterization of calcine and leaching residue, 225
Characterization of the original galvanizing dross, 470
Chemical and physical properties of sintering, 195
Chemicals and materials, 322
Chemistry of sinter, 196
Chlorine impurity, 681–684, 687–690
Chromite, 117, 257–259, 350, 351, 353–356
Chromite fines, 257, 258, 266
Circular economy, 269, 270, 273, 275
Circulation, 15, 16, 21–23, 59, 147, 166, 231, 277
Clogging, 309, 310, 313–316, 513, 519
Cohesive zone, 65, 72, 524–529
Coke, 3–8, 10–12, 15–17, 21, 22, 65, 66, 71–73, 118, 165–173, 191, 192, 194, 212, 215, 217, 218, 262, 271, 273–277, 354, 491–493, 512, 525, 528, 591–593, 595, 596, 649, 651, 652, 654–656, 667, 668
Coke breeze, 86, 87, 191, 257–263, 266, 514, 649–651, 653–656
Coke ratio, 15, 16, 19, 21–23, 521, 592
Coke saving, 65
Cold-bonded pellet, 157, 159–162
Cold-bonded pellet experiments, 159
Cold-rolling sludge, 657–662
Colour change rate feature extraction with flame interference, 202
Combustion, 5–7, 11, 20, 21, 86, 88, 191, 207, 261, 262, 273, 277, 453–455, 458, 460, 461, 464, 592, 595, 596, 649, 655, 656
Combustion efficiency, 649, 655, 656
Composition, 4, 20, 66, 86, 97, 110–112, 119, 120, 122, 128, 138, 149–153, 158, 177, 190, 192, 196, 200, 212, 214, 222, 244–246, 261, 264, 265, 275, 280, 290, 292, 305, 313, 323, 354, 371, 374, 375, 383, 405–408, 415–418, 420, 421, 443–447, 490–494, 496, 502, 511, 512, 514–518, 520, 524, 525, 532, 540, 541, 546, 554, 580, 581, 583, 585, 599, 600, 632–637, 643, 646, 649, 658, 659, 667, 668, 671, 672, 713, 714, 716, 742, 743, 745, 749, 754, 755, 760, 761, 770, 772, 773, 778, 779, 788, 796
Composition and flow of offgas, 581
Composition of the nucleation inclusions in tested steels, 418
Comprehensive process conditions, 81
Comprehensive recovery, 741, 742, 748, 756
Compressive strength, 4, 5, 68, 511–516, 518–521
Computational Fluid Dynamics (CFD), 40, 453–455, 459–464
Concentration ratio of $Sc_2O_3$to Al_2O_3, 325
Concepts of acid and base, 298
Condensation, 393, 394, 397–403
Consolidation behavior, 263
Consolidation characteristic, 257
Construction of full process energy flow network, 272
Converter slag, 115, 693–696, 698, 702, 770
Cooling gas, 779
Copper, 102, 103, 221, 232, 236, 238, 290, 435–439, 666, 705, 707, 720, 741–744, 746–750, 754–756
Copper alloy, 626, 627
Copper smelting slag, 741–743, 745–748, 750, 751, 756
Co-precipitation, 733, 734, 737, 739
Core diameter, 511, 515, 516, 518–520
Crystallization behavior, 443, 444, 451, 599
Cu (II) ion removal, 168, 290, 354, 601, 602, 619–628, 705–712, 741–743, 749, 750, 755, 756, 760

D

Decarburization, 313–316, 352, 579, 580, 584, 585, 587–589, 792
Dechlorination, 658, 661
Decomposition, 118, 185, 192, 236, 660, 670, 693, 694, 696–699, 702, 713–716, 720, 770, 773, 774
Dehydration rate, 501–507
Dendrite, 279–283, 286, 801

Density, 764
Dephosphorization, 277, 352, 769–772, 774, 775
Dezincification, 157, 158, 160–162
Dezincification experiment, 160
Differential Scanning Calorimetry (DSC), 570–573, 575
Directional solidification, 279–281, 283, 284, 286
Direct reduction, 3, 5–12, 21, 511–513, 516, 631–633, 777, 778, 782
Discussion on chemical analysis of experiment products, 749
Dissolution, 118, 139–141, 143, 200, 281, 299, 490, 528, 545, 546, 548–551, 579, 580, 582, 589, 693, 694, 698–702, 769, 770
Dissolution kinetics, 545, 546
Dissolution rate, 139, 140, 142, 546–549, 693, 694, 698, 701, 702, 773–775
Distribution, 5, 15, 16, 51, 102, 134, 168–170, 195, 257, 258, 264, 266, 279, 292, 295, 304, 306, 313, 314, 321, 322, 325, 326, 350, 352, 405–407, 410, 418, 420, 431, 443, 446, 470, 472, 473, 475, 477, 512, 543, 559, 560, 564, 601–603, 619, 621, 623–625, 628, 645, 650, 655, 677, 695, 718, 723, 727, 729–731, 778
Distribution of magnetic field intensity in mould flux, 102
Distribution of oxygen types and Q^nspecies, 603
Distributions of main chemical elements, 558
DOP, 599, 606, 607
Drop method, 569, 571–575
DSC method, 572

E

EAF dust, 117–123
Effective application of high-carbon metallic briquettes to BF, 72
Effect of adding La on the formation of inclusions in steel, 796
Effect of adsorption temperature, 708
Effect of $Al_2O_3 \leftrightarrow SiO_2$on the Structure, 763
Effect of alkali on coke ratio, 19
Effect of Al on the composition, 407
Effect of Al on the inclusion, 407
Effect of Al on the microstructure, 411
Effect of Al on the size of effective inclusions, 408
Effect of CaO/SiO_2ratio on apparent viscosity of slags, 115
Effect of Cr_2O_3 content on the viscosity of CaO-SiO_2-Cr_2O_3slag under different basicity, 130
Effect of Cr_2O_3content on the viscosity of CaO-SiO_2-10%Al_2O_3-Cr_2O_3slag under different temperature program, 131
Effect of Cr_2O_3on apparent viscosity of slags, 112
Effect of distillation temperature, 364
Effect of factors on the aluminum content in liquid phase, 347
Effect of factors on the zinc content in gas phase, 345
Effect of feeding materials(thickness), 367
Effect of FeO on apparent viscosity of slags, 113
Effect of final pH, 737
Effect of gas flow rate, 726
Effect of H_2/CO ratio on the energy utilization rate, 644
Effect of H_2/CO ratio on the gas consumption, 643
Effect of initial Cu (II) ion concentration, 708
Effect of initial pH, 707
Effect of LGPIO dosage, 710
Effect of moisture on sintering performance, 194
Effect of nozzle size, 725
Effect of oxygen blast on antimony volatilization, 676
Effect of particle size, 709
Effect of pressure, 736
Effect of reaction time, 737
Effect of reducing temperature on the gas consumption, 644
Effect of reduction degree on the cohesive zone and permeability, 525
Effect of roasting temperature, 215
Effect of roasting temperature on Cl removal rate, 660
Effect of sintering parameters on microstructure, 180
Effect of soaking time, 365
Effect of the ratio of coal addition, 217
Effect of the size of pellets and the roasting time, 216
Effect on carbon concentration, 611
Effects of bentonite dosage on sintering indexes, 261
Effects of CeO_2on Melting Temperature, 384
Effects of CeO_2on structure, 387
Effects of CeO_2on Viscosity, 385

Effects of coke breeze rate on sintering indexes, 261
Effects of moisture content on sintering indexes, 261
Effects of sintering conditions on sintering indexes, 260
Effects of sintering suction on sintering indexes, 262
Effect on nitrogen, 613
Effect on oxygen, 613
Effect on sulphur, 612
Electrical conductivity, 765
Electric arc furnace, 117, 118, 453, 491, 512, 765
Electric furnace smelting, 777, 782–784
Electrolysis, 199, 200, 207, 283, 321–324, 326, 327
Electrolysis cell, 199
Electrolysis temperature, 323
Electrolytic parameters, 321
Electron-silicothermic, 351
Electrothermal volatilization, 665, 666, 672, 678
Emission profile, 85
Emission property of $PM_{10}$and $PM_{2.5}$, 89
Energy flow network, 269, 272, 278
Energy Flow Network Design, 272
Energy saving, 232, 271, 273–278, 353
Energy saving and energy recycling practice, 274
Energy spectrum analysis of mineral elements, 556
Energy utilization analysis of different system, 782
Engineering concept, 270
Engineering objectives, 271
Enrichment, 3, 8, 9, 11, 12, 19, 22, 232, 274, 275, 431, 433, 468–470, 472, 473, 475–477, 619, 620, 622, 623, 625, 626, 628, 795, 802
Evaluation standard, 77
Exhaust gas circulation, 231

F
Features extraction, 200–203
Features extraction method of fire hole with flame interference, 201
Features extraction of amplitude, 202
FeCrAl alloys, 427–429, 431, 433
Ferrite, 118, 145, 149, 153, 196, 197, 266, 280, 290, 294, 405–407, 411, 412, 415, 416, 421–424, 428, 433, 743, 744
Field-Assisted Sintering Technology (FAST), 25, 176–179, 183, 186, 352, 356, 655, 770
Fine bubble, 725, 726
Fine bubbles generation, 725
Fire hole, 199–201, 203, 204, 206, 207
Flame interference, 199, 201, 202, 206, 207
Flow field in the tundish, 55
Flow field simulation of mold fluxes, 102
Fluidization, 75, 77–82
Fluidizing velocity test, 78
Fluxes and coke, 191
Formation mechanism of As-bearing Inclusions, 799
Formation of molybdates, 236
Formation of sulphates, 237
Free running temperature, 531, 532, 535, 537
Fundamental equation, 100
Fuzzy gray relational analysis, 25, 27, 34, 36

G
Galvanizing dross, 467–472, 477
Gas-based shaft furnace, 632, 777–783
Gas-based shaft furnace and calculating model, 780
Gas-based shaft furnace reduction, 777
Gas consumption, 631, 643, 644, 646
Gasification, 4, 5, 15, 65–67, 70–73, 166
Gasification of high-carbon metallic briquette, 70
Gasification tests, 67
Geometrical model and conditions, 460
Geometric model, 29, 330
Granularity, 356
Granulation, 85, 191, 198, 259, 261, 650, 655, 656
Grinding fineness tests, 81
Growth rate, 279–284, 286

H
Harmless treatment, 741, 746, 753
Heat affected zone, 175, 177, 289, 293, 416
Heat balance, 19–21, 632, 638, 639, 641, 642, 646, 777, 778, 780–782
Heat capacity, 5, 52, 54, 458, 569–575, 639
Heat transfer, 52, 53, 95, 244, 261, 277, 382, 453–455, 457, 461, 463, 464, 640, 694, 696, 782
Heat treatment, 25, 26, 161, 172, 279
High-carbon metallic briquette, 65, 66, 70–72
High efficiency energy conversion and highlight energy saving technology, 276

High heat input welding, 289, 290, 293, 295, 296
High phosphorus oolitic hematite, 777–779, 781–784
High-temperature calorimetric technology, 571
High-temperature wettability, 145, 146, 149, 153
Hot metal, 5–7, 21, 65, 275–277, 454, 494, 540, 769–775
Hybridization, 176

I
Inclusion, 39, 40, 42, 43, 47–49, 52, 54, 58–60, 96, 289–294, 296, 313–316, 405–412, 415, 416, 418, 420–423, 429, 724, 754, 755, 795, 797–799
Inclusion removal, 39, 40, 47–49, 52, 59, 60, 723, 724
Inclusions in the EH40 steel, 290
Inclusions tracking model, 43
Inclusion trajectory, 58
Indium, 733–739
Induction heating, 25–27, 29–31, 36
Industrial application of PRO, 564
Influence factors, 26, 27, 349
Influence mechanism of reduction degree on changes of cohesive zone, 528
Influence of carbon deposition reaction, 583
Influence of CO_2dissolution, 582
Influence of $CO_2$from air, 583
Influence of coke quality on carbon loss, 10
Influence of different factors on carbon loss, 8
Influence of flow rate, 789
Influence of gas injection on flow fields in tundish, 728
Influence of gas injection on temperature fields in tundish, 729
Influence of internal diameter, 46
Influence of low melting point slag on the volatilization of antimony, 675
Influence of nozzle length, 45
Influence of raw materials, 687
Influence of stage blowing, 791
Influence of titanium sponge structure, 685
Influence of twice vacuum distillation, 689
Influence of vacuum distillation process system pressure, 688
Influence of volatile time, 674
Influence of volatilization temperature, 673
Influences of adding organic binder on emission property of $PM_{10}$and $PM_{2.5}$, 91
Iodine, 435, 437–439, 620
Iron, 5–9, 11, 16, 17, 20, 22, 23, 66, 68–72, 85, 86, 114, 117–120, 122, 123, 145, 146, 149, 157, 165, 166, 190, 196, 197, 212, 232, 236, 237, 240, 244, 272, 304, 323, 350–353, 355, 356, 406, 418, 429, 431, 435–437, 443, 454, 467, 468, 472, 474, 475, 477, 489–494, 497, 502, 511–514, 516, 520, 524, 525, 528, 529, 531, 539–543, 545–551, 555, 556, 558, 560, 564, 580, 631–633, 635, 641, 642, 644, 657, 661, 668, 706, 716, 719, 720, 741, 742, 744, 750, 754–756, 770, 771, 777–780, 783, 784, 787, 788, 790, 791, 793, 795, 801
Iron and steel, 95, 269–275, 278, 352, 512, 531, 632, 657, 705, 711, 712, 720, 788
Iron ore, 15, 20, 66, 72, 86, 145, 189, 190, 196, 435, 443, 451, 511–514, 516, 520, 523, 553, 554, 633, 636, 666, 705, 706, 711, 759, 766, 777
Iron ore sintering, 85, 649
Iron removal, 490

K
Kinetics, 5, 70, 71, 238, 458, 459, 495, 545, 548, 550, 592, 675, 676, 696, 769
Kroll process, 681, 687

L
La addition, 427–429, 431, 433, 434
Large pot sintering test, 195
Leaching, 118, 222–228, 231, 468, 490–493, 497, 620
Lead vapor, 393, 398
Limestone, 191, 490, 491, 493, 495, 513, 514, 693–700, 702, 769–775
Limestone decomposition, 693–698, 701, 702, 769, 770, 773–775
Limestone dissolution in converter slag, 698
Limonite, 189, 190, 192, 197, 198
Liquid steel, 54, 309
Low frequency, 95, 96, 103, 104
Low Grade Phosphorus-containing Iron Ore (LGPIO), 705–712
Low-temperature calorimetric technology, 570

M
Magnetic separation, 75–82
Magnetizing roasting, 78
Manganese Sulphide (MnS), 279–281, 283–286, 405, 406, 408, 411, 412, 415, 416, 418, 421–424, 795, 797, 802
Mass and energy balance, 633
Mass balance, 631–633, 638, 645, 646, 777, 780–782
Mathematical model, 29, 724

Mechanical properties, 26, 39, 175–177, 179, 183, 186, 290, 330, 416, 427–429, 431, 433, 511
Mechanical properties of hybrid disks, 183
Medium and low carbon ferrochrome, 352, 355, 356
Medium-low carbon ferrochrome alloys preparation methods, 351
Melting Temperature, 381, 382, 384, 385, 389, 494, 633, 673, 800
Metal, 3, 15, 23, 93, 109, 117, 118, 123, 125, 189, 221–223, 225, 227, 231, 233, 234, 236, 240, 293, 297, 298, 300, 302–304, 306, 321, 350, 354–356, 371, 372, 417, 427, 435, 468, 469, 476, 479–482, 491, 493–495, 511, 528, 546, 548, 560, 620, 623, 626, 657, 658, 666, 671, 706, 713, 714, 716, 718–720, 742–744, 747, 748, 753, 755, 756, 771
Metallurgical performance, 196
Microstructure, 26, 67, 68, 142, 143, 175–177, 180, 186, 214, 223, 264, 266, 280, 281, 283, 290, 323, 354, 406, 411, 412, 416–418, 421, 427–429, 431, 432, 472, 554, 599–601, 607, 624, 693, 694, 714, 742, 743, 753–755
Microstructures of PRO, 555
Microwave drying, 501–503, 507, 508
Mineral composition analysis of limonite, 192
Mineral composition of rare earth slag, 446
Mineral compositions, 196, 245, 248, 553, 554, 659
Mineralogy analysis of PRO, 556
Mineralogy of sinter, 196
Mineral reconstructions, 245, 247, 248
Mineral reconstructions of rare-earth concentrate, 245, 248
Model description, 40
Model hypothesis, 52
Models and prediction of viscosity, 536
Model verification, 31, 55
Mold flux, 95, 96, 98–100, 102–104, 110, 313, 382, 385, 387, 600
Molecular dynamics, 599, 600
Morphological analysis, 172
Morphology, 69, 70, 133, 225, 226, 244, 279, 281, 283, 310, 312, 407, 418, 422, 472, 542, 543, 554, 600, 619, 621, 716, 719
Morphology of iron granules, 542

N
Na_3AlF_6-K_3AlF_6-AlF_3 melt, 321, 322, 326
Ni-based superalloys, 176, 177
Nickel, 175, 176, 221–223, 227, 228, 257, 501, 632, 713, 714, 720
Nucleation, 393, 403
Nucleation inclusion, 415, 418, 423
The nucleation mechanism of inclusion in experimental steels, 421
Numerical simulation, 25–27, 39, 41, 48, 51, 52, 54–56, 60, 95, 96, 100, 454, 461
Numerical solution, 332

O
Offshore steel, 289, 290
Operating line of blast furnace, 19
Operating practice, 274
Operation line calculation results, 21
Oxidation of sulfides, 234
Oxidative roasting, 231, 232, 234–236, 240, 660
Oxygen blowing, 352

P
Parameter selection, 102
Partial RDFs and CNs, 601
Particle size, 47–49, 119, 158, 167–171, 258, 259, 356, 525, 542, 543, 546, 547, 549, 621, 649, 651, 656, 706–712, 714
Pelletizing, 157, 158, 162, 514, 779
Pelletizing experiments, 158
Pellets, 15, 16, 158, 211, 212, 214–217, 219, 233, 491, 493, 511–516, 518–521, 523–529, 533, 592, 632, 633, 635, 639, 640, 658, 760, 777, 779–784
Permeability, 3, 4, 11, 15, 72, 73, 102, 166, 258, 261, 513, 519, 523–529, 649, 654–656
Peru raw ore, 553
Phase diagram, 149–151, 153, 421, 444, 495
Physical analysis of typical copper sulfide, 755
Physical analysis of typical final slag, 753
PM_{10} and $PM_{2.5}$, 85, 86, 88–93
Population balance model, 723
Porosity, 165, 169, 172, 173, 175, 177, 180–183, 186, 456, 457, 511–515, 519–521, 695, 702
Potential analysis of energy conservation, 783
Practice and effect of energy-saving technology innovation, 277
Prediction, 206, 429, 505, 536, 579, 580, 587
Preparation of coke samples, 167
Preparation of Fe–Cu alloy, 436
Preparation of high-carbon metallic briquette, 66
Preparation of pre-reduced pellets, 524

The process mineralogy of limonite, 192
Process optimization, 506
Properties, 26, 29, 52, 85–87, 96, 109, 151, 165, 173, 175, 176, 183–186, 195, 243, 258, 297, 298, 300, 302, 305, 306, 321, 353, 356, 371, 372, 381, 428, 443, 479, 512, 515, 516, 518, 520, 523, 524, 529, 531, 599, 600, 628, 759, 760, 766
Properties of raw materials, 258
Pure metal boiling temperature, 396
Pure metal saturated vapor pressure, 396
Pyrometallurgy, 228, 297, 620, 665

R
Raman spectra analysis, 387
Raman spectroscopy, 298, 303, 381, 600, 604, 605, 714, 760, 762
Rare-earth concentrate, 243–249, 252
Rare earths, 243, 247, 249, 381, 382, 385, 387, 389, 443, 446, 448, 497
Rare earth slag, 443, 444, 446–448, 451
Raw material, 76, 212, 222, 546, 778
Raw materials and equipment, 397
Raw materials and preparation, 468
Reaction mechanism, 221, 224, 372, 455, 459, 657, 658, 661, 713, 714, 769
Reaction mechanism of dechlorination process, 661
Reaction of pure $FeCl_24H_2O$ with air, 660
Recovery, 118, 212, 216–219, 231, 243–245, 247–249, 251–253, 273, 275, 352, 353, 356, 428, 443, 444, 451, 468, 474, 481, 483, 484, 486, 487, 489–493, 495, 497, 620, 621, 658, 677, 741, 742, 746, 756
Recovery of rare earths under super-gravity field, 247, 249
Recycling, 118, 271, 274, 275, 435, 619, 620, 657, 795
Red mud, 489–495, 497
Reducing gas, 779
Reduction agent, 355
Reduction degree, 5, 6, 523–529
Reduction temperature, 354
Reduction time, 355
Removal of copper form Fe–Cu alloy, 437
Removal of inclusion particles, 47
Residence time distribution curve, 55
Residual elements, 416, 795, 796, 799
Response analysis, 505
Response surface, 341–343
Response surface methodology, 501, 502
RE Steel, 382
Results based on fuzzy grey relational analysis, 34
Results based on single-factor experimental design, 31
Roasting, 66, 68, 212, 214–218, 221–227, 231–236, 238, 240, 493, 501, 657–662, 713–720, 777, 779–784
Roasting temperature tests, 80
Roasting time tests, 79
Roman, 601
Rotary disc atomizer, 539–543
Rotary kiln furnace, 591–593
Rotation speed, 541, 546

S
Saturated vapor pressure, 669, 670
Second phase particles of welding joints, 294
SEM-EDS analysis of rare earth slag, 446
Semisteel, 539–541, 543
Separation, 361, 362
Siderite, 75–77, 80–82
Simulating method, 600
Simulation scheme, 54
Single-factor experimental design, 27
Sintering, 85–93, 145, 146, 175–184, 186, 189–191, 193–195, 198, 226, 257–263, 266, 275, 276, 512, 518, 564, 649–651, 654–656
Sintering pot tests, 651
SiO_2–CaO–CaF_2–Ce_2O_3, 443
Size distribution, 195
Slag chemistry, 297–299
Slag composition, 20, 133, 138, 140, 142, 352, 444, 528, 694, 769
Slag preparation, 445
Slag properties, 303
Slag structure, 301, 305, 531, 533, 535, 602, 604–607, 760
Slag structure relationship to physical properties/thermodynamics, 305
Small pot sintering test results and analysis, 194
Smelting, 19, 21, 26, 109, 190, 191, 222, 257, 258, 277, 349–354, 356, 357, 406, 435, 489–494, 497, 512, 539, 545, 633, 637, 638, 665, 666, 668, 674, 675, 677, 741, 742, 749–753, 778, 779, 784
Smelting studies for recovery of iron, 491
Sn-Sb alloy, 361, 362, 364, 365, 367–369
Sodium sulfate, 221–223, 225–228
Softening, melting and dropping experiments of pre-reduced pellets, 525
Solidified structures, 431
Solution loss, 4–7, 66, 71, 72, 166
Species conservation equation, 458
$SrCO_3$, 591–596

Steel, 15, 26, 39–47, 49, 51, 52, 54, 57, 59, 66, 85, 86, 97, 102, 118, 127, 137, 138, 140, 142, 166, 232, 257, 258, 269–273, 276, 277, 279–281, 283, 285, 286, 289–291, 293, 309–311, 313, 315, 323, 349–352, 381, 382, 406, 408, 411, 415–419, 422, 423, 428, 429, 431, 435, 453, 454, 458–462, 467, 468, 480, 481, 491, 496, 511–513, 524, 531, 540, 541, 543, 546, 564, 579–581, 585, 631–633, 657, 658, 693, 713, 723–726, 728–731, 759, 771, 795–799, 801
Steel scrap phase, 459
Stibnite, 665, 667, 672–674, 676, 678
Strength tests, 67
Structural studies of silicates, 303
Structure, 4, 25, 27, 39–41, 43, 46, 47, 52, 69, 99, 102, 104, 140, 145, 165, 166, 168, 169, 171, 177, 193, 249, 261, 264, 271, 272, 274, 275, 289, 293, 300, 301, 304, 306, 321, 353, 354, 371, 381–384, 387, 389, 406, 408, 411, 412, 433, 443, 446–448, 533, 535, 554, 592, 593, 599, 601–607, 670, 714, 716, 718, 725, 747, 750, 759, 760, 762–764, 766, 778
Structure-property relationship, 764
Sulfation roasting, 221, 222, 224, 227, 228
Sulfation roast-water leach process, 225
Sulfide, 222, 223, 227, 228, 279, 290, 310, 553, 556, 558, 559, 561, 562, 564, 669, 672, 673, 676, 713–716, 718–720, 741–744, 751, 752, 754–756, 760
Sulfide minerals, 221, 222, 234, 556, 564
Super-gravity, 244, 247–249, 252, 467–469, 472, 476, 477
Super-gravity enrichment of galvanizing dross, 472
Super-gravity experiments, 469
Super-gravity field, 243, 244, 247, 249, 251–253, 469, 619–623, 625, 628
Superheat degree, 199
Synthesis, 275, 372, 373, 376, 569, 575

T

Technical concept, 274
Technical feature, 273
Technical route and basic principles, 275
Temperature distribution, 25–27, 31, 463
Temperature field in the tundish, 57
Temperature programs, 128, 129, 131, 132, 134
TG analysis, 542
TG-DSC analysis, 224
TG-DSC tests, 651
Theoretical analysis, 746
Theoretical basis, 342
Thermo-analysis of limonite (DTA and TGA), 192
Thermodynamic, 109, 118–120, 122, 123, 231–233, 236, 240, 244, 293, 297, 298, 301, 305–307, 427, 444, 481, 482, 535, 569–571, 573, 575, 646, 672, 673, 678, 769, 774, 775, 796
Thermodynamic analysis, 117–120, 123, 145, 153, 235, 322, 665, 666, 672, 673
Thermodynamic behavior, 17, 22
Thermodynamic calculation, 109, 110, 112, 116, 149, 151, 306, 405, 429, 746
300t converter, 276, 277, 330, 331
Ti–Mg complex deoxidized, 415, 416, 419
Tin-bearing middling, 211–215, 217, 218
TiO_2, 119, 146, 289, 291, 292, 294, 296, 491–496, 517, 531–537, 632, 634, 743, 754, 759–762, 764, 766
Titanium sponge, 681–690
Total energy utilization analysis, 782
Trend analysis of antimony in smelting process, 677
Tundish, 39–46, 48, 49, 51, 52, 54–57, 60, 313, 723–725, 728–730
Tundish model, 40
Tungsten carbide, 371, 372, 374–378
Turbine disk, 176
Turbulence inhibitor, 39, 40, 44, 47, 48
Turbulence model, 42

V

Vacuum, 393, 394, 396–398
Vacuum distillation, 341–343, 345, 347, 361, 362, 369, 681–683, 688–690
Vanadium extraction, 787, 788, 790–794
Viscosity, 41, 43, 52–54, 95–102, 104, 109–111, 113, 116, 125, 126, 128–134, 138, 140, 142, 146, 201, 202, 244, 252, 297, 300, 302, 305, 306, 311, 381–387, 389, 446, 456, 458, 475, 490, 531–537, 599, 600, 741, 746, 747, 750, 752, 759, 760, 764–766, 769, 774, 775, 779
Viscosity measurements, 110, 128
Viscosity models, 531
Volatilization rate, 211, 212, 218, 665, 666, 668, 669, 673–676

W

Wastewater, 705–707, 710–712
Water model, 725
Weight loss of cokes during annealing, 169
Wetting behavior, 151

Wetting experiment, 147

X
XPS spectra analysis, 389
X-ray diffraction, 168
XRD analysis of PRO, 561

Z
Zn-Al-Fe alloy, 341–343, 345, 347
ZrO_2-C refractory, 137, 138, 140, 142, 143

The manufacturer's authorised representative in the EU is Springer Nature Customer Service Centre GmbH, Europaplatz 3, 69115 Heidelberg, Germany. If you have any concerns regarding our products, please contact ProductSafety@springernature.com

Printed and bound by CPI Group (UK) Ltd, Croydon, CR0 4YY
30/09/2026
02225237-0019